S0-BYH-456

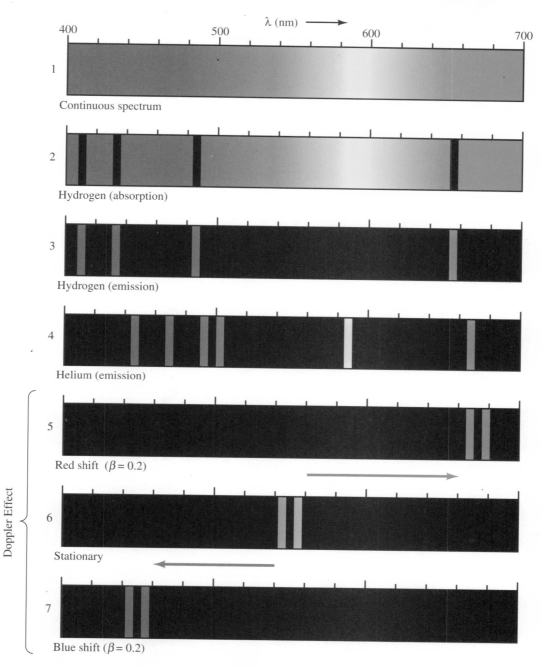

1 — Continuous spectrum

2 — Hydrogen (absorption)

3 — Hydrogen (emission)

4 — Helium (emission)

Doppler Effect

5 — Red shift ($\beta = 0.2$)

6 — Stationary

7 — Blue shift ($\beta = 0.2$)

Atomic spectra like these played a central role in the development of modern physics. (1) The continuous spectrum of white light emitted by a white-hot solid includes all colors from violet through red. (2) The absorption spectrum of hydrogen shows the transmitted light when white light is shone through an assembly of hydrogen atoms (in a discharge tube, for example); the hydrogen atoms absorb visible light at just four discrete wavelengths. (3) When hydrogen atoms are heated they emit light at the same four wavelengths. Spectra like (2) and (3) are called line spectra, and the individual strips of light or dark are called spectral lines. (4) The most prominent lines of the emission spectrum of helium. The final three spectra illustrate the Doppler effect for a hypothetical atom which emits a doublet of green lines, as seen by an observer for whom the atom is at rest (6). If the same atom is moving away from the observer, its spectrum is shifted towards the red (5); if it is approaching the observer, the spectrum is shifted towards the blue (7). Both shifted spectra are for atoms moving at about 0.2 times the speed of light.

Modern Physics
for Scientists and Engineers

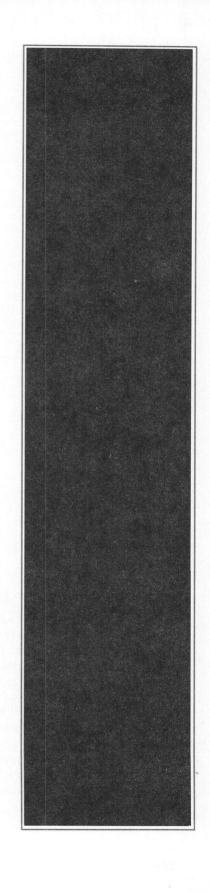

The front cover shows a small part of the four-mile circular tunnel of the particle accelerator at the Fermi National Laboratory outside Chicago. This machine probes the structure of elementary particles by accelerating counter-rotating beams of protons and antiprotons to high energies and then allowing them to collide. The particles are accelerated in vacuum pipes and are guided by magnetic fields as they travel repeatedly round the circular tunnel. They reach about half their final energy guided by the red and blue magnets running across the center of the picture and complete their acceleration in the section just above the floor, where they are guided by the intense fields of superconducting magnets. The machine's operation and the experiments it performs involve many aspects of the modern physics presented in this book. These include relativity, quantum mechanics, atomic physics, particle physics, lasers, and solid state physics.

Modern Physics

for Scientists and Engineers

John R. Taylor • Chris D. Zafiratos

Department of Physics
University of Colorado at Boulder

 Prentice Hall, Englewood Cliffs, New Jersey 07632

Library of Congress Cataloging-in-Publication Data

Taylor, John R. (John Robert)
   Modern physics for scientists and engineers/John R. Taylor and Chris D. Zafiratos.
      p.   cm.
   Includes bibliographical references and index.
   ISBN 0-13-589789-0
   1. Physics.   I. Zafiratos, Chris D.   II. Title.
QC21.2.T393   1991   90-24685
530—dc20   CIP

Designer, Jack Meserole
Production, Nicholas C. Romanelli
Manufacturing buyer, Paula Massenaro
Cover Designer, William Frost Associates
Cover Photo: Fermilab

© 1991 by Prentice-Hall, Inc.
A Paramount Communications Company
Englewood Cliffs, New Jersey 07632

Printed in the United States of America
10 9 8 7 6 5 4 3 2

ISBN 0-13-589789-0

Prentice Hall International (UK) Limited, *London*
Prentice-Hall of Australia Pty. Limited, *Sydney*
Prentice-Hall Canada Inc., *Toronto*
Prentice-Hall Hispanoamericana, S.A., *Mexico*
Prentice-Hall of India Private Limited, *New Delhi*
Prentice-Hall of Japan, Inc., *Tokyo*
Simon & Schuster Asia Pte. Ltd., *Singapore*
Editora Prentice-Hall do Brasil, Ltda, *Rio de Janeiro*

Brief Contents

* Note that the chapters of Parts III and IV are nearly independent of one another. You could omit either part entirely, or even read them in reverse order. You could also study just selected chapters from both parts. In a one-semester course it would normally be necessary to choose to cover either Part III *or* Part IV (or bits of both).

v

Contents

Preface

The name *modern physics* is generally used to describe those parts of physics whose theoretical foundations were laid after the year 1900. In practice, this means that modern physics comprises the two great theories that revolutionized twentieth-century physics—relativity and quantum theory—and those fields, such as atomic and nuclear physics, that can be understood properly only with the help of these two theories. The name *modern physics* is used to distinguish these fields from *classical physics,* which comprises those fields such as Newtonian mechanics and classical electromagnetism that were well established before 1900.

   This book is an introduction to modern physics for students in the physical sciences and engineering. The introductory physics curriculum at American colleges and universities varies widely from institution to institution and is currently undergoing some revisions. Nevertheless, it is probably safe to say that the "normal" procedure is to introduce classical physics first, in a calculus-based course that lasts nearly a year—often two semesters or three quarters. This is followed by an introduction to modern physics in a course that often lasts only half a year—one semester or, perhaps, two quarters. We have designed our book for use in this introductory course in modern physics.

   Given the type of course for which our book is intended, two important features follow immediately. First, we have assumed that our readers already have a reasonable knowledge of classical physics and the way it uses calculus. Second, since there is far more modern physics than can possibly be covered in half a year, some choices have to be made. For the most part we have arranged the book so as to leave these choices to the teacher of the course (or the reader of the book).

   With a few exceptions, we have included everything that seemed a reasonable candidate for inclusion, but have tried to organize the book so that many topics can be omitted without any loss of continuity. First, each chapter is divided into sections, and a number of these sections have been designated as "optional"; all of these optional sections can be omitted without subsequent penalty. More important, the whole book is divided into four parts as follows:

Part I	(Chapters 1 to 3)	Relativity	**THEORIES**
Part II	(Chapters 4 to 11)	Quantum Mechanics	
Part III	(Chapters 12 to 14)	Subatomic Physics	**APPLICATIONS**
Part IV	(Chapters 15 to 17)	Radiation, Molecules, and Solids	

Either Part III or Part IV can be omitted entirely, according to the taste of the teacher or reader. It has been our experience that in one semester one can, with reasonable comfort, cover the "mandatory" Parts I and II (omitting optional sections) and *either* Part III *or* Part IV (again omitting optional sections). Furthermore, the chapters within Parts III and IV are nearly independent. Thus it is possible to pick and choose within Parts III and IV, or even to cover bits of both. Anyone with the luxury of a year to teach modern physics should be able to cover all four parts of the book, including the optional sections. For the majority, who have to make choices, we offer more guidance at the beginning of Parts III and IV.

As with all physics texts, the problem sets at the end of each chapter form an essential part of the book. Anyone who is serious about learning modern physics *must* do several problems for each chapter. To help teachers and students select problems, we have classified them in two ways. We have indicated the approximate difficulty of each problem with a system of dots, ranging from one dot (•), which indicates a straightforward problem involving just one main concept, to three dots (• • •), which indicate a challenging problem that may involve several ideas and lengthy calculations. In addition, we have grouped the problems according to the section to which they relate.*

We have ended each chapter with a list of "ideas that you should now understand" in the hope that this will help in reviewing the chapter. Readers should take a moment to read over this list and ask themselves if they do indeed understand the ideas listed.

Another feature to be aware of is the information given inside the front and back covers. In particular, the front cover contains a list of physical constants given to three or four significant figures. (This is ample accuracy for most purposes; the best known values are given in Appendix A.) This list of constants, and the periodic table inside the back cover, should be valuable references, especially for working the problems. Additional information (more accurate constants, mathematical formulas, further data on atoms and nuclei, picture credits, and suggestions for further reading) is given in the appendices.

In the matter of units and notations we have tried to follow accepted practice and to avoid too many eccentricities. With few exceptions, we use SI units throughout. (For example, we measure wavelengths of light in nanometers, not in angstrom units.) The only non-SI unit used with any regularity is the electron volt (eV), and we use this without apology. The eV and its cousins, the keV, MeV, . . . , are regularly used in almost every branch of modern physics, and a working familiarity with the eV is an essential part of the training of any modern physicist.

As we have already mentioned, we have assumed that our readers will have been exposed to an introductory calculus-based course in classical physics. In particular, we take for granted a familiarity with elementary calculus and its applications. However, we have tried to introduce more advanced ideas, such as partial differentiation and differential equations, in such a way that readers who have never met them before can pick them up as we go along.

Finally, it is a pleasure to express our gratitude to several people. A preliminary version of this book was used for several years at the University of Colorado. The professors who kindly tested the book in this way (besides ourselves) were Professors Ashby, Downs, Miller, Phillipson, Rankin, and Tanttila. Other colleagues at Colorado who helped us in many ways were Dana

---

* This convenient scheme is sometimes criticized for fostering problems that are narrowly focused on a single topic. However, when a problem is listed as relating to a certain section, this does *not* mean that the problem relates *only* to that section; rather, the problem almost certainly involves ideas from earlier sections as well. Thus the designation is only a promise that the problem does not require material from any sections *after* the section indicated.

Anderson, Bob Barkley, Paul Beale, Peter Bender, John Carr, John Cumulat, Bill Ford, Allen Hermann, Carl Lineberger, K. T. Mahanthappa, Uriel Nauenberg, Steve O'Neil, Bill O'Sullivan, Cort Pierpont, Rod Smythe, and Lynn Teets. Colleagues at other universities who read large parts of the manuscript and made many helpful suggestions were Professors Robert Chasson at Denver University, Ed Gibson at Sacramento State, James Ho at Wichita State, Michael Lieber at the University of Arkansas, Ralph Llewellyn at the University of Central Florida, and Mark Semon at Bates College. We are especially indebted to Sanford Kern at Colorado State, who read the entire manuscript and gave us literally hundreds of useful comments, and to Dale Prull at CU, who drew numerous beautiful pictures with his computer and marshalled the data in Appendix D. Harold and Judith Taylor did a painstaking check of all numbers for equations, figures, and so on, and all cross-references. Linda Frueh performed her usual miracles at the word processor, and Frances Kretschmann proofread the entire book at least twice. To all of these people we are truly thankful.

We are deeply grateful to Debby Taylor for her patience and loving encouragement through the years.

J.R.T./C.D.Z.

CHAPTER

1

Relativity in Classical Physics

Two great theories underlie almost all of modern physics, both of them discovered during the first 25 years of the twentieth century. The first of these, relativity, was pioneered mainly by one person, Albert Einstein, and is the subject of Part I of this book (Chapters 1 to 3). The second, quantum theory, was the work of many physicists, including Bohr, Einstein, Heisenberg, Schrödinger, and others; it is the subject of Part II. In Parts III and IV we describe the applications of these great theories to several areas of modern physics.

Part I contains just three chapters. In Chapter 1 we describe how several of the ideas of relativity were already present in the classical physics of Newton and others. In Chapter 2 we describe how Einstein's careful analysis of the relationship between different reference frames, taking account of the observed invariance of the speed of light, changed our whole concept of space and time. Finally, in Chapter 3 we describe how the new ideas about space and time required a radical revision of Newtonian mechanics and a redefinition of the basic ideas — mass, momentum, energy, and force — on which mechanics is built. In the final section of Chapter 3, we briefly describe general relativity, which is the generalization of relativity to include gravity and accelerated reference frames.

1.1 Relativity

Most physical measurements are made *relative* to a chosen reference system. If we measure the time of an event as $t = 5$ seconds, this must mean that t is 5 seconds *relative* to a chosen origin of time, $t = 0$. If we state that the position of a projectile is given by a vector $\mathbf{r} = (x, y, z)$, we must mean that the position vector has components x, y, z *relative* to a system of coordinates with a definite orientation and a definite origin $\mathbf{r} = 0$. If we wish to know the kinetic energy K of a car speeding along a road, it makes a big difference whether we measure K relative to a reference frame fixed on the road or to one fixed on the car. (In the latter case $K = 0$, of course.) A little reflection should convince you that almost every measurement requires the specification of a reference system relative to which the measurement is to be made. We refer to this fact as the *relativity of measurements.*

The theory of relativity is the study of the consequences of this relativity of measurements. It is perhaps surprising that this could be an important subject of study. Nevertheless, Einstein showed, starting with his first paper on relativity in 1905, that a careful analysis of how measurements depend on coordinate systems revolutionizes our whole understanding of space and time, and requires a radical revision of classical, Newtonian mechanics.

In this chapter we discuss briefly some features of relativity as it applies in the classical theories of Newtonian mechanics and electromagnetism. We then describe the Michelson–Morley experiment, which (with the support of numerous other, less direct experiments) shows that something is wrong with the classical ideas of space and time. In Chapter 2 we state the two postulates of Einstein's relativity and show how they lead to a new picture of space and time in which both lengths and time intervals have different values when measured in two reference frames that are moving relative to one another. In Chapter 3 we show how the revised notions of space and time require a revision of classical mechanics. We shall find that the resulting relativistic mechanics is usually indistinguishable from Newtonian mechanics when applied to bodies moving with normal terrestrial speeds, but is entirely different when applied to bodies with speeds that are a substantial fraction of the speed of light, c. In particular, we shall find that no body can be accelerated to a speed greater than c, and that mass is a form of energy, in accordance with the famous relation $E = mc^2$.

Einstein's theory of relativity is really two theories. The first, called the *special* theory of relativity, is "special" in that its primary focus is restricted to unaccelerated frames of reference. This is the theory that we shall be studying in Chapters 2 and 3 and applying to our later discussions of radiation, nuclear, and particle physics.

The second of Einstein's theories is the general theory of relativity, which is "general" in that it includes accelerated frames of reference. Einstein found that the study of accelerated reference frames led naturally to a theory of gravitation, and general relativity turns out to be the relativistic theory of gravity. In practice, general relativity is needed only in areas where its predictions differ significantly from those of Newtonian gravitational theory. These include the study of the intense gravity near black holes, of the large-scale universe, and of the effect the earth's gravity has on extremely accurate time measurements (one part in 10^{12} or so). General relativity is an important part of modern physics; nevertheless, it is an advanced topic and, unlike special relativity, is not required for the other topics we treat in this book. Therefore, we have given only a brief description of general relativity in the final, and optional, section of Chapter 3.

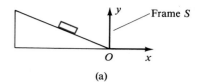

(a)

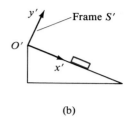

(b)

1.2 The Relativity of Orientation and Origin

In your studies of classical physics you probably did not pay much attention to the relativity of measurements. Nevertheless, the ideas were present and, whether or not you were aware of it, you probably exploited some aspects of relativity theory in solving certain problems. Let us illustrate this claim with two examples.

In problems involving blocks sliding on inclined planes, it is well known that one can choose coordinates in various ways. One could, for example, use a coordinate system S with origin O at the bottom of the slope and with axes Ox horizontal, Oy vertical, and Oz across the slope, as shown in Fig. 1.1(a). Another possibility would be a reference frame S' with origin O' at the top of the slope and axes $O'x'$ parallel to the slope, $O'y'$ perpendicular to the slope, and $O'z'$ across it, as shown in Fig. 1.1(b). The solution of any problem relative to the frame S may look quite different from the solution relative to S', and it often happens that one choice of axes is much more convenient than the other. (For some examples, see Problems 1.1 to 1.3.) On the other hand, the basic laws of motion, Newton's laws, make no reference to the choice of origin and orientation of axes, and are equally true in either coordinate system. In the language of relativity theory, we can say that Newton's laws are **invariant,** or unchanged, as we shift our attention from frame S to S', or vice versa. It is because the laws of motion are the same in either coordinate system that we are free to use whichever system is more convenient.

The invariance of the basic laws when we change the origin or orientation of axes is true in all of classical physics — Newtonian mechanics, electromagnetism, and thermodynamics. It is also true in Einstein's theory of relativity. It means that in any problem in physics one is free to choose the origin of coordinates and the orientation of axes in whatever way is most convenient. This freedom is very useful and we often exploit it. However, it is not especially *interesting* in our study of relativity, and we shall not have much occasion to discuss it further.

FIGURE 1.1 (a) In studying a block on an incline, one could choose axes Ox horizontal and Oy vertical and put O at the bottom of the slope. (b) Another possibility, which is often more convenient, is to use an axis $O'x'$ parallel to the slope with $O'y'$ perpendicular to the slope and to put O' at the top of the slope. (The axes Oz and $O'z'$ point out of the page and are not shown.)

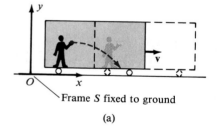

(a)

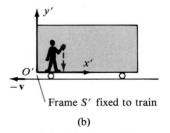

(b)

FIGURE 1.2 (a) As seen from the ground, the train and student move to the right; the ball falls in a parabola and lands at the student's feet. (b) As seen from the train, the ball falls straight down, again landing at the student's feet.

1.3 Moving Reference Frames

As a more important example of relativity, we consider next a question involving two reference frames that are *moving* relative to one another. Our discussion will raise some interesting questions about classical physics, questions that were satisfactorily answered only when Einstein showed that the classical ideas about the relation between moving reference frames needed revision.

Let us imagine a student standing still in a train that is moving with constant velocity **v** along a horizontal track. If the student drops a ball, where will the ball hit the floor of the train? One way to answer this question is to use a reference frame S fixed on the track, as shown in Fig. 1.2(a). In this coordinate system the train and student move with constant velocity **v** to the right. At the moment of release, the ball is traveling with velocity **v** and it moves, under the influence of gravity, in the parabola shown. It therefore lands to the right of its starting point (as measured in the ground-based frame S). However, while the ball is falling the train is moving, and a straightforward calculation shows that the train moves exactly as far to the right as does the ball. Thus the ball hits the floor at the student's feet, vertically below his hand.

Simple as this solution is, one can reach the same conclusion even more simply by using a reference frame S' fixed to the train, as shown in Fig. 1.2(b). In this coordinate system the train and student are at rest (while the track moves to the left with constant velocity $-\mathbf{v}$). At the moment of release the ball is at rest (as measured in the train-based frame S'). It therefore falls straight down and naturally hits the floor vertically below the point of release.

The justification of this second, simpler argument is actually quite subtle. We have taken for granted that an observer on the train (using the coordinates x', y', z') is entitled to use Newton's laws of motion and hence to predict that a ball which is dropped from rest will fall straight down. But is this correct? The question we must answer is this: If we accept as an experimental fact that Newton's laws of motion hold for an observer on the ground (using coordinates x, y, z), does it follow that Newton's laws also hold for an observer in the train (using x', y', z')? Equivalently, are Newton's laws invariant as we pass from the ground-based frame S to the train-based frame S'? Within the framework of classical physics, the answer to this question is "yes," as we now show.

Since Newton's laws refer to velocities and accelerations, let us first consider the velocity of the ball. We let $\mathbf{u}$ denote the ball's velocity relative to the ground-based frame S and $\mathbf{u}'$ the ball's velocity relative to the train-based S'. Since the train moves with constant velocity $\mathbf{v}$ relative to the ground, we naturally expect that

$$\mathbf{u} = \mathbf{u}' + \mathbf{v}. \tag{1.1}$$

We shall refer to this equation as the **classical velocity-addition formula.** It reflects our commonsense ideas about space and time, and asserts that velocities obey ordinary vector addition. Although it is one of the central assumptions of classical physics, equation (1.1) is one of the first victims of Einstein's relativity. In Einstein's relativity the velocities $\mathbf{u}$ and $\mathbf{u}'$ do *not* satisfy (1.1), which is only an approximation (although a very good approximation) that is valid when all speeds are much less than the speed of light, c. Nevertheless, we are for the moment discussing classical physics, and we therefore assume for now that the classical velocity-addition formula is correct.

Now let us examine Newton's three laws, starting with the first: A body on which no external forces act moves with constant velocity. Let us assume that this law holds in the ground-based frame S. This means that if our ball is isolated from all outside forces, its velocity $\mathbf{u}$ is constant. But $\mathbf{u}' = \mathbf{u} - \mathbf{v}$ and the train's velocity $\mathbf{v}$ is constant. It follows at once that $\mathbf{u}'$ is also constant, and Newton's first law also holds in the train-based frame S'. We shall find that this result is also valid in Einstein's relativity; that is, in both classical physics and Einstein's relativity, Newton's first law is invariant as we pass between two frames whose relative velocity is constant.

Newton's second law is a little more complicated. If we assume that it holds in the ground-based frame S, it tells us that

$$\mathbf{F} = m\mathbf{a}$$

where **F** is the sum of the forces on the ball, m its mass, and **a** its acceleration, all measured in the frame S. We now use this assumption to show that $\mathbf{F}' = m'\mathbf{a}'$, where $\mathbf{F}'$, m', $\mathbf{a}'$ are the corresponding quantities measured relative to the train-based frame S'. We shall do this by arguing that each of $\mathbf{F}'$, m', $\mathbf{a}'$ is in fact equal to the corresponding quantity **F**, m, and **a**.

The proof that $\mathbf{F} = \mathbf{F}'$ depends, to some extent, on how one has chosen to define force. Perhaps the simplest procedure is to define forces by their effect on a standard calibrated spring balance. Since observers in the two frames S and S' will certainly agree on the reading of the balance, it follows that any force will have the same value as measured in S and S'; that is, $\mathbf{F} = \mathbf{F}'$.*

Within the domain of classical physics it is an experimental fact that any technique for measuring mass (for example, an inertial balance) will produce the same result in either reference frame; that is, $m = m'$.

Finally, we must look at the acceleration. The acceleration measured in S is

$$\mathbf{a} = \frac{d\mathbf{u}}{dt},$$

where t is the time as measured by ground-based observers. Similarly, the acceleration measured in S' is

$$\mathbf{a}' = \frac{d\mathbf{u}'}{dt'}, \tag{1.2}$$

where t' is the time measured by observers on the train. Now, it is a central assumption of classical physics that time is a single universal quantity, the same for all observers; that is, the times t and t' are the same, or $t = t'$. Therefore, we can replace (1.2) by

$$\mathbf{a}' = \frac{d\mathbf{u}'}{dt}.$$

Since

$$\mathbf{u}' = \mathbf{u} - \mathbf{v}$$

we can simply differentiate with respect to t and find that

$$\mathbf{a}' = \mathbf{a} - \frac{d\mathbf{v}}{dt} \tag{1.3}$$

or, since **v** is constant, $\mathbf{a}' = \mathbf{a}$.

We have now argued that $\mathbf{F}' = \mathbf{F}$, $m' = m$, and $\mathbf{a}' = \mathbf{a}$. Substituting into the equation $\mathbf{F} = m\mathbf{a}$, we immediately find that

$$\mathbf{F}' = m'\mathbf{a}'.$$

That is, Newton's second law is also true for observers using the train-based coordinate frame S'.

The third law,

$$\text{action force} = -(\text{reaction force}),$$

---

* Of course, the same result holds whatever our definition of force, but with some definitions the proof is a little more roundabout. For example, many texts define force by the equation $\mathbf{F} = m\mathbf{a}$. Superficially, at least, this means that Newton's second law is true *by definition* in both frames. Then since $m = m'$ and $\mathbf{a} = \mathbf{a}'$ (as we shall show shortly) it follows that $\mathbf{F} = \mathbf{F}'$.

is easily treated. Since any given force has the same value as measured in S or S', the truth of Newton's third law in S immediately implies its truth in S'.

We have now established that if Newton's laws are valid in one reference frame, they are also valid in any second frame that moves with constant velocity relative to the first. This shows why we could use the normal rules of projectile motion in a coordinate system fixed to the moving train. More generally, in the context of our newfound interest in relativity, it establishes an important property of Newton's laws: If space and time have the usual properties assumed in classical physics, then Newton's laws are invariant as we transfer our attention from one coordinate frame to a second one moving with constant velocity relative to the first.

Newton's laws would *not* still hold in a coordinate system that was *accelerating.* Physically, this is easy to understand. If our train were accelerating forward, then just to keep the ball at rest (relative to the train) would require a force; that is, Newton's first law does not hold in the accelerating train. To see the same thing mathematically, note that if $\mathbf{u}' = \mathbf{u} - \mathbf{v}$ and $\mathbf{v}$ is *changing,* then $\mathbf{u}'$ is not constant even if $\mathbf{u}$ is. Further, the acceleration $\mathbf{a}'$ as given by (1.3) is not equal to $\mathbf{a}$, since $d\mathbf{v}/dt$ is not zero; so our proof of the second law for the train's frame S' also breaks down. In classical physics, the unaccelerated frames in which Newton's laws hold are often called **inertial frames,** and an accelerated frame is then **noninertial.**

1.4 Classical Relativity and the Speed of Light

Although Newton's laws are invariant as we change from one unaccelerated frame to another (if we accept the classical view of space and time), the same is not true of the laws of electromagnetism. We can show this by separately examining each law — Gauss's law, Faraday's law, and so on — but the required calculations are complicated. A simpler procedure is to recall that the laws of electromagnetism imply that, in a vacuum, light signals and all other electromagnetic waves travel in any direction with speed*

$$c = \frac{1}{\sqrt{\epsilon_0 \mu_0}} = 3.00 \times 10^8 \text{ m/s,}$$

where ϵ_0 and μ_0 are the permittivity and permeability of the vacuum. Thus if the electromagnetic laws hold in a frame S, light must travel with the same speed c in all directions, as seen in S.

Let us now consider a second frame S' traveling relative to S and imagine a pulse of light moving in the same direction as S', as shown on the left of Fig. 1.3. The pulse has speed c relative to S. Therefore, by the classical velocity-addition formula, it should have speed $c - v$ as seen from S'. Similarly, a pulse traveling in the opposite direction would have speed $c + v$ as seen from S', and a pulse traveling in any other, oblique direction would have a different speed, intermediate between $c - v$ and $c + v$. We see that in the frame S' the speed of light should vary between $c - v$ and $c + v$ according to its direction of propaga-

---

* More precisely, $c = 299,792,458$ m/s. In fact, the determination of c has become so accurate that, since 1984, the meter has been defined in terms of c, as the fraction $1/299,792,458$ of the distance traveled by light in 1 second. This means that, by definition, c is 299,792,458 m/s *exactly.*

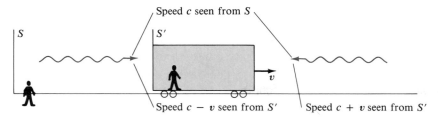

Speed c seen from S

S

S'

v

Speed $c - v$ seen from S' Speed $c + v$ seen from S'

FIGURE **1.3** Frame S' travels with velocity **v** relative to S. If light travels with the same speed c in all directions relative to S, then (according to the classical velocity-addition formula) it should have different speeds as seen from S'.

tion. Therefore, in classical physics, the laws of electromagnetism — unlike those of mechanics — could not be valid in the frame S'.

The situation just described was well understood by physicists toward the end of the nineteenth century. In particular, it was accepted as entirely obvious that there could be only one frame, called the **ether frame,** in which light traveled at the same speed, c, in all directions. The name "ether frame" derived from the belief that light waves must propagate through a medium, in much the same way that sound waves were known to propagate in the air. Since light propagates through a vacuum, physicists recognized that this medium, which no one had ever seen or felt, must have unusual properties. Borrowing the ancient name for the substance of the heavens, they called it the "ether." The unique reference frame in which light traveled at speed c was assumed to be the frame in which the ether was at rest. As we shall see, Einstein's relativity implies that neither the ether, nor the ether frame, actually exist.

1.5 The Michelson–Morley Experiment

Our picture of classical relativity can be quickly summarized. In classical physics we take for granted certain ideas about space and time, all based on our everyday experiences. For example, we assume that relative velocities add like vectors, in accordance with the classical velocity-addition formula; also, that time is a universal quantity, concerning which all observers agree. Accepting these ideas we have seen that Newton's laws should be valid in a whole family of reference frames, any one of which moves uniformly relative to any other. On the other hand, we have seen that there should be a unique reference frame, called the ether frame, relative to which the electromagnetic laws hold, and in which light travels through the vacuum with speed c in all directions.

It should perhaps be emphasized that although this view of nature turned out to be wrong, it was nevertheless perfectly logical and internally consistent. One might argue on philosophical or aesthetic grounds (as Einstein did) that the difference between classical mechanics and classical electromagnetism is surprising and even unpleasing, but theoretical arguments alone could not decide whether or not the classical view is correct. This question could only be decided by experiment. In particular, since classical physics implied that there was a unique ether frame where light travels at speed c in all directions, there had to be some experiment that showed whether or not this was so. This was exactly the experiment that Michelson, later assisted by Morley, performed between the years 1880 and 1887, as we now describe.

If one assumed the existence of a unique ether frame, it seemed clear that as the earth orbits around the sun, it must be moving relative to the ether frame. In principle, this motion relative to the ether frame should be easy to detect. One would simply have to measure the speed (relative to the earth) of light

ISAAC NEWTON (1642–1727, English). Possibly the greatest scientific genius of all time. In addition to his laws of motion and his theory of gravity, his contributions included the invention of calculus and important discoveries in optics. Although he believed in "absolute space" (what we would call the ether frame), Newton was well aware that his laws of motion hold in all unaccelerated frames of reference.

traveling in various directions. If one found different speeds in different directions, one would conclude that the earth is moving relative to the ether frame, and a simple calculation would give the speed of this motion. If, instead, one found the speed of light to be exactly the same in all directions, one would have to conclude that at the time of the measurements the earth happened to be at rest relative to the ether frame. In this case one should probably repeat the experiment a few months later, by which time the earth's velocity relative to the ether frame should surely be nonzero.

In practice, this experiment is extremely difficult because of the enormous speed of light:

$$c = 3 \times 10^8 \text{ m/s}.$$

If our speed relative to the ether is v, then the observed speed of light should vary between $c - v$ and $c + v$. Although the value of v is unknown, it should on average be of the same order as the earth's orbital velocity around the sun,

$$v \sim 3 \times 10^4 \text{ m/s}$$

(or possibly more if the sun is also moving relative to the ether frame). Thus the expected change in the observed speed of light due to the earth's motion is about 1 part in 10^4. This was too small a change to be detected by direct measurement of the speed of light at that time.

To avoid the need for such direct measurements, Michelson devised an interferometer in which a beam of light was split into two beams by a partially reflecting surface; the two beams traveled along perpendicular paths and were then reunited to form an interference pattern; this pattern was sensitive to differences in the speed of light in the two perpendicular directions and so could be used to detect any such differences.

Figure 1.4 is a simplified diagram of Michelson's interferometer. Light from the source hits the half-silvered mirror M and splits, part traveling to the mirror M_1 and part to M_2. The two beams are reflected at M_1 and M_2, and return to M, which sends part of each beam on to the observer. In this way the observer receives two signals, which can interfere constructively or destructively depending on their phase difference.

To calculate this phase difference, suppose for a moment that the two arms of the interferometer, from M to M_1 and M to M_2, have exactly the same length l, as shown. In this case any phase difference must be due to the different

FIGURE 1.4 (a) Schematic diagram of the Michelson interferometer. M is a half-silvered mirror, M_1 and M_2 are mirrors. The vector $\mathbf{v}$ indicates the earth's velocity relative to the supposed ether frame. (b) The vector-addition diagram that gives the light's velocity $\mathbf{u}$, relative to the earth, as it travels from M to M_2. The velocity $\mathbf{c}$ relative to the ether is the vector sum of $\mathbf{v}$ and $\mathbf{u}$.

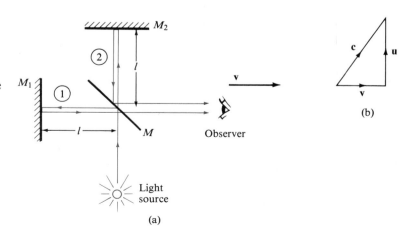

speeds of the two beams as they travel along the two arms. For simplicity, let us assume that arm 1 is exactly parallel to the earth's velocity **v**. In this case the light travels from M to M_1 with speed $c + v$ (relative to the interferometer) and back from M_1 to M with speed $c - v$. Thus the total time for the round trip on path 1 is

$$t_1 = \frac{l}{c + v} + \frac{l}{c - v} = \frac{2lc}{c^2 - v^2}. \tag{1.4}$$

It is convenient to rewrite this in terms of the ratio

$$\beta = \frac{v}{c},$$

which we have seen is expected to be very small, $\beta \sim 10^{-4}$. In terms of β, (1.4) becomes

$$t_1 = \frac{2l}{c} \frac{1}{1 - \beta^2} \approx \frac{2l}{c} (1 + \beta^2). \tag{1.5}$$

In the last step we have used the **binomial approximation** (discussed in Appendix B),

$$(1 - x)^n \approx 1 - nx, \tag{1.6}$$

which holds for any number n and any x much smaller than 1. (In the present case, $n = -1$ and $x = \beta^2$.)

The speed of light traveling from M to M_2 is given by the velocity-addition diagram in Fig. 1.4(b). (Relative to the earth the light has velocity **u** perpendicular to **v**; relative to the ether it travels with speed c in the direction shown.) This speed is

$$u = \sqrt{c^2 - v^2}.$$

Since the speed is the same on the return journey, the total time for the round trip on path 2 is

$$t_2 = \frac{2l}{\sqrt{c^2 - v^2}} = \frac{2l}{c\sqrt{1 - \beta^2}} \approx \frac{2l}{c} \left(1 + \frac{1}{2} \beta^2 \right), \tag{1.7}$$

where we have again used the binomial approximation (1.6), this time with $n = -\frac{1}{2}$.

Comparing (1.5) and (1.7), we see that the waves traveling along the two arms take slightly different times to return to M, the difference being

$$\Delta t = t_1 - t_2 \approx \frac{l}{c} \beta^2. \tag{1.8}$$

If this difference Δt were zero, the two waves would arrive in step and interfere constructively, giving a bright resultant signal. Similarly, if Δt were any integer multiple of the light's period, $T = \lambda/c$ (where λ is the wavelength), they would interfere constructively. If Δt were equal to half the period, $\Delta t = 0.5T$ (or $1.5T$, or $2.5T$, . . .), the two waves would be exactly out of step and would interfere destructively. We can express these ideas more compactly if we consider the ratio

$$N = \frac{\Delta t}{T} = \frac{l\beta^2/c}{\lambda/c} = \frac{l\beta^2}{\lambda}. \tag{1.9}$$

This is the number of complete cycles by which the two waves arrive out of step; in other words, N is the phase difference, expressed in cycles. If N is an integer, the waves interfere constructively; if N is a half-odd integer ($N = \frac{1}{2}, \frac{3}{2}, \frac{5}{2}, \ldots$), the waves interfere destructively.

The phase difference N in (1.9) is the phase difference due to the earth's motion relative to the supposed ether frame. In practice it is impossible to be sure that the two interferometer arms have exactly equal lengths, and there will be an additional phase difference due to the unknown difference in lengths. To circumvent this complication, Michelson and Morley rotated their interferometer through 90°, observing the interference as they did so. This rotation would not change the phase difference due to the different arm lengths, but it should *reverse* the phase difference due to the earth's motion (since arm 2 would now be along **v** and arm 1 across it). Thus, as a result of the rotation, the phase difference N should change by *twice* the amount (1.9),

$$\Delta N = \frac{2l\beta^2}{\lambda}. \qquad (1.10)$$

ALBERT MICHELSON (1852–1931, US). Michelson devoted much of his career to increasingly accurate measurements of the speed of light, and in 1907 he won the Nobel Prize for his contributions to optics. His failure to detect the earth's motion relative to the supposed ether is probably the most famous unsuccessful experiment in the history of science.

This implies that the observed interference should shift from bright to dark and back to bright again ΔN times. Observation of this shift would confirm that the earth is moving relative to the ether frame, and measurement of ΔN would give the value of β, and hence the earth's velocity $v = \beta c$.

In their experiment of 1887 Michelson and Morley had an arm length $l \approx 11$ m. (This was accomplished by having the light bounce back and forth between several mirrors.) The wavelength of their light was $\lambda = 590$ nm; and as we have seen, $\beta = v/c$ was expected to be of order 10^{-4}. Thus the shift should have been at least

$$\Delta N = \frac{2l\beta^2}{\lambda} \approx \frac{2 \times (11 \text{ m}) \times (10^{-4})^2}{590 \times 10^{-9} \text{ m}} \approx 0.4. \qquad (1.11)$$

Although they could detect a shift as small as 0.01, Michelson and Morley observed no significant shift when they rotated their interferometer.

The Michelson–Morley experiment has been repeated many times, at different times of year and with ever-increasing precision, but always with the same final result: There is no shift in the interference when the interferometer is rotated.* With hindsight it is easy to draw the right conclusion from their experiment: Contrary to all our expectations, light always travels with the same speed in all directions relative to an earth-based reference frame, even though the earth has different velocities at different times of year. In other words, there is no unique ether frame in which light has the same speed in all directions.

This conclusion is so surprising that it was not taken seriously for nearly 20 years. Rather, several ingenious alternative theories were advanced that explained the Michelson–Morley result but managed to preserve the notion of an ether. For example, in the "ether-drag" theory, it was suggested that the ether, the medium through which light was supposed to propagate, was dragged along by the earth as it moved through space (in much the same way that the

---

* From time to time experimenters have reported observing nonzero shifts, but closer examination has shown that these are probably due to spurious effects such as expansion and contraction of the interferometer arms resulting from temperature variations. For a careful modern analysis of Michelson and Morley's results and many further references, see M. Handschy, *American Journal of Physics*, vol. 50, p. 987 (1982).

earth does drag its atmosphere with it). If this were the case, an earth-bound observer would automatically be at rest relative to the ether, and Michelson and Morley would naturally have found that light had the same speed in all directions at all times of year. Unfortunately, this neat explanation of the Michelson–Morley result requires that light from the stars would be bent as it entered the earth's envelope of ether. Instead, astronomical observations show that light from any star continues to move in a straight line as it arrives at the earth.*

The ether-drag theory, like all other alternative explanations of the Michelson–Morley result, has been abandoned because it fails to fit all the facts. Today, nearly all physicists agree that Michelson and Morley's failure to detect our motion relative to the ether frame was because there is no ether frame. The first person to accept this surprising conclusion and to develop its consequences into a complete theory was Einstein, as we describe in Chapters 2 and 3.

IDEAS YOU SHOULD NOW UNDERSTAND FROM CHAPTER 1

Relativity of measurements
The classical velocity-addition formula
Invariance of Newton's laws in classical physics

Noninvariance of the speed of light in classical physics
The Michelson–Morley experiment

PROBLEMS FOR CHAPTER 1

The problems for each chapter are arranged according to section number. A problem listed for a given section requires an understanding of that section and earlier sections, but not of later sections. Within each section problems are listed in approximate order of difficulty. A single dot (•) indicates straightforward problems involving just one main concept and sometimes requiring no more than substitution of numbers in the appropriate formula. Two dots (••) identify problems that are slightly more challenging and usually involve more than one concept. Three dots (•••) indicate problems that are distinctly more challenging, either because they are intrinsically difficult or involve lengthy calculations. Needless to say, these distinctions are hard to draw and are only approximate.

Answers to odd-numbered problems are given at the back of the book.

SECTION **1.2** (THE RELATIVITY OF ORIENTATION AND ORIGIN)

1.1 •• At time $t = 0$ a block is released from the Point O on the slope shown in Fig. 1.5. The block accelerates down the slope, overcoming the sliding friction (coefficient μ). (a) Choose axes Oxy as shown and resolve the equation $\Sigma \mathbf{F} = m\mathbf{a}$ into its x and y components. Hence find the block's position (x, y) as a function of time, and the time it takes to reach the bottom.

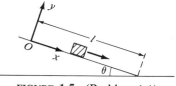

FIGURE **1.5** (Problem 1.1)

* Because of the earth's motion around the sun, the apparent direction of any one star undergoes a slight annual variation—an effect called stellar aberration. This effect is consistent with the claim that light travels in a straight line from the star to the earth's surface, but contradicts the ether drag theory.

(b) Carry out the solution using axes $Ox'y'$, with Ox' horizontal and Oy' vertical, and show that you get the same final answer. Explain why the solution using these axes is less convenient.

1.2 •• A block slides down the slope of Fig. 1.6 from O with initial speed v_0. The sliding friction (coefficient

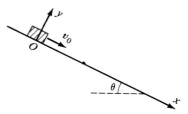

FIGURE 1.6 (Problem 1.2)

μ) brings the block to rest in a time T. (a) Using the axes shown, find T. (b) Solve this problem using axes $Ox'y'$ with Ox' horizontal and Oy' vertical and explain why the solution in this frame is less convenient (although it produces the same final answer, of course).

1.3 ••• At time $t = 0$ a ball is thrown with speed v_0 at an angle θ above a slope that is itself inclined at an angle ϕ above the horizontal (Fig. 1.7). (a) Choosing axes

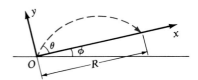

FIGURE 1.7 (Problem 1.3)

as shown, write down the components of the ball's initial velocity $\mathbf{v_0}$ and its acceleration $\mathbf{g}$. Hence find the ball's position (x, y) as a function of time and show that its range up the slope is

$$R = \frac{2v_0^2 \sin \theta \cos(\theta + \phi)}{g \cos^2\phi}.$$

Note that when the ball lands, $y = 0$. (b) Show that you get the same final answer if you use axes $Ox'y'$ with Ox' horizontal and Oy' vertical. Discuss briefly the merits of the two choices of axes. (You can find several useful trig identities in Appendix B.)

SECTION **1.3** (MOVING REFERENCE FRAMES)

1.4 • Consider a head-on, elastic collision between two bodies whose masses are m and M, with $m \ll M$. It is well known that if m has speed v_0 and M is initially at rest, m will bounce straight back with its speed un-

changed, while M will remain at rest (to an excellent approximation). Use this fact to predict the final velocities if M approaches with speed v_0 and m is initially at rest. (*Hint:* Consider the reference frame attached to M.)

1.5 • Use the method of Problem 1.4 to predict the final velocities if two bodies of masses m and M, with $m \ll M$, approach one another both traveling at speed v_0 and undergo a head-on, elastic collision.

1.6 •• A policeman is chasing a robber. Both are in cars traveling at speed v and the distance between them is l. The policeman wishes to shoot the robber with a gun whose muzzle velocity is u_0. At what angle θ above the horizontal should he point his gun? First solve this problem using coordinates traveling with the policeman, as shown in Fig. 1.8. Then sketch the

FIGURE **1.8** (Problem 1.6)

solution using coordinates fixed to the ground; is the angle of the gun the same as the angle of the bullet's initial velocity in this frame? (The advantages of the first frame are not overwhelming; nevertheless, it is clearly the natural choice for the problem.)

SECTION **1.4** (CLASSICAL RELATIVITY AND THE SPEED OF LIGHT)

1.7 • According to the classical ideas of space and time there could only be one frame, the "ether frame," in which light traveled at the same speed c in all directions. It seemed unlikely that the earth would be exactly at rest in this frame and one might reasonably have guessed that the earth's speed v relative to the ether frame would be at least of the order of our orbital speed around the sun ($v \approx 3 \times 10^4$ m/s). (a) What would be the observed speed (on earth) of a light wave traveling parallel to $\mathbf{v}$? (Give your answer in terms of c and v, and then substitute numerical values.) (b) What if it were traveling antiparallel to $\mathbf{v}$? (c) What if it were traveling perpendicular to $\mathbf{v}$ (as measured on earth)? The accepted value of c is 2.9979×10^8 m/s (to five significant figures).

1.8 •• At standard temperature and pressure sound travels at speed $u = 330$ m/s relative to the air through which it propagates. Four students, A, B, C, D, position themselves as shown in Fig. 1.9, with A, B, C in a straight line and D vertically above B. A steady wind is blowing with speed $v = 30$ m/s along the line

FIGURE 1.9 (Problem 1.8)

ABC. If *B* fires a revolver, what are the speeds with which the sound will travel to *A*, *C*, and *D* (in the reference frame of the observers)? Discuss whether the differences in your answers could be detected.

1.9 • • • It is well known that the speed of sound in air is $u = 330$ m/s at standard temperature and pressure. What this means is that sound travels at speed u in all directions *in the frame S where the air is at rest*. In any other frame S', moving relative to S, its speed is *not u* in all directions. To verify this, some students set up a loudspeaker L and receiver R on an open flat car as in Fig. 1.10; by connecting the electrical signals from L

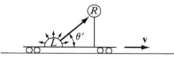

FIGURE 1.10 (Problem 1.9)

and R to an oscilloscope they can measure the time for a sound to travel from L to R and hence find its speed u' (relative to the car). (*a*) Derive an expression for u' in terms of u, v, and θ', where v is the car's speed through the air and θ' is the angle between **v** and LR. (We call this θ' since it is the angle between **v** and **u′**, the velocity of the sound measured in the frame of the car.) (*Hint:* Draw a velocity-addition triangle to represent the relation **u** = **u′** + **v**. The law of cosines should give you a quadratic equation for u'.) (*b*) If the students vary the angle θ' from 0 to 180°, what are the largest and smallest values of u'? (*c*) If v is about 15 m/s (roughly 30 mi/h), what will be the approximate percent variation in u'? Would this be detectable?

SECTION **1.5** (THE MICHELSON–MORLEY EXPERIMENT)

1.10 • In the discussion of the Michelson–Morley experiment we twice used the binomial approximation

$$(1 - x)^n \approx 1 - nx \qquad (1.12)$$

which holds for any number n and any x much smaller than 1 (that is, $|x| \ll 1$). (In the examples, n was -1 and $-\frac{1}{2}$, and $x = \beta^2$ was of order 10^{-8}.) The binomial approximation is frequently useful in rela-

tivity, where one often encounters expressions of the form $(1 - x)^n$ with x small. Make a table showing $(1 - x)^n$ and its approximation $1 - nx$ for $n = -\frac{1}{2}$ and $x = 0.5$, 0.1, 0.01, and 0.001. In each case find the percentage by which the approximation fails.

1.11 • Do the same tasks as in Problem 1.10 but for the case $n = 2$. In this case give an exact expression for the difference between the exact and approximate forms. Explain why the approximation gets better and better as $x \rightarrow 0$.

1.12 • Use the binomial approximation (1.12) (Problem 1.10) to evaluate $(1 - 10^{-20})^{-1} - 1$. Can you evaluate this directly on your calculator?

1.13 • Tom Sawyer and Huck Finn can each row a boat at 5 ft/s in still water. Tom challenges Huck to a race in which Tom is to row the 2000 ft across the Mississippi to a point exactly opposite their starting point and back again, while Huck rows to a point 2000 ft directly downstream and back up again. If the Mississippi flows at 3 ft/s, which boy wins and by how long?

1.14 • An airline, all of whose planes fly with an airspeed of 200 mi/h, serves three cities, A, B, and C, where B is 320 mi due east of A, and C is the same distance due north of A. On a certain day there is a steady wind of 120 mi/h from the east. (*a*) What is the time needed for a round trip from A to B and back? (*b*) What is it from A to C and back?

1.15 • In one of the early (1881) versions of Michelson's interferometer the arms were about 50 cm long. What would be the expected shift ΔN when he rotated the apparatus through 90°, assuming that $\lambda = 590$ nm and that the expected speed of the earth relative to the ether was 3×10^4 m/s? (In this case the expected shift was so small that no one regarded his failure to observe a shift as conclusive.)

1.16 • • One of the difficulties with the Michelson–Morley experiment is that several extraneous effects (mechanical vibrations, variations in temperature, etc.) can produce unwanted shifts in the interference pattern, masking the shift of interest. Suppose, for example, that during the experiment the temperature of one arm of the interferometer were to rise by ΔT. This would increase the arm's length by $\Delta l = \alpha l \, \Delta T$, where α is the arm's coefficient of expansion. Prove that this temperature change would, by itself, cause a shift $\Delta N = 2\alpha l \, \Delta T / \lambda$. For the dimensions given in Problem 1.15 and taking $\alpha \approx 10^{-5}/°C$ (the coefficient for steel) and $\Delta T \approx 0.01°C$, show that the resulting shift is $\Delta N \approx 0.2$, much larger than the expected shift due to the earth's motion. Obviously a successful experiment requires careful temperature control.

CHAPTER

2

The Space and Time of Relativity

2.1 The Postulates of Relativity

We have seen that the classical ideas of space and time had led to two conclusions:

1. The laws of Newtonian mechanics hold in an entire family of reference frames, any one of which moves uniformly relative to any other.

and

2. There can be only one reference frame in which light travels at the same speed c in all directions (and, more generally, in which all laws of electromagnetism are valid).

The Michelson–Morley experiment and numerous other experiments in the succeeding hundred years have shown that the second conclusion is false. Light travels with speed c in all directions in many different reference frames.

Einstein's special theory of relativity is based on the acceptance of this fact.* Einstein proposed two postulates, or axioms, expressing his conviction that *all* physical laws, including mechanics *and* electromagnetism, should be valid in an entire family of reference frames. From these two postulates he developed his special theory of relativity.

Before we state the two postulates of relativity it is convenient to expand the definition of an **inertial frame** to be any reference frame in which *all* the laws of physics hold:

> An inertial frame is any reference frame (that is, system of coordinates x, y, z and time t) where all the laws of physics hold in their simplest form.

Notice that we have not yet said what "all the laws of physics" are; to a large extent Einstein used his postulates to *deduce* what the correct laws of physics could be. It turns out that one of the laws that survives from classical physics into relativity is Newton's first law. Thus our newly defined inertial frames are in fact the familiar "unaccelerated" frames where a body on which no forces act moves with constant velocity. As before, a reference frame anchored to the earth is an inertial frame (to the extent that we ignore the small accelerations due to the earth's rotation and orbital motion); a reference frame fixed to a rapidly rotating turntable is not an inertial frame.

Notice also that in defining an inertial frame we have specified that the laws of physics must hold "in their simplest form." This is because one can sometimes modify physical laws so that they hold in noninertial frames as well. For example, by introducing a "fictitious" centrifugal force, one can arrange that the laws of statics are valid in a rotating frame. It is to exclude this kind of modification that we have added the qualification "in their simplest form."

The first postulate of relativity asserts that there is a whole family of inertial frames:

> ### FIRST POSTULATE OF RELATIVITY
> If S is an inertial frame and if a second frame S' moves with constant velocity relative to S, then S' is also an inertial frame.

We can reword this postulate to say that the laws of physics are *invariant* as we change from one reference frame to a second frame moving uniformly relative

ALBERT EINSTEIN (1879–1955, German–Swiss–US). Like all scientific theories, relativity was the work of many people. Nevertheless, Einstein's contributions outweigh those of anyone else by so much that the theory is quite properly regarded as his. As we shall see in Chapter 5, he also made fundamental contributions to quantum theory, and it was for these that he was awarded the 1921 Nobel Prize.

---

* Whether Einstein actually knew about the Michelson–Morley result when he developed relativity has been a subject of controversy. Until recently the majority view has been that he probably did not, and that he was motivated by purely theoretical arguments. This view was based on the absence of any explicit reference to Michelson and Morley in his paper of 1905 and on his statement in 1954 that "I even do not remember if I knew of it at all when I wrote my first paper." However, the text of a speech made in 1922 has recently been published, in which he said: "While I was thinking of this problem in my student years, I came to know the strange result of Michelson's experiment. Soon I came to the conclusion that our idea about the motion of the earth with respect to the ether is incorrect, if we admit Michelson's null result as a fact. This was the first path which led me to the special theory of relativity." (See *Physics Today,* August 1982, p. 45.) It now seems probable that Einstein was motivated both by the Michelson–Morley result and by the philosophical reasonableness of his theory. Of course, none of these considerations affect the validity and usefulness of relativity as we use it today; nor do they affect the importance of the Michelson–Morley experiment as clear evidence (for us, today) in favor of Einstein's theory.

to the first. This property is familiar from classical mechanics, but in relativity it is postulated for *all* the laws of physics.

The first postulate is often paraphrased as follows: "There is no such thing as absolute motion." To understand what this means, consider a frame S' attached to a rocket moving at constant velocity relative to a frame S anchored to the earth. The question we wish to ask is this: Is there any scientific sense in which we can say that S' is really moving and that S is really stationary (or, perhaps, the other way around)? If the answer were "yes," we could say that S is absolutely at rest and that anything moving relative to S is in absolute motion. However, the first postulate of relativity guarantees that this is impossible: All laws observable by an earth-bound scientist in S are equally observable by a scientist in the rocket S'; any experiment that can be performed in S can be performed equally in S'. Thus no experiment can possibly show which frame is *really* moving. Relative to the earth, the rocket is moving; relative to the rocket, the earth is moving; and this is as much as we can say.

Yet another way to express the first postulate is to say that among the family of inertial frames, all moving relative to one another, there is no *preferred frame*. That is, physics singles out no particular inertial frame as being in any way more special than any other frame.

The second postulate identifies one of the laws that holds in all inertial frames:

> **SECOND POSTULATE OF RELATIVITY**
> In all inertial frames, light travels through the vacuum with the same speed, $c = 299{,}792{,}458$ m/s, in any direction.

This postulate is, of course, the formal expression of the Michelson–Morley result. We can say briefly that it asserts the *universality of the speed of light c.*

The second postulate flies in the face of our normal experience. Nevertheless, it is now a firmly established experimental fact. As we explore the consequences of the two postulates of relativity, we are going to encounter several unexpected effects that may be difficult to accept at first. All of these effects (including the second postulate itself) have the subtle property that they become important only when bodies travel at speeds reasonably close to the speed of light. Under ordinary conditions, at normal terrestrial speeds, these effects simply do not show up. In this sense, none of the surprising consequences of Einstein's relativity really contradict our everyday experience.

2.2 Measurement of Time

Before we begin exploring the consequences of the relativity postulates we need to say a word about the measurement of time. We are going to find that the time of an event may be different when measured from different frames of reference. This being the case, we must first be quite sure we know what we mean by measurement of time in a single frame.

It is implicit in the second postulate of relativity, with its reference to the speed of light, that we can measure distances and times. In particular, we take for granted that we have access to several accurate clocks. These clocks need not all be the same; but when they are all brought to the same point in the same inertial frame and are properly synchronized, they must of course agree.

Consider, now, a single inertial frame S, with origin O and axes x, y, z. We imagine an observer sitting at O and equipped with one of our clocks. Using her clock the observer can easily time any event, such as a small explosion, in the immediate proximity of O, since she will see (or hear) the event the moment it occurs. To time an event far away from O is harder, since the light (or sound) from the event has to travel to O before our observer can sense it. To avoid this complication, we let our observer hire a large number of helpers, each of whom she equips with an accurate clock and assigns to a fixed, known position in the coordinate system S, as shown in Fig. 2.1. Once the helpers are in position, she can check that their clocks are still synchronized by having each helper send a flash of light at a predetermined time; since light travels with the known speed c (second postulate), she can calculate the time for the light to reach O and hence check the setting of the helper's clock.

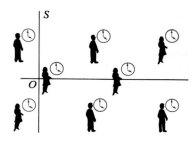

FIGURE 2.1 The chief observer at O distributes her helpers, each with an identical clock, throughout S.

With enough helpers, stationed closely enough together, we can be sure there is a helper sufficiently close to any event to time it effectively instantaneously. Once he has timed it, he can, at his leisure, inform everyone else of the result by any convenient means (by telephone, for example). In this way any event can be assigned a time t, as measured in the frame S.

When we speak of an inertial frame S, we shall always have in mind a system of axes $Oxyz$ and a team of observers who are stationed at rest throughout S and equipped with synchronized clocks. This allows us to speak of the position $\mathbf{r} = (x, y, z)$ and the time t of any event, relative to the frame S.

2.3   The Relativity of Time; Time Dilation

We are now ready to compare measurements of times made by observers in two different inertial frames. To this end we imagine the familiar two frames, S anchored to the ground and S' anchored to a train moving at constant velocity $\mathbf{v}$ relative to the ground. We consider a "thought experiment" in which an observer at rest on the train ignites a flash bulb on the floor of the train, vertically below a mirror mounted on the roof, a height h above. As seen in the frame S' (fixed in the train) a pulse of light travels straight up to the mirror, is reflected straight back, and returns to its starting point on the floor. We can imagine a photocell arranged to give an audible "beep" as the light returns. Our object is to find the time, as measured in either frame, between the two events—the "flash" as the light leaves the floor and the "beep" as it returns.

Our experiment, as seen in the frame S', is shown in Fig. 2.2(a). Since S' is an inertial frame, light travels the total distance $2h$ at speed c. Therefore, the time for the entire trip is

$$\Delta t' = \frac{2h}{c}. \tag{2.1}$$

This is the time that an observer in frame S' will measure between the flash and the beep, provided, of course, that his clock is reliable.

The same experiment, as seen from the inertial frame S, is shown in Fig. 2.2(b). In this frame the light travels along the two sides AB and BC of the triangle shown. If we denote by Δt the time for the entire journey, as measured in S, the time to go from A to B is $\Delta t/2$. During this time the train travels a distance $v\,\Delta t/2$, and the light, moving with speed c, travels a distance $c\,\Delta t/2$. (Note that here is where the postulates of relativity come in; we have taken the

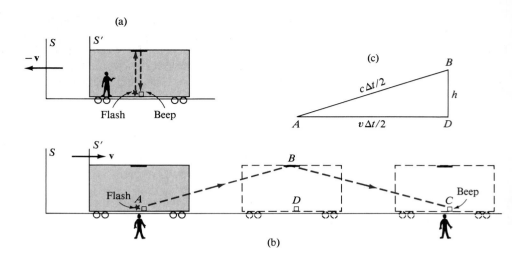

FIGURE 2.2 **(a)** The thought experiment as seen in the train-based frame S'. **(b)** The same experiment as seen from the ground-based frame S. Notice that two observers are needed in this frame. **(c)** The dimensions of the triangle ABD.

speed of light to be c in both S and S'.) The dimensions of the right triangle ABD are therefore as shown in Fig. 2.2(c). Applying Pythogoras's theorem, we see that*

$$\left(\frac{c\,\Delta t}{2}\right)^2 = h^2 + \left(\frac{v\,\Delta t}{2}\right)^2$$

or, solving for Δt,

$$\Delta t = \frac{2h}{\sqrt{c^2 - v^2}} = \frac{2h}{c}\frac{1}{\sqrt{1 - \beta^2}} \tag{2.2}$$

where we have again introduced the ratio

$$\beta = \frac{v}{c}$$

of the speed v to the speed of light c. The time Δt is the time that observers in S will measure between the flash and the beep (provided, again, that their clocks are reliable).

The most important and surprising thing about the two answers (2.1) and (2.2) is that they are not the same. The time between the two events, the flash and the beep, is different as measured in the frames S and S'. Specifically,

$$\Delta t = \frac{\Delta t'}{\sqrt{1 - \beta^2}}. \tag{2.3}$$

We have derived this result for an imagined thought experiment involving a flash of light reflected back to a photocell. However, the conclusion applies to *any* two events that occur at the same place on the train: Suppose, for instance, that we drop a knife on the table and a moment later drop a fork. In principle, at least, we could arrange for a flash of light to occur at the moment the knife lands, and we could position a mirror to reflect the light back to arrive just as the fork lands. The relation (2.3) must, then, apply to these two events (the landing of

---

* Here we are taking for granted, that the height h of the train is the same as measured in either frame, S or S'. We shall prove that this is correct in Section 2.5.

the knife and the landing of the fork). Now, the falling of the knife and fork cannot be affected by the presence or absence of a flashbulb and photocell; thus neither of the times Δt nor $\Delta t'$ can depend on whether or not we actually did the experiment with the light and the photocell. Therefore, the relation (2.3) holds for *any* two events that occur at the same place on board the train.

The difference between the measured times Δt and $\Delta t'$ is a direct consequence of the second postulate of relativity. (In classical physics $\Delta t = \Delta t'$, of course.) You should avoid thinking that the clocks in one of our frames must somehow be running wrong; quite the contrary, it was an essential part of our argument that all the clocks were running right. Moreover, our argument made no reference to the kind of clocks used (apart from requiring that they be correct). Thus the difference (2.3) applies to *all* clocks. In other words, *time itself* as measured in the two frames is different. We shall discuss the experimental evidence for this surprising conclusion shortly.

Several properties of the relationship (2.3) deserve comment. First, if our train is actually at rest ($v = 0$), then $\beta = 0$ and (2.3) tells us that $\Delta t = \Delta t'$. That is, there is no difference unless the two frames are in relative motion. Further, at normal terrestrial speeds, $v \ll c$ and $\beta \ll 1$; thus the difference between Δt and $\Delta t'$ is very small.

EXAMPLE 2.1   The pilot of a jet traveling at a steady 300 m/s sets a buzzer in the cockpit to go off at intervals of exactly 1 hour (as measured on the plane). What would be the interval between two successive buzzes as measured by two observers suitably positioned on the ground? (Ignore effects of the earth's motion; that is, consider the ground to be an inertial frame.)

The required interval between two buzzes is given by (2.3), with $\Delta t' = 1$ hour and $\beta = v/c = 10^{-6}$. Thus

$$\Delta t = \frac{\Delta t'}{\sqrt{1 - \beta^2}} = \frac{1 \text{ hour}}{\sqrt{1 - 10^{-12}}}$$

$$\approx (1 \text{ hour}) \times (1 + \tfrac{1}{2} \times 10^{-12})$$

$$= 1.0000000000005 \text{ hours.}$$

The difference between the two measured times is 5×10^{-13} hour or 1.8 nanoseconds. (A nanosecond, or ns, is 10^{-9} s.) It is easy to see why classical physicists had failed to notice this kind of difference!

The difference between Δt and $\Delta t'$ gets bigger as v increases. In modern particle accelerators it is common to have electrons and other particles with speeds of $0.99c$ and more. If we imagine repeating our thought experiment with the frame S' attached to an electron with $\beta = 0.99$, then (2.3) gives

$$\Delta t = \frac{\Delta t'}{\sqrt{1 - (0.99)^2}} \approx 7\Delta t'.$$

Differences as large as this are routinely observed by particle physicists, as we discuss in the next section.

If we were to put $v = c$ (that is, $\beta = 1$) in Eq. (2.3), we would get the absurd result $\Delta t = \Delta t'/0$; and if we put $v > c$ (that is, $\beta > 1$), we would get an imaginary answer. These ridiculous results indicate that v must always be less than c:

$$v < c.$$

This is one of the most profound results of Einstein's relativity: *The speed of any inertial frame relative to any other inertial frame must always be less than c.* In other words, the speed of light, in addition to being the same in all inertial frames, emerges as the universal speed limit for the relative motion of inertial frames.

The factor $1/\sqrt{1 - \beta^2}$ that appears in Eq. (2.3) crops up in so many relativistic formulas that it is traditionally given its own symbol, γ:

$$\gamma = \frac{1}{\sqrt{1 - \beta^2}} = \frac{1}{\sqrt{1 - (v/c)^2}}. \tag{2.4}$$

Since v is always smaller than c, the denominator in (2.4) is always less than or equal to 1 and hence

$$\gamma \geq 1. \tag{2.5}$$

The factor γ equals 1 only if $v = 0$. The larger we make v, the larger γ becomes; and as v approaches c, the value of γ increases without limit.

In terms of γ, Eq. (2.3) can be rewritten

$$\Delta t = \gamma \Delta t' \geq \Delta t'; \tag{2.6}$$

that is, Δt is always greater than or equal to $\Delta t'$. This asymmetry may seem surprising, and even to violate the postulates of relativity, since it suggests a special role for the frame S'. In fact, however, this is just as it should be. In our experiment the frame S' *is* special, since it is the unique inertial frame where the two events—the flash and the beep—occurred *at the same place*. This symmetry was implicit in Fig. 2.2, which showed *one* observer measuring $\Delta t'$ (since both events occurred at the same place in S') but *two* observers measuring Δt (since the two events were at different places in S). To emphasize this asymmetry, the time $\Delta t'$ can be renamed Δt_0 and (2.6) rewritten as

$$\Delta t = \gamma \Delta t_0 \geq \Delta t_0. \tag{2.7}$$

The subscript 0 on Δt_0 indicates that Δt_0 is the time indicated by a clock that is at rest in the special frame where the two events occurred at the same place. This time is often called the **proper time** between the events. The time Δt is measured in *any* frame and is always greater than or equal to the proper time Δt_0. For this reason, the effect embodied in (2.7) is often called **time dilation.**

The proper time Δt_0 is the time indicated by the clock on the moving train (moving relative to S, that is); Δt is the time shown by the clocks at rest on the ground in frame S. Since $\Delta t_0 \leq \Delta t$, the relation (2.7) can be loosely paraphrased to say that *"a moving clock is observed to run slow."*

Finally, we should reemphasize the fundamental symmetry between any two inertial frames. We chose to conduct our thought experiment with the flash and beep at one spot on the train (frame S'), and we found that $\Delta t > \Delta t'$.

However, we could have done things the other way around: If a ground-based observer (at rest in S) had performed the same experiment with a flash of light and a mirror, the flash and beep would have occurred in the same spot on the ground; and we would have found that $\Delta t' \geq \Delta t$. The great merit of writing the time-dilation formula in the form (2.7), $\Delta t = \gamma \Delta t_0$, is that it avoids the problem of remembering which is frame S and which S'; the subscript 0 always identifies the proper time, as measured in the frame in which the two events were at the same spot.

2.4   Evidence for Time Dilation

In his original paper on relativity Einstein predicted the effect that is now called time dilation. At that time there was no evidence to support the prediction, and many years were to pass before any was forthcoming. In fact, it is only recently, with the advent of atomic clocks, that direct verification using man-made clocks has become possible.

The first such test was carried out in 1971. Four portable atomic clocks were synchronized with a reference clock at the U.S. Naval Observatory in Washington, D.C., and all four clocks were then flown around the world on a jet plane and returned to the Naval Observatory. The discrepancy between the reference clock and the portable clocks after their journey was predicted (using relativity) to be

$$275 \pm 21 \text{ ns} \tag{2.8}$$

while the observed discrepancy (averaged over the four portable clocks) was*

$$273 \pm 7 \text{ ns}. \tag{2.9}$$

We should mention that the excellent agreement between (2.8) and (2.9) is more than a test of the time difference (2.7), predicted by special relativity. Gravitational effects, which require general relativity, contribute a large part of the predicted discrepancy (2.8). Thus this beautiful experiment is a confirmation of general, as well as special, relativity.

Much simpler tests of time dilation, and tests involving much larger dilations, are possible if one is prepared to use the natural clocks provided by unstable subatomic particles. For example, the charged π meson, or pion, is a particle that is formed in collisions between rapidly moving atomic nuclei (as we discuss in detail in Chapter 14). The pion has a definite average lifetime, after which it "decays" or disintegrates into other subatomic particles, and one can use this average life as a kind of natural clock.

One way to characterize the life span of an unstable particle is the **half-life**† $t_{1/2}$, the time after which half of a large sample of the particles in question will have decayed. For example, the half-life of the pion is measured to be

$$t_{1/2} = 1.8 \times 10^{-8} \text{ s}. \tag{2.10}$$

---

* The test was actually carried out twice—once flying east and once west—with satisfactory agreement in both cases. The results quoted here are from the more decisive westward flight. For more details, see J. C. Hafele and R. E. Keating, *Science,* vol. 177, p. 166 (1972). Since the accuracy of this original experiment has been questioned, we should emphasize that the experiment has been repeated many times, with improved accuracy, and there is now no doubt at all that the observations support the predictions of relativity.

† An alternative characterization is the mean life τ, which differs from $t_{1/2}$ by a constant factor. We shall define both of these more carefully in Chapter 13.

This means that if one starts at $t = 0$ with N_0 pions, then after 1.8×10^{-8} s half of them will have decayed and only $N_0/2$ will remain. After a further 1.8×10^{-8} s, half of those $N_0/2$ will have decayed and only $N_0/4$ will remain. After another 1.8×10^{-8} s, only $N_0/8$ will remain. And so on. In general, after n half-lives, $t = nt_{1/2}$, the number of particles remaining will be $N_0/2^n$.

At particle-physics laboratories pions are produced in large numbers in collisions between protons (the nuclei of hydrogen atoms) and various other nuclei. It is usually convenient to conduct experiments with the pions at a good distance from where they are produced, and the pions are therefore allowed to fly down an evacuated pipe to the experimental area. At the Fermilab near Chicago the pions are produced traveling very close to the speed of light, a typical value being

$$v = 0.9999995c,$$

and the distance they must travel to the experimental area is about $L = 1$ km. Let us consider the flight of these pions, first from the (incorrect) classical view with no time dilation, and then from the (correct) relativistic view.

As seen in the laboratory, the pions' time of flight is

$$T = \frac{L}{v} \approx \frac{10^3 \text{m}}{3 \times 10^8 \text{ m/s}} = 3.3 \times 10^{-6} \text{ s.} \qquad (2.11)$$

A classical physicist, untroubled by any notions of relativity of time, would compare this with the half-life (2.10) and calculate that

$$T \approx 183 t_{1/2}.$$

That is, the time needed for the pions to reach the experimental area is 183 half-lives. Therefore, if N_0 is the original number of pions, the number to survive the journey would be

$$N = \frac{N_0}{2^{183}} \approx (8.2 \times 10^{-56})N_0$$

and for all practical purposes, *no* pions would reach the experimental area. This would obviously be an absurd way to do experiments with pions, and is not what actually happens.

In relativity, we now know, times depend on the frame in which they are measured, and we must consider carefully the frames to which the times T and $t_{1/2}$ refer. The time T in (2.11) is, of course, the time of flight of the pions as measured in a frame fixed in the laboratory, the *lab frame.* To emphasize this we rewrite (2.11) as

$$T(\text{lab frame}) = 3.3 \times 10^{-6} \text{ s.} \qquad (2.12)$$

On the other hand, the half-life $t_{1/2} = 1.8 \times 10^{-8}$ s refers to time as *"seen" by the pions;* that is, $t_{1/2}$ is the half-life measured in a frame anchored to the pions, the pions' **rest frame.** (This is an experimental fact: The half-lives quoted by physicists are the proper half-lives, measured in the frame where the particles are at rest.) To emphasize this we write (temporarily)

$$t_{1/2}(\pi \text{ rest frame}) = 1.8 \times 10^{-8} \text{ s.} \qquad (2.13)$$

We see that the classical argument above used two times, T and $t_{1/2}$, measured in *different* inertial frames. A correct argument must work consistently in one frame, for example the lab frame. The half-life measured in the lab frame is given by the time-dilation formula as γ times the half-life (2.13). With $\beta = 0.9999995$ it is easy to see that

$$\gamma = 1000$$

and hence that

$$t_{1/2}(\text{lab frame}) = \gamma t_{1/2}(\pi \text{ rest frame})$$
$$= 1000 \times (1.8 \times 10^{-8} \text{ s})$$
$$= 1.8 \times 10^{-5} \text{ s}. \tag{2.14}$$

Comparing (2.12) and (2.14) we see that

$$T(\text{lab frame}) \approx 0.2 t_{1/2}(\text{lab frame}).$$

That is, the pions' flight down the pipe lasts only one-fifth of the relevant half-life. In this time very few of the pions decay, and almost all reach the experimental area. (The number that survive is $N = N_0/2^{0.2} \approx 0.9N_0$.) That this is exactly what actually happens in all particle-physics laboratories is strong evidence for the relativity of time, as first predicted by Einstein in 1905.

EXAMPLE 2.2 The lambda particle (Λ) is an unstable subatomic particle that decays into a proton and a pion ($\Lambda \rightarrow p + \pi$) with a half-life of $t_{1/2} = 1.7 \times 10^{-10}$ s. If several lambdas are created in a nuclear collision, all with speed $v = 0.6c$, how far will they travel before half of them decay?

The half-life as measured in the laboratory is $\gamma t_{1/2}$ (since $t_{1/2}$ is the proper half-life, as measured in the Λ rest frame). Therefore, the desired distance is $v\gamma t_{1/2}$. With $\beta = 0.6$,

$$\gamma = \frac{1}{\sqrt{1 - \beta^2}} = 1.25$$

and the required distance is

distance $= v\gamma t_{1/2} = (1.8 \times 10^8 \text{ m/s}) \times 1.25 \times (1.7 \times 10^{-10} \text{ s}) = 3.8$ cm.

Notice how even with speeds as large as $0.6c$, the factor γ is not much larger than 1, and the effect of time dilation is not dramatic. Notice also that a distance of a few centimeters is much easier to measure than a time of order 10^{-10} s; thus measurement of the range of an unstable particle is often the easiest way to find its half-life.

2.5 Length Contraction

The postulates of relativity have led us to conclude that time depends on the reference frame in which it is measured. We can now use this fact to show that the same must apply to distances: The measured distance between two events depends on the frame relative to which it is measured. We shall show this with another thought experiment. In the analysis of this thought experiment it will be important to recognize that, even in relativity, the familiar kinematic relation

$$\text{distance} = \text{velocity} \times \text{time}$$

FIGURE 2.3 **(a)** As seen in S the train moves a distance $v \, \Delta t$ to the right. **(b)** As seen in S', the frame S and observer Q move a distance $v \, \Delta t'$ to the left.

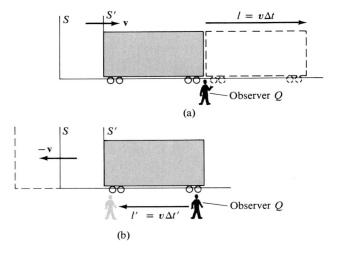

(a)

(b)

is valid in any given inertial frame (with all quantities measured in that frame), since it is just the definition of velocity in that frame.

We imagine again our two frames, S fixed to the ground and S' fixed to a train traveling at velocity $\mathbf{v}$ relative to the ground; and we now imagine observers in S and S' measuring the length of the train. For an observer in S' this measurement is easy since he sees the train at rest and can take all the time he needs to measure the length l' with an accurate ruler. For an observer Q on the ground the measurement is harder since the train is moving. Perhaps the simplest procedure is to time the train as it passes Q [Fig. 2.3(a)]. If t_1 and t_2 are the times at which the front and back of the train pass Q and if $\Delta t = t_2 - t_1$, then Q can calculate the length l (measured in S) as

$$l = v \, \Delta t. \tag{2.15}$$

To compare this answer with l' we note that observers on the train could have measured l' by a similar procedure. As seen from the train the observer Q on the ground is moving to the left with speed* v, and observers on the train can measure the time for Q to move from the front to the back of the train as in Fig. 2.3(b). (This would require two observers on the train, one at the front and one at the back.) If this time is $\Delta t'$, then

$$l' = v \, \Delta t'. \tag{2.16}$$

Comparing (2.15) and (2.16), we see immediately that since the times Δt and $\Delta t'$ are different, the same must be true of the lengths l and l'. To calculate the difference we need to relate Δt and $\Delta t'$ using the time-dilation formula. In the present experiment the two events of interest, "Q opposite the train's front" and "Q opposite the train's back," occur at the same place in S (where Q is at rest). Therefore, the time-dilation formula implies that

$$\Delta t' = \gamma \, \Delta t.$$

Comparing (2.15) and (2.16) we see that

$$l = \frac{l'}{\gamma} \le l'. \tag{2.17}$$

---

* We are taking for granted that the speed of S relative to S' is the same as that of S' relative to S. This follows from the basic symmetry between S and S' as required by the postulates of relativity.

The length of the train as measured in S is less than (or equal to) that measured in S'.

Like time dilation, this result is asymmetric, reflecting the asymmetry of our experiment: The frame S' is special since it is the unique frame where the measured object (the train) is at rest. [We could, of course, have done the experiment the other way around; if we had measured the length of a house at rest in S, the roles of l and l' in (2.17) would have been reversed.] To emphasize this asymmetry, and to avoid confusion as to which frame is which, it is a good idea to rewrite (2.17) as

$$l = \frac{l_0}{\gamma} \leq l_0, \qquad\qquad (2.18)$$

where the subscript 0 indicates that l_0 is the length of an object *measured in its rest frame,* while l refers to the length measured in *any* frame. The length l_0 can be called the object's **proper length.** Since $l \leq l_0$, the effect implied by (2.18) is often called **length contraction** (or Lorentz contraction, or Lorentz–Fitzgerald contraction, after the two physicists who first suggested that there must be some such effect). The effect can be loosely described by saying that *a moving object is observed to be contracted.*

EVIDENCE FOR LENGTH CONTRACTION

Like time dilation, length contraction is a real effect that is well established experimentally. Perhaps the simplest evidence comes from the same experiment as that discussed in connection with time dilation, in which unstable pions fly down a pipe from the collision that produces them to the experimental area. As viewed from the lab frame, we saw that time dilation increases the pions' half-life by a factor of γ, from $t_{1/2}$ to $\gamma t_{1/2}$. In the example discussed it was this increase that allowed most of the pions to complete the journey to the experimental area before they decayed.

Suppose, however, that we viewed the same experiment from the pions' rest frame. In this frame the pions are stationary and there is no time dilation to increase their half-life. So how do they reach the experimental area? The answer is that in this frame the *pipe* is moving, and length contraction reduces its length by the same factor γ, from L to L/γ. Thus observers in this frame would say it is length contraction that allows the pions to reach the experimental area. Naturally, the number of pions completing the journey is the same whichever frame we use for the calculation.

EXAMPLE 2.3 A space explorer of a future era travels to the nearest star, Alpha Centauri, in a rocket with speed $v = 0.9c$. The distance from earth to the star, as measured from earth, is $L = 4$ light years (or $4\,c \cdot$ years). What is this distance as seen by the explorer, and how long will she say the journey to the star lasts?

The distance $L = 4\,c \cdot$ years is the proper distance between earth and the star (which we assume are relatively at rest). Thus the distance as seen from the rocket is given by the length-contraction formula as

$$L(\text{rocket frame}) = \frac{L(\text{earth frame})}{\gamma}.$$

If $\beta = 0.9$, then $\gamma = 2.3$, so

$$L(\text{rocket frame}) = \frac{4\,c \cdot \text{years}}{2.3} = 1.7\ c \cdot \text{years}.$$

We can calculate the time T for the journey in two ways: As seen from the rocket, the star is initially $1.7\ c \cdot$ years away and is approaching with speed $v = 0.9c$. Therefore,

$$T(\text{rocket frame}) = \frac{L(\text{rocket frame})}{v}$$

$$= \frac{1.7\ c \cdot \text{years}}{0.9c} = 1.9 \text{ years}. \tag{2.19}$$

(Notice how the factors of c conveniently cancel when we use $c \cdot$ years and measure speeds as multiples of c.)

Alternatively, as measured from the earth frame the journey lasts for a time

$$T(\text{earth frame}) = \frac{L(\text{earth frame})}{v} = \frac{4\,c \cdot \text{years}}{0.9c} = 4.4 \text{ years};$$

but because of time dilation this is γ times $T(\text{rocket frame})$, which is therefore

$$T(\text{rocket frame}) = \frac{T(\text{earth frame})}{\gamma} = 1.9 \text{ years},$$

in agreement with (2.19), of course.

Notice how time dilation (or length contraction) allows an appreciable saving to the pilot of the rocket. If she returns promptly to earth, then as a result of the complete round trip she will have aged only 3.8 years, while her twin who stayed behind will have aged 8.8 years. This surprising result, sometimes known as the **twin paradox,** is amply verified by the experiments discussed in Section 2.4. In principle, time dilation would allow explorers to make in one lifetime trips that would require hundreds of years as viewed from earth. Since this requires rockets that travel very close to the speed of light, it is not likely to happen soon! See Problem 2.4 for further discussion of this effect.

LENGTHS PERPENDICULAR TO THE RELATIVE MOTION

We have so far discussed lengths that are parallel to the relative velocity, such as the length of a train in its direction of motion. What happens to lengths perpendicular to the relative velocity, such as the height of the train? It is fairly easy to show that for such lengths there is no contraction or expansion. To see this, consider two observers, Q at rest in S and Q' at rest in S', and suppose that Q and Q' are equally tall when at rest. Now, let us assume for a moment that there *is* a contraction of heights analogous to the length contraction (2.18). If this is so, then as seen by Q, Q' will be shorter as he rushes by. We can test this hypothesis by having Q' hold up a sharp knife exactly level with the top of his head; if Q' is shorter, Q will find himself scalped (or worse) as the knife goes by.

This experiment is completely symmetric between the two frames S and

S': There is one observer at rest in each frame and the only difference is the direction in which each sees the other moving.* Therefore, it must also be true that as seen by Q', it is Q who is shorter. But this implies that the knife will *miss* Q. Since it cannot be true that Q is both scalped and not scalped, we have arrived at a contradiction, and there can be no contraction. By a similar argument there can be no expansion and, in fact, the knife held by Q' simply grazes past Q's scalp, as seen in either frame. We conclude that lengths perpendicular to the relative motion are unchanged; and the Lorentz-contraction formula (2.18) applies only to lengths parallel to the relative motion.

2.6 The Lorentz Transformation

We are now ready to answer an important general question: If we know the coordinates x, y, z and time t of an event, as measured in a frame S, how can we find the coordinates x', y', z', and t' of the same event as measured in a second frame S'? Before we derive the correct relativistic answer to this question, we examine briefly the classical answer.

We consider our usual two frames, S anchored to the ground and S' anchored to a train traveling with velocity $\mathbf{v}$ relative to S, as shown in Fig. 2.4. Because the laws of physics are all independent of our choice of origin and orientation, we are free to choose both axes Ox and $O'x'$ along the same line, parallel to $\mathbf{v}$, as shown. We can further choose the origins of time so that $t = t' = 0$ at the moment when O' passes O. We shall sometimes refer to this arrangement of systems S and S' as the *standard configuration*.

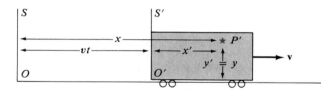

FIGURE 2.4 In classical physics the coordinates of an event are related as shown.

Now consider an event, such as the explosion of a small firecracker, that occurs at position x, y, z and time t as measured in S. Our problem is to calculate, in terms of x, y, z, t, the coordinates x', y', z', t' of the same event, as measured in S' —accepting at first the classical ideas of space and time. First, since time is a universal quantity in classical physics, we know that $t' = t$. Next, from Fig. 2.4 it is easily seen that $x' = x - vt$ and $y' = y$ (and similarly, $z' = z$, although the z coordinate is not shown in the figure). Thus, according to the ideas of classical physics,

$$
\begin{aligned}
x' &= x - vt \\
y' &= y \\
z' &= z \\
t' &= t.
\end{aligned}
\tag{2.20}
$$

---

* Note that our previous two thought experiments were asymmetric, requiring two observers in one of the frames, but only one in the other.

These four equations are often called the **Galilean transformation.** They *transform* the coordinates x, y, z, t of any event as observed in S into the corresponding coordinates x', y', z', t' as observed in S'.

If we had been given the coordinates x', y', z', t' and wanted to find x, y, z, t, we could solve the equations (2.20) to give

$$\begin{aligned} x &= x' + vt' \\ y &= y' \\ z &= z' \\ t &= t'. \end{aligned} \qquad (2.21)$$

Notice that the equations (2.21) can be obtained directly from (2.20) by exchanging x, y, z, t with x', y', z', t' and replacing v by $-v$. This is because the relation of S to S' is the same as that of S' to S except for a change in the sign of the relative velocity.

The Galilean transformation (2.20) cannot be the correct relativistic relation between x, y, z, t, and x', y', z', t'. (For instance, we know from time dilation that the equation $t' = t$ cannot possibly be correct.) On the other hand, the Galilean transformation agrees perfectly with our everyday experience and so must be correct (to an excellent approximation) when the speed v is small compared to c. Thus the correct relation between x, y, z, t and x', y', z', t' will have to reduce to the Galilean relation (2.20) when v/c is small.

To find the correct relation between x, y, z, t and x', y', z', t', we consider the same experiment as before, which is shown again in Fig. 2.5. We have noted before that distances perpendicular to $\mathbf{v}$ are the same whether measured in S or S'. Thus

$$y' = y \quad \text{and} \quad z' = z \qquad (2.22)$$

exactly as in the Galilean transformation. In finding x' it is important to keep careful track of the frames in which the various quantities are measured; in addition, it is helpful to arrange that the explosion whose coordinates we are discussing produces a small burn mark on the wall of the train at the point P' where it occurs. The horizontal distance from the origin O' to the mark at P', as measured in S', is precisely the desired coordinate x'. Meanwhile, the same distance, as measured in S, is $x - vt$ (since x and vt are the horizontal distances from O to P' and O to O' at the instant t, as measured in S). Thus according to the length-contraction formula (2.18),

$$x - vt = \frac{x'}{\gamma}$$

or

$$x' = \gamma(x - vt). \qquad (2.23)$$

This gives x' in terms of x and t and is the third of our four required equations. Notice that if v is small, then $\gamma \approx 1$ and the relation (2.23) reduces to the first of the Galilean relations (2.20), as required.

FIGURE 2.5 The coordinate x' is measured in S'. The distances x and vt are measured at the same time t in the frame S.

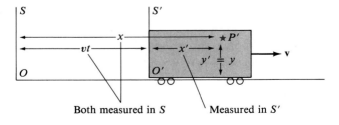

Both measured in S Measured in S'

Finally, to find t' in terms of x, y, z, t, we use a simple trick. We can repeat the argument leading to (2.23) but with the roles of S and S' reversed. That is, we let the explosion burn a mark at the point P on a wall fixed in S and, arguing as before, we find that

$$x = \gamma(x' + vt'). \qquad (2.24)$$

[This can be obtained directly from (2.23) by exchanging x, t with x', t' and replacing v by $-v$.] Equation (2.24) is not yet the desired result, but we can combine it with (2.23) to eliminate x' and find t'. Inserting (2.23) in (2.24), we get

$$x = \gamma[\gamma(x - vt) + vt'].$$

Solving for t' we find that

$$t' = \gamma t - \frac{\gamma^2 - 1}{\gamma v}\, x,$$

or, after some algebra (Problem 2.19),

$$t' = \gamma\left(t - \frac{vx}{c^2}\right). \qquad (2.25)$$

This is the required expression for t' in terms of x and t. When v/c is much smaller than 1, we can neglect the second term and since $\gamma \approx 1$, we get $t' \approx t$, in agreement with the Galilean transformation, as required.

Collecting together (2.22), (2.23), and (2.25), we obtain our required four equations:

$$
\begin{aligned}
x' &= \gamma(x - vt) \\
y' &= y \\
z' &= z \\
t' &= \gamma\left(t - \frac{vx}{c^2}\right).
\end{aligned}
\qquad (2.26)
$$

These equations are called the **Lorentz transformation,** or **Lorentz–Einstein transformation,** in honor of the Dutch physicist Lorentz, who first proposed them, and Einstein, who first interpreted them correctly. The Lorentz transformation is the correct relativistic modification of the Galilean transformation (2.20).

If one wants to know x, y, z, t in terms of x', y', z', t' one can simply exchange the primed and unprimed variables and replace v by $-v$, in the now familiar way, to give

$$
\begin{aligned}
x &= \gamma(x' + vt') \\
y &= y' \\
z &= z' \\
t &= \gamma\left(t' + \frac{vx'}{c^2}\right).
\end{aligned}
\qquad (2.27)
$$

These equations are sometimes called the *inverse* Lorentz transformation.

The Lorentz transformation expresses all of the properties of space and time that follow from the postulates of relativity. From it one can calculate all of the kinematic relations between measurements made in different inertial frames. In the next two sections we give some examples of such calculations.

HENDRIK LORENTZ (1853–1928, Dutch). Lorentz was the first to write down the equations we now call the Lorentz transformation, although Einstein was the first to interpret them correctly. He also preceded Einstein with the length contraction formula (though, again, he did not interpret it correctly). He was one of the first to suggest that electrons are present in atoms, and his theory of electrons earned him the 1902 Nobel Prize.

2.7   Applications of the Lorentz Transformation

In this section we give three examples of problems that can easily be analyzed using the Lorentz transformation. In the first two we rederive two familiar results; in the third we analyze one of the many "paradoxes" of relativity.

EXAMPLE 2.4   Starting with the equations (2.26) of the Lorentz transformation, derive the length-contraction formula (2.18).

Notice that the length-contraction formula was used in our derivation of the Lorentz transformation. Thus this example will not give a new proof of length contraction; it will, rather, be a consistency check on the Lorentz transformation, to verify that it gives back the result from which it was derived. However, one can also take the view that the Lorentz transformation is itself a well-established experimental fact, from which one can legitimately derive the length-contraction formula.

Let us imagine, as before, measuring the length of a train (frame S') traveling at speed v relative to the ground (frame S). If the coordinates of the back and front of the train are x_1' and x_2', as measured in S', then the train's proper length (its length as measured in its rest frame) is

$$l_0 = l' = x_2' - x_1'. \tag{2.28}$$

To find the length l as measured in S, we carefully position two observers on the ground to observe the coordinates x_1 and x_2 of the back and front of the train at some convenient time t. (These two measurements must, of course, be made at the same time t.) In terms of these coordinates, the length l as measured in S is (Fig. 2.6)

$$l = x_2 - x_1.$$

Now, consider the following two events, with their coordinates as measured in S:

Event	Description	Coordinates in S
1	Back of train passes first observer	x_1, t_1
2	Front of train passes second observer	$x_2, t_2 = t_1$

FIGURE 2.6 If the two observers measure x_1 and x_2 at the same time ($t_1 = t_2$), then $l = x_2 - x_1$.

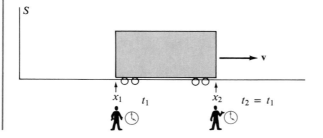

We can use the Lorentz transformation to calculate the coordinates of each event as observed in S':

Event	Coordinates in S'
1	$x_1' = \gamma(x_1 - vt_1)$
2	$x_2' = \gamma(x_2 - vt_2)$

(We have not listed the times t_1' and t_2' since they don't concern us here.) The difference of these coordinates is

$$x_2' - x_1' = \gamma(x_2 - x_1). \qquad (2.29)$$

(Notice how the times t_1 and t_2 cancel out since they are equal.) Since the two differences in (2.29) are respectively $l' = l_0$ and l, we conclude that $l_0 = \gamma l$ or

$$l = \frac{l_0}{\gamma},$$

as required.

EXAMPLE 2.5 Use the Lorentz transformation to rederive the time-dilation formula (2.7).

In our discussion of time dilation we considered two events, a flash and a beep, that occurred at the same place in frame S',

$$x_{\text{flash}}' = x_{\text{beep}}'.$$

The proper time between the two events was the time as measured in S',

$$\Delta t_0 = \Delta t' = t_{\text{beep}}' - t_{\text{flash}}'.$$

To relate this to the time

$$\Delta t = t_{\text{beep}} - t_{\text{flash}}$$

as measured in S, it is convenient to use the inverse Lorentz transformation (2.27), which gives

$$t_{\text{beep}} = \gamma\left(t_{\text{beep}}' + \frac{vx_{\text{beep}}'}{c^2}\right)$$

and

$$t_{\text{flash}} = \gamma\left(t_{\text{flash}}' + \frac{vx_{\text{flash}}'}{c^2}\right).$$

If we take the difference of these two equations, the coordinates x_{beep}' and x_{flash}' drop out (since they are equal) and we get the desired result,

$$\Delta t = t_{\text{beep}} - t_{\text{flash}} = \gamma(t_{\text{beep}}' - t_{\text{flash}}') = \gamma\,\Delta t_0.$$

EXAMPLE 2.6 A relativistic snake of proper length 100 cm is moving at speed $v = 0.6c$ to the right across a table. A mischievous boy, wishing to tease the snake, holds two hatchets 100 cm apart and plans to bounce them simultaneously on the table so that the left hatchet lands immediately behind the snake's tail. The boy argues as follows: "The snake is moving with $\beta = 0.6$. Therefore, its length is contracted by a factor

$$\gamma = \frac{1}{\sqrt{1 - \beta^2}} = \frac{1}{\sqrt{1 - 0.36}} = \frac{5}{4},$$

and its length (as measured in my rest frame) is 80 cm. This implies that the right hatchet will fall 20 cm in front of the snake, and the snake will be unharmed." (The boy's view of the experiment is shown in Fig. 2.7.) On the other hand, the snake argues thus: "The hatchets are approaching me with $\beta = 0.6$, and the distance between them is contracted to 80 cm. Since I am 100 cm long, I shall be cut in pieces when they fall." Use the Lorentz transformation to resolve this paradox.

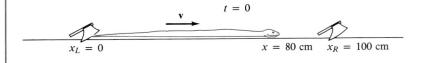

FIGURE 2.7 As seen in the boy's frame S, the two hatchets bounce simultaneously (at $t = 0$) 100 cm apart; since the snake is 80 cm long, it escapes injury.

Let us choose two coordinate frames as follows: The snake is at rest in frame S' with its tail at the origin $x' = 0$ and its head at $x' = 100$ cm. The two hatchets are at rest in frame S, the left one at the origin $x = 0$ and the right one at $x = 100$ cm.

As observed in frame S, the two hatchets bounce simultaneously at $t = 0$. At this time the snake's tail is at $x = 0$ and his head must therefore be at $x = 80$ cm. [You can check this using the transformation $x' = \gamma(x - vt)$; with $x = 80$ cm and $t = 0$ you will find that $x' = 100$ cm, as required.] Thus, as observed in S, the experiment is as shown in Fig. 2.7. In particular, the boy's prediction is correct and the snake is unharmed. Therefore, the snake's argument must be wrong.

To understand what is wrong with the snake's argument we must examine the coordinates, especially the times, at which the two hatchets bounce, as observed in the frame S'. The left hatchet falls at $t_L = 0$ and $x_L = 0$. According to the Lorentz transformation (2.26), the coordinates of this event, as seen in S', are

$$t'_L = \gamma\left(t_L - \frac{vx_L}{c^2}\right) = 0$$

and

$$x'_L = \gamma(x_L - vt_L) = 0.$$

As expected, the left hatchet falls immediately beside the snake's tail, at time $t'_L = 0$, as shown in Fig. 2.8(a).

FIGURE 2.8 As observed in S', both hatchets are moving to the left. The right hatchet falls before the left one; even though the hatchets are only 80 cm apart, this lets them fall at positions that are 125 cm apart.

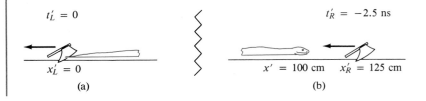

$t'_L = 0$

$x'_L = 0$

(a)

$t'_R = -2.5$ ns

$x' = 100$ cm $x'_R = 125$ cm

(b)

On the other hand, the right hatchet falls at $t_R = 0$ and $x_R = 100$ cm. Thus as seen in S' it falls at a time given by the Lorentz transformation as

$$t'_R = \gamma\left(t_R - \frac{vx_R}{c^2}\right) = \frac{5}{4}\left(0 - \frac{(0.6c)\times(100\text{ cm})}{c^2}\right) = -2.5\text{ ns}.$$

We see that, as measured in S', the two hatchets *do not fall simultaneously.* Since the right hatchet falls before the left one, it does not necessarily have to hit the snake, even though they were only 80 cm apart (in this frame). In fact, the position at which the right hatchet falls is given by the Lorentz transformation as

$$x'_R = \gamma(x_R - vt_R) = \tfrac{5}{4}(100\text{ cm} - 0) = 125\text{ cm},$$

and indeed the hatchet misses the snake, as shown in Fig. 2.8(b).

The resolution of this paradox, and many similar paradoxes, is seen to be that two events which are simultaneous as observed in one frame are not necessarily simultaneous when observed in a different frame. As soon as one recognizes that the two hatchets fall at different times in the snake's rest frame, there is no longer any problem understanding how they can both miss the snake.

2.8   The Velocity-Addition Formula

In Chapter 1 we discussed the classical velocity-addition formula. This relates the velocity **u** of some body or signal, relative to a frame S, and its value **u**′ relative to a second frame S': $\mathbf{u} = \mathbf{u}' + \mathbf{v}$, or equivalently,

$$\mathbf{u}' = \mathbf{u} - \mathbf{v} \tag{2.30}$$

Here **v** is the velocity of S' relative to S, and the formula asserts that in classical physics, relative velocities add and subtract like vectors. Notice that here, as elsewhere, we use **u** and **u**′ for the velocities of a body or signal relative to the two frames, while **v** denotes the relative velocity of the frames themselves.

As we saw in Chapter 1, the classical formula (2.30) cannot be correct, since it contradicts the universality of the speed of light. In this section we use the Lorentz transformation to derive the correct relativistic velocity-addition formula.

Let us imagine some moving object whose velocity we wish to discuss. (For example, this object could be a space rocket, a subatomic particle, or a signal of light.) We consider two neighboring points on its path, as in Fig. 2.9. We denote by $\mathbf{r}_1 = (x_1, y_1, z_1)$ and $\mathbf{r}_2 = (x_2, y_2, z_2)$ the coordinates of these two points, as measured in S, and by t_1 and t_2 the times at which the object passes them. The velocity $\mathbf{u} = (u_x, u_y, u_z)$ as measured in S is then given by

$$u_x = \frac{\Delta x}{\Delta t}, \qquad u_y = \frac{\Delta y}{\Delta t}, \qquad u_z = \frac{\Delta z}{\Delta t} \tag{2.31}$$

where $\Delta x = x_2 - x_1$, and so on (and these equations may, strictly speaking, be valid only in the limit that the two points are close together, $\Delta t \to 0$). The velocity **u**′ relative to S' is defined in the same way using the coordinates and times measured in S'.

We can now use the Lorentz transformation to relate the coordinates and

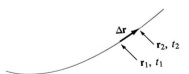

FIGURE 2.9 The velocity of an object is $\mathbf{u} = \Delta\mathbf{r}/\Delta t$.

times of S to those of S', and then, using definition (2.31), relate the corresponding velocities. First, according to the Lorentz transformation (2.26)

$$x_2' = \gamma(x_2 - vt_2), \qquad y_2' = y_2, \qquad t_2' = \gamma\left(t_2 - \frac{vx_2}{c^2}\right)$$

and

$$x_1' = \gamma(x_1 - vt_1), \qquad y_1' = y_1, \qquad t_1' = \gamma\left(t_1 - \frac{vx_1}{c^2}\right).$$

(We omit the equations for z, which transforms just like y.) Subtracting these equations, we find that

$$\Delta x' = \gamma(\Delta x - v\,\Delta t), \qquad \Delta y' = \Delta y, \qquad \Delta t' = \gamma\left(\Delta t - \frac{v\,\Delta x}{c^2}\right).$$

From these we can calculate the components of $\mathbf{u}'$. First

$$u_x' = \frac{\Delta x'}{\Delta t'} = \frac{\gamma(\Delta x - v\,\Delta t)}{\gamma(\Delta t - v\,\Delta x/c^2)}$$

or, canceling the factors of γ and dividing top and bottom by Δt,

$$u_x' = \frac{u_x - v}{1 - u_x v/c^2}. \tag{2.32}$$

Similarly,

$$u_y' = \frac{\Delta y'}{\Delta t'} = \frac{\Delta y}{\gamma(\Delta t - v\,\Delta x/c^2)}$$

or, dividing the top and bottom by Δt,

$$u_y' = \frac{u_y}{\gamma(1 - u_x v/c^2)}. \tag{2.33}$$

Notice that u_y' is not equal to u_y, even though $\Delta y' = \Delta y$; this is because the times $\Delta t'$ and Δt are unequal.

Equations (2.32) and (2.33), with a corresponding equation for u_z', are the **relativistic velocity-addition formulas,** or velocity transformation. Notice that if both u and v are much less than c, we can ignore the term $u_x v/c^2$ in both denominators and put $\gamma \approx 1$ to give

$$u_x' \approx u_x - v$$

and

$$u_y' \approx u_y.$$

These are, of course, the components of the classical addition formula $\mathbf{u}' = \mathbf{u} - \mathbf{v}$.

The inverse velocity transformation, giving $\mathbf{u}$ in terms of $\mathbf{u}'$, can be obtained from (2.32) and (2.33) by exchanging primed and unprimed variables and replacing v by $-v$, in the familiar way.

EXAMPLE 2.7 A rocket traveling at speed $0.8c$ relative to the earth shoots forward a beam of particles with speed $0.9c$ relative to the rocket. What is the particles' speed relative to the earth?

Let S be the rest frame of the earth and S' that of the rocket, with x and x' axes both aligned along the rocket's velocity. The relative speed of the two frames is $v = 0.8c$. We are given that the particles are traveling along the x' axis with speed $u' = 0.9c$ (relative to S'), and we want to find their speed u relative to S. The classical answer is, of course, that $u = u' + v = 1.7c$; that is, because the two velocities are collinear, u' and v simply add in classical physics.

The correct relativistic answer is given by the inverse of (2.32) (from which we omit the subscripts x, since all velocities are along the x axis):

$$u = \frac{u' + v}{1 + u'v/c^2} \tag{2.34}$$

$$= \frac{0.9c + 0.8c}{1 + (0.9 \times 0.8)} = \frac{1.7}{1.72} c$$

$$\approx 0.99c.$$

The striking feature of this answer is that when we "add" $u' = 0.9c$ to $v = 0.8c$ relativistically we get an answer that is less than c. In fact, it is fairly easy to show that for any value of u' which is less than c, the speed u is also less than c (see Problem 2.27); that is, a particle whose speed is less than c in one frame has speed less than c in any other frame.

EXAMPLE 2.8 The rocket of Example 2.7 shoots forward a signal (for example, a pulse of light) with speed c relative to the rocket. What is the signal's speed relative to the earth?

In this case $u' = c$. Thus according to (2.34),

$$u = \frac{u' + v}{1 + u'v/c^2} = \frac{c + v}{1 + v/c} = c. \tag{2.35}$$

That is, anything that travels at the speed of light in one frame does the same as observed from any other frame. (We have proved this here only for the case that **u** is in the same direction as **v**. However, the result is true for any direction; for another example, see Problem 2.28.) We can paraphrase this to say that the speed of light is invariant as we pass from one inertial frame to another. This is, of course, just the second postulate of relativity, which led us to the Lorentz transformation in the first place.

2.9 The Doppler Effect (optional)

It is well known in classical physics that the frequency (and hence pitch) of sound changes if either the source or receiver is put into motion, a phenomenon known as the **Doppler effect**. There is a similar Doppler effect for light: The frequency (and hence color) of light is changed if the source and receiver are put into relative motion. However, the Doppler effect for light differs from that for sound in two important ways. First, since light travels at speed c, one should treat the phenomenon relativistically, and the relativistic Doppler formula differs from its nonrelativistic form because of time dilation. Second, you will probably recall that the nonrelativistic Doppler formula for sound has different forms according to whether the source or receiver is moving. This difference is perfectly reasonable for sound that propagates in air, so that the Doppler shift can legitimately depend on whether it is the source or receiver that moves through the air. On the other hand, we know that light (in the vacuum) has no medium in which it propagates; and in fact relativity has shown us that it can make no difference whether it is the source or receiver that is "really" moving. Thus the Doppler formula for light must be the same for a moving source as for a moving receiver. In this respect the Doppler formula for light is simpler than its nonrelativistic counterpart for sound.

The derivation of the relativistic Doppler formula is very similar to the nonrelativistic argument. Let us consider a train (frame S') that is traveling at speed v relative to the ground (frame S) and whose headlamp is a source of light with frequency f_{source}, as measured in the rest frame of the source. We now imagine an observer Q at rest on the ground in front of the train and wish to find the frequency f_{obs} of the light that Q observes.

We first consider the experiment as seen in frame S, in which it is the source that is moving. In Fig. 2.10(a) and (b) we have shown two successive wave crests (numbered 1 and 2) as they leave the train's headlight. If the time (as measured in S) between the emission of these crests is Δt, then during this time the first crest will move a distance $c \, \Delta t$ to the right. During this same time the train will advance a distance $v \, \Delta t$, and the distance λ between successive crests is therefore

$$\lambda = c \, \Delta t - v \, \Delta t. \tag{2.36}$$

As seen from S, the distance between successive crests is shortened by $v \, \Delta t$ as a result of the train's motion. Since the crests are all approaching with speed c and are a distance λ apart, the frequency f_{obs} with which the observer Q receives them is

$$f_{obs} = \frac{c}{\lambda} = \frac{c}{(c - v) \, \Delta t} = \frac{1}{(1 - \beta) \, \Delta t} \tag{2.37}$$

where, as usual, $\beta = v/c$.

FIGURE 2.10 As seen in frame S the source moves with speed v and the receiver Q is at rest.

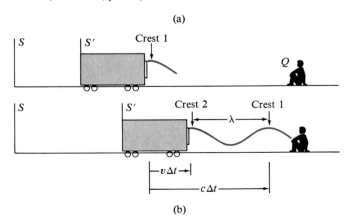

Now, Δt is the time between emission of successive crests *as measured in* S. The corresponding time $\Delta t'$ as measured in S' is the proper time between the two events, since they occur at the same place in S'. Therefore,

$$\Delta t = \gamma \, \Delta t' \tag{2.38}$$

and, from (2.37),

$$f_{obs} = \frac{1}{(1 - \beta)\gamma \, \Delta t'}. \tag{2.39}$$

Finally, the frequency f_{source} measured at the source is just $f_{source} = 1/\Delta t'$. Thus (2.39) implies that

$$f_{obs} = \frac{f_{source}}{(1 - \beta)\gamma} \qquad \text{(approaching)} \tag{2.40}$$

where we have added the parenthesis "(approaching)" to emphasize that this formula applies when the source is approaching the observer.

The relativistic formula (2.40) differs from its nonrelativistic counterpart only by the factor of $1/\gamma$, which arose from the time dilation (2.38). In particular, for slow speeds, with $\gamma \approx 1$, we can use the nonrelativistic formula $f_{obs} = f_{source}/(1 - \beta)$, as one might expect.

It is often convenient to rewrite (2.40), replacing the factor of $1/\gamma$ by

$$\frac{1}{\gamma} = \sqrt{1 - \beta^2} = \sqrt{(1 - \beta)(1 + \beta)}, \tag{2.41}$$

to give

$$f_{obs} = \sqrt{\frac{1 + \beta}{1 - \beta}} f_{source} \qquad \text{(approaching)}. \tag{2.42}$$

We shall see shortly that this formula, which we have so far derived only for a moving source, in fact holds whether it is the source or observer that is moving.

The formula (2.42) applies to a source that is approaching the observer. If the source is moving away from the observer, we have only to change the sign of v to give

$$f_{obs} = \sqrt{\frac{1 - \beta}{1 + \beta}} f_{source} \qquad \text{(receding)}. \tag{2.43}$$

The formulas (2.42) and (2.43) are easy to memorize. In particular, it is easy to remember which is which, since both numerator and denominator lead to the expected *rise* in frequency when source and observer are approaching each other, and the expected *drop* when they are receding.

We have here analyzed only the cases that the source moves directly toward or away from the observer. The case that the source moves obliquely to the observer is more complicated and is discussed in Problem 2.33.

An important example of the Doppler effect is the famous "redshift" of the light from distant stars. A star emits and absorbs light at certain frequencies that are characteristic of the elements in the star. Thus by analyzing the spectrum of light from a star one can identify which elements it contains. Once these elements are identified one can go further. By seeing whether the characteristic frequencies are shifted up or down (as compared to those from a source at rest in the observatory) one can tell whether the star is moving toward or away from us.

The American astronomer Hubble found that the light from distant galaxies is shifted *down* in frequency, or "redshifted" (since red is at the low frequency end of the visible spectrum), indicating that most galaxies are moving away from us. Hubble also found that the speeds of recession of galaxies are roughly proportional to their distances from us. This implied that the universe is expanding uniformly; it also provided a convenient way to find the distance of many galaxies, since measurement of a Doppler shift is usually much easier than the direct measurement of a distance.

EXAMPLE 2.9 It is found that light from a distant galaxy is shifted down in frequency (redshifted) by a factor of 3; that is, $f_{obs}/f_{source} = \frac{1}{3}$. Is the galaxy approaching us or receding? And what is its speed?

Since the observed frequency f_{obs} is less than f_{source}, the galaxy is receding, and we use (2.43) to give

$$\sqrt{\frac{1-\beta}{1+\beta}} = \frac{1}{3}$$

or, solving for β,

$$\beta = 0.8.$$

That is, the galaxy is receding from us at $0.8c$.

EXAMPLE 2.10 The atoms in hot sodium vapor give out light of wavelength $\lambda_{source} = 589$ nm (measured in the atoms' rest frame). Since atoms in a vapor move randomly with speeds up to 300 m/s and even higher, this light is observed with various different Doppler shifts, depending on the atoms' speeds and directions. Taking 300 m/s as the atoms' maximum speed, find the range of wavelengths observed.

The minimum and maximum frequencies observed come from atoms moving directly away from and toward the observer with speed $v = 300$ m/s or $\beta = 10^{-6}$. Since β is so small we can ignore the factor of γ in (2.40) and the extreme frequencies are given by

$$f_{obs} = \frac{f_{source}}{1 \pm \beta}.$$

Since $\lambda = c/f$ (both for source and observer), this implies maximum and minimum wavelengths given by

$$\lambda_{obs} = \lambda_{source}(1 \pm \beta).$$

We can write this as

$$\lambda_{obs} = \lambda_{source} \pm \Delta\lambda$$

where the maximum shift $\Delta\lambda$ in the wavelength is

$$\Delta\lambda = \beta\lambda_{source} = (10^{-6}) \times (589 \text{ nm}) \approx 6 \times 10^{-4} \text{ nm}.$$

This is a very small shift of wavelength, as we should have expected since v

is so small compared to c. Nevertheless, such a shift is easily observed with a good spectrometer. It means that what would otherwise be observed as a sharp spectral line with wavelength λ_{source} is smeared out between $\lambda_{source} \pm \Delta\lambda$. This phenomenon is called **Doppler broadening** and is one of the problems that has to be overcome in precise measurement of wavelengths.

We mentioned earlier that the relativistic Doppler shift for light must be the same whether we view the source as moving and the observer at rest, or vice versa. We check this in our final example:

EXAMPLE 2.11 Rederive the Doppler formula (2.42) working in the rest frame S' of the source (that is, taking the view that the observer is moving).

We consider again two successive wave crests, but examine their *reception* by the observer Q, as shown in Fig. 2.11. As measured in S', the distance between the two crests is the wavelength $\lambda' = c/f_{source}$ (since f_{source} is the frequency measured in S'), and the time between Q's meeting the crests we denote by $\Delta t'$. During the time $\Delta t'$, crest 2 moves a distance $c\,\Delta t'$ to the right and the observer Q moves a distance $v\,\Delta t'$ to the left. The sum of these two distances is just λ',

$$c\,\Delta t' + v\,\Delta t' = \lambda' = \frac{c}{f_{source}},$$

from which we find that

$$\Delta t' = \frac{c}{(c+v)f_{source}} = \frac{1}{(1+\beta)f_{source}}. \qquad (2.44)$$

Now, the frequency with which Q observes wave crests is

$$f_{obs} = \frac{1}{\Delta t} \qquad (2.45)$$

where Δt is the time between arrival of the two successive crests *as measured by the observer Q*. Since these two events occur at the same place in the

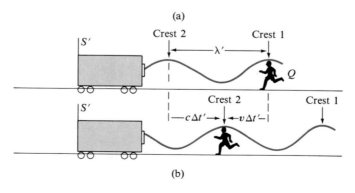

(a)

Crest 2

Crest 1

S'

—λ'—

Q

Crest 2

Crest 1

S'

$|$—$c\,\Delta t'$—$|$—$v\,\Delta t'$—$|$

(b)

FIGURE 2.11 As seen in S', the source is stationary with frequency f_{source} and the observer Q is moving at speed v to the left.

observer's frame S, Δt is the proper time and $\Delta t' = \gamma \, \Delta t$. Substituting into (2.45), and then using (2.44), we find that

$$f_{obs} = \frac{\gamma}{\Delta t'} = \gamma(1 + \beta)f_{source}. \tag{2.46}$$

Apart from the factor γ this is the nonrelativistic formula for a moving receiver. For our present purposes, the important point is that we can replace γ using (2.41) and, after a little algebra, we obtain exactly our previous answer (2.42), as you should check for yourself. As anticipated, the relativistic Doppler shift for light is the same for a moving observer as for a moving source, and depends only on their relative velocity v.

IDEAS YOU SHOULD NOW UNDERSTAND FROM CHAPTER 2

Inertial frames
The postulates of relativity
The universality of the speed of light
Time dilation
The speed limit for relative motion of inertial frames is c
Proper time

Length contraction
Proper length
The Galilean transformation
The Lorentz transformation
The relativistic velocity transformation
[Optional section: the Doppler effect]

PROBLEMS FOR CHAPTER 2

If you have not already done so, read the note preceding the Problems for Chapter 1. This explains the organization of problems and their classification according to difficulty.

SECTIONS 2.3 and 2.4 (THE RELATIVITY OF TIME AND EVIDENCE FOR TIME DILATION)

2.1 • An athlete runs the 100-meter dash at 10 m/s. How much will her watch gain or lose, as compared to ground-based clocks, during the race? [*Hint:* You will need to use the binomial approximation (1.6).]

2.2 • A space vehicle travels at 100,000 m/s (about 200,000 mi/h) relative to the earth. How much time will its clocks gain or lose, as compared to earth-based clocks, in a day?

2.3 • (*a*) What must be one's speed, relative to a frame S, in order that one's clocks will lose 1 second per day as observed from S? (*b*) What if they are to lose 1 minute per day?

2.4 • A space explorer sets off at a steady $v = 0.95c$ to a distant star. After exploring the star for a short time he returns at the same speed and gets home after a total absence of 80 years (as measured by earth-bound observers). How long do his clocks say that he was gone, and by how much has he aged? *Note:* This is the "twin paradox" discussed in Example 2.3. It is easy to get the right answer by judicious insertion of a factor γ in the right place, but to *understand* the result you need to recognize that it involves *three* inertial frames: the earth-bound frame S, the frame S' of the outward-bound rocket, and the frame S'' of the returning rocket. You can write down the time dilation formula (2.3) for each of the two halves of the journey, and add these to give the desired relation. (Notice that the experiment is not symmetrical between the explorer and his friends who stay behind on earth — the earth-bound clocks stay at rest in a single inertial frame, but the rocket's clock and crew occupy at least two different frames. This is what allows the result to be unsymmetrical.)

2.5 • When he returns his Hertz rent-a-rocket after one week's cruising in the galaxy, Spock is shocked to be billed for three weeks' rental. Assuming that he traveled straight out and then straight back, always at the same speed, how fast was he traveling? (See the note in Problem 2.4.)

2.6 •• (a) Use the binomial approximation (1.6) to prove the following useful approximation:

$$\gamma \approx 1 + \tfrac{1}{2}\beta^2$$

when $\beta \ll 1$. (b) Derive a corresponding approximation for $1/\gamma$. (c) When β is close to 1 (v close to c) these approximations are, of course, useless; in this case show that if $\beta = 1 - \varepsilon$, with $\varepsilon \ll 1$, then $\gamma \approx 1/\sqrt{2\varepsilon}$.

2.7 •• Two perfectly synchronized clocks A and B are at rest in S, a distance d apart. If we wanted to verify that they really are synchronized, we might try using a third clock, C. We could bring C close to A and check that A and C agree, then move C over to B and check the agreement of B and C. Unfortunately, this procedure is suspect since clock C will run differently while it is being moved. (a) Suppose that A and C are found to be in perfect agreement and that C is then moved at constant speed v to B. Derive an expression for the disagreement τ between B and C, in terms of v and d. What is τ if $v = 300$ m/s and $d = 1000$ km? (b) Show that the method can nevertheless be made satisfactory *to any desired accuracy* by moving clock C slowly enough; that is, we can make τ as small as we please by choosing v sufficiently small.

2.8 •• A group of π mesons is observed traveling at speed $0.8c$ in a particle-physics laboratory. (a) What is the factor γ for the pions? (b) If the pions' proper half-life is 1.8×10^{-8} s, what is their half-life as observed in the lab frame? (c) If there were initially 32,000 pions, how many will be left after they have traveled 36 m? (d) What would be the answer to (c) if one ignored time dilation?

2.9 •• Muons are subatomic particles that are produced several miles above the earth's surface as a result of collisions of cosmic rays (charged particles, such as protons, that enter the earth's atmosphere from space) with atoms in the atmosphere. These muons rain down more-or-less uniformly on the ground, although some of them decay on the way since the muon is unstable with a proper half-life of about 1.5 μs. (1 μs $= 10^{-6}$ s.) In a certain experiment a muon detector is carried in a balloon to an altitude of 2000 m, and in the course of 1 hour it registers 650 muons traveling at $0.99c$ toward the earth. If an identical detector remains at sea level, how many muons would you expect it to register in 1 hour? (Remember that after n half-lives the number of muons surviving from an initial sample of N_0 is $N_0/2^n$.)

2.10 ••• Time dilation implies that if a clock moves relative to a frame S, then careful measurements made by observers in S [as in Fig. 2.12(a), for example] will

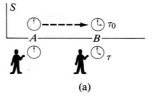

(a)

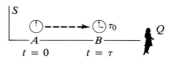

(b)

FIGURE **2.12** (Problem 2.10) (a) Two observers at rest in frame S at A and B time the moving clock as it passes them; they find the dilated time $\tau = \gamma\tau_0$. (b) The single observer Q sees the moving clock at A and B by means of light that has traveled different distances, AQ and BQ.

find that the clock runs slow. This is not at all the same thing as saying that a single observer in S will *see* the clock running slow; and the latter statement is, in fact, not always true. To understand this, remember that what we *see* is determined by the light as it arrives at our eyes. Consider the observer Q in Fig. 2.12(b) and suppose that as the clock moves from A to B it registers the passage of a time τ_0. As measured in S, the time between these two events ("clock at A" and "clock at B") is of course $\tau = \gamma\tau_0$. However, B is closer to Q than A is; thus light from the clock when at B will reach Q more quickly than did light from the clock when at A. Therefore, the time τ_{see} between Q's *seeing* the clock at A and *seeing* it at B is less than τ. (a) Prove that in fact

$$\tau_{\text{see}} = \tau(1 - \beta) = \tau_0 \sqrt{\frac{1 - \beta}{1 + \beta}}.$$

(Prove both equalities.) Since τ_{see} is less than τ_0, the observer Q actually sees the clock running fast. (b) What will Q see once the clock has passed her? That is, find the new value of τ_{see} when the clock is moving away from Q.

Your answers here are closely related to the Doppler effect discussed in Section 2.9. The moral of this problem is that one must be very careful how one states (and thinks about) time dilation. It is safe to say "moving clocks are observed to run slow" [where to "observe" means to "measure carefully" as in Fig. 2.12(a)], but it is certainly wrong to say "moving clocks are seen to run slow."

SECTION 5 (LENGTH CONTRACTION)

2.11 • A rocket of proper length 40 m is observed to be 32 m long as it rushes past the earth. What is its speed relative to the earth?

2.12 • A relativistic conveyor belt is moving at speed $0.5c$ relative to frame S. Two observers standing beside the belt, 10 ft apart as measured in S, arrange that each will paint a mark on the belt at exactly the same instant (as measured in S). How far apart will the marks be as measured by observers on the belt?

2.13 • A rigid spherical ball (rest frame S) is observed from a frame S' which travels with speed $0.5c$ relative to frame S. Describe the ball's shape as measured by observers in S'.

2.14 • Consider the experiment of Problem 2.8 from the point of view of the pions' rest frame. In part (c) how far (as "seen" by the pions) does the laboratory move, and how long does this take? How many pions remain at the end of this time?

2.15 •• A meterstick is moving with speed $0.8c$ relative to a frame S. (*a*) What is the stick's length, as measured by observers in S, if the stick is parallel to its velocity **v**? (*b*) What if the stick is perpendicular to **v**? (*c*) What if the stick is at $60°$ to **v**, as seen in the stick's rest frame? (*Hint:* You can imagine that the meterstick is the hypotenuse of a 30–60–90 triangle of plywood.) (*d*) What if the stick is at $60°$ to **v**, as measured in S?

2.16 •• Like time dilation, the Lorentz contraction cannot be *seen* directly (that is, perceived by the normal process of vision). To understand this claim, consider a rod of proper length l_0 moving relative to S. Careful measurements made by observers in S [as in Fig. 2.13(a), for example] will show that the rod has the contracted length $l = l_0/\gamma$. But now consider what is seen by observer Q in Fig. 2.13(b) (with Q to the right of points A and B). What Q sees at any one instant is determined by the light entering her eyes *at that instant*. Now, consider the light reaching Q at one instant from the front and back of the rod. (*a*) Explain why these two rays must have left the rod (from points A and B) at *different times*. If the x axis has a graduated scale as shown, Q sees (and a photograph would record) a rod extending from A to B; that is, Q sees a rod of length AB. (*b*) Prove that Q sees a rod that is *longer* than l. (In fact, at certain speeds it is even seen to be longer than l_0, and the Lorentz contraction is distorted into an expansion.) (*c*) Prove that once it has passed her, Q will see the rod to be shorter than l.

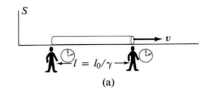

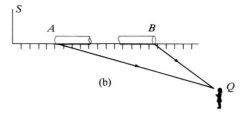

FIGURE 2.13 (Problem 2.16) **(a)** One can measure the Lorentz-contracted length $l = l_0/\gamma$ using two observers to record the positions of the front and back at the same instant. **(b)** What a single observer *sees* is determined by light that left the rod at different times.

SECTIONS 6 and 7 (THE LORENTZ TRANSFORMATION AND APPLICATIONS)

2.17 • The two frames S and S' are in the standard configuration (origins coincident at $t = t' = 0$, x and x' axes parallel, and relative velocity along Ox). Their relative speed is $0.5c$. An event occurs on the x axis at $x = 10$ light seconds (that is, $x = 3 \times 10^9$ m) at time $t = 4$ s in the frame S. What are its coordinates x', y', z', t' as measured in S'?

2.18 • The Lorentz transformation (2.26) consists of four equations giving x', y', z', t' in terms of x, y, z, t. Solve these equations to give x, y, z, t in terms of x', y', z', t'. Show that you get the same result by interchanging primed and unprimed variables and replacing v by $-v$.

2.19 • Give in detail the derivation of the Lorentz transformation (2.25) for t', starting from equations (2.23) and (2.24).

2.20 • Two inertial frames S and S' are in the standard configuration, with relative velocity **v** along the line of the x and x' axes. Consider any two events, 1 and 2. (*a*) From the Lorentz transformation (2.26) derive expressions for the separations $\Delta x'$, $\Delta y'$, $\Delta z'$, $\Delta t'$ (where $\Delta x' = x_2' - x_1'$, etc.) in terms of Δx, Δy, Δz, Δt. (Notice how the transformation of Δx, Δy, Δz, Δt is identical to that of x, y, z, t.) (*b*) If $\Delta x = 0$ and $\Delta t = 4$ s, whereas $\Delta t' = 5$ s, what is the relative speed v, and what is $\Delta x'$?

2.21 •• The frames S and S' are in the standard configuration with relative velocity $0.8c$ along Ox. (*a*) What are the coordinates (x_1, y_1, z_1, t_1) in S of an event that occurs on the x' axis with $x_1' = 1500$ m, $t_1' = 5$ μs? (b) Answer the same for a second event on the x' axis with $x_2' = -1500$ m, $t_2' = 10$ μs. (c) What are the time intervals (Δt and $\Delta t'$) between the two events, as measured in S and S'?

2.22 •• In a frame S two events have spatial separation $\Delta x = 600$ m, $\Delta y = \Delta z = 0$, and temporal separation $\Delta t = 1$ μs. A second frame S' is moving along Ox with nonzero speed v and $O'x'$ parallel to Ox. In S' it is found that the spatial separation $\Delta x'$ is also 600 m. What are v and $\Delta t'$?

2.23 •• Observers in a frame S arrange for two simultaneous explosions at time $t = 0$. The first explosion is at the origin ($x_1 = y_1 = z_1 = 0$) while the second is on the positive x axis 4 light years away ($x_2 = 4$ $c \cdot$years, $y_2 = z_2 = 0$). (a) Use the Lorentz transformation to find the coordinates of these two events as observed in a frame S' traveling in the standard configuration at speed $0.6c$ relative to S. (b) How far apart are the two events as measured in S'? (c) Are the events simultaneous as observed in S'?

2.24 •• A traveler in a rocket of length $2d$ sets up a coordinate system S' with origin O' anchored at the exact middle of the rocket and the x' axis along the rocket's length. At $t' = 0$ she ignites a flashbulb at O'. (a) Write down the coordinates x_F', t_F' and x_B', t_B' for the arrival of the light at the front and back of the rocket. (b) Now consider the same experiment as observed in a frame S relative to which the rocket is traveling at speed v (with S and S' arranged in the standard configuration). Use the Lorentz transformation to find the coordinates x_F, t_F and x_B, t_B of the arrival of the two signals. Explain clearly why the two times are not equal in frame S, although they were in S'. (This illustrates how two events that are simultaneous in S' are not necessarily simultaneous in S.)

SECTION 8 (THE VELOCITY-ADDITION FORMULA)

2.25 • A rocket (rest frame S') traveling at speed $v = 0.5c$ relative to the earth (rest frame S) shoots forward bullets traveling at speed $u' = 0.6c$ relative to the rocket. What is the bullets' speed u relative to the earth?

2.26 • As seen from earth (rest frame S) two rockets A and B are approaching in opposite directions, each with speed $0.9c$ relative to S. Find the velocity of rocket B as measured by the pilot of rocket A. (*Hint:* Consider a coordinate system S' traveling with rocket A; your problem is then to find the velocity $\mathbf{u}'$ of rocket B relative to S', knowing its velocity $\mathbf{u}$ relative to S.)

2.27 •• Using the velocity-addition formula one can prove the following important theorem: If a body's

speed u relative to an inertial frame S is less than c, its speed u' relative to any other inertial frame S' is also less than c. In this problem you will prove this result for the case that all velocities are in the x direction.

Suppose that S' is moving along the x axis of frame S with speed v. Suppose that a body is traveling along the x axis with velocity u relative to S. (We can let u be positive or negative, so that the body can be traveling either way.) (a) Write down the body's velocity u' relative to S'. For a fixed positive v (less than c, of course) sketch a graph of u' as a function of u in the range $-c < u < c$. (b) Hence prove that for any u with $-c < u < c$, it is necessarily true that $-c < u' < c$.

2.28 •• Suppose that as seen in a frame S, a signal (a pulse of light, for example) has velocity c along the y axis (that is, $u_x = u_z = 0$, $u_y = c$). (a) Write down the components of its velocity $\mathbf{u}'$ relative to a frame S' traveling in the standard configuration with speed v along the x axis of frame S. (b) In what direction is the signal traveling relative to S'? (c) Using your answer to part (a), calculate the magnitude of $\mathbf{u}'$.

SECTION 2.9 (THE DOPPLER EFFECT)

2.29 • It is found that the light from a nearby star is blue-shifted by 1%; that is, $f_{obs} = 1.01 f_{source}$. Is the star receding or approaching, and how fast is it traveling? (Assume that it is moving directly toward or away from us.)

2.30 • A star is receding from us at $0.5c$. What is the percent shift in the frequency of light received from the star?

2.31 • Consider the tale of the physicist who is ticketed for running a red light and argues that because he was approaching the intersection, the red light was Doppler shifted and appeared green. How fast would he have been going? ($\lambda_{red} \approx 650$ nm, $\lambda_{green} \approx 530$ nm.)

2.32 • In our discussion of the Doppler shift we found three superficially different expressions for the received frequency f_{obs}: (2.40), (2.42), and (2.46). Show in detail that all three are equal.

2.33 •• Consider a source of light of frequency f_{source} moving obliquely to an observer Q as in Fig. 2.14(a). (a) Prove that Q receives the light with frequency f_{obs} given by the general Doppler formula

$$f_{obs} = \frac{f_{source}}{(1 - \beta \cos \theta)\gamma}.$$

(b) Check that this formula reduces to our previous result (2.40) when the source is approaching Q head-on.

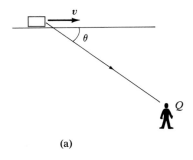

(a)

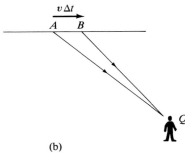

(b)

FIGURE 2.14 (Problem 2.33) (a) Light from the moving source to the observer Q makes an angle θ with velocity **v**. (b) If two successive wave crests are emitted at A and B, a time Δt apart, then AB is $v\,\Delta t$.

The analysis in part (a) is quite similar to that leading to (2.40) but the geometry is more complicated. Consider two successive wave crests emitted at points A and B as in Fig. 2.14(b). Since A and B are in practice very close together, the rays AQ and BQ are effectively parallel. Show that the difference between AQ and BQ is approximately $v\,\Delta t\cos\theta$ and hence that the distance between successive crests as they approach Q is $(c - v\cos\theta)\Delta t$. This is the appropriate generalization of (2.36), and from here the discussion is closely parallel.

CHAPTER
3
Relativistic Mechanics

3.1 Introduction

We have seen that the laws of classical mechanics held in a family of inertial frames that were related to one another by the classical, Galilean transformation; in other words, classical mechanics was invariant under the Galilean transformation. But we now know that the correct, relativistic transformation between inertial frames is the Lorentz, not the Galilean, transformation. It follows that the laws of classical mechanics cannot be correct and that we must find a new, relativistic mechanics that is invariant as we pass from one inertial frame to another using the correct Lorentz transformation.

We shall find that relativistic mechanics, like classical mechanics, is built around the concepts of mass, momentum, energy, and force. However, the relativistic definitions of these concepts are all a little different. In seeking these new definitions and the laws that connect them, we must be guided by three principles: First, a correct relativistic law must be valid in all inertial frames; it must be invariant under the Lorentz transformation. Second, we would expect the relativistic definitions and laws to reduce to their nonrelativistic counterparts when applied to systems moving much slower than the speed of light. Third, and most important, our relativistic laws must agree with experiment.

3.2 Mass in Relativity

We start by considering the mass m of an object—an electron, a space rocket, or a star. The most satisfactory definition of m turns out to be remarkably simple. We know that at slow speeds a suitable definition is the classical one (for example, $m = F/a$, where a is the acceleration produced by a standard force F). In relativity we simply agree to use the same, classical definition of m with the proviso that *before measuring any mass we bring the object concerned to rest.* To emphasize this qualification we shall sometimes refer to m as the **rest mass.** It can also be called the **proper mass** since it is the mass measured in the frame where the object is at rest.

Observers in different inertial frames all agree on the rest mass of an object. Suppose that observers in a frame S take some object, bring it to rest (in S), and measure its mass m. If they then pass the object to observers in a different frame S', who bring it to rest in S' and measure its mass m', it will be found that $m' = m$. That is, rest mass is *invariant* as we pass from S to S'. In fact, this is required by the postulates of relativity: If m' were different from m, we could define a preferred frame (the frame where the rest mass of an object was minimum, for example).

As we shall describe shortly, some physicists use a different definition of mass, called the *variable mass.* We shall not use this concept, and whenever we use the word "mass" without qualification we shall mean the invariant rest mass defined here.

3.3 Relativistic Momentum

The classical definition of the momentum of a single body is

$$\mathbf{p} = m\mathbf{u} \tag{3.1}$$

where m and $\mathbf{u}$ are the body's mass and velocity. Since we know how both m and $\mathbf{u}$ are measured in relativity, it is natural to start by asking the question: Is it perhaps the case that the classical definition (3.1) is the correct definition in relativity as well?

Strictly speaking, this question has no answer. There can be no such thing as a "correct," or "incorrect," definition of $\mathbf{p}$, since one is at liberty to *define* things however one pleases. The proper question is rather: Is the definition $\mathbf{p} = m\mathbf{u}$ a *useful* definition in relativity?

In classical mechanics the concept of momentum has many uses, but its single most useful property is probably the law of conservation of momentum. If we consider n bodies with momenta $\mathbf{p}_1 \ldots, \mathbf{p}_n$, then, in the absence of external forces, the total momentum

$$\sum \mathbf{p} = \mathbf{p}_1 + \cdots + \mathbf{p}_n$$

can never change. It would certainly be useful if we could find a definition of momentum such that this important law carried over into relativity. Accordingly, we shall try seeking a definition of relativistic momentum $\mathbf{p}$ with the following two properties:

1. $\mathbf{p} \approx m\mathbf{u}$ when $u \ll c$

and

2. The total momentum $\Sigma \mathbf{p}$ of an isolated system is constant, as measured in all inertial frames.

Requirement 1 is just that the relativistic definition should agree with the nonrelativistic one in the nonrelativistic domain. Requirement 2 is that the law of conservation of momentum must hold in all inertial frames if it holds at all.

If we were to adopt the classical definition $\mathbf{p} = m\mathbf{u}$, requirement 1 would be satisfied automatically. However, it is fairly easy to construct a thought experiment which illustrates that the classical definition $\mathbf{p} = m\mathbf{u}$ does not meet requirement 2.

Consider two identical billiard balls that collide as shown in Fig. 3.1. Relative to the frame S of Fig. 3.1(a) the initial velocities of the two balls are equal and opposite. Further, the collision is arranged symmetrically, such that the x components of the velocities are unchanged by the impact, whereas the y components reverse. It is an experimental fact that such collisions between particles of equal mass do occur. It is easy to check that if we adopt the classical definition, $\mathbf{p} = m\mathbf{u}$, then as seen in S this collision conserves momentum,

$$\sum m\mathbf{u}(\text{before}) = \sum m\mathbf{u}(\text{after}).$$

The proof that this is so is shown in Table 3.1, where we have denoted by a and b the x and y components of the initial velocity of ball 1. Notice that, as seen in S, the total classical momentum is actually zero before and after the collision; so it is certainly conserved.

Let us now consider the same collision from a second frame S' traveling in the positive x direction of S, at the same rate as ball 1 (that is, the speed v of S' relative to S is equal to the x component of the velocity of ball 1). Figure 3.1(b) shows the collision as seen in S', with ball 1 traveling straight up the y' axis and bouncing straight down again. If the frames S and S' were related by the classical, Galilean transformation (which we know is actually incorrect), the velocities measured in S' could be found from those in Table 3.1 using the classical velocity-addition formula, with the results shown in Table 3.2. From the last column we see that the total classical momentum has the same values before and after impact; that is, the total classical momentum is conserved in

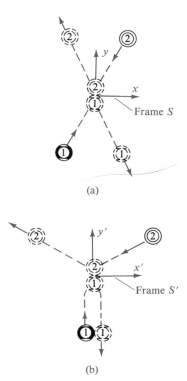

(a)

(b)

FIGURE 3.1 Two different views of a collision between two identical balls. **(a)** In frame S the velocities of the two balls are equal and opposite, before and after the impact. **(b)** The same experiment as seen in frame S', which travels along the x axis at the same rate as ball 1. In this frame ball 1 travels straight up the y' axis and back down again.

TABLE 3.1
The experiment of Fig. 3.1 as seen in S. Each pair of numbers represents the x and y components of the vector indicated.

	First Ball ($\mathbf{u}_1$)	Second Ball ($\mathbf{u}_2$)	$m\mathbf{u}_1 + m\mathbf{u}_2$
Before:	(a, b)	$(-a, -b)$	$(0, 0)$
After:	$(a, -b)$	$(-a, b)$	$(0, 0)$

TABLE 3.2
The experiment of Fig. 3.1 as seen in frame S', *assuming that S and S' are related by the Galilean transformation.* Each velocity here is obtained from that of Table 3.1 by subtracting a from its x component.

	First Ball ($\mathbf{u}_1'$)	Second Ball ($\mathbf{u}_2'$)	$m\mathbf{u}_1' + m\mathbf{u}_2'$
Before:	$(0, b)$	$(-2a, -b)$	$(-2ma, 0)$
After:	$(0, -b)$	$(-2a, b)$	$(-2ma, 0)$

the frame S' also. That is just what we should have expected, since we know that the laws of classical mechanics are invariant under the Galilean transformation.

In fact, however, the Galilean transformation is *not* the correct transformation between frames S and S', and we must compute the velocities in S' using the relativistic transformations (2.32) and (2.33). The results of this rather tedious calculation (Problem 3.3) are shown in Table 3.3. The details of this table are not especially interesting, but there are two important points: First, because the transformation of u_y depends on u_x, the y components of the two balls' velocities transform differently and, as seen in S', they are no longer equal in magnitude. (Compare the y components in the first and second columns.) Consequently, the y component of the total classical momentum (final column of Table 3.3) is positive before the collision and negative after. Thus, even though the total classical momentum *is* conserved in frame S, it is *not* conserved in frame S'. Therefore, the law of conservation of classical momentum (defined as $\mathbf{p} = m\mathbf{u}$ for each body) is incompatible with the postulates of relativity, and the classical definition of momentum does not satisfy our requirement 2 above.

TABLE 3.3
The experiment of Fig. 3.1 as seen in S', based on the correct relativistic velocity transformation from S to S'. The relative velocity of the two frames is $v = a$; accordingly, β denotes a/c and $\gamma = (1 - a^2/c^2)^{-1/2}$.

	First Ball ($\mathbf{u}_1'$)	**Second Ball ($\mathbf{u}_2'$)**	$m\mathbf{u}_1' + m\mathbf{u}_2'$
Before:	$\left(0, \dfrac{b}{\gamma(1-\beta^2)}\right)$	$\left(\dfrac{-2a}{1+\beta^2}, \dfrac{-b}{\gamma(1+\beta^2)}\right)$	$\left(\dfrac{-2ma}{1+\beta^2}, \dfrac{2mb\beta^2}{\gamma(1-\beta^4)}\right)$
After:	$\left(0, \dfrac{-b}{\gamma(1-\beta^2)}\right)$	$\left(\dfrac{-2a}{1+\beta^2}, \dfrac{b}{\gamma(1+\beta^2)}\right)$	$\left(\dfrac{-2ma}{1+\beta^2}, \dfrac{-2mb\beta^2}{\gamma(1-\beta^4)}\right)$

If there is a law of momentum conservation in relativity, the relativistic definition of momentum must be different from the classical one, $\mathbf{p} = m\mathbf{u}$. If we rewrite this classical definition as

$$\mathbf{p} = m\frac{d\mathbf{r}}{dt} \qquad \text{[classical]} \tag{3.2}$$

we get a useful clue for a better definition. The difficulty in our thought experiment originated in the complicated transformation of the velocity $d\mathbf{r}/dt$ (particularly the y component). These complications arose because both $d\mathbf{r}$ and dt change as we transform from S to S'. We can avoid some of this problem if we replace the derivative $d\mathbf{r}/dt$ in (3.2) by the derivative $d\mathbf{r}/dt_0$ with respect to the *proper time, t_0*, of the moving body:

$$\mathbf{p} = m\frac{d\mathbf{r}}{dt_0} \qquad \text{[relativistic].} \tag{3.3}$$

In (3.2), $d\mathbf{r}$ is the vector joining two neighboring points on the body's path, and dt is the time for the body to move between them — both as measured in any one inertial frame S. In (3.3), $d\mathbf{r}$ is the same as before, but dt_0 is the *proper time* between the two points; that is, the time as measured in the body's rest frame. From its definition, the proper time dt_0 (just like the proper mass m) has the same value for all observers in all frames. Thus the vector defined in (3.3) transforms more simply than the classical momentum (3.2), since only the numerator $d\mathbf{r}$ changes as we move from one frame to another. In particular, the y component, $p_y = m\, dy/dt_0$, does not change at all as we pass from S to S', and the difficulty encountered in our thought experiment would not occur if we were to adopt the definition (3.3). (For details, see Problem 3.4.)

At slow speeds dt and dt_0 are indistinguishable, and the new definition (3.3) agrees with the classical one; that is, the definition (3.3) meets requirement 1 above. Further, one can show that it always meets requirement 2; specifically, if the total momentum $\Sigma\, \mathbf{p}$, as defined by (3.3), is constant in one inertial frame S, the same is true in *all* inertial frames.* Since the proof is rather long, although reasonably straightforward, we leave it as a problem (Problem 3.10) at the end of this chapter.

With the definition (3.3) of momentum, the law of conservation of momentum would be logically consistent with the postulates of relativity. Whether or not momentum defined in this way *is* conserved must be decided by experiment. The unanimous verdict of innumerable experiments involving collisions of atomic and subatomic particles is that it *is:* If we adopt the definition (3.3) for the momentum of a body, the total momentum $\Sigma\, \mathbf{p}$ of an isolated system *is* conserved. Under the circumstances we naturally adopt (3.3) as our definition of momentum.

It is convenient to express the definition (3.3) a little differently: The time-dilation formula implies that $dt = \gamma\, dt_0$, where $\gamma = (1 - u^2/c^2)^{-1/2}$. Therefore, (3.3) is the same as

$$\mathbf{p} = m\,\frac{d\mathbf{r}}{dt_0} = m\gamma\,\frac{d\mathbf{r}}{dt} = \gamma m\mathbf{u}.$$

Thus we adopt as the final form of our definition:

> The momentum of a single body
> of mass m and velocity $\mathbf{u}$ is
>
> $$\mathbf{p} = \frac{m\mathbf{u}}{\sqrt{1 - u^2/c^2}} = \gamma m\mathbf{u}.$$

(3.4)

An important consequence of the factor γ in the momentum (3.4) is that no object can be accelerated past the speed of light: We shall find that in relativity, just as in classical mechanics, a constant force on a body increases its momentum p at a constant rate, and if the force acts for long enough, we can make p as large as we please. In classical mechanics, where $p = mu$, this means that a constant force steadily increases u and can eventually make u as large as we please. In relativity, an increase in $p = \gamma mu$ is reflected by increases in u *and* γ. Now, as $u \to c$ we know that γ increases without limit. Thus, as $u \to c$ the constant force keeps increasing γ without u ever reaching c. This difference between classical and relativistic mechanics is illustrated in Fig. 3.2, which shows a plot of u against p for both cases.

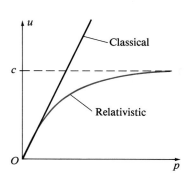

FIGURE 3.2 The speed of a body as a function of its momentum in classical and relativistic mechanics. At low speeds the two curves coincide. In classical mechanics u grows indefinitely as p increases; in relativity u never exceeds c, however large p becomes.

* The proposition as stated is a little oversimplified. What one can actually prove is this: If momentum, as defined by (3.3), and energy (whose definition we give in Section 3.4) are *both* conserved in one inertial frame, they are both automatically conserved in all inertial frames.

EXAMPLE 3.1 A 1-kg lump of metal is observed traveling with speed $0.4c$. What is its momentum? What would its momentum be if we doubled its speed? Compare with the corresponding classical values.

When $\beta = 0.4$, the factor $\gamma = (1 - \beta^2)^{-1/2}$ is easily calculated to be $\gamma = 1.09$, and

$$p = \gamma m u = 1.09 \times (1 \text{ kg}) \times (0.4 \times 3 \times 10^8 \text{ m/s})$$
$$= 1.31 \times 10^8 \text{ kg} \cdot \text{m/s}. \tag{3.5}$$

If we double the speed, then $\beta = 0.8$ and $\gamma = 1.67$. Thus the momentum becomes

$$p = \gamma m u = 1.67 \times (1 \text{ kg}) \times (0.8 \times 3 \times 10^8 \text{ m/s})$$
$$= 4.01 \times 10^8 \text{ kg} \cdot \text{m/s}, \tag{3.6}$$

which is more than three times the previous answer. The classical answers are found by omitting the factors of γ: If $\beta = 0.4$, then $p = 1.20 \times 10^8 \text{ kg} \cdot \text{m/s}$, just a little less than the correct answer (3.5); if $\beta = 0.8$, then $p = 2.40 \times 10^8$ kg·m/s, significantly less than the correct answer (3.6).

Some physicists like to think of the relativistic momentum as the product of γm and $\mathbf{u}$, which they write as

$$\mathbf{p} = m_{\text{var}} \mathbf{u} \tag{3.7}$$

where

$$m_{\text{var}} = \gamma m = \frac{m}{\sqrt{1 - u^2/c^2}}. \tag{3.8}$$

The quantity m_{var} is called the *variable mass* since, unlike the rest mass m, it varies with the body's speed u. The form (3.7) has the advantage of making the relativistic momentum *look* more like its nonrelativistic counterpart $\mathbf{p} = m\mathbf{u}$. On the other hand, it is not always a good idea to give two ideas the appearance of similarity when they are in truth different. Further, the introduction of the variable mass does not achieve a complete parallel with classical mechanics. For example, we shall see that the quantity $\frac{1}{2} m_{\text{var}} u^2$ is *not* the correct expression for the relativistic kinetic energy and the equation $\mathbf{F} = m_{\text{var}} \mathbf{a}$ is *not* the correct relativistic form of Newton's second law (Problems 3.9 and 3.31). For these reasons we shall not use the notion of variable mass in this book.

3.4 Relativistic Energy

Having found a suitable relativistic definition for momentum $\mathbf{p}$, our next task is to do the same for the energy E of a body. Just as with momentum, one is in principle free to define E however one pleases. But the hope of finding a *useful* definition suggests two requirements analogous to those used for momentum:

1. When applied to slowly moving bodies, the new defininition of E should reproduce as closely as possible the classical definition

and

2. The total energy ΣE of an isolated system of bodies should be conserved in all inertial frames.

The definition that fits these requirements turns out to be this:

> The energy of a single body of mass m, moving with speed u, is
> $$E = \frac{mc^2}{\sqrt{1 - u^2/c^2}} = \gamma mc^2.$$

(3.9)

It is important to note that this applies to any single body—an elementary particle, like an electron; an assembly of particles, like an atom; or an assembly of atoms, like a baseball, a space rocket, or a star.

Although we shall not do so here (but see Problem 3.10), one can prove that with the definition (3.9), a law of conservation of energy would be logically consistent with the postulates of relativity: If ΣE were constant as measured in one inertial frame, the same would be true in all inertial frames. Furthermore, experiment shows that the quantity ΣE *is* conserved for any isolated system. Thus the definition (3.9) meets requirement 2.

Just because the quantity ΣE is conserved, we are not yet justified in giving E, as defined by (3.9), the name *energy*. The main reason for doing so will emerge when we establish the connection of (3.9) with the classical definition of energy, that is, when we check requirement 1. Before we do so, we mention two other important points. First, since γ is dimensionless and mc^2 has the dimensions of energy, our definition $E = \gamma mc^2$ at least has the correct dimensions for an energy. Second, although we have not yet defined the concept of force in relativity, when we do so in Section 3.7, we shall prove the following important theorem: If a total force $\mathbf{F}$ acts on a body as it moves through a small displacement $d\mathbf{r}$ the resulting change in the energy E, as defined by (3.9), is

$$dE = \mathbf{F} \cdot d\mathbf{r}.$$

(3.10)

You should recognize the product on the right as the work done by the force $\mathbf{F}$, and the equation (3.10) as the work-energy theorem: The change in a body's energy is the work done on it. The fact that this theorem applies to our new definition of E is strong reason for regarding E as the relativistic generalization of the classical notion of energy.

Let us now evaluate the relativistic energy (3.9) for a slowly moving body. With $u \ll c$ we can use the binomial approximation to write the factor γ as

$$\gamma = \left(1 - \frac{u^2}{c^2}\right)^{-1/2} \approx 1 + \frac{1}{2}\frac{u^2}{c^2}.$$

Therefore,

$$E = \gamma mc^2 \approx \left(1 + \frac{1}{2}\frac{u^2}{c^2}\right)mc^2 = mc^2 + \tfrac{1}{2}mu^2 \qquad [\text{when } u \ll c]. \quad (3.11)$$

We see that for a slowly moving body the relativistic energy is the sum of two terms: a constant term mc^2 that is independent of u, and a second term $\tfrac{1}{2}mu^2$ that is precisely the classical kinetic energy of a body of mass m and speed u.

In classical physics it was believed that mass was always conserved, and the term mc^2 in (3.11) would therefore have been an immutable constant. Further, you will recall that one was always at liberty to add or subtract an overall constant from the energy, since the zero of energy was arbitrary. Thus, in the classical context (3.11) implies that when $u \ll c$ the relativistic energy E of a body is just the classical kinetic energy, plus an irrelevant constant mc^2. Therefore, the relativistic definition (3.9) meets both our requirements 1 and 2, and our identification of (3.9) as the appropriate generalization of the classical notion of energy is complete.

We shall find that in relativity the "irrelevant constant" mc^2 in (3.11) is actually extremely important. The reason is that the classical law of conservation of mass turns out to be wrong. This law was based mainly on nineteenth-century measurements of masses in chemical reactions, where no change of mass was ever detected. In this century, however, nuclear processes have been discovered in which large changes of mass occur, and even where the rest mass of certain particles disappears entirely. Further, we now know that even in chemical reactions the total rest mass of the participating atoms and molecules *does* change, although the changes are much too small (1 part in 10^9 or so) to be detected directly.

Given that rest masses *can* change, it should be clear that the term mc^2 in (3.11) is important. In Section 3.6 we describe some processes in which the rest mass of a system does change, and shall see just how important the term mc^2 is. In the remainder of this section we give two more definitions connected with the relativistic energy (3.9) and describe an application of the laws of energy and momentum conservation to processes in which rest masses do not change — the so-called *elastic* processes.

It is clear from either the exact equation $E = \gamma mc^2$ or the nonrelativistic approximation (3.11) that even when a body is at rest, its energy is not zero but is given, instead, by the famous equation

$$E = mc^2 \quad \text{[when } u = 0\text{]}. \tag{3.12}$$

This energy is called the **rest energy** of the mass m, and we shall see in Section 3.6 how it can be converted into other forms, such as the kinetic energy of other bodies.

EXAMPLE 3.2 What is the rest energy of a 1-kg lump of metal?

Substituting into (3.12), we find that

$$E = mc^2 = (1 \text{ kg}) \times (3 \times 10^8 \text{ m/s})^2 = 9 \times 10^{16} \text{ joules}.$$

This incredible amount of energy would be of no interest if it could not be converted into other forms of energy. In fact, however, such conversion *is* possible. For example, if the metal is uranium 235, about 1 part in 1000 of the rest energy can be converted into heat by the process of nuclear fission. Thus 1 kg of ^{235}U can yield a fantastic 9×10^{13} joules of heat.

When a body is not at rest, we can think of its total energy $E = \gamma mc^2$ as the sum of its rest energy mc^2 *plus* the additional energy $E - mc^2$ that it has by virtue of its motion. This second term we naturally call the **kinetic energy** K, and we write

$$E = mc^2 + K$$

where

$$K = E - mc^2 = (\gamma - 1) \, mc^2 \qquad (3.13)$$

At slow speeds we have seen that $K \approx \frac{1}{2}mu^2$, but in general the relativistic kinetic energy is different from $\frac{1}{2}mu^2$. In particular, as $u \rightarrow c$, the kinetic energy approaches infinity. However much energy we give a body, its speed can therefore never reach c—a conclusion we reached before by considering the relativistic momentum. Notice that since $\gamma \geq 1$, the relativistic kinetic energy is always positive (like its nonrelativistic counterpart $\frac{1}{2}mu^2$).

In classical mechanics a surprising number of interesting problems can be solved using just the laws of energy and momentum conservation. In relativity there are even more such problems, and we conclude this section with an example of one.

EXAMPLE 3.3 Two particles with rest masses m_1 and m_2 collide head-on as shown in Fig. 3.3. Particle 1 has initial velocity u_1, while particle 2 is at rest ($u_2 = 0$). Assuming that the collision is elastic (that is, the rest masses are unchanged), use conservation of energy and momentum to find the velocity u_3 of particle 1 after the collision. Apply the result to the case that a pion (a subatomic particle with mass $m_1 = 2.49 \times 10^{-28}$ kg) traveling at $0.9c$ makes an elastic head-on collision with a stationary proton ($m_2 = 1.67 \times 10^{-27}$ kg).

Since the solution of this problem is very similar to that of the corresponding nonrelativistic problem, let us first review the latter. Conservation of energy implies that

$$E_1 + E_2 = E_3 + E_4.$$

If the rest masses of the two particles are unchanged, this can be rewritten as

$$(m_1c^2 + K_1) + (m_2c^2 + K_2) = (m_1c^2 + K_3) + (m_2c^2 + K_4)$$

and canceling the mass terms, we see that kinetic energy is conserved,

$$K_1 + K_2 = K_3 + K_4,$$

which is the usual definition of an elastic collision in classical mechanics. The

FIGURE 3.3 An elastic, head-on collision.

conservation of momentum and kinetic energy imply (in nonrelativistic mechanics)

$$m_1 u_1 = m_1 u_3 + m_2 u_4 \tag{3.14}$$

and

$$\tfrac{1}{2} m_1 u_1^2 = \tfrac{1}{2} m_1 u_3^2 + \tfrac{1}{2} m_2 u_4^2. \tag{3.15}$$

These two equations can be solved for the two unknowns u_3 and u_4. Since one of the equations is quadratic, we get two solutions. The first solution, $u_3 = u_1$ and $u_4 = 0$, gives the initial velocities before the collision occurred. The second solution is the interesting one and gives

$$u_3 = \frac{m_1 - m_2}{m_1 + m_2} u_1 \tag{3.16}$$

and

$$u_4 = \frac{2 m_1}{m_1 + m_2} u_1. \tag{3.17}$$

Several features of these answers deserve comment. If $m_1 > m_2$, then u_3 is positive and particle 1 continues in its original direction; if $m_1 < m_2$, then u_3 is negative and particle 1 bounces back in the opposite direction. If $m_1 = m_2$, then $u_3 = 0$ and $u_4 = u_1$; that is, particle 1 comes to a dead stop, giving all of its momentum and kinetic energy to particle 2. If $m_1 \ll m_2$ (1 is much lighter than 2), then $u_3 \approx -u_1$ and particle 1 bounces back off the much heavier target with its speed barely changed.

The solution of the corresponding relativistic problem is very similar, although considerably messier because of the square roots involved in the factors γ. Because there are several different velocities in the problem, we must be careful to distinguish the corresponding factors of γ. We therefore write γ_1 for $(1 - u_1^2/c^2)^{-1/2}$, and γ_2 for $(1 - u_2^2/c^2)^{-1/2}$ (which is equal to 1 in this problem), and so on. The conservation of relativistic momentum implies that

$$\gamma_1 m_1 u_1 = \gamma_3 m_1 u_3 + \gamma_4 m_2 u_4,$$

while conservation of relativistic energy implies that

$$\gamma_1 m_1 c^2 + m_2 c^2 = \gamma_3 m_1 c^2 + \gamma_4 m_2 c^2. \tag{3.18}$$

These are two equations for the two unknowns, u_3 and u_4. If, for example, we eliminate u_4, some fairly messy algebra leads us to a quadratic equation for u_3. One solution of this equation is $u_3 = u_1$ (the original velocity before the collision) and the other is

$$u_3 = \frac{m_1^2 - m_2^2}{m_1^2 + m_2^2 + 2 m_1 m_2 \sqrt{1 - u_1^2/c^2}} u_1. \tag{3.19}$$

The answer (3.19) has much in common with the nonrelativistic (3.16). When $m_1 > m_2$, particle 1 continues in its original direction (u_3 positive); if $m_1 < m_2$, particle 1 bounces back (u_3 negative). If $m_1 = m_2$, particle 1 comes to a dead stop. If particle 1 is moving nonrelativistically, the square root in the denominator can be replaced by 1 to give

$$u_3 \approx \frac{m_1^2 - m_2^2}{(m_1 + m_2)^2} u_1 = \frac{m_1 - m_2}{m_1 + m_2} u_1,$$

which is precisely the nonrelativistic answer (3.16).

For the case of a pion, with $u_1 = 0.9c$, colliding with a stationary proton we can substitute the given numbers into (3.19) to give

$$u_3 = -0.76c \quad \text{[relativistic]}.$$

(Notice that this is negative, indicating that the light pion bounces backwards.) If we were to put the same numbers into the nonrelativistic result (3.16), we would find that

$$u_3 = -0.62c \quad \text{[nonrelativistic]}.$$

The difference between these two answers is large enough to be easily detected, and it is the relativistic answer that proves to be correct. More generally, in all collisions between atomic and subatomic particles, one finds perfect agreement between the experimental observations and the predictions based on conservation of relativistic energy and momentum. This is, in fact, the principal evidence that these quantities are conserved.

3.5 Two Useful Relations

We have introduced four parameters, m, $\mathbf{u}$, $\mathbf{p}$, and E, that characterize the motion of a body. Only two of these are independent since $\mathbf{p}$ and E were defined in terms of m and $\mathbf{u}$ as

$$\mathbf{p} = \frac{m\mathbf{u}}{\sqrt{1 - u^2/c^2}} \tag{3.20}$$

and

$$E = \frac{mc^2}{\sqrt{1 - u^2/c^2}} \tag{3.21}$$

We can, of course, rearrange these definitions to give an expression for any one of our parameters in terms of any two others. There are two such expressions that are especially useful, as we discuss now.

First, dividing (3.20) by (3.21) we find that

$$\frac{\mathbf{p}}{E} = \frac{\mathbf{u}}{c^2}$$

or

$$\beta \equiv \frac{\mathbf{u}}{c} = \frac{\mathbf{p}c}{E} \tag{3.22}$$

which gives the dimensionless "velocity" $\beta = \mathbf{u}/c$ in terms of $\mathbf{p}$ and E. (Since $\mathbf{p}c$ has the dimensions of energy, the right side is dimensionless, as it must be.)

Second, by squaring both (3.20) and (3.21) it is easy to verify that (Problem 3.12)

$$E^2 = (pc)^2 + (mc^2)^2. \tag{3.23}$$

This useful expression for E in terms of p and m shows that the three quantities

E, pc, and mc^2 are related like the sides of a right-angled triangle, with E as the hypotenuse. [At this stage there is no deep geometrical significance to this statement; we mention it only as a convenient way to remember the relation (3.23).] In our applications of relativistic mechanics we shall frequently use both the result (3.22) and the "Pythagorean relation" (3.23).

Now is a convenient point to mention some of the units used to measure the parameters u, E, m, and p. First, as we have seen repeatedly, relativistic velocities are best expressed as fractions of c, or by using the dimensionless $\beta = u/c$. The SI unit of energy is the joule (J). However, most of our applications of relativity will be in atomic and subatomic physics, where the joule is an inconveniently large unit of energy, and a much more popular unit is the **electron volt** or eV. This is defined as the work done on an electron (of charge $q = -e = -1.60 \times 10^{-19}$ coulomb) when it is carried through 1 volt ($\Delta V = -1$ volt); thus

$$1 \text{ eV} = q\,\Delta V = (-1.60 \times 10^{-19} \text{ coulomb}) \times (-1 \text{ volt})$$
$$= 1.60 \times 10^{-19} \text{ J}. \tag{3.24}$$

We shall find that typical energies in atomic physics are of the order of 1 eV or so; those in nuclear physics are of order 10^6 eV, or 1 MeV.

Like the joule, the kilogram is inconveniently large as a unit for atomic and subatomic physics. For example, the mass of the electron is 9.11×10^{-31} kg. In fact, in most applications of relativity one is not so much concerned with the mass m as with the corresponding rest energy mc^2. [For instance, in (3.23) it is mc^2, rather than m itself, that appears.] Thus in many relativistic problems, masses are given implicitly by stating the rest energy mc^2, usually in eV. Another way to say this is that mass is measured in units of eV/c^2, as we discuss in the following example.

EXAMPLE 3.4 Given that the electron has mass $m = 9.109 \times 10^{-31}$ kg, what is its rest energy in eV, and what is its mass in eV/c^2?

The rest energy is, of course,

$$mc^2 = (9.109 \times 10^{-31} \text{ kg}) \times (2.998 \times 10^8 \text{ m/s})^2$$
$$= (81.87 \times 10^{-15} \text{ J}) \times \frac{1 \text{ eV}}{1.602 \times 10^{-19} \text{ J}}$$
$$= 5.11 \times 10^5 \text{ eV}$$

or*

$$mc^2 = 0.511 \text{ MeV}. \tag{3.25}$$

Thus a convenient way to specify (and remember) the electron's mass is to say that its rest energy mc^2 is roughly half an MeV.

An alternative way to put this is to divide both sides of (3.25) by c^2 and say that

$$m = 0.511 \text{ MeV}/c^2.$$

---

* Note that to get this answer, correct to three significant figures, we used four significant figures in all input numbers. These were taken from Appendix A, which lists the best known values of the fundamental constants.

If you have not met MeV/c^2 before as a unit of mass, it will probably seem a bit odd at first. The important thing to remember is that the statement "$m = 0.5$ MeV/c^2" is precisely equivalent to the more transparent "$mc^2 = 0.5$ MeV."

In most applications of relativity we shall be less interested in the momentum p than in the product pc. [For example, both of the relations (3.22) and (3.23) involve pc rather than p.] The quantity pc has the dimensions of energy and so is often measured in eV or MeV. This is equivalent to measuring p itself in eV/c or MeV/c.

EXAMPLE 3.5 An electron (rest mass about 0.5 MeV/c^2) is moving with total energy $E = 1.3$ MeV. Find its momentum (in MeV/c and in SI units) and its speed.

Given the energy and mass, we can immediately find the momentum from the "Pythagorean relation" (3.23)

$$pc = \sqrt{E^2 - (mc^2)^2} = \sqrt{(1.3 \text{ MeV})^2 - (0.5 \text{ MeV})^2}$$
$$= 1.2 \text{ MeV}$$

or

$$p = 1.2 \text{ MeV}/c.$$

(Notice how easily the units work out when we measure E in MeV, p in MeV/c, and m in MeV/c^2.) This is easily converted into SI units if necessary: The required conversion is

$$1 \frac{\text{MeV}}{c} = \frac{1.602 \times 10^{-13} \text{ J}}{2.998 \times 10^8 \text{ m/s}}$$
$$= 5.34 \times 10^{-22} \text{ kg} \cdot \text{m/s}. \qquad (3.26)$$

(This and several other conversion factors are listed inside the front cover; more exact values are given in Appendix A.) Therefore,

$$p = (1.2 \text{ MeV}/c) \times \frac{5.3 \times 10^{-22} \text{ kg} \cdot \text{m/s}}{1 \text{ MeV}/c}$$
$$\approx 6 \times 10^{-22} \text{ kg} \cdot \text{m/s}.$$

Once we know the energy and momentum, the speed follows immediately from (3.22):

$$\beta = \frac{pc}{E} = \frac{1.2 \text{ MeV}}{1.3 \text{ MeV}} = 0.92 \qquad \text{or} \qquad u \approx 0.9c.$$

3.6 Conversion of Mass to Energy

According to the relativistic definition of energy, even a mass at rest has an energy, equal to mc^2. As we have emphasized, this statement is meaningless unless there is some way in which the supposed rest energy mc^2 can be converted, or at least partly converted, into other more familiar forms of energy, such as kinetic energy — a process that is often loosely called conversion of mass to energy. Such conversion would require the classical law of conservation of mass to be violated. In this section we argue that if the relativistic mechanics we have developed is correct, the nonconservation of mass is logically necessary; and we describe some processes in which mass is not conserved.

Let us consider two bodies that can come together to form a single composite body. The two bodies could be two atoms that can bind together to form a molecule, or two atomic nuclei that can fuse to form a larger nucleus. Because it is probably easier to think about everyday objects we shall imagine two macroscopic (that is, nonmicroscopic) blocks. When the bodies are far apart we take for granted that the forces between them are negligible. But when they are close together there will be forces, and we distinguish two cases: If the forces are predominantly repulsive, we shall have to supply energy to push the two bodies together; if the forces are predominantly attractive, energy will be released as they move together.

As a model of the repulsive case we imagine two blocks with a compressible spring attached to one of them, as shown in Fig. 3.4(a). To hold the blocks together we attach a pivoted catch to the second block, as shown. Suppose now that we push the two blocks together until the catch closes as in Fig. 3.4(b). This will require us to do work, which is stored by the system as potential energy of the spring. Because of this potential energy the bound state of the system is unstable: If we release the catch, the blocks will fly apart and the stored energy will reappear as kinetic energy, as shown in Fig. 3.4(c). What we now wish to argue is that the potential energy stored in the bound system manifests itself as an increase in the mass of the system; that is, the rest mass M of the unstable bound state is greater than the sum of the separate rest masses m_1 and m_2 of the blocks when far apart. (In classical mechanics $M = m_1 + m_2$, of course.)

As long as the two blocks remain together, we can treat them as a single body, whose rest mass we denote by M. If the body is at rest, then according to the definition (3.9) of relativistic energy its total energy is

$$\text{total energy} = Mc^2. \tag{3.27}$$

[Notice that we are assuming that the definition (3.9) can be applied to a composite system like our two blocks locked together; in the final analysis this assumption must be—and is—justified by experiment.] If we now gently release the catch the two blocks will fly apart. Once they are well separated, we

FIGURE 3.4 (a) Two blocks repel one another at close range because of the spring attached to block 1. (b) If the blocks are pushed together, the catch on block 2 can hold them in a bound state of rest mass M. (c) The bound state is unstable in the sense that the blocks fly apart when the catch is released.

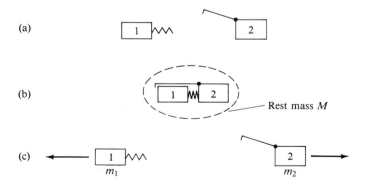

can treat them as two separate bodies with rest masses m_1 and m_2 and total energy,

$$\text{total energy} = E_1 + E_2$$
$$= K_1 + m_1 c^2 + K_2 + m_2 c^2. \qquad (3.28)$$

By conservation of energy, (3.27) and (3.28) must be equal. Therefore,

$$Mc^2 = (K_1 + K_2) + (m_1 + m_2)c^2. \qquad (3.29)$$

Since both K_1 and K_2 are greater than zero, we are forced to the conclusion that M is greater than the sum of the separate masses, $m_1 + m_2$. This conclusion is independent of the mechanical details of our example and applies to any unstable system of rest mass M that can fly apart into two (or more) pieces.

We can solve (3.29) to give the amount by which the rest mass decreases as the blocks fly apart. If we denote this by

$$\Delta M = M - (m_1 + m_2),$$

then (3.29) implies that

$$\Delta M c^2 = K_1 + K_2$$
$$= \text{energy released as bodies fly apart.} \qquad (3.30)$$

We can say that a mass ΔM has been converted into kinetic energy, the rate of exchange between mass and energy being given by the familiar "energy = mass $\times c^2$."

The kinetic energy released as the bodies move apart is the same as the work done to bring them together in the first place. Thus we can rephrase our result to say that when we push our two blocks together, the work done results in an increase in rest mass, ΔM, given by

$$\Delta M c^2 = \text{work done to push bodies together.} \qquad (3.31)$$

EXAMPLE 3.6 The nuclei of certain atoms are naturally unstable, or radioactive, and spontaneously fly apart, tearing the whole atom into two pieces. For example, the atom called thorium 232 splits spontaneously into two "offspring" atoms, radium 228 and helium 4,

$$^{232}\text{Th} \rightarrow {}^{228}\text{Ra} + {}^{4}\text{He}. \qquad (3.32)$$

The combined kinetic energy of the two offspring is 4 MeV. By how much should the rest mass of the "parent" ^{232}Th differ from the combined rest mass of its offspring? Compare this with the difference in the measured masses listed in Appendix D.

The reaction (3.32) is analogous to the example of the two blocks just discussed. (In fact, the analogy is remarkably good. The repulsive force is the electrostatic repulsion of the offspring nuclei, both of which are positively charged; the "catch" that holds the offspring together is their nuclear attraction, which is not quite strong enough and eventually "releases," letting the offspring fly apart.) The required mass difference ΔM is given by (3.30) as

$$\Delta M = \frac{K_1 + K_2}{c^2} = 4 \text{ MeV}/c^2.$$

Therefore, ^{232}Th should be heavier than the combined mass of ^{228}Ra and ^{4}He by 4 MeV/c^2. We can convert this to kilograms if we wish:

$$\Delta M = 4 \frac{\text{MeV}}{c^2} = 4 \times \frac{10^6 \times 1.6 \times 10^{-19} \text{ J}}{(3 \times 10^8 \text{ m/s})^2}$$

$$= 7 \times 10^{-30} \text{ kg.} \qquad (3.33)$$

This mass difference is very small, even compared to the masses of the atoms involved. (^{232}Th has a mass of about 4×10^{-25} kg; so the mass difference is of order 1 part in 10^5 of the total mass.) Nevertheless, it is large enough to be measured directly.

To check our predicted mass difference against the measured masses, we refer to Appendix D, which lists the masses concerned as follows:

Atom	Mass (in u)	
^{232}Th	232.038	Initial
^{228}Ra $\left.\begin{array}{c} 228.031 \\ 4.003 \end{array}\right\}$ ^{4}He	232.034	Final total
	0.004	Difference

These masses are given in atomic mass units (denoted u). We shall discuss this unit in Chapter 4, but for now we need to know only that

$$1 \text{ u} = 1.66 \times 10^{-27} \text{ kg.}$$

Thus the measured mass difference is

$$\Delta M = 0.004 \text{ u} \times \frac{1.7 \times 10^{-27} \text{ kg}}{1 \text{ u}}$$

$$= 7 \times 10^{-30} \text{ kg}$$

in agreement with our prediction (3.33).

We next consider briefly the case that the force between the bodies is predominantly attractive, so that work is required to pull them apart. For example, if our two bodies are an electron and a proton, they attract one another because of their opposite electric charges; if they come close together they can form the stable bound state that we call the hydrogen atom, and an external agent must do work (13.6 eV, in fact) to pull them apart again. The work needed to pull a bound state apart (leaving the pieces well separated and at rest) is called the **binding energy** and is denoted by B. As before, we denote by M the rest mass of the bound state and by m_1 and m_2 the separate masses of the two bodies. By conservation of energy

$$Mc^2 + B = m_1 c^2 + m_2 c^2. \qquad (3.34)$$

In this case, we see that M is less than $m_1 + m_2$, the difference $\Delta M = m_1 + m_2 - M$ being given by

$$\Delta M c^2 = \text{binding energy, } B$$

$$= \text{work to pull the bodies apart.} \qquad (3.35)$$

If we now released our two bodies, they would accelerate back together and could reenter their bound state, with the release of energy B. (In the example of the electron and proton, the energy is released mostly as light when they form into a hydrogen atom.) Thus we can rephrase (3.35) to say that as the two bodies come together and form the stable bound state, there is a release of energy and corresponding loss* of mass, ΔM, given by

$$\Delta Mc^2 = \text{energy released as bodies come together to form}$$
$$\text{bound state.} \qquad (3.36)$$

EXAMPLE 3.7 It is known that two oxygen atoms attract one another and can unite to form an O_2 molecule, with the release of energy $E_{out} \approx 5$ eV (mostly in the form of light if the reaction takes place in isolation). By how much is the O_2 molecule lighter than two O atoms? If one formed 1 gram of O_2 in this way, what would be the total loss of rest mass and what the total energy released? (The O_2 molecule has a mass of about 5.3×10^{-26} kg.)

By (3.36) the mass of one O_2 molecule is less than that of two O atoms by an amount

$$\Delta M = \frac{E_{out}}{c^2} \approx 5 \text{ eV}/c^2$$

$$\approx 5 \times \frac{1.6 \times 10^{-19} \text{ J}}{(3 \times 10^8 \text{ m/s})^2}$$

$$\approx 9 \times 10^{-36} \text{ kg.}$$

Dividing this by the mass 5.3×10^{-26} of a single O_2 molecule, we see that the *fractional* loss of mass is about 2 parts in 10^{10}. Thus if we were to form 1 gram (g) of O_2 this way, the total loss of mass would be

$$\Delta M \approx 2 \times 10^{-10} \text{ g} \qquad (3.37)$$

which is much too small to be measured directly. This is fairly typical of the mass changes in chemical reactions and explains why nonconservation of mass does not show up in chemistry.

While the mass change (3.37) is exceedingly small, the total energy released,

$$E_{out} = \Delta Mc^2 \approx (2 \times 10^{-13} \text{ kg}) \times (3 \times 10^8 \text{ m/s})^2$$

$$\approx 2 \times 10^4 \text{ J,}$$

is large. This is because the conversion factor, c^2, from mass to energy is so large.

So far in this section we have focused on the conservation of relativistic energy and the concomitant nonconservation of mass. For an isolated system

---

* It is quite easy to confuse the direction of the mass change (gain or loss) in a given process. If this happens, just go back to the fact of *energy conservation,* which you can easily write in a form like (3.29) or (3.34). From this you can see immediately which mass is greater.

momentum is also conserved, and in many problems conservation of energy and momentum give enough information to determine all that one needs to know. We conclude this section with an example.

EXAMPLE 3.8   The Λ particle is a subatomic particle which (as mentioned in Example 2.2) can decay spontaneously into a proton and a negatively charged pion:

$$\Lambda \rightarrow p + \pi^-.$$

(This immediately tells us that the rest mass of the Λ is greater than the total rest mass of the proton and pion.) In a certain experiment the outgoing proton and pion were observed both traveling in the same direction along the x axis with momenta

$$p_p = 581 \text{ MeV}/c \quad \text{and} \quad p_\pi = 256 \text{ MeV}/c.$$

Given that their rest masses are known to be

$$m_p = 938 \text{ MeV}/c^2 \quad \text{and} \quad m_\pi = 140 \text{ MeV}/c^2,$$

find the rest mass m_Λ of the Λ.

We can solve this problem in three steps: First, knowing p and m for the proton and pion, we can calculate their energies using the "Pythagorean relation"

$$E^2 = (pc)^2 + (mc^2)^2. \tag{3.38}$$

This gives

$$E_p = 1103 \text{ MeV} \quad \text{and} \quad E_\pi = 292 \text{ MeV}.$$

Second, using conservation of energy and momentum we can reconstruct the energy and momentum of the original Λ:

$$E_\Lambda = E_p + E_\pi = 1395 \text{ MeV} \tag{3.39}$$

and

$$\mathbf{p}_\Lambda = \mathbf{p}_p + \mathbf{p}_\pi = 837 \text{ MeV}/c \quad \text{(along the x axis).} \tag{3.40}$$

Finally, knowing E and p for the Λ we can use the relation (3.38) again to give the mass

$$m_\Lambda = \frac{\sqrt{E_\Lambda^2 - (p_\Lambda c)^2}}{c^2} = 1116 \text{ MeV}/c^2.$$

This is, in fact, how the masses of many unstable subatomic particles are measured. As we discuss in the next section, it is fairly easy to measure the momenta of the decay products as long as they are charged. If the decay masses are known, we can then calculate their energies and hence reconstruct the parent particle's energy and momentum. From these one can calculate its mass.

3.7   Force in Relativity

We have come a surprisingly long way in relativistic mechanics without defining the notion of force. This reflects correctly the comparative unimportance of forces in relativity. Nonetheless, forces are what change the momentum of a body and we must now discuss them.

Just as with momentum and energy, our first task is to decide on a suitable definition of force. In classical mechanics two equivalent definitions of the force acting on a body are

$$\mathbf{F} = m\mathbf{a} \tag{3.41}$$

and

$$\mathbf{F} = \frac{d\mathbf{p}}{dt}. \tag{3.42}$$

These are equivalent since $\mathbf{p} = m\mathbf{u}$, with m constant, so $d\mathbf{p}/dt = m\mathbf{a}$. In relativity, we have defined $\mathbf{p}$ as $\gamma m\mathbf{u}$ and it is no longer true that $d\mathbf{p}/dt = m\mathbf{a}$. Therefore, we certainly cannot carry over *both* definitions, (3.41) and (3.42), into relativity. In fact, for most purposes the convenient definition of force in relativity is the second of the classical definitions, (3.42):

> The total force $\mathbf{F}$ acting on a body with momentum $\mathbf{p}$ is defined as
> $$\mathbf{F} = \frac{d\mathbf{p}}{dt}. \tag{3.43}$$

Evidently, the definition (3.43) reduces to the nonrelativistic definition if the body concerned is moving nonrelativistically. Further, with this definition of $\mathbf{F}$, the work-energy theorem carries over into relativity, as we prove in the following example.

EXAMPLE 3.9 Prove that if a mass m, acted on by a total force $\mathbf{F}$, moves a small distance $d\mathbf{r}$, the change in its energy, dE, equals the work done by $\mathbf{F}$:

$$dE = \mathbf{F} \cdot d\mathbf{r}. \tag{3.44}$$

To prove this we replace $\mathbf{F}$ by $d\mathbf{p}/dt$ and $d\mathbf{r}$ by $\mathbf{u}\,dt$ on the right to give

$$\mathbf{F} \cdot d\mathbf{r} = \frac{d\mathbf{p}}{dt} \cdot \mathbf{u}\,dt. \tag{3.45}$$

Now, we shall prove in a moment that

$$\frac{d\mathbf{p}}{dt} \cdot \mathbf{u} = \frac{dE}{dt}. \tag{3.46}$$

Thus from (3.45) it follows that

$$\mathbf{F} \cdot d\mathbf{r} = \frac{dE}{dt}\,dt = dE,$$

which is the work-energy theorem. It remains only to prove the identity (3.46), as follows: From the "Pythagorean relation" we know that

$$E = (p^2c^2 + m^2c^4)^{1/2}.$$

Using the chain rule to differentiate this, we find (Problem 3.32)

$$\frac{dE}{dt} = \frac{1}{2}(p^2c^2 + m^2c^4)^{-1/2}2\mathbf{p}c^2 \cdot \frac{d\mathbf{p}}{dt} = \frac{\mathbf{p}c^2}{E} \cdot \frac{d\mathbf{p}}{dt} = \mathbf{u} \cdot \frac{d\mathbf{p}}{dt}, \quad (3.47)$$

which is (3.46).

The most important force in most applications of relativistic mechanics is the electromagnetic force. It is found experimentally that when a charge q is placed in electric and magnetic fields $\mathbf{E}$ and $\mathbf{B}$, its relativistic momentum changes at a rate equal to $q(\mathbf{E} + \mathbf{u} \times \mathbf{B})$. Thus, having defined $\mathbf{F}$ as $d\mathbf{p}/dt$, we find that the electromagnetic force is given by the classical formula, often called the **Lorentz force,**

$$\mathbf{F} = q(\mathbf{E} + \mathbf{u} \times \mathbf{B}).$$

Even when the force $\mathbf{F}$ is known, the equation $d\mathbf{p}/dt = \mathbf{F}$ is usually hard to solve for a body's position as a function of time. One important case where it *is* easily solved is for a charged body in a uniform magnetic field $\mathbf{B}$. In this case, the force $\mathbf{F} = q\mathbf{u} \times \mathbf{B}$ is perpendicular to $\mathbf{u}$ and the body's energy is therefore constant. [This follows from the work-energy theorem (3.44), because $\mathbf{F}$ is perpendicular to $d\mathbf{r}$ as the body moves along.] Therefore, the velocity is constant in magnitude and changes only its direction. In particular, if the body is moving in a plane perpendicular to $\mathbf{B}$, it moves in a circular path, whose radius R can be found as follows: Since γ is constant, the equation

$$\frac{d\mathbf{p}}{dt} = q\mathbf{u} \times \mathbf{B}$$

becomes

$$\gamma m \frac{d\mathbf{u}}{dt} = q\mathbf{u} \times \mathbf{B}.$$

Since $d\mathbf{u}/dt = \mathbf{a}$ and $\mathbf{u}$ is perpendicular to $\mathbf{B}$, this implies that

$$\gamma ma = quB.$$

For motion in a circle we know that a is the centripetal acceleration u^2/R. (This purely kinematic result is true in relativity for exactly the same reasons as in classical mechanics.) Therefore,

$$\gamma m \frac{u^2}{R} = quB;$$

whence, since $\gamma mu = p$,

$$R = \frac{p}{qB}. \quad (3.48)$$

This result provides a convenient way to measure the momentum of a particle of known charge q. If we send the particle into a known magnetic field B, then by measuring the radius R of its curved path we can find its momentum p from (3.48).

EXAMPLE 3.10   A proton of unknown momentum p is sent through a uniform magnetic field $B = 1.0$ tesla (T), perpendicular to $\mathbf{p}$, and is found to move in a circle of radius $R = 1.4$ m. What are the proton's momentum in MeV/c and its energy in MeV?

The proton's charge is known to be $q = e = 1.6 \times 10^{-19}$ coulomb (C). Thus from (3.48) its momentum is

$$p = qBR = (1.6 \times 10^{-19} \text{ C}) \times (1.0 \text{ T}) \times (1.4 \text{ m})$$

$$= 2.24 \times 10^{-19} \text{ kg} \cdot \text{m/s}.$$

From the list of conversion factors inside the front cover, we find 1 MeV/c = 5.34×10^{-22} kg·m/s. [This was derived in (3.26).] Therefore,

$$p = (2.24 \times 10^{-19} \text{ kg} \cdot \text{m/s}) \times \frac{1 \text{ MeV}/c}{5.34 \times 10^{-22} \text{ kg} \cdot \text{m/s}} = 420 \text{ MeV}/c.$$

Since the proton's rest mass is known to be $m = 938$ MeV/c^2, we can find its energy from the "Pythagorean relation" (3.38) to be

$$E = \sqrt{(pc)^2 + (mc^2)^2} = 1030 \text{ MeV}.$$

3.8   Massless Particles

In this section we consider a question that will probably strike you as peculiar if you have never met it before: In the framework of relativistic mechanics, is it possible that there are particles of zero mass, $m = 0$?

In classical mechanics the answer to this question is undoubtedly "no." The classical momentum and kinetic energy of a particle are $m\mathbf{u}$ and $\frac{1}{2}mu^2$. If $m = 0$, both of these are identically zero, and a particle whose momentum and kinetic energy are always zero is presumably nothing at all.

In relativity the answer is not as clear cut. Our definitions of energy and momentum were

$$\mathbf{p} = \gamma m\mathbf{u} \tag{3.49}$$

and

$$E = \gamma mc^2, \tag{3.50}$$

and from these we derived the two important relations

$$E^2 = (pc)^2 + (mc^2)^2 \tag{3.51}$$

and

$$\beta = \frac{u}{c} = \frac{pc}{E}. \tag{3.52}$$

Let us consider the last two relations first. If there were a particle with $m = 0$ [and if the relations (3.51) and (3.52) applied to this particle], then (3.51) would imply that

$$E = pc \quad (\text{if } m = 0) \tag{3.53}$$

and (3.52) would then tell us that

$$\beta = 1 \quad \text{or} \quad u = c \quad (\text{if } m = 0);$$

that is, the massless particle would always have speed $u = c$. The converse of this statement is also true. If we discovered a particle that traveled with speed c, then by (3.52) it would satisfy $pc = E$, and (3.51) would then require that $m = 0$.

If we turn to our original definitions of $\mathbf{p}$ and E, (3.49) and (3.50), a superficial glance suggests the same result as in classical mechanics. With $m = 0$, (3.49) and (3.50) appear at first to imply that $\mathbf{p}$ and E are zero. However, a closer look shows that this is not so. We have already seen that a massless particle would have to travel at speed c; and if $u = c$, then $\gamma = \infty$. Thus both (3.49) and (3.50) have the form $\infty \times 0$, which is undefined. Thus our original definitions fail to define $\mathbf{p}$ and E if $m = 0$, but they do not actually contradict the possible existence of particles with $m = 0$.

Apparently, it is logically possible in relativity to have particles with $m = 0$; and in fact experiment shows that such particles do exist, the most important example being the photon, or particle of light. In classical physics, light (like all other forms of electromagnetic radiation) was assumed to be a wave in which the energy and momentum were distributed continuously through space. As we shall describe in Chapter 5, it was discovered at the beginning of this century that the energy and momentum in an electromagnetic wave are actually confined to many tiny localized bundles. These tiny bundles, which have come to be called *photons,* display many of the properties of ordinary particles like the electron; in particular, they have energy and momentum. But unlike the more familiar particles, they travel at the speed of light and, by what we have already said, must therefore have $m = 0$.

Since the definitions (3.49) and (3.50) for momentum and energy cannot be used for massless particles, the question naturally arises how we can define and measure these quantities. One simple answer is to consider a process that involves only one massless particle. For example, light shining on an atom can tear loose one of the atom's electrons, a process called photoionization. According to the photon theory of light, photoionization occurs when one of the photons that make up the light collides with the atom and ejects one of the atom's electrons. In the process the photon is absorbed and disappears. For example, the photoionization of the simplest atom, hydrogen (made up of one electron and a proton), can be represented as

$$\gamma + H \rightarrow e + p \tag{3.54}$$

where γ is the traditional symbol for a photon and H, e, and p stand for the hydrogen atom, the electron, and the proton. The important point is that the three bodies H, e, and p all have mass, and hence have well-defined momenta and energies. Thus if we assume that momentum and energy are conserved in the process (3.54), we can calculate the momentum and energy of the photon from the known values for the other three particles.

If we define $\mathbf{p}$ and E for the photon as just described, we find two essential properties. First, the definition is consistent: One obtains the same values for $\mathbf{p}$ and E when identical photons are observed in two or more different processes. Second, the values obtained satisfy (3.53), $pc = E$, which we saw was an essential characteristic of a particle with $m = 0$ and $u = c$.

We shall review much more of the evidence for the existence of photons and their zero rest mass in Chapter 5. Today it is almost universally accepted that the photon is a particle, which carries energy and momentum and can be treated much like other particles, except that it travels at speed c and has mass exactly equal to zero.

There is considerable evidence that a second particle, the neutrino, has zero rest mass. This is a particle observed in the decay of many radioactive nuclei, as we shall describe in Chapter 13. However, there is some recent evidence that the neutrino's mass, although very small (possibly of order 10^{-5} times the electron mass), may not be exactly zero. There are theoretical reasons for believing that there is a massless particle called the graviton, which is related to gravity in the same way that the photon is related to light; but there is, as yet, no experimental evidence for the graviton.

EXAMPLE 3.11 As we shall discuss in Chapters 13 and 14, there is a subatomic particle called the *positron,* or *antielectron,* with exactly the same mass as the electron (0.511 MeV/c^2) but the opposite charge. The most remarkable property of the positron is that when it collides with an electron the two particles can annihilate one another, converting themselves into two or more photons. Consider the case that the electron and positron are both at rest and that just two photons are produced, with energies E_1 and E_2 and momenta $\mathbf{p}_1$ and $\mathbf{p}_2$ (Fig. 3.5). Use conservation of energy and momentum to find the energies, E_1 and E_2, of the two photons.

CARL ANDERSON (born 1905, US). Positrons (or anti-electrons) were discovered by Anderson in 1932, and their annihilation with electrons was the first example of the total conversion of mass energy to other forms of energy. Anderson won the 1936 Nobel Prize for this discovery.

Before the annihilation the total momentum is zero and the total energy is $2mc^2$, where m is the rest mass of the electron. By conservation of momentum and energy, it follows that

$$\mathbf{p}_1 + \mathbf{p}_2 = 0 \tag{3.55}$$

and

$$E_1 + E_2 = 2mc^2. \tag{3.56}$$

From (3.55) we see that the photons have equal and opposite momenta ($\mathbf{p}_1 = -\mathbf{p}_2$, as was suggested in Fig. 3.5). This means that the photons' energies are equal, $E_1 = E_2$ (since $E_1 = p_1 c$ and $E_2 = p_2 c$). Thus from (3.56)

$$E_1 = E_2 = mc^2 = 0.511 \text{ MeV}.$$

Each photon carries away exactly the rest energy of one electron.

The most remarkable thing about this process is this: In the initial state the total energy is the rest energy of the two particles, $2mc^2$; that is, all the energy is mass energy. In the final state the two photons have no rest mass and there is therefore no mass energy. Thus the process involves 100% conversion of mass energy into another form of energy (electromagnetic, in this case). The observation of such processes is triumphant justification for the claim that the rest

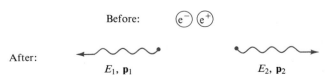

FIGURE 3.5 The annihilation of an electron–positron pair into two photons.

energy mc^2 of a mass m is a real energy, not just the result of a whimsical choice for the zero of energy.

This process of electron–positron annihilation with production of two 0.511-MeV photons is observed routinely in particle-physics laboratories using artificially produced positrons. It also occurs naturally in the earth's atmosphere where the positrons are produced by cosmic radiation. Recently, astronomers have observed photons with exactly 0.511 MeV coming from the center of our galaxy; the conclusion is almost inescapable that somewhere in the galaxy's center positrons are being created and then annihilating with electrons to produce these photons.

3.9　When Is Nonrelativistic Mechanics Good Enough?

In most problems—although certainly not all—the equations of classical mechanics are easier to use than their relativistic counterparts. It is therefore important to be able to recognize situations where nonrelativistic mechanics is a good enough approximation and the complications of relativity can be ignored. Loosely speaking, the rule is clear: If a body has speed much less than c at all times, it can be treated nonrelativistically. Unfortunately, this gives us no guidance as to how small the speed must be before it can be considered "much less than c," and in fact there is no clear-cut answer. The speed at which relativistic effects must be considered depends on the desired accuracy.

We can illustrate these ideas with two simple examples: First, we saw in Section 2.3 (Example 2.1) that in an airplane traveling at 300 m/s for an hour, relativistic time dilation can affect the plane's clock by a few nanoseconds (1 ns $= 10^{-9}$ s). In almost any situation this effect is utterly insignificant, and if this is the case, we can ignore relativity. Nevertheless, certain proposed navigational systems depend on timings with an accuracy of a few nanoseconds, and if we wish to use these systems, we must take account of relativity.

As a second example, suppose that we want to know the kinetic energy K of a mass m at speed u. The correct relativistic answer is $K_{rel} = (\gamma - 1)mc^2$, while its nonrelativistic approximation is $K_{NR} = \frac{1}{2}mu^2$. We have tabulated both of these, with their percent discrepancy, for various speeds, in Table 3.4. We see that at a speed $u = 0.01c$ (some 300 times the speed of an orbiting satellite) K_{NR} is within 0.01% of the correct K_{rel}. Thus for most purposes the nonrelativistic answer would be quite good enough when $u = 0.01c$; nevertheless, if for some reason we needed an accuracy better than 0.01%, we would have to work relativistically. At $u = 0.1c$, K_{NR} is still within 1% of the correct K_{rel}, which for some purposes would still be acceptable. At $u = 0.5c$, K_{NR} differs from K_{rel} by 20%, which in most cases would *not* be acceptable (although, even here, K_{NR} still has the right order of magnitude).

TABLE 3.4

The relativistic and nonrelativistic kinetic energy of a mass m at various speeds u, in units of mc^2.

u:	0.01c	0.1c	0.5c
$K_{rel} = (\gamma - 1)mc^2$:	5.0004×10^{-5}	5.038×10^{-3}	0.155
$K_{NR} = \frac{1}{2}mu^2$:	5.0000×10^{-5}	5.000×10^{-3}	0.125
% difference:	0.01%	1%	20%

These examples show clearly that whether or not nonrelativistic mechanics gives an acceptable approximation depends on what one wishes to calculate and with what accuracy. Nevertheless, as a very rough guide for the user of this book, we can state the following three rules:

1. If a particle has speed much less than $0.1c$, it is usually satisfactory to apply nonrelativistic mechanics. For example, we shall find that the electron in a hydrogen atom has $u \approx 0.01c$ and we shall get excellent results for its energy treating it nonrelativistically.

2. If a particle has a speed of $0.1c$ or more, then, except in very approximate calculations, one should usually use relativistic mechanics.

3. Since they always travel at speed c, massless particles can never be treated nonrelativistically. For example, in considering the photons of light emitted by an atom we can frequently treat the atom nonrelativistically, but the photons themselves must always be treated relativistically.

Finally, since we often know a particle's energy, rather than its speed, it is convenient to rephrase these rough rules in terms of energy. To decide whether we must use relativity, we compare the particle's kinetic energy K with its rest energy mc^2. We see from Table 3.4 that if $u \approx 0.1c$, then (very roughly) $K \approx 0.01\,mc^2$. Thus the first two rules above can be rephrased as follows:

1. If a particle's kinetic energy is much less that 1% of its rest energy, it is usually satisfactory to use nonrelativistic mechanics.

2. If the particle has kinetic energy equal to 1% of its rest energy or more, one should usually use relativistic mechanics.

3.10 General Relativity (optional)

Einstein's general relativity, completed in 1915, is the extension of special relativity to include the effects of gravity. We shall see that the examination of gravity led Einstein to consider noninertial reference frames. Thus general relativity is more general than special relativity both because it includes gravity *and* because it focuses on noninertial, as well as inertial, reference frames.* Because it is mathematically more complicated than special relativity, and because we shall have no occasion to use it again in this book, we content ourselves here with a brief and descriptive introduction to general relativity.

INERTIAL FORCES

We can learn much about general relativity — just as we did about special relativity — by first examining certain questions within classical physics. In particular, let us consider two classical frames, an inertial frame S and a *noninertial* frame S', accelerating relative to S with acceleration $\mathbf{A}$. In the inertial frame S the equation of motion for a mass m is Newton's second law, $m\mathbf{a} = \Sigma \mathbf{F}$, where $\Sigma \mathbf{F}$ is the sum of all forces on the mass. To find the corresponding

---

* Experts may object that even special relativity *can* handle noninertial frames; nevertheless, its primary focus is inertial frames. We should also mention that when we speak of inertial frames here, we mean the inertial frames of special relativity (or of Newtonian mechanics when our discussion is classical).

equation in the noninertial frame S', we recall the classical velocity addition formula, $\mathbf{u}' = \mathbf{u} - \mathbf{v}$. Differentiating, we find that the mass's acceleration as measured in S' is

$$\mathbf{a}' = \mathbf{a} - \mathbf{A}, \tag{3.57}$$

since $d\mathbf{v}/dt = \mathbf{A}$, the acceleration of the frame S'. Multiplying (3.57) by m, and putting $m\mathbf{a} = \Sigma \mathbf{F}$, we find that

$$m\mathbf{a}' = \sum \mathbf{F} - m\mathbf{A}.$$

This equation has exactly the form of Newton's second law *except* that in addition to all the forces identified in S, there is an extra force term equal to $-m\mathbf{A}$. Thus Newton's second law is also valid in the noninertial frame S', *provided* that we recognize that in a noninertial frame every mass must experience an additional **inertial force**

$$\mathbf{F}_{in} = -m\mathbf{A}. \tag{3.58}$$

This inertial force experienced in noninertial frames is familiar in several everyday situations: If we sit in an aircraft accelerating rapidly toward takeoff, then, from our point of view, there is a force that pushes us back into our seat. If we are standing in a bus that brakes suddenly ($\mathbf{A}$ backward), the inertial force $-m\mathbf{A}$ is forward and can make us fall on our faces if we aren't properly braced. As a car goes rapidly around a sharp curve, the inertial force experienced by its occupants is the so-called centrifugal force that pushes them outward. One can take the view that the inertial force is a "fictitious" force, introduced simply to preserve the form of Newton's second law in noninertial frames. Nevertheless, for an observer in an accelerating frame, it is entirely real.

THE EQUIVALENCE PRINCIPLE

The starting point of general relativity is called the equivalence principle and arises from the following observation: The inertial force $\mathbf{F}_{in} = -m\mathbf{A}$ in (3.58) is proportional to the mass m of the object under consideration, and the same is true of the gravitational force $\mathbf{F}_{gr} = m\mathbf{g}$ on any mass. That F_{gr} is proportional to m follows from experiments in which two different masses are dropped simultaneously and fall at the same rate — as supposedly tested by Galileo off the leaning tower of Pisa. All objects fall at the same rate only if F_{gr} is proportional to m, since only then does m cancel out of the equation of motion. The proportionality of F_{gr} and m has now been verified to about 1 part in 10^{11}.

That both inertial and gravitational forces are proportional to mass causes a remarkable ambiguity. Suppose that we are in an enclosed cabin, at rest on the surface of a planet with gravitational acceleration $\mathbf{g}$. Then one of the forces on any mass m is the gravitational attraction of the planet, $m\mathbf{g}$. We could try to verify this, for example, by dropping the mass, or by weighing it [Fig. 3.6(a)]. However, a little reflection should convince you that there is no mechanical experiment that can unambiguously confirm the presence of the gravitational force $m\mathbf{g}$. The trouble is that either of the experiments suggested in Fig. 3.6(a) (or any other mechanical experiment) can equally be explained by assuming that our cabin is in gravity-free space but is accelerating upward with $\mathbf{A} = -\mathbf{g}$, as in Fig. 3.6(b). Any effect that we attributed to the gravitational force $m\mathbf{g}$ can equally be explained by the inertial force $-m\mathbf{A}$. The impossibility of distinguishing between a gravitational force $m\mathbf{g}$ and the equivalent inertial force

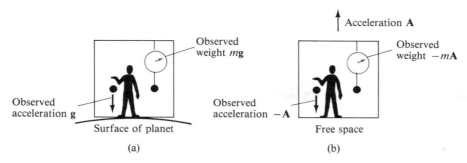

FIGURE 3.6 (a) Two experiments designed to verify the existence of the gravitational force $m\mathbf{g}$ on the surface of a planet. (b) The same experiments would produce the same results, if we were *not* near any gravitating body, but were instead accelerating with an acceleration $\mathbf{A} = -\mathbf{g}$.

$-m\mathbf{A}$ (with $\mathbf{A}$ equal but opposite to $\mathbf{g}$) is called the principle of equivalence. In classical physics, the equivalence principle applies only to mechanical experiments, but Einstein's general relativity starts from the assumption that the equivalence principle applies to *all* the laws of physics, mechanical and otherwise:*

> ### EINSTEIN'S EQUIVALENCE PRINCIPLE
> No experiment, mechanical or otherwise, can distinguish between a uniform gravitational field ($\mathbf{g}$) and the equivalent uniform acceleration ($\mathbf{A} = -\mathbf{g}$).

The general theory of relativity is built on this postulate in much the same way that special relativity was built on its two postulates (Section 2.1). As we shall describe briefly, the experimental evidence is such as to convince most physicists that general relativity and the equivalance principle on which it is based are correct.

As one would expect, general relativity agrees with the Newtonian theory of gravity under those conditions where the latter was already known to work well. Specifically, as the gravitating masses get smaller, the differences between Einstein's and Newton's theories approach zero. In fact, even with a mass as large as the sun's, the difference is usually very small. General relativity is of practical importance only for systems that include very large, dense masses and in situations requiring high precision, where very small differences may be important.

One branch of physics where general relativity has always been important is cosmology, the study of the structure and evolution of the whole universe. Here the effects of gravity are paramount and *have* to be treated properly (that is, using general relativity). In the last three decades there have been several developments relevant to general relativity, and there has been a burgeoning of interest in the subject. One such development is the discovery of the probable existence of black holes, whose behavior can only be analyzed with the help of general relativity. Another is the success of several experiments with sufficient precision to distinguish between Newton's and Einstein's (and other) theories

---

* Notice the striking parallel between the general and special theories. In both cases, Einstein took a principle that applied to classical mechanics and made the bold assumption that it applied to all physical laws.

of gravity. (And, incidentally, all such experiments seem to favor Einstein's theory.) Finally, navigational systems using satellites, such as the Global Positioning System of the U.S. Air Force, have become so precise that tiny corrections for the effect of gravity on time have to be made using general relativity.

A remarkable feature of general relativity is that the effects of gravity are incorporated into the *geometry* of space. Thus, instead of saying that the sun exerts forces on the planets causing them to follow their curved orbits, we say that the gravitational field of the sun causes a curvature of space, and it is this curvature that is responsible for the curved orbits of the planets. These ideas are expressed mathematically in the language of *differential geometry,* but to describe how this works would take us too far afield. Instead, we shall just describe a few of the theory's simpler consequences.

THE BENDING OF LIGHT BY GRAVITY

We can use the equivalence principle to predict the effects of a gravitational field in several experiments. Suppose, for example, that we shine a beam of light horizontally across the gravitational field of a massive star. The equivalence principle guarantees that the outcome of this experiment must be the same as if we were to observe the beam of light from the equivalent accelerated frame in gravity-free space. Accordingly, we consider first a flash of light that is shone through the window of a cabin S' that accelerates upward relative to an inertial frame S, in free space, as in Fig. 3.7. In part (a) we see the experiment as viewed in the inertial frame S, where the light travels in a horizontal path, but the cabin moves upward with increasing speed; evidently, the light hits the far wall at a point lower than the point at which it entered. Part (b) shows the same experiment as seen inside the accelerating cabin S'; in this frame the path of the light is curved downward. Finally, the equivalence principle guarantees that the behavior observed in the upwardly accelerating frame of part (b) must be the same as would be observed in the downward gravitational field of part (c). We conclude that light traveling across the gravitational field of a massive body must be deflected downward, toward the body.

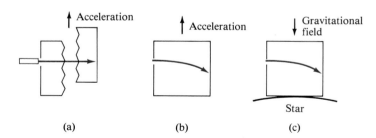

| (a) | (b) | (c) |

FIGURE **3.7** **(a)** A pulse of light is shone into the window of an accelerating cabin. In the inertial frame of the light source, the light travels in a horizontal line. By the time the pulse reaches the far wall, the cabin has moved upward. **(b)** This means that as seen in the accelerating cabin the light's path angles downward. Because the cabin moves upward with increasing speed, the observed path is actually *curved* down. **(c)** The equivalence principle guarantees that the same experiment performed in the equivalent gravitational field must appear the same as in (b); that is, the gravitational field must bend the light downward.

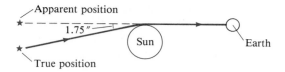

FIGURE 3.8 Light from the star on the left is deflected by 1.75 arcseconds as it skims past the sun. The star's apparent position is therefore 1.75 arcseconds above its true position.

Using general relativity, Einstein calculated that a light ray skimming past the surface of the sun should be deflected by 1.75 arcseconds.* This means that the apparent position of a star closely aligned with the sun's rim should be shifted by 1.75 arcseconds from its true position, as shown in Fig. 3.8. This shift is very hard to observe, since a star closely aligned with the sun is invisible because of the sun's own, brighter light. However, it was observed during the solar eclipse of 1919, and although the measurement was only about 30% accurate, it provided early experimental support for general relativity. In the early 1970s, the deflection was measured much more accurately using radio telescopes, and the results agreed with general relativity within about 1%. More recently, experiments using radar reflected from planets and spacecraft have provided results that agree within about 0.1%.

THE GRAVITATIONAL REDSHIFT

A second effect of gravity on light is the gravitational redshift. Imagine a beam of light fired upward from the floor to the ceiling of a cabin in the gravitational field of the earth (or other massive body), as in Fig. 3.9(a). The question we consider is this: If the frequency of the source on the floor is f_0, what is the frequency of the light received by a detector on the ceiling? To answer this question, we have only to imagine the cabin to be in gravity-free space, with acceleration $\mathbf{A} = -\mathbf{g}$, as in Fig. 3.9(b). We consider two inertial frames, S, in

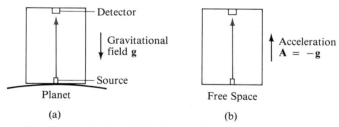

FIGURE 3.9 (a) In the gravitational redshift experiment, light is shone upward in a gravitational field. (b) The experiment of (a) is equivalent to this experiment, in which the cabin is accelerating upward in free space. The rest frame of the detector (at the time of detection) moves upward relative to the rest frame of the source (at the time of emission). Therefore, the light received is redshifted.

* This calculation is actually quite complicated: The simple argument based on the equivalence principle gives only half of the deflection, and the other half comes from a subtle geometrical effect of general relativity.

which the source is at rest when the light emitted, and S' the frame in which the detector is at rest when the light is received. The emitted frequency f_0 is the frequency measured in S, and the received frequency f is that in S'. During the time that the light travels from source to detector, the whole cabin is accelerating upward. This means that frame S' is moving upward relative to S, and the frequency f measured in S' is therefore *redshifted* compared to the frequency f_0 measured in S. By the equivalence principle the same conclusion applies in the gravitational field of Fig. 3.9(a). Therefore, light is redshifted as it travels upward in a gravitational field. By a similar argument, it is blueshifted if it travels downward. The shifts concerned are usually very small, and this prediction was not accurately confirmed until 1960, when Pound and Rebka verified it using high-frequency radiation that traveled up and down a tower at Harvard. It has subsequently been checked with even greater precision (about 0.02%) using radiation sent from high-altitude rockets down to the earth's surface.

THE PRECESSION OF MERCURY'S ORBIT

The first test of general relativity was based on the discovery made in 1859 that the planet Mercury does not move in the perfect, fixed ellipse predicted by Newtonian theory [Fig. 3.10(a)]; instead, the orbit precesses in such a way that its axis rotates slowly (by 43 arcseconds per century) as in Fig. 3.10(b).* This anomalous motion had suggested to some astronomers that Mercury's orbit was being disturbed by a hitherto undetected planet, which was even given the name Vulcan. However, careful searches failed to find Vulcan. Some 50 years later, when he completed his general theory of relativity, Einstein found that the theory predicted a slow precession of all planetary orbits. For Mercury the theory predicted a precession in almost perfect agreement with the (present) observed 42.98 arcseconds per century. For the other planets, the predicted values were much smaller, but several of these have now been measured and agree well with the predictions.

BLACK HOLES

The prediction of general relativity that has most caught the public imagination is the black hole. This is a very heavy and dense star, whose gravitational field is so strong that no light — nor anything else — can escape from its interior. A black hole is formed when a star of several solar masses exhausts its supply of nuclear fuel. Without the fuel to maintain the pressure needed to support it, the star begins to collapse and continues to do so until its radius approaches a value called the Schwarzschild, or gravitational, radius, R_s. As $r \rightarrow R_s$, the rate of collapse slows down (as observed from far away), the redshift of light from the star grows indefinitely, and light from inside R_s can no longer escape at all. Since light from inside a black hole cannot escape, direct evidence for black holes is hard to come by. Nevertheless, there is now strong circumstantial evidence for their existence. For example, if an ordinary star were caught in orbit around a black hole, matter would be torn from the star by the black hole.

---

* The situation is actually more complicated: The other planets distort Mercury's orbit much more than this tiny effect, but when all these well-understood distortions were subtracted out, there remained an unexplained advance of 43 arcseconds per century. Because an orbit's axis is conveniently specified by its perihelion (the point of closest approach to the sun), this effect is usually referred to as the advance of Mercury's perihelion.

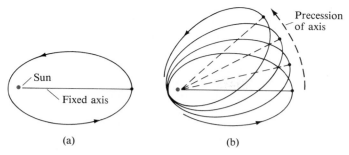

FIGURE 3.10 (a) According to Newtonian theory, a planet should move round an ellipse whose axis is fixed. (b) The axis of the ellipse actually rotates slowly, in agreement with general relativity. The effect is greatly exaggerated in this picture; the actual precession is greatest for Mercury, whose axis advances by just 43 arcseconds per century. (Courtesy Dale Prull.)

Before it disappears inside the black hole, this falling matter should reach such high energies that its collisions should produce X rays. Certain astronomical X-ray sources have been observed that fit the predictions of this model well. In addition, it is thought that extremely massive black holes may be responsible for the intense radiation from the superenergetic and distant objects called quasars.

GRAVITATIONAL WAVES

Another prediction of general relativity that has attracted much attention in the last two decades is the possibility of gravitational waves. Unlike the Newtonian theory, general relativity predicts that accelerating masses should radiate gravitational waves, just as accelerating electric charges radiate electromagnetic waves (like radio waves or light). However, even the most violent cosmic events produce very feeble gravitational waves, and although there have been several heroic attempts at detection, there have so far been no reproducible, direct observations of gravitational waves. On the other hand, there is strong indirect evidence, in that certain rapidly rotating star systems have been observed to be losing energy at precisely the rate that should result from their gravitational radiation. Intense efforts are under way to detect gravitational waves, and astronomers hope that they may soon give us an entirely new tool for studying the universe.

OVERVIEW

The observation in 1919 of the gravitational deflection of light passing the sun has been described as the first scientific media event. It drew public attention to relativity theory and helped make Einstein the best known scientific celebrity of all time. Nevertheless, for the next 40 years general relativity attracted less and less attention, and it remained a specialized, and even obscure, branch of physics. Then, sometime in the 1950s, it entered a renaissance, with the new generation of experiments that were sensitive enough to test it, and with the discovery of phenomena, like black holes, that absolutely require general relativity for their understanding. Today, it is, once again, undeniably an important part of modern physics.

Rest mass
Relativistic momentum, $\mathbf{p} = \gamma m \mathbf{u}$
Relativistic energy, $E = \gamma m c^2$
Rest energy, $E = mc^2$
Kinetic energy, $K = E - mc^2$
The "Pythagorean relation",
$$E^2 = (pc)^2 + (mc^2)^2$$
MeV, MeV/c, and MeV/c^2 as units of energy, momentum, and mass

Conversion of mass to other forms of energy
Binding energy
Circular motion of a charge in a magnetic field
Massless particles
[Optional section: inertial forces; principle of equivalence; bending of light by gravity; gravitational redshift; precession of Mercury's orbit]

PROBLEMS FOR CHAPTER 3

SECTION 3.3 (RELATIVISTIC MOMENTUM)

3.1 • The mass of an electron is about 9.11×10^{-31} kg. Make a table showing an electron's momentum, both the correct relativistic momentum and the nonrelativistic form, at speeds with $\beta = 0.1, 0.5, 0.9, 0.99$.

3.2 • Consider the collision between the two billiard balls of Fig. 3.1. Table 3.1 gives the four velocities measured in frame S. (a) Use the classical velocity-addition formula, Eq. (2.30), to verify that the velocities measured in S' would be as given in Table 3.2, *if S and S' were related classically.* (b) Verify that the correct relativistic answers given in Table 3.3 reduce to the values of Table 3.2 if all speeds are much less than c.

3.3 •• Consider the collision between the two billiard balls of Fig. 3.1. The four velocities involved, as measured in S, are given in Table 3.1. Use the velocity transformations (2.32) and (2.33) to verify that the four velocities measured in S' are as given in Table 3.3. Verify that the total classical momentum as measured in S' (that is, Σ $m\mathbf{u}'$) is *not* conserved.

3.4 ••• (a) Consider the collision of the two billiard balls of Fig. 3.1. Make a table similar to Table 3.1 but showing the relativistic momenta of the balls before and after the collision, as measured in S. Verify that, as seen in S, the total relativistic momentum is conserved. (b) Now, using the velocities of Table 3.3, make a table showing the four relativistic momenta as measured in S'. Verify that the total relativistic momentum is also conserved in S'. (Be careful! Remember that the factor γ in the definition $p = \gamma m u$ depends on u. Therefore with four different velocities there are in general four different factors γ. It takes some courage to wade through the algebra here, but it is rewarding to see the momenta come out equal.)

SECTION 3.4 (RELATIVISTIC ENERGY)

3.5 • At what speed would a body's relativistic energy E be twice its rest energy mc^2?

3.6 • An electron (rest mass 9×10^{-31} kg) is moving at $0.6c$. What is its energy E? At this speed, what fraction of its energy is rest energy?

3.7 • At what speed is a body's kinetic energy equal to twice its rest energy?

3.8 • We saw that the relativistic momentum $\mathbf{p} = \gamma m \mathbf{u}$ of a mass m can also be expressed as

$$\mathbf{p} = m \frac{d\mathbf{r}}{dt_0} \qquad (3.59)$$

where dt_0 denotes the *proper* time between two neighboring points on the body's path and has the same value for all observers. Show that the relativistic energy $E = \gamma m c^2$ can similarly be rewritten as

$$E = mc^2 \frac{dt}{dt_0}. \qquad (3.60)$$

The relations (3.59) and (3.60) make it easy to see how $\mathbf{p}$ and E transform from one inertial frame to another. (See Problem 3.10.)

3.9 •• If one defines a variable mass $m_{var} = \gamma m$, the relativistic momentum $\gamma m \mathbf{u}$ becomes $m_{var} \mathbf{u}$, which looks more like the classical definition. Show, however, that the relativistic kinetic energy $(\gamma - 1)mc^2$ is *not* equal to $\frac{1}{2} m_{var} u^2$.

3.10 ••• (a) Suppose that a mass m has momentum $\mathbf{p}$ and energy E as measured in a frame S. Use the relations (3.59) and (3.60) (Problem 3.8) and the known transformation of $d\mathbf{r}$ and dt to find the values of $\mathbf{p}'$ and E' as measured in a second frame S' traveling with speed v along Ox. (Notice that apart from some factors of c, the quantities $\mathbf{p}$ and E transform just like $\mathbf{r}$ and t. Remember that dt_0 has the same value for all observers.) (b) Use the results of part (a) to prove the following important result: If the total momentum and energy of a system are conserved as measured in one inertial frame S, the same is true in any other inertial frame S'.

3.11 • • • Consider the elastic, head-on collision of Example 3.3 (Section 3.4). The algebra leading to Eq. (3.19) for u_3 is surprisingly messy and not very illuminating. To simplify it, consider the case that the two masses are equal. Write down the equations for energy and momentum conservation, then prove that the final velocities are $u_3 = 0$ and $u_4 = u_1$. (In this special case, with equal rest masses, the relativistic result agrees with the nonrelativistic result.) [*Hints:* Remember that the factor γ that appears in the definitions of $\mathbf{p}$ and E depends on velocity; thus, with four different velocities one has four different values of γ. The identity $\gamma^2(1 - \beta^2) = 1$ will be useful.]

SECTION **3.5** (TWO USEFUL RELATIONS)

3.12 • By squaring the definitions (3.20) and (3.21) of $\mathbf{p}$ and E, verify the "Pythagorean relation" $E^2 = (pc)^2 + (mc^2)^2$.

3.13 • A nuclear particle has mass $3 \text{ GeV}/c^2$ and momentum $4 \text{ GeV}/c$. (a) What is its energy? (b) What is its speed? ($1 \text{ GeV} = 10^9 \text{ eV}$.)

3.14 • A particle is observed with momentum $500 \text{ MeV}/c$ and energy 1746 MeV. What is its speed? What is its mass (in MeV/c^2 and in kg)?

3.15 • A proton (rest mass $938 \text{ MeV}/c^2$) has kinetic energy 500 MeV. What is its momentum (in MeV/c and in $\text{kg} \cdot \text{m/s}$)? How fast is it traveling?

3.16 • At the Stanford Linear Accelerator electrons are accelerated to energies of 50 GeV ($1 \text{ GeV} = 10^9 \text{ eV}$). (a) If this energy were classical kinetic energy, what would be the electrons' speed? (Take the electrons' mass to be $0.5 \text{ MeV}/c^2$.) (b) Calculate γ and hence find the electrons' actual speed.

3.17 • The rest energy of a certain nuclear particle is 5 GeV ($1 \text{ GeV} = 10^9 \text{ eV}$) and its kinetic energy is found to be 8 GeV. What is its momentum (in GeV/c), and what is its speed?

3.18 • (a) Using the two identities (3.22) and (3.23) prove that the speed of any particle with $m > 0$ is always less than c. (b) If there could be particles with $m = 0$ (as in fact there can), prove from the same two identities that such particles would always have speed equal to c.

SECTION **3.6** (CONVERSION OF MASS TO ENERGY)

3.19 • When the radioactive nucleus of astatine 215 decays, it tears the whole atom into two atoms, bismuth 211 and helium 4:

$$^{215}\text{At} \rightarrow {}^{211}\text{Bi} + {}^{4}\text{He}.$$

(This type of decay is called α decay, because the helium 4 nucleus is often called an alpha particle.) The masses of the three atoms are

$$^{215}\text{At}: \quad 3.57019 \times 10^{-25} \text{ kg}$$
$$^{211}\text{Bi}: \quad 3.50358 \times 10^{-25} \text{ kg}$$
$$^{4}\text{He}: \quad 0.06647 \times 10^{-25} \text{ kg}.$$

What is the kinetic energy released in the decay (in joules and in MeV)?

3.20 • If an electron and proton (both initially at rest and far apart) come together to form a hydrogen atom, 13.6 eV of energy is released (mostly as light). By how much does the mass of an H atom differ from the sum of the electron and proton masses? What is the fractional difference $\Delta M/(m_e + m_p)$?

3.21 • When two molecules of hydrogen combine with one molecule of oxygen to form two water molecules,

$$2\text{H}_2 + \text{O}_2 \rightarrow 2\text{H}_2\text{O},$$

the energy released is 5 eV. (a) What is the mass difference between the three original molecules and the two final ones? (b) Given that the water molecule has mass about 3×10^{-26} kg, what is the fractional change in mass, $\Delta M/(\text{total mass})$? (Does it matter significantly whether you use the initial or final total mass?) (c) If one were to form 10 g of water by this process, what would be the total change in rest mass?

3.22 • A subatomic particle A decays into two identical particles B; that is, $\text{A} \rightarrow \text{B} + \text{B}$. The two B particles are observed to have exactly equal and opposite momenta of magnitude p. (a) What can you deduce about the velocity of A just before the decay? (b) Derive an expression for the mass, m_A, of A in terms of m_B and p.

3.23 • • A lambda particle (Λ) decays into a proton and a pion, $\Lambda \rightarrow \text{p} + \pi$, and it is observed that the proton is left at rest. (a) What is the energy of the pion? (b) What was the energy of the original Λ? (The masses involved are $m_\Lambda = 1116$, $m_p = 938$, and $m_\pi = 140$, all in MeV/c^2. As is almost always the case, your best procedure is to solve the problem algebraically, in terms of the symbols m_Λ, m_p, m_π and, only at the end, to put in numbers.)

3.24 • • A particle A moving with momentum $\mathbf{p}_A \neq 0$ decays into two particles B and C as in Fig. 3.11. (a) Prove that the three momenta $\mathbf{p}_A$, $\mathbf{p}_B$, $\mathbf{p}_C$ lie in a plane. (b) If $m_B = m_C$ and if it is found that $\theta_B = \theta_C$,

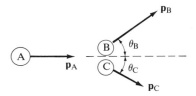

FIGURE 3.11 (Problems 3.24 and 3.35)

prove that particles B and C must have equal energies. (c) If it is found that $\theta_B = 0$, prove that $\theta_C = 0$ or 180°.

3.25 •• The K° meson is a subatomic particle of rest mass $m_K = 498$ MeV/c^2 that decays into two charged pions, K° → $\pi^+ + \pi^-$. (The π^+ and π^- have opposite charges but exactly the same mass, $m_\pi = 140$ MeV/c^2.) A K° at rest decays into two pions. Use conservation of energy and momentum, to find the energies, momenta and speeds of the two pions. (Give algebraic answers, in terms of the symbols m_K and m_π, first; then put in numbers.)

3.26 •• Many problems in relativity are best solved by viewing them first in a cleverly chosen reference frame. Here is an example: A K° meson (see Problem 3.25) traveling at $0.9c$ decays into a π^+ and π^- sending the π^+ exactly forward and the π^- exactly backward. Using the results of Problem 3.25, find the velocities of the two pions. (*Hint:* Let S be the frame of this problem and S' the K° rest frame. You can use the results of Problem 3.25 in S' and then use the velocity transformation to find the velocities in S.)

3.27 •• A particle of unknown mass M decays into two particles of known masses $m_1 = 0.5$ GeV/c^2 and $m_2 = 1$ GeV/c^2, whose momenta are measured to be $\mathbf{p}_1 = 2$ GeV/c along the y axis and $\mathbf{p}_2 = 1.5$ GeV/c along the x axis. (1 GeV = 10^9 eV.) Find the unknown mass M and its speed.

3.28 •• A mad scientist claims to have observed the decay of a particle of mass M into two identical particles of mass m with $M < 2m$. In response to the objection that this violates conservation of energy, he retorts that if M was traveling fast enough it could easily have energy greater than $2mc^2$ and hence could decay into the two particles of mass m. Show that he is wrong. (He has forgotten that momentum, as well as energy, must be conserved. You can analyze this problem in terms of these two conservation laws, but it is much easier to view the proposed reaction from the rest frame of the particle M.)

SECTION 3.7 (FORCE IN RELATIVITY)

3.29 • An electron with kinetic energy 1 MeV enters a uniform magnetic field B = 0.1 T (perpendicular to

the electron's velocity). What is the radius of the resulting circular orbit? [Take the electron's rest mass to be 0.5 MeV/c^2. Don't forget that to use Eq. (3.48) you must express p in the proper SI units, which are kg·m/s. The needed conversion factor can be found inside the front cover.]

3.30 • An electron (rest mass 0.5 MeV/c^2) traveling at $0.7c$ enters a magnetic field of strength of 0.02 T and moves on a circular path of radius R. (a) What would be the value of R according to classical mechanics? (b) What is R according to relativity? (The fact that the observed radius agrees with the relativistic answer is good evidence in favor of relativistic mechanics.)

3.31 •• Consider the relativistic form of Newton's second law, $\mathbf{F} = d\mathbf{p}/dt$, for a single mass m with velocity $\mathbf{u}$ and momentum $\mathbf{p} = \gamma m\mathbf{u}$. (a) Prove that if $\mathbf{F}$ is always perpendicular to $\mathbf{u}$ (as is the case if $\mathbf{F}$ is the magnetic force on a charged particle), then γ is constant and we can write $\mathbf{F} = \gamma m\mathbf{a}$, where $\mathbf{a}$ is the acceleration $d\mathbf{u}/dt$. (b) If, instead, $\mathbf{F}$ is parallel to $\mathbf{u}$, show that $\mathbf{F} = \gamma^3 m\mathbf{a}$. (c) Rewrite your results in terms of the variable mass $m_{var} = \gamma m$. Note that in case (a) the result ($\mathbf{F} = m_{var}\mathbf{a}$) looks just like the nonrelativistic form of Newton's second law, but in case (b) it does not.

3.32 •• In deriving the work-energy theorem (Example 3.9 in Section 3.7) we used the chain rule to find the derivative dE/dt in Eq. (3.47). This is actually a rather subtle point because E is a function of three variables, $E(p_x, p_y, p_z)$, and the relevant form of the chain rule is

$$dE = \frac{\partial E}{\partial p_x} dp_x + \frac{\partial E}{\partial p_y} dp_y + \frac{\partial E}{\partial p_z} dp_z,$$

where the three derivatives are partial derivatives. Use this equation to fill in the details in Eq. (3.47). (If you don't know about partial derivatives, don't trouble with this problem now. We shall describe partial differentiation carefully in Chapter 9.)

SECTION 3.8 (MASSLESS PARTICLES)

3.33 • The uncharged pion, $\pi°$, is a subatomic particle that is closely related to the charged pions discussed before. The $\pi°$ has mass 135 MeV/c^2 (very nearly the same as the charged pions, whose mass is 140 MeV/c^2) and decays into two photons, $\pi° \rightarrow \gamma + \gamma$. Assuming that the pion was at rest, what is the energy of each photon?

3.34 •• A neutral pion traveling along the x axis decays into two photons, one being ejected exactly forward, the other exactly backward. The first photon has three times the energy of the second. Prove that the original pion had speed $0.5c$.

3.35 •• Consider the decay $A \rightarrow B + C$ shown in Fig. 3.11 (Problem 3.24). Suppose that both B and C are massless and that $\theta_B = \theta_C$. Use conservation of energy and momentum to prove that $\cos \theta_B = \beta$, where β is the dimensionless "velocity" (u/c) of particle A.

3.36 •• The positive pion decays into a muon and a neutrino, $\pi^+ \rightarrow \mu^+ + \nu$. The pion has rest mass $m_\pi = 140$ MeV/c^2, the muon has $m_\mu = 106$ MeV/c^2, while the neutrino apparently has $m_\nu = 0$. (Strictly speaking, there are several kinds of neutrino; the one discussed here is denoted ν_μ to indicate that it is the one produced with a muon.) Assuming the original pion was at rest, use conservation of energy and momentum to show that the speed of the muon is given by

$$\frac{u}{c} = \frac{(m_\pi/m_\mu)^2 - 1}{(m_\pi/m_\mu)^2 + 1}.$$

Evaluate this numerically.

3.37 ••• An atom (or a nucleus) is ordinarily found in its *ground state,* or state of lowest energy. However, by supplying some energy one can lift it to an *excited state.* An excited state of an atom X is sometimes denoted X* and, because of the additional energy ΔE, has a mass m^* slightly greater than that of the ground state (m):

$$m^* = m + \Delta m,$$

where $\Delta m = \Delta E/c^2$. If left in isolation, the excited state X* usually drops back to the ground state emitting a single photon:

$$X^* \rightarrow X + \gamma.$$

If the atom were immovable, the photon would carry off all the additional energy:

$$E_\gamma = \Delta mc^2.$$

In reality, conservation of momentum requires the atom to recoil, so that a little of the energy goes to kinetic energy of the recoiling atom. (*a*) Using conservation of momentum and energy and assuming that the excited atom X* was at rest, show that

$$E_\gamma = \Delta mc^2 \left(1 - \frac{\Delta m}{2m^*}\right).$$

(*b*) The energy needed to lift a hydrogen atom to its lowest excited state is 10.2 eV. Evaluate the fraction $\Delta m/m^*$ for this state. (Does it make a significant difference whether you use m or m^* in the denominator?) What percentage of the available energy Δmc^2 goes to the photon in the decay of this excited state? (Your result illustrates what a good approximation it usually is to assume that all the energy goes to the photon in an atomic transition.)

SECTION 3.9 (WHEN IS NONRELATIVISTIC MECHANICS GOOD ENOUGH?)

3.38 • If u is much less than c, the binomial approximation ($\gamma \approx 1 + \beta^2/2$) shows that the relativistic kinetic energy, $K_{rel} = (\gamma - 1)mc^2$ is about equal to the nonrelativistic $K_{NR} = \frac{1}{2}mu^2$. An even better approximation is to keep the first *three* terms in the binomial series for γ. (The binomial series can be found in Appendix B.) Show in this way that the difference between K_{rel} and K_{NR} is about $3\beta^4 mc^2/8$. Express this difference as a fraction of K_{NR} and find the maximum value of β for which the difference is less than 1%.

3.39 • Consider a relativistic particle of mass m and kinetic energy K. Derive an expression for the particle's speed u in terms of K and m. Make a table showing both the correct value of u and the value you would get if you assumed that $K = \frac{1}{2}mu^2$, for $K = 0.001mc^2, 0.01mc^2, 0.1mc^2$, and mc^2. Calculate the percent discrepancy for each case. Approximately what is the largest value of K for which a nonrelativistic calculation of u is within 1% of the correct value?

3.40 •• The cyclotron is a device for accelerating protons (or other charged particles) to high energies. The protons are held in a circular orbit of radius $R = p/eB$ by a uniform magnetic field B. Twice in each orbit they are subjected to an accelerating electric field. As they speed up, R increases and they spiral slowly outward. (*a*) Show that the period of each orbit is $T = 2\pi\gamma m/(eB)$. (*b*) As long as the motion is nonrelativistic, $\gamma \approx 1$ and the period is constant. This greatly simplifies the design of the cyclotron, since the accelerating field can be applied at a constant frequency. Assuming that a cyclotron can tolerate no more than a 2% increase in the period, what is the highest kinetic energy of the protons it can produce?

Atoms

In Part II (Chapters 4 to 11) we describe the second of the two theories that transformed twentieth-century physics — quantum theory. Just as relativity can be roughly characterized as the study of phenomena involving high speeds ($v \sim c$), so quantum theory can be described as the study of phenomena involving *small* objects — generally of atomic size or smaller.

We begin our account of quantum theory, in Chapter 4, with a descriptive survey of the main properties of atoms (their size, mass, constituents, and so on). In Chapters 5 and 6 we describe some puzzling properties of microscopic systems that began to emerge in the late nineteenth century — some properties of light in Chapter 5, and of atoms in Chapter 6. All of these puzzles pointed up the need for a new mechanics — quantum mechanics, as we now say — to replace classical mechanics in the treatment of microscopic systems. In Chapters 8 to 10 we describe the basic ideas of the new quantum mechanics, which began to develop around 1900 and was practically complete by 1925. In particular, we introduce the Schrödinger equation, which is the basic equation of quantum mechanics, just as Newton's second law is the basic equation of Newtonian mechanics.

Armed with the Schrödinger equation, we can calculate most of the important properties of the simplest of all atoms, the hydrogen atom (with its one electron), as we describe in Chapter 9. In Chapter 10 we introduce one more important idea of the new quantum mechanics—the electron's spin angular momentum. Then in Chapter 11 we are ready to apply all these ideas to the general multielectron atom. We shall see that quantum theory gives a remarkably complete account of all the 100 or so known different atoms, and that, in principle at least, it explains all of chemistry.

Although quantum mechanics was first developed to explain the properties of atoms, it has also been applied with extraordinary success to other systems, some smaller than atoms, such as subatomic particles, and some larger, such as molecules and solids. We shall describe these other applications in Parts III and IV.

4.1 Introduction

Quantum physics is primarily the physics of microscopic systems. Historically, the most important such system was the atom, and even today the greatest triumph of quantum theory is the complete and accurate account that it gives us of atomic properties. For the next several chapters we shall be discussing atoms, and in this chapter, therefore, we give a brief description of the atom and its constituents, the electron, proton, and neutron.

Much of what we describe in this chapter is by now almost general knowledge. Certainly, if you have taken a course in high-school physics or chemistry, you will have met most of it before. If this is the case, you can treat this chapter as a review, or even just glance through it and proceed to Chapter 5.

4.2 Elements, Atoms, and Molecules

The concept of the atom arose in scientists' search to identify the basic constituents of matter. More than 2000 years ago Greek philosophers had recognized that this search requires answers to two questions: First, among the thousands of different substances we find around us—sand, air, water, soil, gold, diamonds—how many are the basic substances, or **elements,** from which all the rest are formed? Second, if one took a sample of one of these elements and subdivided it over and over again, could this process of repeated subdivision go on forever, or would one eventually arrive at some smallest, indivisible unit, or **atom** (from a Greek word meaning "indivisible")? With almost no experimental data, the Greeks could not find satisfactory answers to these questions; but this in no way lessens their achievement in identifying the right questions to ask.

Serious experimental efforts to identify the elements began in the eighteenth century with the work of Lavoisier, Priestley, and other chemists. By the end of the nineteenth century, about 80 of the elements had been correctly identified, including all of the examples listed in Table 4.1. Today we know that there are 90 elements that occur naturally on earth. All of the matter on earth is made up from these 90 elements, occasionally in the form of a pure element, but usually as a chemical compound of two or more elements, or as a mixture of such compounds.

TABLE 4.1

A few of the elements with their chemical symbols. For a complete list of the elements, see the periodic table inside the back cover or the alphabetical list in Appendix C.

Hydrogen (H)	Oxygen (O)	Iron (Fe)
Helium (He)	Sodium (Na)	Copper (Cu)
Carbon (C)	Aluminum (Al)	Lead (Pb)
Nitrogen (N)	Chlorine (Cl)	Uranium (U)

In addition to the 90 naturally occurring elements, there are about a dozen elements that can be created artificially in nuclear collisions. All of these artificial elements are unstable and disintegrate with half-lives much less than the age of the earth; this means that even if any of them were present when the earth was formed, they have long since decayed and are not found naturally in appreciable amounts.*

The evidence that the elements are composed of characteristic smallest units, or atoms, began to emerge about the year 1800. Chemists discovered the law of definite proportions: When two elements combine to form a pure chemical compound, they always combine in a definite proportion by mass. For example, when carbon (C) and oxygen (O) combine to form carbon monoxide (CO), they do so in the proportion 3:4; three grams of C combine with four of O to form seven grams of CO:

$$(3 \text{ g of C}) + (4 \text{ g of O}) \rightarrow (7 \text{ g of CO}).$$

If we were to add some extra carbon, we would not get any additional CO; rather, we would get the same 7 g of CO, with all the extra carbon remaining unreacted.

The law of definite proportions was correctly interpreted by the English chemist Dalton as evidence for the existence of atoms. Dalton argued that if we assume that carbon and oxygen are composed of atoms whose masses are in the ratio† 3:4, and if CO is the result of an exact pairing of these atoms (one atom of C paired with each atom of O), the law of definite proportions would immediately follow: The total masses of C and O that combine to form CO would be in the same ratio as the masses of the individual atoms, namely 3:4. If we add extra C (without any additional O), the extra C atoms have no O atoms with which to pair; thus we get no additional CO and the extra C remains unreacted.

We now know that this interpretation of the law of definite proportions was exactly correct. However, its general acceptance took 60 years or more, mainly because the situation was considerably more complicated than our single example suggests. For example, carbon and oxygen can also combine in the ratio 3:8 to form carbon dioxide, CO_2. Dalton was aware of this particular complication and argued (correctly) that the carbon atom must be able to combine with one O atom (to form CO) *or* two O atoms (to form CO_2). Nevertheless, this kind of ambiguity clouded the issue for many years.

JOHN DALTON (1766–1844, English chemist). In his book, *New System of Chemical Philosophy,* Dalton argued that the chemists' law of definite proportions is evidence for the existence of atoms. Using this idea, he was the first to draw up a table of relative atomic masses.

* A few elements with half-lives much less than the earth's age are observed to occur naturally, but this is because some natural nuclear process (radioactivity or cosmic-ray collisions, for example) is creating a fresh supply all the time.

† In stating Dalton's arguments we have replaced his terminology and measured masses with their modern counterparts. In particular, Dalton had measured the ratio of the C and O masses to be 5:7, not 3:4, as we now know it.

TABLE 4.2

A few simple molecules.

Carbon monoxide, CO	Oxygen, O_2	Glucose, $C_6H_{12}O_6$
Water, H_2O	Nitrogen, N_2	Ethyl alcohol, C_2H_6O
Ammonia, NH_3	Sulfur, S_8	Urea, CON_2H_4

A stable group of atoms, such as CO or CO_2, is called a **molecule**. A few examples of molecules are listed in Table 4.2, where we follow the convention that a subscript on any symbol indicates the number of atoms of that kind (and single atoms carry no subscript — thus H_2O has two H atoms and one O atom). As the list shows, a molecule may contain two or more different atoms, for example, the CO and $C_6H_{12}O_6$ molecules; in this case we say that the two or more elements have combined to form a compound, of which the molecule is the smallest unit. Molecules can also contain atoms of just one kind, for example, O_2; in this case, we do not speak of a new compound; we say simply that the element normally occurs (or sometimes occurs) not as separate atoms, but as groups of atoms clustered together, that is, as molecules.* Molecules can contain small numbers of atoms, like the examples in Table 4.2; but certain organic molecules, such as proteins, can contain tens of thousands of atoms.

Although most of the credit for establishing the existence of atoms and molecules goes to the chemists, a second strand of evidence came from the kinetic theory of gases — which we would regard today as a part of physics. This theory assumed, correctly, that a gas consists of many tiny molecules in rapid motion. By applying Newton's laws of motion to the molecules, one could explain several important properties of gases (Boyle's law, viscosity, Brownian motion, and more). These successes gave strong support to the atomic and molecular hypotheses. In addition, kinetic theory, unlike the chemical line of reasoning, gave information on the actual size and mass of molecules. (The law of definite proportions implied that the C and O atoms have masses in the proportion 3:4 but gave no clue as to the actual magnitude of either.) Kinetic theory gave, for example, an expression for the viscosity of a gas in terms of the size of the individual molecules. Thus measurement of viscosity allowed one to determine the actual size of the molecules — a method first exploited by Loschmidt in 1885.

By the beginning of the twentieth century it was fairly generally accepted that all matter was made up of elements, the smallest units of which were atoms. Atoms could group together into molecules, the formation of which explained the chemical compounds. The relative masses of many atoms and molecules were known quite accurately, and there were already reasonably reliable estimates of their actual masses.

4.3 Electrons, Protons, and Neutrons

Our story so far has the atom, true to its name, as the indivisible smallest unit of matter, and until the late nineteenth century there was, in fact, no direct evidence that atoms could be subdivided. The discovery of the first subatomic

* This possibility was probably the source of greatest confusion for the nineteenth-century chemists; if one wrongly identified the O_2 molecule as an atom, then one's interpretation of all other molecules containing oxygen — CO, CO_2, and so on — would also be incorrect.

J. J. THOMSON (1856–1940, English). As head of the famous Cavendish Laboratory in Cambridge for 35 years, Thomson was one of the most influential figures in the development of modern physics. Seven of his students and assistants went on to win the Nobel Prize, and he, himself, won it for his discovery of the electron.

particle, the negatively charged **electron,** is generally attributed to J. J. Thomson (1897), whose experiments we describe briefly in Section 4.7. On the basis of his experiments, Thomson argued that electrons must be contained inside atoms, and hence that the atom is in fact divisible.

Fourteen years later (1911) Rutherford argued convincingly for the now familiar picture of the atom as a tiny planetary system, in which the negative electrons orbit about a central positive **nucleus.** Rutherford's conclusion was based on an experiment that was the forerunner of many modern experiments in atomic and subatomic physics. In these experiments, called **scattering experiments,** one fires a subatomic projectile, such as an electron, at an atom or nucleus; by observing how the projectile is deflected, or *scattered,* one can deduce the properties of the target atom or nucleus. In the Rutherford experiment, the projectiles were alpha (α) particles — positively charged, subatomic particles ejected by certain radioactive substances. When these were directed at a thin metal foil, Rutherford found that almost all of them passed straight through, but that a few were deflected through large angles. In terms of his planetary model, Rutherford argued that the great majority of α particles never came close to any nuclei and encountered only a few electrons, which were too light to deflect them appreciably. On the other hand, a few of the α particles would pass close to a nucleus and would be deflected by the strong electrostatic force between the α particle and the nucleus; these, Rutherford argued, were the α particles that scattered through large angles. As we shall describe in Section 4.9, Rutherford used his model to predict the number of α particles that should be scattered as a function of scattering angle and energy. The beautiful agreement between Rutherford's predictions and the experimental observations was strong evidence for the planetary model of the atom, with its electrons orbiting around a tiny nucleus.

Within another eight years (1919) Rutherford had shown that the atomic nucleus can itself be subdivided, by establishing that nuclear collisions can break up a nitrogen nucleus. He identified one of the ejected particles as a hydrogen nucleus, for which he proposed the new name **proton** ("first one") in honor of its role in other nuclei. In 1932, Chadwick showed that nuclei contain a second kind of particle, the neutral **neutron.** (The experiments that identified the proton and neutron as constituents of nuclei will be described in Chapter 13.) With this discovery, the modern picture of the constituents of atoms was complete. Every atom contains a definite number of electrons, each with charge $-e$, in orbit around a central nucleus; and nuclei consist of two kinds of nuclear particles, or **nucleons,** the proton, with charge $+e$, and the uncharged neutron.

The constituents of five common atoms are shown in Table 4.3. In every

ERNEST RUTHERFORD (1871–1937, British). Rutherford succeeded J. J. Thomson as Cavendish Professor in 1919. In 1908, he had won the Nobel Prize in chemistry for showing that radioactivity transforms one element into another. Within another four years, he had established the existence of the atomic nucleus, and in 1917 he identified the first man-made nuclear reaction and proved that the proton is one of the constituents of nuclei.

TABLE 4.3
The constituents of five representative atoms.

Atom	Electrons	Nucleons	
		Protons	**Neutrons**
Hydrogen, H	1	1	0
Helium, He	2	2	2
Carbon, C	6	6	6
Iron, Fe	26	26	30
Uranium, U	92	92	146

case the number of electrons is equal to the number of protons, reflecting that the atom is neutral (in its normal state). Hydrogen is the only atom that has no neutrons. In all other atoms the numbers of neutrons and protons are roughly equal. (We shall see the reason for this in Chapter 12.) In many lighter atoms (helium or carbon, for example) the two numbers are exactly the same, while in most medium atoms the number of neutrons is a little larger; in the heaviest atoms there are about 50% more neutrons than protons.

In addition to their role as the building blocks of atoms, the electron and proton also define the smallest observed unit of charge. Their charges are exactly equal and opposite, with magnitude e,

$$q_p = -q_e = e = 1.60 \times 10^{-19} \text{ C}.$$

No charge smaller than e has been observed, and all known charges are integral multiples of e,*

$$0, \pm e, \pm 2e, \pm 3e, \ldots$$

Because e is so small, typical macroscopic charges are very large multiples of e, and the restriction to integral multiples is usually unimportant. On the atomic level, the existence of a smallest unit of charge is obviously very important.

By 1932 it appeared that all matter was made from just three subatomic particles, the electron, proton, and neutron. This picture of matter was a distinct simplification compared to its predecessor, with its 100 or so elements, each with a characteristic atom. As we shall describe in Chapter 14, it now appears that at least some of the subatomic particles themselves have an internal structure, being made of sub-subatomic particles called quarks. However, for the purposes of atomic physics, and much of nuclear physics, the picture of matter as made of electrons, protons, and neutrons seems to be quite sufficient, and this is where we shall stop the story for now.

4.4   Some Atomic Parameters

The distribution of electrons in their atomic orbits, and of protons and neutrons inside the nucleus, are two of the major concerns of quantum theory, as we shall describe in several later chapters. Indeed, they are still the subjects of current research. Nevertheless, a surprising number of atomic and nuclear properties can be understood just from an approximate knowledge of a few parameters, such as the size of the electron orbits and the masses of the electron, proton, and neutron. In this section we discuss some of these parameters, with which you should become familiar.

The size of an atom is not a precisely defined quantity, but can be characterized roughly as the radius of the outermost electron's orbit. This radius varies surprisingly little among the atoms, ranging from about 0.05 nm (helium) to about 0.3 nm (cesium, for example). In the majority of atoms it is between 0.1 and 0.2 nm. Thus for all atoms we can say that

---

* As we shall discuss in Chapter 14, there are reasons for believing that certain "sub-subatomic" particles, called quarks, have charge $e/3$. However, there have been no authenticated direct observations of such fractional charges.

$$\boxed{\begin{array}{c} \text{atomic radius} \approx \text{radius of outer electron orbits} \\ \sim 0.1 \text{ nm} = 10^{-10} \text{ m.} \end{array}} \quad (4.1)$$

The radii of nuclei are all much smaller than atomic radii and are usually measured in terms of the *femtometer* (fm) or fermi, defined as

$$1 \text{ fm} = 10^{-15} \text{ m.} \quad (4.2)$$

Nuclear radii increase steadily from about 1 fm for the lightest nucleus, hydrogen (which is a single proton, of course), to about 8 fm for the heaviest nuclei (for example, uranium). Thus we can say that

$$\boxed{\text{nuclear radius} \approx \text{a few fm} = \text{a few} \times 10^{-15} \text{ m.}} \quad (4.3)$$

The single most important thing about the two sizes (4.1) and (4.3) is their great difference. The atomic radius is at least 10^4 times larger than the nuclear radius. Thus if we made a scale model in which the nucleus was represented by a pea, the atom would be the size of a football stadium. Now it turns out that in chemical reactions atoms approach one another only close enough for their outer electron orbits to overlap. This means that the chemical properties of atoms are determined almost entirely by the distribution of electrons, and are substantially independent of what is happening inside the nucleus. Further, quantum mechanics predicts (as we shall find in Chapter 11) that the distribution of electrons is almost completely determined just by the *number* of electrons. Thus the chemical properties of an atom are determined almost entirely by the number of its electrons. This important number is called the **atomic number** and is denoted by the letter Z:

$$\boxed{\begin{array}{l} \text{atomic number, } Z, \text{ of an atom} \\ \quad = \text{number of electrons in neutral atom} \\ \quad = \text{number of protons in nucleus.} \end{array}} \quad (4.4)$$

That Z is also the number of protons in the nucleus follows because the numbers of electrons and protons are equal in a neutral atom.

We should mention that many atoms can gain or lose a few electrons fairly easily. When this happens we say that the atom is *ionized* and refer to the charged atom as an **ion**. (For example, the Ca^{2+} ion is a calcium atom that has lost two electrons; the Cl^- ion is a chlorine atom that has gained one electron.) Since the number of electrons in an atom can vary in this way, one must specify that Z is the number of electrons in the *neutral* atom, as was done in (4.4).

Since chemical properties are determined mainly by the atomic number Z, one might expect that each chemical element could be identified by the atomic number of its atoms, and this proves to be so. All the atoms of a given element have the same atomic number Z, and, conversely, for every number Z between 1 and 107 (the highest value to be officially recognized with a name) there is exactly one element. Table 4.4 lists a few elements with their atomic numbers. We see that the hydrogen atom has one electron ($Z = 1$), helium two ($Z = 2$), carbon six, oxygen eight, and so on through uranium with 92, and on to the artificial unnilseptium with 107. An alphabetical list of all the elements

TABLE 4.4

A few elements listed by atomic number, Z.

1 Hydrogen	6 Carbon	82 Lead
2 Helium	7 Nitrogen	⋮
3 Lithium	8 Oxygen	92 Uranium
4 Beryllium	⋮	⋮
5 Boron	26 Iron	107 Unnilseptium

can be found in Appendix C; a table of their properties — the periodic table — is inside the back cover.

The mass of an atom depends on the masses of the electron, proton, and neutron, which are as follows:

$$\text{electron:} \quad m_e = 0.511 \text{ MeV}/c^2 = 9.11 \times 10^{-31} \text{ kg}$$

$$\text{proton:} \quad m_p = 938.3 \text{ MeV}/c^2 = 1.673 \times 10^{-27} \text{ kg} \qquad (4.5)$$

$$\text{neutron:} \quad m_n = 939.6 \text{ MeV}/c^2 = 1.675 \times 10^{-27} \text{ kg}.$$

To an excellent approximation (1 part in 1000) the proton and neutron masses are equal, and, by comparison, the electron mass is negligible (1 part in 2000). Thus we can say that

$$m_p \approx m_n \approx m_H \approx 940 \text{ MeV}/c^2, \qquad (4.6)$$

where m_H denotes the mass of the hydrogen atom (a proton plus an electron). For many purposes it is sufficient to approximate the mass (4.6) as roughly 1 GeV/c^2.

The result (4.6) makes the approximate calculation of atomic masses extremely simple, since one has only to count the total number of nucleons (that is, the number of protons plus the number of neutrons). For example, the helium atom has two protons and two neutrons and so is four times as massive as hydrogen:

$$m_{He} \approx 4m_H;$$

the carbon atom has six protons and six neutrons and so

$$m_C \approx 12m_H.$$

Since its number of nucleons determines an atom's mass, this number is called the **mass number** of the atom. It is denoted by A:

$$\text{mass number, } A, \text{ of an atom} = \text{number of nucleons in atom}$$
$$= (\text{number of protons}) + (\text{number of neutrons}). \qquad (4.7)$$

An atom with mass number A has mass approximately equal to Am_H.

It often happens that two atoms with the same atomic number Z (and hence the same numbers of electrons and protons) have different numbers of neutrons in their nuclei. Such atoms are said to be **isotopes** of one another. Since the two isotopes have the same number of electrons, they have almost identical chemical properties and so belong to the same chemical element. But

TABLE 4.5

The stable isotopes of four elements. The percent of each element that occurs naturally in each isotope is shown in parentheses. A complete list, including all stable isotopes of all the elements and their abundances, is given in Appendix D.

Element	Atomic Number	Isotopes
Carbon	6	^{12}C(98.9%), ^{13}C(1.1%)
Magnesium	12	^{24}Mg(79.0%), ^{25}Mg(10.0%), ^{26}Mg(11.0%)
Chlorine	17	^{35}Cl(75.8%), ^{37}Cl(24.2%)
Iron	26	^{54}Fe(5.8%), ^{56}Fe(91.7%), ^{57}Fe(2.2%), ^{58}Fe(0.3%)

since they have different numbers of neutrons, their mass numbers, and hence masses, are different. For example, the commonest carbon atom has 6 protons and 6 neutrons in its nucleus, but there is also an isotope with 6 protons but 7 neutrons. These two atoms have the same chemical properties, but have different masses, about $12m_H$ and $13m_H$, respectively. To distinguish isotopes we sometimes write the mass number A as a superscript before the chemical symbol. Thus the two isotopes of carbon just mentioned are denoted ^{12}C and ^{13}C (usually read as "carbon 12" and "carbon 13").

While some elements have only one type of stable atom, the majority have two or more stable isotopes. The maximum number belongs to tin with 10 stable isotopes, and on average there are about 2.5 stable isotopes for each element. Table 4.5 lists all of the stable isotopes of four representative elements.

Since the chemical properties of isotopes are so similar, the proportion of isotopes that occur in nature does not change in normal chemical processes. This means that the atomic mass of an element, as measured by chemists, is the weighted average of the masses of its various natural isotopes. This explains why chemical atomic masses are not always close to integer multiples of m_H. For example, we see from Table 4.5 that natural chlorine is $\frac{3}{4}$ the isotope ^{35}Cl and $\frac{1}{4}$ the isotope ^{37}Cl. Thus the chemical atomic mass of Cl is the weighted average $35.5m_H$. This complication caused some confusion in the historical development of atomic theory. The English physician Prout had pointed out as early as 1815 that the masses of atoms appeared to be integral multiples of m_H, suggesting that all atoms were made from hydrogen atoms. As more atomic masses were measured, examples of nonintegral masses (such as chlorine) were found, and Prout's hypothesis was rejected. Not until a hundred years later was it seen to be very nearly correct.

4.5 The Atomic Mass Unit

Since all atomic masses are approximately integer multiples of the hydrogen atom's mass, m_H, it would be natural to use a mass scale with m_H as the unit of mass, so that all atomic masses would be close to integers. In fact, this is approximately (although not exactly) how the atomic mass scale is defined, and the so-called **atomic mass unit,** or u, is to a good approximation just m_H.

To understand the exact definition of the atomic mass unit, we must examine why atomic masses are only approximately integral multiples of m_H. Let us consider an atom with atomic number Z and mass number A. The number of neutrons we denote by N, so that

$$A = Z + N.$$

If we ignore for a moment the requirements of relativity, the mass of our atom would be just the sum of the masses of Z electrons, Z protons, and N neutrons:

$$m = Zm_e + Zm_p + Nm_n = Zm_H + Nm_n. \qquad (4.8)$$

To the extent that $m_n \approx m_H$ we can write this as

$$m \approx Zm_H + Nm_H \qquad (4.9)$$

or

$$m \approx Am_H. \qquad (4.10)$$

In fact, of course, m_H and m_n are not exactly the same, the difference being of order 1 part in 1000. Thus the step from (4.8) to (4.9) is only an approximation, and our conclusion (4.10) can be in error by about 1 part in 1000.

There is a second, and more important, reason why (4.10) is only an approximation. We have learned from relativity that when bodies come together to form a stable bound system, the bound system has less mass than its separate constituents by an amount

$$\Delta m = \frac{B}{c^2} \qquad (4.11)$$

where B is the binding energy, or energy required to pull the bound system apart into its separate pieces. For example, the helium nucleus (2 protons + 2 neutrons) has a binding energy of about 28 MeV; thus the helium atom has 28 MeV/c^2 less mass than the nearly 4000 MeV/c^2 predicted by (4.10). This is a correction of about 7 parts in 1000 and is appreciably more important than the correction discussed previously.*

For all atoms (except hydrogen itself) there is a similar adjustment due to the nuclear binding energy. Further, it turns out that in almost all atoms the correction is in the same proportion, namely 7 or 8 parts in 1000. (We shall see why the correction is always about the same in Chapter 12.) The only significant exception to this statement is the hydrogen atom itself, which has *no* nuclear binding energy and hence no correction. Thus if we want a mass scale on which most atomic masses are as close as possible to integers, the mass of the H atom is certainly not the best choice for the unit. If we chose instead ^{4}He or ^{12}C and defined the unit as $\frac{1}{4}$ the mass of ^{4}He or $\frac{1}{12}$ the mass of ^{12}C, all atoms (except hydrogen) would have masses closer to integral multiples of the atomic mass unit. Which atom we actually choose is a matter of convenience, and by international agreement we use ^{12}C, defining the u as follows:†

$$
\begin{aligned}
1 \text{ atomic mass unit} = 1 \text{ u} &= \tfrac{1}{12} \text{ (mass of one neutral } ^{12}\text{C atom)} \\
&= 1.661 \times 10^{-27} \text{ kg} = 931.5 \text{ MeV}/c^2.
\end{aligned} \qquad (4.12)
$$

With the definition (4.12) we can write the mass of an atom whose mass number is A as

---

* There is also a correction for the binding energy of the electrons, but this is much smaller (1 part in 10,000 at the most) and is almost always completely negligible.

† The atomic mass unit defined here, sometimes called the unified mass unit, replaces two older definitions, one based on ^{16}O and the other on the natural mixture of the three oxygen isotopes.

TABLE 4.6

Atomic masses in *u* are always close to an integer.

^{1}H	1.008	^{35}Cl	34.969
^{4}He	4.003	^{56}Fe	55.939
^{12}C	12 exactly	^{208}Pb	207.977
^{16}O	15.995	^{238}U	238.049

$$\text{(mass of atom with mass number } A) \approx A \text{ u.} \qquad (4.13)$$

For most atoms this approximation is good to 1 part in 1000, and even in the worst case, hydrogen, it is better than 1 part in 100. Both of these claims are illustrated by the examples in Table 4.6.

EXAMPLE 4.1   Using the natural abundances in Table 4.5, find the chemical atomic mass of magnesium in u to three significant figures.

The required mass is the weighted average of the masses of the isotopes ^{24}Mg, ^{25}Mg, ^{26}Mg, as listed in Table 4.5. To three significant figures we can use the approximation (4.13) to give

average atomic mass of Mg
$$= [(0.79 \times 24) + (0.10 \times 25) + (0.11 \times 26)] \text{ u} = 24.3 \text{ u}, \quad (4.14)$$

in agreement with the observed value given inside the back cover. On those rare occasions when one needs greater accuracy, one can use the more precise atomic masses given in Appendix D.

4.6   Avogadro's Number and the Mole

From a fundamental point of view the natural way to measure a quantity of matter is to count the number of molecules (or atoms). In practice, a convenient amount of matter usually has a very large number of molecules, and it is sometimes better to have a larger unit than the individual molecule. The usual choice for that larger unit is the *mole,* which is defined as follows.

We first define **Avogadro's number,** N_A, as the number of atoms in 12 grams of ^{12}C:

$$\text{Avogadro's number, } N_A$$
$$= \text{(number of atoms in 12 grams of } ^{12}\text{C)} = 6.022 \times 10^{23}. \qquad (4.15)$$

In this definition there is nothing sacred or fundamental about 12 grams; it is simply a reasonable macroscopic amount. (There *is* something *convenient* about the number 12 when used in conjunction with ^{12}C, as we shall see directly.) Since N_A atoms of ^{12}C (each of mass 12 u) have total mass 12 g we see that

$$N_A \times (12 \text{ u}) = 12 \text{ grams.}$$

Canceling the 12 and dividing, we see that

$$N_A = \frac{1 \text{ gram}}{1 \text{ u}}; \qquad (4.16)$$

that is, N_A is just the number of u in a gram.

We now define a **mole** of objects (carbon atoms, water molecules, elephants) as N_A objects. Thus a mole of carbon is N_A carbon atoms, a mole of water is N_A water molecules (and a mole of elephants is N_A elephants). Since a mole always contains the same number of objects, the mass of a mole is proportional to the mass of the object concerned. From its definition, a mole of ^{12}C has mass exactly 12 grams. Thus a mole of ^{4}He has mass approximately 4 grams (more exactly 4.003 grams); a mole of hydrogen atoms, about 1 gram; a mole of H_2 molecules, about 2 grams; a mole of water, about 18 grams (since the H_2O molecule has mass $2 + 16 = 18$ u); and so on.

In chemistry the masses of atoms and molecules are usually specified by giving, not the mass of an individual atom or molecule, but the mass of a mole. Thus instead of saying that

$$\text{mass of } ^{12}\text{C atom} = 12 \text{ u}$$

one says that

$$\text{mass of 1 mole of } ^{12}\text{C} = 12 \text{ grams.}$$

Masses expressed in this way, in grams per mole, are often misleadingly called atomic or molecular weights. Evidently, the mass of an atom or molecule in u has the same numerical value as its atomic or molecular weight in grams per mole.

The mole is not a fundamental unit. It is an arbitrary unit, which is convenient for people who deal with large numbers of atoms and molecules. (It can be compared with the "dozen", another arbitrary unit, which is convenient for people who deal with large numbers of eggs or doughnuts.) A purely theoretical physicist has little call to work with either moles or Avogadro's number, but for most of us it is important to be able to translate between the fundamental language of molecules and the practical language of moles.

To conclude this brief discussion of the mole we should mention that the mole is officially considered to have an independent dimension called the "amount of substance" (in the same way that meters have the dimension called "length"). For this reason Avogadro's number,

$$N_A = 6.022 \times 10^{23} \text{ objects/mole,}$$

is often called Avogadro's *constant,* since it is viewed as a dimensional quantity with the units mole^{-1}.

4.7 Thomson's Discovery of the Electron (optional)

Our main purpose in this chapter has been to introduce the principal characters in the story of atomic physics—atoms and molecules, electrons, protons, and neutrons. The history of their discovery is an interesting and important subject, with which the educated scientist should be acquainted. However, the early part of that history (prior to about 1890) can no longer be regarded as "modern physics," and in this book we shall content ourselves with the brief sketch already given.* On the other hand, the recent history of atomic physics is very much a part of modern physics, and from time to time we shall give more detailed descriptions of some of its experimental highlights. In particular, we conclude this chapter by describing three key experiments: the experiments of J. J. Thomson and Millikan, which together identified the electron and its

---

* The interested reader can find an authoritative but highly readable history of atomic theory from about 600 B.C. to 1960 in *The World of the Atom* by H. A. Boorse and L. Motz (New York: Basic Books 1966).

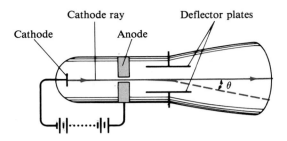

FIGURE **4.1** Thomson's cathode ray tube. When opposite charges were placed on the deflector plates the "rays" were deflected through an angle θ as shown.

principal properties, and the so-called Rutherford scattering experiment, which established the existence of the atomic nucleus.

The discovery of the electron is generally attributed to J. J. Thomson (1897) for a series of experiments in which he showed that "cathode rays" were in fact a stream of the negative particles that we now call electrons. Cathode rays, which had been discovered some 30 years earlier, were the "rays" emitted from the cathode, or negative electrode, of a cathode ray tube—a sealed glass tube containing two electrodes and low-pressure gas (Fig. 4.1). When a large potential is applied between the cathode and anode some of the gas atoms ionize and an electric discharge occurs. Positive ions hitting the cathode eject electrons, which are then accelerated toward the anode. With the arrangement shown in Fig. 4.1 some of the electrons pass through the hole in the anode and coast on to the far end of the tube. At certain pressures the "rays" can be seen by the glow that they produce in the gas, while at lower pressures they produce a fluorescent patch where they strike the end of the tube.

It had been shown by Crookes (1879) and others that cathode rays normally travel in straight lines and that they carry momentum. (When directed at the mica vanes of a tiny "windmill" they caused the vanes to rotate.) Crookes had also shown that the rays can be bent by a magnetic field, the direction of deflection being what would be expected for negative charges. All of this suggested that the rays were actually material particles carrying negative charge. Unluckily, attempts to deflect the rays by a transverse electric field had failed, and this failure had lead to the suggestion that cathode rays were not particles at all, but were instead some totally new phenomenon—possibly some kind of disturbance of the ether.

J. J. Thomson carried out a series of experiments that settled these questions beyond any reasonable doubt. He showed that if the rays were deflected into an insulated metal cup, the cup became negatively charged; and that as soon as the rays were deflected away from the cup, the charging stopped. He showed, by putting opposite charges on the two deflecting plates in Fig. 4.1, that cathode rays *were* deflected by a transverse electric field. (He was also able to explain the earlier failures to observe this effect. Previous experimenters had been unable to achieve as good a vacuum as Thomson's. The remaining gas in the tube was ionized by the cathode rays, and the ions were attracted to the deflecting plates, neutralizing the charge on the plates and canceling the electric field.) After further experiments with magnetic deflection of cathode rays, Thomson concluded: "I can see no escape from the conclusion that they are charges of negative electricity carried by particles of matter." Accepting this hypothesis, he next measured some of the particles' properties. By making each measurement in several ways he was able to demonstrate the consistency of his measurements and to add weight to his identification of cathode rays as negatively charged particles.

Thomson measured the speed of the cathode rays by applying electric and magnetic fields. The force on each electron in fields **E** and **B** is the well-known Lorentz force

$$\mathbf{F} = -e(\mathbf{E} + \mathbf{u} \times \mathbf{B}). \tag{4.17}$$

Using an E field alone he measured the deflection of the electrons. After removing E and switching on a transverse magnetic field B, he adjusted B until the magnetic deflection was equal to the previous electric deflection. Under these conditions $\mathbf{E} = \mathbf{u} \times \mathbf{B}$, or since **u** and **B** were perpendicular,

$$u = \frac{E}{B}. \tag{4.18}$$

(In Thomson's experiments u was of order 0.1 c. Given his experimental uncertainties, this means that his experiments can be analyzed nonrelativistically.) By measuring the heating of a solid body on to which he directed the electrons, he got a second estimate of u, which agreed with the value given by (4.18) given his fairly large uncertainties. (See Problem 4.21.)

Knowing the electrons' speed he could next find their "mass-to-charge" ratio, m/e. For example, as we saw in Chapter 3 — Eq. (3.48) — a B field causes electrons to move in a circular path of radius

$$R = \frac{mu}{eB}. \tag{4.19}$$

Thus measurement of R (plus knowledge of u and B), gave the ratio m/e. Similarly, it is easy to show (Problem 4.19) that an E field alone deflects the electrons through an angle

$$\theta = \frac{eEl}{mu^2} \tag{4.20}$$

where l denotes the length of the plates that produce E. Thus by measuring the deflection θ produced by a known E field, he could again determine the ratio m/e.

Even though Thomson used several different gases in his tube and different metals for his cathode, he always found the same value for m/e (within his experimental uncertainties). From this observation he argued, correctly, that there was apparently just one kind of electron, which must be contained in all atoms. He could also compare his value of m/e for electrons with the known values of m/e for ionized atoms. (The mass-to-charge ratio for ionized atoms had been known for some time from experiments in electrolysis — the conduction of currents through liquids by transport of ionized atoms.) Thomson found that even for the lightest atom (hydrogen) the value of m/e was about 2000 times greater than its value for the electron. The smallness of m/e for electrons had to be due to the smallness of their mass or the largeness of their charge, or some combination of both. Thomson argued (again correctly, as we now know) that it was probably due to the smallness of the electrons' mass.

4.8 Millikan's Oil-Drop Experiment (optional)

A frustrating shortcoming of Thomson's measurements was that they allowed him to calculate the ratio m/e, but not m or e separately. The reason for this is easy to see: Newton's second law for an electron in E and B fields states that

$$m\mathbf{a} = -e(\mathbf{E} + \mathbf{u} \times \mathbf{B}).$$

Evidently, any calculation of the electron's motion is bound to involve just the ratio m/e and not m or e separately. One way around this difficulty was to study the motion of some larger body, such as a droplet of water, whose mass M could be measured and which had been charged by the gain or loss of an electron. In this way one could measure the ratio M/e. Knowing the mass M of the droplet, one could find e, and thence the mass m of the electron from the known value of m/e.

The need to measure e (or m) separately was recognized by Thomson and his associates, who quickly set about measuring e using drops of water. However, the method proved difficult (for example, the water drops evaporated rapidly), and Thomson could not reduce his uncertainties below about 50%.

In an effort to avoid the problem of evaporation, the American physicist Millikan tried using oil drops and quickly developed an extremely effective technique. His apparatus is sketched in Fig. 4.2. Droplets of oil from a fine spray were allowed to drift into the space between two horizontal plates. The plates were connected to an adjustable voltage, which produced a vertical electric field E between the plates. With the field off, the droplets all drifted downward and quickly acquired their terminal velocity, with their weight balanced by the viscous drag of the air. When the E field was switched on, Millikan found that some of the drops moved down more rapidly, while others started moving upward. This showed that the drops had already acquired electric charges of both signs, presumably as a result of friction in the sprayer. The method by which Millikan measured these charges was ingenious and intricate, but the essential point can be understood from the following simplified account. (For some more details, see Problem 4.23.)

It proved possible to adjust the electric field E between the plates so as to hold any chosen droplet stationary. When this was the case the upward electric force must exactly have balanced the downward force of gravity:

$$qE = Mg, \qquad (4.21)$$

where q was the charge on the droplet. Since E and g were known, it remained only to find M in order to determine the charge q.

To measure M Millikan switched off the E field and observed the terminal

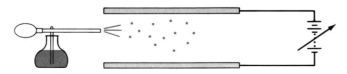

FIGURE 4.2 Millikan's oil-drop experiment. The potential difference between the plates is adjustable, both in magnitude and direction.

velocity v with which the droplet fell. (This velocity was very small and so easily measured.) From classical fluid mechanics it was known that the terminal velocity of a small sphere acted on by a constant force F is

$$v = \frac{F}{6\pi r \eta}, \tag{4.22}$$

where r is the radius of the sphere and η the viscosity of the gas in which it moves. In the present case F is just the weight of the droplet,

$$F = Mg = \tfrac{4}{3}\pi r^3 \rho g, \tag{4.23}$$

where ρ is the density of the oil. Substitution of (4.23) in (4.22) gives the speed of the falling droplet as

$$v = \frac{2r^2 \rho g}{9\eta}. \tag{4.24}$$

Since ρ, g, and η were known, measurement of v allowed Millikan to calculate the droplet's radius r and thence, from (4.23), its mass M. Using (4.21), he could then find the charge q.

From time to time it was found that a balanced droplet would suddenly move up or down, indicating that it had picked up an extra ion from the air. Millikan quickly learned to change the charge of the droplet at will by ionizing the air (by passing X rays through the apparatus, for example). In this way he could measure, not only the initial charge on the droplet, but also the *change* in the charge as it acquired extra positive and negative ions.

In the course of several years, Millikan observed thousands of droplets, sometimes watching a single droplet for several hours as it changed its charge a score or more of times. He used several different liquids for his droplets and various different procedures for changing the droplets' charges. In every case the original charge q, and all subsequent changes in q, were found to be integer multiples (positive and negative) of a single basic charge,

$$e = 1.6 \times 10^{-19} \text{ C}.$$

That the basic unit of charge was the same as the electron's charge was checked by pumping most of the air out of the apparatus. With few air molecules, there could be few ions for the droplet to pick up. Nevertheless, Millikan found that X rays could still change the charge on the droplets, but only in the direction of increasing positive charge. He interpreted this, correctly, to mean that X rays were knocking electrons out of the droplet. Since the changes in charge were still multiples of e (including sometimes e itself), it was clear that the electron's charge was itself equal to the unit of charge e (or rather $-e$, to be precise).

It is worth emphasizing the double importance of Millikan's beautiful experiments. First, he had measured the charge of the electron, $-e$. Combined with the measurement of m/e by Thomson and others, this also determined the mass of the electron (as about 1/2000 of the mass of the hydrogen atom, as Thomson had guessed). Second, and possibly even more important, he had established that *all* charges, positive and negative, come in multiples of the same basic unit e.

4.9 Rutherford and the Nuclear Atom (optional)

As soon as the electron had been identified by Thomson, physicists began considering its role in the structure of atoms. In particular, Kelvin and Thomson developed a model of the atom (usually called the Thomson model) in which the electrons were supposed to be embedded in a uniform sphere of positive charge, somewhat like the berries in a blueberry muffin. The nature of the positive charge was unknown, but the small mass of the electron suggested that whatever carried the positive charge must account for most of the atom's mass.

To investigate these ideas it was necessary to find an experimental probe of the atom, and the energetic charged particles ejected by radioactive substances proved suitable for this purpose. Especially suitable was the α particle, which is emitted by many radioactive substances and which Rutherford had identified (1909) as a positively ionized helium atom. (In fact, it is a helium atom that has lost both of its two electrons; that is, it is a helium nucleus.) It was found that if α particles were directed at a thin layer of matter, such as a metal foil, the great majority passed almost straight through, suffering only small deflections. It seemed clear that the α particles must be passing through the atoms in their path, and that the deflections must be caused by the electric fields inside the atoms. All of this was consistent with the Thomson model, which necessarily predicted only small deflections, since the electrons are too light, and the field of the uniform positive charge too small, to produce large changes in the velocity of a massive α particle.

However, it was observed by two of Rutherford's assistants, Geiger and Marsden, that even with the thinnest of metal foils (of order 1 μm) a few α particles were deflected through very large angles — 90° and even more. Since a single encounter with Thomson's atom could not possibly cause such a deflection, it was necessary to assume (if one wished to retain the Thomson model) that these large deflections were the result of many encounters, each causing a small deflection. But Rutherford was able to show that the probability of such multiple encounters was much too small to explain the observations.

Rutherford argued that the atom must contain electric fields far greater than predicted by the Thomson model, and large enough to produce the observed large deflections in a single encounter. To account for these enormous electric fields he proposed his famous nuclear atom, with the positive charge concentrated in a tiny massive nucleus. (Initially, he did not exclude the possibility of a negative nucleus with positive charges outside, but it was quickly established that the nucleus was positive.)

According to Rutherford's model, the majority of α particles going through a thin foil would not pass close enough to any nuclei to be appreciably deflected. On the other hand, a few would come close to a nucleus, and these would be the ones scattered through a large angle. A simplifying feature of the model is that the large deflections occur close to the nucleus, well inside the electron orbits, and are unaffected by the electrons. That is, one can analyze the large deflections in terms of the Coulomb force of the nucleus, ignoring all the atomic electrons.

Rutherford was able to calculate the trajectory of an α particle in the Coulomb field of a nucleus and hence to predict the number of deflections through different angles; he also predicted how this number should vary with the particle's energy, the foil's thickness, and other variables. These predictions

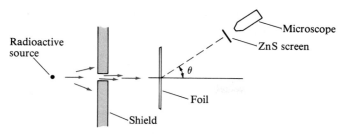

FIGURE 4.3 The "Rutherford scattering" experiment of Geiger and Marsden. Alpha particles from a radioactive source pass through a narrow opening in a thick metal shield and impinge on the thin foil. The number scattered through the angle θ is counted by observing the scintillations they cause on the zinc sulfide screen.

were published in 1911, and all were subsequently verified by Geiger and Marsden (1913), whose results provided the most convincing support for the nuclear atom.

The experiment of Geiger and Marsden is shown schematically in Fig. 4.3. A narrow stream of α particles from a radioactive source was directed toward a thin metal foil. The number of particles deflected through an angle θ was observed with the help of a zinc sulfide screen, which gave off a visible scintillation when hit by an α particle; these scintillations could be observed through a microscope and counted.

To calculate the number of deflections expected on the basis of Rutherford's model of the atom, let us consider a single α particle of mass m, charge $q = 2e$, and energy E. Since we are interested in large deflections (θ more than a couple of degrees, say) we suppose that the particle passes reasonably close to the nucleus of an atom and well inside the atomic electrons, whose presence we can therefore ignore. We denote the nuclear charge by $Q = Ze$ and, for simplicity, suppose the nucleus to be fixed. (This is a good approximation since most nuclei are much heavier than the α particle.) The only force acting on the α particle is the Coulomb repulsion of the nucleus,*

$$F = \frac{kqQ}{r^2} = \frac{2Zke^2}{r^2}, \tag{4.25}$$

where k is the Coulomb force constant, $k = 8.99 \times 10^9 \text{ N} \cdot \text{m}^2/\text{C}^2$. Under the influence of this inverse-square force the α particle follows a hyperbolic path as shown in Fig. 4.4.

We can characterize the path followed by any particular α particle by its *impact parameter b,* defined as the perpendicular distance from the nucleus to the α particle's original line of approach (Fig. 4.4). Our first task is to relate the angle of deflection θ to the impact parameter b. We defer this rather tedious exercise in mechanics to Section 4.10, where we shall prove that

$$b = \frac{Zke^2}{E} \cot \frac{\theta}{2}, \tag{4.26}$$

where E denotes the energy of the incident α particles. Notice that a large

* The Coulomb force constant k is often written in the form $k = 1/(4\pi\epsilon_0)$, where ϵ_0 is called the permittivity of the vacuum.

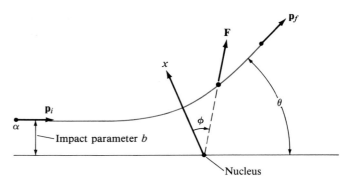

FIGURE 4.4 Trajectory of the α particle in the Coulomb field of a nucleus. The initial and final momenta are labeled $\mathbf{p}_i$ and $\mathbf{p}_f$; the direction labeled x is used as x axis in Section 4.10.

scattering angle θ corresponds to a small impact parameter b, and vice versa, just as one would expect.

Let us now focus on a particular value of the impact parameter b and the corresponding angle θ. All particles whose impact parameter is less than b will be deflected by more than θ [Fig. 4.5(a)]. Thus all those particles that impinge on a circle of radius b (and area πb^2) are scattered by θ or more. If the original beam of particles has cross-sectional area A [Fig. 4.5(b)], the proportion of particles scattered by θ or more is $\pi b^2/A$. If the total number of particles is N, the number scattered through θ or more by any one atom in the foil is

$$\text{number scattered through } \theta \text{ or more by one atom} = N\frac{\pi b^2}{A}. \quad (4.27)$$

This must now be multiplied by the number of target atoms that the beam of α particles encounters. If the target foil has thickness t and contains n atoms in unit volume, the number of atoms that the beam meets is [Fig. 4.5(b)]

$$\text{number of target atoms encountered} = nAt. \quad (4.28)$$

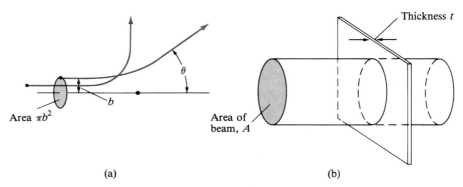

(a)

(b)

FIGURE 4.5 (**a**) A particle with impact parameter b is deflected by an angle θ; all those particles that impinge on the circle of area πb^2 are deflected by more than θ. (**b**) The cross-sectional area of the whole beam is A; the volume of target intersected by the beam is At.

Combining (4.27) and (4.28) we find that the total number of α particles scattered through θ or more is

$$N_{sc}(\theta \text{ or more}) = \frac{N\pi b^2}{A} \times nAt = \pi Nntb^2. \qquad (4.29)$$

By differentiating (4.29) we can find the number of particles emerging between θ and $\theta + d\theta$; and finally, by elementary geometry we can find the number that hit unit area on the zinc sulfide screen at angle θ and distance R from the foil. We again defer the details of the calculation to Section 4.10. The result is that

$$\boxed{\begin{aligned} n_{sc}(\theta) &= \text{number of particles per unit area at } \theta \\ &= \frac{Nnt}{4R^2} \cdot \left(\frac{Zke^2}{E}\right)^2 \cdot \frac{1}{\sin^4(\theta/2)}. \end{aligned}} \qquad (4.30)$$

This important result is called the **Rutherford formula.**

Some features of the Rutherford formula would be expected of almost any reasonable atomic model. For example, it is almost inevitable that $n_{sc}(\theta)$ should be proportional to N, the original number of α particles, and inversely proportional to R^2, the square of the distance to the detector. On the other hand, several features are specific to Rutherford's assumption that each large angle deflection results from an encounter with the tiny, but massive, charged nucleus of a single target atom. Among the features specific to Rutherford's model are:

1. $n_{sc}(\theta)$ is proportional to the thickness t of the target foil.
2. $n_{sc}(\theta)$ is proportional to Z^2, the nuclear charge squared.
3. $n_{sc}(\theta)$ is inversely proportional to E^2, the incident energy squared.
4. $n_{sc}(\theta)$ is inversely proportional to the fourth power of $\sin(\theta/2)$.

Geiger and Marsden were able to check each of these predictions separately and found excellent agreement in all cases. Their beautiful experiments established Rutherford's nuclear atom beyond reasonable doubt and paved the way for the modern quantum theory of the atom.

One final consequence of Rutherford's analysis deserves mention. The Rutherford formula (4.30) was derived by assuming that the force of the nucleus on the α particles is the Coulomb force

$$F = \frac{kqQ}{r^2}. \qquad (4.31)$$

This is true provided that the α particles remain outside the nucleus at all times. If the α particles penetrated the nucleus, the force would not be given by (4.31), and the Rutherford formula would presumably not hold. Thus the fact that Geiger and Marsden found the Rutherford formula to be correct in all cases indicated that no α particles were penetrating into the target nuclei. Since it was easy to calculate the minimum distance of the α particle from the center of any nucleus, this gave an upper limit on the nuclear radius, as we see in the following example.

EXAMPLE 4.2 When 7.7-MeV alpha particles are fired at a gold foil ($Z = 79$), the Rutherford formula agrees with the observations at all angles. Use this fact to obtain an upper limit on the radius R of the gold nucleus.

Since the Rutherford formula holds at all angles, none of the α particles penetrate inside the nucleus; that is, for all trajectories the distance r from the α to the nuclear center is always greater than R:

$$r > R. \tag{4.32}$$

(See Fig. 4.6.) Now, the minimum value of r occurs for the case of a head-on collision, in which the α particle comes instantaneously to rest. At this point its kinetic energy is zero and its potential energy $2Zke^2/r_{min}$ is equal to its total energy, $E = 7.7$ MeV. Thus

$$\frac{2Zke^2}{r_{min}} = E, \tag{4.33}$$

whence

$$r_{min} = \frac{2Zke^2}{E}. \tag{4.34}$$

Since $R < r$ for *all* orbits it follows that $R < r_{min}$ and hence

$$R \lesssim \frac{2Zke^2}{E}. \tag{4.35}$$

We have used the sign $\lesssim$ to emphasize that this is only an approximate result. In the first place, the nuclear radius is itself only an approximate notion. Second, we are ignoring the size of the α particle. [We could improve on (4.35) by replacing R by $R + R_\alpha$.] Finally, it is possible that non-Coulombic, nuclear forces may have an appreciable effect even a little before the α particle penetrates the nucleus.

Before we substitute numbers into the inequality (4.35), it is convenient to note that the Coulomb constant, $k = 9 \times 10^9$ N·m²/C², almost always appears in atomic and nuclear calculations in the combination ke^2. Since ke^2/r is an energy, ke^2 has the dimensions of energy × length and is conveniently expressed in the units eV·nm or, what is the same thing, MeV·fm. To this end, we detach one factor of e and write

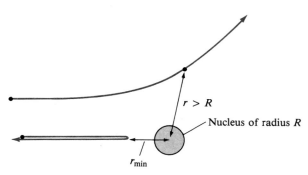

$r > R$

Nucleus of radius R

r_{min}

FIGURE 4.6 If none of the α particles penetrate into the nucleus, $r > R$ for all points on all orbits. The closest approach r_{min} occurs in a head-on collision.

$$ke^2 = (8.99 \times 10^9 \text{ N} \cdot \text{m}^2/\text{C}^2) \times (1.60 \times 10^{-19} \text{ C}) \times e$$
$$= 1.44 \times 10^{-9} e \cdot \text{N} \cdot \text{m}^2/\text{C}.$$

Now the unit $\text{N} \cdot \text{m/C} = \text{J/C}$ is the same as a volt. Multiplied by e this gives an electron volt; therefore, $ke^2 = 1.44 \times 10^{-9} \text{ eV} \cdot \text{m}$ or

$$\boxed{ke^2 = 1.44 \text{ eV} \cdot \text{nm} = 1.44 \text{ MeV} \cdot \text{fm}.} \tag{4.36}$$

Returning to the inequality (4.35), we find that

$$R \lesssim \frac{2Zke^2}{E} = \frac{2 \times 79 \times (1.44 \text{ MeV} \cdot \text{fm})}{7.7 \text{ MeV}}$$

or

$$R \lesssim 30 \text{ fm}. \tag{4.37}$$

That is, the fact that Geiger and Marsden found the Rutherford formula valid for 7.7-MeV alpha particles on gold implied that the gold nucleus has radius less than 30 fm. Since the radius of the gold nucleus is in fact about 8 fm, this was perfectly correct.

If the incident energy E were steadily increased, some of the α particles would eventually penetrate the target nucleus, and the Rutherford formula would cease to hold, first at $\theta = 180°$ (that is, for head-on collisions), then as the energy increased, at smaller angles as well. From (4.35) we see that the energy at which the Rutherford formula first breaks down is

$$E \approx \frac{2Zke^2}{R}. \tag{4.38}$$

For gold this energy is about 30 MeV, an energy that was not available to Rutherford until much later. However, for aluminum, with Z only 13, it is about 6 MeV (Problem 4.27), and Rutherford was able to detect some departure from his formula at large angles.

4.10 Derivation of Rutherford's Formula (optional)

In deriving the Rutherford formula (4.30) we omitted two slightly tedious calculations. First, the relation (4.26) between impact parameter b and scattering angle θ: If we call the initial and final momenta of the α particle $\mathbf{p}_i$ and $\mathbf{p}_f$, the angle between $\mathbf{p}_i$ and $\mathbf{p}_f$ is the scattering angle θ, as shown in Fig. 4.7. Since the α particle's energy is unchanged, $\mathbf{p}_i$ and $\mathbf{p}_f$ are equal in magnitude and the triangle on the left of Fig. 4.7 is isosceles. If we denote the change in momentum by $\Delta\mathbf{p} = \mathbf{p}_f - \mathbf{p}_i$, it is clear from the construction shown that

$$\Delta p = 2p_i \sin\frac{\theta}{2}. \tag{4.39}$$

Now, we know from Newton's second law $(d\mathbf{p}/dt = \mathbf{F})$ that

$$\Delta\mathbf{p} = \int_{-\infty}^{\infty} \mathbf{F} \, dt.$$

To evaluate this integral we choose our x axis in the direction of $\Delta\mathbf{p}$. This

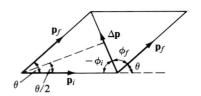

FIGURE 4.7 Geometry of the initial and final momenta in Rutherford scattering.

direction was indicated in Fig. 4.4, where we labeled the polar angle of the α particle as ϕ. With these notations

$$\Delta p = \int_{-\infty}^{\infty} F_x \, dt = \int_{-\infty}^{\infty} \frac{2Zke^2}{r^2} \cos \phi \, dt. \tag{4.40}$$

This integral can be easily evaluated by a trick that exploits conservation of angular momentum. (Since the force on the α is radially outward from the nucleus, the angular momentum L about the nucleus is constant.) Long before the collision $L = bp_i$, whereas we know that at all times

$$L = mr^2\omega = mr^2 \frac{d\phi}{dt}.$$

Equating these two expressions for L, we find that

$$\frac{d\phi}{dt} = \frac{bp_i}{mr^2}.$$

Thus we can rewrite (4.40) as

$$\Delta p = 2Zke^2 \int \frac{\cos \phi}{r^2} \cdot \frac{dt}{d\phi} \, d\phi$$

$$= 2Zke^2 \int \frac{\cos \phi}{r^2} \cdot \frac{mr^2}{bp_i} \, d\phi$$

$$= \frac{2Zke^2 m}{bp_i} \int_{\phi_i}^{\phi_f} \cos \phi \, d\phi$$

$$= \frac{2Zke^2 m}{bp_i} (\sin \phi_f - \sin \phi_i). \tag{4.41}$$

Comparing Figs. 4.4 and 4.7, we see that $\phi_f = -\phi_i$ and that $\theta + 2\phi_f = 180°$. Therefore, $\phi_f = 90° - \theta/2$ and $\sin \phi_f = \cos(\theta/2)$. Thus (4.41) becomes

$$\Delta p = \frac{4Zke^2 m}{bp_i} \cos \frac{\theta}{2}. \tag{4.42}$$

Equating the two expressions (4.39) and (4.42) for Δp, we see that

$$2p_i \sin \frac{\theta}{2} = \frac{4Zke^2 m}{bp_i} \cos \frac{\theta}{2},$$

which we can solve for b:

$$b = \frac{Zke^2}{E} \cot \frac{\theta}{2}, \tag{4.43}$$

since $E = p_i^2/2m$. This completes the proof of the relation (4.26).

The other step that was omitted in Section 4.9 was the derivation of the final result, the Rutherford formula (4.30), from Equation (4.29),

$$N_{sc}(\theta \text{ or more}) = \pi N n t b^2.$$

Substituting (4.43), we can rewrite the latter as

$$N_{sc}(\theta \text{ or more}) = \pi N n t \left(\frac{Zke^2}{E} \right)^2 \cot^2 \frac{\theta}{2}. \tag{4.44}$$

This gives the number of α particles deflected through θ or more. The number that emerge between θ and $\theta + d\theta$ is found by differentiating (4.44) to give (see Problem 4.29)

$$N_{sc}(\theta \text{ to } \theta + d\theta) = \pi Nnt \left(\frac{Zke^2}{E}\right)^2 \frac{\cos(\theta/2)}{\sin^3(\theta/2)} \, d\theta. \qquad (4.45)$$

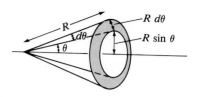

FIGURE 4.8 Particles whose deflection is between θ and $\theta + d\theta$ pass through the ring-shaped surface shown.

Now, at a distance R from the target (where the detector is placed) the particles emerging between θ and $\theta + d\theta$ are distributed uniformly over the ring-shaped surface shown in Fig. 4.8. This ring has area

$$\text{area of ring} = (2\pi R \sin \theta) \times (R \, d\theta). \qquad (4.46)$$

To find the number of particles per unit area at distance R, we must divide the number (4.45) by this area, to give

$$n_{sc}(\theta) = \frac{N_{sc}(\theta \text{ to } \theta + d\theta)}{2\pi R^2 \sin \theta \, d\theta}$$

$$= \frac{Nnt}{4R^2} \cdot \left(\frac{Zke^2}{E}\right)^2 \cdot \frac{1}{\sin^4(\theta/2)},$$

where we have used the identity $\sin \theta = 2 \sin(\theta/2) \cos(\theta/2)$. This completes our derivation of the Rutherford formula.

IDEAS YOU SHOULD NOW UNDERSTAND FROM CHAPTER 4

Elements and compounds
Atoms, molecules, and their chemical formulas
Electrons
Nuclei
Rutherford scattering
Protons and neutrons
Nucleon is the name for protons *and* neutrons
Atomic radii are about 0.1 nm
Nuclear radii are a few fm
Atomic number, Z

Ions
Mass number, A
Isotopes
Atomic mass unit, u
Avogadro's number, N_A
The mole
[Optional sections: measurements of m/e and e by Thomson and Millikan; the Thomson and Rutherford models of the atom, the Rutherford formula.]

PROBLEMS FOR CHAPTER 4

4.1 • The elements are generally listed in order of their "atomic number" Z (which is just the number of electrons in a neutral atom of the element). Hydrogen is first with $Z = 1$, helium next with $Z = 2$, and so on. There are 90 elements that occur naturally on earth, but they are not just numbers 1 through 90. What are

they? What are the atomic numbers of the known artificial elements? (You can find this information in the periodic table inside the back cover.)

4.2 • With what mass of oxygen would 2 g of hydrogen combine to form water (H_2O)? (All you need to know is the ratio of the masses of the atoms concerned; you can find these masses, listed in atomic mass units, inside the back cover.)

4.3 • With what mass of nitrogen would 1 g of hydrogen combine to form ammonia (NH_3)? (Read the hint to Problem 4.2.)

SECTION 4.4 (SOME ATOMIC PARAMETERS)

4.4 • Tabulate the numbers of electrons, protons, neutrons, and nucleons in the most common form of the following neutral atoms: hydrogen, carbon, nitrogen, oxygen, aluminum, iron, lead. (The necessary information can be found in Appendix D.) On the basis of this information, can you suggest why lead is much denser than aluminum?

4.5 • Make a table similar to Table 4.5 showing all of the stable isotopes for hydrogen, helium, oxygen, and aluminum, including their percentage abundances. (Use the information in Appendix D.)

4.6 • Use Appendix D to find four elements that have only one type of stable atom (that is, an atom with no stable isotopes).

4.7 • Two atoms that have the same mass number A but different atomic numbers Z are called *isobars*. (Isobar means "of equal mass.") For example, ^3He and ^3H are isobars. Use the data in Appendix D to find three examples of pairs of isobars. Find an example of a triplet of isobars (that is, three different atoms with the same mass number).

4.8 • • It is found that the radius R of any nucleus is given approximately by

$$R = R_o A^{1/3}$$

where A is the mass number of the nucleus and R_o is a constant whose value depends a little on how R is defined, but is about 1.1 fm. (a) What are the radii of the nuclei of helium, carbon, iron, lead, and lawrencium? (b) How does the volume of a nucleus (assumed spherical) depend on A? What does your answer tell you about the average density of nuclei?

SECTION 4.5 (THE ATOMIC MASS UNIT)

4.9 • The mass of a carbon 12 atom is 1.992648×10^{-26} kg (with an uncertainty of 1 in the final digit). Calculate the value of the atomic mass unit to six significant figures in kg and in MeV/c^2.

4.10 • Find the mass of a ^{12}C nucleus in u correct to five significant figures. (The mass of an electron is 0.000549 u. Ignore the binding energy of the electrons.)

4.11 • Using the natural abundances listed in Table 4.5 (or Appendix D) calculate the average atomic masses of naturally occurring carbon, chlorine, and iron. (Your answers should be in u and correct to three significant figures.)

4.12 • What is the mass (correct to the nearest u) of the following molecules: (a) water, H_2O; (b) laughing gas, N_2O; (c) ozone, O_3; (d) glucose, $C_6H_{12}O_6$; (e) ammonia, NH_3; (f) limestone, $CaCO_3$

4.13 • • (a) Calculate the mass of the ^4He *nucleus* in atomic mass units by subtracting the mass of two electrons from that of a neutral ^4He atom. (See Appendix D and give four decimal places.) (b) The procedure suggested for part (a) is theoretically incorrect because we have neglected the binding energy of the electrons. Given that the energy needed to pull both electrons far away from the nucleus is 80 eV, is the correct answer larger or smaller than implied by part (a), and by how much (in u)? In stating the mass of the nucleus, about how many significant figures can you give before this effect would show up?

SECTION 4.6 (AVOGADRO'S NUMBER AND THE MOLE)

4.14 • How many moles are in (a) 1 g of carbon? (b) 1 g of hydrogen molecules? (c) 10 g of water? (d) 1 ounce of gold?

4.15 • How many moles are in (a) 10 g of NaCl? (b) 2 kg of NH_3? (c) 10 cm³ of Hg (density 13.6 g/cm³)? (d) 1 pound of sugar ($C_{12}H_{22}O_{11}$)?

4.16 • A mole of O_2 molecules has mass 32 g (that is, the molecular "weight" of O_2 is 32 g/mole). What is the mass of a single O_2 molecule (in u and in grams)?

4.17 • What is the total charge of 1 mole of Cl^- ions? (The Cl^- ion is a chlorine atom that has acquired one extra electron.) This important quantity is called the faraday; it is easily measured since it is the charge that must pass to release 1 mole of chlorine in electrolysis of NaCl solution, (or, more generally, 1 mole of any monovalent ion in electrolysis of an appropriate solution). Once the electron charge was known, knowledge of the faraday let one calculate Avogadro's number.

4.18 • • How many C atoms are there in (a) 1 g of CO_2? (b) 1 mole CH_4? (c) 1 kmole of C_2H_6? In each case what fraction of the atoms are C atoms, and what fraction of the mass is C?

SECTION 4.7 (THOMSON'S DISCOVERY OF THE ELECTRON)

4.19 • • In Thomson's experiment electrons travel with velocity u in the x direction. They enter a uniform electric field E, which points in the y direction and has total width l (Fig. 4.9). Find the time for an electron to cross the field and the y component of its velocity when it leaves the field. Hence show that its velocity is deflected through an angle $\theta \approx eEl/(mu^2)$

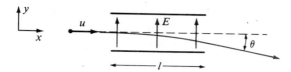

FIGURE 4.9 (Problem 4.19)

(provided that θ is small). Assume that the electrons are nonrelativistic, as was the case for Thomson.

4.20 •• Suppose that the electrons in Thomson's experiment enter a uniform magnetic field B, which is in the z direction (with axes defined as in Fig. 4.9) and has total width l. Show that they are deflected through an angle $\theta \approx eBl/(mu)$ (provided that θ is small). Assume that the electrons are nonrelativistic.

4.21 •• Thomson directed his cathode rays at a metal body and measured the total charge Q that it acquired and its rise in temperature ΔT. From ΔT and the body's known thermal capacity, he could find the heat given to the body. Show that this heat should be

$$\text{heat} = \frac{Qmu^2}{2e}$$

where m, u, and e are the electron's mass, speed (nonrelativistic), and charge. Combine this with the result of Problem 4.20 to give expressions for u and m/e in terms of measured quantities.

4.22 •• In one version of Thomson's measurement of m/e, nonrelativistic electrons are accelerated through a measured potential difference V_0. This means that their kinetic energy is $K = eV_0$. They are then passed into a known magnetic field B and the radius $R = mu/(eB)$ of their circular orbit is measured. Eliminate the speed u from these two equations and derive an expression for m/e in terms of the measured quantities V_0, B, and R.

SECTION 4.8 (MILLIKAN'S OIL-DROP EXPERIMENT)

4.23 ••• In order to find the electron's charge e Millikan needed to know the mass M of his oil drops, and this was actually the source of his greatest uncertainty in determining e. (His value of e was about 0.5% low as a result.) However, to show that all charges are multiples of some basic unit charge, it was *not* necessary to know M, which cancels out of the charge ratios. The

following problem illustrates this point and gives some more details of Millikan's experiment.

By switching the E field off and on, Millikan could time an oil drop as it fell and rose through a measured distance l (falling under the influence of gravity, rising under that of E and gravity). The downward and upward speeds are given by (4.22) as

$$v_d = \frac{Mg}{6\pi r\eta} \quad \text{and} \quad v_u = \frac{qE - Mg}{6\pi r\eta}. \quad (4.47)$$

Both speeds were measured in the form $v = l/t$, where t is the time for the droplet to traverse l.

(a) By adding the two equations (4.47) and rearranging, show that

$$\frac{1}{t_d} + \frac{1}{t_u} = \left(\frac{E}{6\pi r\eta l}\right) q = Kq \quad (4.48)$$

where the quantity $K = E/(6\pi r\eta l)$ was constant as long as Millikan watched a single droplet and did not vary E.

From (4.48) we see that the charge q is proportional to the quantity $(1/t_d) + (1/t_u)$. Thus if it is true that q is always an integer multiple of e, it must also be true that $(1/t_d) + (1/t_u)$ is always an integer multiple of some fixed quantity. Table 4.7 shows a series of timings for a single droplet (made with a commerical version of the Millikan experiment used in a teaching laboratory). That t_u changes abruptly from time to time indicates that the charge on the droplet has changed as described in Section 4.8. (The times t_d should theoretically all be the same, of course. The small variations give you a good indication of the uncertainties in all timings. For t_d you should use the average of all measurements of t.)

(b) Calculate $(1/t_d) + (1/t_u)$ and show that (within the uncertainties) this quantity is always an integer multiple of one fixed number, and hence that the charge is always a multiple of one fixed charge.

TABLE 4.7

A series of measurements of t_d and t_u, the times for a single droplet to travel down and up a fixed distance l. All times are in seconds. (Problem 4.23)

t_d:	15.2	15.0	15.1	15.0	14.9	15.1	15.1	15.0	15.2	15.2
t_u:	6.4	6.3	6.1	24.4	24.2	3.7	3.6	1.8	2.0	1.9

TABLE 4.8

Additional data for the oil-drop experiment. (Problem 4.23)

$l = 8.30 \times 10^{-4}$ m	(distance traveled by droplet)
$\rho = 839$ kg/m³	(density of oil)
$g = 9.80$ m/s²	(acceleration of gravity)
$\eta = 1.60 \times 10^{-5}$ N·s/m²	(viscosity of air, adjusted*)
$E = 1.21 \times 10^5$ N/C	(electric field)

* For very small droplets there is a small correction to the formula (4.22) for the terminal velocity. For the experiment reported here this correction amounts to a 12% reduction in the viscosity. We have included this correction in the value given.

(c) Use Equation (4.24) for v_d and the additional data in Table 4.8 to find the radius r of the droplet.

(d) Finally, use (4.48) to find the four different charges q on the droplet. What is the best estimate of e based on this experiment?

SECTION 4.9 (RUTHERFORD AND THE NUCLEAR ATOM)

4.24 • A student doing the Rutherford scattering experiment arranges matters so that she gets 80 counts/min at a scattering angle of $\theta = 10°$. If she now moves her detector around to $\theta = 150°$, keeping it at the same distance from the target, how many counts would she expect to observe in a minute? (This illustrates an awkward feature of the experiment, especially before the days of automatic counters. An arrangement that gives a reasonable counting rate at small θ gives far too few counts at large θ; and one that gives a reasonable rate for large θ will overwhelm the counter at small θ.)

4.25 •• The Rutherford model of the atom could explain the large-angle scattering of α particles because it led to very large electric fields compared to the Thomson model. To see this, note that in the Thomson model the positive charge of an atom was uniformly distributed through a sphere of the same size as the atom itself. According to this model, what would be the maximum E field (in volts/meter) produced by the positive charge in a gold atom ($Z = 79$, atomic radius ≈ 0.18 nm)? What is the corresponding maximum field in Rutherford's model of the gold atom, with the positive charge confined to a sphere of radius about 8 fm? (In the Thomson model the actual field would be even less because of the electrons. Note that since $ke^2 = 1.44$ eV·nm it follows that $ke = 1.44$ V·nm.)

4.26 •• Consider several different α particles approaching a nucleus, with various different impact parameters, b, but all with the same energy, E. Prove that an α particle that approaches the nucleus head on ($b = 0$) gets closer to the nucleus than any other.

4.27 •• (a) If the Rutherford formula is found to be correct at all angles when 15-MeV alpha particles are fired at a silver foil ($Z = 47$), what can you say about the radius of the silver nucleus? (b) Aluminum has atomic number $Z = 13$ and a nuclear radius about $R_{Al} \approx 4$ fm. If one were to bombard an aluminum foil with α particles and slowly increase their energy, at about what energy would you expect the Rutherford formula to break down? You can make the estimate (4.38) a bit more realistic by taking R to be $R_{Al} + R_{He}$, where R_{He} is the α particle's radius (about 2 fm).

4.28 •• In a student version of the Rutherford experiment ^{210}Po is used as a source of 5.2-MeV alpha particles, which are directed at a gold foil (thickness 2 μm) at a rate of 10^5 particles per minute. The scattered particles are detected on a screen of area 1 cm² at a distance of 12 cm. Use the Rutherford formula (4.30) to predict the number of α particles observed in 10 minutes at $\theta = 15°$, $30°$, and $45°$, given the following data:

$N = 10^6$	(number of incident particles in 10 min)
$\rho = 19.3$ g/cm³	(density of gold)
$m_{Au} = 197$ u	(mass of gold atom)
$t = 2 \times 10^{-6}$ m	(thickness of foil)
$Z = 79$	(atomic number of gold)
$E = 5.2$ MeV	(energy of α particles)

(Note that the number density n of gold nuclei is ρ/m_{Au}, and don't forget that $ke^2 = 1.44$ MeV·fm.)

SECTION 4.10 (DERIVATION OF RUTHERFORD'S FORMULA)

4.29 • (a) If $N_{sc}(\theta$ or more) denotes the number of α particles scattered by θ or more, show that the number whose deflection is between θ and $\theta + d\theta$ is

$$N_{sc}(\theta \text{ to } \theta + d\theta) = -\frac{dN_{sc}(\theta \text{ or more})}{d\theta} d\theta.$$

(b) Differentiate the expression (4.44) for N_{sc} (θ or more) and verify the result (4.45).

CHAPTER

5

Quantization of Light

5.1 Quantization

Quantum theory replaced classical physics in the microscopic domain because classical physics proved incapable of explaining a wide range of microscopic phenomena. In this chapter we describe the first clear evidence for this failure of classical physics, the evidence that light and all other forms of electromagnetic radiation are *quantized.*

We say that something is **quantized** if it can occur only in certain discrete amounts. Thus "quantized" is the opposite of "continuous." (The word comes from the Latin "quantum" meaning "how much.") In our everyday experience at the store, eggs are quantized (since one can take only integer numbers of eggs), whereas gasoline is not (since one can take any amount of gasoline). On the microscopic level, we have seen that matter is quantized, the smallest unit, or *quantum,* of normal matter being the atom. Similarly, we have seen that electric charge is quantized, the quantum of charge being $e = 1.6 \times 10^{-19}$ C.

During the first quarter of the twentieth century it was found that electromagnetic radiation is quantized. The energy contained in light* of a given

---

* When there is no serious risk of confusion we use "light" to mean any form of electromagnetic radiation. When necessary, we use "visible light" to emphasize that we are speaking only of the radiation to which the human eye is sensitive.

MAX PLANCK (1858–1947, German). Planck is best known for suggesting (in 1900) the quantization of radiant energy — an idea that won him the 1918 Nobel Prize. He was quick to take up Einstein's relativity in 1905, and was the first to propose the correct relativistic expression for the momentum of a massive particle (in 1906).

frequency is not a continuous variable; instead, it can only be an integer multiple of a certain basic quantum of energy. In fact, it was found that an electromagnetic wave consists of tiny localized bundles of energy, which also carry momentum and have most of the properties of ordinary particles, except that their mass is zero. These bundles, or quanta of light, have come to be called **photons.**

The quantization of matter and of electric charge had not contradicted any basic principles of classical physics. On the other hand, the quantization of radiation was definitely inconsistent with classical electromagnetic theory, which predicted unambiguously that the energy of radiation should be a continuous variable. Thus the discovery that light is quantized required the development of a new theory, the quantum theory of radiation.

5.2 Planck and Blackbody Radiation

The first person to propose that electromagnetic radiation must be quantized was the German physicist Planck, in connection with his studies of blackbody radiation (1900). A blackbody is any body that is a perfect absorber of radiation, and blackbody radiation is the radiation given off by such a body when heated. Using classical electromagnetic theory it was possible to calculate how much energy should be radiated at any given frequency, and the result of this calculation is called the Rayleigh–Jeans formula. This formula agreed well with the observed energy distribution at low (infrared) frequencies, but was badly wrong at high (ultraviolet) frequencies. In fact, the Rayleigh–Jeans formula made the disastrous prediction that the total energy emitted at all frequencies should be infinite, a result known as the "ultraviolet catastrophe."

Planck found that he could avoid this unfortunate result by assuming that the radiation emitted by a body is quantized. Specifically, he proposed that radiation of frequency f can be emitted only in integral multiples of a basic quantum hf,

$$E = 0, hf, 2hf, 3hf, \ldots, \qquad (5.1)$$

where h was an unknown constant, now called **Planck's constant.** Notice that Planck's quantum of energy, hf, varies with the frequency. This contrasts with the quantum of charge e, which is the same for *all* charges.

Planck found that the assumption (5.1) led to a different formula for blackbody radiation. This formula depended on the unknown constant h, which Planck chose so as to give the best fit to the experimental data. With h chosen in this way, Planck's formula fits the data perfectly at all frequencies and all temperatures. The modern value of Planck's constant h is

$$h = 6.63 \times 10^{-34} \text{ J} \cdot \text{s.} \qquad (5.2)$$

Notice that since hf is an energy, h has the dimensions of energy × time.

Planck did not claim to understand *why* the radiation from a body was quantized in multiples of hf. In fact, it appears that he regarded the idea as an *ad hoc,* temporary hypothesis that would ultimately prove dispensible. As we shall see, it was in fact the first indication of a fundamental and universal property of all electromagnetic radiation.

5.3   The Photoelectric Effect

Planck's ideas were taken up and extended by Einstein (1905), who showed that they explain several phenomena, of which the most important was the **photoelectric effect.** In this effect, discovered by Hertz in 1887, light is shone on a metal and is found to eject electrons from the surface. At first sight, this process, which is the basis of many modern light-detecting devices, appeared perfectly consistent with classical electromagnetic theory. Light waves were known to carry energy, in the form of oscillating electric and magnetic fields, and it seemed perfectly reasonable that some electrons in the metal could absorb enough of this energy to be ejected. However, closer investigation showed that several features of the process were incompatible with classical electromagnetic theory.

An apparatus for investigating the photoelectric effect is sketched in Fig. 5.1. Light is shone on one of the two electrodes in an evacuated glass tube and electrons are ejected. If the other electrode is kept at a higher potential, it attracts these electrons and a current flows, whose magnitude indicates the number of electrons being ejected. If, instead, the second electrode is kept at a lower potential, it repels the electrons and only those electrons with enough kinetic energy to overcome the retarding potential V reach the second electrode. As one increases the retarding potential, the current drops until at a certain *stopping potential V_s* all current ceases. Evidently V_s is given by*

$$V_s e = K_{max}, \qquad (5.3)$$

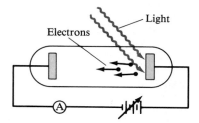

FIGURE 5.1   A photoelectric cell. The applied voltage can be adjusted in magnitude and sign.

where K_{max} denotes the maximum kinetic energy of the ejected electrons. Thus by measuring the stopping potential V_s, one can find K_{max}.

When the apparatus of Fig. 5.1 is used to investigate the numbers and kinetic energies of the electrons ejected, two important facts emerge:

1. If the intensity of the incident light is increased, the number of ejected electrons increases (as one might expect), but quite unexpectedly, *their kinetic energy does not change at all.*

2. If the frequency f of the incident light is reduced, then below a certain critical frequency f_0, *no electrons are ejected* however intense the light may be.

Neither of these results is consistent with the classical view of electromagnetic waves as a continuous distribution of oscillating electric and magnetic fields. According to this view, an increased intensity means increased field strengths, which should surely eject some electrons with increased kinetic energy; and (again in the classical view) there is no reasonable way to explain why low-frequency fields (however strong) should be unable to eject any electrons.

Einstein proposed that as a natural extension of Planck's ideas, one should assume that "the energy in a beam of light is not distributed continuously through space, but consists of a finite number of energy quanta, which are localized at points, which cannot be subdivided, and which are absorbed and emitted only as whole units." The energy of a single quantum, or photon as

---

* If the two electrodes are made of different metals, there is a small complication that arises because the two metals attract electrons differently. (This causes the so-called contact potential of the two metals.) To avoid this complication we assume that the two electrodes are made of the same metal.

we would now say, he took to be hf. (These proposals go beyond — but include — Planck's assumption that light is emitted in multiples of hf.) Since it seemed unlikely that two photons would strike one electron, Einstein argued that each ejected electron must be the result of a single photon giving up its energy hf to the electron. With these assumptions, both of the properties mentioned above are easily explained, as follows:

If the intensity of light is increased, then, according to Einstein's assumptions, the number of photons is increased, but the energy hf of an individual photon is unchanged. With more photons, more electrons will be ejected. But since each photon has the same energy, each ejected electron will be given the same energy. Therefore, K_{max} will not change, and point 1 is explained.

Point 2 is equally easy: For any given metal, there is a definite minimum energy needed to remove an electron. This minimum energy is called the **work function** for the metal and is denoted ϕ. If the photon energy hf is less than ϕ, no photons will be able to eject any electrons; that is, if f is less than a critical frequency f_0 given by

$$hf_0 = \phi, \tag{5.4}$$

no electrons will be ejected, as observed.

Einstein carried this reasoning further to make a quantitative prediction. If the frequency f is greater than f_0, each ejected electron should have gained energy hf from a photon but lost ϕ, or more, in escaping from the metal. Thus its kinetic energy on emerging should be $hf - \phi$, or less. Therefore,

$$\boxed{K_{max} = hf - \phi;} \tag{5.5}$$

that is, the observed maximum energy of the ejected electrons should be a linear function of the frequency of the light, and the slope of this function should be Planck's constant, h. The test of this prediction proved difficult but was successfully carried out by Millikan, one of whose trials (1916) is shown in Fig. 5.2. It will be seen that Millikan's data are a beautiful fit to the expected straight line. In particular, by measuring the slope Millikan was able to determine h and obtained a value in agreement with that found by Planck.

If Planck and Einstein were right that light is quantized (as they were), the question naturally arises why this quantization had not been observed sooner. The answer is that, by everyday standards, the energy hf of a single photon is very small. Thus the number of photons in a normal beam of light is enormous, and the restriction of the energy to integer multiples of hf is correspondingly unimportant, as the following example illustrates.

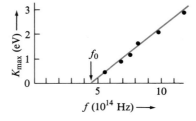

FIGURE 5.2 Millikan's data for K_{max} as a function of frequency f for the photoelectric effect in sodium.

ROBERT MILLIKAN (1868–1953, US). Although Millikan won the 1923 Nobel Prize mainly for his beautiful measurement of the electron's charge (around 1910), the citation also mentioned his verification of Einstein's equation for the photoelectric effect.

EXAMPLE 5.1 What is the energy of a typical visible photon, and about how many photons enter the eye per second when one looks at a weak source of light such as the moon (whose intensity is about 3×10^{-4} watts/m²)?

The wavelength of visible light is between 400 and 700 nm. Thus we can take a typical visible wavelength to be

$$\lambda \approx 550 \text{ nm.} \tag{5.6}$$

The energy of a single photon is

$$E = hf = \frac{hc}{\lambda}. \tag{5.7}$$

Before we evaluate this, it is useful to note that the product hc enters into many calculations and is a useful combination to remember. Since h has the dimension energy $\times$ time, hc has the dimension energy $\times$ length and is conveniently expressed in eV·nm, as follows:

$$hc = (6.63 \times 10^{-34} \text{ J·s}) \times (3.00 \times 10^8 \text{ m/s}) \times \frac{1 \text{ eV}}{1.60 \times 10^{-19} \text{ J}}$$
$$= 1.24 \times 10^{-6} \text{ eV·m}$$

or

$$\boxed{hc = 1240 \text{ eV·nm.}} \tag{5.8}$$

Putting numbers into (5.7) we find the energy for a typical visible photon to be

$$E = \frac{hc}{\lambda} \approx \frac{1240 \text{ eV·nm}}{550 \text{ nm}} \approx 2.3 \text{ eV.} \tag{5.9}$$

On the atomic level this energy is significant, but by everyday standards it is extremely small.

When we look at the moon, the energy entering our eye per second is given by IA, where I is the intensity ($I \approx 3 \times 10^{-4}$ W/m²) and A is the area of the pupil ($A \approx 3 \times 10^{-5}$ m², if we take the diameter of the pupil to be about 6 mm). Thus the number of photons entering our eye per second is

$$\text{number of photons per second} = \frac{IA}{E}$$
$$\approx \frac{(3 \times 10^{-4} \text{ W/m}^2) \times (3 \times 10^{-5} \text{ m}^2)}{(2.3 \times 1.6 \times 10^{-19} \text{ J})}$$
$$\approx 2.5 \times 10^{10} \text{ photons per second.}$$

This is such a large number that the restriction to integer numbers of photons is quite unimportant even for this weak source.*

EXAMPLE 5.2 The work function of silver is $\phi = 4.7$ eV. What is the potential V_s needed to stop all electrons when ultraviolet light of wavelength $\lambda = 200$ nm shines on silver?

The energy of the UV photon is

$$hf = \frac{hc}{\lambda} = \frac{1240 \text{ eV·nm}}{200 \text{ nm}} = 6.2 \text{ eV.}$$

* Note, however, that the individual receptors in the eye are extraordinarily sensitive, being able to detect just a few photons per second.

The stopping potential V_s is given by $V_s e = K_{max}$, where K_{max} is given by the Einstein equation (5.5). Thus

$$V_s e = K_{max} = hf - \phi = (6.2 - 4.7)\ eV = 1.5\ eV$$

or

$$V_s = 1.5\ \text{volts}.$$

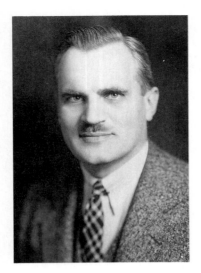

ARTHUR COMPTON (1892–1962, US). Compton's observations (1923) of the scattering of electromagnetic radiation by atoms showed clearly that photons should be regarded as particles that carry energy and momentum and earned him the 1927 Nobel Prize.

5.4 X Rays and Bragg Diffraction

The ideas that light was quantized, and that the quantum of light should be regarded as a particle, were slow to gain wide acceptance. In fact, the name "photon," which recognizes these ideas, was not coined until 1926 (by the chemist G. N. Lewis). Probably the decisive event was the experiment (1923) in which A. H. Compton showed that photons carry momentum as well as energy, and are subject to the same conservation laws of energy and momentum as other, more familiar particles. Compton's experiment used X-ray photons and we therefore use the next two sections to describe X rays, before taking up the Compton effect itself.

X rays are electromagnetic radiation whose wavelength is in or near the range from 0.1 nm down to 0.001 nm, more than 1000 times shorter than visible wavelengths. Since shorter wavelengths mean higher photon energies, X-ray photons are much more energetic than visible photons, with energies of many thousands of electron volts. X rays were discovered in 1895 by Roentgen, who found that when energetic electrons were fired into a solid target, a very penetrating radiation was produced. Unable to identify this radiation, he gave it the name "X rays."

Figure 5.3 is a sketch of a modern medical X-ray tube (whose basic principles are actually quite similar to those of Roentgen's original arrangement). Two electrodes are enclosed in an evacuated glass tube. Electrons are ejected from the heated cathode on the left. A potential difference of several thousand volts between the cathode and anode accelerates the electrons, which therefore acquire several keV of kinetic energy (and hence speeds of $0.1c$ and more). X rays are produced when the electrons crash into the anode and are brought to an abrupt stop. It is found that most of the rays are emitted near 90° to the electrons' path, and for this reason the anode is tilted to encourage the X rays to exit in one direction, as shown.

The ability of X rays to penetrate solids of low density was put to medical use within a few months of their discovery, but the problem of identifying what X rays really were took longer. It was known that an accelerating electric charge produces electromagnetic waves. (For example, the oscillating charges in a

FIGURE 5.3 An X-ray tube.

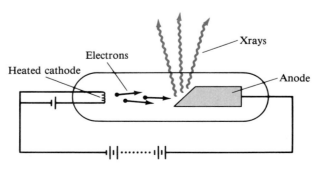

radio antenna produce the long-wavelength radiation that we call radio waves.) Thus it was perfectly reasonable to suppose that X rays were electromagnetic waves produced by the enormous deceleration of the electrons stopping in the anode. (Radiation produced in this way is called **bremsstrahlung,** the German for "braking radiation.") The problem was to verify that X rays really were waves, and the difficulty (as we now know) was that their wavelength is so very short.

Probably the most effective way to show that something is a wave, and to measure its wavelength, is to pass it through a diffraction grating and to observe the resulting interference pattern. In an efficient grating the slits must be spaced regularly with a separation of the same order as the wavelength. Thus a good grating for visible light has spacing of order 1000 nm, but a diffraction grating for X rays would require spacing of order 0.1 nm — not something that could be easily made. In 1912 von Laue suggested that since the atoms in a crystal are spaced regularly with separations of order 0.1 nm, it should be possible to use a crystal as a kind of three-dimensional grating for X rays. This proved correct, and Laue and his assistants quickly established that X rays were waves, with wavelengths of order 0.1 nm.

The use of crystals as X-ray diffraction gratings was developed by the English physicists W. L. Bragg and his father W. H. Bragg, and is often called **Bragg diffraction** (or Bragg scattering or Bragg reflection). The technique was historically important in the study of X rays and is even more important today in the study of crystal structures. To understand the Braggs' analysis, we first note that we can think of a crystal as a large number of regularly spaced, identical, parallel planes, each containing many regularly spaced atoms as in Fig. 5.4(a). [These crystal planes can be chosen in several ways; another possibility is shown in Fig. 5.4(b).] Let us consider an electromagnetic wave ap-

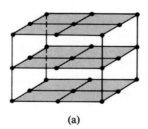
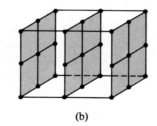

(a)   (b)

FIGURE 5.4 The atoms of a crystal define sets of identical, parallel planes, each containing many atoms. Two such sets of planes are shown.

proaching the planes of Fig. 5.4(a), with glancing angle θ, as shown in Fig. 5.5(a). (In Bragg diffraction, the incident direction is traditionally specified by the glancing angle measured up from the plane, rather than the angle down from the normal as in optics.) When the wave strikes the crystal, each atom will scatter some of the radiation and we shall observe diffraction maxima in those directions where all of the scattered waves are in phase. If we consider first waves scattered by atoms in a single plane, all of the scattered waves will be in phase in the direction given by the familiar law of reflection [Fig. 5.5(a)]

$$\theta = \theta'. \tag{5.10}$$

(This is why Bragg diffraction is often called Bragg reflection.)

Let us next consider the waves scattered by atoms in two adjacent planes, a perpendicular distance d apart. It is easy to see, as in Fig. 5.5(b), that the path difference for these two waves is $2d \sin \theta$. Thus the waves from adjacent planes

FIGURE 5.5 Side view of waves striking the crystal planes of Fig. 5.4(a). **(a)** Waves scattered by atoms in a single plane are in phase if the path lengths AA' and BB' are equal; this requires that $\theta = \theta'$. **(b)** The path difference for waves scattered from two successive planes is the distance $LMN = 2d \sin \theta$.

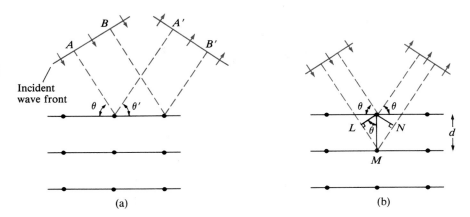

(a)   (b)

will be in phase provided that $2d \sin \theta$ is an integral multiple of the wavelength λ:

$$2d \sin \theta = n\lambda, \tag{5.11}$$

where n is any integer, 1, 2, 3, . . . , and is known as the *order* of the diffraction maximum. In many applications the maxima with $n > 1$ are relatively weak, so that only $n = 1$ is important.

The condition (5.11) is called the **Bragg law.** In any direction for which both (5.10) and (5.11) are satisfied, the waves from all atoms in the crystal will be in phase, and a strong maximum will be observed. This result can be used in several ways. For many simple crystals the spacing, d, of the planes can be calculated from knowledge of the density of the crystal and the mass of the atoms. If monochromatic X rays (that is, X rays of a single wavelength λ) are fired at a crystal of known spacing, the resulting pattern can be used, in conjunction with the Bragg law, to measure λ. If the X rays contain a spread of wavelengths ("white" X rays), the different wavelengths will give maxima in different directions, and one can use the crystal as an **X-ray spectrometer,** to find out what wavelengths are present and with what intensity. (A spectrometer is any device — like the familiar diffraction grating — that sorts and measures the different wavelengths in radiation.)

A simple X-ray spectrometer of the type used by the Braggs is shown in Fig. 5.6. The X rays under study pass through a collimator (a small aperture to define their direction) and are reflected off the face of a crystal whose plane spacing is known. The intensity I of the reflected rays is measured by a detector, such as an ionization chamber. (In this device, the X rays pass through a gas between two plates at different potentials. The X rays ionize the gas allowing a current to flow, whose magnitude is proportional to the X-ray intensity.) By

FIGURE 5.6 An X-ray spectrometer. X rays are reflected off the crystal and detected by the ionization chamber. The crystal and chamber can both rotate in such a way that the two angles θ are always equal.

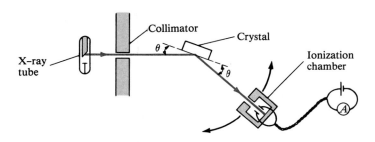

rotating the crystal and detector, one can find the intensity I as a function of θ. By the Bragg law (5.11), this is equivalent to finding I as a function of λ, which is the required X-ray spectrum. By selecting just the X rays scattered at one angle, one can obtain a monochromatic beam of X rays for use in some other experiment (in which case we would call the arrangement an X-ray monochromer).

Once the wavelength of X rays is known, one can use the Bragg law to investigate unknown crystal structures. This use of X rays is called X-ray crystallography and is an important tool in solid-state physics and molecular biology. Figure 5.7 shows a short section of the biological molecule DNA, whose structure was found from X-ray studies.

The diffraction pattern implied by the Bragg law (5.11) is often fairly complicated. The main reason is that as mentioned in connection with Fig. 5.4, there are many different sets of crystal planes, with different orientations and different spacings, d. For each such set of planes, the Bragg condition implies certain definite directions of maximum intensity. The resulting pattern in a typical X-ray diffraction experiment is shown in Fig. 5.8.

A second complication in Bragg diffraction is that many solids are not a single monolithic crystal, but are instead a jumbled array of many microcrystals. When X rays pass into such a polycrystal, only those microcrystals that are oriented correctly for the Bragg condition will produce constructive interference. The locus of these constructive directions for all such microcrystals is a cone, and the resulting pattern is a series of rings as illustrated in Fig. 5.9. These rings can be thought of as the result of rotating a single-crystal pattern like Fig. 5.8 about its center.

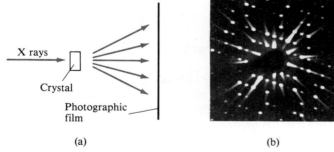

(a)   (b)

FIGURE 5.8 **(a)** One possible arrangement for observing X-ray diffraction with a single crystal. **(b)** The X-ray diffraction pattern produced by a crystal of the compound $P_2S_2(NC_6H_5)_2$ $(NHC_6H_5)_2$. (Courtesy Curtis Haltiwanger.)

FIGURE 5.7   The double helix of the DNA molecule was inferred by Crick and Watson from X-ray studies. Each circle represents a *base,* comprising about 30 atoms, and the order of these bases carries genetic information. (The bases in the two strands are shaded differently only to emphasize the helical structure.)

5.5   The Duane–Hunt Law

With an X-ray spectrometer such as the one described in Section 5.4, it was possible to analyze the distribution of wavelengths produced in an X-ray tube. This kind of distribution is generally recorded as a *spectrum;* that is, as a graph of intensity as a function of wavelength or frequency. Two typical spectra, made with the same accelerating potential but with different anode metals, are sketched in Fig. 5.10, where the intensity is shown as a function of frequency.

According to the classical theory of bremsstrahlung (the "braking radiation" produced by a decelerating charge), the X rays should be produced in a broad spread of frequencies, such that the intensity varies smoothly with f and

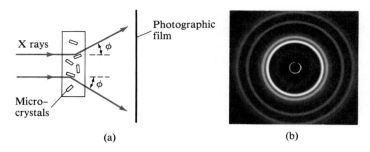

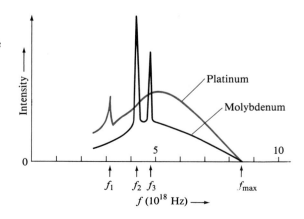

FIGURE 5.9 X-ray diffraction with a polycrystal. **(a)** Constructive interference occurs only for those microcrystals that are oriented correctly for the Bragg condition; the locus of all constructive directions is a cone whose angle is shown as ϕ. **(b)** The resulting pattern for X rays with $\lambda = 0.07$ nm fired at an aluminum foil. (Courtesy Educational Development Center.)

drops slowly to zero at high frequencies. The observed spectra in Fig. 5.10 contradict this prediction in two important ways. First, each curve shows one or more tall, sharp spikes superposed on an otherwise smooth background. These spikes indicate that an appreciable fraction of the radiation produced in either metal is produced at certain isolated frequencies (marked f_1 for platinum and f_2, f_3 for molybdenum in the picture). Similar sharp spikes appear whatever metal is used for the anode, but they occur at different frequencies that are characteristic of the metal concerned. For this reason the X rays at these frequencies are called **characteristic X rays.**

Classical physics has no explanation for the characteristic X rays, which are produced by a mechanism quite different from the bremsstrahlung of the smooth background curves in Fig. 5.10. We shall return to the characteristic X rays in Section 6.9, where we shall see that they are the result of the quantization of atomic energy levels.

The second important feature of the spectra shown in Fig. 5.10 is that both drop abruptly to zero at a certain maximum frequency f_{max}, which is the same for both metals. This phenomenon, which lacks any classical explanation, is easily explained if we recognize, first, that X rays are quantized with energy hf, and second, that the energy of each quantum is supplied by one of the electrons striking the anode in the X-ray tube. These electrons have kinetic energy $K = V_0 e$, where V_0 is the accelerating voltage of the tube. Since the most energy an

FIGURE 5.10 X-ray spectra produced by platinum and molybdenum anodes, both made with an accelerating potential of 35 kV. Note how both spectra terminate abruptly at the same frequency, f_{max}.

electron can possibly give up is K, no X-ray photons can be produced with energy hf greater than K. Thus the maximum frequency produced is

$$f_{max} = \frac{K}{h} = \frac{V_0 e}{h}.$$

(5.12)

This result shows that f_{max} varies in proportion to the accelerating voltage V_0, but is the same for all anode materials. It is called the **Duane–Hunt law,** after its discoverers Duane and Hunt, who used it to obtain a value for Planck's constant h. Their result was in excellent agreement with Planck's value, furnishing "strong evidence in favor of the fundamental principle of the quantum hypothesis."

EXAMPLE 5.3 The spacing of one set of crystal planes in common salt (NaCl) is $d = 0.282$ nm. A monochromatic beam of X rays produces a Bragg maximum when its glancing angle with these planes is $\theta = 7°$. Assuming that this is the first-order maximum ($n = 1$), find the wavelength of the X rays. What is the minimum possible accelerating voltage, V_0, that produced the X rays?

From the Bragg law (5.11), with $n = 1$, we find that

$$\lambda = 2d \sin \theta = 2 \times (0.282 \text{ nm}) \times \sin 7°$$
$$= 0.069 \text{ nm}.$$

The Duane–Hunt law requires that the electrons' kinetic energy, $V_0 e$, in the X-ray tube must be at least equal to the energy, hf, of the X-ray photons. Therefore,

$$V_0 e \geq hf = \frac{hc}{\lambda} = \frac{1240 \text{ eV·nm}}{0.069 \text{ nm}} = 18,000 \text{ eV}$$

or

$$V_0 \geq 18,000 \text{ Volts.}$$

5.6 The Compton Effect

When a beam of light is fired at a system of charges, like an atom or a single electron, some of the beam is scattered in various directions. The classical theory of such scattering was straightforward: The oscillating electric field of the incident light caused the charges to oscillate, and the oscillating charges then radiated secondary waves in various directions. The angular distribution of these scattered waves depended on the detailed arrangement of the target charges, but one prediction of the theory was common to all targets: The frequency f of the scattered waves must be the same as that of the oscillating charges, which in turn must be the same as the incident frequency f_0. Thus the scattered and incident frequencies were necessarily the same,

$$f = f_0.$$

(5.13)

Numerous experiments with visible light and preliminary observations with X rays had all seemed to confirm this prediction.

Starting in 1912, however, there appeared various reports that when high frequency X rays were scattered off electrons, the scattered frequency f was less than f_0,

$$f < f_0. \tag{5.14}$$

This claim was so surprising that it was not taken seriously at first. But in 1923 the American physicist A. H. Compton published two papers in which he argued that if light is quantized, then one should *expect* to find that $f < f_0$; and he reported experiments which showed that for X rays scattering off electrons f *is* less than f_0.

Compton argued that if photons carry energy, they should also carry momentum. This momentum, p, should be related to the energy, E, by the "Pythagorean relation" (3.23)

$$E^2 = (pc)^2 + (mc^2)^2, \tag{5.15}$$

except that since photons travel with speed c, they must have $m = 0$, and hence satisfy

$$E = pc, \tag{5.16}$$

as discussed in Section 3.8. Since $E = hf$, this implies that

$$p = \frac{E}{c} = \frac{hf}{c} = \frac{h}{\lambda}. \tag{5.17}$$

He proposed further that when radiation is scattered by electrons, each scattered photon results from a collision with a single electron, and that the ordinary rules of conservation of energy and momentum hold in this collision.

Compton's assumptions immediately explain the observed shift in frequency of the scattered X rays: When a photon (of frequency f_0) strikes a stationary, free electron, the electron must recoil. Since the electron gains energy in this process, the photon must lose energy. Therefore, its final energy hf is less than the original hf_0, and $f < f_0$, as observed.

In fact, Compton used conservation of energy and momentum to predict the scattered frequency f as a function of the scattering angle θ of the X rays. Before we describe this calculation, we should emphasize that Compton did not, of course, have a target of stationary, free electrons. In fact, his target electrons were the electrons in the carbon atoms of a graphite block. Thus his electrons were moving (in their atomic orbits), and because of the binding forces in the atoms, they were not perfectly free to recoil when struck by the photons. Fortunately, neither of these complications is important. X rays have wavelengths of order 0.1 nm or less (Compton's had $\lambda = 0.07$ nm) and energies of order

$$E = hf = \frac{hc}{\lambda} = \frac{1240 \text{ eV} \cdot \text{nm}}{0.1 \text{ nm}} \approx 10^4 \text{ eV}$$

or more. Now, the kinetic energies of the outer electrons in an atom are a few eV, and the energy needed to remove them is of the same order. These energies are negligible compared to the incident photons' energies and we can, therefore, treat the target electrons as if they were at rest and free.

Let us now consider a photon of energy E_0 and momentum $\mathbf{p_0}$ approaching a stationary electron, whose energy is mc^2 and momentum zero. After the encounter we suppose that the photon has an energy $E = hf$ and a momentum $\mathbf{p}$ that makes an angle θ with $\mathbf{p_0}$ as in Fig. 5.11. The electron recoils with total energy E_e and momentum $\mathbf{p_e}$. By conservation of energy and momentum, we expect that

$$E_e + E = mc^2 + E_0 \tag{5.18}$$

and

$$\mathbf{p_e} + \mathbf{p} = \mathbf{p_0}. \tag{5.19}$$

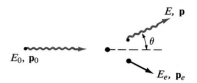

FIGURE 5.11 Compton scattering, in which a photon scatters off a free electron.

(We treat the electron relativistically so that our analysis will apply even at very high energies.) We can eliminate the electron's energy and momentum from these two equations and find the photon's final energy E in terms of the initial E_0, and hence f in terms of f_0.

We first solve (5.18) for E_e to give

$$E_e = mc^2 + E_0 - E. \tag{5.20}$$

The photon energies can be rewritten, using (5.16), as $E_0 = p_0 c$ and $E = pc$; and the electron's energy E_e is given by the "Pythagorean relation" (5.15). Canceling a common factor of c we find that (5.20) implies that

$$\sqrt{p_e^2 + (mc)^2} = mc + p_0 - p. \tag{5.21}$$

We can solve (5.19) to give $\mathbf{p_e} = \mathbf{p_0} - \mathbf{p}$ and hence

$$\begin{aligned}
p_e^2 &= \mathbf{p_e} \cdot \mathbf{p_e} = (\mathbf{p_0} - \mathbf{p}) \cdot (\mathbf{p_0} - \mathbf{p}) \\
&= p_0^2 + p^2 - 2\mathbf{p_0} \cdot \mathbf{p} \\
&= p_0^2 + p^2 - 2p_0 p \cos\theta.
\end{aligned} \tag{5.22}$$

We can next substitute (5.22) for p_e^2 into (5.21). After squaring both sides and canceling several terms, we find that

$$mc(p_0 - p) = p_0 p(1 - \cos\theta) \tag{5.23}$$

or

$$\frac{1}{p} - \frac{1}{p_0} = \frac{1}{mc}(1 - \cos\theta). \tag{5.24}$$

From (5.24) we can find the scattered photon's momentum p; and from this we can calculate its energy $E = pc$, its frequency $f = E/h$, or its wavelength $\lambda = c/f$. The result is most simply expressed in terms of λ, since according to (5.17) the photon momentum is $p = h/\lambda$, so that $1/p = \lambda/h$ (and similarly for p_0). Thus (5.24) implies that

$$\Delta\lambda \equiv \lambda - \lambda_0 = \frac{h}{mc}(1 - \cos\theta). \tag{5.25}$$

This result, first derived by Compton, gives the increase in a photon's wavelength when it is scattered through an angle θ. Since $\Delta\lambda \geq 0$ the wavelength is always *increased,* and the frequency *decreased,* as anticipated. The shift in wavelength, $\Delta\lambda$, is zero at $\theta = 0$ and increases as a function of θ, to a maximum

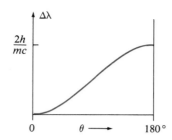

FIGURE 5.12 Increase, $\Delta\lambda$, in the wavelength of photons in Compton scattering. Note that $\Delta\lambda$ is zero at $\theta = 0$ and rises to a maximum at $\theta = 180°$.

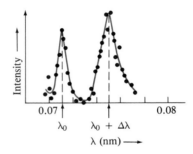

FIGURE 5.13 The spectrum of X rays scattered at $\theta = 135°$ off graphite, as measured by Compton; λ_0 is the incident wavelength (0.0711 nm) and $\Delta\lambda$ is the shift predicted by Compton's formula (5.25).

of $2h/(mc)$ at $\theta = 180°$, as sketched in Fig. 5.12. The magnitude of the shift is determined by the quantity*

$$\frac{h}{mc} = \frac{hc}{mc^2} = \frac{1240 \text{ eV} \cdot \text{nm}}{0.511 \text{ MeV}} = 0.00243 \text{ nm}. \qquad (5.26)$$

In Compton's experiment the incident wavelength was $\lambda_0 = 0.07$ nm. Thus the predicted shift ranged up to 7% of the incident λ_0, a shift that Compton's spectrometer could certainly detect.

Compton measured the scattered wavelength at four different angles θ and found excellent agreement with his prediction (5.25). This agreement gave "very convincing" support to the assumptions on which (5.25) was based, that photons carry momentum as well as energy, and can be treated like particles, subject to the ordinary laws of conservation of energy and momentum.

Two final comments on Compton's formula and his experimental results: First, it is clear from (5.25) that the change of wavelength, $\Delta\lambda$, is independent of the wavelength itself. Thus we should see the same change $\Delta\lambda$ for visible light as for X rays. Why, then, is no shift observed with visible light? The answer is simple: The wavelength of visible light (400 to 700 nm) is about 5000 times larger than that of X rays. Thus the fractional change $\Delta\lambda/\lambda_0$ is about 5000 times smaller for visible light than for X rays and is in practice unobservable.† Quite generally, the longer the incident wavelength, the less important Compton's predicted shift will be.

Second, Compton found that at each angle *some* of the scattered X rays had the wavelength given by his formula (5.25), but that some had λ equal to the incident wavelength λ_0. This is clearly visible in Fig. 5.13, which shows Compton's spectrum of X rays scattered at $\theta = 135°$. The explanation of these unshifted X rays is also simple, at least qualitatively. The derivation of (5.25) assumed a collision between the incident photon and a free electron (free, that is, to recoil when struck by the photon). As already discussed, this assumption is certainly good for the outer atomic electrons that are weakly bound. But the inner electrons are very tightly bound to the atom. (Several hundred eV is needed to remove an inner electron from a carbon atom.) Thus if a photon interacts with either an inner electron or the atomic nucleus, it may not detach an electron at all, and the whole atom may recoil as a unit. In this case we should treat the process as a collision between the photon and the whole atom, and the same analysis would lead to the same formula (5.25), but with the electron mass m replaced by the mass of the whole atom. This implies a change $\Delta\lambda$ which is about 20,000 times smaller (in the case of carbon). For these collisions the change of wavelength will appear to be zero.

5.7 Particle–Wave Duality

Today almost all physicists accept that the photoelectric effect, the Compton effect, and numerous other experiments demonstrate beyond doubt the particle nature of light. But what about the many experiments that had established the

---

* This quantity is rather misleadingly called the *electron's Compton wavelength*. It is not, in any obvious sense, the wavelength of the electron; it is simply a quantity, with the dimensions of length, that determines the shift in wavelength of photons scattered off electrons.

† Even if this small shift were detectable, the Compton effect for visible light would be more complicated, because the target electron cannot be considered as stationary and free if it is struck by a visible photon whose energy is only a few eV.

wave nature of light? Were these experiments somehow wrong? The answer is that both kinds of experiment are right, and that light exhibits wave properties *and* particle properties. In fact, both aspects of light are inextricably mixed in the two basic equations

$$E = hf \quad \text{and} \quad p = \frac{h}{\lambda} \qquad (5.27)$$

since the energy E and momentum p refer to the particle nature of the photon, whereas the frequency f and wavelength λ are both wave properties.

We shall see in Chapter 7 that ordinary particles like electrons and protons also show this **particle–wave duality,** and the first task of quantum theory is to reconcile these seemingly contradictory aspects of all the particles of modern physics. It turns out that the quantum theory of electrons, protons, and other massive particles is much simpler than the quantum theory of the massless photon. This is because massive particles can move at nonrelativistic speeds, whereas photons always move at speed c and are intrinsically relativistic. Thus for massive particles we can, and shall, develop a nonrelativistic quantum mechanics, which is able to explain a wide range of phenomena in atomic, nuclear, and condensed-matter physics. Relativistic quantum theory—in particular, the quantum theory of electromagnetic radiation, or *quantum electrodynamics*—is beyond the scope of this book. Fortunately, we can understand a large part of modern physics, armed with just the basic facts of quantum radiation theory, as summarized in the two equations (5.27).

IDEAS YOU SHOULD NOW UNDERSTAND FROM CHAPTER 5

Quantization
Planck's constant, h
Photons; $E_{photon} = hf$
The photoelectric effect; work function ϕ
X rays

Bragg diffraction
The Duane–Hunt law
The Compton effect
Photons have momentum $p_{photon} = h/\lambda$

PROBLEMS FOR CHAPTER 5

SECTION 5.3 (THE PHOTOELECTRIC EFFECT)

5.1 • Given that visible light has $400 < \lambda < 700$ nm, what is the range of energies of visible photons (in eV)? Ultraviolet (UV) radiation has wavelengths *shorter* than visible, while infrared (IR) wavelengths are *longer* than visible. What can you say about the energies of UV and IR photons?

5.2 • (*a*) X rays are electromagnetic radiation with wavelengths much shorter than visible—of order 0.1 nm

or less. What are the energies of X-ray photons? (*b*) Electromagnetic waves with wavelengths even shorter than X rays are called γ rays and are produced in many nuclear processes. A typical γ-ray wavelength is 10^{-4} nm (or 100 fm); what is the corresponding photon energy? Give your answers in keV or MeV, as appropriate.

5.3 • Microwaves (as used in microwave ovens, telephone transmission, etc.) are electromagnetic waves with wavelength of order 1 cm. What is the energy of a typical microwave photon in eV?

5.4 • The work function of cesium is 1.9 eV. (*a*) What is the maximum wavelength of light that can eject photoelectrons from cesium? (*b*) If light with $\lambda = 500$ nm strikes cesium, what is the maximum kinetic energy of the ejected electrons?

5.5 • The longest wavelength of light that can eject electrons from potassium is $\lambda_0 = 560$ nm. (*a*) What is the work function of potassium? (*b*) If ultraviolet radiation with $\lambda = 300$ nm is shone onto potassium, what will be the stopping potential V_s (the potential that just stops all the ejected electrons)?

5.6 •• A light bulb that is rated at 60 W actually produces only about 3 W of visible light, most of the rest of the energy being infrared (or heat). (*a*) About how many visible photons does such a light bulb produce each second? Use the average value $\lambda \approx 550$ nm. (*b*) If a person looks at such a bulb from about 1 ft away about how many visible photons enter the eye per second? (When looking at a bright light, the pupil has a diameter of about 1 mm.) (*c*) By how many powers of 10 does this exceed the minimum detectable intensity, which is about 100 photons entering the eye per second?

5.7 •• Use Millikan's data from the photoelectric effect in sodium (Fig. 5.2) to get a rough value of Planck's constant h in eV·s and in J·s.

5.8 •• The work function for tungsten is $\phi = 4.6$ eV. (*a*) If light is incident on tungsten, find the critical frequency f_0, below which no electrons will be ejected, and the corresponding wavelength λ_0. Use the equation $K_{max} = hf - \phi$ to find the maximum kinetic energy of ejected electrons if tungsten is irradiated with light with (*b*) $\lambda = 200$ nm and (*c*) 300 nm. Explain your answer to part (*c*).

SECTION 5.4 (X RAYS AND BRAGG DIFFRACTION)

5.9 • Potassium chloride (KCl) has a set of crystal planes separated by a distance $d = 0.31$ nm. At what glancing angle, θ, to these planes would the first order Bragg maximum occur for X rays of wavelength 0.05 nm?

5.10 • When X rays of wavelength $\lambda = 0.20$ nm are reflected off the face of a crystal, a Bragg maximum is observed at a glancing angle of $\theta = 17.5°$, with sufficient intensity that it is judged to be first order. (*a*) What is the spacing d of the planes that are parallel to the face in question? (*b*) What are the glancing angles of all higher-order maxima?

5.11 •• A student is told to analyze a crystal using Bragg diffraction. She finds that the elderly equipment has seized up and cannot turn to glancing angles below $\theta = 30°$. She bravely persists and, using X rays with $\lambda = 0.0438$ nm, finds three weak maxima at $\theta = 36.7°$, $52.8°$, and $84.5°$. What are the orders of these maxima, and what is the spacing of the crystal planes?

SECTION 5.5 (THE DUANE–HUNT LAW)

5.12 • What is the shortest wavelength of X rays that can be produced by an X-ray tube whose accelerating voltage is 10 kV?

5.13 • What is the voltage of an X-ray tube that produces X rays with wavelengths down to 0.01 nm but no shorter?

5.14 •• A monochromatic beam of X rays produces a first-order Bragg maximum when reflected off the face of an NaCl crystal with glancing angle $\theta = 20°$. The spacing of the relevant planes is $d = 0.28$ nm. What is the minimum possible voltage of the tube that produced the X rays?

SECTION 5.6 (THE COMPTON EFFECT)

5.15 • Use Compton's formula (5.25) to calculate the predicted shift $\Delta\lambda$ of wavelength at $\theta = 135°$ for Compton's data shown in Fig. 5.13. What percent shift in wavelength was this?

5.16 • Find the change in wavelength for photons scattered through 180° by free protons. Compare with the corresponding shift for electrons.

5.17 • A photon of 1 MeV collides with a free electron and scatters through 90°. What are the energy of the scattered photon and the kinetic energy of the recoiling electron?

5.18 •• Consider a head-on, elastic collision between a massless photon (momentum $\mathbf{p}_0$ and energy E_0) and a stationary free electron. (*a*) Assuming that the photon bounces directly back with momentum $\mathbf{p}$ (in the direction of $-\mathbf{p}_0$) and energy E, use conservation of energy and momentum to find p. (*b*) Verify that your answer agrees with that given by Compton's formula (5.25) with $\theta = \pi$.

5.19 •• Compton showed that an individual photon carries momentum, $p = E/c$, as well as energy, E. This momentum manifests itself in the radiation pressure felt by bodies exposed to bright light, as the following problem illustrates: A 100-W beam of light shines for 1000 s on a 1-g black body initially at rest in a frictionless environment. (*a*) Calculate the total energy and momentum of the photons absorbed by the black body. (*b*) Use conservation of momentum to find the body's final velocity. (*c*) Calculate the body's final kinetic energy; explain how this can be less than the original energy of the photons.

5.20 •• A student wants to measure the shift in wavelength predicted by Compton's formula (5.25). She finds that she cannot measure a shift of less than 5% and that she cannot conveniently measure at angles greater than $\theta = 150°$. What is the longest X-ray wavelength that she can use and still observe the shift?

5.21 •• If the maximum kinetic energy given to the electrons in a Compton scattering experiment is 10 keV, what is the wavelength of the incident X rays?

CHAPTER

6

Quantization of Atomic Energy Levels

6.1 Introduction

In Chapter 5 we described the discovery of the quantization of light, which made clear that classical electromagnetic theory is incorrect on the microscopic level. In the present chapter we describe the corresponding failure of classical *mechanics* when applied to microscopic systems. With hindsight, we can see that the evidence for this breakdown of classical mechanics goes back to the discovery of atomic spectra in the middle of the nineteenth century, as we describe in Sections 6.2 and 6.3. However, it was not until 1913 that any satisfactory explanation of atomic spectra was found — by the Danish physicist Niels Bohr — and it became clear that a substantial revision of classical mechanics was required.

Bohr's work was originally prompted by a problem concerning the stability of Rutherford's model of the atom (a problem we discuss briefly in Section 6.4), but he quickly found that many properties of atomic spectra could easily be explained if one assumed that the total energy of the electrons in an atom is quantized. This quantization of atomic energies has no classical explanation, and to account for it Bohr developed a mechanics that is now called the **Bohr model** or the old quantum theory. As Bohr was well aware, his ideas were not really a complete theory, and they have now been superseded by modern quantum mechanics. Nevertheless, Bohr's ideas were a crucial step in the development of modern quantum mechanics and were correct in several important respects. For these reasons, we describe the Bohr theory and some of its successes in the last six sections of this chapter.

6.2   Atomic Spectra

Perhaps the most famous spectrum of all time was the one discovered in 1666 by Newton, who shone a narrow beam of white light through a glass prism, producing the well-known ribbon of rainbow colors, as shown in the first spectrum at the front of this book. This established that what we perceive as white light is a mixture of different colors, or different wavelengths as we would now say.

In 1814, the German physicist Fraunhofer discovered that when viewed more closely, the spectrum of sunlight is crossed by dark lines, like those in the second spectrum at the front of this book (though narrower and much more numerous). This showed that certain colors, or wavelengths, are missing from the light that reaches us from the sun. Today, we know that this is because the gases in the sun's outer atmosphere absorb light at certain discrete wavelengths. The light with these wavelengths is therefore removed from the white light that leaves the sun's surface, and this causes the dark lines observed by Fraunhofer.

By the middle of the nineteenth century it was known that all gases absorb light at wavelengths that are characteristic of the atoms and molecules they contain. For example, if white light is shone through a gas containing just one kind of atom, the gas will absorb certain wavelengths characteristic of that atom; if the transmitted light is then passed through a prism or diffraction grating, it will produce an **absorption spectrum,** consisting of a bright ribbon of rainbow colors crossed with dark **absorption lines,** like Fraunhofer's. Furthermore, it is found that if the same gas is heated sufficiently, it will *emit* light. Moreover, the wavelengths of this emitted light are the same as those which the gas absorbed when illuminated with white light. If this emitted light is passed through a prism it will produce an **emission spectrum,** consisting of bright **emission lines** against a dark background.

The absorption and emission spectra produced by atomic hydrogen are shown in color at the front of this book. (At ordinary temperatures hydrogen gas consists of H_2 *molecules;* to produce the atomic spectra shown, one must use gas that is heated enough — by an electric discharge, for example — to dissociate the molecules into atoms.) The same spectra are shown schematically in black and white in Fig. 6.1, where the pictures (a) and (c) show the emission and absorption spectra themselves, while (b) and (d) are the corresponding graphs of intensity against wavelength.

Each atom, and likewise each molecule, emits and absorbs at its own characteristic wavelengths. Thus emission and absorption spectra act like fingerprints, uniquely identifying the atom or molecule that produced them. By about 1870 spectroscopy had become a powerful tool of chemical analysis and

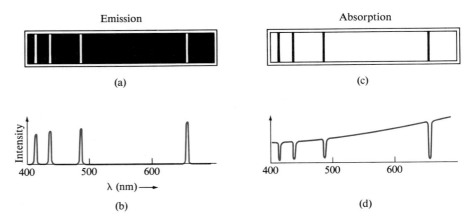

Emission

(a)

Absorption

(c)

Intensity

400   500   600

λ (nm) →

(b)

400   500   600

(d)

FIGURE 6.1 Emission and absorption spectra of atomic hydrogen. **(a)** The emission spectrum; the white stripes represent bright lines against a dark background. **(b)** The corresponding graph of intensity against wavelength, on which the spikes correspond to the bright lines of the spectrum itself. (The relative intensities of the four lines depend on the temperature.) **(c)** The absorption spectrum, with dark lines against a bright background. **(d)** The corresponding graph, on which the dips correspond to the dark lines of the spectrum.

had led to the discovery of several previously unknown elements. In particular, it was, and still is, the only way to determine the chemical composition of the sun, other stars, and interstellar matter.

Despite the many successful applications of spectroscopy in the nineteenth century, there was no satisfactory theory of atomic spectra. Classically, emission and absorption were easy to understand: An atom would be expected to emit light if some source of energy, such as collisions with other atoms, caused its electrons to vibrate and produce the oscillating electric fields that constitute light. Conversely, if light were incident on an atom, the oscillating electric field would cause the electrons to start vibrating and hence to absorb energy from the incident light. The observation that light is emitted and absorbed only at certain characteristic frequencies was understood to imply that the atomic electrons could vibrate only at these same frequencies; but no completely satisfactory classical model was ever found that could explain (let alone *predict*) these characteristic frequencies of vibration.

As we shall see, Bohr's ideas (and likewise modern quantum mechanics) explain the characteristic spectra in a quite different way. First, the characteristic frequencies, $f_\alpha, f_\beta, \ldots$ of light emitted by an atom imply that atoms emit photons with characteristic energies, $hf_\alpha, hf_\beta, \ldots$. (This connection between frequency and energy was, of course, completely unknown in the nineteenth century.) These characteristic energies are explained by establishing that the total energy of the electrons in an atom is *quantized,* with discrete values $E_1, E_2, E_3, \ldots$. An atom emits or absorbs light by making an abrupt jump from one energy state to another — for example, by changing from energy E_2 to E_1, or vice versa. If $E_2 > E_1$, when the atom changes from E_2 to E_1 it must release the excess energy $E_2 - E_1$, and it does so in the form of a photon of energy $hf = E_2 - E_1$; similarly, it can only change from E_1 to E_2 if it is supplied with energy $E_2 - E_1$, and one way this can happen is by absorption of a photon of energy $hf = E_2 - E_1$. The characteristic energies of the photons emitted and absorbed by an atom are thus explained as the differences in the characteristic quantized energies of the atom. Before we explore Bohr's explanation of atomic spectra further, we relate a little more history.

6.3   The Balmer–Rydberg Formula

The simplest of all atoms is hydrogen and it is therefore not surprising that the spectrum of atomic hydrogen was the first to be thoroughly analyzed. By 1885, the four visible lines shown in Fig. 6.1 had been measured very accurately by Ångstrom. These measurements were examined by a Swiss school teacher, Balmer, who found (in 1885) that the observed wavelengths fitted the formula

$$\frac{1}{\lambda} = R\left(\frac{1}{4} - \frac{1}{n^2}\right), \tag{6.1}$$

where R was a constant (with the dimension length^{-1}), which Balmer determined as

$$R = 0.0110 \text{ nm}^{-1}, \tag{6.2}$$

and n was an integer equal to 3, 4, 5, and 6 for the four lines in question. Ångstrom had measured these wavelengths to four significant figures, and Balmer's formula fitted them to the same accuracy. Balmer guessed (correctly, as we now know) that such an excellent fit could not be a coincidence and that there were probably other lines given by other values of the integer n in the formula (6.1). For example, if we take $n = 7$, then (6.1) gives

$$\lambda = 397 \text{ nm},$$

a wavelength near the violet edge of the visible spectrum; with $n = 8, 9, \ldots$, Eq. (6.1) predicts shorter wavelengths in the ultraviolet. One can imagine Balmer's delight when he discovered that several more lines had already been observed in the spectrum of hydrogen and that they were indeed given by his formula with $n = 7, 8, 9, \ldots$.

We can rewrite Balmer's formula (6.1) in the form

$$\frac{1}{\lambda} = R\left(\frac{1}{2^2} - \frac{1}{n^2}\right) \qquad (n = 3, 4, 5, \ldots). \tag{6.3}$$

It is tempting to guess that this is just a special case of the more general formula

$$\frac{1}{\lambda} = R\left(\frac{1}{n'^2} - \frac{1}{n^2}\right) \qquad (n > n', \text{ both integers}) \tag{6.4}$$

and that the spectrum of atomic hydrogen should contain all wavelengths given by all integer values of n' and n. Balmer himself had guessed that some such generalization might be possible, but the form (6.4) was apparently first written down by Rydberg, for whom (6.4) is usually called the **Rydberg formula** and R, the **Rydberg constant**. If we take $n' = 1$ and $n = 2$, for example, Rydberg's formula predicts

$$\lambda = 121 \text{ nm},$$

a wavelength well into the ultraviolet; with $n' = 3$, $n = 4$, we get

$$\lambda = 1870 \text{ nm}$$

in the infrared. In fact, all of the additional wavelengths predicted by (6.4) (with n' any integer other than Balmer's original value of 2) are either in the ultraviolet or the infrared. It was several years before any of these additional lines were observed, but in 1908 Paschen found some of the infrared lines with $n' = 3$, and in 1914 Lyman found some of the ultraviolet lines with $n' = 1$. Today, it is well established that the Rydberg formula (6.4) accurately describes all of the wavelengths in the spectrum of atomic hydrogen.

EXAMPLE 6.1   Conventional spectrometers with glass components do not transmit ultraviolet light ($\lambda \lesssim 380$ nm). Explain why none of the lines predicted by (6.4) with $n' = 1$ could be observed with a conventional spectrometer.

For the case $n' = 1$, $n = 2$, Eq. (6.4) predicts that

$$\frac{1}{\lambda} = R\left(\frac{1}{1} - \frac{1}{4}\right) = \frac{3}{4}R$$

and hence

$$\lambda = \frac{4}{3R} = \frac{4}{3 \times (0.0110 \text{ nm}^{-1})} = 121 \text{ nm},$$

as stated earlier. Similarly, for $n' = 1$ and $n = 3$, one finds that $\lambda = 102$ nm, and inspection of (6.4) shows that the larger we take n, the *smaller* the corresponding wavelength. Therefore, all lines with $n' = 1$ lie well into the ultraviolet and are unobservable with a conventional spectrometer.

NIELS BOHR (1885–1962, Danish). Bohr's model of the hydrogen atom (1913) was the first reasonably successful explanation of atomic spectra in terms of the atom's internal structure and was an essential step on the path to modern quantum theory. It also earned him the 1922 Nobel Prize. Bohr continued active in the development of modern physics and set up an institute in Copenhagen that was one of the world's great centers for theoretical physics, both before and after World War II.

It is often convenient to rewrite the Rydberg formula (6.4) in terms of photon energies rather than wavelengths. Since the energy of a photon is $E_{\text{ph}} = hc/\lambda$, we have only to multiply (6.4) by hc to give

$$E_{\text{ph}} = hcR\left(\frac{1}{n'^2} - \frac{1}{n^2}\right) \qquad (n > n', \text{ both integers}). \qquad (6.5)$$

These are the energies of the photons emitted or absorbed by a hydrogen atom.

It is important to recognize that neither Balmer, Rydberg, nor anyone else prior to 1913, could explain *why* the formula (6.5) gave the spectrum of hydrogen. It was perhaps Bohr's greatest triumph that his theory *predicted* the Rydberg formula (6.5), including the correct value of the Rydberg constant R.

6.4   The Problem of Atomic Stability

A satisfactory theory of atomic spectra obviously required a correct knowledge of the structure of the atom. It is therefore not surprising that atomic spectra lacked any explanation until Rutherford had proposed his planetary model of the atom in 1911, nor that Bohr's explanation came soon after that proposal, in 1913. Curiously enough, the planetary model posed a serious problem of its own, and it was in solving this problem that Bohr succeeded in explaining atomic spectra as well.

As Rutherford himself was aware, his model of the atom raises an awkward problem of stability. Superficially, Rutherford's atom resembles the solar system, and electrons can orbit around the nucleus just as planets orbit around the sun. According to classical mechanics the planets' orbits are circles or ellipses and are *stable:* Once a planet is placed in a given orbit it will remain there indefinitely (if we ignore small effects like tidal friction). Unfortunately, the same is not true in an atom. The electron carries an electric charge and should radiate electromagnetic waves as it moves in its orbit. This means that the orbiting electron should gradually lose energy, and hence that its orbital radius should steadily shrink and its orbital frequency increase. The steady increase in orbital frequency means that the frequency of emitted light should keep changing—in sharp contrast to the observed spectrum, with its discrete, fixed frequencies. Worse still, one can estimate the rate at which the radius will shrink, and one finds that all electrons should collapse into the nucleus in a time of order 10^{-11} s. (See Section 15.2.) That is, stable atoms, as we know them, could not exist.* This was the problem that Bohr originally set out to solve.

6.5   Bohr's Explanation of Atomic Spectra

To solve the problem of atomic stability, Bohr proposed that the laws of classical mechanics must be modified and that among the continuum of electron orbits that classical mechanics predicts, only a certain discrete set are actually possible. He gave these allowed orbits the name **stationary orbits.** Since the possible orbits were discrete, their energies would also be discrete; that is, the energies of the electrons in an atom would be quantized, and the only possible energies of the whole atom would be a discrete set $E_1, E_2, E_3, \ldots$. If this were true, it would be impossible for the atom to lose energy steadily and continuously, as required by classical electromagnetic theory. Therefore, Bohr simply postulated that an electron in one of the allowed, stationary orbits does not radiate energy and remains in exactly the same orbit as long as it is not disturbed.

Bohr could not show *why* the electrons in his stationary orbits do not radiate, and one cannot really say that his ideas *explain* the stability of atoms. Nevertheless, his ideas come very close to being correct, and his name "stationary orbit" has proved remarkably apt. In modern quantum mechanics, we shall find that the electron does not have a classical orbit at all; rather, it is distributed continuously through the atom and can be visualized as a cloud of charge surrounding the nucleus. The stable states of the atom—corresponding to Bohr's stationary orbits—are states in which this charge cloud *actually is stationary* and does not radiate.†

Having solved the problem of atomic stability (simply by postulating his quantized, stationary orbits, in which electrons did not radiate) Bohr realized that his theory gave a beautiful explanation of atomic spectra. As we have

* In theory, we should consider the same problem with respect to the solar system. As the planets move in their orbits they should radiate *gravitational* waves, analogous to the electromagnetic waves radiated by electrons. However, gravitational radiation is so small that it has not yet (1991) been detected directly, and its effect on the planetary orbits is certainly unimportant.

† We shall discuss this further in Section 8.3.

already described, if the total energy of the electrons in an atom is quantized, with allowed values $E_1, E_2, E_3, \ldots$, the energy can change only by making a discontinuous transition from one value E_n to another $E_{n'}$,

$$E_n \rightarrow E_{n'}.$$

Bohr did not try to explain the detailed mechanisms by which such a transition can occur, but one method is certainly the emission or absorption of a photon. If $E_n > E_{n'}$, a photon of energy $E_n - E_{n'}$ must be *emitted;* if $E_n < E_{n'}$, a photon of energy $E_{n'} - E_n$ must be *absorbed.* Either way, the photon's energy must be the difference of two of the allowed energies, E_n and $E_{n'}$, of the atom. This immediately explains why the energies (and hence frequencies) of the *emitted* and *absorbed* photons are the same. Further, one would naturally expect the allowed energies $E_1, E_2, E_3, \ldots$ to be different for different atoms. Thus the same should be true of the differences $E_n - E_{n'}$, and this would explain why each atom emits and absorbs with its own characteristic spectrum.

EXAMPLE 6.2 The helium atom has two stationary orbits, designated $3p$ and $2s$, with energies

$$E_{3p} = 23.1 \text{ eV} \quad \text{and} \quad E_{2s} = 20.6 \text{ eV}.$$

(We shall see the significance of the designations $3p$ and $2s$ later.) What will be the wavelength of a photon emitted when the atom makes a transition from the $3p$ orbit to the $2s$?

In moving from the $3p$ orbit to the $2s$, the atom loses energy $E_{3p} - E_{2s}$. This is therefore the energy of the emitted photon

$$E_{\text{ph}} = E_{3p} - E_{2s} = 2.5 \text{ eV},$$

so the wavelength is

$$\lambda = \frac{hc}{E_{\text{ph}}} = \frac{1240 \text{ eV} \cdot \text{nm}}{2.5 \text{ eV}} \approx 500 \text{ nm}.$$

Light with this wavelength is blue-green, and the $3p \rightarrow 2s$ transition is, in fact, responsible for the blue-green line that is visible in the helium spectrum inside the front cover.

6.6 The Bohr Model of the Hydrogen Atom

It was obviously desirable that Bohr find a way to predict the allowed energies, $E_1, E_2, \ldots$, of an atom, and in the case of hydrogen he was able to do so. In fact, he produced several different arguments, all of which gave the same answer for the allowed energies. All these arguments were, as Bohr himself acknowledged, quite tentative, and their main justification was that they produced the right answer; that is, they predicted certain energy levels, which in turn led to the Rydberg formula for the spectrum of hydrogen.

We shall describe one of Bohr's arguments, which is the simplest and, in many ways, the closest to modern quantum mechanics. Since Bohr assumed that the possible orbits of the electron were a subset of the classical orbits, we begin by reviewing the classical mechanics of an orbiting electron.

Our system consists of an electron of mass m and charge $-e$, which orbits around a proton of charge $+e$, as shown in Fig. 6.2. For simplicity, we shall treat the case where the electron moves in a circular orbit, and because the proton is so much heavier than the electron we shall make the approximation that the proton is fixed in position. (In reality, the proton moves a little, and this requires a very small correction to our answers, as we discuss later.)

The electron's acceleration is the centripetal acceleration, $a = v^2/r$, and the only force acting on the electron is the Coulomb attraction of the proton,

$$F = \frac{ke^2}{r^2}$$

where k is the Coulomb force constant, $k = 1/(4\pi\epsilon_0) = 8.99 \times 10^9$ N·m²/C². Thus Newton's second law implies that

$$(\text{mass}) \times (\text{centripetal acceleration}) = \text{Coulomb force}$$

or

$$m\frac{v^2}{r} = \frac{ke^2}{r^2}. \tag{6.6}$$

This condition is a relation between v and r, which can be solved to give v in terms of r, or vice versa.

In classical mechanics, (6.6) is the only constraint between v and r; so neither v nor r is fixed. On the contrary, the possible values of v and r range continuously from 0 to ∞, and this means that the energy of the electron is *not* quantized. To see this explicitly, we note that (6.6) implies that

$$mv^2 = \frac{ke^2}{r}. \tag{6.7}$$

Now, the electron's kinetic energy K is $K = \frac{1}{2}mv^2$, while the potential energy of an electron (charge $-e$) in the field of a proton (charge $+e$) is

$$U = -\frac{ke^2}{r}, \tag{6.8}$$

if we define U to be zero when the electron is far from the proton (that is, when $r = \infty$). Thus (6.7) implies that for an electron in a circular orbit

$$K = -\tfrac{1}{2}U. \tag{6.9}$$

(This result was well known in classical mechanics and was called the virial theorem.) The total energy is therefore

$$E = K + U = \frac{1}{2}U = -\frac{1}{2}\frac{ke^2}{r}. \tag{6.10}$$

Notice that the total energy is negative, as it has to be since the electron is bound to the proton and cannot escape to infinity. Since r can have any value in the range $0 < r < \infty$, it is clear from (6.10) that the energy of our bound electron can have any value in the range $-\infty < E < 0$.

Our analysis so far has been purely classical. Some new hypothesis was needed if the allowed energies were to be quantized and if one was to find what

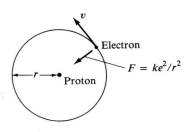

FIGURE 6.2 The hydrogen atom consists of an electron in orbit around a proton. The centripetal acceleration, $a = v^2/r$, is supplied by the Coulomb attraction, $F = ke^2/r^2$.

those allowed energies were. To understand the hypothesis that Bohr proposed, we note that Planck's constant has the same dimensions as angular momentum:

$$[h] = \text{energy} \times \text{time} = \frac{ML^2}{T^2} \times T = \frac{ML^2}{T}$$

and

$$[\text{angular momentum}] = [mvr] = M \times \frac{L}{T} \times L = \frac{ML^2}{T}. \qquad (6.11)$$

This suggests that the electron's angular momentum could be quantized in multiples of h; and Bohr proposed specifically that its allowed values are integer multiples of $h/(2\pi)$:

$$L = \frac{h}{2\pi}, \quad 2\frac{h}{2\pi}, \quad 3\frac{h}{2\pi}, \quad \ldots, \qquad (6.12)$$

where L denotes the electron's angular momentum. Bohr was led to propose these values for L by what he called the correspondence principle, which we shall describe briefly in Problem 15.10. For the moment, we simply accept (6.12) as a judicious guess which, like Planck's hypothesis that light is quantized in multiples of hf, was principally justified by the fact that it led to the correct answers. When we go on to discuss modern quantum mechanics we shall be able to *prove* that **Bohr's quantization condition** (6.12) is essentially correct.*

The combination $h/2\pi$ in (6.12) appears so frequently that it is often given its own symbol:

$$\hbar = \frac{h}{2\pi} = 1.054 \times 10^{-34} \text{ J} \cdot \text{s}, \qquad (6.13)$$

where $\hbar$ is read as "h cross" or "h bar." Thus we can rewrite the quantization condition (6.12) as

$$L = n\hbar \qquad (n = 1, 2, 3, \ldots). \qquad (6.14)$$

For the circular orbits, which we are discussing, the angular momentum is $L = mvr$, and (6.14) can be rewritten as

$$mvr = n\hbar \qquad (n = 1, 2, 3, \ldots). \qquad (6.15)$$

This condition is a second relation between r and v. [The first was (6.7), which expressed Newton's second law.] With two equations for two unknowns we can now solve to find the allowed values of r (or v). If we solve (6.15) to give $v = n\hbar/(mr)$ and then substitute into (6.7), we find

---

* The hypothesis (6.12) is not exactly correct. We shall prove, rather, that any *component* of the vector **L** is an integer multiple of $h/2\pi$. However, for the present this comes to the same thing, because we can take the electron's orbit to lie in the xy plane, in which case the total angular momentum is the same as its z component.

$$m\left(\frac{n\hbar}{mr}\right)^2 = \frac{ke^2}{r};$$

whence

$$r = \frac{n^2\hbar^2}{ke^2m}. \tag{6.16}$$

That is, the values of r are quantized, with values given by (6.16), which we write as

$$\boxed{r = n^2 a_B \qquad (n = 1, 2, 3, \ldots).} \tag{6.17}$$

Here we have defined the **Bohr radius** a_B, which is easily evaluated to be (Problem 6.5)

$$\boxed{a_B = \frac{\hbar^2}{ke^2m} = 0.0529 \text{ nm.}} \tag{6.18}$$

Knowing the possible radii of the electron's orbits, we can immediately find the possible energies from (6.10):

$$E = -\frac{ke^2}{2r} = -\frac{ke^2}{2a_B}\frac{1}{n^2} \qquad (n = 1, 2, 3, \ldots). \tag{6.19}$$

We see that the possible energies of the hydrogen atom *are* quantized. If we denote the energy (6.19) by E_n, then, as argued in Section 6.5, the energy of a photon emitted or absorbed by hydrogen must have the form

$$E_{ph} = E_n - E_{n'} = \frac{ke^2}{2a_B}\left(\frac{1}{n'^2} - \frac{1}{n^2}\right), \tag{6.20}$$

which has precisely the form of the Rydberg formula (6.5)

$$E_{ph} = hcR\left(\frac{1}{n'^2} - \frac{1}{n^2}\right). \tag{6.21}$$

Comparing (6.20) and (6.21) we see that Bohr's theory predicts both the Rydberg formula and the value of the Rydberg constant,*

$$R = \frac{ke^2}{2a_B(hc)} = \frac{1.44 \text{ eV} \cdot \text{nm}}{2 \times (0.0529 \text{ nm}) \times (1240 \text{ eV} \cdot \text{nm})}$$
$$= 0.0110 \text{ nm}^{-1}$$

in perfect agreement with the observed value (6.2).

Because of its close connection with the Rydberg constant, the energy $hcR = ke^2/(2a_B)$ in (6.21) and (6.20) is called the **Rydberg energy** and is denoted E_R. Its value, and several equivalent expressions for it, are (as you should check for yourself—Problem 6.6)

---

* In evaluating this we have used the two useful combinations $ke^2 = 1.44$ eV·nm and $hc = 1240$ eV·nm. (See Problem 6.3.) These are listed, along with many other physical constants, inside the front cover.

$$E_R = hcR = \frac{ke^2}{2a_B} = \frac{m(ke^2)^2}{2\hbar^2} = 13.6 \text{ eV}. \qquad (6.22)$$

In terms of E_R, the allowed energies (6.19) of the electron in a hydrogen atom are

$$E_n = -\frac{E_R}{n^2}. \qquad (6.23)$$

This is the most important result of the Bohr model, and we take up its implications in the next section.

6.7 Properties of the Bohr Atom

As compared to the modern quantum-mechanical view, Bohr's model of the hydrogen atom is not completely correct. Nevertheless, it is correct in several important features, and is often easier to visualize and remember than its modern counterpart. For these reasons we review some of its properties.

Bohr's model predicts (and modern quantum mechanics agrees) that the possible energies of an electron in a hydrogen atom are quantized, their allowed values being $E_n = -E_R/n^2$, where $n = 1, 2, \ldots$. The lowest possible energy is that with $n = 1$ and is

$$E_1 = -E_R = -13.6 \text{ eV}. \qquad (6.24)$$

This state of lowest energy is called the **ground state.** It is the most stable state of the atom, and is the state into which an isolated atom will eventually find its way. The significance of the energy $E_1 = -13.6$ eV is that an energy $+13.6$ eV must be supplied to remove the electron entirely from the proton. That is, the Bohr theory predicts that the binding energy of the hydrogen atom is 13.6 eV, in excellent agreement with its observed value.

According to (6.17) the radius of the $n = 1$ orbit is just the Bohr radius a_B:

$$r = a_B = 0.05 \text{ nm}. \qquad (6.25)$$

This agrees well with the observed size of the hydrogen atom and was regarded by Bohr as an important accomplishment of his theory.* The primary significance of a_B is that it gives the radius of the ground state of hydrogen. However, we shall find that it also gives the order of magnitude of the outer radius of all atoms in their ground states. For this reason, the Bohr radius a_B is often used as a unit of distance in atomic physics.

The orbits with energies greater than the ground-state energy are called *excited states.* Their energies are given by $E_n = -E_R/n^2$, with $n = 2, 3, \ldots$; that is,

$$E_2 = -\frac{E_R}{4} = -3.4 \text{ eV}, \qquad E_3 = -\frac{E_R}{9} = -1.5 \text{ eV},$$

---

* We shall see later that in modern quantum mechanics there is no uniquely defined radius of the electron's orbit. However, the average radius is about a_B; thus in this sense, Bohr's theory agrees with modern quantum mechanics.

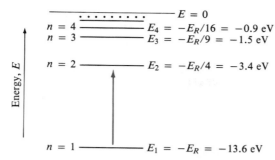

FIGURE 6.3 Energy levels of the hydrogen atom. The lowest level is called the ground state and the higher levels, excited states. There are infinitely many levels, $n = 5, 6, 7, \ldots$, all squeezed below $E = 0$. The upward arrow represents a transition in which the atom is excited from the ground state to the first excited level, by the absorption of an energy 10.2 eV. (Note that the diagram is not exactly to scale; the interval between E_1 and E_2 is really larger than shown.)

and so on. The allowed energies are often called **energy levels** and are traditionally displayed graphically as in Fig. 6.3. In these **energy-level diagrams** (also called term diagrams or Grotrian diagrams) the energy is plotted vertically upward, and the allowed energies are shown as horizontal lines, somewhat like the rungs of a ladder.

Energy-level diagrams provide a convenient way to represent transitions between the energy levels. For example, if the atom is in the lowest state, $n = 1$, the only possible change is an upward transition, which will require the supply of some energy. If, for instance, we were to shine photons of energy 10.2 eV on a hydrogen atom, they would have exactly the right energy to lift the electron to the $n = 2$ level, and the atom could make the transition, absorbing one photon in the process. This transition is indicated in Fig. 6.3 by an arrow between the levels concerned. The energy 10.2 eV is called the **first excitation energy** of hydrogen, since it is the energy required to raise the atom to its first excited state; similarly, the second excitation energy is $E_3 - E_1 = 12.1$ eV, and so on.

If the atom is in an excited state n (with $n > 1$), it can drop to a lower state n' (with $n' < n$) by emitting a photon of energy $E_n - E_{n'}$. If, for example, the original level is $n = 3, 4, 5, \ldots$ and the electron drops to the $n = 2$ orbit, the photon will have energy

$$E_{\text{ph}} = E_n - E_2 = E_R\left(\frac{1}{2^2} - \frac{1}{n^2}\right). \tag{6.26}$$

These are, of course, the photon energies implied by Balmer's original formula (6.1). For this reason, the spectral lines given by this formula (with the lower level being $n' = 2$) are often called the **Balmer series**. The transitions in which the lower level is the ground state ($n' = 1$) are called the Lyman series, and those in which the lower level is $n' = 3$, the Paschen series (after their respective discoverers). These three series are illustrated in Fig. 6.4.

According to (6.17) the radius of the nth circular orbit is proportional to n^2:

$$r = n^2 a_B.$$

Thus the radii of the Bohr orbits increase rapidly with n, as indicated in Fig. 6.5.

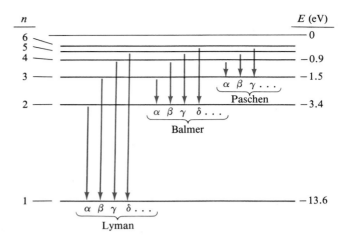

FIGURE 6.4 Some of the transitions of the Lyman, Balmer, and Paschen series in atomic hydrogen. The lines of each series are labeled α, β, γ, . . . , starting with the line of longest wavelength, or least energy. In principle, each series has infinitely many lines. (Not to scale.)

This agrees qualitatively with modern quantum theory and with experiment: In the excited states of hydrogen atoms (and all other atoms, in fact) the electrons tend to be much farther away from the nucleus than in their ground states.

In addition to the circular orbits that we have discussed, Bohr's theory also allowed certain elliptical orbits. However, Bohr was able to show that the allowed energies of these elliptical orbits were the same as those of the circular orbits. Thus, for our present purposes the elliptical orbits do not add any important further information. Since the precise details of the Bohr orbits are not correct anyway, we shall not discuss the elliptical orbits any further.

EXAMPLE 6.3 Modern atomic physicists have observed hydrogen atoms in states with $n > 100$. What is the diameter of a hydrogen atom with $n = 100$?

The diameter is

$$d = 2r = 2n^2 a_B = 2 \times 10^4 \times (0.05 \text{ nm}) = 1 \text{ } \mu\text{m}.$$

By atomic standards this is an enormous size — 10^4 times the diameter in the ground state. For comparison, we note that a quartz fiber with this diameter is visible to the naked eye. Atoms with these high values of n can only exist in a good vacuum since the interatomic spacing at normal pressures is of order 3 nm and leaves no room for atoms this large. (See Problem 6.12.)

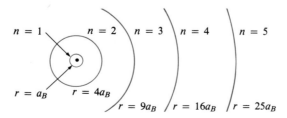

FIGURE 6.5 The radius of the nth Bohr orbit is $n^2 a_B$.

6.8 Hydrogen-Like Ions

For several years, at least, it appeared that Bohr's theory gave a perfect account of the hydrogen atom. The important problem was to generalize the theory to atoms with more than one electron, and in this no one succeeded. In fact, the Bohr theory was never successfully generalized to explain multielectron atoms, and a satisfactory quantitative theory had to await the development of modern quantum mechanics around 1925. Nevertheless, the Bohr theory did give a successful quantitative account of some atomic problems besides the spectrum of atomic hydrogen. In this and the next section we describe two of these successes.

In Section 6.6 we found the allowed radii and energies of a hydrogen atom, that is, a single electron in orbit around a proton of charge $+e$. The arguments given there started with Bohr's quantization condition, that the allowed values of angular momentum are integer multiples of $\hbar$. Starting from this same assumption, we can modify those arguments to apply to any **hydrogen-like ion;** that is, any atom that has lost all but one of its electrons and therefore comprises a single electron in orbit around a nucleus of charge $+Ze$. We might consider, for example, the He$^+$ ion (an electron and a helium nucleus of charge $2e$) or the Li^{2+} ion (an electron and a lithium nucleus of charge $3e$).

To adapt the arguments of Section 6.6 to hydrogen-like ions, we have only to note that the force ke^2/r^2 on the electron in hydrogen must be replaced by

$$F = \frac{Zke^2}{r^2}.$$

In other words, wherever ke^2 appears in Section 6.6, it must be replaced by Zke^2. For example, the allowed orbits of hydrogen had radii given by (6.16) as

$$r = n^2 \frac{\hbar^2}{ke^2 m} = n^2 a_B;$$

therefore, the orbits of an electron orbiting around a charge Ze are

$$r = n^2 \frac{\hbar^2}{Zke^2 m} = n^2 \frac{a_B}{Z}. \qquad (6.27)$$

We see that the radius of any given orbit is inversely proportional to Z. The larger the nuclear charge Z, the closer the electron is pulled in toward the nucleus, just as one might expect.

The potential energy of the hydrogen-like ion is $U = -Zke^2/r$. The total energy is, according to (6.10), $E = K + U = U/2$ or

$$E = -\frac{Zke^2}{2r}. \qquad (6.28)$$

Inserting (6.27) for the radius of the nth orbit, we find that

$$E_n = -Z^2 \frac{ke^2}{2a_B} \frac{1}{n^2}$$

or

$$E_n = -Z^2 \frac{E_R}{n^2}. \qquad (6.29)$$

That is, the allowed energies of the hydrogen-like ion with nuclear charge Ze are Z^2 times the corresponding energies in hydrogen. [The two factors of Z are easy to understand: One is the Z in the expression (6.28) for the energy, the other comes from the $1/Z$ in the allowed radii.]

The result (6.29) implies that the energy levels of the He^+ ion should be four times those in hydrogen. Thus the energies of the photons emitted and absorbed by He^+ should be

$$E_{ph} = 4E_R\left(\frac{1}{n'^2} - \frac{1}{n^2}\right) \tag{6.30}$$

(that is, four times those of the hydrogen atom). This formula looks so like the Rydberg formula for hydrogen that when the spectrum of He^+ had been observed in 1896 in light from the star Zeta Puppis, it had been wrongly interpreted as a new series of lines for hydrogen. It was another of the triumphs for Bohr's theory that he could explain these lines as belonging to the spectrum of once-ionized helium. Today, the spectra of hydrogen-like ions ranging from He^+ and Li^{2+} to Fe^{25+} (iron with 25 of its 26 electrons removed) have been observed and are all in excellent agreement with the Bohr formula (6.29).

There is a small but interesting correction to (6.29) that we should mention. So far we have supposed that our single electron orbits around a fixed nucleus; in reality, the electron and nucleus both orbit around their common center of mass. Because the electron is so light compared to the nucleus, the center of mass is very close to the nucleus. Thus the nucleus is very nearly stationary and our approximation is very good. Nonetheless, the nucleus *does* move, and this motion is fairly easily taken into account. In particular, it can be shown (Problem 6.17) that the allowed energies are still $E = -Z^2E_R/n^2$ and that the Rydberg energy is still given by (6.22) as

$$E_R = \frac{m(ke^2)^2}{2\hbar^2}, \tag{6.31}$$

provided that the mass m of the electron is replaced by the so-called **reduced mass:**

$$\text{reduced mass} = \frac{m}{1 + m/m_{nuc}}, \tag{6.32}$$

where m is the electron mass and m_{nuc} the mass of the nucleus. In the case of hydrogen, the nucleus is a proton and $m/m_{nuc} \approx 1/1800$. Therefore, the energy levels in hydrogen are all reduced by about 1 part in 1800 as a result of this correction. This is a rather small change, but one that can be easily detected by the careful spectroscopist.

The interesting thing (for the present discussion) is that the correction represented by (6.32) is different for different nuclei. In He^+, the nucleus is four times heavier than in hydrogen, and the factor m/m_{nuc} is four times smaller. Because of this difference the ratio of the He^+ frequencies to those of hydrogen is not exactly 4 but is about 4.002. This small difference was observed and added further weight to Bohr's interpretation.

EXAMPLE 6.4 Hydrogen has an isotope, ^2H, called deuterium or heavy hydrogen, whose nucleus is almost exactly twice as heavy as that of ordinary hydrogen, since it contains a proton *and* a neutron. It was discovered because its spectrum is not exactly the same as that of ordinary hydrogen. Calculate the Balmer α wavelengths of ordinary hydrogen and of deuterium, both to five significant figures, and compare.

Since ^{1}H and ^{2}H have the same nuclear charge, they would have identical spectra, if it were not for the motion of the nuclei. In particular, the Balmer α line would be given by the Rydberg formula (6.4) with $n = 3$, $n' = 2$, and $R = 0.0109737$ nm^{-1} (to six significant figures); this would give

$$\lambda = 656.114 \text{ nm}.$$

However, all energy levels of each atom must be corrected, in accordance with (6.32) and (6.31), by dividing by the factor $(1 + m/m_{\text{nuc}})$. Since wavelength is inversely proportional to energy, the correct wavelengths are found by multiplying by this same factor. Therefore, for ordinary hydrogen, the Balmer α wavelength is really

$$\lambda(^1\text{H}) = (656.114 \text{ nm})\left(1 + \frac{m}{m(^1\text{H})}\right)$$

$$= (656.114 \text{ nm})\left(1 + \frac{1}{1800}\right) = 656.48 \text{ nm}.$$

For deuterium the corresponding wavelength is

$$\lambda(^2\text{H}) = (656.114 \text{ nm})\left(1 + \frac{1}{3600}\right) = 656.30 \text{ nm},$$

a difference of about 1 part in 4000.

Since natural hydrogen contains 0.015% deuterium, its spectrum has a very faint component with these slightly shorter wavelengths. It was by observing these lines that the American chemist Urey proved the existence of deuterium in 1931.

6.9 X-Ray Spectra

The quantitative successes of Bohr's theory all concerned systems in which a single electron moves in the field of a single positive charge. The most obvious example is the hydrogen-like ion discussed in Section 6.8, but another system that fits the description, at least approximately, is the innermost electron of a multielectron atom. To the extent that the charge distribution of the other, outer electrons is spherical (which is actually true to a fair approximation) the outer electrons exert no net force on the innermost electron.* Therefore, the latter feels only the force of the nuclear charge Ze, and its allowed energies should be given by (6.29) as about

$$E_n = -Z^2 \frac{E_R}{n^2}. \tag{6.33}$$

The factor Z^2 means that for medium and heavy atoms the inner electron is

---

* Remember that the field due to a spherically symmetric shell of charge is zero inside the shell.

HENRY MOSELEY (1887–1915, English). Moseley exploited the newly discovered X-ray diffraction to measure the wavelengths of X rays emitted by atoms. He showed that the wavelengths depend on the atomic number exactly as predicted by the Bohr model, and used this dependence to identify unambiguously several previously uncertain atomic numbers. He was killed at age 27 in World War I.

TABLE 6.1

The labeling of transitions of an inner atomic electron. The lower level is identified by a capital letter $K, L, M, \ldots$, and the upper by a Greek subscript, $\alpha, \beta, \gamma, \ldots$.

K Series		L Series	
Label	Transition	Label	Transition
K_α	$n = 1 \leftrightarrow n = 2$	L_α	$n = 2 \leftrightarrow n = 3$
K_β	$n = 1 \leftrightarrow n = 3$	L_β	$n = 2 \leftrightarrow n = 4$
K_γ	$n = 1 \leftrightarrow n = 4$	L_γ	$n = 2 \leftrightarrow n = 5$

very tightly bound. For example, in zinc, with $Z = 30$, the energy needed to remove the innermost electron from the $n = 1$ orbit is about

$$Z^2 E_R = (30)^2 \times (13.6 \text{ eV}) \approx 12,000 \text{ eV}.$$

Thus atomic transitions that involve the inner electrons would be expected to involve energies of order several thousand eV; in particular, a photon emitted or absorbed in such a transition should be an X-ray photon. This fact was recognized by the young British physicist Moseley, who, within a few months of Bohr's paper, had shown that the Bohr theory gave a beautiful explanation of the characteristic X rays that were produced at discrete frequencies, characteristic of the anode material of an X-ray tube, as discussed in Section 5.5.

Moseley's explanation of the characteristic X rays was very simple: In an X-ray tube, the anode is struck by high-energy electrons, which can eject one or more of the electrons in the anode. If an electron in the $n = 1$ orbit is ejected, this will create a vacancy in the $n = 1$ level, into which an outer atomic electron can now fall.* If, for example, an $n = 2$ electron falls into this $n = 1$ vacancy, a photon will be emitted with energy given by (6.33) as

$$E_{\text{ph}} = E_2 - E_1 = Z^2 E_R \left(1 - \frac{1}{4} \right) = \frac{3}{4} Z^2 E_R. \tag{6.34}$$

The traditional terminology for the transitions of inner atomic electrons is shown briefly in Table 6.1; in particular, transitions between $n = 2$ and $n = 1$ are traditionally identified as the K_α transitions. Using this terminology we can say that the Bohr theory predicts the K_α photons emitted or absorbed by an atom should have energy $E_{\text{ph}} = 3Z^2 E_R/4$. If one observes several different elements and measures the frequencies of their K_α X rays (or any other definite X-ray line), then the photon energies, and hence frequencies, should vary like the square of the atomic number Z; that is, we should find $f \propto Z^2$, or equivalently

$$\sqrt{f} \propto Z. \tag{6.35}$$

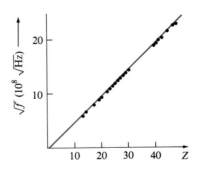

FIGURE 6.6 Moseley measured the frequencies f of K_α X rays, using several different elements for the anode of his X-ray tube. The graph shows clearly that $\sqrt{f}$ is a linear function of the atomic number Z of the anode material.

Moseley measured the K_α lines of some 20 elements. By plotting $\sqrt{f}$ against the known values of Z and showing that the data fitted a straight line (Fig. 6.6) he verified the prediction (6.35) and gave strong support to the Bohr theory. At that time (1913) the significance of the atomic number Z as the number of positive charges on the nucleus was only just becoming apparent, and Moseley's work settled this point conclusively. The atomic numbers of

* Implicit in this argument is the idea that each orbit can hold only a certain number of electrons. We shall see later that this is correct.

several elements were still in doubt, and Moseley's data, plotted as in Fig. 6.6, allowed these numbers to be determined unambiguously. Moseley was also able to identify three atomic numbers for which the corresponding elements had not yet been found—for example, $Z = 43$, technetium, which does not occur naturally and was first produced artificially in 1937.

A close look at Fig. 6.6 shows that the data do not confirm the prediction (6.35) exactly. If $\sqrt{f} \propto Z$, the line in Fig. 6.6 should pass through the origin, which it does not quite do. The line shown (which is a least-squares best fit) meets the Z axis close to $Z = 1$. That is, the data show that $\sqrt{f} \propto (Z - 1)$, or equivalently,

$$E_{\text{ph}} \propto (Z - 1)^2. \tag{6.36}$$

This small discrepancy was explained (and, in fact, anticipated) by Moseley, as follows: The prediction that the X-ray frequencies of a given line should be proportional to Z^2 was based on the assumption that the inner electron feels only the force of the nuclear charge Ze and is completely unaffected by any of the other electrons. This is a fair approximation, but certainly not perfect. An inner electron does experience some repulsion by the other electrons, and this slightly offsets, or *screens*, the attraction of the nucleus. This amounts to a small reduction in the nuclear charge, which we can represent by replacing Ze with $(Z - \delta)e$, where δ is some (unknown) small number. In this case the energy levels of the inner electron, and hence the X-ray energies, should be proportional to $(Z - \delta)^2$, rather than Z^2; specifically, (6.34) should be replaced by

$$E_{\text{ph}} = \frac{3}{4} (Z - \delta)^2 E_R. \tag{6.37}$$

According to (6.36) the observed data fit this prediction perfectly, with a screening factor δ close to $\delta = 1$.

As we have seen, the dependence of the characteristic X-ray frequencies on atomic number is very simple (namely, f approximately proportional to Z^2). Also, because the transitions involve the inner electrons, the frequencies are independent of the external conditions of the atom (for example, whether or not it is bound to other atoms in a molecule or a solid). Further, Moseley found that with an impure anode he could easily detect the X-ray lines of the impurities. For all these reasons, he predicted that X-ray spectroscopy would "prove a powerful method of chemical analysis." This prediction has proved correct. In modern X-ray spectroscopy, a sample (a biological tissue, for example) is put in a beam of electrons, protons, or X rays. The beam ejects inner electrons of many of the atoms in the sample, which then emit X rays. By measuring the wavelengths emitted one can identify all elements in the sample, down to the "trace" level of 1 part per million or even less.

EXAMPLE 6.5 Most X-ray spectrometers have a thin window through which the X rays must enter. Although high-energy X rays pass easily through such windows, low-energy X rays are severely attenuated and cannot be analyzed. A certain spectrometer, which is used for chemical analysis, cannot detect X rays with $E_{\text{ph}} \lesssim 2.4$ keV. If an unambiguous identification requires that one observe the K_α line of an element, what is the lightest element that can be identified using this spectrometer?

The energy of a K_α photon emitted by an element Z is given by (6.37) as $3(Z - \delta)^2 E_R/4$ (with $\delta \approx 1$). The lowest detectable element is found by equating this energy to 2.4 keV and solving for Z, to give

$$Z = \sqrt{\frac{4 \times (2.4 \times 10^3 \text{ eV})}{3 \times (13.6 \text{ eV})}} + 1 = 16.3.$$

We see from the table inside the back cover that the element with $Z = 17$ is chlorine and that with $Z = 16$ is sulfur. Therefore, our spectrometer can just detect chlorine, but cannot detect sulfur.

6.10 Other Evidence for Atomic Energy Levels (optional)

Although atomic spectroscopy gives abundant evidence for the quantization of atomic energy levels, one might hope to find other types of evidence as well. And, in fact, almost any process that transfers energy to or from an atom provides us with such evidence.

Imagine, for example, we fire a stream of electrons, all with the same kinetic energy K_0, at a target of stationary atoms; and suppose, for simplicity, that all the atoms are in their ground state, with energy E_1. The possible collisions between any one electron and an atom can be divided into two classes, the elastic and the inelastic. An *elastic* collision is defined as one in which the atom's internal state of motion is unaltered; this means that the total kinetic energy (of the incident electron plus the atom) does not change. Since the atom can recoil as a whole, it can gain some kinetic energy; but because the atom is so heavy compared to the electron this recoil energy is very small (Problem 6.23). Therefore, for most purposes, an elastic collision can be characterized as one in which the scattered electron is deflected by the atom but suffers no appreciable loss of kinetic energy.

An *inelastic* collision is one in which the atom is excited to a different energy level and there is a corresponding loss of kinetic energy. Because the atomic energy levels are quantized, the same must be true of the energy lost by the electron. Specifically, the electron can lose kinetic energy only in amounts equal to $E_n - E_1$, where E_n is an allowed energy of the atom. In particular, if the original kinetic energy K_0 is less than the first excitation energy, $E_2 - E_1$, of the atom, the electrons cannot excite the atoms at all, and only elastic scattering is possible. Obviously, by studying electron scattering at various incident energies and by measuring how much energy the electrons lose, one should be able to demonstrate and measure the allowed energies of the atoms.

THE FRANCK–HERTZ EXPERIMENT

The first experiment along these lines was carried out by Franck and Hertz in 1914 and is duplicated in many undergraduate teaching laboratories today. In this experiment a stream of electrons is passed through a tube of mercury vapor, as shown schematically in Fig. 6.7(a). The electrons leave a heated cathode and are attracted toward the grid by an adjustable accelerating potential V_0. Those electrons that pass through the grid will reach the anode, provided that they have enough energy to overcome the small retarding poten-

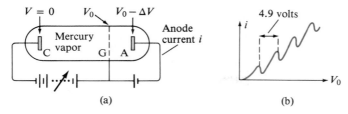

FIGURE 6.7 The Franck–Hertz experiment. **(a)** Electrons leave the heated cathode C and pass through mercury vapor to the anode A. The grid G is kept at a higher potential, V_0, than C and attracts the electrons; the anode is kept at a slightly lower potential $V_0 - \Delta V$. **(b)** The anode current i as a function of accelerating potential V_0.

tial ΔV. The current i reaching the anode is measured, and the observed behavior of i as a function of the accelerating potential V_0 is shown in Fig. 6.7(b). This behavior is easily explained in terms of the quantized energy levels of the mercury atom; in particular, the abrupt drop in the current each time the accelerating potential reaches a multiple of 4.9 V shows that the first excitation energy of mercury is 4.9 eV, as we now argue.

Even when the accelerating potential V_0 is zero, the heated cathode emits some electrons, but these collect in a cloud around the cathode, and the resulting field prevents the emission of any more electrons; so no steady current flows. When V_0 is slowly increased some of the electrons in this cloud are drawn away, which allows more electrons to be emitted, and a current now flows. When an electron is accelerated by a potential difference V_0, it acquires an energy $V_0 e$. As long as this energy is less than the first excitation energy of the mercury atom, only elastic collisions are possible, and there is no way for the electrons to lose any energy to the vapor. (See Problem 6.23.) Thus until $V_0 e$ reaches the first excitation energy, the anode current increases steadily as we increase V_0. Once $V_0 e$ reaches the excitation energy, some electrons can excite the mercury atoms and lose most of their energy as a result. When these electrons reach the grid, they have insufficient energy to overcome the retarding potential ΔV and cannot reach the anode. Thus when $V_0 e$ reaches the first excitation energy, we expect a drop in the current i. The fact that the drop is observed when $V_0 = 4.9$ V shows that the first excitation energy of mercury is 4.9 eV.

When V_0 increases beyond 4.9 V the current increases again. But when $V_0 e$ reaches twice the excitation energy, some electrons can undergo two inelastic collisions between the cathode and grid, and the current drops again. This process continues, and it is possible under favorable conditions to observe 10 or more drops in the current, regularly spaced at intervals of 4.9 V.*

This interpretation of the Franck–Hertz experiment can be confirmed by examining the optical spectrum of mercury. If 4.9 eV is the first excitation energy, the mercury atom should be able to emit and absorb photons of energy

---

* One can sometimes observe drops in i at voltages corresponding to the excitation of higher levels as well. However, the experiment is usually arranged in such a way that the probability for exciting the first excited state is very large and hence that very few electrons acquire enough energy to excite any higher levels.

4.9 eV. That is, the spectrum of mercury should include a line whose wavelength is

$$\lambda = \frac{hc}{E_{ph}} = \frac{1240 \text{ eV} \cdot \text{nm}}{4.9 \text{ eV}} = 250 \text{ nm},$$

and this is, indeed, the wavelength of a prominent line in the spectrum of mercury. Better still, one finds that the mercury vapor in the Franck–Hertz experiment begins emitting this line just as soon as V_0 passes 4.9 V and excitation of the first excited state becomes possible.

ENERGY-LOSS SPECTRA

Today, energy levels of atoms (and, even more, nuclei) are routinely measured by finding the energy lost by inelastically scattered particles. Figure 6.8(a) is a schematic diagram of an arrangement for measuring the energy levels of the helium atom using inelastically scattered electrons. A beam of electrons, all with the same incident kinetic energy, is fired through a container of helium gas. Those electrons scattered at some convenient angle are sent through a magnetic field, which bends them into circular paths, whose radii depend on their energies. [See Equation (3.48).] Therefore, a photographic film placed as shown lets one determine how many electrons are scattered at each different energy.

Figure 6.8(b) is a schematic plot of the number of scattered electrons as a function of their energy lost in the collision. This kind of plot is called an *energy-loss spectrum* — a natural generalization of the word "spectrum," which in the context of light refers to numbers of photons as a function of their energy. The large peak on the left, which occurs at zero energy loss, corresponds to the many electrons that scatter elastically and hence lose no energy. The next peak

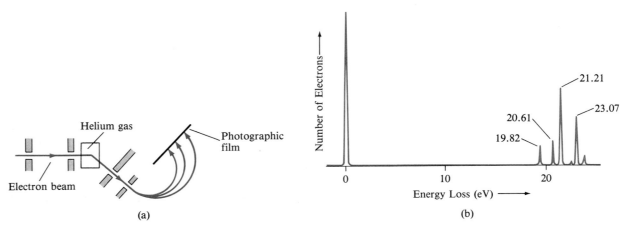

(a)   (b)

FIGURE 6.8 Measurement of the energy levels of helium by inelastic scattering of electrons. **(a)** Electrons are scattered in helium gas. Those electrons scattered at one convenient angle are passed into a magnetic field, which bends them onto different paths according to their energy. **(b)** A typical energy-loss spectrum, showing the numbers of scattered electrons as a function of the energy which they lost in the collision. Each spike corresponds to the excitation of a particular atomic energy level. The different heights of the spikes depend on how easily the different levels can be excited. (Based on data of Trajmar, Rice, and Kuppermann taken at 25° with incident energy 34 eV.)

occurs at an energy loss of 19.8 eV, indicating that the first excited state of helium is 19.8 eV above the ground state. The subsequent peaks indicate further excited states at 20.6 eV, 21.2 eV, and so on. All of the energies shown in Fig. 6.8 agree with the energy levels deduced from the optical spectrum, within the accuracy of the measurements.

Experiments like the Franck–Hertz experiment and the energy-loss measurements just described add confirmation (if any is needed) to Bohr's hypothesis that atomic energy levels are quantized. In particular, they demonstrate clearly that the quantized energies of atomic spectra are more than just a property of the light emitted and absorbed by atoms; rather, as suggested by Bohr, they reflect the quantization of the atomic energy levels themselves.

IDEAS YOU SHOULD NOW UNDERSTAND FROM CHAPTER 6

Emission and absorption spectra
The Balmer–Rydberg formula
The Bohr model of hydrogen:
    quantization of angular momentum
    allowed radii
    energy levels

Energy-level diagrams
Hydrogen-like ions
X-ray spectra of the elements
[Optional section: Franck–Hertz experiment;
    energy-loss spectra]

PROBLEMS FOR CHAPTER 6

SECTION 6.3 (THE BALMER–RYDBERG FORMULA)

6.1 • Find the wavelength of the light emitted by hydrogen as predicted by the Rydberg formula (6.4) with $n = 4$ and $n' = 3$. What is the nature of this radiation? (Visible? X-ray? . . .)

6.2 • Use Equation (6.5) to calculate the upper limit of the energies of photons that can be emitted by a hydrogen atom (in eV). (This energy is called the Rydberg energy, as discussed in Section 6.6.) What is the lower limit?

6.3 • Starting from the SI values of k, e, h, and c, find the values of ke^2 and hc in eV·nm. Prove, in particular, that

$$ke^2 = 1.44 \text{ eV·nm}.$$

6.4 •• The spectral lines of atomic hydrogen are given by the Rydberg formula (6.4). Those lines for which $n' = 1$ are called the Lyman series. Since n can be any integer greater than 1, there are (in principle, at least)

infinitely many lines in the Lyman series. (a) Calculate the five longest wavelengths of the Lyman series. Mark the positions of these five lines along a linear scale of wavelength. (b) Prove that the successive lines in the Lyman series get closer and closer together, approaching a definite limit (the *series limit*) as $n \to \infty$. Show this limit on your plot. What kind of radiation is the Lyman series? (Visible? X-ray? . . .)

SECTION 6.6 (THE BOHR MODEL OF THE HYDROGEN ATOM)

6.5 • (a) Find the value of the Bohr radius $a_B = \hbar^2/(ke^2m)$ (where m is the electron's mass) by substituting the SI values of the constants concerned. (b) It is usually easier to do such calculations by using common combinations of constants, which can be memorized in convenient units ($ke^2 = 1.44$ eV·nm, for example). Find the value of the convenient combination $\hbar c$ in eV·nm from your knowledge of hc. [The value of hc was given in (5.8). Both hc and $\hbar c$ are worth remembering in eV·nm.] Now calculate a_B by writing it as $(\hbar c)^2/(ke^2 mc^2)$ and using known values of $\hbar c$, ke^2, and mc^2.

6.6 • Two equivalent definitions of the Rydberg energy E_R are

$$E_R = \frac{ke^2}{2a_B} = \frac{m(ke^2)^2}{2\hbar^2}.$$

(a) Using the definition (6.18) of a_B, verify that these two definitions are equivalent. (b) Find the value of E_R from each of these expressions. (In the first case use $ke^2 = 1.44\ \text{eV} \cdot \text{nm}$ and the known value of a_B; in the second, multiply top and bottom by c^2 and then use the known values of mc^2, ke^2, and $\hbar c$.)

6.7 • Consider a charge q_1 with mass m in a circular orbit around a fixed charge q_2, with q_1 and q_2 of opposite sign. Show that the kinetic energy K is $-\frac{1}{2}$ times the potential energy U and hence that $E = K + U = U/2$. [Your arguments can parallel those leading to (6.10). The point is for you to make sure you understand those arguments, and to check that the conclusion is true for any two charges of opposite sign.]

6.8 • • (a) Derive an expression for the electron's speed in the nth Bohr orbit. (b) Prove that the orbit with highest speed is the $n = 1$ orbit, with $v_1 = ke^2/\hbar$. Compare this with the speed of light and comment on the validity of ignoring relativity (as we did) in discussing the hydrogen atom. (c) The ratio

$$\alpha = \frac{v_1}{c} = \frac{ke^2}{\hbar c} \tag{6.38}$$

is called the *fine-structure constant* (for reasons that are discussed in Problems 10.20 and 10.21) and is generally quoted as $\alpha \approx 1/137$. Verify this value.

SECTION **6.7** (PROPERTIES OF THE BOHR ATOM)

6.9 • Find the range of wavelengths in the Balmer series of hydrogen. Does the Balmer series lie completely in the visible region of the spectrum? If not, what other regions does it include?

6.10 • • Find the range of wavelengths in each of the Lyman, Balmer, and Paschen series of hydrogen. Show that the lines in the Lyman series are all in the UV, those of the Pashchen are all in the IR, while the Balmer series is in the visible and UV. (Note that visible light ranges from violet at about 400 nm to deep red at about 700 nm.) Show that these three series do not overlap one another, but that the next series, in which the lower level is the $n = 4$ level, overlaps the Paschen series.

6.11 • • The negative muon is a subatomic particle with the same charge as the electron but a mass that is about 207 times greater: $m_\mu \approx 207 m_e$. A muon can be captured by a proton to form a "muonic hydrogen atom," with energy and radius given by the Bohr model, except that m_e must be replaced by m_μ. (a) What are the radius and energy of the first Bohr orbit in a muonic hydrogen atom? (b) What is the wavelength of the Lyman α line in muonic hydrogen? What sort of electromagnetic radiation is this? (Visible? IR? . . .)

6.12 • • • The average distance D between the atoms or molecules in a gas is of order $D \sim 3$ nm at atmospheric pressure and room temperature. This distance is much larger than typical atomic sizes, and it is therefore reasonable to treat an atom as an isolated system, as we did in our discussion of hydrogen. However, the Bohr theory predicts that the radius of the nth orbit is $n^2 a_B$. Thus for sufficiently large n, the atoms would be larger than the spaces between them and our simple theory would surely not apply. Therefore, one would not expect to observe energy levels for which the atomic diameter $2n^2 a_B$ is of order D or more. (a) At normal densities (with $D \sim 3$ nm) what is the largest n that you would expect to observe? (b) If one reduced the pressure to 1/1000 of atmospheric, what would be the largest n? [Remember that the spacing D is proportional to $(\text{pressure})^{-1/3}$ for constant temperature.] (c) Modern experiments have found hydrogen atoms in levels with $n \approx 100$. What must be the pressure in these experiments?

SECTION **6.8** (HYDROGEN-LIKE IONS)

6.13 • What are the energy and wavelength of photons in the Lyman α line of Fe^{25+} (an iron nucleus with all but one of its 26 electrons removed)? What kind of electromagnetic radiation is this? (Visible? UV? . . .)

6.14 • What is the radius of the $n = 1$ orbit in the O^{7+} ion? What are the wavelength and energy of photons in the Lyman α line of O^{7+}?

6.15 • • When the spectrum of once-ionized helium, He^+, was first observed it was interpreted as a newly discovered part of the hydrogen spectrum. The following two questions illustrate this confusion: (a) Show that alternate lines in the Balmer series of He^+ — that is, those lines given by the Rydberg formula (6.30) with the lower level $n' = 2$ — coincide with the lines of the Lyman series of hydrogen. (b) Show that *all* lines of He^+ could be interpreted (incorrectly) as belonging to hydrogen, if one supposed that the numbers n and n' in the Rydberg formula for hydrogen could be *half-integers* as well as integers.

6.16 • • The negative pion, π^-, is a subatomic particle with the same charge as the electron but $m_\pi = 273 m_e$. A π^- can be captured into Bohr orbits around an atomic nucleus, with radius given by the Bohr formula (6.27), except that m_e must be replaced by m_π. (a) What is the orbital radius for a π^- captured in the $n = 1$ orbit by a carbon nucleus? (b) Given that the carbon nucleus has radius $R \approx 3 \times 10^{-15}$ m, can this orbit be formed? (c) Repeat parts (a) and (b) for a lead nucleus (nuclear radius $\approx 7 \times 10^{-15}$ m). (The required atomic numbers can be found in Appendix C.)

6.17 • • • In Section 6.6 we treated the H atom as if the electron moves around a fixed proton. In reality both the electron and proton orbit around their center of mass as shown in Fig. 6.9. Using this figure you can repeat the analysis of Section 6.6, including the small effects of the proton's motion, as follows:

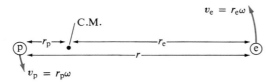

FIGURE **6.9** (Problem 6.17)

(a) Write down the distances r_e and r_p in terms of r, m_e, and m_p.
(b) Because both e and p move, it is easiest to work with the angular velocity ω, in terms of which v_e and v_p are as given in Fig. 6.9. Write down the total kinetic energy $K = K_e + K_p$ and prove that

$$K = \tfrac{1}{2}\mu r^2 \omega^2, \qquad (6.39)$$

where μ is the *reduced mass*,

$$\mu = \frac{m_p m_e}{m_p + m_e} = \frac{m_e}{1 + (m_e/m_p)} \approx 0.9995 \, m_e. \quad (6.40)$$

Notice that the expression (6.39) for K differs from its fixed-proton counterpart, $K = \tfrac{1}{2}m_e r^2 \omega^2$, only in the replacement of m_e by μ.
(c) Show that Newton's law, $F = ma$, applied to either the electron or proton gives

$$\frac{ke^2}{r^2} = \mu \omega^2 r. \qquad (6.41)$$

(Again this differs from the fixed-proton equivalent only in that μ has replaced m_e.)
(d) Use (6.39) and (6.41) to show that $K = -U/2$ and $E = U/2$.
(e) Show that the total angular momentum is

$$L = L_e + L_p = \mu r^2 \omega \qquad (6.42)$$

(in place of $L = m_e r^2 \omega$ if the proton is fixed).
(f) Assuming that the allowed values of L are $L = n\hbar$, where $n = 1, 2, 3, \ldots$, use (6.41) and (6.42) to find the allowed radii r, and prove that the allowed

energies are given by the usual formula $E = -E_R/n^2$, except that

$$E_R = \frac{\mu(ke^2)^2}{2\hbar^2}. \qquad (6.43)$$

This is the result quoted without proof in (6.31) and (6.32).
(g) Calculate the energy of the ground state of hydrogen using (6.43) and compare with the result of using the fixed-proton result $E_R = m_e(ke^2)^2/(2\hbar^2)$. (Give five significant figures in both answers.) The difference in your answers is small enough that we are usually justified in ignoring the proton's motion. Nevertheless, the difference can be detected, and the result (6.43) is found to be correct.

SECTION **6.9** (X-RAY SPECTRA)

6.18 • Use Equation (6.37) to predict the slope of a graph of $\sqrt{f}$ against Z for the K_α frequencies. Do the data in Fig. 6.6 bear out your prediction?

6.19 • The K_α line from a certain element is found to have wavelength 0.475 nm. Use Equation (6.37), with $\delta \approx 1$, to determine what the element is.

6.20 • The K series of X rays consists of photons emitted when an electron drops from the nth Bohr orbit to the first ($n \rightarrow 1$). (a) Use (6.33) to derive an expression for the wavelengths of the K series. [This will be approximate, since (6.33) ignores effects of screening.] (b) Find the wavelengths of the K_α, K_β, and K_γ lines ($n = 2, 3, 4$) of uranium. (For the atomic numbers of uranium and other elements, see the periodic table inside the back cover or the alphabetical lists in Appendix C.)

6.21 • What is the approximate radius of the $n = 1$ orbit of the innermost electron in the lead atom? Compare with the radius of the lead nucleus, $R \approx 7 \times 10^{-15}$ m.

6.22 • • Suppose that a negative muon (see Problem 6.11) penetrates the electrons of a silver atom and is captured in the first Bohr orbit around the nucleus. (a) What is the radius (in fm) of the muon's orbit? Is it a good approximation to ignore the atomic electrons when considering the muon? (The muon's orbital radius is very close to the nuclear radius. For this reason, the details of the muon's orbit are sensitive to the charge distribution of the nucleus, and the study of muonic atoms is a useful probe of nuclear properties.) (b) What are the energy and wavelength of a photon emitted when a muon drops from the $n = 2$ to the $n = 1$ orbit?

SECTION **6.10** (OTHER EVIDENCE FOR ATOMIC ENERGY LEVELS)

6.23 • • • When an electron, with initial kinetic energy K_0, scatters elastically from a stationary atom, there is no loss of total kinetic energy. Nevertheless, the electron loses a little kinetic energy to the recoil of the atom. (*a*) Use conservation of momentum and kinetic energy to prove that the maximum kinetic energy of the recoiling atom is approximately $(4m/M)K_0$, where m and M are the masses of the electron and atom. (*Hints:* The maximum recoil energy is in a head-on collision. Remember that $m \ll M$ and use nonrelativistic mechanics.)

(*b*) If a 3-eV electron collides elastically with a mercury atom, what is its maximum possible loss of kinetic energy? (Your answer should convince you that it is a good approximation to say that the electrons in the Franck–Hertz experiment lose no kinetic energy in elastic collisions with atoms.)

7

Matter Waves

7.1 Introduction

Bohr published his model of the hydrogen atom in 1913. Although it was studied intensely during the next 10 years, there was little progress toward a complete theory that could explain the model's success with hydrogen, or could give a satisfactory account of multielectron atoms. Then, in 1923, a French doctoral student named de Broglie proposed an idea that gave a new understanding of the Bohr model and proved to be the essential step in the development of modern quantum mechanics.

Appealing to the hope that nature is symmetric, de Broglie reasoned that if light has both wave-like and particle-like properties, material objects such as electrons might also exhibit this dual character. At that time, there was no known evidence for wave-like properties of any material particles. Nevertheless, de Broglie showed that if electrons were assumed to behave like waves, Bohr's stationary orbits could be explained as standing waves inside the hydrogen atom. These proposed waves, whose exact nature did not become clear for another two or three years, came to be called *matter waves*.

De Broglie's idea that wave–particle duality applied to electrons as well as photons appealed to many physicists. It was taken up by the Austrian physicist

Schrödinger, whose four papers published in 1926 mark the birth of modern *wave mechanics,* or *quantum mechanics* as we usually say today.* The next year, 1927, saw direct experimental verification of de Broglie's matter waves, with the observation of interference patterns (the hallmark of any wave phenomenon) produced by electrons.

In this chapter we discuss the properties of de Broglie's matter waves and their experimental verification. We write down de Broglie's relations for the frequency and wavelength of matter waves, and we discuss their physical interpretation — suggested by the German physicist Max Born in 1926 — as waves whose intensity at any point gives the *probability* of finding the particle at that point. In Chapter 8 we introduce the Schrödinger equation, which is the equation of motion for matter waves and plays the same role in quantum mechanics as Newton's second law plays in classical mechanics.

7.2 De Broglie's Hypothesis

We saw in Chapter 5 that photons display the properties of both waves and particles, and that the two kinds of properties are related by the equations (5.27)

$$E = hf \quad \text{and} \quad p = \frac{h}{\lambda}. \tag{7.1}$$

De Broglie proposed that material particles like electrons should show a similar particle–wave duality. He did not know the precise nature of the proposed "matter waves," but he argued that they should satisfy the same two relations (7.1) as apply to light waves.† For this reason, the relations (7.1), as applied to matter waves, are often called the **de Broglie relations.**

At the time there was no known experimental evidence for de Broglie's proposed matter waves. However, one can see (at least in retrospect) that electron waves are a plausible way to try explaining the quantization of atomic energy levels. If we accept the de Broglie relations (7.1), quantization of the electron's energy E is equivalent to quantization of the frequency f of the electron wave. Now, it is a familiar fact from classical physics that waves which are confined in some region (sound waves in an organ pipe, waves on a stretched string) can vibrate only at certain discrete, quantized, frequencies. This suggests that quantization of atomic energy levels might be explained as the quantization of the frequency of the electron waves confined inside the atom.

In fact, de Broglie was able to argue that if electrons were some kind of wave satisfying the relations (7.1), then the angular momentum of an electron in a hydrogen atom would be quantized in multiples of $\hbar$, exactly as required by the Bohr model. He pictured the electron wave as somehow vibrating around

---

* At about the same time, Heisenberg developed an independent form of quantum mechanics, which was later shown to be equivalent to Schrödinger's theory. We have chosen to describe Schrödinger's approach because it is more easily understood.

† This suggestion was not as obvious as we may have made it seem. In the case of light waves, whose speed is c, the relations (7.1) can be rewritten in various equivalent ways. For example, since $f = c/\lambda$ (for light) the relation $E = hf$ can be rewritten as $E = hc/\lambda$. Since matter waves do not have speed c, this second form is *incorrect* for matter waves. De Broglie was led to the correct relations (7.1) mainly by considerations of relativistic invariance.

the Bohr orbit, as shown in Fig. 7.1. It is evident that a wave can fit onto a circular path as shown only if the circumference can accommodate an integral number of wavelengths; that is,

$$2\pi r = n\lambda, \qquad n = 1, 2, 3, \ldots \ . \tag{7.2}$$

Now, according to (7.1) $\lambda = h/p$. Thus (7.2) implies that $2\pi r = nh/p$ or

$$rp = \frac{nh}{2\pi}. \tag{7.3}$$

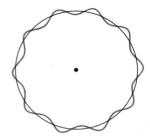

FIGURE 7.1 If an electron wave is pictured as circling around the atomic nucleus, its wavelength must fit an integer number of times into the circumference.

But for a circular orbit (which we are considering) rp is just the angular momentum L, and we conclude that

$$L = \frac{nh}{2\pi} = n\hbar, \qquad n = 1, 2, 3, \ldots ,$$

which is just the Bohr quantization condition.

We should emphasize that this explanation of the quantization of angular momentum is no longer considered completely satisfactory. For instance, the wave shown in Fig. 7.1 is some kind of one-dimensional wave, constrained to move around a circular path with a definite radius r, whereas modern quantum mechanics envisions a three-dimensional wave that is spread throughout the whole atom. Nevertheless, the central idea — that quantization of angular momentum was explained by the notion of electron waves satisfying the de Broglie relations — was absolutely correct.

7.3 Experimental Verification

If electrons and other material particles have wave properties, as de Broglie suggested, the question naturally arose why these wave properties had never been observed. The answer lies in the extremely short wavelength of most matter waves. You will recall that the wave nature of light was only firmly established in the nineteenth century by the observation of interference patterns in Young's double-slit experiment and other similar experiments. What made those experiments difficult was that the wavelength of light ($\lambda \approx 400$ to 700 nm) is very small by everyday, macroscopic standards. To obtain an interference pattern that is easily observed, one must make the distance between the slits of the same general order as the wavelength, and this was not easy to do. In the case of de Broglie's matter waves, the wavelengths are usually much shorter than those of visible light, as we see in Example 7.1 below. Thus the wave nature of material particles was even more difficult to observe than was that of light.

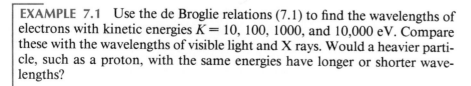

EXAMPLE 7.1 Use the de Broglie relations (7.1) to find the wavelengths of electrons with kinetic energies $K = 10, 100, 1000$, and $10,000$ eV. Compare these with the wavelengths of visible light and X rays. Would a heavier particle, such as a proton, with the same energies have longer or shorter wavelengths?

The wavelength is given by the relation $\lambda = h/p$. Thus our first task is to express the momentum p in terms of the kinetic energy K.* Since all energies concerned are small compared to the electron's rest energy, $mc^2 \approx 0.5$ MeV, we can use the nonrelativistic expression

$$K = \frac{1}{2}mv^2 = \frac{1}{2}\frac{p^2}{m}$$

which implies that

$$p = \sqrt{2mK}.$$

Therefore,

$$\lambda = \frac{\hbar}{p} = \frac{h}{\sqrt{2mK}}. \tag{7.4}$$

For an electron with $K = 10$ eV and $mc^2 = 0.51$ MeV, this gives

$$\lambda = \frac{hc}{\sqrt{2mc^2 K}} = \frac{1240 \text{ eV} \cdot \text{nm}}{\sqrt{2 \times (0.51 \times 10^6 \text{ eV}) \times (10 \text{ eV})}} = 0.39 \text{ nm}.$$

Since λ is inversely proportional to $\sqrt{K}$, we can immediately write down λ for all four energies as follows:

K (eV):	10	100	1000	10,000
λ (nm):	0.39	0.12	0.039	0.012

All of these wavelengths are very much shorter than those of visible light (400 to 700 nm), but they span the range of X-ray wavelengths. From (7.4) it is clear that for a given energy K, the wavelength λ is inversely proportional to $\sqrt{m}$. Thus the heavier the particle, the shorter will be the wavelength; and for a given energy, the best chance of observing matter waves is with electrons.

Since the wavelength of electrons in the range 100 to 1000 eV was comparable with that of the X rays being used in X-ray diffraction from crystals, de Broglie suggested that it might be possible to observe diffraction of electron waves by crystals. Unfortunately, a beam of electrons with kinetic energies of only a few hundred eV requires an extremely good vacuum to avoid severe scattering of the electrons by the remaining gas in the electron tube. For this reason, several attempts to observe electron diffraction failed, and it was not until 1927 that the American physicists Davisson and Germer published conclusive evidence for the diffraction of electron waves. Working in the Western Electric Laboratories in New York, they directed a beam of 54-eV electrons at a crystal of nickel. They found numerous maxima and minima in the scattered intensity at angles that were consistent with the diffraction of waves with wavelength $\lambda = h/p$. A photograph of Davisson and Germer's electron tube is shown in Fig. 7.2. Their experiment left no doubt of the existence of electron waves and of the correctness of the de Broglie relation $\lambda = h/p$.

* An alternative might seem to be to find the frequency from the relation $E = hf$ and then the wavelength from $\lambda = v_{\text{wave}}/f$, where v_{wave} denotes the speed of the matter wave. Unfortunately, we do not know v_{wave}, which cannot be assumed to be the same as the speed of the particle. See Section 7.10.

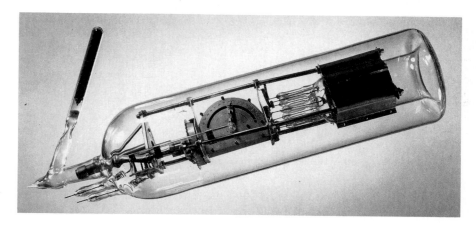

FIGURE 7.2 The electron tube in which Davisson and Germer observed the diffraction of electron waves. Notice the graduated turntable for rotating the target crystal near the center of the tube. (Courtesy AT&T Archives.)

In the same year, G. P. Thomson (son of J. J. Thomson, discoverer of the electron) demonstrated diffraction of electrons transmitted through thin metal foils. (Thomson and Davisson shared the 1937 Nobel Prize for their discovery of matter waves.) Within a few years diffraction of several other particles—hydrogen atoms, helium atoms, and later neutrons—had been observed, and it was reasonably clear that de Broglie's ideas applied to all material particles. Figure 7.3 shows diffraction patterns made by X rays, electrons, and neutrons, transmitted through polycrystalline metals. The first picture is the same picture shown in Fig. 5.9 as evidence for the wave nature of X rays. The similarity of the three patterns is unmistakable evidence that electrons and neutrons are also wave phenomena.

Since one of the first and best known demonstrations that light is a wave was Young's two-slit experiment, it is interesting that this same experiment has now been carried out with electron waves. This requires extremely narrow slits, very close together, and the pattern has to be enlarged many times to be discern-

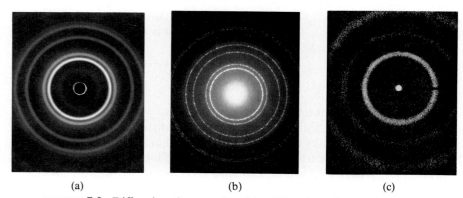

| (a) | (b) | (c) |

FIGURE 7.3 Diffraction rings produced by diffraction of waves in polycrystalline metal samples with (a) X rays, (b) electrons, (c) neutrons. (Courtesy Educational Development Center and Prof. C. G. Shull.)

(a) Light   (b) Electrons

FIGURE 7.4 Two-slit interference patterns produced by light and electrons. (Courtesy Addison-Wesley and Prof. C. Jönsson.)

ible. Figure 7.4 shows photographs of two-slit patterns made with beams of light and electrons. Here again the similarity is unmistakable.

Today we take for granted that all material particles have wave properties with wavelength and frequency given by the de Broglie relations, and these wave properties have found many applications. The short wavelength of electron waves is exploited in the electron microscope. An ordinary microscope uses light (focused by glass lenses) and cannot resolve objects much smaller than 10^{-6} m. An electron microscope uses electrons, focused by magnetic fields; because the electrons have wavelengths thousands of times smaller than light, the electron microscope can resolve objects down to about 10^{-10} m.

The diffraction phenomena that originally established the wave properties of matter are now used as probes of the structure of solids. As we saw, electrons with suitable wavelength have energies of 100 eV or so — much lower than the energy of X rays with the same wavelength. These low-energy electrons do not penetrate as deeply into matter as do the X rays, and low-energy electron diffraction (or LEED) is therefore used to study surface properties of solids. Since these surface properties are important in electronic and catalytic devices, LEED has become a widely used research tool.

It has also proved possible to work with neutrons of very low energy — a few hundredths of an eV. Even though the neutron is so much heavier than the electron, its wavelength at these low energies is comparable with that of X rays, and neutron diffraction is another important probe of the structure of solids. One advantage of using neutrons is that they are scattered appreciably by hydrogen, since they are subject to the nuclear force of the proton; X rays and electrons are scattered only weakly by hydrogen, since they interact mainly with electric charge (of which the hydrogen atom contains relatively little — one electron and one proton). Thus neutron diffraction is often the most effective way to study crystals containing hydrogen.

7.4   The Quantum Wave Function

For a complete description of any wave we must discuss its **wave function.** This is the mathematical function that specifies the wave disturbance at each point of space and time. For waves on a taut string, aligned along the x axis, the wave function $y(x, t)$ gives the transverse displacement of the string, for all positions x and times t. For sound waves, the wave function is the pressure change p resulting from the wave. Since sound waves usually travel outward in all directions, p depends on all three spatial coordinates and time,

$$p = p(x, y, z, t) = p(\mathbf{r}, t).$$

For light waves, the wave function is the electric field strength $\mathscr{E}(\mathbf{r}, t)$ at the point $\mathbf{r} = (x, y, z)$ and time t. De Broglie had no clear idea what his matter waves really *were;* in other words, he did not know the nature of their wave function. The interpretation of the matter wave function that is generally accepted today was proposed by Born in 1926. Born's ideas were taken up and extended by Bohr and his associates in Copenhagen and, for this reason, are often called the Copenhagen interpretation of quantum mechanics. The distinctive feature of Born's proposal is that the matter wave function specifies only *probabilities* — rather than specific values — of a particle's properties, as we now describe.

To understand Born's proposal it is helpful to consider, as Born did, the connection between the electromagnetic wave function $\mathscr{E}(\mathbf{r}, t)$ and the photon. It was known in classical electromagnetism that the electric field strength determines the energy carried by an electromagnetic wave. Specifically, the energy E in any small volume dV at a point $\mathbf{r}$ (and time t) is

MAX BORN (1882–1970, German–British). Born is best known for his work on the mathematical structure of quantum mechanics and the interpretation of the wave function. When Hitler came to power, Born left Germany and became a professor at Edinburgh University. He won the Nobel Prize in 1954 for his contributions to quantum theory.

$$E(\text{in volume } dV \text{ at } \mathbf{r}) = \epsilon_0 [\mathscr{E}(\mathbf{r}, t)]^2 \, dV, \qquad (7.5)$$

where ϵ_0 is the constant called the permittivity of the vacuum. To avoid having to worry about constants like ϵ_0 we shall rewrite (7.5) as a proportion:

$$E(\text{in volume } dV \text{ at } \mathbf{r}) \propto [\mathscr{E}(\mathbf{r}, t)]^2 \, dV. \qquad (7.6)$$

From a quantum point of view, we know that the energy of an electromagnetic wave is carried by discrete photons. If, for simplicity, we consider a wave with a single fixed frequency f, each photon has energy hf. Therefore, we can divide (7.6) by hf to give

$$(\text{number of photons in } dV \text{ at } \mathbf{r}) = \frac{E(\text{in } dV \text{ at } \mathbf{r})}{hf} \propto [\mathscr{E}(\mathbf{r}, t)]^2 \, dV. \quad (7.7)$$

(Note that the proportionality sign lets us omit the constants h and f.) Since the square of any wave function is often called the *intensity,* we can paraphrase (7.7) to say that the number of photons in a small volume dV is proportional to the intensity $[\mathscr{E}(\mathbf{r}, t)]^2$ of the light.

The result (7.7) can not be exactly true as written. If we were to choose a small enough volume dV, then (7.7) would predict a fractional number of photons, and this is impossible. A correct statement is that (7.7) gives the *probable number* of photons in the volume dV:

$$(\text{probable number of photons in } dV \text{ at } \mathbf{r}) \propto [\mathscr{E}(\mathbf{r}, t)]^2 \, dV. \qquad (7.8)$$

To illustrate what this means, suppose that we shine a steady beam of light across a room. We select some definite small volume dV and imagine somehow counting the number of photons in dV (at any instant t). The "probable number" given by (7.8) is the *average result* expected if we repeat this same counting experiment many times. Suppose, for example, the number predicted by (7.8) for a certain small volume dV is 1.5. Of course, we shall not find 1.5 photons in dV in any one observation. On the contrary, each observation will yield some whole number, and we might get 1 photon, then 2, then 0, and so on. After many repeated observations, our *average* result will be 1.5.

Born proposed that a relation similar to (7.8) should apply to electron waves (and any other matter waves); that is, there must be an electron wave function, usually denoted by the Greek capital letter psi, $\Psi(\mathbf{r}, t)$, whose square gives the probable number of electrons in a small volume dV:

$$\text{(probable number of electrons in } dV \text{ at } \mathbf{r}) \propto [\Psi(\mathbf{r}, t)]^2 \, dV. \qquad (7.9)$$

A small but important complication with matter waves is that the wave function Ψ has two parts, which can be conveniently expressed by use of complex numbers: The value of Ψ (at any point) can be written as a complex number, with real and imaginary parts:

$$\Psi = \Psi_{\text{real}} + i\Psi_{\text{imag}}, \qquad (7.10)$$

where i is the imaginary number, $i = \sqrt{-1}$. We shall see later, in Section 8.3, that this unexpected feature of matter waves is central to the modern explanation of Bohr's stationary orbits. With Ψ complex, (7.9) is modified to read

$$\text{(probable number of electrons in } dV \text{ at } \mathbf{r}) \propto |\Psi(\mathbf{r}, t)|^2 \, dV, \qquad (7.11)$$

where $|\Psi|$ denotes the absolute value of the complex number (7.10),

$$|\Psi| = \sqrt{\Psi_{\text{real}}^2 + \Psi_{\text{imag}}^2}. \qquad (7.12)$$

We shall refer to $|\Psi|^2$ as the intensity of the wave function. Since $|\Psi|$ is a real number, the intensity is real and positive, and (7.11) guarantees that the probable number of electrons is positive, as it has to be.

To understand better the significance of the matter (or quantum) wave function Ψ, let us consider in detail an interference experiment such as the two-slit experiment. As we have seen, this can be done with electrons or photons, but we shall consider the case of electrons. A beam of electrons is directed at a metal foil with two narrow slits, as shown in Fig. 7.5. The electron waves that pass through the slits can interfere constructively and destructively on the far side. The resulting interference pattern can be recorded by placing a photographic film as shown, since each electron striking the film will leave a small dark spot. The directions in which constructive interference occurs are given by the well-known condition

$$d \sin \theta = n\lambda, \qquad n = 0, \pm 1, \pm 2, \ldots$$

where d is the separation of the slits and λ the wavelength of the wave. The resulting intensity, as a function of position along the screen, is plotted at the bottom of Fig. 7.5.

According to (7.11) we expect many electrons to arrive at those points where the intensity $|\Psi|^2$ is large, and very few to be seen where $|\Psi|^2$ is small. This is exactly what is found, as illustrated in Fig. 7.6, which shows a typical two-slit pattern after 40, 200, and 2000 electrons have arrived. After just 40 electrons no obvious pattern stands out, but it is clear that each electron, as represented by a black dot on the film, arrives at a definite position. After 200

FIGURE 7.5 Schematic diagram of the two-slit experiment. The graph shows the intensity as a function of position along the film.

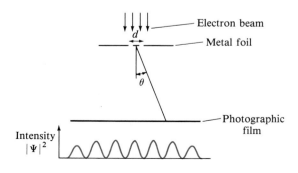

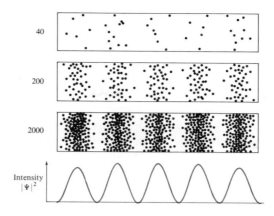

FIGURE 7.6 Development of a two-slit interference pattern. The three pictures show the pattern after 40, 200, and 2000 electrons (or photons) have arrived. The graph shows the intensity $|\Psi|^2$ of the wave as a function of position. Note that each particle arrives at a definite position, but more particles arrive where $|\Psi|^2$ is larger.

electrons, the characteristic two-slit pattern is becoming discernible; after 2000, the pattern is obvious and we see clearly that most of the electrons arrive near those points where $|\Psi|^2$ is maximum.

If we use a weak beam of electrons, the patterns of Fig. 7.6 will develop slowly. But even if the beam is so weak that only one electron is in the apparatus at a time, the interference pattern still appears. This implies that there is a quantum wave associated with *each individual* electron. This wave passes through the two slits, and the resulting interference determines where the electron is most likely to be found. Each electron arrives at a definite spot, but after many electrons have arrived we can identify the regions where $|\Psi|^2$ is greatest as the regions where more electrons have arrived.

These considerations let us sharpen our interpretation of the quantum wave function: Associated with each individual quantum particle there is a wave function $\Psi(\mathbf{r}, t)$, whose intensity at any position $\mathbf{r}$ determines the probability P of finding the particle at $\mathbf{r}$ (at time t):

$$P(\text{finding particle in } dV \text{ at } \mathbf{r}) \propto |\Psi(\mathbf{r}, t)|^2 \, dV. \qquad (7.13)$$

Since P is proportional to $|\Psi|^2 \, dV$, we can choose the scale of Ψ so that P *equals* $|\Psi|^2 \, dV$; that is, we choose the units of Ψ so that the constant of proportionality in (7.13) has the value 1. With this choice we can rewrite (7.13) as follows: If $\Psi(\mathbf{r}, t)$ is the wave function associated with a quantum particle, then

$$|\Psi(\mathbf{r}, t)|^2 \, dV = P(\text{finding particle in } dV \text{ at } \mathbf{r}) \qquad (7.14)$$

This is Born's interpretation of the quantum wave function $\Psi(\mathbf{r}, t)$: It is a wave whose intensity gives the probability of finding the particle at $\mathbf{r}$. Another way to express (7.14) is to divide both sides by the volume dV, in which case the right-hand side becomes the **probability density** (that is, the probability per unit volume). We can then say that

$$|\Psi(\mathbf{r}, t)|^2 = \text{probability density for finding particle at } \mathbf{r}. \qquad (7.15)$$

It is important to understand what (7.14) and (7.15) assert. Suppose that we can somehow arrange an experiment with a succession of quantum particles all with the same wave function Ψ. If we measure *one* particle's position (by letting it run into a photographic film, for example), we will find some definite result (in the form of a single dark spot on the film). However, after measuring several particles, all with exactly the same wave function, we will generally get several different answers. What the quantum wave function Ψ tells us is the frequency of occurrence of the various possible positions. The particles will be found most often at those points where $|\Psi|^2$ is largest, and least often where $|\Psi|^2$ is smallest, in accordance with (7.14) or (7.15).

Even when we know a particle's wave function Ψ exactly, we cannot specify a unique position of the particle; rather, we can give only the respective probabilities that the particle will be found at the various possible positions. For this reason, modern quantum mechanics is often described as a probabilistic theory. This is perhaps the most profound difference between classical and quantum mechanics. In classical mechanics every experiment has an outcome that can, in principle, be predicted unambiguously; in quantum mechanics, the same experiment, repeated under the same conditions, can produce different outcomes. Nonetheless, the probabilities of the various results *can* be predicted, and it is these probabilities that are measured in an experiment such as the two-slit experiment.

Equation (7.14) gives the answer to the question: What *is* the wave function Ψ of a quantum particle? According to (7.14), Ψ is a function whose intensity $|\Psi|^2$ gives the probability of finding the particle at any particular position. The quantum wave function is much harder to visualize than classical wave functions, like the displacement $y(x, t)$ produced by a wave on a string. Nevertheless, the probabilistic nature of (7.14) is an essential characteristic of the quantum wave function, and is one of the properties on which modern quantum mechanics is built.

The probabilistic character of quantum mechanics appears in almost all quantum processes. Consider, for example, the scattering of X rays by electrons in the Compton effect. Quantum mechanics allows one to calculate the *probability* that an X-ray photon will be scattered by an electron. Thus, if we send in many identical photons, quantum mechanics predicts unambiguously the fraction of photons scattered, but it *cannot* predict the fate of any single photon. This can be likened to the situation in an everyday toss of a coin: Probability theory tells us that if we toss many identical coins, then 50% will land "heads," but the theory *cannot* tell us the fate of any individual coin.*

7.5 Which Slit Does the Electron Go Through?

When an electron passes through the two-slit apparatus, its wave must pass through both slits in order to produce interference. Thus if we observe interference, we can be sure that the wave certainly passed through *both* slits. Nevertheless, it is natural to ask which slit the *electron itself* went through. Reasonable as this question seems, it is a question without an answer. Since the wave had the same intensity in each slit, there is an equal probability that we would have found it in either slit if we had placed suitable detectors in the slits; but if we conduct the experiment in the usual way, without any such detectors, it is impossible to say which slit the electron actually went through. Furthermore, if

---

* This analogy is not perfect. The probabilistic character of the coin toss arises from our ignorance of the details of the toss. If we knew the precise conditions of the coin's launch, then — in principle — we *could* predict the fate of a single coin. By contrast, the probabilistic character of quantum mechanics is an inherent feature of the theory.

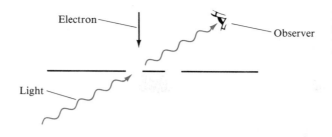

Electron

Light

Observer

FIGURE 7.7 To find out which slit the electron passes through, we shine a narrow beam of light through one of the slits.

we *do* use detectors to monitor which slit the electron uses, the detectors will disturb the electron waves as they pass through, and the two-slit pattern will disappear.

A general proof of this last statement is beyond the scope of this book, and we shall have to be content with illustrating it by means of a simple thought experiment. Let us suppose that we decide to find out which slit each electron passes through by shining a narrow beam of light through one slit, as in Fig. 7.7. Each time an electron passes through that slit, it will scatter a few photons and can be detected by a brief reduction in the light transmitted through the slit. In this way, we can determine (in principle, at least) which slit each electron passes through. Unfortunately, we acquire this knowledge at the cost of having the electron collide with at least one photon; and this collision disturbs the electron's motion enough to destroy completely the two-slit interference pattern, as we now argue.

We have already noted that to observe a two-slit pattern it is convenient to use electrons with wavelength of the same order as the slit separation d. We can write this requirement as*

$$\lambda_{el} \sim d,$$

which implies that

$$p_{el} = \frac{h}{\lambda_{el}} \sim \frac{h}{d}. \tag{7.16}$$

Now, to distinguish between the two slits we must make our beam of light appreciably narrower than d, and it is a well-known result that this is possible only if the wavelength of the light is of order d or less:

$$\lambda_{ph} \lesssim d.$$

This requires that

$$p_{ph} = \frac{h}{\lambda_{ph}} \gtrsim \frac{h}{d}. \tag{7.17}$$

Comparing (7.16) and (7.17) we see that the photons used to detect the electrons must have momentum at least of the same order as that of the electrons themselves. Under these conditions, a single collision with a photon is enough to change, completely and randomly, the electron's momentum. After the passage of many electrons, each of which has been disturbed in this random fashion, there will be no interference pattern. By using a beam of light to see which slit each electron traversed, we have destroyed the two-slit interference pattern.

---

* For the sake of brevity, we consider only the case that λ_{el} is of order d. If λ_{el} is much less than d (which *is* possible), the argument becomes more complicated. However, the conclusion is the same.

This conclusion is independent of the details of the experiment (see Problem 7.15) and illustrates an important difference between classical and quantum mechanics. According to classical mechanics, the electron follows a definite path and hence must pass through one slit or the other, in a predictable manner. In quantum mechanics, precise statements *can* be made about the probabilities of finding the electron at various positions, but one unique position *cannot* be predicted. Indeed, when one uses a detector to measure the electron's position, subsequent predictions of its position are radically altered. The need to analyze carefully the process of measurement and the disturbances it produces was first pointed out by Heinsenberg, who was led by these considerations to his famous uncertainty principle, which we describe in Sections 7.8 and 7.9.

7.6 Sinusoidal Waves

The de Broglie relations, $E = hf$ and $p = h/\lambda$, imply that if a particle has definite values for its energy and momentum, its wave function has corresponding definite values for its frequency and wavelength. A wave with definite frequency and wavelength is called a *sinusoidal* or *harmonic wave*. In this section we review briefly the properties of these waves. As we shall argue later, an *exactly* sinusoidal wave is an idealization that never really occurs in practice. Nevertheless, many real waves are well approximated by sinusoidal waves, and as we shall describe in Section 7.7, *any* wave can be built up from sinusoidal waves. Thus it is important to be familiar with their properties.

To simplify our discussion we shall consider the simplest of classical waves, namely waves on a string. However, most of the ideas of this and the next section apply with only small changes to any wave, classical or quantum.

Let us consider first a sinusoidal wave traveling to the right on a taut string. The wave function for such a wave has the form

$$y(x, t) = A \sin 2\pi \left(\frac{x}{\lambda} - \frac{t}{T} \right). \tag{7.18}$$

[Note that we shall use the word "sinusoidal" for a wave given by the sine function (7.18) or a cosine function, since both have the same general characteristics.] In (7.18), $y(x, t)$ is the transverse displacement of the string at position x along the string and time t. The constants A, λ, and T are called the amplitude, wavelength, and period.

A function like (7.18) that depends on two variables x and t is hard to visualize. One way to describe it is by a sequence of snapshots taken at equally spaced times as in Fig. 7.8. Each picture shows the displacement y as a function of x for one definite time t, and for the wave (7.18) each snapshot has the shape of a sine function. As time goes by, the whole wave moves steadily to the right, as can be seen by focusing on the surfer shown riding on one wave crest.* If, instead, we focus on one particular position x, the string oscillates up and down as a sinusoidal function of time. Thus a graph of y against t for any fixed x would have the same general shape as any one of the pictures in Fig. 7.8.

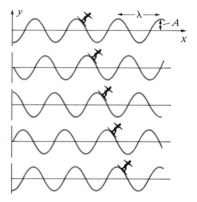

FIGURE 7.8 Five successive snapshots of the sinusoidal wave (7.18). Any definite point on the wave, like the crest on which the surfer is riding, moves steadily to the right. At any fixed position x, the string bobs up and down sinusoidally in time.

---

* To ride on the "surf" of a string would require good balance and a skate board in place of a surf board.

The wave (7.18), with a minus sign in its argument, travels to the right. If we replace the minus sign with a plus, we get a wave,

$$y(x, t) = A \sin 2\pi\left(\frac{x}{\lambda} + \frac{t}{T}\right),$$

that travels to the left. (See Problem 7.24.)

The significance of the constants A and λ is clear in Fig. 7.8. The amplitude A is the maximum displacement of the string from its mean position. The wavelength λ is the distance one must move (at one fixed time t) before the wave repeats itself. The period T is the time one must wait (at one fixed point x) for the wave to repeat itself. The frequency f is the number of oscillations in unit time at one fixed point and is given by $f = 1/T$. The frequency is usually measured in s^{-1} or hertz (1 hertz $\equiv$ 1 Hz $\equiv$ 1 s^{-1}).

It is often convenient to rewrite the wave function (7.18) as

$$y(x, t) = A \sin(kx - \omega t) \tag{7.19}$$

where k is called the **wave number** and ω the **angular frequency**. Comparing (7.18) and (7.19) we can express k and ω in terms of λ and T (or f). The various parameters characterizing a wave can be summarized as follows: The spatial parameters are

$$
\begin{array}{c}
\text{wavelength, } \lambda \\[1mm]
\text{wave number, } k = \dfrac{2\pi}{\lambda}
\end{array}
\tag{7.20}
$$

and the parameters related to time,

$$
\begin{array}{c}
\text{period, } T \\[1mm]
\text{frequency, } f = \dfrac{1}{T} \\[2mm]
\text{angular frequency, } \omega = 2\pi f = \dfrac{2\pi}{T}.
\end{array}
\tag{7.21}
$$

The speed with which any particular crest (or any other definite point on the wave) moves is the wave speed

$$v = \frac{\lambda}{T} = \lambda f = \frac{\omega}{k}.$$

In quantum mechanics the parameters k and ω are used frequently, and it is a good idea to be familiar with the de Broglie relations in terms of them:

$$E = hf = \hbar \omega \tag{7.22}$$

and

$$p = \frac{h}{\lambda} = \hbar k. \tag{7.23}$$

7.7 Wave Packets

A sinusoidal wave, with definite frequency and wavelength, is a mathematical idealization that never occurs in practice. For example, we speak of a pure musical tone as a harmonic wave with one precise frequency, but a careful analysis shows that any real musical note is a mixture of many different frequencies. Similarly, we speak of monochromatic light as light with a single well-defined frequency, but even light from the most monochromatic laser is found on close inspection to have a small spread of frequencies.

It is easy to see why an exact sine wave cannot occur. A pure sine wave like (7.19) is perfectly periodic, repeating itself endlessly in time and in space. A real wave may repeat itself for a long time and over a large distance, but certainly not *forever.* This is especially clear in the case of matter waves. A wave function that was a pure sine wave would extend indefinitely and would represent a particle that was equally likely to be found *anywhere.* This is obviously absurd. Imagine, for example, an electron in a TV tube. At some time t_0 it is ejected from the cathode, and at some later time t_1 it hits the screen, creating a small flash of light. At any time between t_0 and t_1, the electron—being a wave—cannot have a precisely defined position. Nevertheless, we know from experience that the electron is definitely within some region that is tiny compared to the size of the TV tube. When t is close to t_0, this region is near the cathode; and as t advances, the region moves toward the screen. Thus at any instant during its flight, the electron's wave function might look something like Fig. 7.9. The important points about this wave function are that (1) there is a region, labeled $\pm\Delta x$, where the wave differs from zero and the electron may be found, and that (2) outside this region the wave is zero (or very small) and the electron will not be found (or is very unlikely to be found). We refer to this kind of wave function that is *localized* within some region as a **wave packet** or **wave pulse.**

Our aim in this section is to describe the mathematical properties of localized wave packets like that shown in Fig. 7.9. We shall find that they can be built up as superpositions of many different sinusoidal waves. The analysis of complicated waves as sums of sinusoidal waves is called **harmonic analysis** or **Fourier analysis** (after the French mathematician Fourier, who discovered the technique). As an important preliminary to discussing wave packets, we shall first describe the harmonic analysis of a periodic wave, such as a musical tone. We shall then go on to the harmonic analysis of a localized wave packet like that in Fig. 7.9.

Let us consider a typical sound wave, such as a tone produced by a musical instrument. We can consider the pressure variation p as a function of time at one fixed position, or as a function of position at one fixed time. Most of what we shall say applies equally to either function, but to be definite we shall consider the former. This has the advantage that it can be measured easily by

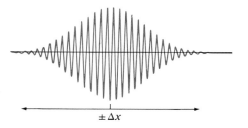

$\pm\Delta x$

FIGURE 7.9 A localized wave packet, which is nonzero in an interval $\pm\Delta x$ and zero elsewhere. A particle with this wavefunction would not be found outside the interval $\pm\Delta x$. (Courtesy Dale Prull.)

FIGURE 7.10 The sound waves produced by a violin are not pure sine functions. The basic period of this wave is 5 ms, but more rapid variations are also present.

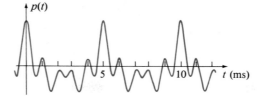

placing a microphone at a fixed position and connecting it to an oscilloscope to display the pressure as a function of time.

In Figure 7.10 we show the pressure variations of a musical tone, such as could be produced by a violin. Although the pressure variation is clearly not sinusoidal, it is nonetheless periodic. In fact, the pattern obviously repeats itself every 5 ms, as indicated by the identical peaks at $t = 0$, 5, and 10 ms. Thus the wave has period $T = 5$ ms, or frequency $f = 1/T = 200$ Hz.

Unlike a pure sine curve, the graph of Fig. 7.10 shows more rapid variations superposed on the basic 200-Hz oscillation. The most obvious of these are the sharp peaks that occur once every 1 ms, corresponding to a frequency of 1000 Hz. Since it is known that a violin string can vibrate at a series of different *harmonic frequencies* of the form

$$f = f_1, \ 2f_1, \ 3f_1, \ \ldots \tag{7.24}$$

we might reasonably guess that the complicated structure of Fig. 7.10 is the result of the string vibrating at several of these different frequencies at once.

One can test this hypothesis by letting the sound from the violin strike a large number of resonant cavities, each of which can oscillate at just one definite frequency. The result of this experiment is that only a few of the cavities are set into oscillation, and that the frequencies of these cavities form a harmonic series like (7.24). The strengths of the cavities' responses to the signal of Fig. 7.10 are plotted in Fig. 7.11. Notice that the lowest, or *fundamental,* frequency is 200 Hz, the frequency of the overall pattern in Fig. 7.10. The strongest of the higher harmonics is the fifth harmonic with $f = 1000$ Hz, which explains the prominent 1000-Hz variations in Fig. 7.10.

We conclude that the sound wave shown in Fig. 7.10 is a superposition of harmonic waves with frequencies of the form (7.24). As a function of time, its mathematical form could be*

$$p(t) = A_{f_1} \cos 2\pi f_1 t + A_{f_2} \cos 2\pi f_2 t + A_{f_3} \cos 2\pi f_3 t + \cdots$$
$$= \sum_f A_f \cos 2\pi f t \tag{7.25}$$

where $f_1, f_2, f_3, \ldots$ denote the various harmonic frequencies (7.24). The coefficients $A_{f_1}, A_{f_2}, A_{f_3}, \ldots$ indicate the amplitudes of the corresponding harmonics in the overall signal, and are what were plotted in Fig. 7.11. A graph of A_f against frequency f, like that of Fig. 7.11, is called the **spectrum** of the signal. The series (7.25) that represents $p(t)$ as a sum of harmonic waves is called a **Fourier series.**

Any periodic function, with period T, can be synthesized as a Fourier

---

* For simplicity we do not give the most general possible form, which contains a cosine and sine function for each harmonic frequency. This added complication is unimportant here. In particular, the waveform of Fig. 7.10 was chosen to contain only cosines.

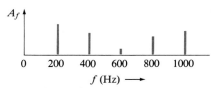

FIGURE 7.11 The waveform of Fig. 7.10 is a superposition of several different sinusoidal waves, whose frequencies, 200, 400, 600, . . . Hz, form a harmonic series. This bar graph, or spectrum, shows the amplitudes of the first five harmonics.

series similar to (7.25), consisting of sinusoidal functions with frequencies f_1, $f_2 = 2f_1, f_3 = 3f_1, . . .$, where the fundamental frequency is $f_1 = 1/T$. In Fig. 7.12 we show two more examples of musical tones, one produced by a piano and the other by a clarinet. On the left are the graphs of pressure as a function of time. On the right are the spectra obtained when these tones are analyzed as sums of sinusoidal waves.

So far we have discussed the Fourier analysis of a wave that is not itself sinusoidal, but *is* periodic. Our real objective is to discuss the kind of wave packet that could represent a single isolated electron. To understand how an isolated, and hence nonperiodic, wave can be synthesized with sinusoidal waves, let us consider a signal $F(t)$ of the kind shown in Fig. 7.13. This signal is periodic, with period T, and each cycle consists of a single rectangular pulse of duration τ. We choose this rectangular waveform mainly for mathematical convenience. However, it is interesting to note that it is the kind of signal used to transmit digital information, and we can imagine that $F(t)$ represents the voltage in a telephone line connected to a computer. We shall Fourier analyze this periodic series of pulses and then allow the period T to increase. In the limit $T \to \infty$ we shall be left with a single isolated (and nonperiodic) pulse of width τ.

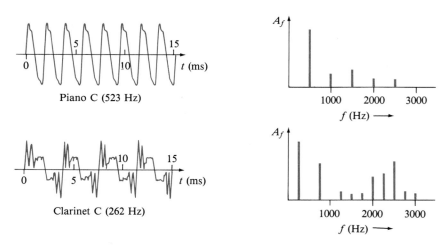

FIGURE 7.12 The harmonic analysis of tones produced by a piano and a clarinet. The two graphs on the left show pressure as a function of time; those on the right are the corresponding spectra when these tones are analyzed as sums of sinusoidal waves. The presence of more high harmonics in the clarinet is responsible for the greater complexity of the clarinet waveform. (Courtesy McGraw-Hill.)

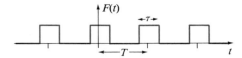

FIGURE 7.13 A periodic series of rectangular pulses of width τ and period T.

For any given value of T we can express the signal $F(t)$ as a Fourier series like (7.25):

$$F(t) = \sum_f A_f \cos 2\pi f t \tag{7.26}$$

where the frequencies in this series are the multiples of the fundamental frequency $f_1 = 1/T$,

$$f = 0, \ \frac{1}{T}, \ \frac{2}{T}, \ \frac{3}{T}, \ \cdots \tag{7.27}$$

(Note that a Fourier series can contain a zero-frequency term, that is, a constant. This is because the general periodic function is made up of oscillating sinusoidal terms *plus* a nonoscillating constant.) For the case that $T = 2.5\tau$, for example, it can be shown that the series has the form (Problems 7.30 and 7.33)

$$F(t) = 0.40 + 0.61 \cos \frac{2\pi t}{T} + 0.19 \cos \frac{4\pi t}{T} - 0.12 \cos \frac{6\pi t}{T}$$

$$- 0.15 \cos \frac{8\pi t}{T} + \cdots \tag{7.28}$$

To obtain the exact signal $F(t)$ one has to sum this whole infinite series. However, one obtains a remarkably good approximation using just the first few terms. This is illustrated in Fig. 7.14, which shows the sum of the five terms given explicitly in (7.28). Although this finite sum of sinusoidal functions cannot reproduce exactly the abrupt jumps of $F(t)$, it manages to approximate them surprisingly well.

Having analyzed the signal for any fixed period T, we can now let T begin to grow so that the pulses move farther and farther apart. In Fig. 7.15 we show our signal and its spectrum for the cases $T = 3\tau, 6\tau$, and 9τ. As T increases, the fundamental frequency $f_1 = 1/T$ gets smaller and the harmonic frequencies

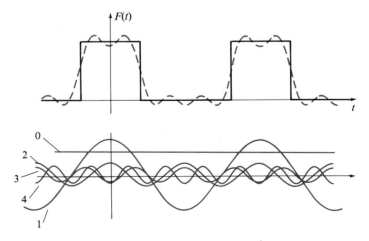

FIGURE 7.14 The series of rectangular pulses $F(t)$ (solid line in the upper picture) is approximated surprisingly well by the sum of the first five terms in its Fourier series (dashed curve). The graphs in the lower picture show the five separate terms of (7.28), whose sum is the dashed curve above; the constant term is labeled 0, the first harmonic 1, and so on.

FIGURE 7.15 The series of
rectangular pulses $F(t)$ is shown
on the left for the cases $T = 3\tau$,
6τ, and 9τ. On the right are the
corresponding spectra when $F(t)$
is analyzed as a sum of
sinusoidal functions. When
$T \to \infty$, all that remains of $F(t)$ is
a single isolated pulse, and the
spectrum approaches the dashed
curve superposed on each graph.

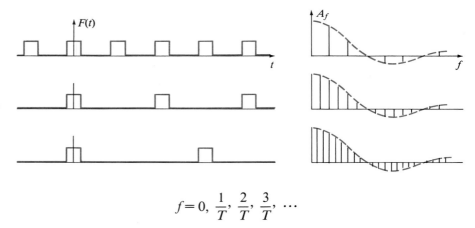

$$f = 0, \ \frac{1}{T}, \ \frac{2}{T}, \ \frac{3}{T}, \ \cdots$$

all get closer together. Thus the spectrum involves more and more frequencies,
which are closer and closer together. Looking at the three spectra on the right of
Fig. 7.15, it is not hard to guess that as $T \to \infty$ the spectrum will approach the
smooth continuous function shown as a dashed curve above each discrete
spectrum. In the limit, the sum over harmonic frequencies in (7.26) will become
an integral (called a Fourier integral) over a continuum of frequencies.

The discussion above suggests what is in fact true, that one can build up
any isolated pulse out of sinusoidal waves, provided that one is prepared to use a
continuous spread of different frequencies. This is illustrated in Fig. 7.16, which
shows two different isolated pulses and their corresponding spectra. The upper
pulse is the rectangular pulse discussed above, while the lower pulse has a
similar shape but lasts for half as long; that is, the durations of the two pulses are
τ and $\tau/2$, as shown. Notice that the spectrum of the narrower pulse contains a
wider range of frequencies. In fact, if we define the spread of frequencies Δf
as suggested in the figure, the spread Δf for the second pulse is exactly twice
that of the first; that is, the spread of frequencies Δf needed to synthesize
these pulses is inversely proportional to their duration.

The rectangular pulses of Fig. 7.16 are constant during the time when they
are nonzero. A more typical wave packet would oscillate during that time, like
either of the pulses shown in Fig. 7.17. These pulses can also be synthesized
from sinusoidal waves, and the corresponding spectra are shown on the right of
the figure. Notice that each of the spectra is centered on a definite nonzero

FIGURE 7.16 Two different
isolated rectangular pulses are
shown on the left; the lower
pulse lasts half as long as the
upper. On the right are the
corresponding spectra of
frequencies required when the
pulses are built up from
sinusoidal waves. Note that the
second, narrower pulse requires a
wider spread of frequencies.

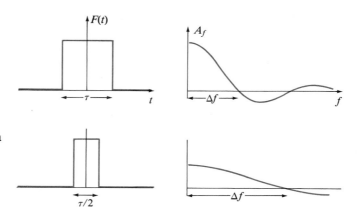

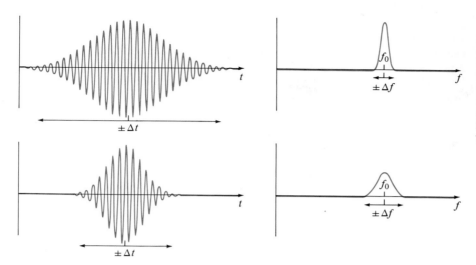

FIGURE 7.17 Two oscillating pulses, the second of which lasts half as long as the first, are shown on the left. On the right are the spectra of frequencies contained in the pulses. The frequency f_0 on which both spectra are centered is the mean frequency of the wave crests in the pulses. (Courtesy Dale Prull.)

frequency, marked f_0, which we can describe as the dominant, or central, frequency of the pulse. This frequency is the average frequency of the individual wave crests in the pulse itself. Notice also that we have characterized the widths of the graphs by quantities Δt and Δf, defined so that each graph is spread out from its central value by an amount $\pm \Delta t$ or $\pm \Delta f$.

The pulses in Fig. 7.17 have the same property noted in connection with Fig. 7.16: The pulse of shorter duration (Δt smaller) requires a wider spread of frequencies (Δf larger), and vice versa. This is a property of all wave packets and follows from a theorem in Fourier analysis, which we quote without proof: If a wave pulse has a duration characterized by Δt, its Fourier synthesis requires a spread in frequency of order Δf, where

$$\Delta t \, \Delta f \gtrsim 1. \tag{7.29}$$

Here the sign $\gtrsim$ means "greater than or of the same order as." This inequality shows clearly that if Δt is made smaller, Δf must become larger, and vice versa.

We have used the sign $\gtrsim$ in (7.29) because we have not yet given a precise definition of the quantities Δt and Δf. In fact, several definitions, differing by assorted factors of 2 and π, are possible, and the precise form of (7.29) depends on the definition used. In most advanced work Δt and Δf are defined as root-mean-square deviations,* and with this definition one can prove that (7.29) takes the exact form

$$\Delta t \, \Delta f \geq \frac{1}{4\pi} \tag{7.30}$$

or, since $f = \omega/2\pi$,

$$\Delta t \, \Delta \omega \geq \tfrac{1}{2}. \tag{7.31}$$

For uniformity we shall always use the form (7.31). Nevertheless, you should be aware that the exact form of the inequality depends on the definitions of Δt and

---

* The root-mean-square, or rms, deviation of f, for example, is defined as the square root of the average of the squared difference between f and the central value f_0. In this book we shall not require any explicit definition.

$\Delta\omega$, and you will see it written in different forms. For example, many authors write $\Delta t \, \Delta\omega \gtrsim 1$. In practice, the inequality (7.31) is used mainly to give rough estimates of various quantities, and in this context the odd factor of 2 is not especially important anyway.

The inequality (7.31) has important implications for the transmission of information, as the following example illustrates.

EXAMPLE 7.2 A TV picture is composed of 525 horizontal lines, each of which is drawn across the screen in 10^{-4} s. What is the approximate range of frequencies Δf at which a TV transmitter must be able to broadcast, if the horizontal and vertical resolutions in the TV picture are to be about the same?

The vertical resolution of detail is 1/525 of the screen height. Therefore, to achieve a comparable horizontal resolution, each individual line should be capable of becoming brighter or darker in about 1/525 of a screen width. Since the beam travels across the screen in 10^{-4} s, the transmitter must be able to send bright or dark pulses of total duration

$$2\Delta t \approx \frac{1}{525} \times 10^{-4}\text{ s} \approx 2 \times 10^{-7}\text{ s.}$$

According to (7.31), this requires a spread of frequencies

$$\Delta\omega \gtrsim \frac{1}{2\Delta t} \approx 5 \times 10^6\text{ s}^{-1}$$

or

$$\Delta f = \frac{\Delta\omega}{2\pi} \gtrsim 8 \times 10^5\text{ Hz.}$$

The range of frequencies, $\pm\Delta f$, contained in a signal is called the *band-width*. Thus we can say that TV video signals require a total bandwidth of at least $2\Delta f = 1.6 \times 10^6$ Hz. By comparison, high-fidelity sound systems require a frequency range of 2×10^4 Hz. The much higher bandwidth required for video signals delayed the development of videotape recorders many years after that of audiotape recorders.

We have so far discussed a wave as a function of time t at one fixed position x. We can equally consider the wave as a function of x at one fixed time. Any wave pulse considered as a function of x can be expressed as a superposition of sinusoidal functions $\cos kx$ and $\sin kx$, with properties analogous to those discussed above. The synthesis of a localized pulse requires a continuous spread of wave numbers k (or wavelengths $\lambda = 2\pi/k$). The smaller the interval Δx in which the wave is localized, the larger must be the spread Δk of wave numbers used in the superposition. Specifically, the spreads Δx in position and Δk in wave number satisfy an inequality exactly analogous to (7.31):

$$\Delta x \, \Delta k \gtrsim \tfrac{1}{2}. \tag{7.32}$$

In the next two sections we shall see that the two inequalities (7.31) and (7.32) are especially important in the case of matter waves, for which they imply the celebrated Heisenberg uncertainty principle. In Section 7.8 we discuss the consequences of the inequality (7.32) and in Section 7.9 those of (7.31).

To conclude this section we note that one can apply the inequality (7.32) to the special case of a perfectly sinusoidal wave like $A \sin(kx - \omega t)$. This wave has an exactly defined wave number k and hence has $\Delta k = 0$. According to (7.32), if Δk approaches zero, then Δx must approach ∞, and this is exactly what we knew already: A perfectly sinusoidal wave must extend periodically through all of space, and therefore has $\Delta x = \infty$. Of course, the values $\Delta k = 0$ and $\Delta x = \infty$ are an extremely special case of the inequality (7.32). Nevertheless, it is worth recognizing that (7.32) does cover even this extreme case.

7.8 The Uncertainty Relation for Position and Momentum

We have seen that the wave function of a single particle is spread out over some interval. This means that a measurement of the particle's position x may yield any value within this interval. (To simplify our discussion we suppose that our particle moves in one dimension and so has a single coordinate x.) Therefore, the particle's position is *uncertain* by an amount $\pm \Delta x$, and we refer to Δx as the *uncertainty in the position.* Most physicists take the view that this uncertainty is not just a reflection of our ignorance of the particle's position. Rather, the particle *does not have a definite position.* The uncertainty Δx can be smaller in some states than in others; but for any given state, specified by a wave function $\Psi(x, t)$, there is some nonzero interval within which the particle may be found, and the particle's position is simply not defined any more precisely than that.

In Section 7.7 we saw that the wave function that describes a particle can be built up from sinusoidal waves, but that this requires a spread of different wave numbers k (or wavelengths λ). From the de Broglie relation,

$$p = \frac{h}{\lambda} = \hbar k$$

it follows that a spread of wave numbers implies a spread of momenta; that is, the particle's momentum p, like its position x, is uncertain. A measurement of the momentum may yield any of several values in a range given by

$$\Delta p = \hbar \, \Delta k. \tag{7.33}$$

We have seen that the spreads Δx and Δk are not independent, but always satisfy the inequality (7.32),

$$\Delta x \, \Delta k \geq \tfrac{1}{2}.$$

If we multiply this relation by $\hbar$, we find that

$$\Delta x \, \Delta p \geq \frac{\hbar}{2}. \tag{7.34}$$

This is one of several inequalities called the **Heisenberg uncertainty relations** and known collectively as the **Heisenberg uncertainty principle.** It implies that both the position and momentum of a particle have uncertainties, in the sense just described. One can find states for which Δx is small, but then (7.34) tells us

WERNER HEISENBERG (1901– 1976, German). After earning his PhD at Munich, Heisenberg worked with Born and then Bohr. His many contributions to modern physics include an early formulation of quantum mechanics in terms of matrices, several ideas in nuclear physics, and the famous uncertainty principle, for which he won the 1932 Nobel Prize.

that Δp will be large; one can also find states for which Δp is small, but then Δx will be large. In all cases their product, $\Delta x \, \Delta p$, will never be less than $\hbar/2$.

In classical physics it was taken for granted that particles have definite values of their position x and momentum p. It was recognized, of course, that x and p could not be measured with perfect accuracy. But it was assumed that with enough care, one could make both experimental uncertainties as small as one pleased. Heisenberg's uncertainty relation (7.34) shows that these assumptions were incorrect. There are intrinsic uncertainties, or spreads, Δx and Δp in the position and momentum of any particle. Whereas either one of Δx and Δp can be made as small as one pleases, their product can never be less than $\hbar/2$.

We now know that the uncertainty principle applies to all particles. On the macroscopic level, however, it is seldom important, as the following example shows.

EXAMPLE 7.3 The position x of a 0.01-g pellet has been carefully measured and is known within $\pm 0.5 \ \mu$m. According to the uncertainty principle, what are the minimum uncertainties in its momentum and velocity, consistent with our knowledge of x?

If x is known within $\pm 0.5 \ \mu$m, the spread $\pm \Delta x$ in the position is certainly no larger than 0.5 μm:

$$\Delta x \leq 0.5 \ \mu\text{m}.$$

According to the uncertainty relation (7.34), this implies that the momentum is uncertain by an amount

$$\Delta p \geq \frac{\hbar}{2\Delta x} \geq \frac{10^{-34} \ \text{J} \cdot \text{s}}{10^{-6} \ \text{m}} = 10^{-28} \ \text{kg} \cdot \text{m/s}.$$

Therefore, the velocity $v = p/m$ is uncertain by*

$$\Delta v = \frac{\Delta p}{m} \geq \frac{10^{-28} \ \text{kg} \cdot \text{m/s}}{10^{-5} \ \text{kg}} = 10^{-23} \ \text{m/s}.$$

Clearly, the inevitable uncertainties in p and v required by the uncertainty principle are of no practical importance in this case. (To appreciate how small 10^{-23} m/s is, notice that at this speed our pellet would take about a million years to cross an atomic diameter.)

Although the uncertainty principle is seldom important on the macroscopic level, it is frequently very important on the quantum level, as the next example illustrates.

EXAMPLE 7.4 An electron is known to be somewhere in an interval of total width $a \approx 0.1$ nm (the size of a small atom). What is the minimum uncertainty in its velocity, consistent with this knowledge?

---

* The mass of a stable particle has no uncertainty, so we can treat m as a constant in the relation $v = p/m$.

If we know the electron is certainly inside an interval of total width a, then

$$\Delta x \le \frac{a}{2}. \tag{7.35}$$

(Remember that Δx is the spread from the central value out to either side.) According to the uncertainty relation (7.34), this implies that

$$\Delta p \ge \frac{\hbar}{2\Delta x} \ge \frac{\hbar}{a}. \tag{7.36}$$

This implies that $\Delta v = \Delta p/m \ge \hbar/(am)$ or

$$\Delta v \ge \frac{\hbar c^2}{amc^2} = \frac{200 \text{ eV} \cdot \text{nm}}{(0.1 \text{ nm}) \times (0.5 \times 10^6 \text{ eV})} c = \frac{c}{250} \approx 10^6 \text{ m/s}$$

(where we multiplied numerator and denominator by c^2 to take advantage of the useful combinations $\hbar c$ and mc^2). This large uncertainty in v shows the great importance of the uncertainty principle for systems with atomic dimensions.

Perhaps the most dramatic consequence of the uncertainty principle is that a particle confined in a small region cannot be exactly at rest, since if it were, its momentum would be precisely zero, which would violate (7.36). Since its momentum cannot be precisely zero, the same is true of its kinetic energy. Therefore, the particle has a minimum kinetic energy, which we can estimate as follows: Since the momentum is spread out by an amount given by (7.36) as

$$\Delta p \ge \frac{\hbar}{a}, \tag{7.37}$$

the magnitude of p must be, on average, at least of this same order. Thus the kinetic energy, whether it has a definite value or not, must on average have magnitude

$$K_{av} = \left(\frac{p^2}{2m}\right)_{av} \gtrsim \frac{(\Delta p)^2}{2m} \tag{7.38}$$

or, by (7.37),

$$K_{av} \gtrsim \frac{\hbar^2}{2ma^2}. \tag{7.39}$$

The energy (7.39) is called the **zero-point energy.** It is the minimum possible kinetic energy for a quantum particle confined inside a region of width a. The kinetic energy can, of course, be larger than this, but it cannot be any smaller.

EXAMPLE 7.5   What is the minimum kinetic energy of an electron confined in a region of width $a \approx 0.1$ nm, the size of a small atom?

According to (7.39),

$$K_{av} \gtrsim \frac{\hbar^2}{2ma^2} = \frac{(\hbar c)^2}{(2mc^2)a^2} = \frac{(200 \text{ eV} \cdot \text{nm})^2}{(10^6 \text{ eV}) \times (0.1 \text{ nm})^2}$$
$$= 4 \text{ eV}.$$

This lower bound is satisfactorily consistent with the known kinetic energy, 13.6 eV, of an electron in the ground state of a hydrogen atom.*

The bound (7.39) gives a useful estimate of the minimum kinetic energy of several other systems. For more examples, see Problems 7.37, 7.40, and 7.41.

We have so far written the uncertainty relation only for the case of a particle moving in one dimension. In three dimensions there is a corresponding inequality for each dimension separately:

$$\Delta x \, \Delta p_x \geq \frac{\hbar}{2}, \qquad \Delta y \, \Delta p_y \geq \frac{\hbar}{2}, \qquad \Delta z \, \Delta p_z \geq \frac{\hbar}{2} \qquad (7.40)$$

where x, y, z are the particle's three coordinates and p_x, p_y, p_z the three components of its momentum. (See Problem 7.40 for an application.)

The uncertainty principle can be illustrated by several thought experiments, the best known of which is sometimes called the Heisenberg microscope. In this thought experiment, a classical physicist — reluctant to accept the uncertainty principle — tries to disprove it by showing that he can measure the position and momentum of a particle with uncertainties smaller than are allowed by the uncertainty relation (7.34).

To find the position x of the particle, our classical physicist observes it with a microscope, as shown in Fig. 7.18. Now, it is a fact — well known in classical physics — that the resolution of any microscope is limited by the diffraction of light. Specifically, the angular resolution θ_{min} (the minimum angle at which two points can be told apart) is given by the so-called Rayleigh criterion,

$$\theta_{min} \approx \frac{\lambda}{d} \qquad (7.41)$$

where λ is the wavelength of light used and d the diameter of the objective lens. If the particle is a distance l below the lens, the minimum uncertainty in x is [see Fig. 7.18(a)]

$$\Delta x \approx l\theta_{min} \approx \frac{l\lambda}{d} \qquad (7.42)$$

(where we assume for simplicity that all angles are small, so that $\sin \theta \approx \theta$). Our classical physicist is aware of this limitation, but points out that he can make Δx as small as he pleases, for example by using light of very short wavelength λ.

Simply to pin down the particle's position with arbitrarily small Δx does not itself conflict with the uncertainty principle. Our classical physicist must show that he can also know the momentum with a suitably small uncertainty; and if we recall that light is quantized, we can quickly show that this is impossible: In order to observe the particle, he must allow at least one photon to strike

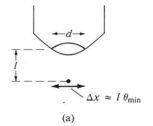

(a)

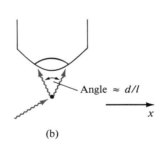

(b)

FIGURE 7.18 The Heisenberg microscope. (a) The minimum experimental uncertainty in the particle's position x is determined by the microscope's resolution as $\Delta x \approx l\theta_{min} \approx l\lambda/d$. (b) The direction of a photon entering the microscope is uncertain by an angle of order d/l; therefore, the photon gives the particle a momentum (in the x direction) which is uncertain by $\Delta p_x \approx p_{ph} \, d/l$.

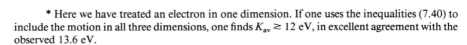

* Here we have treated an electron in one dimension. If one uses the inequalities (7.40) to include the motion in all three dimensions, one finds $K_{av} \gtrsim 12$ eV, in excellent agreement with the observed 13.6 eV.

it, and this collision will change the particle's momentum. Now, he has no way of knowing which part of the lens the photon passed through, since the lens sends *any* light from the object through the same image point. Therefore, the direction in which the photon approached the lens is uncertain by an angle of order d/l. [See Fig. 7.18(b).] This means that the x component of the photon's momentum is uncertain by an amount of order $p_{ph}\, d/l$. Since the particle was struck by the photon, the x component of the electron's momentum is now uncertain by at least this same amount; that is,

$$\Delta p_x \gtrsim p_{ph} \frac{d}{l} = \frac{h}{\lambda} \cdot \frac{d}{l}. \tag{7.43}$$

Our classical physicist can make this uncertainty in p_x as small as he pleases, for example by making λ large. But comparing (7.42) and (7.43) we see that whatever he does to reduce Δp_x will increase Δx, and vice versa. In particular, multiplying (7.42) by (7.43) we find that

$$\Delta x\, \Delta p_x \gtrsim h, \tag{7.44}$$

and our classical physicist has failed in his attempt to disprove the uncertainty principle.*

The uncertainty principle is a general result which follows from the wave–particle duality of nature. We should emphasize that our analysis of the Heisenberg microscope is not an alternative proof of this general result; it serves only to illustrate the inevitable appearance of the uncertainty principle in the context of one particular experiment.

7.9   The Uncertainty Relation for Time and Energy

Just as the inequality $\Delta x\, \Delta k \geq \frac{1}{2}$ implies the position-momentum uncertainty relation, $\Delta x\, \Delta p \geq \hbar/2$, so the inequality (7.31)

$$\Delta t\, \Delta \omega \geq \tfrac{1}{2} \tag{7.45}$$

implies a corresponding relation for time and energy. Specifically, if we multiply by $\hbar$, we find the **time-energy uncertainty relation**

$$\Delta t\, \Delta E \geq \frac{\hbar}{2}. \tag{7.46}$$

Here ΔE is the uncertainty in the particle's energy: A quantum particle generally does not have a definite energy, and measurement of its energy can yield any answer within a range $\pm \Delta E$. To understand the significance of Δt, recall that the inequality (7.45) arose when we considered a wave pulse as a function of time t, at one fixed position x. The time Δt characterizes the duration of the pulse at that position. Thus, for a quantum wave, Δt characterizes the time for which the particle is likely to be found at the position x. According to (7.46), if Δt is small, then the particle must have a large uncertainty ΔE in its energy, and vice versa.

---

* The fact that we have found $\Delta x\, \Delta p_x \gtrsim h$, rather than $\hbar/2$, is not significant, since the arguments leading to (7.44) were only order-of-magnitude arguments.

If a particle has a definite energy, then $\Delta E = 0$, and (7.46) tells us that Δt must be infinite. That is, a quantum particle with definite energy stays localized in the same region (and in the same state, in fact) *for all time.* States with this property are the quantum analog of Bohr's stationary orbits and are called stationary states, as we discuss in Chapter 8.

If a particle (or, more generally, any quantum system) does *not* remain in the same state for ever, then Δt is finite and (7.46) tells us that ΔE cannot be zero; that is, the energy must be uncertain. For example, any unstable state of an atom or nucleus lives for a certain finite time Δt, after which it decays by emitting a particle (an electron, photon, or α particle, for example). This means that the energy of any unstable atom or nucleus has a minimum uncertainty *

$$\Delta E \approx \frac{\hbar}{2\Delta t}. \tag{7.47}$$

Since the energy of the original unstable state is uncertain, the same is true of the ejected particle. In some cases one can measure both the spread of energies of the ejected particles (from many decays of identical unstable systems) and the lifetime Δt; one can then confirm the relation (7.47). In many applications one measures one of the quantities ΔE or Δt, then uses (7.47) to estimate the other.

EXAMPLE 7.6 Many excited states of atoms are unstable and decay by emission of a photon in a time of order $\Delta t \approx 10^{-8}$ s. What is the minimum uncertainty in the energy of such an atomic state?

According to (7.47), the minimum uncertainty in energy is

$$\Delta E \approx \frac{\hbar}{2\Delta t} = \frac{\hbar c}{2c\Delta t} \approx \frac{200 \text{ eV} \cdot \text{nm}}{2 \times (3 \times 10^{17} \text{ nm/s}) \times (10^{-8} \text{ s})} \approx 3 \times 10^{-8} \text{ eV}.$$

Compared to the several eV between typical atomic energy levels, this uncertainty ΔE is very small. Nevertheless, the resulting spread in the energy, and hence frequency, of the ejected photon is easily measurable with a modern laser spectrometer.

Nowadays, the frequencies of photons ejected in atomic transitions are used as standards for the definition and calibration of frequency and time. Because of the uncertainty principle, the frequency of any such photon is uncertain by an amount

$$\Delta \omega = \frac{\Delta E}{\hbar} \approx \frac{1}{2\Delta t},$$

where Δt is the lifetime of the emitting state. Therefore, it is important to choose atomic states with very long lifetimes Δt to use as standards.

---

* In (7.47) we have used the symbol $\approx$ because several different definitions of ΔE and Δt are commonly used. For example, Δt can be defined as the half-life (discussed in Section 2.4) or the mean life (to be discussed in Chapter 13), and the precise form of the relation depends on which definition we adopt. For the case of an unstable particle, (7.47) is exact if we take ΔE to be the so-called half-width at half-height and Δt to be the mean life.

7.10 Velocity of a Wave Packet (optional)

In this section we discuss a puzzle concerning the speed of matter waves. This is one of those curious problems that are important in principle, but relatively unimportant in practice. In particular, this material will not be needed again in this book, and you can therefore omit this section if you wish.

The puzzle in question is this: When we compute the speed of a matter wave, we find that the wave's speed is not equal to the speed of the particle that the wave describes. To see this, consider a single particle of mass m moving nonrelativistically with momentum p, in a region free of all forces. The particle's energy is just its kinetic energy:

$$E = K = \frac{p^2}{2m}.$$

From the de Broglie relations (7.1) we can find the frequency and wavelength:

$$f = \frac{E}{h} = \frac{p^2}{2mh} \quad \text{and} \quad \lambda = \frac{h}{p}.$$

For any wave, the wave speed is just the frequency times the wavelength. Thus for our matter wave

$$v_{\text{wave}} = f\lambda = \frac{p^2}{2mh}\frac{h}{p} = \frac{p}{2m}. \tag{7.48}$$

However, for a nonrelativistic particle with velocity v_{part}, we know that $p = mv_{\text{part}}$ and hence $v_{\text{part}} = p/m$. Thus (7.48) implies that

$$v_{\text{wave}} = \frac{v_{\text{part}}}{2}. \tag{7.49}$$

At first sight this result is very surprising, since it is natural to assume that the velocity, v_{wave}, of a matter wave should be the same as the velocity, v_{part}, of the particle to which it corresponds.

To understand why (7.49) is perfectly acceptable (and correct) we must consider carefully the significance of the wave velocity, v_{wave}, namely, that v_{wave} is the velocity with which any crest of a sinusoidal wave moves, as was shown in Fig. 7.8. We argued in Section 7.7 that a sinusoidal wave is an idealization that cannot occur in practice. The wave representing a particle is necessarily a *wave packet* like that in Fig. 7.9, and as we shall see shortly, the velocity v_{pack} with which a wave packet moves as a whole is not necessarily the same as the velocity, v_{wave}, of its individual crests. Since the particle is represented by the whole wave packet, rather than any single wave crest, it is v_{pack} that should equal the velocity of the corresponding particle. We shall prove in a moment that indeed $v_{\text{pack}} = v_{\text{part}}$, but let us first consider how it is that v_{pack} and v_{wave} can be different.

To understand the difference between v_{pack} and v_{wave}, consider Fig. 7.19, which shows the observed behavior of a wave packet on deep water. We see that the packet as a whole is moving to the right at a constant velocity v_{pack} (often called the **group velocity**). On closer inspection we see that any given wave crest, like the one carrying the surfer, is traveling to the right faster than the packet as a whole. In the first picture the surfer's crest is at the center of the packet, but in each successive picture it is closer to the front of the packet, until in the sixth picture the surfer's crest disappears entirely and the surfer sinks ignominiously into the water. As a packet of this type advances, wave crests steadily move

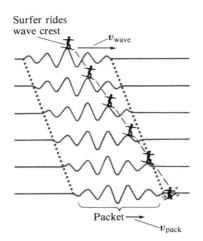

Surfer rides
wave crest

v_{wave}

Packet

v_{pack}

FIGURE 7.19 Six successive views of a wave packet, or group, moving to the right. The velocity of the whole packet is v_{pack}, often called the *group velocity*. The velocity of an individual crest, like the one carrying the surfer, is v_{wave}, the *wave velocity* (or phase velocity). For this wave $v_{\text{wave}} > v_{\text{pack}}$, and the surfer moves steadily toward the front of the packet. The two sloping dotted lines indicate the motion of the front and back of the packet; the sloping dashed line indicates the motion of a single wave crest.

forward and disappear at the front of the packet, while others appear at the rear, allowing the packet to maintain its overall shape. The behavior of the packet in Fig. 7.19 is typical of waves on deep water and can, in fact, be observed by carefully watching the bow waves of a boat on a calm lake.*

To understand mathematically how the behavior shown in Fig. 7.19 occurs, we must recognize that a wave packet is the result of the interference of many different sinusoidal waves, as discussed in Section 7.7. When one superposes sinusoidal waves of different frequencies, there are two main possibilities: For some waves (light waves in vacuum, for example) the wave speed v_{wave} is the same for all frequencies. This means that the various sine waves all travel at the same speed, and their interference pattern (that is, the packet) is carried along at the same speed as the individual waves. In this case, $v_{\text{pack}} = v_{\text{wave}}$.

For many waves, including waves on deep water and matter waves, the wave speed, v_{wave}, is different for different frequencies. In this case, the various sine waves that make up the packet move at different speeds, and the resulting interference pattern shifts steadily, relative to the component sine waves. This means that the interference pattern (or wave packet) moves at a speed different from that of the component sine waves: $v_{\text{pack}} \neq v_{\text{wave}}$.

A realistic wave packet, like that of Fig. 7.19, is a superposition of infinitely many different sine waves, and the mathematical analysis of such a wave is beyond our scope here. Instead, we consider a superposition of just two sine waves. Although this does not produce a realistic wave packet, it illustrates the main ideas and gives us the correct expression for the velocity, v_{pack}, of the wave packet.

Let us consider, then, the superposition of two sine waves with equal amplitudes, but with wave numbers that differ by Δk and frequencies that differ by $\Delta \omega$:

$$\Psi(x, t) = A \sin[(k + \Delta k)x - (\omega + \Delta \omega)t] + A \sin[kx - \omega t]. \quad (7.50)$$

Using the identity (Appendix B)

$$\sin \theta + \sin \phi = 2 \sin \frac{\theta + \phi}{2} \cos \frac{\theta - \phi}{2}$$

* On *shallow* water it turns out that $v_{\text{wave}} = v_{\text{pack}}$ (Problem 7.47), so a surfer *can* ride an individual crest until it breaks as it nears the shore.

we can rewrite Ψ as (Problem 7.49)

$$\Psi(x, t) = 2A \sin(\bar{k}x - \bar{\omega}t) \cos\left(\frac{\Delta k}{2}x - \frac{\Delta \omega}{2}t\right) \qquad (7.51)$$

where $\bar{k}$ and $\bar{\omega}$ denote the mean wave number and mean frequency ($\bar{k} = k + \Delta k/2$ and $\bar{\omega} = \omega + \Delta\omega/2$). The important property of (7.51) is that it is the product of two terms. The first term is a sine wave with wave number and frequency corresponding to the average of the original two waves; the second, cosine, term has wave number and frequency corresponding to half the difference of the original two waves. If the original two waves were close together in frequency ($\Delta\omega$ and Δk small), this second term oscillates much more slowly than the first, and forms the envelope for the more rapid oscillations of the first term, as in Fig. 7.20(b).

Since the individual crests in Fig. 7.20(b) are defined by the rapidly oscillating first term in (7.51), they travel to the right with speed (Problem 7.25)

$$v_{\text{wave}} = \frac{\bar{\omega}}{\bar{k}}. \qquad (7.52)$$

When $\Delta\omega$ and Δk are small, this is close to the wave speed of either of the original two waves. To find the velocity of the interference pattern we must look at the second term in (7.51), which travels to the right with speed (again see Problem 7.25)

$$v_{\text{pack}} = \frac{\Delta\omega/2}{\Delta k/2} \approx \frac{d\omega}{dk}. \qquad (7.53)$$

We can relate ω to k as follows: $\omega = 2\pi f = 2\pi(v_{\text{wave}}/\lambda) = v_{\text{wave}}k$. Thus

$$\omega = v_{\text{wave}}k. \qquad (7.54)$$

If the wave speed is the same for all frequencies, v_{wave} is a constant, and differentiation of (7.54) gives

$$v_{\text{pack}} = \frac{d\omega}{dk} = v_{\text{wave}} \qquad [\text{if } v_{\text{wave}} = \text{constant}].$$

(a)

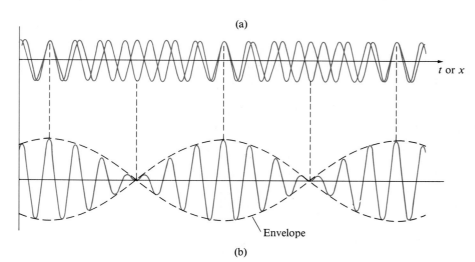

t or x

Envelope

(b)

FIGURE 7.20 (a) If two waves with slightly different frequencies ω_1 and ω_2 are superposed, they are alternately in and out of step. (b) The resultant wave shows the phenomenon of beats, in which a wave of frequency $\bar{\omega} = (\omega_1 + \omega_2)/2$ is modulated by an envelope, which oscillates at the difference frequency. (Courtesy Dale Prull.)

This result agrees with our previous argument that if v_{wave} is the same for all frequencies, the interference pattern will be carried along at the speed v_{wave}. If, however, v_{wave} is not constant, differentiation of (7.54) yields two terms (v_{wave} *plus* $k dv_{\text{wave}}/dk$). Therefore, $d\omega/dk$ is not the same as v_{wave}, and the speed with which the envelope moves is different from v_{wave}.

The results just derived were for the superposition of two sine waves. As long as the range of different frequencies is small (as is usually the case) similar results can be proved for superpositions of any number of sine waves. In particular, for a wave packet like that in Fig. 7.19, one can prove that the individual crests move with velocity $v_{\text{wave}} = \overline{\omega}/\overline{k}$, where $\overline{\omega}$ and $\overline{k}$ are the average frequency and wave number, but that the whole packet moves with the group velocity

$$v_{\text{pack}} = \frac{d\omega}{dk}. \qquad (7.55)$$

Only when v_{wave} is independent of frequency are v_{pack} and v_{wave} equal; otherwise, they are not.

We can now apply these ideas to matter waves, whose frequency and wave number are determined by the de Broglie relations

$$E = \hbar\omega \qquad \text{and} \qquad p = \hbar k.$$

In this case (7.55) gives for the group velocity, v_{pack},

$$v_{\text{pack}} = \frac{d\omega}{dk} = \frac{dE}{dp}. \qquad (7.56)$$

For a nonrelativistic particle moving freely, $E = p^2/2m$ and (7.56) implies that

$$v_{\text{pack}} = \frac{d}{dp}\left(\frac{p^2}{2m}\right) = \frac{p}{m} = v_{\text{part}}.$$

That is, the velocity of the wave packet is the same as the particle's velocity, and the de Broglie wave packet does indeed move with the speed of the particle it represents.

IDEAS YOU SHOULD NOW UNDERSTAND FROM CHAPTER 7

The de Broglie relations, $E = hf$ and $p = h/\lambda$

De Broglie's explanation of quantization of angular momentum

The Davisson–Germer experiment

The wave nature of all matter

Probabilistic significance of the quantum wave function:

$|\Psi|^2$ = probability density

relevance to two-slit experiment

Wave parameters: λ, k, f, ω, T

Wave packets

Fourier analysis

The Heisenberg uncertainty relations:

$\Delta x \, \Delta p \geq \hbar/2$

$\Delta t \, \Delta E \geq \hbar/2$

[Optional section: wave velocity and group velocity]

SECTIONS 7.2 AND 7.3 (DeBROGLIE'S HYPOTHESIS AND
EXPERIMENTAL VERIFICATION)

7.1 • Use the de Broglie relation $\lambda = h/p$ to find the wavelength of electrons with kinetic energy 500 eV.

7.2 • Find the kinetic energy of an electron with the same wavelength as blue light ($\lambda \approx 450$ nm).

7.3 • Find the kinetic energy of a neutron with the same wavelength as blue light ($\lambda \approx 450$ nm).

7.4 • Compare the wavelengths of electrons and neutrons, both with $K = 3$ eV.

7.5 • Find and compare the wavelengths of an electron and a muon ($m_\mu = 207 m_e$) each with kinetic energy 15 keV.

7.6 • In order to investigate a certain crystal we need a wave with $\lambda = 0.05$ nm. If we wish to use neutrons, what should be their kinetic energy? What if we use electrons? What for photons?

7.7 • • Find the wavelength of an electron with energy $E = 2$ MeV. [*Hint:* The de Broglie relation $\lambda = h/p$ is correct at all energies, but since this energy is relativistic, you will have to use the relation $E^2 = (pc)^2 + (mc^2)^2$ to find p.]

7.8 • • Find the wavelength of an electron with *kinetic* energy $K = 2$ MeV. (See the hint in Problem 7.7.)

7.9 • • Using the appropriate relativistic relations between energy and momentum, find and compare the wavelengths of electrons and photons at the three different kinetic energies: 1 keV, 1 MeV, 1 GeV.

7.10 • • (*a*) Use the relativistic relation between E and p to show that electrons and photons with the same energy E have different wavelengths. (*Note:* Even at relativistic energies the de Broglie relation $\lambda = h/p$ is correct.) (*b*) Show that their wavelengths approach equality as their common energy E gets much larger than $m_e c^2$.

7.11 • • At what common energy E do the wavelengths of electrons and photons differ by a factor of (*a*) 2, (*b*) 1.1, (*c*) 1.01? (See Problem 7.10.)

SECTIONS 7.4 AND 7.5 (THE QUANTUM WAVE FUNCTION
AND WHICH SLIT DOES THE ELECTRON GO THROUGH?)

7.12 • Electrons with $K = 100$ eV are directed at two narrow slits a distance d apart. If the angle between the central maximum of the resulting interference pattern and the next maximum is to be 1°, what should d be?

7.13 • An experimenter wishes to arrange a two-slit experiment with 3-eV electrons so that the $n = 1$ maximum occurs at 15°. What will his slit separation, d, have to be?

7.14 • What are the dimensions of a matter wave $\Psi(\mathbf{r}, t)$ describing an electron in three dimensions? What are its SI units? [*Hint:* Probabilities are dimensionless. (Why?)]

7.15 • • A classical physicist is determined to find out which slit each electron passes through in the two-slit experiment (without disrupting the interference pattern). To this end he places a molecule near one slit, in the hope that electrons passing through this slit will excite the molecule, causing it to give out a characteristic pulse of light. Show that this arrangement fares no better than the thought experiment using light that was described in Section 7.5.

SECTION 7.6 (SINUSOIDAL WAVES)

7.16 • A wave described by (7.19) has $k = 6$ rad/m and $\omega = 22$ rad/s. Find λ, f, and v.

7.17 • A traveling wave is given by $y(x, t) = A \sin(kx - \omega t)$ with $A = 4$ cm, $k = 12$ rad/cm, and $\omega = 2 \times 10^3$ rad/s. Find the wave speed v, wavelength λ, and frequency f.

7.18 • What are the SI units of λ, T, f, k, ω, and v?

7.19 • For green light ($\lambda \approx 550$ nm) find k and ω.

7.20 • Find k and ω for X rays with $\lambda = 0.05$ nm.

7.21 • Use the de Broglie relation (7.23) to find λ and k for electrons with kinetic energy 300 eV.

7.22 • If we observe a point on a string with a fixed value of x, it will oscillate up and down as the wave (7.18) travels past it. Show that it oscillates with simple harmonic motion of frequency $f = 1/T$.

7.23 • • (*a*) At any fixed point $x = x_0$ the traveling wave $y(x, t) = A \sin(kx - \omega t)$ can be expressed as $y(x_0, t) = A \sin(\omega t + \phi)$. Find ϕ. (*b*) For $x_0 = 0$, what is ϕ? (*c*) By how much must one change x_0 [from its value in part (*b*)] such that ϕ is π larger than in part (*b*)? Express your answer in terms of λ.

7.24 • • (*a*) Prove that a crest of the wave (7.18) moves with speed $v = \lambda/T$ to the right. [*Hint:* Focus attention on one wave crest P (for example, the crest for which the argument of the sine function is $\pi/2$) in order to find an expression for x_p in terms of t.] (*b*) Show that if the minus sign in (7.18) is replaced by a plus sign, then the wave moves to the left.

7.25 • • Show that the crests of a wave $\Psi = A \sin(ax - bt)$ move to the right at a speed v equal to the ratio of the coefficients of t and x; $v = b/a$. (See the hint to Problem 7.24.)

7.26 • A telephone line can transmit a range of frequencies $\Delta f \approx 2500$ Hz. Roughly what is the duration of the shortest pulse that can be sent over this line?

7.27 • A space probe sends a picture containing 500×500 elements each containing a brightness scale with 256 possible levels. This scale requires eight binary digits. Thus altogether $8 \times 500 \times 500 = 2 \times 10^6$ pulses are required to encode the picture for transmission. Suppose that the transmitter uses a bandwidth of 1000 Hz. (For the faint signals from distant space probes the bandwidth must be kept small to reduce the effects of electronic noise—which is present at all frequencies.) Roughly how long is needed to send one picture? Note that the center-to-center separation of adjacent pulses must be at least $2\Delta t$ (the total width of any one pulse).

7.28 • (a) By inspection of Fig. 7.21 deduce the fundamental frequency in the spectrum of $F(t)$. (b) What

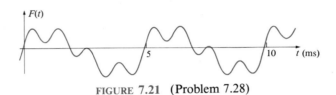

FIGURE 7.21 (Problem 7.28)

would you suggest is the highest harmonic that has a substantial amplitude? What is its frequency?

7.29 • • Figure 7.22 shows a snapshot of a stretched garden hose that has been given an impulse at one end, so that a pulse is propagating along it with speed $v = 6$

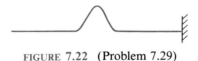

FIGURE 7.22 (Problem 7.29)

m/s. The width of the pulse is $\pm \Delta x$, with $\Delta x = 30$ cm. (a) Find the approximate range Δk of wave numbers needed to build up this pulse. (b) When the pulse passes a certain point it produces a brief vertical deflection $y = F(t)$. Find the width Δt of this pulse and the spread $\Delta \omega$ of its spectrum.

7.30 • • For a given periodic function $F(t)$, the coefficients A_n of its Fourier expansion can be found using the formulas (7.60) and (7.61) in Problem 7.33. [This is for the case of an even function, for which only cosine terms appear in the Fourier series. The general case involves sine terms as well, with coefficients given by (7.62) in Problem 7.34, but these do not appear in this problem.] Use (7.60) and (7.61) to verify that the

Fourier expansion for the pulse function of Fig. 7.13 (with period $T = 2.5\tau$) is as given in Eq. (7.28). Let the pulse height be 1.

7.31 • • • Consider the "sawtooth" wave shown in Fig. 7.23 with period 2 and maximum value 1. It can be

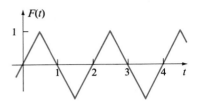

FIGURE 7.23 (Problems 7.31, 7.32, and 7.34)

shown this wave has a spectrum consisting only of odd harmonics whose amplitude alternates in sign and decreases like $1/n^2$, where n is the harmonic number. Specifically:

$$F(t) = \sum_{n=1}^{\infty} B_n \sin n\pi t \qquad (7.57)$$

where B_n is zero for n even, whereas

$$B_n = (-1)^{(n-1)/2} \frac{8}{\pi^2 n^2} \qquad \text{for } n \text{ odd.} \quad (7.58)$$

Sketch the spectrum, showing B_n as a function of n for the first few harmonics.

Calculate the first two nonzero terms ($n = 1$ and 3) in the series (7.57) for the points $t = 0, 0.1, 0.2, \ldots, 2.0$. Make graphs of these two terms and of their sum. On the last graph draw the triangular wave itself for comparison. (If you make rough plots first, you will notice symmetries that allow you to calculate considerably fewer points.)

7.32 • • • Calculate the first three terms in the Fourier series of the "sawtooth" function $F(t)$ in Problem 7.31 for $t = 0, 0.1, 0.2, 0.3, 0.4,$ and 0.5. To compare the quality of the fit using the first one, or two, or three terms of the series, calculate the rms difference between each approximation and $F(t)$ itself. [The rms difference is the square root of the mean of the squares of the differences between the approximation and $F(t)$ at each calculated point; it indicates the quality of the fit, approaching zero as the approximation approaches $F(t)$.]

7.33 • • • The Fourier expansion theorem proves that any periodic function* $F(t)$ can be expanded in terms of sines and cosines. If the function happens to be even

---

* $F(t)$ must satisfy some conditions of "reasonableness." For example, the theorem is certainly true if $F(t)$ is continuous, although it is also true for many discontinuous functions as well.

$[F(t) = F(-t)]$, only cosines are needed and the expansion has the form

$$F(t) = \sum_{n=0}^{\infty} A_n \cos \frac{2n\pi t}{T} \qquad (7.59)$$

where T is the period of the function. In this problem you will see how to find the Fourier coefficients A_n. (a) Prove that

$$A_0 = \frac{1}{T} \int_0^T F(t)\, dt. \qquad (7.60)$$

[Hint: Integrate Eq. (7.59) from $t = 0$ to T.] (b) Prove that for $m > 0$,

$$A_m = \frac{2}{T} \int_0^T F(t) \cos \frac{2m\pi t}{T}\, dt, \qquad (7.61)$$

where we have labeled the coefficient as A_m (rather than A_n) for reasons that will become apparent in your proof. (Hint: Multiply both sides of (7.59) by $\cos(2m\pi t/T)$ and integrate from 0 to T. Using the trig identities in Appendix B, you can prove that $\int_0^T \cos(2m\pi t/T) \cos(2n\pi t/T)\, dt$ is zero if $m \ne n$ and equals $T/2$ if $m = n$. In both parts of this problem you may assume that the integral of an infinite series, $\int [\sum g_n(t)]\, dt$, is the same as the series of integrals $\sum [\int g_n(t)\, dt]$.)

7.34 • • • (a) Let $F(t)$ be a periodic function which is odd — $F(t) = -F(-t)$. The Fourier expansion of such a function requires only sine functions:

$$F(t) = \sum_{n=1}^{\infty} B_n \sin \frac{2n\pi t}{T}.$$

Following the suggestions in Problem 7.33, prove that

$$B_m = \frac{2}{T} \int_0^T F(t) \sin \frac{2m\pi t}{T}\, dt. \qquad (7.62)$$

(Note that the sine series has no $n = 0$ term, since $\sin 0 = 0$.) (b) Use this result to verify that the Fourier coefficients of the "sawtooth" function in Fig. 7.23 are as given in Problem 7.31.

SECTION 7.8 (THE UNCERTAINTY RELATION FOR POSITION AND MOMENTUM)

7.35 • A proton is known to be within an interval ± 6 fm (the radius of a large nucleus). Roughly what is the minimum uncertainty in its velocity? Treat this problem as one-dimensional and express your answer as a fraction of c.

7.36 • An air rifle pellet has a mass of 20 g and a velocity of 100 m/s. If its velocity is known to an accuracy of $\pm 0.1\%$, what is the minimum possible uncertainty in the pellet's position?

7.37 • The position of a 60-gram golf ball sitting on a tee is determined within $\pm 1\ \mu$m. What is its minimum possible energy? Moving at the speed corresponding to this kinetic energy, how far would the ball move in a year?

7.38 • A classical physicist wishes to use the Heisenberg microscope to disprove the uncertainty principle. To reduce the unknown momentum imparted to the electron, he reduces the lens diameter to one-third of its original value. How does this change Δp_x? Δx? Their product?

7.39 • Having failed to disprove the uncertainty principle in Problem 7.38, the physicist tries to reduce Δx by halving the object distance l. How does this change Δx? Δp_x? Their product?

7.40 • • Consider a proton confined to a region of typical nuclear dimensions, about 5 fm. (a) Use the uncertainty principle to estimate its minimum possible kinetic energy in MeV, assuming that it moves in only one dimension. (b) How would your result be modified if the proton were confined in a three-dimensional cube of side 5 fm? [See Eq. (7.40).] The actual kinetic energy of protons in nuclei is somewhat larger than this estimated minimum, being of order 10 MeV.

7.41 • • Consider an electron confined in a region of nuclear dimensions (about 5 fm). Find its minimum possible kinetic energy in MeV. Treat this problem as one-dimensional and use the relativistic relation between E and p. (The large value you will find is a strong argument against the presence of electrons inside nuclei, since no known mechanism could contain an electron with this much energy.)

SECTION 7.9 (THE UNCERTAINTY RELATION FOR TIME AND ENERGY)

7.42 • An excited state of a certain nucleus has a lifetime of 5×10^{-18} s. Find the minimum possible uncertainty in its energy.

7.43 • The subatomic particle called the $\Delta(1232)$ is an excited state of the proton, as we describe in Chapter 14. It decays into a pion and a proton with a mean life of order 10^{-23} s. What is the approximate uncertainty in its energy?

7.44 • The lifetime of the radioactive nucleus ^{235}U is about 1 billion years. Roughly what is the uncertainty in its energy?

7.45 • • An unusually long-lived unstable atomic state has a lifetime of 1 ms. (a) Roughly what is the minimum uncertainty in its energy? (b) Assuming that the photon emitted when this state decays is visible ($\lambda \approx 550$ nm), what are the uncertainty and fractional uncertainty in its wavelength?

7.46 • We have seen that for a nonrelativistic free particle, the particle's velocity v_{part} equals the velocity v_{pack} of the corresponding wave packet [as given by (7.56)]. Prove that the same is true for a relativistic free particle. (Remember the Pythagorean relation between E and p.)

7.47 • (a) For waves on deep water (depth h much greater than wavelength λ) the wave velocity is given by* $v_{wave} = \sqrt{g/k}$. Prove that for these waves the packet velocity v_{pack} is half the wave velocity v_{wave}. (b) In

* Both of the formulas given here for v_{wave} ignore surface tension, which becomes important when λ is of order 1 cm or less.

shallow water ($h \ll \lambda$), $v_{wave} = \sqrt{gh}$. Prove that, in this case, $v_{pack} = v_{wave}$.

7.48 • (a) Use Eqs. (7.54) and (7.55) to prove that

$$v_{pack} = v_{wave} - \lambda \frac{dv_{wave}}{d\lambda}.$$

(b) When monochromatic light passes from a vacuum into glass it is refracted, with blue light bending more than red. Use this observation to decide for which color v_{wave} is greater in glass — red or blue? (c) Use your results in parts (a) and (b) to prove that $v_{pack} < v_{wave}$ for light in glass.

7.49 • Prove that Eq. (7.51) follows from Eq. (7.50).

C H A P T E R

8

The Schrödinger Equation in One Dimension

8.1 Introduction

In classical mechanics, the state of motion of a particle is specified by giving the particle's position and velocity. In quantum mechanics, the state of motion of a particle is specified by giving the wave function. In either case, the fundamental question is to predict how the state of motion will evolve as time goes by, and in each the answer is given by an *equation of motion*. The classical equation of motion is Newton's second law, $\mathbf{F} = m\mathbf{a}$; if we know the particle's position and velocity at time $t = 0$, Newton's second law determines the position and velocity at any other time. In quantum mechanics, the equation of motion is called the *time-dependent Schrödinger equation*. If we know a particle's wave function at $t = 0$, the time-dependent Schrödinger equation determines the wave function at any other time.

The time-dependent Schrödinger equation is a partial differential equation, a complete understanding of which requires more mathematical preparation than we are assuming here. Fortunately, the majority of interesting problems in quantum mechanics do not require use of the equation in its full generality. By far the most interesting states of any quantum system are those states in which the system has a definite total energy, and it turns out that for these states the wave function is a *standing wave,* analogous to the familiar standing waves on a string. When the time-dependent Schrödinger equation is applied to these standing waves, it reduces to a simpler equation called the *time-independent Schrödinger equation.* We shall need only this time-independent equation, which will let us find the wave functions of the standing waves and the corresponding allowed energies. Because we shall be using only the time-independent Schrödinger equation we shall often refer to it as just "the Schrödinger equation." Nevertheless, you should know that there are really two Schrödinger equations (the time-dependent and the time-independent). Unfortunately, it is almost universal to refer to either as "the Schrödinger equation" and to let the context decide which is being discussed. In this book, however, "the Schrödinger equation" will always mean the simpler, time-independent equation.

In Section 8.2 we review some properties of classical standing waves, using waves on a uniform, stretched string as our example. In Section 8.3 we discuss quantum standing waves. Then, in Section 8.4, we show how the familiar properties of classical standing waves let one find the allowed energies of one simple quantum system, namely a particle that moves freely inside a perfectly rigid box.

Using our experience with the wave functions of a particle in a rigid box we next write down the time-independent Schrödinger equation, with which one can, in principle, find the allowed energies and wave functions for any system. Then in Sections 8.6 to 8.9 we use the Schrödinger equation to find the allowed energies of various simple systems.

Throughout this chapter we treat particles that move nonrelativistically in one dimension. All real systems are, of course, three-dimensional. Nevertheless, just as is the case in classical mechanics, it is a good idea to start with the simpler problem of a particle confined to move in just one dimension. In the classical case it is easy to find examples of systems that are at least approximately one-dimensional — a railroad car on a straight track, a bead threaded on a taut string. In quantum mechanics there are fewer examples of one-dimensional systems. However, we can for the moment imagine an electron moving along a very narrow wire.* The main importance of one-dimensional systems is that they provide a good introduction to three-dimensional systems, and that several one-dimensional solutions find direct application in three-dimensional problems.

8.2 Classical Standing Waves

We start with a review of some properties of classical standing waves. We could discuss waves on a string, for which the wave function is the string's transverse displacement $y(x, t)$; or we might consider sound waves, for which the wave

---

* More realistic examples include the motion of electrons along one axis in certain crystals and in some linear molecules.

function is the pressure variation, $p(x, t)$. If we considered electromagnetic waves, then the wave function would be the electric field strength, $\mathcal{E}(x, t)$. In this section we choose to discuss waves on a string, but since our considerations apply equally to all waves, we shall use the general notation $\Psi(x, t)$ to represent the wave function.

Let us consider first two sinusoidal traveling waves, one moving to the right,

$$\Psi_1(x, t) = B \sin(kx - \omega t)$$

(this is the wave sketched in Fig. 7.8) and the other moving to the left with the same amplitude,

$$\Psi_2(x, t) = B \sin(kx + \omega t).$$

If we imagine both of these waves traveling simultaneously in the same string, then the resultant wave is their sum*:

$$\Psi(x, t) = \Psi_1(x, t) + \Psi_2(x, t) = B[\sin(kx - \omega t) + \sin(kx + \omega t)]. \quad (8.1)$$

If we recall the important trigonometric identity (Appendix B)

$$\sin a + \sin b = 2 \sin \frac{a + b}{2} \cos \frac{a - b}{2},$$

we can rewrite the wave (8.1) as

$$\Psi(x, t) = 2B \sin kx \cos \omega t$$

or if we set $2B = A$,

$$\Psi(x, t) = A \sin kx \cos \omega t. \quad (8.2)$$

A series of snapshots of the resultant wave (8.2) is sketched in Fig. 8.1. The important point to observe is that the resultant wave is not traveling to the right or left. At certain fixed points called **nodes,** where $\sin kx$ is zero, $\Psi(x, t)$ is always zero and the string is stationary. At any other point the string simply oscillates up and down in proportion to $\cos \omega t$, with amplitude $A \sin kx$. By superposing two traveling waves we have formed a **standing wave.**

Because the string never moves at the nodes, we could clamp it at two nodes and remove the string outside the clamps, leaving a standing wave on a finite length of string as in Fig. 8.2. This is the kind of wave produced on a piano or guitar string when it sounds a pure musical tone.

If we now imagine a string clamped between two fixed points separated by a distance a, we can ask: What are the possible standing waves that can fit on the string? The answer is that a standing wave is possible, *provided* that it has nodes at the two fixed ends of the string. Since the distances between the nodes of a wave are $\lambda/2$, λ, $3\lambda/2$, and so on, we conclude that a standing wave is possible if it has

$$\frac{\lambda}{2} = a \quad \text{or} \quad \lambda = a \quad \text{or} \quad \frac{3\lambda}{2} = a \quad \text{or} \quad \cdots;$$

that is, if

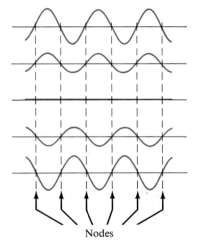

FIGURE 8.1 Five successive snapshots of the standing wave of Eq. (8.2). The nodes, where the string is stationary, are a distance $\lambda/2$ apart.

Nodes

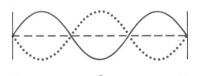

FIGURE 8.2 Three successive snapshots of a standing wave on a finite string, of length a, clamped at its two ends. The solid curve shows the string at maximum displacement; the dashed and dotted curves show it after successive quarter-cycle intervals.

* We are assuming here that whenever Ψ_1 and Ψ_2 are possible waves, the same is true of $\Psi_1 + \Psi_2$. This important property, called the superposition principle, is true of many waves, including those considered here.

$$\lambda = 2a \qquad\qquad \lambda = a \qquad\qquad \lambda = 2a/3$$

FIGURE 8.3 The first three possible standing waves on a string of length a, fixed at both ends.

$$\lambda = \frac{2a}{n}, \qquad \text{where } n = 1, 2, 3, \ldots . \qquad (8.3)$$

We see that the possible wavelengths of a standing wave on a string of length a are *quantized,* the allowed values being $2a$ divided by any positive integer. The first three of these allowed waves are sketched in Fig. 8.3.

It is important to recognize that the quantization of wavelengths arises from the requirement that the wave function must always be zero at the two fixed ends of the string. We refer to this kind of requirement as a **boundary condition,** since it relates to the boundaries of the system. We shall find that for quantum waves, just as for classical waves, it is the boundary conditions that lead to quantization.

8.3 Standing Waves in Quantum Mechanics; Stationary States

Before we discuss quantum standing waves we need to examine more closely the form of the classical standing wave (8.2):

$$\Psi(x, t) = A \sin kx \cos \omega t. \qquad (8.4)$$

This function is a *product* of one function of x (namely $A \sin kx$) and one function of t (namely $\cos \omega t$). We can emphasize this by rewriting (8.4) as

$$\Psi(x, t) = \psi(x) \cos \omega t, \qquad (8.5)$$

where we have used the capital letter Ψ for the full wave function $\Psi(x, t)$ and the lower case letter ψ for its spatial part $\psi(x)$. The spatial function $\psi(x)$ gives the full wave function $\Psi(x, t)$ at time $t = 0$ (since $\cos \omega t = 1$ when $t = 0$); more generally, at any time t the full wave function $\Psi(x, t)$ is $\psi(x)$ times the oscillatory factor $\cos \omega t$.

In our particular example (a wave on a uniform string) the spatial function $\psi(x)$ was a sine function

$$\psi(x) = A \sin kx, \qquad (8.6)$$

but in more general problems, such as waves on a nonuniform string, $\psi(x)$ can be a more complicated function of x. On the other hand, even in these more complicated problems the time dependence is still sinusoidal; that is, it is given by a sine or cosine function of t. The difference between the sine and the cosine is just a difference in the choice of origin of time. Thus either function is possible, and the general sinusoidal standing wave is a combination of both:

$$\Psi(x, t) = \psi(x)(a \cos \omega t + b \sin \omega t). \qquad (8.7)$$

Different choices for the ratio of the coefficients a and b correspond to different choices of the origin of time. (See Problem 8.11.)

The standing waves of a quantum system have the same form (8.7), but with one important difference. For a classical wave, the function $\Psi(x, t)$ is, of course, a real number. (It would make no sense to say that the displacement of a string, or the pressure of a sound wave, had an imaginary part.) Therefore, the function $\psi(x)$ and the coefficients a and b in (8.7) are always real for any classical wave.* In quantum mechanics, on the other hand, the wave function can be a complex number; and for quantum standing waves it usually *is* complex. Specifically, the time-dependent part of the wave function (8.7) always occurs in precisely the combination

$$\cos \omega t - i \sin \omega t, \qquad (8.8)$$

where i is the imaginary number $i = \sqrt{-1}$. That is, the standing waves of a quantum particle have the form

$$\Psi(x, t) = \psi(x)(\cos \omega t - i \sin \omega t). \qquad (8.9)$$

In more advanced texts, this specific time dependence is derived from the time-dependent Schrödinger equation. Here we simply state it as a fact that quantum standing waves have the sinusoidal time dependence of the particular combination of $\cos \omega t$ and $\sin \omega t$ in (8.9).

The form (8.9) can be simplified if we use Euler's formula from the theory of complex numbers (Problem 8.12),

$$\cos \theta + i \sin \theta = e^{i\theta}. \qquad (8.10)$$

This identity can be illustrated in an Argand diagram, as in Fig. 8.4, where the complex number $z = x + iy$ is represented by a point with coordinates x and y

FIGURE 8.4 The complex number $\cos \theta + i \sin \theta$ is represented by a point with coordinates $(\cos \theta, \sin \theta)$ in the complex plane. The absolute value of any complex number $z = x + iy$ is defined as $|z| = \sqrt{x^2 + y^2}$. Since $\cos^2\theta + \sin^2\theta = 1$, we see that $|e^{i\theta}| = 1$.

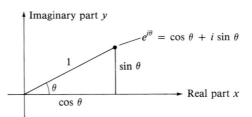

in the complex plane. Since the number $e^{i\theta}$ (with θ any real number) has coordinates $\cos \theta$ and $\sin \theta$, we see from Pythagoras's theorem that its absolute value is 1:

$$|e^{i\theta}| = \sqrt{(\cos \theta)^2 + (\sin \theta)^2} = 1.$$

Thus the complex number $e^{i\theta}$ lies on a circle of radius 1, with polar angle θ as shown. Notice that since $\cos(-\theta) = \cos \theta$ and $\sin(-\theta) = -\sin \theta$,

$$\cos \theta - i \sin \theta = e^{-i\theta}.$$

Returning to (8.9) and using the identity (8.10), we can write for the general standing wave of a quantum system

$$\boxed{\Psi(x, t) = \psi(x)e^{-i\omega t}.} \qquad (8.11)$$

Since this function has a definite angular frequency, ω, any quantum system with this wave function has a definite energy given by the de Broglie relation

---

* As some readers may know, it is sometimes a mathematical convenience to introduce a certain *complex* wave function. Nonetheless, in classical physics the actual wave function is always the real part of this complex function.

$E = \hbar\omega$. Conversely (although we shall not prove this), any quantum system that has a definite energy has a wave function of the form (8.11).

We saw in Chapter 7 that the probability density associated with a quantum wave function $\Psi(x, t)$ is the absolute value squared, $|\Psi(x, t)|^2$. For the complex standing wave (8.11) this has a remarkable property:

$$|\Psi(x, t)|^2 = |\psi(x)|^2 |e^{-i\omega t}|^2$$

or, since $|e^{-i\omega t}| = 1$,

$$|\Psi(x, t)|^2 = |\psi(x)|^2 \qquad \text{(for quantum standing waves).} \qquad (8.12)$$

That is, for a quantum standing wave, the probability density is *independent of time*. This is possible because the time-dependent part of the wave function,

$$e^{-i\omega t} = \cos \omega t - i \sin \omega t,$$

is complex, with two parts that oscillate 90° out of phase; when one is growing the other is shrinking, in such a way that the sum of their squares is constant. Thus for a quantum standing wave the distribution of matter (of electrons in an atom, or nucleons in a nucleus, for example) is time independent or *stationary*. For this reason a quantum standing wave is often called a **stationary state.** The stationary states are the modern counterpart of Bohr's stationary orbits and are precisely the states of definite energy. Because their charge distribution is static, atoms in stationary states do not radiate.*

An important practical consequence of (8.12) is that in most problems the only interesting part of the wave function $\Psi(x, t)$ is its spatial part $\psi(x)$. We shall see that a large part of quantum mechanics is devoted to finding the possible spatial function $\psi(x)$ and their corresponding energies. Our principal tool in finding these will be the time-independent Schrödinger equation.

8.4   The Particle in a Rigid Box

Before we write down the Schrödinger equation, we consider a simple example that we can solve using just our experience with standing waves on a string. We consider a particle that is confined to some finite interval on the x axis, and moves freely inside that interval — a situation we describe as a **one-dimensional rigid box.** For example, in classical mechanics we could consider a bead on a frictionless straight thread between two rigid knots; the bead can move freely between the knots, but cannot escape outside them. In quantum mechanics we can imagine an electron inside a length of very thin conducting wire; to a fair approximation the electron would move freely back and forth inside the wire, but could not escape from it.

Let us consider, then, a quantum particle of mass m moving nonrelativistically in a one-dimensional rigid box of length a, with no forces acting on it between $x = 0$ and $x = a$. The absence of forces means that the potential energy is constant inside the box, and we are free to choose that constant to be zero. Therefore, its total energy is just its kinetic energy. In quantum mechanics, it is almost always more convenient to think of the kinetic energy as $p^2/2m$, rather

---

* Of course, atoms *do* radiate from excited states, but as we discuss in Chapter 15, this is always because some external influence disturbs the stationary state.

than $\frac{1}{2}mv^2$, because of the de Broglie relation, $\lambda = h/p$, between the momentum and wavelength. Therefore, we write the energy as

$$E = K = \frac{p^2}{2m}. \qquad (8.13)$$

As we have said, the states of definite energy are the standing waves. Therefore, to find the allowed energies we must find the possible standing waves for the particle's wave function $\Psi(x, t)$. We know that the standing waves have the form

$$\Psi(x, t) = \psi(x)e^{-i\omega t}. \qquad (8.14)$$

By analogy with waves on a string, one might guess that the spatial function $\psi(x)$ will be a sinusoidal function inside the box; that is, $\psi(x)$ should have the form $\sin kx$ or $\cos kx$ or a combination of both:

$$\psi(x) = A \sin kx + B \cos kx \qquad (8.15)$$

for $0 \leq x \leq a$. (We make no claim to have proved this; but it is certainly a reasonable guess, and we shall prove in Section 8.6 that it is correct.)

Since it is impossible for the particle to escape from the box, the wave function must be zero outside; that is, $\psi(x) = 0$ when $x < 0$ and when $x > a$. If we make the reasonable (and, again, correct) assumption that $\psi(x)$ is continuous, then it must also vanish *at* $x = 0$ and $x = a$:

$$\psi(0) = \psi(a) = 0. \qquad (8.16)$$

These are the boundary conditions that the wave function (8.15) must satisfy. Notice that these boundary conditions are identical to those for a classical wave on a string clamped at $x = 0$ and $x = a$.

From (8.15) we see that $\psi(0) = B$. Thus the wave function (8.15) can satisfy the boundary condition (8.16) only if the coefficient B is zero; that is, the condition $\psi(0) = 0$ restricts $\psi(x)$ to have the form

$$\psi(x) = A \sin kx. \qquad (8.17)$$

Next, the boundary condition that $\psi(a) = 0$ requires that

$$A \sin ka = 0, \qquad (8.18)$$

which implies that*

$$ka = \pi, \quad \text{or } 2\pi, \quad \text{or } 3\pi, \quad \ldots \qquad (8.19)$$

or

$$k = \frac{n\pi}{a}, \qquad n = 1, 2, 3, \ldots . \qquad (8.20)$$

We conclude that the only standing waves that satisfy the boundary conditions (8.16) have the form $\psi(x) = A \sin kx$ with k given by (8.20). In terms of wavelength, this condition implies that

$$\lambda = \frac{2\pi}{k} = \frac{2a}{n}, \qquad n = 1, 2, 3, \ldots , \qquad (8.21)$$

---

* Strictly speaking, (8.18) implies *either* that k satisfies (8.19) *or* that $A = 0$; but if $A = 0$, then $\psi = 0$ for all x, and we get no wave at all. Thus only the solution (8.19) corresponds to a particle in a box.

which is precisely the condition (8.3) for standing waves on a string. This is, of course, not an accident. In both cases, the quantization of wavelengths arose from the boundary condition that the wave function must be zero at $x = 0$ and $x = a$.

For our present discussion the important point is that quantization of wavelength λ implies quantization of momentum, and hence also of energy. Specifically, substituting (8.21) into the de Broglie relation $p = h/\lambda$, we find that

$$p = \frac{nh}{2a} = \frac{n\pi\hbar}{a}, \qquad n = 1, 2, 3, \ldots \qquad (8.22)$$

Since $E = K + U$, and $U = 0$ in this case, we have $E = p^2/2m$. Therefore, (8.22) means that the allowed energies for a particle in a one-dimensional rigid box are

$$E_n = n^2 \frac{\pi^2 \hbar^2}{2ma^2}, \qquad n = 1, 2, 3, \ldots \qquad (8.23)$$

The lowest energy for our particle is obtained when $n = 1$ and is

$$E_1 = \frac{\pi^2 \hbar^2}{2ma^2}. \qquad (8.24)$$

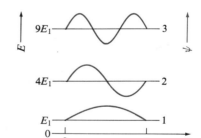

FIGURE 8.5 The first four energy levels and wave functions for a particle in a rigid box. Each horizontal line indicates an energy level, and is also used as the axis for a plot of the corresponding wave function.

This is consistent with the lower bound derived from the Heisenberg uncertainty principle in Chapter 7, where we argued that for a particle confined in a region of length a,

$$E \geq \frac{\hbar^2}{2ma^2} \qquad (8.25)$$

For our particle in a rigid box the actual minimum energy (8.24) is larger than the lower bound (8.25) by a factor of π^2.

In terms of the ground-state energy E_1 the energy of the nth level (8.23) is

$$E_n = n^2 E_1, \qquad n = 1, 2, 3, \ldots \qquad (8.26)$$

These energy levels are sketched in Fig. 8.5. Notice that (quite unlike those of the hydrogen atom) the energy levels are farther and farther apart as n increases, and that E_n increases without limit as $n \to \infty$. The corresponding wave functions $\psi(x)$ (which look exactly like the standing waves on a string) have been superimposed on the same picture, the wave function for each level being plotted on the line that represents its energy. Notice how the number of nodes of the wave functions increases steadily with energy; this is what one should expect, since more nodes mean shorter wavelength and hence larger momentum and kinetic energy.

The complete wave function $\Psi(x, t)$ for any of our standing waves has the form

$$\Psi(x, t) = \psi(x)e^{-i\omega t} = A \sin(kx)e^{-i\omega t}.$$

We can rewrite this, using the identity (Problem 8.13)

$$\sin \theta = \frac{e^{i\theta} - e^{-i\theta}}{2i}, \qquad (8.27)$$

to give

$$\Psi(x, t) = \frac{A}{2i} \left(e^{i(kx - \omega t)} - e^{-i(kx + \omega t)} \right). \qquad (8.28)$$

We see that our quantum standing wave (just like the classical standing wave of Section 8.2) can be expressed as the sum of two traveling waves, one moving to the right and one to the left. The wave moving to the right represents a particle with momentum $\hbar k$ *directed to the right,* and that moving to the left, a particle with momentum of the same magnitude $\hbar k$ but *directed to the left.* Thus a particle in one of our stationary states has a definite magnitude, $\hbar k$, for its momentum but is a 50:50 superposition of momenta in either direction. To some extent this situation is analogous to the result that *on average* a classical particle is equally likely to be moving in either direction as it bounces back and forth inside a rigid box.

8.5 The Time-Independent Schrödinger Equation

Our discussion of the particle in a rigid box depended on some guessing as to the form of the spatial wave function $\psi(x)$. There are very few problems where this kind of guesswork is possible, and no problems where it is entirely satisfying. What we need is the equation that determines $\psi(x)$ in *any* problem, and this equation is the time-independent Schrödinger equation. Like all basic laws of physics, the Schrödinger equation cannot be *derived.* It is simply a relation, like Newton's second law, that experience has shown us is true. Thus a legitimate procedure would be simply to state the equation and to start using it. Nevertheless, it may be helpful to offer some arguments that *suggest* the equation, and this is what we shall try to do.

Almost all laws of physics can be expressed as *differential equations,* that is, as equations that involve the variable of interest and some of its derivatives. The most familiar example is Newton's second law for a single particle, which we can write as

$$m \frac{d^2x}{dt^2} = \sum F. \tag{8.29}$$

If, for example, the particle in question were immersed in a viscous fluid that exerted a drag force $-bv$, and attached to a spring that exerted a restoring force $-kx$, then (8.29) would read

$$m \frac{d^2x}{dt^2} = -b \frac{dx}{dt} - kx. \tag{8.30}$$

This is a differential equation for the particle's position x as a function of time t, and since the highest derivative involved is the second derivative, the equation is called a *second-order differential equation.*

The equation of motion for classical waves (which is often not discussed in an introductory physics course) is a differential equation. It is therefore natural to expect the equation that determines the possible standing waves of a quantum system to be a differential equation. Since we already know the form of the wave functions for a particle in a rigid box, what we shall do is examine these wave functions and try to spot a simple differential equation which they satisfy and which we can generalize to more complicated systems.

We saw in Section 8.4 that the spatial wave functions for a particle in a rigid box have the form

$$\psi(x) = A \sin kx. \tag{8.31}$$

To find a differential equation that this function satisfies, we naturally differentiate it, to give

$$\frac{d\psi}{dx} = kA \cos kx. \tag{8.32}$$

There are several ways in which we could relate $\cos kx$ in (8.32) to $\sin kx$ in (8.31) and hence obtain an equation connecting $d\psi/dx$ with ψ. However, a simpler course is to differentiate a second time to give

$$\frac{d^2\psi}{dx^2} = -k^2 A \sin kx. \tag{8.33}$$

Comparing (8.33) and (8.31) we see at once that $d^2\psi/dx^2$ is proportional to ψ; specifically,

$$\frac{d^2\psi}{dx^2} = -k^2\psi. \tag{8.34}$$

We can rewrite k^2 in (8.34) in terms of the particle's kinetic energy, K. We know that $p = \hbar k$. Therefore,

$$K = \frac{p^2}{2m} = \frac{\hbar^2 k^2}{2m}; \tag{8.35}$$

hence

$$k^2 = \frac{2mK}{\hbar^2}. \tag{8.36}$$

Thus we can write (8.34) as

$$\frac{d^2\psi}{dx^2} = -\frac{2mK}{\hbar^2}\,\psi, \tag{8.37}$$

which gives us a second-order differential equation satisfied by the wave function $\psi(x)$ of a particle in a rigid box.

The particle in a rigid box is an especially simple system, with potential energy equal to zero throughout the region where the particle moves. It is not at all obvious how the equation (8.37) should be generalized to include the possibility of a nonzero potential energy, $U(x)$, which may vary from point to point. However, since the kinetic energy K is the difference between the total energy E and the potential energy $U(x)$, it is perhaps natural to replace K in (8.37) by

$$K = E - U(x). \tag{8.38}$$

This gives us the differential equation*

$$\frac{d^2\psi}{dx^2} = \frac{2m}{\hbar^2}\,[U(x) - E]\psi. \tag{8.39}$$

* In more advanced texts this equation is usually written in the form

$$-\frac{\hbar^2}{2m}\frac{d^2\psi}{dx^2} + U(x)\psi = E\psi$$

because the differential operator $-(\hbar^2/2m)d^2/dx^2$ is intimately connected with the kinetic energy. Nevertheless, for the applications in this book the form (8.39) is the most convenient, and we shall always write it this way.

This differential equation is called the **Schrödinger equation** (time-independent Schrödinger equation, in full), in honor of the Austrian physicist who first published it in 1926. Like us, Schrödinger had no way to *prove* that his equation was correct. All he could do was argue that the equation seemed reasonable and that its predictions should be tested against experiment. In the 60 years since then, it has passed this test repeatedly. In particular, Schrödinger himself showed that it predicts correctly the energy levels of the hydrogen atom, as we describe in Chapter 9. Today, it is generally accepted that the Schrödinger equation is the correct basis of nonrelativistic quantum mechanics, in just the same way that Newton's second law is accepted as the basis of nonrelativistic classical mechanics.

The Schrödinger equation as written in (8.39) applies to one particle moving in one dimension. We shall need to generalize it later to cover systems of several particles, in two or three dimensions. Nevertheless, the general procedure for using the equation is the same in all cases. Given a system whose stationary states and energies we wish to know, we must first find the potential energy function $U(x)$. For example, a particle held in equilibrium at $x = 0$ by a force obeying Hooke's law ($F = -kx$) has potential energy

ERWIN SCHRÖDINGER (1887 – 1961, Austrian). After learning of de Broglie's matter waves, Schrödinger proposed the equation — the Schrödinger equation — that governs the waves' behavior and earned him the 1933 Nobel Prize. He left Austria after Hitler's invasion and became a professor in Dublin.

$$U(x) = \tfrac{1}{2}kx^2. \tag{8.40}$$

An electron in a hydrogen atom has

$$U(r) = -\frac{ke^2}{r}. \tag{8.41}$$

(We shall return to this three-dimensional example in Chapter 9.) Once we have identified $U(x)$, the Schrödinger equation (8.39) becomes a well-defined equation that we can try to solve.* In most cases, it turns out that for many values of the energy E the Schrödinger equation *has no solutions* (no acceptable solutions, satisfying the particular conditions of the problem, that is). This is exactly what leads to the quantization of energies. Those values of E for which the Schrödinger equation has no solution are not allowed energies of the system. Conversely, those values of E for which there is a solution *are* allowed energies, and the corresponding solutions $\psi(x)$ give the spatial wave functions of these stationary states.

As we hinted in the last paragraph, there are usually certain conditions that the wave function $\psi(x)$ must satisfy to be an acceptable solution of the Schrödinger equation. First, there may be boundary conditions on $\psi(x)$, for example, the condition that $\psi(x)$ must vanish at the walls of a perfectly rigid box. In addition, there are certain general restrictions on $\psi(x)$; for example, as we anticipated in Section 8.4, $\psi(x)$ must always be *continuous,* and in most problems its first derivative must also be continuous. When we speak of an acceptable solution of the Schrödinger equation, we shall mean a solution that satisfies all the conditions appropriate to the problem at hand.

In this section you may have noticed that in quantum mechanics it is the potential energy $U(x)$ that appears in the basic equation, whereas in classical mechanics it is the *force F.* Of course, U and F are closely related, U being the integral of F. Nevertheless, it is an important difference of emphasis that quan-

---

* The necessity of identifying U before one can solve the Schrödinger equation corresponds to the necessity of identifying the total force F on a classical particle before one can solve Newton's second law, $F = ma$.

tum mechanics focuses primarily on potential energies, whereas Newtonian mechanics focuses on forces.

8.6 The Rigid Box Again

As a first application of the Schrödinger equation, we use it to rederive the allowed energies of a particle in a rigid box and check that we get the same answers as before.* The first step in applying the Schrödinger equation to any system is to identify the potential energy function $U(x)$. Inside the box the particle has zero potential energy (if we choose our zero correctly), and outside the box the potential energy is infinite. This is the mathematical expression of our idealized *perfectly* rigid box — no finite amount of energy can remove the particle from it. Thus

$$U(x) = \begin{cases} 0, & \text{for } 0 \le x \le a \\ \infty, & \text{for } x < 0 \text{ and } x > a \end{cases} \tag{8.42}$$

That $U(x) = \infty$ outside the box implies that the particle can never be found there, and hence that the wave function $\psi(x)$ must be zero when $x < 0$ and when $x > a$. The continuity of $\psi(x)$ then requires that

$$\psi(0) = \psi(a) = 0 \tag{8.43}$$

(all of which we had argued in Section 8.4). Inside the box, where $U(x) = 0$, the Schrödinger equation (8.39) reduces to

$$\frac{d^2\psi}{dx^2} = -\frac{2mE}{\hbar^2}\psi \qquad \text{for } 0 \le x \le a. \tag{8.44}$$

This is the differential equation whose solutions we must investigate. In particular, we want to find those values of E for which it has a solution satisfying the boundary conditions (8.43).

Before solving (8.44) we remark that it is a nuisance, both for the printer of a book and for the student taking notes, to keep writing the symbols $d\psi/dx$ and $d^2\psi/dx^2$. For this reason we introduce the shorthand

$$\psi' \equiv \frac{d\psi}{dx} \qquad \text{and} \qquad \psi'' \equiv \frac{d^2\psi}{dx^2}.$$

From now on we shall use this notation whenever convenient. In particular, we rewrite (8.44) as

$$\psi''(x) = -\frac{2mE}{\hbar^2}\psi(x). \tag{8.45}$$

We now consider whether there is an acceptable solution of (8.45) for any particular value of E, starting with the case that E is negative. (We do not *expect* any states with $E < 0$, since then E would be less than the minimum potential energy. But we have already encountered several unexpected consequences of quantum mechanics, and we should check this possibility.) If E were negative,

---

* You may reasonably object that it is circular to apply the Schrödinger equation to a particle in a box, when we used the latter to derive the former. Nevertheless, it is a legitimate consistency check, as well as an instructive exercise, to see how the Schrödinger equation gives back the known energies and wave functions.

then the coefficient $-2mE/\hbar^2$ on the right of (8.45) would be *positive* and we could call it α^2, where

$$\alpha = \frac{\sqrt{-2mE}}{\hbar}. \tag{8.46}$$

With this notation, (8.45) becomes

$$\psi''(x) = \alpha^2 \psi(x). \tag{8.47}$$

The simplification of rewriting (8.45) in the form (8.47) has the disadvantage of requiring a new symbol (namely α); but it has the important advantage of letting us focus on the mathematical structure of the equation.

Equation (8.47) is a second-order differential equation, which has the solutions (Problem 8.19) $e^{\alpha x}$ and $e^{-\alpha x}$ or any combination of these,

$$\psi(x) = Ae^{\alpha x} + Be^{-\alpha x} \tag{8.48}$$

where A and B are any constants, real or complex.

It is important in what follows that (8.48) is the most general solution of (8.47), that is, that *every* solution of (8.47) has the form (8.48). This follows from a theorem about second-order differential equations of the same type as the one-dimensional Schrödinger equation.* This theorem states three facts: First, these equations always have two independent solutions. For example, $e^{\alpha x}$ and $e^{-\alpha x}$ are two independent solutions of (8.47).† Second, if $\psi_1(x)$ and $\psi_2(x)$ denote two such independent solutions, then the linear combination

$$A\psi_1(x) + B\psi_2(x) \tag{8.49}$$

is also a solution, for any constants A and B. Third, given two independent solutions $\psi_1(x)$ and $\psi_2(x)$, *every* solution can be expressed as a linear combination of the form (8.49). These three properties are illustrated in Problems 8.19 to 8.24.

That the general solution of a second-order differential equation contains two arbitrary constants is easy to understand: A second-order differential equation amounts to a statement of the second derivative ψ''; to find ψ, one must somehow accomplish two integrations, which should introduce two constants of integration; and this is what the two arbitrary constants A and B in (8.49) are. The theorem above is very useful in seeking solutions of such differential equations. If, by any means, we can spot two independent solutions, we are assured that *every* solution is a combination of these two. Since $e^{\alpha x}$ and $e^{-\alpha x}$ are independent solutions of (8.47), it follows from the theorem that the most general solution is (8.48).

Equation (8.48) gives all solutions of the Schrödinger equation for negative values of E. The important question is now whether any of these solutions could satisfy the required boundary conditions (8.43), and the answer is "no." With $\psi(x)$ given by (8.48), the condition that $\psi(0) = 0$ implies that

$$A + B = 0,$$

---

* To be precise, ordinary second-order differential equations that are linear and homogeneous.

† When we say that two functions are *independent,* we mean that neither function is just a constant multiple of the other. For example, $e^{\alpha x}$ and $e^{-\alpha x}$ are independent, but $e^{\alpha x}$ and $2e^{\alpha x}$ are not; similarly $\sin x$ and $\cos x$ are independent, but $5\cos x$ and $\cos x$ are not.

while the requirement that $\psi(a) = 0$ implies that

$$Ae^{\alpha a} + Be^{-\alpha a} = 0.$$

It is easy to check that the only values of A and B that satisfy these two simultaneous equations are $A = B = 0$. That is, if $E < 0$, the only solution of the Schrödinger equation that satisfies the boundary conditions is the zero function. In other words, with $E < 0$ there can be no standing waves, so negative values of E are not allowed.

Let us next see if the Schrödinger equation (8.45) has any acceptable solutions for positive energies (as we expect it does). With $E > 0$, the coefficient $-2mE/\hbar^2$ on the right of (8.45) is negative and can conveniently be called $-k^2$, where

$$k = \frac{\sqrt{2mE}}{\hbar}. \tag{8.50}$$

With this notation, the Schrödinger equation reads

$$\psi''(x) = -k^2\psi(x). \tag{8.51}$$

This differential equation has the solutions $\sin kx$ and $\cos kx$, or any combination of both:

$$\psi(x) = A \sin kx + B \cos kx \tag{8.52}$$

(see Example 8.1 below). This is exactly the form of the wave function that we assumed at the beginning of Section 8.4. The important difference is that in Section 8.4 we could only guess the form (8.52), whereas we have now *derived* it from the Schrödinger equation. From here on the argument follows precisely the argument given before. As we saw, the boundary condition $\psi(0) = 0$ requires that the coefficient B in (8.52) be zero, whereas the condition that $\psi(a) = 0$ can be satisfied without A being zero, provided that ka is an integer multiple of π (so that $\sin ka = 0$); that is,

$$k = \frac{n\pi}{a}$$

or, from (8.50),

$$E = \frac{\hbar^2 k^2}{2m} = n^2 \frac{\pi^2 \hbar^2}{2ma^2}$$

exactly as before.

EXAMPLE 8.1   Verify explicitly that the function (8.52) is a solution of the Schrödinger equation (8.51) for any values of the constants A and B. [This illustrates part of the theorem stated in connection with (8.49).]

To verify that a given function satisfies an equation, one must substitute the function into one side of the equation and then manipulate it until one arrives at the other side. Thus, for the proposed solution (8.52),

$$\psi''(x) = \frac{d^2}{dx^2}(A \sin kx + B \cos kx)$$

$$= \frac{d}{dx}(kA \cos kx - kB \sin kx)$$

$$= -k^2A \sin kx - k^2B \cos kx$$

$$= -k^2(A \sin kx + B \cos kx)$$

$$= -k^2\psi(x)$$

and we conclude that the proposed solution does satisfy the desired equation.

There is one loose end in our discussion of the particle in a rigid box that we can now dispose of. We have seen that the stationary states have wave functions

$$\psi(x) = A \sin \frac{n\pi x}{a}, \tag{8.53}$$

but we have not yet found the constant A. Whatever the value of A, the function (8.53) satisfies the Schrödinger equation and the boundary conditions. Clearly, therefore, neither the Schrödinger equation nor the boundary conditions fix the value of A.

To see what does fix A, recall that $|\psi(x)|^2$ is the probability density for finding the particle at x. This means, in the case of a one-dimensional system, that $|\psi(x)|^2 dx$ is the probability P of finding the particle between x and $x + dx$:

$$P(\text{between } x \text{ and } x + dx) = |\psi(x)|^2 dx. \tag{8.54}$$

Since the total probability of finding the particle *anywhere* must be 1, it follows that

$$\int_{-\infty}^{\infty} |\psi(x)|^2 dx = 1. \tag{8.55}$$

This relation is called the **normalization condition** and a wave function that satisfies it is said to be **normalized.** It is the condition (8.55) that fixes the value of the constant A, which is therefore called the **normalization constant.**

In the case of the rigid box, $\psi(x)$ is zero outside the box; therefore, (8.55) can be rewritten as

$$\int_0^a |\psi(x)|^2 dx = 1 \tag{8.56}$$

or, with the explicit form (8.53) for $\psi(x)$,

$$A^2 \int_0^a \sin^2\left(\frac{n\pi x}{a}\right) dx = 1. \tag{8.57}$$

The integral here is easily seen to be $a/2$ (Problem 8.25). Therefore, (8.57) implies that

$$\frac{A^2 a}{2} = 1 \tag{8.58}$$

and hence that*

$$A = \sqrt{\frac{2}{a}}. \tag{8.59}$$

We conclude that the normalized wave functions for the particle in a rigid box are given by

$$\psi(x) = \sqrt{\frac{2}{a}} \sin \frac{n\pi x}{a}. \tag{8.60}$$

EXAMPLE 8.2 Consider a particle in the ground state of a rigid box of length a. (a) Find the probability density $|\psi|^2$. (b) Where is the particle most likely to be found? (c) What is the probability of finding the particle in the interval between $x = 0.50a$ and $x = 0.51a$? (d) What is it for the interval [0.75a, 0.76a]? (e) What would be the average result if the position of a particle in the ground state were measured many times?

(*a*) The probability density is just $|\psi(x)|^2$, where $\psi(x)$ is given by (8.60) with $n = 1$. Therefore, it is

$$|\psi(x)|^2 = \frac{2}{a} \sin^2\left(\frac{\pi x}{a}\right), \tag{8.61}$$

which is sketched in Fig. 8.6.

(*b*) The most probable value x_{mp} is the value of x for which $|\psi(x)|^2$ is maximum. From Fig. 8.6 this is clearly seen to be

$$x_{\text{mp}} = \frac{a}{2}. \tag{8.62}$$

(*c*) The probability of finding the particle in any small interval from x to $x + \Delta x$ is given by (8.54) as

$$P(\text{between } x \text{ and } x + \Delta x) \approx |\psi(x)|^2 \, \Delta x. \tag{8.63}$$

(This is exact in the limit $\Delta x \to 0$ and is therefore a good approximation for any small interval Δx.) Thus the two probabilities are

$$P(0.50a \leq x \leq 0.51a) \approx |\psi(0.50a)|^2 \, \Delta x = \frac{2}{a} \sin^2\left(\frac{\pi}{2}\right) \times 0.01a$$

$$= 0.02 = 2\%.$$

(*d*) Similarly,

$$P(0.75a \leq x \leq 0.76a) \approx \frac{2}{a} \sin^2\left(\frac{3\pi}{4}\right) \times 0.01a = 0.01 = 1\%.$$

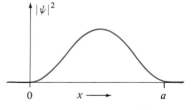

FIGURE 8.6 The probability density $|\psi(x)|^2$ for a particle in the ground state of a rigid box. Inside the box $|\psi|^2$ is given by (8.61); outside, it is zero.

* You may have noticed that, strictly speaking, the argument leading from (8.56) to (8.59) implies only that the *absolute value* of A is $\sqrt{2/a}$. This is because (8.57) and (8.58) should both contain $|A|$ rather than A. However, since the probability density depends only on the absolute value of ψ, we are free to choose any value of A satisfying $|A| = \sqrt{2/a}$ (for example, $A = -\sqrt{2/a}$ or $i\sqrt{2/a}$); the choice (8.59) is convenient and customary.

(e) The average result if we measure the position many times (always with the particle in the same state) is the integral, over all possible positions, of x times the probability of finding the particle at x:

$$x_{av} = \int_0^a x|\psi(x)|^2 \, dx. \tag{8.64}$$

This average value x_{av} is also denoted $\bar{x}$ or $\langle x \rangle$ and is often called the **expectation value** of x. (But note that it is not the value we expect in any one measurement; it is rather the average value expected after many measurements.) In the present case

$$x_{av} = \frac{2}{a} \int_0^a x \sin^2\left(\frac{\pi x}{a}\right) dx. \tag{8.65}$$

This integral can be evaluated to give (Problem 8.29)

$$x_{av} = \frac{a}{2}, \tag{8.66}$$

an answer that is easily understood from Fig. 8.6; since $|\psi(x)|^2$ is symmetric about $x = a/2$, any two points an equal distance on either side of $a/2$ are equally likely, and the average value must be $a/2$. We see from (8.62) and (8.66) that for the ground state of a rigid box, the most probable position x_{mp} and the mean position x_{av} are the same. We shall see in the next example that x_{mp} and x_{av} are not always equal.

EXAMPLE 8.3   Answer the same questions as in Example 8.2 but for the first excited state of the rigid box.

The wave function is given by (8.60) with $n = 2$, so

$$|\psi(x)|^2 = \frac{2}{a} \sin^2\left(\frac{2\pi x}{a}\right).$$

This is plotted in Fig. 8.7, where it is clear that $|\psi(x)|^2$ has two equal maxima at

$$x_{mp} = \frac{a}{4} \quad \text{and} \quad \frac{3a}{4}.$$

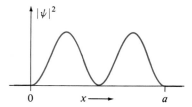

FIGURE 8.7   The probability density $|\psi(x)|^2$ for a particle in the first excited state ($n = 2$) of a rigid box.

Since $|\psi(x)|^2$ is symmetric about $x = a/2$, the average value x_{av} is the same as for the ground state,

$$x_{av} = \frac{a}{2}.$$

The probabilities of finding the particle in any small intervals are given by (8.63) as

$$P(0.50a \le x \le 0.51a) \approx |\psi(0.50a)|^2 \, \Delta x = 0 \tag{8.67}$$

since* $\psi(0.50a) = 0$; and

---

* Note that the probability for the interval $[0.50a, 0.51a]$ is not *exactly* zero, since the probability density $|\psi(x)|^2$ is zero only at the one point $0.50a$. The significance of (8.67) is really that the probability for this interval is *very small* compared to the probability for intervals of the same width elsewhere.

$$P(0.75a \le x \le 0.76a) \approx |\psi(0.75a)|^2 \, \Delta x = \frac{2}{a} \sin^2\left(\frac{3\pi}{2}\right) \times (0.01a) = 0.02.$$

In particular notice that although $x = a/2$ is the average value of x, the probability of finding the particle in the immediate neighborhood of $x = a/2$ is zero. This result, although a little surprising at first, is easily understood by reference to Fig. 8.7.

8.7 The Free Particle

As a second application of the Schrödinger equation, we investigate the possible energies of a free particle; that is, a particle subject to no forces and completely unconfined (still in one dimension, of course). The potential energy of a free particle is constant and can be chosen to be zero. With this choice, we shall show that the energy of the particle can have any positive value, $E \ge 0$. That is, the energy of a free particle is not quantized, and its allowed values are the same as those of a classical free particle.

To prove these assertions we must write down the Schrödinger equation and find those E for which it has acceptable solutions. With $U(x) = 0$, the Schrödinger equation is

$$\psi''(x) = -\left(\frac{2mE}{\hbar^2}\right)\psi(x). \tag{8.68}$$

This is the same equation that we solved for a particle in a rigid box. However, there is an important difference, since the free particle can be anywhere in the range

$$-\infty < x < \infty.$$

Thus we must look for solutions of (8.68) for all x rather than just those x between 0 and a.

If we consider first the possibility of states with $E < 0$, the coefficient $-2mE/\hbar^2$ in front of ψ in (8.68) is positive and we can write (8.68) as

$$\psi''(x) = \alpha^2\psi(x),$$

where $\alpha = \sqrt{-2mE}/\hbar$. Just as with the rigid box, this equation has the solutions $e^{\alpha x}$ and $e^{-\alpha x}$, or any combination of both:

$$\psi(x) = Ae^{\alpha x} + Be^{-\alpha x}. \tag{8.69}$$

But in the present case we can immediately see that none of these solutions can possibly be physically acceptable. The point is that (8.69) is the solution in the whole range $-\infty < x < \infty$. Now, as $x \to \infty$, the exponential $e^{\alpha x}$ grows without limit or "blows up," and it obviously makes no sense to have a wave function $\psi(x)$ that gets bigger and bigger without limit as we move farther and farther away from the origin. The only way out of this difficulty is to have the coefficient A of $e^{\alpha x}$ in (8.69) equal to zero. Similarly, as $x \to -\infty$, the exponential $e^{-\alpha x}$ blows up; thus by the same argument the coefficient B must also be zero, and we are left with just the zero solution $\psi(x) \equiv 0$. That is, there are no acceptable states with $E < 0$, just as we expected.

The argument just given crops up surprisingly often in solving the Schrödinger equation. If a solution of the equation blows up as $x \to \infty$, or as $x \to -\infty$, that solution is obviously not acceptable. Thus we can add to our list of conditions that must be satisfied by an acceptable wave function $\psi(x)$ the requirement that $\psi(x)$ must not blow up as $x \to \pm\infty$. We speak of a function that satisfies this requirement as being "well behaved" as $x \to \pm\infty$. This requirement is actually another example of a boundary condition, since the "points" $x = \pm\infty$ are the boundaries of our system.

Let us next examine the possibility of states of our free particle with $E \geq 0$. In this case the Schrödinger equation can be written as

$$\psi''(x) = -\left(\frac{2mE}{\hbar^2}\right)\psi(x) = -k^2\psi(x) \tag{8.70}$$

where

$$k = \frac{\sqrt{2mE}}{\hbar}. \tag{8.71}$$

As before, the general solution of this equation is

$$\psi(x) = A \sin kx + B \cos kx. \tag{8.72}$$

The important point about this solution is that neither $\sin kx$ nor $\cos kx$ blows up as $x \to \pm\infty$. Thus neither function suffers the difficulty that we encountered with negative energies. Thus for any value of k the function (8.72) is an acceptable solution, for any constants A and B. According to (8.71), this means that all energies in the continuous range $0 \leq E < \infty$ are allowed. In particular, the energy of a free particle is not quantized. Evidently, it is only when a particle is confined in some way that its energy is quantized.

To understand what the positive-energy wave functions (8.72) represent, it is helpful to recall the identities (Problem 8.13)

$$\sin kx = \frac{e^{ikx} - e^{-ikx}}{2i} \quad \text{and} \quad \cos kx = \frac{e^{ikx} + e^{-ikx}}{2}. \tag{8.73}$$

Substituting these expansions into the wave function (8.72), we can write

$$\psi(x) = Ce^{ikx} + De^{-ikx}, \tag{8.74}$$

where you can easily find C and D in terms of the original coefficients A and B. It is important to note that since A and B were arbitrary, the same is true of C and D; that is, (8.74) is an acceptable solution for any values of C and D.

The full, time-dependent wave function $\Psi(x, t)$ for the spatial function (8.74) is

$$\Psi(x, t) = \psi(x)e^{-i\omega t} = Ce^{i(kx-\omega t)} + De^{-i(kx+\omega t)}. \tag{8.75}$$

This is a superposition of two traveling waves, one moving to the right (with coefficient C) and the other moving to the left (with coefficient D). If we choose the coefficient $D = 0$, then (8.75) represents a particle with definite momentum $\hbar k$ to the right; if we choose $C = 0$, then (8.75) represents a particle with momentum of the same magnitude $\hbar k$ but directed to the left. If both C and D are nonzero, then (8.75) represents a superposition of both momenta.

8.8 The Nonrigid Box

So far our only example of a particle that is confined, or "bound," is the rather unrealistic case of a particle in a perfectly rigid box. In this section we apply the Schrödinger equation to a particle in the more realistic nonrigid box. This is a

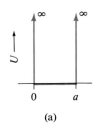

(a)

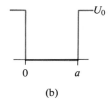

(b)

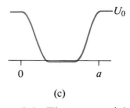

(c)

FIGURE 8.8 Three potential wells: **(a)** the infinite well (8.76); **(b)** the finite square well (8.77); **(c)** the finite rounded well.

rather long section; but the ideas it contains are all fairly simple and are central to an understanding of many quantum systems.

The first step in applying the Schrödinger equation to any system is to determine the potential energy function. Therefore, we must first decide what is the potential energy, $U(x)$, of a particle in a nonrigid box. For a *rigid* box we know that

$$U(x) = \begin{cases} 0, & 0 \le x \le a \\ \infty, & x < 0 \quad \text{and} \quad x > a. \end{cases} \tag{8.76}$$

This is infinite outside the box because no finite amount of energy can remove the particle from a perfectly rigid box.* For most systems a more realistic assumption would be that there is a *finite* minimum energy needed to remove a stationary particle from the box. If we call this minimum energy U_0, the potential energy function would be

$$U(x) = \begin{cases} 0, & 0 \le x \le a \\ U_0, & x < 0 \quad \text{and} \quad x > a. \end{cases} \tag{8.77}$$

In Fig. 8.8(a) and (b) we plot the potential energy functions (8.76) and (8.77). For obvious reasons these functions are often called *potential wells,* and we speak of a particle "moving in a well" when its potential energy is given by one of these functions. The rigid box is an infinitely deep well and the nonrigid box a finite well.

Even the finite well of Fig. 8.8(b) is unrealistic in that the potential energy jumps abruptly from 0 to U_0 at $x = 0$ and $x = a$. For a real particle in a box (for example, an electron in a conductor) the potential energy changes continuously near the walls, more like the well shown in Fig. 8.8(c). This well is sometimes called a *rounded well,* while those of Fig. 8.8(a) and (b) are called *square wells.* To simplify our discussion we shall suppose that the rounded well has $U(x)$ exactly constant, $U(x) = U_0$, for $x < 0$ and $x > a$, as shown in Fig. 8.8(c).

In this section we wish to investigate the energy levels of a particle confined in a nonrigid box such as either Fig. 8.8(b) or (c). As one might expect, the properties of both wells are qualitatively similar. To be definite we shall consider mostly the rounded well of Fig. 8.8(c).

Like the infinite well, the finite wells allow no states with $E < 0$, if we define the zero of U at the bottom of the well. (See Problem 8.34.) An important difference is that for $E > U_0$ a particle in the finite well is not confined; that is, it has enough energy to escape to $x = \pm\infty$. This means that the wave functions for $E > U_0$ are quite similar to those of a free particle. In particular, the possible energies for $E > U_0$ are not quantized; but we shall not pursue this point here since our main interest is in the bound states, whose energies lie in the interval $0 < E < U_0$.

A classical particle moving in a finite well with energy in the interval $0 < E < U_0$ would simply bounce back and forth indefinitely. In the square

---

* Until a few years ago one would have said that in this repect the perfectly rigid box is totally unrealistic—a real bound system might require a *large* energy to pull it apart, but surely not an infinite amount. As we discuss in Chapter 14, it now appears that subatomic particles like neutrons and protons are made up of sub-subatomic particles called quarks, and that an infinite energy *is* needed to pull them apart (that is, they cannot be pulled apart). Thus a potential energy function like (8.76) may be more realistic than we had formerly appreciated.

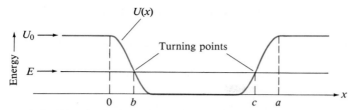

FIGURE 8.9 The classical turning points. A classical particle trapped in the potential well oscillates back and forth, turning around at the points $x = b$ and $x = c$ where the kinetic energy is zero and hence $E = U(x)$.

well of Fig. 8.8(b) it would bounce between the points $x = 0$ and $x = a$. For the rounded well, the points at which the particle turns around are determined by the condition $E = U(x)$ (since the kinetic energy must be zero at the turning point where the particle comes instantaneously to rest). These points can be found graphically as in Fig. 8.9, by drawing a horizontal line at the height representing the energy E. The points, $x = b$ and $x = c$, at which this line meets the potential-energy curve are the two classical turning points, and a classical particle with energy E simply bounces back and forth between these points.

Let us now consider the Schrödinger equation,

$$\psi''(x) = \frac{2m}{\hbar^2}[U(x) - E]\psi(x),$$

for a quantum particle with the rounded potential well for its potential energy. We focus attention on a particular energy E in the range $0 < E < U_0$ and consider whether or not the Schrödinger equation has an acceptable solution for this energy.

WAVE FUNCTIONS OUTSIDE THE WELL

We first note that in the regions $x < 0$ and $x > a$, the potential energy is constant, $U(x) = U_0$, and the Schrödinger equation is easily solved. If we set

$$\frac{2m}{\hbar^2}[U_0 - E] = \alpha^2, \tag{8.78}$$

then in the region $x < 0$ the general solution has the form

$$\psi(x) = Ae^{\alpha x} + Be^{-\alpha x}, \qquad x < 0, \tag{8.79}$$

with A and B arbitrary. Now, as $x \to -\infty$, the exponential $e^{-\alpha x}$ blows up and is physically unacceptable. Thus the form (8.79) is physically acceptable only if the coefficient $B = 0$.

Similarly, in the region $x > a$, the general solution has the form

$$\psi(x) = Ce^{\alpha x} + De^{-\alpha x}, \qquad x > a \tag{8.80}$$

and this is acceptable only if the coefficient $C = 0$.

NUMERICAL SOLUTION OF THE SCHRÖDINGER EQUATION

Equations (8.79) and (8.80) give the form of the general solution of the Schrödinger equation in the particular regions $x < 0$ and $x > a$. We must now find a single solution for the entire interval $-\infty < x < \infty$. For most potential energy functions $U(x)$ the solution cannot be expressed in terms of known elementary functions, such as $e^{\alpha x}$ or $\sin kx$, in the interval $0 < x < a$. When this *is* possible we say that the equation can be *solved analytically,* but there are very few real problems where there are analytic solutions of the Schrödinger equation (or Newton's second law, or any other differential equation). Usually, one has to find the solution numerically using a calculator or computer. It is easy to understand the general principles of such numerical calculations, and once you do, we can discuss and understand the way in which acceptable solutions of the Schrödinger equation exist for some energies and not for others.

To start solving the Schrödinger equation, or any other second-order differential equation, one needs to know the values of ψ and its derivative ψ' at one point $x = x_0$. (This is exactly analogous to the familiar statement that to solve Newton's second law one has to know the particle's position and velocity at one time t_0.) If the potential energy function is known, the Schrödinger equation,

$$\psi''(x) = \frac{2m}{\hbar^2}[U(x) - E]\psi(x), \tag{8.81}$$

tells us $\psi''(x_0)$ in terms of the known $\psi(x_0)$. We therefore know ψ and its first two derivatives at $x = x_0$.

Suppose now that we want to find ψ at some other point x. Our first step is to divide the interval between x_0 and x into n small intervals at points

$$x_0 < x_1 < x_2 < \cdots < x_{n-1} < x_n = x,$$

a distance Δx apart. Knowing ψ, ψ', and ψ'' at x_0, we can now set up a procedure for finding the same three functions at x_1, then at x_2, and so on, all the way to $x_n = x$.

To find $\psi(x_1)$ we use the approximation

$$f(x + \Delta x) = f(x) + f'(x)\,\Delta x, \tag{8.82}$$

which gives the value of f at $x + \Delta x$, by approximating the curve of f, between x and $x + \Delta x$, with a straight line of slope $f'(x)$, as shown in Fig. 8.10. Applying this approximation to the function $\psi(x)$, we find that

$$\psi(x_1) = \psi(x_0) + \psi'(x_0)\,\Delta x, \tag{8.83}$$

which gives $\psi(x_1)$ in terms of known quantities. This is, of course, only an approximation, but by making Δx smaller we can achieve any desired accuracy — at the expense of a longer overall computation. Applying the same approximation (8.82) to $\psi'(x)$, we find similarly that

$$\psi'(x_1) = \psi'(x_0) + \psi''(x_0)\,\Delta x,$$

which gives $\psi'(x_1)$ in terms of known quantities. Knowing $\psi(x_1)$ and $\psi'(x_1)$, we can use the Schrödinger equation to find $\psi''(x_1)$, and we are now ready to start again and move from x_1 on to x_2, and so on. After n steps this procedure gives us the wave function at the desired point $x_n = x$, as illustrated in Fig. 8.11.

In the problem of the particle in a nonrigid box we know that for $x < 0$ the wave function must have the form $\psi(x) = Ae^{\alpha x}$ if it is to be physically acceptable. We can now imagine using the numerical procedure just outlined to follow $\psi(x)$ across the well from $x = 0$ to $x = a$ and beyond. In the region $x > a$ we already know that any solution has the form

$$Ce^{\alpha x} + De^{-\alpha x}. \tag{8.84}$$

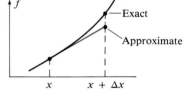

FIGURE 8.10 Starting from known values at x, we can approximate the value of f at $x + \Delta x$ using a straight line through $f(x)$ with slope $f'(x)$. The smaller we choose Δx, the better this approximation becomes.

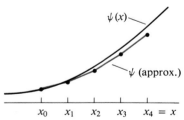

FIGURE 8.11 The numerical solution described in the text replaces the curve $\psi(x)$ by the polygon labeled ψ(approx.). If we use smaller and smaller subdivisions (and hence more and more steps), then ψ(approx.) approaches the exact $\psi(x)$. A typical computer calculation might use several thousand steps and give an accuracy of many significant figures.

By comparing our numerical solution with this form we can identify coefficients C and D for our solution. If we find that $C \neq 0$, our solution blows up as $x \to \infty$ and is not physically acceptable. In this case the energy E is not an allowed energy of the particle, and we must try a different value of E. If we find that C *is* zero, our solution is well behaved both as $x \to -\infty$ and as $x \to \infty$, and E *is* an allowed energy.

SOME GENERAL PROPERTIES OF WAVE FUNCTIONS

To understand when we should expect to find allowed energies, it is useful to explore a little farther the general behavior of solutions of the Schrödinger equation

$$\psi''(x) = \frac{2m}{\hbar^2}[U(x) - E]\psi(x). \tag{8.85}$$

If, as usual, we focus attention on a particular value of E (with $0 < E < U_0$) we can distinguish two important ranges of x: those x where the factor $[U(x) - E]$ in (8.85) is positive, and those x where it is negative. The dividing points between these regions are the classical turning points $x = b$ and $x = c$, where $U(x) = E$. (These were defined in Fig. 8.9 and are shown again in Fig. 8.12.) The region where $[U(x) - E]$ is positive is outside these turning points ($x < b$ and $x > c$) and is often called the classically forbidden region, since a classical particle with energy E cannot penetrate there. The region where $[U(x) - E]$ is negative is the interval $b < x < c$ and is called the classically allowed region. We shall see that the behavior of the wave function $\psi(x)$ is quite different in these two regions.

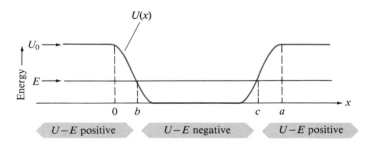

FIGURE **8.12** The factor $[U(x) - E]$, which appears on the right side of the Schrödinger equation, is positive for $x < b$ and $x > c$, and is negative for $b < x < c$.

In the region where $[U(x) - E]$ is positive, the Schrödinger equation has the form

$$\psi''(x) = (\text{positive function}) \times \psi(x). \tag{8.86}$$

In an interval where $\psi(x)$ is positive, this implies that $\psi''(x)$ is also positive and hence that $\psi(x)$ curves *upward,* as in Fig. 8.13(a) and (b). If $\psi(x)$ is negative, then (8.86) implies that $\psi''(x)$ is negative and hence that $\psi(x)$ curves *downward,* as in Fig. 8.13(c) and (d). In either case, we see that $\psi(x)$ curves *away from the axis.* In particular, if the Schrödinger equation has the form (8.86) as $x \to \infty$, as it does in our case, then either $\psi(x)$ will blow up as $x \to \infty$ [as when $\psi(x) = Ce^{\alpha x}$—see Fig. 8.13(a) and (c)], or $\psi(x)$ will approach zero as $x \to \infty$ [as when $\psi(x) = De^{-\alpha x}$—see Fig. 8.13(b) and (d)].

In the region $b < x < c$, the Schrödinger equation has the form

$$\psi''(x) = (\text{negative function}) \times \psi(x), \tag{8.87}$$

and we can argue that $\psi(x)$ curves *toward* the axis and tends to oscillate, as

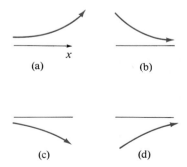

FIGURE **8.13** If $\psi(x)$ satisfies an equation of the form (8.86), it curves away from the axis.

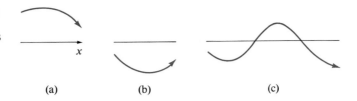

x

(a)   (b)   (c)

follows: If $\psi(x)$ is positive, then $\psi''(x)$ is negative and $\psi(x)$ curves *downward,* as in Fig. 8.14(a); if $\psi(x)$ is negative, the argument reverses and $\psi(x)$ bends *upward,* as in Fig. 8.14(b). In either case, $\psi(x)$ curves toward the axis. If the interval $b < x < c$ is sufficiently wide, a function bending toward the axis will cross the axis and immediately start bending the other way, as in Fig. 8.14(c). Thus we can say that in the region $b < x < c$ the wave function tends to oscillate.

If the "negative function" in (8.87) has a large magnitude, then $\psi''(x)$ tends to be large and $\psi(x)$ curves and oscillates rapidly. Conversely, when the "negative function" is small, $\psi(x)$ bends gradually and oscillates slowly. This is all physically reasonable: The "negative function" in (8.87) is proportional to $[U(x) - E]$, which is just minus the kinetic energy; according to de Broglie, large kinetic energy means short wavelength and hence rapid oscillation, and vice versa.

Now that we understand the qualitative behavior of solutions of the Schrödinger equation for all x, let us return to our hunt for an acceptable solution. We can start in the region $x < 0$, where $U(x)$ is constant (Fig. 8.12), and begin with the known, acceptable form

$$\psi(x) = Ae^{\alpha x} \qquad (x < 0).$$

When we move to the right, our solution will cease to have this explicit form once $U(x)$ starts to vary, but it will continue to bend away from the axis until x reaches the point b. At $x = b$, it will start oscillating and continue to do so until $x = c$, where it will start curving away from the axis again. Thus the general appearance of $\psi(x)$ will be as shown in Fig. 8.15, with two regions where $\psi(x)$ bends away from the axis, separated by one region where $\psi(x)$ oscillates. Figure 8.15 shows a solution that blows up on the right. We must now find out if there are any values of E for which the solution is well behaved both on the left and right.

FIGURE 8.15 The wave function oscillates between the two turning points $x = b$ and $x = c$, and curves away from the axis outside them. The example shown is well behaved as $x \rightarrow -\infty$, but blows up as $x \rightarrow \infty$.

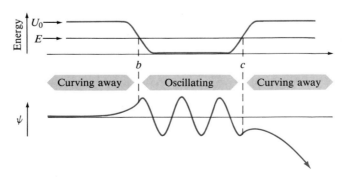

Let us begin a systematic search for allowed energies, starting with E close to zero. We shall show first that with E sufficiently close to zero, an acceptable wave function is impossible. We start with the wave function $Ae^{\alpha x}$ in the region $x < 0$, and follow $\psi(x)$ to the right. When we reach the classical turning point $x = b$, $\psi(x)$ has a positive slope and starts to bend toward the axis. But with E very small, $\psi(x)$ bends very slowly. Thus when we reach the second turning point $x = c$, the slope is *still* positive. With ψ and ψ' both positive, $\psi(x)$ continues to increase without limit, as shown in Fig. 8.16. Therefore, the wave function which is well behaved as $x \to -\infty$ blows up as $x \to +\infty$, and we conclude that there can be no acceptable wave function with E very close to zero.

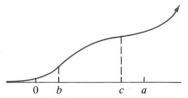

FIGURE 8.16 When E is very small the wave function bends too slowly inside the well. The function that has the form $Ae^{\alpha x}$ when $x < 0$ blows up as $x \to +\infty$.

Suppose now that we slowly increase E, continuing to hunt for an allowed energy. For larger values of E the kinetic energy is larger and, as we have seen, $\psi(x)$ bends more rapidly inside the well. Thus it can bend enough that its slope becomes negative, as shown in Fig. 8.17, curve 2. Eventually, its value and slope at the right of the well will be just right to join onto the solution, which is well behaved as $x \to \infty$, and we have an acceptable wave function (curve 3 in Fig. 8.17). If we increase E any further, $\psi(x)$ will bend over too far inside the well and will now approach $-\infty$ as $x \to \infty$, like curve 4 in Fig. 8.17. Evidently, there is exactly one allowed energy in the range explored so far.

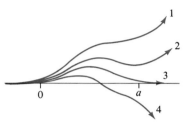

FIGURE 8.17 Solutions of the Schrödinger equation for four successively larger energies. All four solutions have the well-behaved form $Ae^{\alpha x}$ for $x < 0$; only number 3 is also well behaved as $x \to +\infty$.

If E is increased still further, the wave function may bend over *and back* just enough to fit onto the function that is well behaved as $x \to \infty$ as in Fig. 8.18(a). If this happens, we have a second allowed energy. Beyond this, we may find a third acceptable wave function, like that in Fig. 8.18(b), and so on.

Figure 8.19 shows the first three wave functions for a finite square well, beside the corresponding wave functions for an infinitely deep square well of the same width a. Notice the marked similarity of corresponding functions: The first wave function of the infinite well fits exactly half an oscillation into the well, while that of the finite well fits somewhat less than half an oscillation into the well, since it doesn't actually vanish until $x = \pm\infty$. Similarly, the second function of the infinite well makes one complete oscillation, whereas that of the finite well makes just less than one oscillation. For this reason the energy of each level in the finite well is slightly lower than that of the corresponding level in the infinite well.

It is useful to note that the ground-state wave function for any finite well has no nodes, while that for the second level has one node, and that for the nth level has $n - 1$ nodes. This general trend (more nodes for higher energies) could have been anticipated, since higher energy corresponds to a wave function that oscillates more quickly and hence has more nodes.

The wave functions in Fig. 8.19 illustrate two more important points. First, the wave functions of the finite well are nonzero outside the well, in the

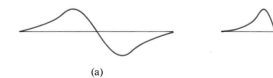

(a)   (b)

FIGURE 8.18 Wave functions for the second and third energy levels of a nonrigid box.

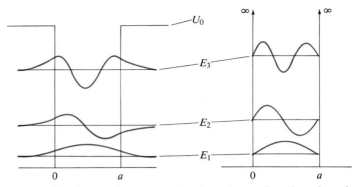

FIGURE 8.19   The lowest three energy levels and wave functions for a finite square well and for an infinite square well of the same width. The horizontal lines that represent each energy level have been used as the axes for drawing the corresponding wave function.

classically forbidden region. (Remember that a classical particle with energy $0 < E < U_0$ cannot escape outside the turning points.) However, since the wave function is largest inside the well, the particle is most likely to be found inside the well; and since $\psi(x)$ approaches zero rapidly as one moves away from the well, we can say that our particle is bound inside, *or close to,* the potential well. Nevertheless, there is a definite, nonzero probability of finding the particle in the classically forbidden regions. This difference between classical and quantum mechanics is due to the wave nature of quantum particles. The ability of the quantum wave function to penetrate classically forbidden regions has important consequences, as we discuss in Section 13.9.

A second important point concerns the number of bound states of the finite well. With the infinite well one can increase E indefinitely and always encounter more bound states. With the finite well, however, the particle is no longer confined when E reaches U_0, and there are no more bound states. The number of bound states depends on the well depth U_0 and width a, but it is always finite.

8.9   The Simple Harmonic Oscillator (optional*)

As another example of a one-dimensional bound particle we consider the simple harmonic oscillator. The name **simple harmonic oscillator** (or **SHO**) is used, in classical and quantum mechanics, for a system that oscillates about a stable equilibrium point to which it is bound by a force obeying Hooke's law.

Familiar classical examples of harmonic oscillators are a mass suspended from an ideal spring and a pendulum oscillating with small amplitude. Before we give any quantum examples, it may be worth recalling why the harmonic oscillator is such an important system. If a particle is in equilibrium at a point x_0, the total force on the particle is zero at x_0; that is, $F(x_0) = 0$. If the particle is displaced to a neighboring point x, the total force will be approximately

$$F(x) = F(x_0) + F'(x_0)(x - x_0), \tag{8.88}$$

* If you choose to omit this section, be aware that we shall use its results in the optional Sections 16.6 and 16.7 on excited states of molecules, and in Section 17.7 on vibrations in solids.

with $F(x_0) = 0$ in this case. [This is just the approximation (8.82) and should be good for any x sufficiently close to x_0.] If x_0 is a point of stable equilibrium, the force is a restoring force; that is, $F(x)$ is negative when $x - x_0$ is positive, and vice versa. Therefore, $F'(x_0)$ must be negative and we denote it by $-k$, where k is called the *force constant*. With $F'(x_0) = -k$ and $F(x_0) = 0$, Eq. (8.88) becomes

$$F(x) = -k(x - x_0),\qquad(8.89)$$

which is Hooke's law. Thus any particle oscillating about a stable equilibrium point will oscillate harmonically, at least for small displacements $(x - x_0)$.

An important example of a quantum harmonic oscillator is the motion of any one atom inside a solid crystal; each atom has a stable equilibrium position relative to its neighboring atoms and can oscillate harmonically about that position. Another important example is a diatomic molecule, such as HCl, whose two atoms can vibrate harmonically, in and out from one another.

In quantum mechanics we work, not with the force F, but with the potential energy U. This is easily found by integrating (8.89) to give

$$U(x) = -\int_{x_0}^{x} F(x)\, dx = \tfrac{1}{2}k(x - x_0)^2\qquad(8.90)$$

if we take U to be zero at x_0. Thus in quantum mechanics the SHO can be characterized as a system whose potential energy has the form (8.90). This function is a parabola, with its minimum at $x = x_0$, as shown in Fig. 8.20(a).

It is important to remember that (8.89) and (8.90) are approximations that are usually valid only for small displacements from x_0. This point is illustrated in Fig. 8.20(b), which shows the potential energy of a typical diatomic molecule, as a function of the distance r between the two atoms. The molecule is in equilibrium at the separation r_0. For r close to r_0, the potential energy is well approximated by a parabola of the form $U(r) = \tfrac{1}{2}k(r - r_0)^2$, but when r is far from r_0, $U(r)$ is quite different. Thus for small displacements the molecule will behave like an SHO, but for large displacements it will not. This same statement can be made about almost any oscillating system. (For example, the simple pendulum is well known to oscillate harmonically for small amplitudes, but not when the amplitude is large.) It is because small displacements from equilibrium are very common that the harmonic oscillator is so important.

There is a close connection between the classical and quantum harmonic oscillators, and we start with a quick review of the former. If we choose our origin at the equilibrium position, then $x_0 = 0$ and the force is $F = -kx$. If the particle has mass m, then Newton's second law reads

$$ma = -kx.$$

If we define the classical angular frequency

$$\omega_c = \sqrt{\frac{k}{m}}\qquad(8.91)$$

this becomes

$$\frac{d^2x}{dt^2} = -\omega_c^2 x,$$

which has the general solution

$$x = a \sin \omega_c t + b \cos \omega_c t.\qquad(8.92)$$

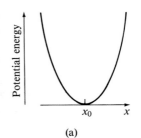

(a)

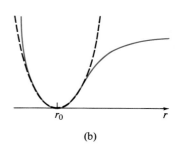

(b)

FIGURE 8.20 (a) The potential energy of an ideal simple harmonic oscillator is a parabola. (b) The potential energy of a typical diatomic molecule (solid curve) is well approximated by that of an SHO (dashed curve) when r is close to its equilibrium value r_0.

Thus the position of the classical SHO varies sinusoidally in time with angular frequency ω_c. If we choose our origin of time, $t = 0$, at the moment when the particle is at $x = 0$ and moving to the right, then (8.92) takes the form

$$x = a \sin \omega_c t. \tag{8.93}$$

The positive number a is the amplitude of the oscillations, and the particle oscillates between $x = a$ and $x = -a$. In other words, the points $x = \pm a$ are the classical turning points. When the particle is at $x = \pm a$, all of its energy is potential energy; thus $E = \frac{1}{2}ka^2$ and hence

$$a = \sqrt{\frac{2E}{k}}. \tag{8.94}$$

As one would expect, the classical amplitude a increases with increasing energy.

To find the allowed energies of a *quantum* harmonic oscillator we must solve the Schrödinger equation with $U(x)$ given by

$$U(x) = \tfrac{1}{2}kx^2, \tag{8.95}$$

where we have again chosen $x_0 = 0$. Qualitatively, the analysis is very similar to that for the finite well discussed in the preceding section. For any choice of E the solutions will bend away from the axis outside the classical turning points $x = a$ and $x = -a$, and will oscillate between these points. As before, there is just one solution that is well behaved as $x \to -\infty$, and for most values of E this solution blows up as $x \to \infty$. Thus the allowed energies are quantized. The potential energy (8.95) increases indefinitely as x moves away from the origin [see Fig. 8.20(a)] and the particle is therefore confined for all energies. In this respect, the SHO resembles the infinitely deep potential well, and we would expect to find infinitely many allowed energies, all of them quantized.

An important difference between the SHO and most other potential wells is that the Schrödinger equation for the SHO can be solved analytically. The solution is quite complicated and will not be given here. On the other hand, the answer for the energy levels is remarkably simple: The allowed energies turn out to be

$$E = \frac{1}{2}\hbar\sqrt{\frac{k}{m}}, \quad \frac{3}{2}\hbar\sqrt{\frac{k}{m}}, \quad \frac{5}{2}\hbar\sqrt{\frac{k}{m}}, \quad \dots$$

The quantity $\sqrt{k/m}$ is the frequency ω_c, defined in (8.91), of a classical oscillator with the same force constant and mass. It is traditional to rewrite the allowed energies in terms of ω_c as

$$E_n = \left(n + \frac{1}{2}\right)\hbar\omega_c, \qquad n = 0, 1, 2, \dots \tag{8.96}$$

As anticipated, the allowed energies are all quantized, and there are levels with arbitrarily large energies. A remarkable feature of (8.96) is that the energy levels of the harmonic oscillator are all equally spaced. This is a property that we could not have anticipated in our qualitative discussion, but which emerges from the exact solution.

The allowed energies of the harmonic oscillator are shown in Fig. 8.21, which also shows the corresponding wave functions (each drawn on the line representing its energy). Notice the similarity of these wave functions to those of

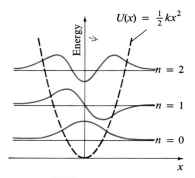

FIGURE 8.21 The first three energy levels and wave functions of the simple harmonic oscillator.

the finite well shown in Fig. 8.19. In particular, just as with the finite well, the lowest wave function has no nodes, the next has one node, and so on. Notice also that the wave functions with higher energy spread out further from $x = 0$, just as the classical turning points $x = \pm a$ move farther out when E increases. Finally, note that the ground state ($n = 0$) has a "zero-point" energy (equal to $\frac{1}{2}\hbar\omega_c$) as required by the uncertainty principle.

The wave functions shown in Fig. 8.21 were plotted using the known analytic solutions, which we list in Table 8.1, where, for convenience, we have introduced the parameter

$$b = \sqrt{\frac{\hbar}{m\omega_c}}. \tag{8.97}$$

(This parameter has the dimensions of a length and is, in fact, the half-width of the SHO well at the ground-state energy — see Problem 8.44.) The factors A_0, A_1, and A_2 in the table are normalization constants*.

TABLE 8.1

The energies and wave functions of the first three levels of a quantum harmonic oscillator. The length b is defined as $\sqrt{\hbar/m\omega_c}$.

n	E_n	$\psi(x)$
0	$\frac{1}{2}\hbar\omega_c$	$A_0 e^{-x^2/2b^2}$
1	$\frac{3}{2}\hbar\omega_c$	$A_1 \dfrac{x}{b} e^{-x^2/2b^2}$
2	$\frac{5}{2}\hbar\omega_c$	$A_2 \left(1 - 2\dfrac{x^2}{b^2}\right) e^{-x^2/2b^2}$

EXAMPLE 8.4   Verify that the $n = 1$ wave function given in Table 8.1 is a solution of the Schrödinger equation with $E = \frac{3}{2}\hbar\omega_c$. (For the cases $n = 0$ and $n = 2$, see Problems 8.45 and 8.48.)

The potential energy is $U = \frac{1}{2}kx^2$ and the Schrödinger equation is therefore

$$\psi'' = \frac{2m}{\hbar^2}\left(\frac{1}{2}kx^2 - E\right)\psi. \tag{8.98}$$

Differentiating the wave function ψ_1 of Table 8.1, we find that

$$\psi_1' = A_1\left(\frac{1}{b} - \frac{x^2}{b^3}\right)e^{-x^2/2b^2}$$

and

$$\psi_1'' = A_1\left(-\frac{3x}{b^3} + \frac{x^3}{b^5}\right)e^{-x^2/2b^2} = \left(\frac{x^2}{b^4} - \frac{3}{b^2}\right)A_1\frac{x}{b}e^{-x^2/2b^2},$$

where the last three factors together are just ψ_1. Using (8.97) to replace b, we find

$$\psi_1'' = \left(\frac{m^2\omega_c^2 x^2}{\hbar^2} - \frac{3m\omega_c}{\hbar}\right)\psi_1.$$

Replacing ω_c^2 by k/m in the first term, we get

$$\psi_1'' = \frac{2m}{\hbar^2}\left(\frac{1}{2}kx^2 - \frac{3}{2}\hbar\omega_c\right)\psi_1,$$

which is precisely the Schrödinger equation (8.98), with $E = \frac{3}{2}\hbar\omega_c$, as required.

---

* The three functions in Table 8.1 illustrate, what is true for all n, that the wave function of the nth level has the form $P_n(x)\exp(-x^2/2b^2)$, where $P_n(x)$ is a certain polynomial of degree n, called a Hermite polynomial.

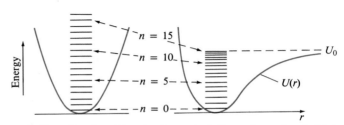

FIGURE 8.22 Vibrational levels of a typical diatomic molecule (right), and the SHO that approximates the molecule for small displacements (left). The first five or so levels correspond very closely; the higher levels of the molecule are somewhat closer together, and the molecule has no levels above the energy U_0.

The properties of the quantum SHO are nicely illustrated by the example of a diatomic molecule, such as HCl. A diatomic molecule is, of course, a complicated, three-dimensional system. However, the energy of the molecule can be expressed as the sum of three terms: an electronic term, corresponding to the motions of the individual electrons; a rotational term, corresponding to rotation of the whole molecule; and a vibrational term, corresponding to the in-and-out, radial vibrations of the two atoms. Careful analysis of molecular spectra lets one disentangle the possible values of these three terms. In particular, the vibrational motion of the two atoms is one-dimensional (along the line joining the atoms) and the potential energy approximates the SHO potential. Therefore, the allowed values of the vibrational energy should approximate those of an SHO, and this is amply borne out by observation, as illustrated in Fig. 8.22.

The molecular potential energy $U(r)$ in Fig. 8.22 deviates from the parabolic shape of the SHO at higher energies. Therefore, we would expect the higher energy levels to depart from the uniform spacing of the SHO. This, too, is what is observed.

The observation of photons emitted when molecules make transitions between different vibrational levels is an important source of information on interatomic forces, as we discuss further in Chapter 16. These same photons can also be used to identify the molecule that emitted them. For example, the H_2 molecule emits infrared photons of frequency 1.2×10^{14} Hz when it drops from one vibrational level to the next. This radiation is used by astronomers to locate clouds of H_2 molecules in our galaxy.

IDEAS YOU SHOULD NOW UNDERSTAND FROM CHAPTER 8

Standing waves
Time dependence of quantum standing waves
Particle in a rigid box:
    boundary conditions
    allowed energies
The Schrödinger equation
Normalization condition

Properties of wave functions:
    in the classically allowed regions ($E > U$)
    in the classically forbidden regions ($U > E$)
Allowed energies and well-behaved wave functions
[Optional section: the SHO has potential energy $\frac{1}{2}kx^2$ and allowed energies $(n + \frac{1}{2})\hbar\omega_c$, where $n = 0$, 1, 2, etc.]

8.1 • A string is oscillating with wave function

$$y(x, t) = A \sin kx \cos \omega t \qquad (8.99)$$

with $A = 3$ cm, $k = 0.2\pi$ rad/cm, and $\omega = 10\pi$ rad/s. For each of the times $t = 0, 0.05,$ and 0.07 s, find the displacement at $x = 0, 1, \ldots, 10$ cm. Sketch the string at each of these times.

8.2 • A standing wave of the form (8.99) has amplitude 2 m, wavelength 10 m, and period $T = 2$ s. Write down the expression for $y(x, t)$.

8.3 • A string is clamped at the points $x = 0$ and $x = 30$ cm. It is oscillating sinusoidally with amplitude 2 cm, wavelength 20 cm, and frequency $f = 40$ Hz. Write an expression for its wave function $y(x, t)$.

8.4 • For the standing wave of Eq. (8.99) calculate the string's transverse velocity at any fixed position x. [That is, differentiate (8.99) with respect to t, treating x as a constant.] At certain times the string's displacement is zero for all x, and an instantaneous snapshot of the string would show no wave at all. Sketch the string at such a moment and indicate, with several small arrows, the velocity of several points on the string at that moment.

8.5 • In Fig. 8.23 are sketched three successive snapshots of a standing wave on a string. The first (solid) curve shows the string at maximum displacement. Were these snapshots taken at equally spaced times? If not, make a sketch in which they were. (Use the same first and third curves.)

FIGURE **8.23**   (Problems 8.5 and 8.6)

8.6 • Figure 8.23 shows a standing wave on a string. In any position other than the dotted one the string has potential energy because it is stretched (compared to the straight, dotted position). Let U_0 denote this potential energy at the position of the solid curve (which shows the maximum displacement). What is the potential energy when the string reaches the position of the dotted line? What is the kinetic energy at this position?

8.7 • Consider a standing wave on a string, clamped at the points $x = 0$ and $x = 40$ cm. It is oscillating with amplitude 3 cm and wavelength 80 cm, and its maxi-

mum transverse velocity is 60 cm/s. Write an expression for its displacement $y(x, t)$ as a function of x and t.

8.8 •• A string is oscillating with wave function of the form (8.99) but with $A = 2.5$ cm, $k = 1$ rad/cm, and $\omega = 10\pi$ rad/s. (a) Sketch two complete wavelengths of the wave at each of the times $t = 0.05$ s and $t = 0.1$ s. (b) For a fixed value of x, differentiate (8.99) with respect to t to give the string's transverse velocity at any position x. Graph this velocity as a function of x for each of the two times in part (a).

8.9 • Prove that any complex number $z = x + iy$ (with x and y real) can be written as $z = re^{i\theta}$ (with r and θ real). Give expressions for x and y in terms of r and θ, and vice versa. [*Hint:* Use Euler's formula, (8.10).]

8.10 • Consider a complex function of time, $z = re^{-i\omega t}$, where r is a real constant. (a) Write z in terms of its real and imaginary parts, x, and y, and show that they oscillate sinusoidally and 90° out of phase. (b) Show that $|z| = \sqrt{x^2 + y^2}$ is constant.

8.11 •• We claimed in connection with Eq. (8.7) that the general (real) sinusoidal wave has time dependence

$$a \cos \omega t + b \sin \omega t. \qquad (8.100)$$

Another way to say this is that the general sinusoidal time dependence is

$$A \sin(\omega t + \phi). \qquad (8.101)$$

(a) Show that these two forms are equivalent; that is, prove that the function (8.100) can be expressed in the form (8.101), and vice versa. Give expressions for a and b in terms of A and ϕ, and vice versa. (Remember the trig identities in Appendix B.) (b) Show that by changing the origin of time (that is, rewriting everything in terms of $t' = t + $ constant, with a suitably chosen constant) you can eliminate the constant ϕ from (8.101) [that is, rewrite (8.101) as $A \sin \omega t'$.]

8.12 •• The function e^z is defined for any z, real or complex, by its *power series*

$$e^z = 1 + z + \frac{z^2}{2!} + \frac{z^3}{3!} + \cdots.$$

Write down this series for the case that z is pure imaginary, $z = i\theta$. Note that the terms in this series are alternately real and imaginary. Group together all the real terms and all the imaginary terms and prove the important identity, called Euler's formula,

$$e^{i\theta} = \cos \theta + i \sin \theta. \qquad (8.102)$$

(*Hint:* You will need to remember the power series for cos θ and sin θ given in Appendix B. You may wonder whether it is legitimate to regroup the terms of an infinite series as recommended here; for power series like those in this problem, it is.)

8.13 • • Use the result (8.102) to prove that

$$\cos\theta = \frac{e^{i\theta} + e^{-i\theta}}{2} \quad \text{and} \quad \sin\theta = \frac{e^{i\theta} - e^{-i\theta}}{2i}.$$

SECTION 8.4 (THE PARTICLE IN A RIGID BOX)

8.14 • Find the lowest three energies, in eV, for an electron in a one-dimensional box of length $a = 0.2$ nm (about the size of an atom).

8.15 • Find the lowest three energies, in MeV, of a proton in a one-dimensional rigid box of length $a = 5$ fm (a typical nuclear size).

8.16 • What is the spacing, in eV, between the lowest two levels of an electron confined in a one-dimensional wire of length 1 cm?

8.17 • Sketch the energy levels and wave functions for the levels $n = 5, 6, 7$ for a particle in a one-dimensional rigid box. (See Fig. 8.5.)

8.18 • For the ground state of a particle in a rigid box we have seen that the momentum has a definite magnitude $\hbar k$, but is equally likely to be in either direction. This means that the uncertainty in p is $\Delta p \approx \hbar k$. The uncertainty in position is $\Delta x \approx a/2$. Verify that these uncertainties satisfy the Heisenberg uncertainty principle (7.34).

SECTION 8.6 (THE RIGID BOX AGAIN)

8.19 • Prove that the function $\psi = Ae^{\alpha x} + Be^{-\alpha x}$ satisfies the equation $\psi'' = \alpha^2\psi$ for any two constants A and B.

8.20 • Prove that the function $\psi = Ae^{ikx} + Be^{-ikx}$ satisfies the equation $\psi'' = -k^2\psi$ for any two constants A and B.

8.21 • • We have seen that second-order differential equations like the Schrödinger equation have two independent solutions. Consider the *fourth*-order equation $d^4\psi/dx^4 = \beta^4\psi$, where β is a positive constant. Prove that each of the four functions $e^{\pm\beta x}$ and $e^{\pm i\beta x}$ is a solution. (It is a fact—though harder to prove—that *any* solution of the equation can be expressed as a linear combination of these four solutions.)

8.22 • • A second-order differential equation like the Schrödinger equation has two independent solutions $\psi_1(x)$ and $\psi_2(x)$. These two solutions can be chosen in many ways, but once they are chosen *any* solution can be expressed as a linear combination $A\psi_1(x) + B\psi_2(x)$ (where A and B are constants, real

or complex). (*a*) To illustrate this property, consider the differential equation $\psi'' = -k^2\psi$ where k is a constant. Prove that each of the three functions sin kx, cos kx, and e^{ikx} is a solution. (*b*) Show that each can be expressed as a combination of the other two.

8.23 • • Many physical problems lead to a differential equation of the form

$$a\psi''(x) + b\psi'(x) + c\psi(x) = 0,$$

where a, b, c are constants. [An example was given in (8.30).] (*a*) Prove that this equation has two solutions of the form $\psi(x) = e^{\gamma x}$, where γ is either solution of the quadratic equation $a\gamma^2 + b\gamma + c = 0$. (*b*) Prove that any linear combination of these two solutions is itself a solution.

8.24 • • Consider the second-order differential equation

$$f(x)\psi''(x) + g(x)\psi'(x) + h(x)\psi(x) = 0,$$

where $f, g,$ and h are known functions of x. Prove that if $\psi_1(x)$ and $\psi_2(x)$ are both solutions of this equation, the linear combination $A\psi_1(x) + B\psi_2(x)$ is also a solution for any two constants A and B.

8.25 • • Show that the integral which appears in the normalization condition for a particle in a rigid box has the value

$$\int_0^a \sin^2\left(\frac{n\pi x}{a}\right) dx = \frac{a}{2}.$$

(*Hint:* Use the identity for $\sin^2\theta$ in terms of cos 2θ given in Appendix B.)

8.26 • • (*a*) Write down and sketch the probability distribution $|\psi(x)|^2$ for the second excited state ($n = 3$) of a particle in a rigid box of length a. (*b*) What are the most probable positions, x_{mp}? (*c*) What are the probabilities of finding the particle in the intervals $[0.50a, 0.51a]$ and $[0.75a, 0.76a]$?

8.27 • • Answer the same questions as in Problem 8.26 but for the third excited state ($n = 4$).

8.28 • • If a particle has wave function $\psi(x)$, the probability of finding the particle between any two points b and c is

$$P(b \leq x \leq c) = \int_b^c |\psi(x)|^2 \, dx. \quad (8.103)$$

For a particle in the ground state of a rigid box calculate the probability of finding it between $x = 0$ and $x = a/3$ (where a is the width of the box). Use the hint in Problem 8.25.

8.29 • • Evaluate the integral (8.65) to give the average result x_{av} found when the position of a particle in the ground state of a rigid box is measured many times.

[*Hint:* Rewrite $\sin^2(\pi x/a)$ in terms of $\cos(2\pi x/a)$ and use integration by parts.]

8.30 •• Do Problem 8.29 for the case of the second excited state ($n = 3$) of a rigid box.

8.31 ••• Consider a particle in the ground state of a rigid box of length a. (*a*) Evaluate the integral (8.103) to give the probability of finding the particle between $x = 0$ and $x = c$ for any $c \leq a$. (*b*) What does your result give when $c = a$? Explain. (*c*) What if $c = a/2$? (*d*) What if $c = a/4$? (*e*) The answer to part (*c*) is half that for part (*b*), whereas that to part (*d*) is *not* half that for part (*c*). Explain.

SECTION 8.8 (THE NONRIGID BOX)

8.32 • Consider the potential-energy function

$$U(x) = \tfrac{1}{2}kx^2.$$

(*a*) Sketch U as a function of x. (*b*) For a classical particle of energy E, find the turning points in terms of E and k. (*c*) If we double the energy, what happens to the length of the classically allowed region?

8.33 •• Consider the potential-energy function

$$U(x) = U_0(1 - e^{-x^2/a^2}),$$

where U_0 and a are positive constants. (*a*) Sketch $U(x)$. (*b*) For $0 < E < U_0$, find the classical turning points (in terms of U_0, E, and a).

8.34 •• Give an argument that a particle moving in either of the finite wells of Figs. 8.8(b) and (c) can have no states with $E < 0$. [*Hint:* Remember that whenever $U(x) - E$ is positive, $\psi(x)$ must bend away from the axis; show that a wave function that is well behaved as $x \to -\infty$ (that is, behaves like $e^{\alpha x}$) necessarily blows up as $x \to \infty$.]

8.35 •• Make sketches of the probability density $|\psi(x)|^2$ for the first four energy levels of a particle in a nonrigid box. Draw next to each graph the corresponding graph for the rigid box.

8.36 •• Consider the potential well shown in Fig. 8.24(a). Sketch the wave function for the eighth excited level of this well, assuming that it has energy $E > U_0$.

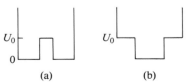

(a)   (b)

FIGURE 8.24 (Problems 8.36 to 8.39)

(*Hint:* The wave function has eight nodes — not counting the nodes at the walls of the well.)

8.37 •• Do the same task as in Problem 8.36, but for the well in Fig. 8.24(b).

8.38 •• Sketch the wave function for the fourth excited level of the well of Fig. 8.24(a), assuming that it has energy in the interval $0 < E < U_0$.

8.39 •• Do Problem 8.38 but for the well of Fig. 8.24(b).

8.40 ••• You do not have to be able to carry out the numerical solution of differential equations yourself to understand the method. Nevertheless, a good way to make sure you *do* understand is to do a simple example like the following: Consider the differential equation

$$\psi''(x) = -\psi(x) \qquad (8.104)$$

(*a*) What is the analytic form of the general solution to this equation? Find the solution $\psi(x)$ that satisfies

$$\psi(0) = 1 \quad \text{and} \quad \psi'(0) = 0; \quad (8.105)$$

what is its value at $x = 0.4$? (*b*) Now imagine trying to find $\psi(x)$ by the numerical technique described in Section 8.8, starting from the given values (8.105). What would you get for $\psi(x)$ at $x = 0.4$ if you used just one step [that is, took $\Delta x = 0.4$ in (8.83)]? What if you used two steps ($\Delta x = 0.2$) and three steps ($\Delta x = 0.133$)? Compare these approximate answers with the answer found analytically in part (*a*). (*c*) If you have access to a programmable calculator or computer, write a program to find $\psi(x)$ using N steps. What does this give for $\psi(0.4)$ with $N = 5$, 10, 50, 100? Compare with the analytic answer.

8.41 ••• Answer the same questions as in Problem 8.40 but with Equation (8.104) replaced by $\psi''(x) = \psi(x)$ and (8.105) replaced by $\psi(0) = \psi'(0) = 1$.

8.42 ••• Answer the same questions as in Problem 8.40 with the same differential equation (8.104) but with the starting conditions (8.105) replaced by $\psi(0) = \psi'(0) = 1$.

8.43 ••• The differential equation $\psi'' - 3\psi' + 2\psi = 0$ has the solutions $\psi_1 = e^{2x}$ and $\psi_2 = e^x$. Verify this claim and then answer the same questions as in Problem 8.40 for this equation, using the same starting conditions (8.105).

SECTION 8.9 (THE SIMPLE HARMONIC OSCILLATOR)

8.44 • Show that the length parameter b defined for the SHO in Eq. (8.97) is equal to the value of x at the classical turning point for a particle with the energy of the quantum ground state.

8.45 • Verify that the $n = 0$ wave function for the SHO, given in Table 8.1, satisfies the Schrödinger equation with $E = \tfrac{1}{2}\hbar\omega_c$. (The potential energy function is $U = \tfrac{1}{2}kx^2$.)

8.46 • The wave function $\psi_0(x)$ for the ground state of a harmonic oscillator is given in Table 8.1. Show that its normalization constant A_0 is

$$A_0 = (\pi b^2)^{-1/4}. \qquad (8.106)$$

You will need to know the integral $\int_{-\infty}^{\infty} e^{-\lambda x^2}\, dx$, which can be found from Appendix B.

8.47 • Find the value of the normalization constant A_1 for the wave function of the first excited state of the SHO. Give your answer in terms of h, m, and ω_c. (The wave function is given in Table 8.1. You will need to know the integral $\int_{-\infty}^{\infty} x^2 e^{-\lambda x^2}\, dx$, which is given in Appendix B.)

8.48 • • Verify that the $n = 2$ wave function for the SHO, given in Table 8.1, satisfies the Schrödinger equation with energy $\frac{5}{2}\hbar\omega$.

8.49 • • • The wave functions of the harmonic oscillator, like those of a particle in a finite well, are nonzero in the classically forbidden regions, outside the classical turning points. In this question you will find the probability that a quantum particle which is in the ground state of an SHO be found outside its classical turning points. The wave function for this state is in Table 8.1 and its normalization constant A_0 is given in Problem 8.46. (a) What are the turning points for a classical particle with the ground-state energy $\frac{1}{2}\hbar\omega_c$ in the SHO, $U = \frac{1}{2}kx^2$? Relate your answer to the constant b in Eq. (8.97). (b) For a quantum particle in the ground state, write down the integral which gives the total probability for finding the particle between the two classical turning points. The form of the required integral is given in Problem 8.28, Equation (8.103). To evaluate it, change variables until you get an integral of the form $\int_{-1}^{1} e^{-y^2}\, dy$; this is a standard integral of mathematical physics (called the error function) with the known value 1.49. What is the probability of finding the particle between the classical turning points? (c) What is the probability of finding it *outside* the classical turning points?

9

The Three-Dimensional Schrödinger Equation

9.1   Introduction

In Chapter 8 we studied the one-dimensional Schrödinger equation and saw how it determines the allowed energies and corresponding wave functions of a particle in one dimension. If the world in which we lived were one-dimensional, we could now proceed to apply these ideas to various real systems: atoms, molecules, nuclei, and so on. However, our world is *three*-dimensional, and we must first describe how the one-dimensional equation is generalized to three dimensions.

We shall find that the three-dimensional equation is appreciably more complicated than its one-dimensional counterpart, involving derivatives with respect to all three coordinates x, y, and z. Nevertheless, its most important properties will be familiar. Specifically, in three dimensions, just as in one dimension (and two), the time-independent Schrödinger equation is a differential equation for the wave function ψ. For most systems this equation has acceptable solutions only for certain particular values of the energy E. Those E for which it has an acceptable solution are the allowed energies of the system, and the solutions ψ are the corresponding wave functions.

In this chapter we write down the three-dimensional Schrödinger equation and describe its solutions for some simple systems, culminating with the hydrogen atom. Since many of the important features of the three-dimensional equation are already present in the simpler case of two dimensions, two of our examples will be two-dimensional.

9.2 The Three-Dimensional Schrödinger Equation and Partial Derivatives

In one dimension the wave function $\psi(x)$ for a particle depends on the one coordinate x, and the time-independent Schrödinger equation has the now familiar form

$$\frac{d^2\psi}{dx^2} = \frac{2M}{\hbar^2}[U - E]\psi \tag{9.1}$$

where U is the particle's potential energy and we temporarily use capital M for the particle's mass.* In this equation, remember that both ψ and U are functions of x, whereas E, although it can take on various values, does not depend on x. In three dimensions we would naturally expect the wave function ψ to depend on all three coordinates x, y, and z; that is,

$$\psi = \psi(x, y, z) = \psi(\mathbf{r})$$

where

$$\mathbf{r} = (x, y, z).$$

Similarly the potential energy U will normally depend on x, y, and z:

$$U = U(x, y, z) = U(\mathbf{r}).$$

How the differential equation (9.1) generalizes to three dimensions is not so obvious. Here we shall simply state that the correct generalization of (9.1) is this:

$$\frac{\partial^2\psi}{\partial x^2} + \frac{\partial^2\psi}{\partial y^2} + \frac{\partial^2\psi}{\partial z^2} = \frac{2M}{\hbar^2}[U - E]\psi, \tag{9.2}$$

where the three derivatives on the left are the so-called *partial derivatives,* whose definition and properties we discuss in a moment. That (9.2) is a possible generalization of (9.1) is perhaps fairly obvious. That it is the *correct* generalization is certainly not obvious. In more advanced texts you will find various arguments that *suggest* the correctness of (9.2). However, the ultimate test of any equation is whether its predictions agree with experiment, and we shall see

* In the context of the three-dimensional Schrödinger equation, the letter m is traditionally used for the integer that labels the allowed values of the components of angular momentum. It is to avoid confusion with this notation that we use M for the mass in the first six sections of this chapter. In Section 9.7 we return to the hydrogen atom, in which the relevant particle is the electron, whose mass we shall call m_e.

that the three-dimensional Schrödinger equation in the form (9.2) has passed this test repeatedly when applied to atomic and subatomic systems.

The three derivatives that appear in (9.2) are called **partial derivatives,** and it is important that you understand what these are. An *ordinary* derivative, like $d\psi/dx$, is defined for a function such as $\psi(x)$ that depends on just one variable (for example, the temperature as a function of position x along a narrow rod). Partial derivatives arise when one considers functions of two or more variables, such as the temperature as a function of position (x, y, z) in a three-dimensional room. If ψ depends on three variables x, y, z, we define the partial derivative $\partial\psi/\partial x$ as the derivative of ψ with respect to x, *obtained when we hold y and z fixed.* Similarly, $\partial\psi/\partial y$ is the derivative with respect to y, when x and z are held constant; and similarly with $\partial\psi/\partial z$. Notice that it is customary to use the symbol ∂ for this new kind of derivative.

The calculation of partial derivatives is very simple in practice, as the following example shows.

EXAMPLE 9.1 Find the three partial derivatives $\partial\psi/\partial x$, $\partial\psi/\partial y$, and $\partial\psi/\partial z$ for $\psi(x, y, z) = x^2 + 2y^3z + z$.

If y and z are held constant, then ψ has the form

$$\psi = x^2 + \text{constant},$$

and the rules of ordinary differentiation give

$$\frac{\partial\psi}{\partial x} = 2x.$$

If instead we hold x and z constant, then ψ has the form

$$\psi = \text{constant} + 2zy^3.$$

Since the coefficient $2z$ is constant, the rules of ordinary differentiation give

$$\frac{\partial\psi}{\partial y} = 6zy^2.$$

Finally, if x and y are constant, then

$$\psi = \text{constant} + 2y^3z + z,$$

and since $2y^3$ is constant,

$$\frac{\partial\psi}{\partial z} = 2y^3 + 1.$$

Higher partial derivatives are defined similarly. For example, $\partial^2\psi/\partial x^2$ is the second derivative of ψ with respect to x, obtained if we hold y and z constant, and so on. Partial derivatives are really no harder to use than ordinary derivatives, once you understand their definition. If you have never worked with them before, you might find it useful to try some of Problems 9.1 to 9.5 right away.

An equation, like the Schrödinger equation (9.2), that involves partial derivatives is called a *partial differential equation.* Now that we know what partial derivatives are, we can discuss methods for solving such equations, and this is what we shall do in the remainder of this chapter.

9.3 The Two-Dimensional Square Box

Before we consider any three-dimensional systems we consider an example in two dimensions. This shares several important features of three-dimensional systems, but is naturally somewhat simpler.

Looking at the Schrödinger equations (9.1) and (9.2) in one and three dimensions, it is easy to guess, correctly, that the Schrödinger equation for a particle of mass M in two dimensions should read:

$$\frac{\partial^2 \psi}{\partial x^2} + \frac{\partial^2 \psi}{\partial y^2} = \frac{2M}{\hbar^2}[U - E]\psi. \tag{9.3}$$

Here ψ is a function of the two-dimensional coordinates, $\psi = \psi(\mathbf{r}) = \psi(x, y)$, and $U = U(\mathbf{r})$ is the particle's potential energy.

The method for solving the Schrödinger equation depends on the potential-energy function $U(\mathbf{r})$. In this section we consider a particle confined in a two-dimensional, rigid square box; that is, a particle for which $U(\mathbf{r})$ is zero inside a square region like that shown in Fig. 9.1, but is infinite outside:

$$U(x, y) = \begin{cases} 0, & \text{when } 0 \le x \le a \quad \text{and} \quad 0 \le y \le a \\ \infty, & \text{otherwise.} \end{cases} \tag{9.4}$$

A classical example of such a system would be a metal puck sliding on a frictionless, square air table with perfectly rigid, elastic bumpers at its edges. As a quantum example we could imagine an electron confined inside a thin square metal sheet.

A classical particle inside a square, rigid box would bounce indefinitely inside the box. Since $U = 0$ inside the box, its energy E would be all kinetic and could have any value in the range

$$0 \le E < \infty.$$

To find the possible energies for the corresponding quantum system we must solve the Schrödinger equation (9.3) with the potential-energy function (9.4).

Since the particle cannot escape from the box, the wave function $\psi(x, y)$ is zero outside the box, and since $\psi(x, y)$ must be continuous, it must also be zero on the boundary:

$$\psi(x, y) = 0, \quad \text{if } x = 0 \text{ or } a, \quad \text{or} \quad y = 0 \text{ or } a. \tag{9.5}$$

Since $U(x, y) = 0$ inside the box, the Schrödinger equation reduces to

$$\frac{\partial^2 \psi}{\partial x^2} + \frac{\partial^2 \psi}{\partial y^2} = -\frac{2ME}{\hbar^2}\psi \tag{9.6}$$

for all x and y inside the box. We must solve this equation, subject to the boundary conditions (9.5).

SEPARATION OF VARIABLES

If we knew nothing at all about partial differential equations, the solution of (9.6) would be a formidable prospect. Fortunately, there is an extensive mathematical theory of partial differential equations, which tells us that equa-

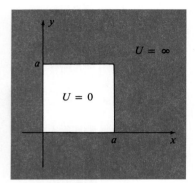

FIGURE 9.1 The two-dimensional, rigid square box. The particle is confined by perfectly rigid walls to the unshaded, square region, within which it moves freely.

tions like (9.6) can be solved by a method called **separation of variables.** In this method one seeks solutions with the form

$$\psi(x, y) = X(x) Y(y) \qquad (9.7)$$

where $X(x)$ is a function of x alone and $Y(y)$ a function of y alone. We describe a function with the form (9.7) as a *separated function.** It is certainly not obvious that there will be any solutions with this separated form. On the other hand, there is nothing to stop us from seeing if there are such solutions, and we shall find that indeed there are. Furthermore, the mathematical theory of equations like (9.6) guarantees that *any* solution of the equation can be expressed as a sum of separated solutions. This means that once we have found all of the solutions with the form (9.7), we have, in effect, found *all* solutions. It is this mathematical theorem that is the ultimate justification for using separation of variables to solve the Schrödinger equation for many two- and three-dimensional systems.

To see whether the Schrödinger equation (9.6) does have separated solutions with the form (9.7), we substitute (9.7) into (9.6). When we do this two simplifications occur. First, the partial derivatives simplify. Consider, for example,

$$\frac{\partial^2 \psi}{\partial x^2} = \frac{\partial^2}{\partial x^2} [X(x) Y(y)].$$

This derivative is evaluated by treating y as fixed. Therefore, the term $Y(y)$ can be brought outside, to give

$$\frac{\partial^2 \psi}{\partial x^2} = Y(y) \frac{\partial^2}{\partial x^2} X(x).$$

Since $X(x)$ depends only on x, the remaining derivative is an ordinary derivative, which we can write as

$$\frac{\partial^2}{\partial x^2} X(x) = \frac{d^2}{dx^2} X(x) = X''(x)$$

where, as usual, the double prime indicates the second derivative of the function concerned. Thus

$$\frac{\partial^2 \psi}{\partial x^2} = Y(y) X''(x). \qquad (9.8)$$

Similarly,

$$\frac{\partial^2 \psi}{\partial y^2} = X(x) Y''(y). \qquad (9.9)$$

If we substitute (9.8) and (9.9) into the Schrödinger equation (9.6) we find that

$$Y(y) X''(x) + X(x) Y''(y) = -\frac{2ME}{\hbar^2} X(x) Y(y).$$

To separate the terms that depend on x from those that depend on y, we divide by $X(x) Y(y)$ to give

---

* Note that by no means can every function of x and y be separated in this way. As simple a function as $x + y$ cannot be expressed as the product of one function of x and one of y.

$$\frac{X''(x)}{X(x)} + \frac{Y''(y)}{Y(y)} = -\frac{2ME}{\hbar^2}. \tag{9.10}$$

The right side of this equation is constant (independent of x and y). Thus (9.10) has the general form

(function of x) + (function of y) = constant

for all x and y (in the box). To see what this implies, we move the function of y over to the right:

(function of x) = constant − (function of y).

This equation asserts that a certain function of x is equal to a quantity that does not depend on x at all. In other words, this function, which can only depend on x, is in fact independent of x. This is possible only if the function in question is a constant. We conclude that the quantity $X''(x)/X(x)$ in (9.10) is a constant:

$$\frac{X''(x)}{X(x)} = \text{constant}. \tag{9.11}$$

If we call this constant $-k_x^2$, then (9.11) can be rewritten as

$$X''(x) = -k_x^2 X(x). \tag{9.12}$$

This equation has exactly the form of the Schrödinger equation (8.51) for a particle in a one-dimensional rigid box,

$$\psi''(x) = -k^2 \psi(x), \tag{9.13}$$

whose solutions we have already discussed in Chapter 8. In particular, we saw there that this equation has acceptable solutions only when the constant on the right is negative, which is why we called the constant in (9.11) $-k_x^2$.

An exactly parallel argument shows that the quantity $Y''(y)/Y(y)$ in (9.10) has to be independent of y; that is,

$$\frac{Y''(y)}{Y(y)} = \text{constant}, \tag{9.14}$$

or, if we call this second constant $-k_y^2$,

$$Y''(y) = -k_y^2 Y(y). \tag{9.15}$$

We see that the method of separation of variables has let us replace the partial differential equation (9.6), involving the two variables x and y, by two ordinary differential equations (9.12) and (9.15), one of which involves only the variable x, and the other only y.

Before we seek the acceptable solutions of these two equations, we must return to the boundary condition that $\psi(x, y)$ is zero at the edges of our box ($x = 0$ or a, and $y = 0$ or a). Since $\psi(x, y) = X(x) Y(y)$ this requires that

$$X(x) = 0 \qquad \text{when } x = 0 \text{ or } a \tag{9.16}$$

and

$$Y(y) = 0 \qquad \text{when } y = 0 \text{ or } a. \tag{9.17}$$

Now, the differential equation (9.12) and boundary conditions (9.16) for $X(x)$ are exactly the equation and boundary conditions for a particle in a one-dimensional rigid box. And we already know the solutions for that problem: The wave function must have the form

$$X(x) = B \sin k_x x$$

where B is a constant. This satisfies the boundary conditions only if k_x is an integral multiple of π/a,

$$k_x = \frac{n_x \pi}{a}, \tag{9.18}$$

where n_x is any positive integer

$$n_x = 1, 2, 3, \ldots$$

Therefore,

$$X(x) = B \sin k_x x = B \sin \frac{n_x \pi x}{a}. \tag{9.19}$$

The equation and boundary conditions for $Y(y)$ are also the same as those for a one-dimensional rigid box, and there are acceptable solutions only if

$$k_y = \frac{n_y \pi}{a} \tag{9.20}$$

(with n_y any positive integer), in which case

$$Y(y) = C \sin k_y y = C \sin \frac{n_y \pi y}{a}. \tag{9.21}$$

Combining (9.19) and (9.21), we find for the complete wave function

$$\psi(x, y) = X(x)Y(y) = BC \sin k_x x \sin k_y y \tag{9.22}$$

$$= A \sin \frac{n_x \pi x}{a} \sin \frac{n_y \pi y}{a} \tag{9.23}$$

where n_x and n_y are any two positive integers.* In writing (9.23) we have renamed the constant BC as A; the value of this constant is fixed by the requirement that the integral of $|\psi|^2$ over the whole box must be 1.

Using the form (9.22) we can see the physical significance of the separation constants k_x and k_y. If we fix y and move in the x direction, then ψ varies sinusoidally in x, with wavelength $\lambda = 2\pi/k_x$. According to de Broglie, this means that the particle has momentum in the x direction of magnitude $h/\lambda = \hbar k_x$. Since a similar argument can be applied in the y direction, we conclude that

$$|p_x| = \hbar k_x \quad \text{and} \quad |p_y| = \hbar k_y. \tag{9.24}$$

By analogy with the one-dimensional wave number k, satisfying $p = \hbar k$, we can call k_x and k_y by the components of a *wave vector*. Note, however, that since

$$\sin k_x x = \frac{e^{ik_x x} - e^{-ik_x x}}{2i},$$

the wave function (9.22) is a superposition of states with $p_x = \pm \hbar k_x$, and similarly with $p_y = \pm \hbar k_y$. This is the same situation that we encountered in

---

* By labeling these two integers n_x and n_y we do not wish to imply that there is necessarily a vector **n** of which n_x and n_y are the components. For the moment, n_x and n_y are simply two integers, one of which characterizes the function $X(x)$ and the other $Y(y)$.

Section 8.4 for the one-dimensional rigid box and explains the absolute value signs in Eq. (9.24).

ALLOWED ENERGIES

In solving for the wave function (9.22) we have temporarily lost sight of the energy E. In fact, the last place that E appeared was in (9.10):

$$\frac{X''(x)}{X(x)} + \frac{Y''(y)}{Y(y)} = -\frac{2ME}{\hbar^2}. \tag{9.25}$$

Now, we know from (9.12) that X''/X is the constant $-k_x^2$, and from (9.18) that $k_x = n_x\pi/a$. Therefore, X''/X is equal to $-n_x^2\pi^2/a^2$. Inserting this, and the corresponding expression for Y''/Y, into (9.25), we obtain

$$-\frac{n_x^2\pi^2}{a^2} - \frac{n_y^2\pi^2}{a^2} = -\frac{2ME}{\hbar^2}.$$

Solving for E, we find that the allowed values of the energy are

$$E = E_{n_x,n_y} = \frac{\hbar^2\pi^2}{2Ma^2}(n_x^2 + n_y^2) \tag{9.26}$$

where n_x and n_y are any two positive integers. This energy is the sum of two terms, each of which has exactly the form of an allowed energy for the one-dimensional box; namely,

$$E = \frac{\hbar^2\pi^2}{2Ma^2}n^2 \qquad [n = 1, 2, 3, \ldots]. \tag{9.27}$$

If we adopt the notation

$$E_0 = \frac{\hbar^2\pi^2}{2Ma^2}, \tag{9.28}$$

where you can think of the subscript 0 as standing for one-dimensional, we can rewrite the allowed energies for a particle in a two-dimensional square box as

$$E = E_{n_x,n_y} = E_0(n_x^2 + n_y^2). \tag{9.29}$$

QUANTUM NUMBERS

Just like the one-dimensional box, the two-dimensional box has energy levels that are quantized. The main difference is that where the one-dimensional energy levels are characterized by a single integer n, the two-dimensional levels are given by two integers, n_x and n_y. We are going to find many more examples of quantities whose allowed values are characterized by integers (and sometimes half integers, such as $\frac{1}{2}$, $1\frac{1}{2}$, . . .). In general, any integer or half integer that gives the allowed values of some physical quantity is called a **quantum number**. With this terminology we can say that the energy levels of a particle in a two-dimensional square box are characterized by two quantum numbers, n_x and n_y.

The lowest possible energy for the two-dimensional box occurs when both quantum numbers are equal to 1,

$$n_x = n_y = 1,$$

and the corresponding, ground-state, energy is given by (9.29) as

$$E_{11} = 2E_0.$$

The first excited energy occurs when $n_x = 1$, $n_y = 2$, or vice versa:

$$E_{12} = E_{21} = 5E_0.$$

In Fig. 9.2 are sketched the lowest four energy levels for the square box.

DEGENERACY

An important new feature of the two-dimensional box is that there can be several different wave functions for which the particle has the same energy. For example, we saw that $E_{12} = E_{21} = 5E_0$. That is, the state with $n_x = 1$, $n_y = 2$ has the same energy as that with $n_x = 2$ and $n_y = 1$. The corresponding wave functions are

$$\psi_{12} = A \sin \frac{\pi x}{a} \sin \frac{2\pi y}{a} \quad \text{and} \quad \psi_{21} = A \sin \frac{2\pi x}{a} \sin \frac{\pi y}{a}. \quad (9.30)$$

Since these correspond to different probability densities $|\psi|^2$, they represent experimentally distinguishable states that happen to have the same energy.

In general, if there are N independent wave functions ($N > 1$) all with the same energy E, we say that the energy level E is **degenerate** and has **degeneracy** N (or is N-fold degenerate). If there is only one wave function with energy E, we say that the energy E is **nondegenerate** (or has degeneracy 1). Looking at Fig. 9.2, we see that the ground state and second excited state of the square box are nondegenerate, while the first and third excited states are both twofold degenerate. In general, most of the levels E_{11}, E_{22}, E_{33}, . . . are nondegenerate, while most of the levels E_{n_x,n_y} with $n_x \neq n_y$ are twofold degenerate, since $E_{n_x,n_y} = E_{n_y,n_x}$. A few of the levels have higher degeneracies; for example, since

$$5^2 + 5^2 = 1^2 + 7^2$$

it follows that $E_{55} = E_{17} = E_{71}$, and this level is threefold degenerate; since

$$1^2 + 8^2 = 4^2 + 7^2$$

it follows that $E_{18} = E_{81} = E_{47} = E_{74}$ and this level is fourfold degenerate.

In Chapter 11 we shall see that degeneracy has an important effect on the structure and chemical properties of atoms. Therefore, it is important not just

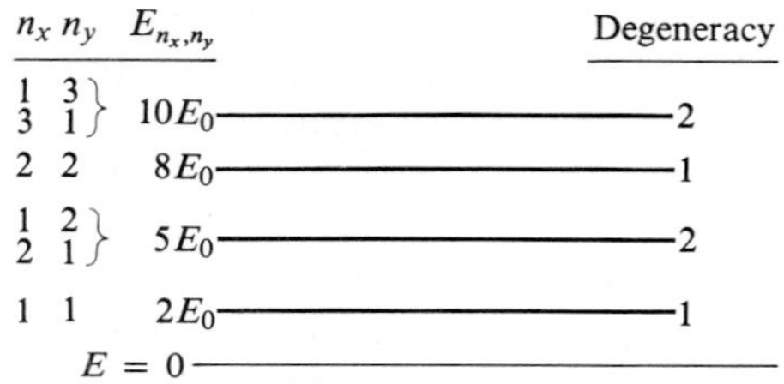

$n_x\ n_y$	E_{n_x,n_y}	Degeneracy
$\left.\begin{matrix}1\ 3\\3\ 1\end{matrix}\right\}$	$10E_0$	2
2 2	$8E_0$	1
$\left.\begin{matrix}1\ 2\\2\ 1\end{matrix}\right\}$	$5E_0$	2
1 1	$2E_0$	1
	$E = 0$	

FIGURE 9.2 The energy levels of a particle in a two-dimensional, square, rigid box. The lowest allowed energy is $2E_0$; the line at $E = 0$ is merely to show the zero of the energy scale. The degeneracies, listed on the right, refer to the number of independent wave functions with the same energy.

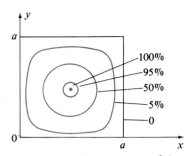

FIGURE 9.3 Contour map of the probability density $|\psi|^2$ for the ground state of the square box. The percentages shown give the value of $|\psi|^2$ as a percentage of its maximum value.

to find the energy levels of a quantum system, but to find the degeneracy of each level.

CONTOUR MAPS OF $|\psi|^2$

It is often important to know how the probability density $|\psi(x, y)|^2$ is distributed in space. Because $|\psi(x, y)|^2$ depends on two variables, it is harder to visualize than in the one-dimensional case. One method that is quite successful is to draw a contour map of $|\psi(x, y)|^2$. Figure 9.3 shows such a contour map for the ground-state density

$$|\psi(x, y)|^2 = A^2 \sin^2\left(\frac{\pi x}{a}\right) \sin^2\left(\frac{\pi y}{a}\right). \qquad (9.31)$$

The density $|\psi|^2$ is maximum at the center of the box $x = y = a/2$. The contours shown are for $|\psi|^2$ equal to 95%, 50%, and 5% of its maximum value. Notice how the contour lines become more square near the edges of the box. The contour line $|\psi|^2 = 0$ is, of course, the square boundary of the box itself.

Figure 9.4 shows the same three contour lines for each of three excited states. Notice how the higher energies correspond to more rapid oscillations of the wave functions and hence to larger numbers of hills and valleys on the map.

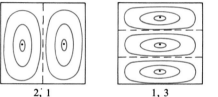

2, 1   1, 3   2, 3

FIGURE 9.4 Contour maps of $|\psi|^2$ for three excited states of the square box. The two numbers under each picture are n_x and n_y. The dashed lines are nodal lines, where $|\psi|^2$ vanishes; these occur where ψ passes through zero as it oscillates from positive to negative values.

EXAMPLE 9.2 Having solved the Schrödinger equation for a particle in the two-dimensional square box, one can solve the corresponding three-dimensional problem very easily (see Problem 9.13). The result is that the allowed energies for a mass M in a rigid cubical box of side a have the form

$$E = E_0(n_x^2 + n_y^2 + n_z^2), \qquad (9.32)$$

where $E_0 = \hbar^2\pi^2/(2Ma^2)$ is the same energy introduced in (9.28), and the quantum numbers n_x, n_y, n_z are any three positives integers. Use this result to find the lowest five energy levels and their degeneracies for a mass M in a rigid, cubical box of side a.

Equation (9.32) shows that the energies of a particle in a three-dimensional box are characterized by three quantum numbers n_x, n_y, n_z. The lowest energy occurs for $n_x = n_y = n_z = 1$ and is

$$E_{111} = 3E_0.$$

The next level corresponds to the three quantum numbers being 2, 1, 1 or 1, 2, 1 or 1, 1, 2:

$$E_{211} = E_{121} = E_{112} = 6E_0.$$

This level is evidently threefold degenerate. The higher levels are easily calculated, and the first five levels are found to be as shown in Fig. 9.5.

(n_x, n_y, n_z)	E	Degeneracy
(222)	$12E_0$	1
(113) or (131) or (311)	$11E_0$	3
(122) or (212) or (221)	$9E_0$	3
(112) or (121) or (211)	$6E_0$	3
(111)	$3E_0$	1
	$E = 0$	

FIGURE 9.5 The first five levels and their degeneracies for a particle in a three-dimensional cubical box.

9.4 The Two-Dimensional Central-Force Problem

Many physical systems involve a particle that moves under the influence of a **central force;** that is, a force that always points exactly toward, or away from, a force center O. The most obvious example is the hydrogen atom, in which the electron is held to the proton by the central Coulomb force. Other examples where the force is at least approximately central include the motion of any one electron in a multielectron atom, and the motion of either atom as it orbits around the other atom in a diatomic molecule.

If the force on a particle is central, it does no work when the particle moves in any direction perpendicular to the radius vector, as shown in Fig. 9.6. This means that the particle's potential energy U is constant in any such displacement. Thus U may depend on the particle's distance, r, from the force center O, but not on its direction. Therefore, instead of writing the potential energy as $U(x, y, z)$, we can write simply $U(r)$ when the force is central. This property of central forces will allow us to solve the Schrödinger equation using separation of variables.

As an introduction to the three-dimensional central-force problem we consider first a two-dimensional particle moving in a central-force field. We shall not present a complete solution of the Schrödinger equation for this system, since it is fairly complicated and we are not really interested in two-dimensional systems anyway. Nevertheless, we shall carry it far enough to see two important facts: First, like the energies for a square box, the allowed energies of a two-dimensional particle in a central-force field are given by two quantum numbers. Second, we shall find that one of the two quantum numbers is closely connected with the angular momentum of the particle.

Since the potential energy U depends only on r, the distance of the particle from the force center O, it is natural to adopt r as one of our coordinates. In two dimensions the simplest way to do this is to use *polar coordinates* (r, ϕ) as defined in Fig. 9.7.* It is a simple trigonometric exercise to express x and y in

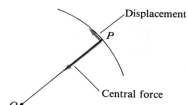

FIGURE 9.6 A central force points exactly toward, or away from, O. If the particle undergoes a displacement perpendicular to the radius vector OP, the force does no work and the potential energy U is therefore constant.

---

* In two dimensions the angle that we are calling ϕ is more often called θ. However, in three dimensions it is usually called ϕ and our primary interest will be in three dimensions.

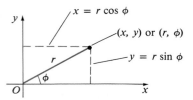

FIGURE 9.7 Definition of the polar coordinates r and ϕ in two dimensions.

$$x = r \cos \phi$$
$$(x, y) \text{ or } (r, \phi)$$
$$y = r \sin \phi$$

terms of r and ϕ as in Fig. 9.7, or vice versa (Problem 9.14). Note that r is defined as the distance from O to the point of interest and is therefore always positive: $0 \leq r < \infty$. If we increase the angle ϕ by 2π (one complete revolution), we come back to our starting direction. Therefore, you must remember that $\phi = \phi_0$ and $\phi = \phi_0 + 2\pi$ represent exactly the same direction.

The wave function ψ, which depends on x and y, can just as well be expressed as a function of r and ϕ:

$$\psi = \psi(r, \phi),$$

and the Schrödinger equation can similarly be rewritten in terms of r and ϕ. When one rewrites the partial derivatives of the Schrödinger equation in terms of r and ϕ, one finds that

$$\frac{\partial^2 \psi}{\partial x^2} + \frac{\partial^2 \psi}{\partial y^2} = \frac{\partial^2 \psi}{\partial r^2} + \frac{1}{r} \frac{\partial \psi}{\partial r} + \frac{1}{r^2} \frac{\partial^2 \psi}{\partial \phi^2}. \tag{9.33}$$

If you have had some experience with handling partial derivatives, you should be able to verify this rather messy identity (Problem 9.17). Otherwise, it is probably simplest for now to accept it without proof. In any case, it is helpful to note that all three terms are dimensonally consistent. Given the identity (9.33), we can rewrite the Schrödinger equation (9.6) in terms of r and ϕ as

$$\frac{\partial^2 \psi}{\partial r^2} + \frac{1}{r} \frac{\partial \psi}{\partial r} + \frac{1}{r^2} \frac{\partial^2 \psi}{\partial \phi^2} = \frac{2M}{\hbar^2} [U(r) - E]\psi. \tag{9.34}$$

SEPARATION OF VARIABLES

The equation (9.34) can be solved by separation of variables, very much as described in Section 9.3, except that we now work with the coordinates r and ϕ instead of x and y. We first seek a solution with the separated form

$$\psi(r, \phi) = R(r)\Phi(\phi). \tag{9.35}$$

Substituting (9.35) into (9.34), we find that

$$\Phi(\phi) \left[R''(r) + \frac{R'(r)}{r} \right] + \frac{R(r)}{r^2} \Phi''(\phi) = \frac{2M}{\hbar^2} [U(r) - E]R(r)\Phi(\phi),$$

where, as before, primes denote differentiation with respect to the argument ($R' = dR/dr$, $\Phi' = d\Phi/d\phi$). If we now multiply both sides by $r^2/(R\Phi)$ and regroup terms, this gives

$$\frac{\Phi''(\phi)}{\Phi(\phi)} = -\frac{r^2 R''(r) + r R'(r)}{R(r)} + \frac{2Mr^2}{\hbar^2} [U(r) - E], \tag{9.36}$$

for all r and ϕ.

The equation (9.36) has the form

$$(\text{function of } \phi) = (\text{function of } r), \tag{9.37}$$

and this will allow us to separate variables. Notice that the separation in (9.37) occurs because the potential energy in (9.36) depends only on r (which follows because the force is central). This explains why we went to the trouble of rewriting the equation in terms of r and ϕ. It would not have separated if we had used x and y, since $U(r)$ depends on both x and y (because $r = \sqrt{x^2 + y^2}$).

The right side of (9.37) is a function of r but is independent of ϕ. Since the two sides are equal for all r and ϕ, it follows that the left side is *also* independent of ϕ, and hence a constant. By a similar argument the right side is likewise constant, and by (9.37) these two constants are equal. It is traditional (and, as we shall see, convenient) to call this constant* $-m^2$. Since each side of (9.36) is equal to the constant $-m^2$, we get two equations:

$$\Phi''(\phi) = -m^2\Phi(\phi) \tag{9.38}$$

and

$$R'' + \frac{R'}{r} - \left[\frac{m^2}{r^2} + \frac{2M}{\hbar^2}(U - E)\right]R = 0. \tag{9.39}$$

Once again, separation of variables has reduced a single partial differential equation in two variables to two separate equations each involving just one variable. The two equations (9.38) and (9.39) are called the ϕ equation and the radial equation. We discuss the ϕ equation (9.38) first.

We already know that the general solution of (9.38) is an arbitrary combination of $\cos m\phi$ and $\sin m\phi$, or, equivalently, of $e^{im\phi}$ and $e^{-im\phi}$. We can economize a little in notation if we use the second pair and if we agree to let m be positive or negative. For example, both $e^{i\phi}$ and $e^{-i\phi}$ can be written as $e^{im\phi}$ if we let m be ± 1. Thus we can say that all solutions of (9.38) can be built up from the functions

$$\Phi(\phi) = e^{im\phi} \tag{9.40}$$

with m positive or negative.

The ϕ equation (9.38) proved easy to solve, but we must now ask whether there are any boundary conditions to be met. In fact, there are. For any given r, the points labeled by ϕ and $\phi + 2\pi$ are the same. Therefore, the wave function $\psi(r, \phi)$ must satisfy

$$\psi(r, \phi) = \psi(r, \phi + 2\pi).$$

For our separated solutions this means that

$$\Phi(\phi) = \Phi(\phi + 2\pi);$$

that is, $\Phi(\phi)$ must be *periodic* and must repeat itself each time ϕ increases by 2π.

Now it is known from trigonometry that the function $\cos m\phi$ is periodic and repeats itself every 2π if m is an integer, but not otherwise (Fig. 9.8). The

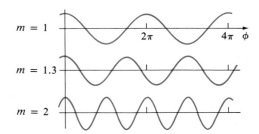

FIGURE 9.8 If m is an integer, the function $\cos m\phi$ repeats itself each time ϕ increases by 2π; for intermediate values it does not.

---

* It is to avoid confusion with this m that we are using M for the mass of the particle.

same is true of $\sin m\phi$ and hence also of $e^{im\phi} = \cos m\phi + i \sin m\phi$. Thus the solution (9.40) of the ϕ equation is acceptable if and only if m is an integer:

$$m = 0, \pm 1, \pm 2, \ldots \tag{9.41}$$

Incidentally, we can now see why it was convenient to use the notation $-m^2$ for the separation constant in (9.38). If we had called it K, for instance, then (9.41) would have read $\sqrt{-K} = 0, \pm 1, \ldots$

QUANTIZATION OF ANGULAR MOMENTUM

Our conclusion so far is that there are solutions of the Schrödinger equation (9.34) with the form

$$\psi(r, \phi) = R(r)e^{im\phi} \tag{9.42}$$

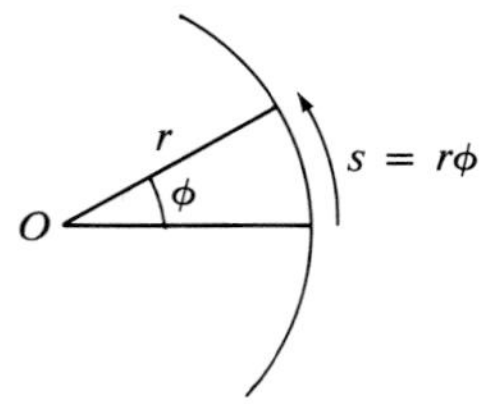

provided the quantum number m is an integer. That m must be an integer indicates that something related to the ϕ dependence of $\psi(r, \phi)$ is quantized. To decide what this is, let us fix r so that we can study just the ϕ dependence. If we let ϕ increase, we move around a circle, as shown in Fig. 9.9, the distance that we travel being $s = r\phi$. According to (9.42), the wave function varies sinusoidally with the distance s:

$$\psi(r, \phi) \propto e^{im\phi} = e^{i(m/r)s}. \tag{9.43}$$

Now, we saw in Chapter 8 that a wave function

$$e^{ikx}, \tag{9.44}$$

where x is the distance along a line, represents a particle with momentum $\hbar k$ along that line.* Comparison of (9.43) and (9.44) suggests that (9.43) represents a particle with momentum

$$p_{\text{tang}} = \hbar \frac{m}{r}$$

in the direction tangential to the circle. If we multiply p_{tang} by r, this gives the angular momentum

$$L = p_{\text{tang}}r = m\hbar;$$

that is, the wave functions (9.42) define states in which the particle has a definite angular momentum

$$L = m\hbar. \tag{9.45}$$

That m has to be an integer shows that L is quantized in multiples of $\hbar$, as originally proposed by Bohr. That is, we have justified Bohr's quantization of angular momentum.

You should bear in mind that we have so far discussed the central-force problem only in two dimensions. For a particle confined to the x-y plane, the angular momentum L is the same thing as its z component L_z. Thus it is not clear whether the result (9.45) will apply to L or L_z when we go on to discuss

FIGURE 9.9 If we move through an angle ϕ on a circle of radius r, the distance traveled is $s = r\phi$.

---

* Recall that the full time-dependent wave function corresponding to (9.44) is $\Psi(x, t) = \exp[i(kx - \omega t)]$. This wave travels to the right and is sinusoidal in x with wavelength $2\pi/k$; that is, it has momentum $p = h/\lambda = \hbar k$ to the right.

three-dimensional motion. In fact, we shall see that in three dimensions it is L_z that is restricted to integer multiples of $\hbar$:

$$L_z = m\hbar \qquad [m = 0, \pm 1, \pm 2, \ldots]. \qquad (9.46)$$

Of course, there is nothing special about the z axis (in three dimensions), and the general statement of (9.46) is that *any* component of the vector $\mathbf{L}$ is restricted to integer multiples of $\hbar$.

THE ENERGY LEVELS

Let us turn now to the radial equation (9.39),

$$R'' + \frac{R'}{r} - \left[\frac{m^2}{r^2} + \frac{2M}{\hbar^2}(U - E)\right]R = 0. \qquad (9.47)$$

Since this equation contains the energy E, its solution will determine the allowed values of E. The details of the solution depend on the particular potential-energy function $U(r)$, which we have not specified. However, we can understand several general features: The radial equation is an ordinary differential equation, which involves the energy E as a parameter. Just as with the one-dimensional Schrödinger equation, one can show that there are acceptable solutions only for certain particular values of E, and these are the allowed energies of our particle. Notice that the equation (9.47) that determines the allowed energies depends on the quantum number m. Thus for each value of m we have a different equation to solve, and will usually get different allowed energies; that is, the allowed energies of our particle will depend on its angular momentum. This is what one would expect classically: The more angular momentum the particle has, the more kinetic energy it will have in its orbital motion; thus, in general, we expect different energies for different angular momenta.

For each value of m we could imagine finding all the allowed energies. We could then label them in increasing order by an integer $n = 1, 2, 3, \ldots$, so that the nth level with angular momentum $m\hbar$ would have energy $E_{n,m}$, as shown in Fig. 9.10. Just as with the square box, each level is then identified by two quantum numbers. In this case, one of the quantum numbers, m, identifies the angular momentum, while the other, n, identifies the energy level for given m.

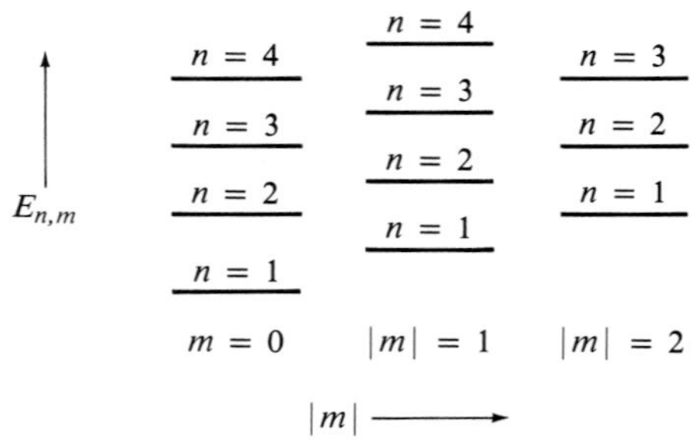

FIGURE 9.10 General appearance of the energy levels for a two-dimensional central force. Energy is plotted upward and the angular-momentum quantum number is plotted to the right. For each value of m, there may be several possible energies, which we label with a quantum number $n = 1, 2, 3, \ldots$. For each pair of values n and m we denote the corresponding energy by $E_{n,m}$. Notice that the variable plotted horizontally is the *absolute value* of m, since $E_{n,m}$ depends only on the magnitude of the angular momentum, not its direction.

The radial equation (9.39) involves m only in the term m^2/r^2. Because this depends on m^2, we get the same equation, and hence the same energies, whether m is positive or negative:

$$E_{n,m} = E_{n,-m}.$$

In other words, except when $m = 0$ there are two states with the same energy, and the level $E_{n,m}$ is twofold degenerate. This is a property that we would also have found in a classical analysis: Two states that differ only by having $L_z = \pm m\hbar$ are different only because the particle is orbiting in opposite directions, and we would expect two such states to have the same energy.

We have not actually *found* the allowed energies $E_{n,m}$ of our two-dimensional particle. These depend on the particular potential-energy function $U(r)$ under consideration. Whatever the form of $U(r)$, the detailed solution of the radial equation (9.39) is fairly complicated. Since our real concern is with three-dimensional systems, we shall not pursue the two-dimensional problem any further here.

9.5 The Three-Dimensional Central-Force Problem

The three-dimensional central-force problem is very similar to the two-dimensional one, but the additional complexity due to the extra dimension means that we shall have to state several results without proof.

Since the potential energy depends only on r (the particle's distance from the force center O) we first choose a coordinate system that includes r as one of the coordinates. For this we use **spherical polar coordinates,** which are defined in Fig. 9.11. Any point P is identified by the three coordinates (r, θ, ϕ) where r is the distance from O to P, θ is the angle between the z axis and OP, and ϕ is the angle between the xz plane and the vertical plane containing OP, as shown. If we imagine P to be a point on the earth's surface and put the origin O at the earth's center, with the z axis pointing to the north pole, then θ is the colatitude of P (the latitude measured down from the north pole) and ϕ is its longitude measured in an easterly direction from the xz plane. The angle θ lies between 0, at the north pole, and π, at the south pole. If ϕ increases from 0 to 2π (with r and θ fixed), then P circles the earth at fixed latitude and returns to its starting point. The rectangular coordinates (x, y, z) are given in terms of (r, θ, ϕ) by

$$x = r \sin\theta \cos\phi, \qquad y = r \sin\theta \sin\phi, \qquad z = r \cos\theta.$$

You should verify these expressions for yourself and derive the corresponding expressions for (r, θ, ϕ) in terms of (x, y, z) (Problem 9.18).

To write down the Schrödinger equation in terms of spherical coordinates we must write the derivatives with respect to x, y, z in terms of r, θ, and ϕ. When this is done, one finds that

$$\frac{\partial^2\psi}{\partial x^2} + \frac{\partial^2\psi}{\partial y^2} + \frac{\partial^2\psi}{\partial z^2} = \frac{1}{r}\frac{\partial^2}{\partial r^2}(r\psi)$$

$$+ \frac{1}{r^2 \sin\theta}\frac{\partial}{\partial\theta}\left(\sin\theta\frac{\partial\psi}{\partial\theta}\right) + \frac{1}{r^2 \sin^2\theta}\frac{\partial^2\psi}{\partial\phi^2}. \qquad (9.48)$$

The proof of this identity is analogous to that of the two-dimensional identity

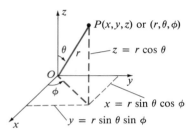

FIGURE 9.11 The spherical polar coordinates of a point P are (r, θ, ϕ), where r is the distance OP, θ is the angle between OP and the z axis, and ϕ is the angle between the xz plane and the vertical plane containing OP.

(9.33) (see Problem 9.17); however, it is appreciably more complicated and is certainly not worth giving here. We shall simply accept the identity (9.48) and use it to write down the three-dimensional Schrödinger equation in spherical coordinates:

$$\frac{1}{r}\frac{\partial^2}{\partial r^2}(r\psi) + \frac{1}{r^2\sin\theta}\frac{\partial}{\partial\theta}\left(\sin\theta\frac{\partial\psi}{\partial\theta}\right)$$

$$+ \frac{1}{r^2\sin^2\theta}\frac{\partial^2\psi}{\partial\phi^2} = \frac{2M}{\hbar^2}[U(r) - E]\psi. \tag{9.49}$$

SEPARATION OF VARIABLES

The three-dimensional Schrödinger equation for a central force, (9.49), can be solved by separation of variables. We start by seeking a solution with the separated form

$$\psi(r, \theta, \phi) = R(r)\Theta(\theta)\Phi(\phi). \tag{9.50}$$

If we substitute (9.50) into (9.49), we can rearrange the resulting equation in the form

$$\frac{\Phi''(\phi)}{\Phi(\phi)} = \text{function of } r \text{ and } \theta. \tag{9.51}$$

(For guidance in checking this and the next few steps, see Problem 9.20.) By the now familiar argument, each side of this equation is equal to the same constant, which we call $-m^2$. This gives us two equations:

$$\Phi''(\phi) = -m^2\Phi(\phi) \tag{9.52}$$

and a second equation involving r and θ. This second equation can next be rearranged in the form

$$(\text{function of } \theta) = (\text{function of } r).$$

Once again each side must be equal to a constant, which we shall temporarily call $-k$. This gives two final equations with the form (Problem 9.20)

$$\frac{1}{\sin\theta}\frac{d}{d\theta}\left(\sin\theta\frac{d\Theta}{d\theta}\right) + \left(k - \frac{m^2}{\sin^2\theta}\right)\Theta = 0 \tag{9.53}$$

and

$$\frac{d^2}{dr^2}(rR) = \frac{2M}{\hbar^2}\left[U(r) + \frac{k\hbar^2}{2Mr^2} - E\right]rR. \tag{9.54}$$

We see that separation of variables has reduced the partial differential equation (9.49) in r, θ, and ϕ to three ordinary differential equations, each involving just one variable. Notice that the ϕ equation (9.52) is exactly the same as Equation (9.38) for the two-dimensional case. Notice also that neither the ϕ equation (9.52) nor the θ equation (9.53) involves the potential-energy function $U(r)$. This means that the solutions for the angular functions $\Theta(\theta)$ and $\Phi(\phi)$ will apply to *any* central-force problem. Finally, note that both $U(r)$ and E appear only in the radial equation (9.54). Therefore, it is the radial equation that determines the allowed values of the energy, and these do depend on the potential-energy function $U(r)$, as one would expect.

9.6   Quantization of Angular Momentum

In this section we discuss the two angular equations (9.52) and (9.53) that resulted from separating the three-dimensional Schrödinger equation. The first of these is exactly the ϕ equation that arose in the two-dimensional central-force problem, and it has the same solutions

$$\Phi(\phi) = e^{im\phi}. \tag{9.55}$$

Since $\Phi(\phi)$ must be periodic with period 2π, it follows, as before, that m must be an integer

$$m = 0, \pm 1, \pm 2, \ldots .$$

The significance of m is essentially the same as in two dimensions: If we fix r and θ and let ϕ vary, then we move around a circle about the z axis. The radius of this circle is $\rho = r \sin \theta$, as shown in Fig. 9.12. In terms of the distance s traveled around this circle, the angle ϕ is

$$\phi = \frac{s}{\rho}$$

and we can temporarily rewrite (9.55) as

$$\Phi(\phi) = e^{im\phi} = e^{i(m/\rho)s}. \tag{9.56}$$

Comparing this with the familiar one-dimensional wave e^{ikx} with momentum $\hbar k$, we see that (9.56) represents a state with tangential momentum

$$p_{\text{tang}} = \frac{\hbar m}{\rho}. \tag{9.57}$$

If we multiply p_{tang} by the radius ρ we obtain the z component of angular momentum, $L_z = p_{\text{tang}}\rho$. Thus, from (9.57) our wave function represents a particle with

$$L_z = m\hbar, \qquad m = 0, \pm 1, \pm 2, \ldots$$

as anticipated in Section 9.4.

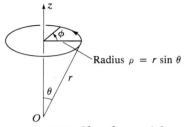

FIGURE 9.12   If we fix r and θ and let ϕ vary, we move around a circle of radius $\rho = r \sin \theta$.

THE θ EQUATION

The second angular equation (9.53), which determines $\Theta(\theta)$, is much harder and we shall have to be satisfied with stating its solutions. The equation is one of the standard equations of mathematical physics and is called Legendre's equation. It has solutions for any value of the separation constant k. However, for most values of k, these solutions are infinite at $\theta = 0$ or at $\theta = \pi$ and are therefore physically unacceptable. It turns out that the equation has one (and only one) acceptable solution for each k of the form

$$k = l(l + 1), \tag{9.58}$$

where l is a positive integer greater than or equal in magnitude to m,

$$l \geq |m|. \tag{9.59}$$

If we denote these acceptable solutions by $\Theta_{lm}(\theta)$, our solutions of the Schrödinger equation have the form

$$\psi(r, \theta, \phi) = R(r)\Theta_{lm}(\theta)e^{im\phi}. \tag{9.60}$$

We shall find the specific form of the function $\Theta_{lm}(\theta)$ for a few values of l and m later. Figure 9.13 shows the function $\Theta_{lm}(\theta)$ for the case $l = 2$ and $m = 0$, as well as one of the unacceptable solutions for the case $l = 1.75$, $m = 0$.

The physical significance of the quantum number m is, as we have seen, that a particle with the wave function (9.60) has a definite value of L_z equal to $m\hbar$. In more advanced texts it is shown that a particle with the wave function (9.60) also has a definite value for the magnitude of $\mathbf{L}$ equal to

$$L = \sqrt{l(l + 1)}\,\hbar. \qquad (9.61)$$

That is, the quantum number l identifies the magnitude of $\mathbf{L}$, according to (9.61).

The quantum number m can be any integer (positive or negative) while, for given m, l can be any integer greater than or equal to the magnitude of m. Turning this around we can say that l can be any positive integer

$$l = 0, 1, 2, \ldots \qquad (9.62)$$

while for given l, m can be any integer less than or equal (in magnitude) to l. That is,

$$L_z = m\hbar, \qquad m = l, l - 1, \ldots, -l. \qquad (9.63)$$

According to (9.61) and (9.62) the possible magnitudes of $\mathbf{L}$ are as follows:

quantum number, l:	0	1	2	3	4	$\cdots$
magnitude, L:	0	$\sqrt{2}\hbar$	$\sqrt{6}\hbar$	$\sqrt{12}\hbar$	$\sqrt{20}\hbar$	$\cdots$

When l is large we can approximate $l(l + 1)$ by l^2 and write

$$L \approx l\hbar.$$

Thus for large l the possible magnitudes of $\mathbf{L}$ are close to those of the Bohr model, but for small l there is an appreciable difference. (See Problems 9.25 and 9.26.)

THE VECTOR MODEL

We have found wave functions for which the magnitude and z component of $\mathbf{L}$ are quantized. This is, of course, a purely quantum result. Nevertheless, it is sometimes useful to try to visualize it classically. The magnitude of $\mathbf{L}$ is $\sqrt{l(l + 1)}\hbar$, so we imagine a vector of length $L = \sqrt{l(l + 1)}\hbar$. Since $L_z = m\hbar$, this vector must be oriented so that its z component is $m\hbar$, and since m can take any of the $(2l + 1)$ different values (9.63) there are $(2l + 1)$ possible orientations, as shown in Fig. 9.14 for the case $l = 2$. We can describe this state of affairs by saying that the spatial orientation of $\mathbf{L}$ is quantized, and in the older literature the quantization of L_z was sometimes called "space quantization."

We should emphasize that there is nothing special about the z axis. When

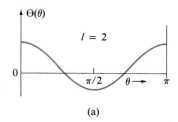

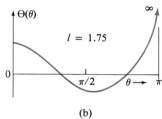

FIGURE 9.13 (a) If the constant k has the form $l(l + 1)$ with l an integer greater than or equal to $|m|$, then the θ equation (9.53) has one acceptable solution, finite for all θ from 0 to π. The picture shows this acceptable solution for the case $m = 0$, $l = 2$. (b) Otherwise, every solution of the θ equation is infinite at $\theta = 0$ or π or both. The picture shows a solution that is finite at $\theta = 0$ but infinite at $\theta = \pi$ for the case $m = 0$ and $l = 1.75$.

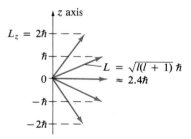

FIGURE 9.14 Classical representation of the quantized values of angular momentum $\mathbf{L}$ for the case $l = 2$. The z component has $(2l + 1) = 5$ possible values, $L_z = m\hbar$ with $m = 2, 1, 0, -1, -2$. The magnitude of $\mathbf{L}$ is $L = \sqrt{l(l + 1)}\hbar = \sqrt{2 \times 3}\,\hbar \approx 2.4\hbar$ in all five cases.

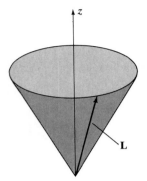

FIGURE 9.15 The quantum properties of angular momentum can be visualized by imagining the vector **L** randomly distributed on the cone shown. This represents the quantum situation, where L and L_z have definite values but L_x and L_y do not.

we defined our spherical coordinates we chose to use the z direction as the polar axis ($\theta = 0$), and when we separated variables the Schrödinger equation led us to wave functions with a definite value for L_z. If we had chosen the x direction as polar axis, the same procedure would have produced states with a definite value for L_x, and so on. For the moment it makes no difference which component of **L** we choose to focus on, and we shall continue to work with states that have definite L_z. However, in more advanced books it is shown that the Heisenberg uncertainty principle extends to angular momentum and implies that no two components of **L** can simultaneously have definite values (except in the special case that all three components are zero). Therefore, states that have definite L_z do not have definite values of L_x and L_y.*

Since our wave functions do not have definite values of L_x and L_y, the vectors shown in Fig. 9.14, with definite components in all directions, are a bit misleading. One must somehow imagine that the components L_x and L_y are random and no longer have definite values. Since the magnitude and z component are fixed, this means that the vector **L** is randomly distributed on a cone as shown in Fig. 9.15. This reflects the quantum situation, where L_x and L_y simply do not have definite values. It is sometimes helpful to use this classical picture — often called the **vector model** — as an aid in thinking about the quantum properties of angular momentum.

The wave functions that we have found for a particle in a central force field have the form

$$\psi(r, \theta, \phi) = R(r)\Theta_{lm}(\theta)e^{im\phi} \tag{9.64}$$

and represent a particle with definite values for the magnitude and z component of **L**:

$$L = \sqrt{l(l + 1)}\,\hbar \qquad \text{and} \qquad L_z = m\hbar.$$

It is sometimes important to know the explicit form of these wave functions. Fortunately, we shall usually be concerned only with states with small values of l — in chemistry, for example, all of the electrons most involved in molecular bonding have $l = 0$ or 1 — and for these the angular wave functions are quite simple, as the following example shows.

EXAMPLE 9.3 Write down the θ equation (9.53) for the cases that $l = 0$ and that $l = 1$, $m = 0$. Find the angular functions $\Theta_{lm}(\theta)e^{im\phi}$ explicitly for these two cases.

The θ equation (9.53) is

$$\frac{1}{\sin\theta}\frac{d}{d\theta}\left(\sin\theta\frac{d\Theta}{d\theta}\right) + \left[l(l + 1) - \frac{m^2}{\sin^2\theta}\right]\Theta = 0. \tag{9.65}$$

[Recall that the separation constant k got renamed $l(l + 1)$.] If $l = 0$, the only allowed value for m is $m = 0$, and the θ equation reduces to

$$\frac{d}{d\theta}\left(\sin\theta\frac{d\Theta}{d\theta}\right) = 0.$$

---

* The wave functions that have definite L_x are different from those with definite L_z. However, any one of the former can be expressed as a sum of the latter, and vice versa. In this sense, it is enough to consider just those with definite L_z. We shall see one example of this in Section 9.8.

By inspection, we see that one solution of this equation is

$$\Theta(\theta) = \text{constant.} \tag{9.66}$$

As stated in connection with (9.58), the θ equation has only one acceptable solution for each value of l and m. (This is illustrated in Problem 9.27.) Therefore, we need look no further for any other solutions. We see that with $l = 0$ the function $\Theta(\theta)$ is independent of θ. Further, with $m = 0$, $e^{im\phi} = 1$ is independent of ϕ. Thus with $l = m = 0$ the wave function (9.64) is actually independent of θ and ϕ, and depends only on r:

$$\psi(r, \theta, \phi) = R(r) \times \text{constant} \qquad [\text{for } l = m = 0].$$

This means that the probability distribution $|\psi(r, \theta, \phi)|^2$ for a particle with zero angular momentum is spherically symmetric. We shall find, for example, that the ground state of the electron in the hydrogen atom has $l = 0$ and hence that the hydrogen atom is spherically symmetric in its ground state.

If $l = 1$, then m can be $m = 1$, 0, or -1. In this example we are asked to consider the case $m = 0$, for which the θ equation (9.65) reads

$$\frac{1}{\sin \theta} \frac{d}{d\theta} \left(\sin \theta \frac{d\Theta}{d\theta} \right) + 2\Theta = 0.$$

By inspection we see that the solution of this equation is

$$\Theta(\theta) = \cos \theta. \tag{9.67}$$

Therefore, with $l = 1$ and $m = 0$, the complete wave function given by (9.64) is

$$\psi(r, \theta, \phi) = R(r) \cos \theta \qquad [l = 1, m = 0]. \tag{9.68}$$

As is obviously the case whenever $m = 0$, this wave function is independent of the angle ϕ. On the other hand, it *does* depend on θ. Since $|\psi|^2 \propto \cos^2\theta$, a particle with $l = 1$ and $m = 0$ is most likely to be found near the polar axes $\theta = 0$ and π (where $\cos^2\theta = 1$), and has zero probability of being found in the x-y plane (where $\theta = \pi/2$ and $\cos \theta = 0$). We shall find that this distribution of electrons in atoms has important implications for the shape of many molecules.

9.7 The Energy Levels of the Hydrogen Atom

Of the three equations that resulted from separating the Schrödinger equation, we have now discussed the two angular ones. It remains to consider the radial equation (9.54), which we rewrite as

$$\frac{d^2}{dr^2} (rR) = \frac{2M}{\hbar^2} \left[U(r) + \frac{l(l+1)\hbar^2}{2Mr^2} - E \right] (rR). \tag{9.69}$$

[Remember that the separation constant k in (9.54) got renamed $l(l + 1)$.] This is the equation that determines the allowed values of the energy E, which will, of course, depend on the precise form of the potential-energy function $U(r)$. Since the equation involves the angular-momentum quantum number l, the allowed values of E will generally depend on l as well. Notice, however, that (9.69) does *not* involve the quantum number m. Thus the allowed values of E will not depend on m; that is, for a given magnitude of $\mathbf{L}$ equal to $\sqrt{l(l + 1)}\hbar$, we shall find the same allowed energies for all $(2l + 1)$ different orientations given by $m = l, l - 1, \ldots, -l$. This is just what we would expect classically: Since the force field is *spherically symmetric,* the energy of the particle cannot depend on the *orientation* of its orbit. Quantum mechanically it means that in any central-force problem, a level with $L = \sqrt{l(l + 1)}\hbar$ will always be at least $(2l + 1)$-fold degenerate. As we shall see shortly, it can sometimes be more degenerate, since two states with different l may happen to have the same energy.

The detailed solution of the differential equation (9.69) depends on the potential-energy function $U(r)$. As a first and very important example, we consider the electron bound to a proton in a hydrogen atom, for which

$$U(r) = \frac{-ke^2}{r}. \tag{9.70}$$

If we substitute (9.70) into (9.69), we obtain the differential equation

$$\frac{d^2}{dr^2}(rR) = \frac{2m_e}{\hbar^2}\left[-\frac{ke^2}{r} + \frac{l(l + 1)\hbar^2}{2m_e r^2} - E\right](rR), \tag{9.71}$$

where we have replaced M by m_e, the mass of the electron. This equation has been studied extensively by mathematical physicists. Here we must be content with simply stating the facts about its solutions (but see Problems 9.33 and 9.34 for some simple special cases): The equation (9.71) has acceptable solutions only if E has the form

$$E = -\frac{m_e(ke^2)^2}{2\hbar^2}\frac{1}{n^2}, \tag{9.72}$$

where n is any integer greater than l. That is, the energy is quantized, and its allowed values are given by (9.72). You may recognize the first factor in (9.72) as the Rydberg energy, originally defined in (6.22),

$$E_R = \frac{m_e(ke^2)^2}{2\hbar^2} = 13.6 \text{ eV}.$$

Thus, solution of the three-dimensional Schrödinger equation for a hydrogen atom has brought us back to exactly the energy levels

$$E = -\frac{E_R}{n^2}$$

predicted by the Bohr model. Since these levels are known to be correct, this is a most satisfactory result.

The possible values of the quantum number l are the integers $l = 0, 1, 2, \ldots$, and for each value of l, we have stated, the radial equation has a solution only if n is an integer greater than l:

$$n > l.$$

Turning these statements around, we can say that the possible values of n are the positive integers,

$$n = 1, 2, 3, \ldots \tag{9.73}$$

and that, for each value of n, l can be any integer less than n,

$$l < n;$$

that is,

$$l = 0, 1, 2, \ldots, (n-1). \tag{9.74}$$

For the ground state, $n = 1$, the only possible value of l is $l = 0$, and the ground state of hydrogen therefore has zero angular momentum. With $l = 0$, the only possible value of m is $m = 0$ and the ground state is characterized by the unique set of quantum numbers

$$\text{ground state:} \quad n = 1, \quad l = 0, \quad m = 0.$$

Notice that although the Schrödinger equation and the Bohr model give the same *energy* for the ground state, there is an important difference: Whereas the Bohr model assumed a magnitude $L = 1\,\hbar$ in the ground state, the Schrödinger equation predicts that $L = 0$, a prediction that is borne out by experiment.

For the first excited level, $n = 2$, there are two possible values of l, namely 0 or 1. If $l = 0$, then m can only be 0; but with $l = 1$ there are three possible orientations of $\mathbf{L}$, given by $m = 1, 0$, or -1. Thus there are four independent wave functions for the first excited level, with quantum numbers:

$$\text{first excited level:} \quad n = 2, \quad l = \begin{cases} 0, & m = 0 \\ \text{or} \\ 1, & m = 1, 0, \text{ or } -1. \end{cases}$$

This means that the first excited level is *fourfold degenerate*.

For the nth level, there are n possible values of L, given by $l = 0, 1, \ldots, (n-1)$. To display this graphically, it is convenient to draw energy-level diagrams in which the energy is plotted upward as usual, but with the different angular momenta $L = \sqrt{l(l+1)}\,\hbar$ shown separately by plotting l horizontally. The first four levels of the hydrogen atom are plotted in this way in Fig. 9.16.

In Figure 9.16 we have introduced the code letters $s, p, d, f, \ldots$, which are traditionally used to identify the magnitude of the angular momentum. These are as follows:

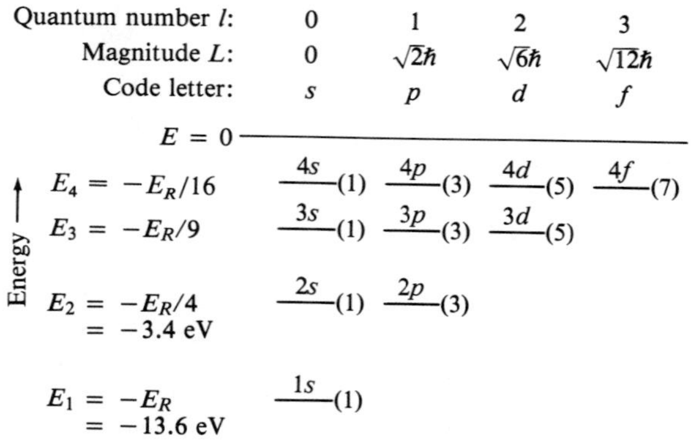

Quantum number l:	0	1	2	3
Magnitude L:	0	$\sqrt{2}\hbar$	$\sqrt{6}\hbar$	$\sqrt{12}\hbar$
Code letter:	s	p	d	f

$$E = 0$$

$E_4 = -E_R/16$ $\underline{\quad}4s$ (1) $\underline{\quad}4p$ (3) $\underline{\quad}4d$ (5) $\underline{\quad}4f$ (7)

$E_3 = -E_R/9$ $\underline{\quad}3s$ (1) $\underline{\quad}3p$ (3) $\underline{\quad}3d$ (5)

$E_2 = -E_R/4$ $\underline{\quad}2s$ (1) $\underline{\quad}2p$ (3)
$\quad\quad = -3.4$ eV

$E_1 = -E_R$ $\underline{\quad}1s$ (1)
$\quad\quad = -13.6$ eV

Energy →

FIGURE 9.16 Energy-level diagram for the hydrogen atom, with energy plotted upward and angular momentum to the right. The letters $s, p, d, f, \ldots$ are code letters traditionally used to indicate $l = 0, 1, 2, 3, \ldots$. (Energy spacing not to scale.)

code letter:	s	p	d	f	g	h	i
quantum number l:	0	1	2	3	4	5	6

These code letters are a survival from early attempts to classify spectral lines; in particular, s, p, d, and f stood for *sharp, principal, diffuse,* and *fundamental.* After f the letters continue alphabetically, although code letters are seldom used for values of l greater than 6. When specifying the values of n and l it is traditional to give the number n followed by the code letter for l. Thus the ground state of hydrogen is called $1s$; the first excited level can be $2s$ or $2p$; and so on. Lower case letters, s, p, d, . . . , are generally used when discussing a single electron, and capitals, S, P, D, . . . , when discussing the total angular momentum of a multielectron atom.

Even when n and l are specified, there are still $(2l + 1)$ distinct states corresponding to the $(2l + 1)$ orientations $m = l, l - 1, . . . , -l$. For s states ($l = 0$) there is just one orientation, for p states ($l = 1$) there are $(2 \times 1) + 1 = 3$, for d states ($l = 2$) there are $(2 \times 2) + 1 = 5$, and so on. These numbers are shown in parentheses on the right of each horizontal bar in Fig. 9.16. The total degeneracy of any level can be found by adding all of these numbers for the level in question. For example, the $n = 1$ level is nondegenerate; the $n = 2$ level has degeneracy 4; the $n = 3$ level 9. The nth level has $l = 0, 1, . . . , (n - 1)$ and hence has degeneracy* (Problem 9.29)

$$1 + 3 + 5 + \cdots + (2n - 1) = n^2. \tag{9.75}$$

In conclusion, the stationary states of hydrogen can be identified by three quantum numbers, n, l, and m. The numbers l and m characterize the magnitude and z component of the angular momentum $\mathbf{L}$. The number n determines the energy as $E_n = -E_R/n^2$ and, for this reason, is often called the **principal quantum number.** It is a peculiarity of the hydrogen atom that the energy depends only on n and is independent of l. We shall see that in other atoms the energy of an electron is determined mainly by n, but does, nonetheless, depend on l as well.

9.8   Hydrogenic Wave Functions

In many applications of atomic physics it is important to know at least the qualitative behavior of the electron wave functions. In this section we discuss the wave functions for the lowest two levels in the hydrogen atom.

THE GROUND STATE

The ground state is the $1s$ state with $n = 1$ and $l = 0$. Since $l = 0$, m has to be zero and, as discussed below (9.66), the wave function is spherically symmetric (that is, is independent of θ and ϕ and depends only on r):

$$\psi_{1s}(r, \theta, \phi) = R_{1s}(r). \tag{9.76}$$

* Actually, the total degeneracy is twice this answer. This is because the electron has another degree of freedom, called spin, which can be thought of as the angular momentum due to its spinning on its own axis (much as the earth spins on its north-south axis). This spin can have two possible orientations, and for each of the states described here, there are really *two* states, one for each orientation of the spin. This will be discussed in Chapter 10.

The radial function $R_{1s}(r)$ is determined by the radial equation (9.71), which we can rewrite (for the particular case that $l = 0$) as

$$\frac{d^2}{dr^2}(rR) = \frac{2m_e}{\hbar^2}\left[-\frac{ke^2}{r} + \frac{E_R}{n^2}\right](rR). \qquad (9.77)$$

If we recall that $\hbar^2/(m_e ke^2)$ is the Bohr radius a_B and that $E_R = ke^2/2a_B$, we can rewrite this equation more simply as

$$\frac{d^2}{dr^2}(rR) = \left(\frac{1}{n^2 a_B^2} - \frac{2}{a_B r}\right)(rR). \qquad (9.78)$$

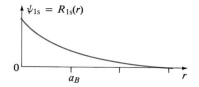

FIGURE 9.17 The wave function (9.79) for the ground state of hydrogen, as a function of r.

For the case $n = 1$, it is easy to verify that the solution of this equation is (Problem 9.33)

$$R_{1s}(r) = Ae^{-r/a_B}. \qquad (9.79)$$

This wave function is plotted in Fig. 9.17. Since $|\psi|^2$ is the probability density for the electron, it is clear from this picture that the probability density is maximum at the origin. In fact, it is characteristic of all s states (states with zero angular momentum) that $|\psi|^2$ is nonzero at the origin; whereas for any state with $l \neq 0$, $|\psi|^2$ is zero at the origin. This situation is easy to understand classically: A classical particle can be found at $r = 0$ only if its angular momentum is zero. This difference between states with $l = 0$ and those with $l > 0$ has important consequences in multielectron atoms, as we discuss in Chapter 11. It also means that the exact energy of s states is slightly dependent on the spatial extent of the nucleus. In fact, careful measurements of energies of atomic electrons in s states have been used to measure nuclear radii.

Since the electron's potential energy depends only on its distance from the nucleus, it is often more important to know the probability of finding the electron at any particular *distance* from the nucleus than to know the probability of its being at any specific *position*. More precisely, we seek the probability of finding it anywhere between the distances r and $r + dr$ from 0, that is, anywhere in a spherical shell between the radii r and $r + dr$. This can be evaluated if we recall that the probability of finding the electron in a small volume dV is $|\psi|^2 dV$. The volume of this spherical shell is the area of the sphere, $4\pi r^2$, times its thickness, dr:

$$(\text{volume between } r \text{ and } r + dr) = 4\pi r^2\, dr. \qquad (9.80)$$

For the ground state of hydrogen the wave function depends on r only and is the same at all points in this thin shell. Therefore, the required probability is just

$$P(\text{between } r \text{ and } r + dr) = |\psi|^2\, dV = |R(r)|^2 4\pi r^2\, dr.$$

We can rewrite this as

$$P(\text{between } r \text{ and } r + dr) = P(r)\, dr \qquad (9.81)$$

if we introduce the **radial probability density** (or radial distribution)

$$P(r) = 4\pi r^2 |R(r)|^2. \qquad (9.82)$$

We have dropped the subscripts $1s$ in these important relations, since they are in fact true for all wave functions.

An important feature of the function (9.82) is the factor of r^2, which comes from the factor $4\pi r^2$ in the volume of the spherical shell (9.80). It means that when we discuss the probability of different distances r (as opposed to different positions), large distances are more heavily weighted, just because larger r corresponds to larger spherical shells, with more volume than those with small r.

For the ground state of hydrogen, with wave function (9.79), the radial probability density is

$$P_{1s}(r) = 4\pi A^2 r^2 e^{-2r/a_B}. \tag{9.83}$$

This is plotted in Fig. 9.18.* Perhaps its most striking property is that its maximum is at $r = a_B$. That is, the most probable distance between the electron and proton in the $1s$ state is the Bohr radius a_B. Thus although quantum mechanics gives a very different picture of the hydrogen atom (with the electron's probability density spread continuously through space), it agrees exactly with the Bohr model as to the electron's most probable radius in the ground state.

Armed with the radial density $P_{1s}(r)$ one can calculate several important properties of the atom. Problems 9.31 and 9.35 to 9.37 contain some examples, and here is another.

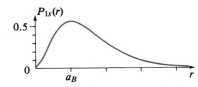

FIGURE 9.18 The probability of finding the electron a distance r from the nucleus is given by the radial probability density $P(r)$. For the $1s$ or ground state of hydrogen $P(r)$ is maximum at $r = a_B$. The density $P(r)$ has the dimensions of inverse length and is shown here in units of $1/a_B$.

EXAMPLE 9.4 Find the constant A in the $1s$ wave function $R_{1s} = Ae^{-r/a_B}$ and the average value of the potential energy for the ground state of hydrogen.

The constant A is determined by the requirement that the total probability of finding the electron at *any* radius must be 1:

$$\int_0^\infty P(r)\, dr = 1. \tag{9.84}$$

Substituting (9.83) we find that

$$4\pi A^2 \int_0^\infty r^2 e^{-2r/a_B}\, dr = 1. \tag{9.85}$$

This integral can be evaluated with two integrations by parts to give $a_B^3/4$ (Problem 9.35). Therefore,

$$\pi A^2 a_B^3 = 1$$

and

$$A = \frac{1}{\sqrt{\pi a_B^3}}. \tag{9.86}$$

To find the average value of the potential energy $U(r)$ we multiply $U(r)$ by the probability $P(r)\, dr$ that the electron be found at distance r, and integrate over all r:

$$U_{av} = \int_0^\infty U(r) P(r)\, dr.$$

---

* Note that the radial density in Fig. 9.18 is zero at $r = 0$ even though the wave function itself is not. This is due to the factor r^2 in (9.82).

If we substitute $U(r) = -ke^2/r$ and replace $P(r)$ by (9.83) this gives

$$U_{av} = -\frac{4ke^2}{a_B^3} \int_0^\infty re^{-2r/a_B}\, dr.$$

The integral can be evaluated by parts as $a_B^2/4$, and we find that

$$U_{av} = -\frac{ke^2}{a_B}.$$

Note that this quantum value for the mean potential energy agrees exactly with the potential energy of the Bohr model.

THE 2s WAVE FUNCTION

In the $n = 2$ level, with $E = -E_R/4$, we have seen that there are four independent wave functions to consider. Of these, the 2s wave function depends only on r:

$$\psi_{2s}(r, \theta, \phi) = R_{2s}(r),$$

where $R_{2s}(r)$ is determined by the radial equation (9.77) to be (Problem 9.38)*

$$R_{2s}(r) = A\left(2 - \frac{r}{a_B}\right)e^{-r/2a_B}. \tag{9.87}$$

The probability of finding the electron between distances r and $r + dr$ from the origin is again given by $P(r)\, dr$ with

$$P_{2s}(r) = 4\pi r^2 |R_{2s}(r)|^2. \tag{9.88}$$

This function is plotted in Fig. 9.19. As we would expect, it is peaked at a much larger radius than the 1s function. Specifically, the most probable radius for the 2s state is $r \approx 5.2a_B$, in approximate (though not exact) agreement with the second Bohr radius, $r = 4a_B$. An important feature of the 2s distribution is the small secondary maximum much closer to $r = 0$ (at $r \approx 0.76a_B$). This means that there is a small (but not negligible) probability of finding the 2s electron close to the nucleus.

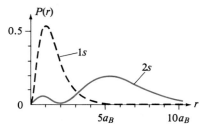

FIGURE 9.19 The radial distribution $P(r)$ for the 2s state (solid curve). The most probable radius is $r \approx 5.2a_B$, with a small secondary maximum at $r \approx 0.76a_B$. For comparison, the dashed curve shows the 1s distribution on the same scale. (Vertical axis in units of $1/a_B$.)

---

* As usual, A denotes a constant, which is determined by the normalization condition (9.84). For simplicity we use the same symbol, A, for all such constants, but we do not wish to imply that they have the same value for all wave functions.

THE $2p_z$ WAVE FUNCTION

There are three $2p$ wave functions, corresponding to the three possible orientations of an $l = 1$ state. In Section 9.6 (Example 9.3) we saw that the angular part of the $m = 0$ wave function is $\Theta(\theta) = \cos\theta$, and the complete wave function is therefore

$$\psi(r, \theta, \phi) = R_{2p}(r) \cos\theta. \tag{9.89}$$

For reasons that we shall see in a moment, this is often called the $2p_z$ wave function. The radial function $R_{2p}(r)$ is found by solving the radial equation for $E = -E_R/4$ and $l = 1$. This gives (Problem 9.39)

$$R_{2p}(r) = Are^{-r/2a_B}. \tag{9.90}$$

Notice that $R_{2p}(r)$ is zero at $r = 0$. Thus the probability density $|\psi|^2$ is zero at the origin, a result that applies (as already mentioned) to any state with nonzero angular momentum.

Substituting (9.90) into (9.89), we find for the complete wave function of the $2p$ state with $m = 0$:

$$\psi(r, \theta, \phi) = Are^{-r/2a_B} \cos\theta \qquad [n = 2, l = 1, m = 0]. \tag{9.91}$$

Since this depends on r and θ it is harder to visualize than the $l = 0$ wave functions (which depend on r only). One way to show its main features is to draw a contour map of the probability density $|\psi|^2$ in the x-z plane, as shown in Fig. 9.20(a). Since $|\psi|^2$ is independent of ϕ, one would find the same picture in any other plane containing the z axis, and one obtains the full three-dimensional distribution simply by rotating Fig. 9.20(a) about the z axis. Figure 9.20(b) shows a perspective view of the 75% contour obtained in this way.

The probability density $|\psi|^2$ for (9.91) is largest on the z axis (where $\cos\theta = \pm 1$) at the points $z = \pm 2a_B$, and is zero in the x-y plane (where $\cos\theta = 0$). The region in which the electron is most likely to be found consists of two approximately spherical volumes centered on the z axis, one above and the other below the x-y plane, as shown in Fig. 9.20(b). It is because the electron is concentrated near the z axis that the $2p$ state with $m = 0$ is called the $2p_z$ state.

FIGURE 9.20 (a) Contour map of $|\psi|^2$ in the x-z plane for the $2p$ $(m = 0)$ state. The density is maximum at the points $z = \pm 2a_B$ on the z axis and zero in the x-y plane. The contours shown are for $|\psi|^2$ equal to 75%, 50%, and 25% of its maximum value. (b) A three-dimensional view of the 75% contour, obtained by rotating the 75% contour of (a) about the z axis.

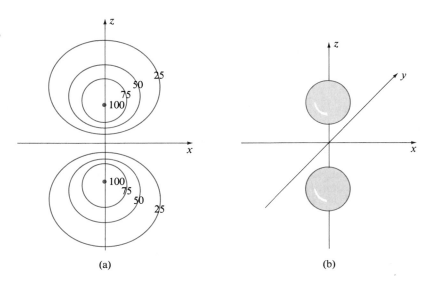

(a)   (b)

THE $2p_x$ AND $2p_y$ WAVE FUNCTIONS

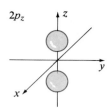

There are still two more $2p$ states to be discussed. An easy way to write these down is to note that the $2p_z$ wave function (9.91) can be rewritten as

$$\psi_{2p_z} = Are^{-r/2a_B}\cos\theta = Aze^{-r/2a_B} \tag{9.92}$$

since $r\cos\theta = z$. Now, the Schrödinger equation, from which this was derived, involves each of the coordinates x, y, z in exactly the same way. Thus if (9.92) is a solution, so must be the two functions obtained from (9.92) by replacing z with x or with y:

$$\psi_{2p_x} = Axe^{-r/2a_B} \tag{9.93}$$

and

$$\psi_{2p_y} = Aye^{-r/2a_B}. \tag{9.94}$$

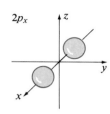

The properties of these two wave functions are very similar to those of ψ_{2p_z}, except that where ψ_{2p_z} is concentrated near the z axis, ψ_{2p_x} is concentrated near the x axis and ψ_{2p_y} near the y axis. Figure 9.21 shows perspective views of all three wave functions. We shall see in Chapter 16 that the concentration of the electron near one of the axes in each of these states has important implications for the shape of some molecules.

As with s waves, it is often important to know the probability of finding the electron at a certain distance from the origin (as opposed to that for finding it at one particular position). Because the $2p$ wave functions depend on θ and ϕ as well as r, the probability of finding the electron between r and $r + dr$ must be calculated by integrating over the angles θ and ϕ. However, the result is exactly the same as (9.82) for s waves

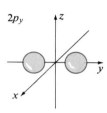

FIGURE 9.21 Perspective views of the 75% contours of $|\psi|^2$ for the $2p_z$, $2p_x$, and $2p_y$ wave functions.

$$P(\text{between } r \text{ and } r + dr) = P(r)\,dr$$

where, for any of the $2p$ states,

$$P_{2p}(r) = 4\pi r^2 |R_{2p}(r)|^2 = 4\pi A^2 r^4 e^{-r/a_B}. \tag{9.95}$$

This function is plotted in Fig. 9.22, where we see that $P_{2p}(r)$ is maximum at $r = 4a_B$ (Problem 9.44); that is, the most probable radius for the $2p$ states agrees exactly with the radius of the second circular Bohr orbit.

Before leaving the $2p$ wave functions, we should mention a final complication. In our general discussion of the central-force problem, we saw that for any p state ($l = 1$) there must be three possible orientations given by $m = 1, 0$, or -1. In the case of the $2p$ states we found, explictly, three independent wave

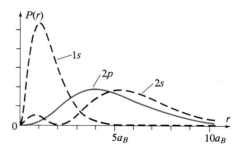

FIGURE 9.22 The radial probability density for the $2p$ states (solid curve). The most probable radius is $r = 4a_B$. For comparison the dashed curves show the $1s$ and $2s$ distributions to the same scale.

functions $2p_x$, $2p_y$, and $2p_z$. It turns out that these latter three wave functions are not exactly the same as the former. Specifically, the $2p_z$ state is precisely the $m = 0$ state. (This was how we derived it.) On the other hand, the $2p_x$ state is *not* the $m = 1$, nor the $m = -1$, state. Instead, the $2p_x$ wave function is the sum of the wave functions for $m = \pm 1$, while the $2p_y$ function is their difference (Problem 9.41).

The important property of the $2p$ states is this: *Any $2p$ wave function can be written as a linear combination of the three wave functions with $m = 1, 0,$ and -1, or as a combination of the wave functions $2p_x$, $2p_y$, and $2p_z$. Which set of three functions we choose to focus on is largely a matter of convenience, and for our purposes the three functions $2p_x$, $2p_y$, and $2p_z$ are usually more suitable.

This situation is very similar to what we saw in Sections 8.6 and 8.7 when solving the differential equation $\psi'' = -k^2\psi$. Any solution of that equation could be expressed as a linear combination,

$$A \sin kx + B \cos kx,$$

of the two solutions $\sin kx$ and $\cos kx$, *or* as a combination,

$$Ce^{ikx} + De^{-ikx},$$

of the two solutions $e^{\pm ikx}$. When seeking the energy levels of a rigid box we found it convenient to use the pair $\sin kx$ and $\cos kx$ to apply the boundary conditions $\psi(0) = \psi(a) = 0$. On the other hand, to interpret the solutions in terms of momentum, it was convenient to re-express them in terms of the pair $e^{\pm ikx}$, as in Eq. (8.75).

9.9 Shells

We have seen that the most probable radius for the $1s$ state of hydrogen is $r = a_B$, while those for the $2s$ and $2p$ states are $r \approx 5.2a_B$ and $r = 4a_B$. For the $3s$, $3p$, and $3d$ states the most probable radii are $13.1a_B$, $12a_B$, and $9a_B$, respectively. These results are illustrated in Fig. 9.23, which shows the radial densities and most probable radii for all of the states concerned.

Figure 9.23 suggests what is found to be true for all of the states with which we shall be concerned: For all the different states with a given value of n, the most probable radii are quite close to one another, and are reasonably well

FIGURE 9.23 The radial distributions for the $n = 1, 2,$ and 3 states in hydrogen. The numbers shown are the most probable radii in units of a_B.

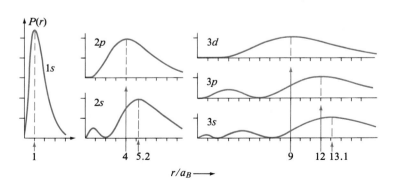

separated from those with any other value of n. This important property is illustrated in a different way in Fig. 9.24, which shows how the most probable radii for the states with quantum number n tend to bunch together in concentric spherical shells with radii close to the Bohr values of $n^2 a_B$. For this reason the word **shell** is often used for the set of all states with a given value of n.

In the hydrogen atom one can characterize a shell in two different ways that are exactly equivalent. As we have just seen, for all states in the nth shell, the most probable distances of the electron from the nucleus are clustered close to the Bohr value $n^2 a_B$. Alternatively, since all states of a given shell have the same value of n, they all have the same energy. Thus the word "shell" can refer either to a clustering in space (what we could call a **spatial shell**) or to a clustering in energy (an **energy shell**).

The notion of shells is very important in atoms with more than one electron, as we shall see in Chapter 11. We shall find that the possible states of any one electron in a multielectron atom can be identified by the same three quantum numbers, n, l, m, that label the states of hydrogen. Furthermore, just as with hydrogen, all states with a given n have radial distributions that peak at about the same radius, and this most probable radius is well separated from the most probable radius for any other value of n. Thus we can speak of spatial shells, as a characteristic clustering of the radial distributions for given n, in just the same sense as in hydrogen.

On the other hand, the allowed energies of any one electron in a multi-electron atom are more complicated than those of hydrogen. In particular, we shall find that states with the same principal quantum number, n, do not necessarily have the same energy. Nevertheless, the states can be grouped into energy shells, such that all levels within one shell are closer to one another than to any level in a neighboring shell. However, these energy shells do not correspond to unique values of n: States with the same n may belong to different shells, and one shell may contain states with different values of n. For example, in many atoms the $3s$ and $3p$ levels are close to one another but are quite well separated from the $3d$ level, which is closer to the $4s$ and $4p$ levels. In this case the $3s$ and $3p$ levels form one energy shell, and the $3d$, $4s$, and $4p$ another.

Unfortunately, the word "shell" is commonly used (without qualification) to denote both what we have called a spatial shell and what we have called an energy shell. We shall discuss all this in more detail in Chapter 11. We mention it here only to emphasize that the simple situation in hydrogen (for which the grouping of states according to energy is exactly the same as the grouping according to distance from the nucleus) is unique to hydrogen.

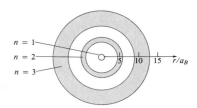

FIGURE 9.24 The most probable radius for any $n = 3$ state in hydrogen is between $9a_B$ and $13.1 a_B$. The corresponding range for the $n = 2$ states is from $4a_B$ to $5.2 a_B$. The most probable radius for the $n = 1$ state is a_B.

9.10 Hydrogen-Like Ions

In Chapter 6 we saw that Bohr's model of the hydrogen atom could be easily generalized to any hydrogen-like ion (that is, a single electron bound to a nucleus of charge Ze). The modern, Schrödinger theory of hydrogen can be generalized in exactly the same way. The potential energy of the electron in hydrogen is $U = -ke^2/r$; that of the electron in a hydrogen-like ion is $U = -Zke^2/r$. Thus the Schrödinger equation for the latter case differs from that of the former only in that ke^2 is replaced by Zke^2 in the potential-energy function.* Therefore, we can convert our hydrogen solutions into solutions for the hydrogen-like ion simply by substituting Zke^2 whenever the term ke^2 appears. This lets us draw three important conclusions with almost no additional labor.

First, the properties of the angular wave functions and the allowed values of angular momentum do not involve the potential energy U at all. Therefore, these angular properties are exactly the same for any hydrogen-like ion as for hydrogen itself.

Second, the Schrödinger equation for hydrogen has acceptable solutions only for the allowed energies,

$$E = -\frac{m_e(ke^2)^2}{2\hbar^2}\frac{1}{n^2} = -\frac{E_R}{n^2}.$$

Replacing ke^2 by Zke^2, we find for the allowed energies of a hydrogen-like ion:

$$E = -\frac{m_e(Zke^2)^2}{2\hbar^2}\frac{1}{n^2} = -Z^2\frac{E_R}{n^2}. \tag{9.96}$$

Third, the spatial extent of the hydrogen wave functions is determined by the Bohr radius,

$$\frac{\hbar^2}{m_e ke^2} = a_B;$$

thus the corresponding parameter for a hydrogen-like ion is

$$\frac{\hbar^2}{m_e Zke^2} = \frac{a_B}{Z}. \tag{9.97}$$

For example, the ground-state wave function of hydrogen is $\psi_{1s} = A\exp[-r/a_B]$; therefore, that for a hydrogen-like ion, with a_B replaced by a_B/Z, is

$$\psi_{1s} = Ae^{-Zr/a_B}.$$

Since all wave functions are modified in the same way, each state of a hydrogen-like ion is pulled inward by a factor $1/Z$, compared to the corresponding state in hydrogen.

The relationship between the quantum properties of the hydrogen atom and the hydrogen-like ion is closely analogous to the corresponding relationship for the Bohr model. This provides the ultimate justification for the several properties of hydrogen-like ions described in Chapter 6 in connection with the Bohr model.

When we discuss multielectron atoms in Chapter 11, we shall make extensive use of the two results (9.96) and (9.97). These are so important, let us close by reiterating them in words: When an electron moves around a total charge Ze, its allowed energies are Z^2 times the allowed energies of a hydrogen atom; and its spatial distribution is scaled inward by a factor of $1/Z$ compared to hydrogen.

---

* Throughout this chapter we are ignoring motion of the nucleus. If we were to include this, there would be a second difference between the hydrogen atom and the hydrogen-like ion, because of the different masses of the nuclei. This small effect can be allowed for by introducing a reduced mass, as described briefly in Section 6.8, but we shall ignore it here.

Partial derivatives
The three-dimensional Schrödinger equation
The square box
Separation of variables
Quantum numbers
Degeneracy
Central-force problems
Polar and spherical polar coordinates

Quantization of angular momentum:
    magnitude — quantum number l
    z component — quantum number m
    vector model
    the code letters $s, p, d, f, \ldots$
The hydrogen atom:
    the quantum numbers n, l, m
    degeneracy
    wave functions
    properties of p states
    radial probability density and shells

PROBLEMS FOR CHAPTER 9

SECTION 9.2 (THE THREE-DIMENSIONAL SCHRÖDINGER EQUATION AND PARTIAL DERIVATIVES)

9.1 • Find the two partial derivatives of (a) $x^2y^3 + x^4y^2$, (b) $(x + y)^3$, (c) $\sin x \cos y$.

9.2 • Find all three partial derivatives of (a) $x^2 + y^2 + z^2$, (b) $(\sin y + \cos z)^2$, (c) $x^2 e^y \sin z$.

9.3 • A mountain can be described by the function $h(x, y)$ which gives the height above sea level of a point that is x east and y north of the origin O. (a) Describe in words the meaning of $\partial h/\partial x$ and $\partial h/\partial y$. (b) What does it mean to a hiker who is walking due north if $\partial h/\partial y$ is positive? (c) What if he is walking due north but $\partial h/\partial y$ is zero and $\partial h/\partial x$ is positive?

9.4 •• Let $h(x, y)$ describe a mountain as in Problem 9.3. If the same mountain is given by the contour map in Fig. 9.25, give estimates for $\partial h/\partial x$ and $\partial h/\partial y$ at points P, Q, R, and the summit S. The scale for x and y (shown by the ruled line) and contours are given in meters.

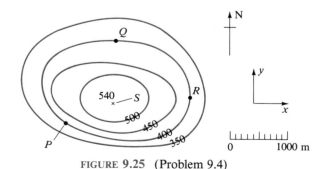

FIGURE 9.25 (Problem 9.4)

9.5 •• If one differentiates a function $f(x, y)$ with respect to x and then differentiates the result with respect to y, one obtains the mixed second derivative

$$\frac{\partial}{\partial y}\left(\frac{\partial f}{\partial x}\right) = \frac{\partial^2 f}{\partial y\, \partial x}.$$

It is a theorem that for any "reasonable" function (which includes any function normally encountered in physics) it makes no difference which differentiation is done first. That is,

$$\frac{\partial^2 f}{\partial y\, \partial x} = \frac{\partial^2 f}{\partial x\, \partial y}. \tag{9.98}$$

A proof of this useful result is beyond the level of this book, but you can *check* the truth of (9.98) for some specific functions. Evaluate both sides of (9.98) for the following functions and verify that they are equal: (a) $f = (x + y^2)^3$, (b) $xe^{(x+y)^2}$, (c) $(x + y) \ln(x - y)$.

SECTION 9.3 (THE TWO-DIMENSIONAL SQUARE BOX)

9.6 • Make an energy-level diagram similar to Fig. 9.2 showing the quantum numbers, energies, and degeneracies of the lowest eight levels for a particle in a two-dimensional, rigid, square box.

9.7 •• Consider a particle of mass M in a two-dimensional, rigid, rectangular box with sides a and b. Using the method of separation of variables, find the allowed energies and wave functions for this particle. In particular, show that the allowed energies are identified by two integers n_x and n_y and have the form

$$E_{n_x, n_y} = \frac{\hbar^2 \pi^2}{2M}\left(\frac{n_x^2}{a^2} + \frac{n_y^2}{b^2}\right). \tag{9.99}$$

Your analysis will be very similar to that given in Section 9.3; the main purposes of this problem are that you go through that analysis yourself and understand how it generalizes to the case of a rectangle with unequal sides.

9.8 •• Consider a particle in a rigid, rectangular box with sides a and $b = a/2$. Using the result (9.99) (Problem 9.7), find the lowest six energy levels with their quantum numbers and degeneracies.

9.9 •• The energy levels for a rectangular rigid box with sides a and b are given by Eq. (9.99) in Problem 9.7. When $a \neq b$ some of the degeneracies noticed for the square box (Fig. 9.2) are no longer present. To illustrate this, find the lowest six levels and their degeneracies for the case $a = 1.1b$. Compare with the levels for the case $a = b$ (Fig. 9.2). Your results will illustrate a general trend: When one reduces the symmetry of a system, its degeneracies usually decrease.

9.10 •• (a) Consider the state with $n_x = 1$ and $n_y = 2$ for a particle in a two-dimensional, rigid, square box. Write down $|\psi|^2$. At what points is the particle most likely to be found? How many such points are there? Sketch a contour map similar to those in Fig. 9.4. (b) Repeat for $n_x = 2$, $n_y = 2$. (c) Repeat for $n_x = 4$, $n_y = 3$.

9.11 •• The energy levels of a particle in a cubical box can be found from Eq. (9.100) (Problem 9.13) by setting $a = b = c$. Find the lowest eight energy levels for a particle in a three-dimensional, rigid, cubical box. Draw an energy-level diagram for these levels showing their quantum numbers, energies, and degeneracies.

9.12 •• In Chapter 8 we claimed that an electron confined inside a thin conducting wire was essentially a one-dimensional system. To illustrate this, take as a model of the wire a long thin rigid box of length a and square cross section $b \times b$ (with $a \gg b$). (a) Using the formula (9.100) (Problem 9.13), write down the ground-state energy for an electron in this box. (b) Write down the energy, measured up from the ground state, of the general excited state. (c) Do the same for an electron in a one-dimensional box of the same length a. (d) Suppose that $a = 1$ m and $b = 1$ mm. Show that the first 1700 (approximately) levels of the electron in the wire are identical to those for the one-dimensional box.

9.13 ••• Show that the allowed energies of a mass M confined in a three-dimensional rectangular rigid box with sides a, b, and c are

$$E = \frac{\hbar^2 \pi^2}{2M} \left(\frac{n_x^2}{a^2} + \frac{n_y^2}{b^2} + \frac{n_z^2}{c^2} \right) \quad (9.100)$$

where the three quantum numbers n_x, n_y, n_z are any

three positive integers $(1, 2, 3, \ldots)$. [*Hint:* Use separation of variables and seek a solution of the form $\psi = X(x)Y(y)Z(z)$. Note that by setting $a = b = c$, one obtains the cubical box of Example 9.2.]

SECTION **9.4** (THE TWO-DIMENSIONAL CENTRAL-FORCE PROBLEM)

9.14 • (a) For the two-dimensional polar coordinates defined in Fig. 9.7 (Section 9.4), prove the relations

$$x = r \cos \phi \quad \text{and} \quad y = r \sin \phi. \quad (9.101)$$

(b) Find corresponding expressions for r and ϕ in terms of x and y.

9.15 • A certain point P in two dimensions has rectangular coordinates (x, y) and polar coordinates (r, ϕ). What are the polar coordinates of the point $Q = (-x, -y)$? Illustrate your answer with a picture.

9.16 •• Changes of coordinates in two dimensions (such as that from x, y to r, ϕ) are much more complicated than in one dimension. In one dimension if we have a function $f(x)$ and choose to regard x as a function of some other variable u, then the derivative of f with respect to u is given by the *chain rule,*

$$\frac{df}{du} = \frac{df}{dx} \frac{dx}{du}.$$

In two dimensions the chain rule reads

$$\frac{\partial f}{\partial r} = \frac{\partial f}{\partial x} \frac{\partial x}{\partial r} + \frac{\partial f}{\partial y} \frac{\partial y}{\partial r} \quad (9.102)$$

and

$$\frac{\partial f}{\partial \phi} = \frac{\partial f}{\partial x} \frac{\partial x}{\partial \phi} + \frac{\partial f}{\partial y} \frac{\partial y}{\partial \phi}.$$

(a) Use the relations (9.101) (Problem 9.14) to evaluate the four derivatives $\partial x/\partial r$, $\partial y/\partial r$, $\partial x/\partial \phi$, and $\partial y/\partial \phi$. (b) If $f = \exp\sqrt{x^2 + y^2}$, use (9.102) to find $\partial f/\partial r$. (c) What is $\partial f/\partial \phi$? (d) By noticing that $\sqrt{x^2 + y^2} = r$ and hence that $f = \exp r$, evaluate $\partial f/\partial r$ and $\partial f/\partial \phi$ directly and check that your answers in parts (b) and (c) are correct.

9.17 ••• A crucial step in solving the Schrödinger equation for the central-force problem was the identity (9.33)

$$\frac{\partial^2 \psi}{\partial x^2} + \frac{\partial^2 \psi}{\partial y^2} = \frac{\partial^2 \psi}{\partial r^2} + \frac{1}{r} \frac{\partial \psi}{\partial r} + \frac{1}{r^2} \frac{\partial^2 \psi}{\partial \phi^2}. \quad (9.103)$$

The purpose of this problem is to prove this identity by showing that the right-hand side is equal to the left. (a) If you have not already done so, do part (a) of Problem 9.16. (b) Use the chain rule (9.102) to show that

$$\frac{\partial \psi}{\partial r} = \frac{\partial \psi}{\partial x} \cos \phi + \frac{\partial \psi}{\partial y} \sin \phi.$$

(c) Use the chain rule on each term in $\partial\psi/\partial r$ to find $\partial^2\psi/\partial r^2$ in terms of $\partial^2\psi/\partial x^2$, $\partial^2\psi/\partial x\,\partial y$, $\partial^2\psi/\partial y^2$. [Remember (9.98), and remember that $\partial f/\partial r$ denotes the derivative with respect to r when ϕ is fixed; therefore, $\partial\phi/\partial r = 0$.] (d) Similarly, find $\partial^2\psi/\partial\phi^2$ in terms of derivatives with respect to x and y. Remember that $\partial r/\partial\phi = 0$. (e) Substitute the results of the previous three parts into the right-hand side of (9.103) and show that you get the left-hand side.

SECTION 9.5 (THE THREE-DIMENSIONAL CENTRAL-FORCE PROBLEM)

9.18 •• The spherical polar coordinates (r, θ, ϕ) are defined in Fig. 9.11. Derive the expressions given there for x, y, and z in terms of (r, θ, ϕ). Find corresponding expression for r, θ, and ϕ in terms of (x, y, z).

9.19 •• A point P on the earth's surface has rectangular coordinates (x, y, z) and spherical polar coordinates (r, θ, ϕ) (with coordinates defined so that the origin is at the earth's center and the z axis points north). What are the coordinates (rectangular and spherical) of the place Q at the opposite end of the earth diameter through P?

9.20 •• Substitute the separated form $\psi = R(r)\Theta(\theta)\Phi(\phi)$ into the Schrödinger equation (9.49). (a) Show that if you multiply through by $r^2\sin^2\theta/(R\Theta\Phi)$ and rearrange, you get an equation of the form $\Phi''/\Phi = $ (function of r and θ). Explain clearly why each side of this equation must be a constant, which we can call $-m^2$. (b) Show that the resulting equation, (function of r and θ) $= -m^2$, can be put in the form

$$\frac{1}{\Theta\sin\theta}\frac{d}{d\theta}\left(\sin\theta\,\frac{d\Theta}{d\theta}\right) - \frac{m^2}{\sin^2\theta} = \text{(function of } r\text{)}.$$

Explain (again) why each side of this equation must be a constant, which we can call $-k$. Derive the r and θ equations (9.54) and (9.53).

SECTION 9.6 (QUANTIZATION OF ANGULAR MOMENTUM)

9.21 • Consider the vector model for the case $l = 2$. Referring to Fig. 9.14, find the minimum possible angle between **L** and the z axis.

9.22 • (a) Draw a vector model diagram similar to Fig. 9.14 for angular momentum of magnitude given by $l = 1$. (b) How many possible orientations are there? (c) What is the minimum angle between **L** and the z axis?

9.23 • Do the same tasks as in Problem 9.22 but for $l = 3$.

9.24 • For a given magnitude $L = \sqrt{l(l+1)}\,\hbar$ of **L**, what is the largest allowed value of L_z? Prove that this largest value of L_z is less than or equal to L (as one would certainly expect classically).

9.25 • The allowed magnitudes of the angular momentum are $L = \sqrt{l(l+1)}\,\hbar$, whereas the Bohr model assumes that $L = l\hbar$. (In both cases, l is restricted to integers.) Compute the ratio $L(\text{correct})/L(\text{Bohr})$ for $l = 1, 2, 3, 4, 10$, and 100. Comment.

9.26 •• The allowed magnitudes of angular momentum are $L = \sqrt{l(l+1)}\,\hbar$. Use the binomial expansion to prove that when l is large, $L \approx (l + \frac{1}{2})\hbar$. (This shows that even for large l, modern quantum mechanics does not quite agree with the Bohr model.)

9.27 •• Write down the θ equation (9.65) for the special case that $l = m = 0$. (a) Verify that $\Theta = $ constant is a solution. (b) Verify that a second solution is $\Theta = \ln[(1 + \cos\theta)/(1 - \cos\theta)]$ and show that this is infinite when $\theta = 0$ or π (and hence is unacceptable). (c) Since the θ equation is a second-order differential equaton, *any* solution must be a linear combination of these two. Write down the general solution and prove that the only acceptable solution is $\Theta = $ constant.

9.28 •• Write down the θ equation (9.65) for the case $l = m = 1$. Verify that $\Theta = \sin\theta$ is a solution. (Any other solution is infinite at $\theta = 0$ or π, so $\sin\theta$ is the only acceptable solution.) Write down the complete wave function (9.64), showing its explicit dependence on θ and ϕ, for $l = 1$, $m = 1$ and for $l = 1$, $m = -1$. [You don't know the radial function, so just leave it as $R(r)$.] With r fixed, in what directions is $|\psi|^2$ a maximum for these states?

SECTIONS 9.7 AND 9.8 (THE ENERGY LEVELS OF THE HYDROGEN ATOM AND HYDROGENIC WAVE FUNCTIONS)

9.29 • Prove that the degeneracy of the nth level in the hydrogen atom is n^2; that is, verify the result (9.75). (But see the footnote below that equation.)

9.30 • It is known that a certain hydrogen atom has a definite value of l. (a) What does this statement tell you about the angular momentum? (b) What are the allowed energies consistent with this information?

9.31 • The mean value of $1/r$ for any state is $(1/r)_{av} = \int_0^\infty (1/r)P(r)\,dr$. Find $(1/r)_{av}$ for the $1s$ state of hydrogen. (*Hint:* See the integrals in Appendix B.)

9.32 •• (a) It is known that a certain hydrogen atom has $n = 5$ and $m = 2$. How many different states are consistent with this information? (b) Answer the same question (in terms of n and m) for arbitrary values of n and m.

9.33 •• The radial equation for $l = 0$ states in hydrogen was given in (9.77). (a) Verify that this can be rewritten as

$$\frac{d^2}{dr^2}(rR) = \left(\frac{1}{n^2 a_B^2} - \frac{2}{a_B r}\right)(rR). \quad (9.104)$$

(b) For the case that $n = 1$, prove that $R_{1s} = e^{-r/a_B}$ is a solution of this equation (that is, calculate the derivative on the left and show that it *is* equal to the right-hand side).

9.34 • • The hydrogenic radial functions $R(r)$ are relatively simple for the case $l = n - 1$ (the maximum allowed value of l for given n):

$$R(r) = Ar^{n-1}e^{-r/na_B} \qquad [l = n - 1] \quad (9.105)$$

(a) Write down the radial Schrödinger equation, (9.71), for this case. (b) Verify that the proposed solution (9.105) does indeed satisfy this equation if and only if $E = -E_R/n^2$.

9.35 • • Use integration by parts to evaluate the integral in (9.85) and hence verify that the normalization constant for the 1s wave function is $A = 1/\sqrt{\pi a_B^3}$.

9.36 • • The mean radius r_{av} for any state is $\int_0^\infty rP(r)\,dr$. Find the mean radius for the 1s state of hydrogen. Referring to Fig. 9.18, explain the difference between the mean and most probable radii.

9.37 • • • The probability of finding the electron in the region $r > a$ is $\int_a^\infty P(r)\,dr$. What is the probability that a 1s electron in hydrogen would be found outside the Bohr radius ($r > a_B$)?

9.38 • • • (a) Write down the radial equation (9.104) for the case that $n = 2$ and $l = 0$ and verify that

$$R_{2s} = A\left(2 - \frac{r}{a_B}\right)e^{-r/2a_B}$$

is a solution. (b) Use the normalization condition (9.84) to find the constant A. (See Appendix B.)

9.39 • • • Write down the radial equation (9.71) for the case that $n = 2$ and $l = 1$. Put in the value $-E_R/4$ for the energy and use the known expressions for a_B and E_R to eliminate all dimensional constants except a_B [as was done in (9.78)]. Verify that $R_{2p} = Are^{-r/2a_B}$ is a solution, and use the normalization condition (9.84) with $P_{2p} = 4\pi r^2|R_{2p}|^2$ to prove that $A = 1/(4\sqrt{6\pi a_B^5})$.

9.40 • • • (a) Use the wave function R_{2p} with the normalization constant A as given in Problem 9.39 to find the mean value of the radius, $r_{av} = \int_0^\infty rP(r)\,dr$, for any of the 2p states of hydrogen. (b) Find the mean potential energy. (c) Compare your results with the values predicted by the Bohr model. (Do they agree exactly? Roughly?)

9.41 • • • (a) Write down the θ equation (9.65) for the 2p states with $m = \pm 1$. Show that the solution is $\Theta(\theta) = \sin\theta$. (There are, of course, two solutions of this second-order equation, but this is the only acceptable one.) This means that the complete wave functions for the 2p states with $m = \pm 1$ are

$$\psi_{2,1,\pm 1} = R_{2p}(r)\sin\theta\, e^{\pm i\phi}.$$

(b) Prove that the sum of these two wave functions is the $2p_x$ wave function (times an uninteresting factor of 2) and that the difference is the $2p_y$ function (times $2i$). (*Hint:* Rewrite $e^{\pm i\phi}$ as $\cos\phi \pm i\sin\phi$ and remember the relations for x and y in terms of r, θ, ϕ in Fig. 9.11.)

SECTION **9.9** (SHELLS)

9.42 • Consider the radial probability density $P(r)$ for the ground state of hydrogen, as given by Eq. (9.83). By finding where $P(r)$ is maximum, find the most probable radius for this state.

9.43 • Using the wave function (9.105) given in Problem 9.34, write down the radial probability density for a hydrogen atom in a state with $l = n - 1$. Find the most probable radius. Notice that in this case — with l equal to its maximum possible value, $l = n - 1$ — the quantum mechanical answer agrees with the Bohr model.

9.44 • • Write down the radial density $P(r)$ for the 2s and 2p states of hydrogen. [See (9.88) and (9.95).] Find the most probable radius for each of these states. [*Hint:* If $P(r)$ is maximum, so is $\sqrt{P(r)}$.]

SECTION **9.10** (HYDROGEN-LIKE IONS)

9.45 • What is the most probable radius for a 1s electron in the hydrogen-like ion Ni^{27+}? What is its binding energy?

9.46 • An inner electron in a heavy atom is affected relatively little by the other electrons, and hence has a wave function very like that for a single electron in orbit around the same nucleus. Approximately what is the most probable radius for a 1s electron in lead? What is this electron's approximate binding energy?

9.47 • A hydrogen-like ion Mg^{11+} drops from its $n = 2$ to its $n = 1$ level. What is the wavelength of the photon emitted?

9.48 • • A hydrogen-like ion of calcium emits a photon with energy $E_{ph} = 756$ eV. What transition was involved?

C H A P T E R

10

Electron Spin

10.1 Introduction

In Chapter 9 we saw how the Schrödinger equation can explain many properties of the hydrogen atom. Our next logical step would be to describe how Schrödinger's theory — unlike the Bohr model — also gives an excellent account of all higher, multielectron atoms. Before we can do this, however, we need to introduce another important property of the electron, its *spin angular momentum,* or *spin.* In this chapter we describe the electron's spin and several of its experimental consequences. Then in Chapter 11 we shall return to the Schrödinger equation and use it to explain the properties of multielectron atoms.

In Section 10.2 we state the observed facts about the electron's spin angular momentum and its quantized values. Most of the more obvious manifestations of the electron's spin concern the magnetic effects associated with a spinning charged particle. Therefore, in Sections 10.3 to 10.5 we describe the magnetic properties of orbiting and spinning charged particles. Then in Sections 10.6 and 10.7 we describe several important experimental consequences of the electron's spin magnetic moment. These effects, all important in their own right, are historically important because they gave evidence for the existence of spin.

10.2 Spin Angular Momentum

As the earth orbits around the sun its total angular momentum $\mathbf{J}$ is the sum of two terms,

$$\mathbf{J} = \mathbf{L} + \mathbf{S}. \qquad (10.1)$$

Here the first term, $\mathbf{L}$, is $\mathbf{r} \times \mathbf{p}$, where $\mathbf{r}$ is the vector from the sun to the earth and $\mathbf{p}$ is the earth's linear momentum; because this term arises from the earth's yearly orbital motion it is called the *orbital angular momentum.* The second term, $\mathbf{S}$, is $I\boldsymbol{\omega}$, where I is the earth's moment of inertia and $\boldsymbol{\omega}$ is the angular velocity of its daily rotation on its own axis; this second term, $\mathbf{S}$, is called the earth's *spin.* In a similar way, the angular momentum of an electron is found to be the sum of two terms with the same form (10.1). The first term, $\mathbf{L}$, is the orbital angular momentum, and this is the angular momentum discussed in Chapter 9, with quantized magnitude $\sqrt{l(l+1)}\hbar$ and components $m\hbar$. The second term, $\mathbf{S}$, is called the electron's **spin.** For most purposes one can visualize the electron's spin as analogous to the earth's spinning motion as it rotates on its own axis.*

We saw in Chapter 9 that the magnitude of $\mathbf{L}$ is quantized, with allowed values

$$L = \sqrt{l(l+1)}\hbar.$$

The magnitude of the spin $\mathbf{S}$ is found to be given by a similar formula,

$$S = \sqrt{s(s+1)}\hbar. \qquad (10.2)$$

Here the spin quantum number s determines the magnitude of the spin $\mathbf{S}$ in just the same way that l determines the magnitude of $\mathbf{L}$. There is, however, an important difference: As we saw in Chapter 9, l can be any integer:

$$l = 0, 1, 2, 3, \ldots .$$

Experiment shows that s always has a fixed value, which is not an integer:

$$s = \tfrac{1}{2}. \qquad (10.3)$$

According to (10.2), this means that the electron's spin $\mathbf{S}$ always has the same magnitude,

$$S = \sqrt{s(s+1)}\hbar = \sqrt{\tfrac{1}{2}(\tfrac{1}{2}+1)}\,\hbar = \frac{\sqrt{3}}{2}\,\hbar, \qquad (10.4)$$

and for this reason the spin is sometimes described as the *intrinsic angular momentum* of the electron. Because the quantum number s is $\tfrac{1}{2}$, one often refers to the electron as having "spin half."

The possible values of the z component (or any other component) of the orbital angular momentum $\mathbf{L}$ have the form

$$L_z = m\hbar,$$

where m runs in integer steps from l to $-l$:

$$m = l, l-1, \ldots, -l.$$

SAMUEL GOUDSMIT (1902–1978, Dutch–US).

GEORGE UHLENBECK (born 1900, Dutch–US). In 1925, while both graduate students at Leiden, Goudsmit and Uhlenbeck showed that several puzzles in atomic spectra could be explained if the electron was assumed to have a spin angular momentum with quantum number $s = 1/2$. Both moved to the US in 1927, and both worked at Michigan and then MIT. Goudsmit became editor of the *Physical Review.*

---

* We should emphasize that this analogy is not exact. For example, if the electron were, like the earth, a spinning ball of matter, its spin angular momentum would be characterized by a quantum number, l, that could take on all integer values, $l = 0, 1, 2, \ldots$. As we shall see shortly, the spin quantum number is *fixed* and *noninteger.*

It is found that a corresponding result holds for spin. The possible values of S_z are

$$S_z = m_s \hbar$$

where m_s is a quantum number that runs in integer steps from s to $-s$. However, since $s = \frac{1}{2}$ this gives only two possible values:

$$m_s = \frac{1}{2} \quad \text{or} \quad -\frac{1}{2}$$

and hence

$$S_z = \frac{1}{2}\hbar \quad \text{or} \quad -\frac{1}{2}\hbar. \tag{10.5}$$

We often describe these two possibilities by saying the electron's spin can be "up" or "down," and represent these two states by arrows, $\uparrow$ and $\downarrow$. However, note that in neither case is $\mathbf{S}$ actually parallel to the z axis, since S_z is smaller than S, as is clear from (10.4) and (10.5). This is the same thing that we saw with respect to $\mathbf{L}$: Even when L_z has its maximum value it is smaller than L, so that $\mathbf{L}$ is never exactly parallel to the z axis.

A complete specification of an electron's state of motion requires that one specify its spin orientation as well as its orbital motion. For example, for an electron in a hydrogen atom the quantum numbers n, l, m specify the orbital motion, but for each choice of n, l, m the spin can be either up or down, corresponding to $m_s = \frac{1}{2}$ or $-\frac{1}{2}$. It turns out that the energy of the H atom is almost completely independent of the spin orientation. Thus the allowed energies calculated in Chapter 9 are still correct, but there are twice as many independent states in each energy level as we had calculated formerly. The ground state, with $n = 1$ and $l = m = 0$, can have $m_s = \pm\frac{1}{2}$ and is therefore twofold degenerate. More generally, we saw that for the nth energy level there are n^2 possible values of l and m; for each of these there are two possible spin orientations. Therefore, the total degeneracy of the nth level is $2n^2$:

$$\boxed{(\text{degeneracy of } n\text{th level in hydrogen}) = 2n^2.} \tag{10.6}$$

When an electron's state of motion is completely specified, we say that the electron is in a definite **quantum state,** or just **state;** for example, we can speak of the quantum state identified by the four quantum numbers n, l, m, and m_s in hydrogen. A specification of an electron's orbital motion, but not its spin orientation, is sometimes called an **orbital;** for example, we can speak of the orbital given by the three quantum numbers n, l, and m. For each orbital, there are evidently two independent quantum states, corresponding to the two possible values of m_s.

10.3 Magnetic Moments

There is an enormous body of evidence for the electron's spin angular momentum. However, most of this evidence is fairly indirect. In particular, much of the evidence for spin relates not to the angular momentum itself, but to the magnetic moment associated with any rotating electric charge. We must, therefore, review the concept of magnetic moment and describe its relation to the rotational motion of a charged particle. In this section and Section 10.4 we confine ourselves to the magnetic properties associated with the orbital motion of an electron. Then in Section 10.5 we describe the additional magnetic moment

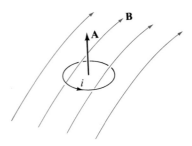

FIGURE 10.1 The current i flows around a loop specified by the vector **A**, whose magnitude gives the loop's area and whose direction is perpendicular to the plane of the loop. A magnetic field **B** exerts a torque that tends to align **A** with **B**.

that results from the electron's spin. Finally, in Sections 10.6 and 10.7 we describe some phenomena in which the spin magnetic moment plays an important role.

We start by considering the magnetic properties of a classical point electron traveling around a nucleus in a circular orbit. An orbiting charge acts like a small current loop, and we know from classical electromagnetic theory that a current loop both produces a magnetic field and responds to an externally applied field. If a current i flowing around a small plane loop of area A is placed in a magnetic field **B**, it experiences a torque **Γ** given by

$$\mathbf{\Gamma} = i\mathbf{A} \times \mathbf{B},$$

where the vector **A** has magnitude equal to the area A and is perpendicular to the plane of the loop as in Fig. 10.1. The sense of the vector **A** is given by the familiar right-hand rule: If you curl the fingers of your right hand around the loop in the direction of i, your thumb will point in the direction of **A**.

It is usual to rewrite the torque $\mathbf{\Gamma} = i\mathbf{A} \times \mathbf{B}$ as

$$\mathbf{\Gamma} = \boldsymbol{\mu} \times \mathbf{B}, \tag{10.7}$$

where the vector $\boldsymbol{\mu}$,

$$\boldsymbol{\mu} = i\mathbf{A}, \tag{10.8}$$

is called the **magnetic moment** of the loop. The torque **Γ** tends to turn the loop so that $\boldsymbol{\mu}$ points in the same direction as the magnetic field **B**.

Because of the torque (10.7), a current loop in a B field has a potential energy U that depends on the loop's orientation. To evaluate this energy we recall that the work done by a torque **Γ** as it turns through an angle $d\theta$ is $\Gamma \, d\theta$. The torque (10.7) has magnitude $\mu B \sin \theta$ and is in a direction to decrease θ. Thus the work done by B when the loop is brought to angle θ is

$$W = -\mu B \int \sin \theta \, d\theta = \mu B \cos \theta + \text{constant}. \tag{10.9}$$

The potential energy U is defined as the negative of this work. Since the definition of potential energy always contains an arbitrary constant, it is customary to set the constant in (10.9) equal to zero, with the result that

$$U = -\mu B \cos \theta = -\boldsymbol{\mu} \cdot \mathbf{B}. \tag{10.10}$$

Notice that this potential energy is minimum when $\theta = 0$, with $\boldsymbol{\mu}$ pointing along **B** in stable equilibrium (see Fig. 10.2).

Let us now consider the current loop produced by an orbiting electron. The current i is equal to the total charge passing any fixed point in unit time. The electron has charge of magnitude e, and the number of times it passes a fixed point in unit time is $v/(2\pi r)$. Therefore, the current has magnitude

$$i = e \frac{v}{2\pi r},$$

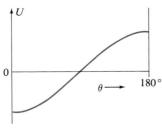

FIGURE 10.2 The potential energy $U = -\mu B \cos \theta$ of a magnetic moment $\boldsymbol{\mu}$ in a field **B** is minimum when $\theta = 0$ and $\boldsymbol{\mu}$ is parallel to **B**.

and the magnetic moment has magnitude

$$\mu = iA = \frac{ev}{2\pi r} \pi r^2 = \frac{1}{2} evr. \tag{10.11}$$

It is convenient to relate the magnetic moment μ to the angular momentum L. (Since μ and L both result from the electron's orbital motion, one might expect some simple relation between them.) Since the angular momentum has magnitude

$$L = m_e vr$$

(where m_e denotes the electron mass), we see from (10.11) that

$$\frac{\mu}{L} = \frac{e}{2m_e}. \tag{10.12}$$

We conclude that the ratio of μ to L—the so-called **gyromagnetic ratio**—is a constant that depends only on the charge and mass of the electron. Because the electron's charge is negative, the current is in a direction opposite to the electron's velocity, so that the vectors μ and $\mathbf{L}$ are antiparallel. Thus we can rewrite (10.12) in vector form as

$$\mu = -\frac{e}{2m_e}\mathbf{L}. \tag{10.13}$$

We have derived the result (10.13) for the magnetic moment of a classical point electron in a circular orbit. In quantum mechanics, it turns out that exactly the same expression correctly predicts the magnetic moment due to the orbital motion of an electron, provided that we use the correct quantum values for the magnitude and components of $\mathbf{L}$. In particular, for a given magnitude $\sqrt{l(l+1)}\,\hbar$ we know that $\mathbf{L}$ has just $2l + 1$ possible orientations. According to (10.13), the same is true of the magnetic moment μ of an orbiting electron: For a given value of l, μ has just $2l + 1$ possible orientations.

Equation (10.13) gives the magnetic moment due to the orbital motion of an electron. As one might expect, there is an additional magnetic moment due to the electron's spinning motion. Before we discuss this spin magnetic moment, we use (10.10) and (10.13) to see how the energy of an atom changes when it is put in a magnetic field.

10.4 The Zeeman Effect

Because of the motion of their electrons most atoms have a magnetic moment μ. Therefore, by applying a magnetic field $\mathbf{B}$ one can change an atom's energy levels by an amount $-\mu\cdot\mathbf{B}$. This means that the energies of photons emitted and absorbed by the atom will change. That is, by putting it in a magnetic field one can change an atom's spectrum. This effect was first observed in 1896 by the Dutch physicist Zeeman and is called the **Zeeman effect.**

To simplify our discussion, we consider at first an atom in which the magnetic moments due to the electrons' spins cancel out. The simplest atom in which this can happen is helium, with its two electrons, and this is the atom that we consider. In several states of helium—known as the **singlet states**—the spins of the two electrons point in opposite directions so that the total magnetic moment due to the spins is zero. Furthermore, in all the bound states of helium one of the electrons has zero orbital angular momentum. Therefore, the total magnetic moment for any of the singlet states of helium is just the moment $\mu = -(e/2m_e)\mathbf{L}$ due to the orbital motion of the second electron.

PIETER ZEEMAN (1865–1943, Dutch). At the suggestion of his teacher, Lorentz, Zeeman investigated the effect of magnetic fields on atomic spectra. The results confirmed Lorentz's suspicion that atomic spectra are somehow connected to the motion of electrons in the atoms. Zeeman and Lorentz shared the 1902 Nobel Prize for this work.

In the absence of a magnetic field, the helium atom has an energy that we call E_0, and its angular momentum (the orbital momentum $\mathbf{L}$ of the second electron) has a magnitude given by the quantum number l. The angular momentum $\mathbf{L}$ has $2l + 1$ possible different orientations, corresponding to the $2l + 1$ possible values of $L_z = m\hbar$, with $m = l, l - 1, \ldots, -l$. In the absence of a magnetic field the energy is the same for all of these states, and the level E_0 is $(2l + 1)$-fold degenerate.

Suppose now that we apply a magnetic field $\mathbf{B}$ to our atom. According to (10.10) this will change the atom's energy by the amount $-\boldsymbol{\mu} \cdot \mathbf{B}$, which depends on the orientation of $\boldsymbol{\mu}$. According to (10.13), $\boldsymbol{\mu}$ is given by

$$\boldsymbol{\mu} = -\frac{e}{2m_e}\mathbf{L}$$

and has $2l + 1$ different possible orientations. Therefore, we can anticipate *not only that the energy will change as a result of the magnetic field,* but that *it will change by a different amount for each of the $2l + 1$ different orientations.* That is, by applying a magnetic field we remove the $(2l + 1)$-fold degeneracy of the original energy level.

The size of the energy shift due to the magnetic field is easily calculated: With the field switched on, we denote the total energy by $E = E_0 + \Delta E$, where the shift ΔE is given by (10.10) and (10.13) as

$$\Delta E = -\boldsymbol{\mu} \cdot \mathbf{B} \tag{10.14}$$

$$= \left(\frac{e}{2m_e}\right)\mathbf{L} \cdot \mathbf{B}. \tag{10.15}$$

If we choose our z axis in the direction of the applied field $\mathbf{B}$, then (10.15) simplifies to

$$\Delta E = \left(\frac{e}{2m_e}\right)L_z B,$$

or, since the possible values of L_z are $m\hbar$,

$$\Delta E = \left(\frac{e\hbar}{2m_e}\right)mB. \tag{10.16}$$

As anticipated, the magnetic field changes the atom's energy by an amount that depends on the quantum number m. This is why m is often called the **magnetic quantum number** and explains the traditional choice of the letter m.

Comparing (10.16) with (10.14) (and remembering that the quantum number m is dimensionless) we see that the quantity in parentheses, $(e\hbar/2m_e)$, must have the dimensions of a magnetic moment. (You should check this directly; see Problem 10.12.) In atomic physics this quantity is a convenient unit for magnetic moments and is called the **Bohr magneton** μ_B, with the value

$$\mu_B = \frac{e\hbar}{2m_e} = 9.27 \times 10^{-24}\ \text{A} \cdot \text{m}^2. \tag{10.17}$$

In terms of μ_B, we can rewrite (10.16) in the compact form

$$\Delta E = m\mu_B B. \tag{10.18}$$

Since m can have the $2l + 1$ values $l, l - 1, \ldots , -l$, we see that the $2l + 1$ states of the original degenerate level now have energies that are equally spaced, an energy $\mu_B B$ apart:

$$\text{separation of adjacent levels} = \mu_B B. \qquad (10.19)$$

The result (10.19) shows that the dimensions of μ_B can also be expressed as energy/magnetic field; that is, the unit $A \cdot m^2$ in (10.17) can be replaced by joules/tesla. If we convert the joules to eV, we get the useful result

$$\mu_B = 5.79 \times 10^{-5} \text{ eV/T}. \qquad (10.20)$$

According to (10.19), this means that a field of 1 tesla leads to a separation of adjacent levels by 5.79×10^{-5}eV — a very small separation on the scale of normal atomic levels, which are typically a few eV apart.

EXAMPLE 10.1   A helium atom is in one of its singlet states (with the two spins antiparallel and hence no spin magnetic moment). One of its electrons is in an s state ($l = 0$) and the other a d state ($l = 2$). The atom is placed in a magnetic field, $B = 2$ T (by normal laboratory standards a fairly strong field). By how much does the magnetic field change the atom's energy?

The shift in energy is given by (10.18) as $\Delta E = m\mu_B B$, where, since $l = 2$, the quantum number m can have any of the five values $m = 2, 1, 0, -1, -2$. If the atom is in the state with $m = 0$, its energy is unaltered. If it is in any of the other four states, its energy is shifted as shown in Fig. 10.3. The five resulting energy levels are said to form a *multiplet* and are evenly spaced above and below the original E_0, with separation $\mu_B B$. With $B = 2$ tesla, the separation of adjacent levels is

$$\mu_B B = \left(5.79 \times 10^{-5} \frac{\text{eV}}{\text{T}} \right) \times (2 \text{ T}) = 1.2 \times 10^{-4} \text{ eV}.$$

We see from this example that even a relatively strong field of a few teslas produces a very small separation of energy levels. Nevertheless, this small splitting of each level into several levels results in an observable splitting of the

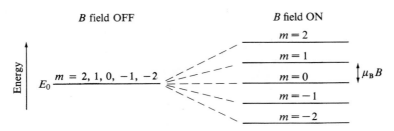

FIGURE 10.3 In the absence of a magnetic field an atomic level with $l = 2$ (and no spin magnetic moment) is fivefold degenerate with energy E_0. When B is switched on, the level splits into a multiplet of five equally spaced levels with separation $\mu_B B$.

spectral lines of the light emitted and absorbed by the atom. To illustrate this effect — the Zeeman effect — we consider again the helium atom. Specifically, we consider the ground state and one of the low-lying excited levels, 21.0 eV above the ground state.

In both of these levels the two electron spins are antiparallel and the resultant spin magnetic moment is zero. Thus the shift in energy produced by a magnetic field is correctly given by (10.18) as

$$\Delta E = m\mu_B B.$$

In the ground state both electrons have $l = 0$. Thus the only possible value of m is $m = 0$, and the shift ΔE is zero. That is, the energy of the ground state is unchanged by a magnetic field. In the excited level, one electron has zero orbital angular momentum while the other has $l = 1$. With $l = 1$, the possible values of m are $m = 1, 0,$ and -1, and the magnetic field splits this level into three equally spaced levels, a distance $\mu_B B$ apart, as shown in Fig. 10.4.

Let us now consider transitions in which a helium atom drops from the excited level just described to the ground state and emits a photon. In the absence of a magnetic field both levels have unique energies separated by 21.0 eV, and the photon has energy $E_{ph} = 21.0$ eV and frequency $f_0 = E_{ph}/h$. Thus a spectrometer would reveal a single spectral line with frequency f_0, as indicated at the bottom left of Fig. 10.4. If we now apply a magnetic field B, the upper level splits into three closely spaced levels. Therefore, there are now three possible transitions with three slightly different energies, as indicated by the three downward arrows on the right side of Fig. 10.4. In a gas of excited helium there would normally be atoms in all three levels. Therefore, all three transitions would occur, and a spectrometer would now reveal a *triplet* of three closely spaced spectral lines with frequencies $f_0 + \mu_B B/h, f_0,$ and $f_0 - \mu_B B/h$, as shown at the bottom right of Fig. 10.4.

The Zeeman effect was discovered in 1896, well before the development of quantum mechanics. Naturally, attempts were made to explain it on the

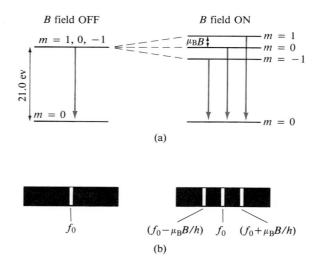

FIGURE 10.4 (a) The ground state ($l = 0$) and one of the excited levels ($l = 1$) of helium. When a magnetic field is applied, the upper level splits into three, while the ground state is unaffected. With the magnetic field on, there are three distinct transitions possible and hence three distinct spectral lines, as shown in (b).

basis of classical mechanics. As it happens, the classical theory of the Zeeman effect gives the correct answers for any two atomic levels whose spin magnetic moments are zero. Thus a Zeeman splitting like that shown in Fig. 10.4(b) agreed with classical predictions and came to be called the **normal Zeeman effect.*** Unfortunately (for classical mechanics), the effect of a magnetic field on many atoms was found to be much more complicated than this and did not agree with the classical theory. As we shall see, these more complicated shifts — which were called **anomalous Zeeman splittings** — involve the spin magnetic moment of the electron and were among the first indications that the electron has a spin.

EXAMPLE 10.2   What is the wavelength λ_0 of the transition shown on the left of Fig. 10.4? If a magnetic field of $2\,T$ is applied to the helium atom, what are the shifts $\Delta\lambda$ of the outer two spectral lines on the right of Fig. 10.4(b)?

With $B = 0$ the two levels are 21.0 eV apart and the emitted photon has $E_{ph} = 21.0$ eV. The corresponding wavelength is

$$\lambda_0 = \frac{hc}{E_{ph}} = \frac{1240 \text{ eV} \cdot \text{nm}}{21.0 \text{ eV}} = 59.0 \text{ nm},$$

which is well into the ultraviolet.

If we switch on a magnetic field, the upper level splits into the three equally spaced levels shown in Fig. 10.4, separated by energy $\mu_B B$:

separation of adjacent levels $= \mu_B B$

$$= \left(5.79 \times 10^{-5} \frac{\text{eV}}{\text{T}}\right) \times (2 \text{ T})$$

$$\approx 1.2 \times 10^{-4} \text{ eV}.$$

The $m = 0$ state has the same energy as when $B = 0$, and the wavelength of photons emitted from this state is unchanged. The energy of photons emitted from the states with $m = \pm 1$ is changed by $\Delta E_{ph} = \pm \mu_B B$. Since this is small, the shift in their wavelength is well approximated as

$$\Delta\lambda \approx \frac{d\lambda}{dE_{ph}} \Delta E_{ph} = -\frac{hc}{E_{ph}^2} \Delta E_{ph}$$

$$= -\frac{1240 \text{ eV} \cdot \text{nm}}{(21.0 \text{ eV})^2} \times (\pm 1.2 \times 10^{-4} \text{ eV})$$

$$= \mp 3.4 \times 10^{-4} \text{ nm}. \tag{10.21}$$

The Zeeman shift of wavelength is so small that the earliest observations could not distinguish the separate spectral lines. At first all that was detected was a broadening of the original single line; but later experiments with better resolution showed that the line was indeed split into several separate lines. Today, spectrometers can resolve splittings of order 10^{-8} nm, and the Zeeman shifts

---

* The example of Fig. 10.4(b) involved a transition from $l = 1$ to $l = 0$ and gave a splitting into three spectral lines. One might imagine that higher l values would lead to more than three lines; in fact, however, the normal Zeeman effect always produces exactly three lines (Problem 10.15).

can be measured very accurately. An important modern application is to measure the splitting of an identified spectral line and hence to find an unknown magnetic field. This is especially useful in astronomy, since the magnetic fields of the sun and stars cannot be measured directly.

10.5 Spin Magnetic Moments

As an atomic electron orbits around the nucleus we have seen that it produces a magnetic moment given by (10.13) as

$$\boldsymbol{\mu}_{orb} = -\frac{e}{2m_e}\mathbf{L}. \tag{10.22}$$

If we visualize the electron as a tiny, rigid ball of charge spinning on its axis, we would expect this spinning motion to produce an additional, spin, magnetic moment. Each piece of the electron would be carried in a circular path around the axis and hence constitute a small current loop. Each such loop would produce a magnetic moment, and the sum of all these moments would be the total spin magnetic moment $\boldsymbol{\mu}_{spin}$. Since $\boldsymbol{\mu}_{spin}$ would be proportional to the angular velocity $\boldsymbol{\omega}_{spin}$, which in turn is proportional to the spin angular momentum $\mathbf{S}$, we would expect to find $\boldsymbol{\mu}_{spin} \propto \mathbf{S}$ or

$$\boldsymbol{\mu}_{spin} = -\gamma\mathbf{S}, \tag{10.23}$$

where γ is a constant called the spin gyromagnetic ratio. [We have put a minus sign in (10.23) because the electron's charge is negative, and $\boldsymbol{\mu}_{spin}$ and $\mathbf{S}$ are in opposite directions.] In the case of the *orbital* motion, the gyromagnetic ratio is seen from (10.22) to be $e/2m_e$. The spin gyromagnetic ratio would not necessarily have this same value, since it would depend on the distributions of charge and mass within the electron. If the charge were concentrated farther out than the mass, then μ_{spin}/S would be relatively large; if the charge were concentrated nearer the center, μ_{spin}/S would be smaller.

The classical picture of the electron as a rigid spinning ball of charge is not strictly correct. For example, if the radius of the ball is taken consistent with modern observations, the equatorial speed turns out to be greater than c, which is impossible (Problem 10.6). Nevertheless, the conclusions that there should be a magnetic moment with the general form (10.23) and that the spin gyromagnetic ratio does not necessarily have the same value as the orbital ratio $e/2m_e$ *are* both correct. Experiment shows that there is a magnetic moment with the form (10.23) and that the spin gyromagnetic ratio γ is e/m_e, just twice* the value of the orbital ratio, $e/2m_e$; that is,

$$\boldsymbol{\mu}_{spin} = -\frac{e}{m_e}\mathbf{S}. \tag{10.24}$$

The total magnetic moment of any electron is just the sum of its orbital and spin moments

* Precise measurements show that the spin ratio is actually not *exactly* twice the orbital value (instead of 2, the factor is 2.0023). The observed value is successfully predicted by the relativistic quantum theory called quantum electrodynamics.

$$\boldsymbol{\mu}_{\text{tot}} = \boldsymbol{\mu}_{\text{orb}} + \boldsymbol{\mu}_{\text{spin}} = -\frac{e}{2m_{\text{e}}}(\mathbf{L} + 2\mathbf{S}). \tag{10.25}$$

As we describe in the next two sections, much of the evidence for the electron's spin comes from the repeated success of the formula (10.25) in explaining a wide variety of experimental results.

The suggestion that the electron has a spin angular momentum and a corresponding magnetic moment [given, as we now know, by (10.24)] is generally credited to the Dutch physicists Goudsmit and Uhlenbeck (1925). Their suggestion was based on an analysis of the anomalous Zeeman effect (which we describe in Section 10.6) and of the fine structure in atomic spectra (Section 10.7). However, it is worth mentioning that similar suggestions had been made by other physicists. In particular, Compton had suggested that a spin magnetic moment for the electron could possibly explain the phenomenon of ferromagnetism, a suggestion that later proved to be correct.

10.6 The Anomalous Zeeman Effect

When an atom is placed in a magnetic field, we have seen that its energy levels undergo small shifts, and individual levels get split into several closely spaced levels. This results in a splitting of the spectral lines into closely spaced "multiplets" of lines — an effect known as the Zeeman effect.

In Section 10.4 we calculated in detail the Zeeman splitting for the so-called singlet states of the helium atom (the states in which the two spin magnetic moments cancel out and can therefore be ignored). The results of those calculations are correct for any atomic state in which the spin magnetic moments cancel. In general, however, the Zeeman effect does *not* agree with the splittings calculated in Section 10.4, but *does* agree with a corresponding calculation using the correct magnetic moment (10.25), including both orbital and spin moments. For historical reasons the splitting of levels and spectral lines is called the *normal Zeeman effect* in those cases where spin has no effect, and the *anomalous Zeeman effect* in those cases where spin does contribute.

The correct calculation of the anomalous Zeeman splitting is quite complicated, depending as it does on both orbital and spin moments. To simplify our discussion we consider here the simple case of a hydrogen atom in a state with no orbital angular momentum; that is, an *s* state, with $l = 0$. If the electron had no spin, then, with $l = 0$, the atom would have no magnetic moment at all, and would be completely unaffected by a magnetic field. In fact, of course, the electron does have spin and, even though $l = 0$, there is a magnetic moment

$$\boldsymbol{\mu} = \boldsymbol{\mu}_{\text{spin}} = -\frac{e}{m_{\text{e}}}\mathbf{S}. \tag{10.26}$$

When a magnetic field $\mathbf{B}$ (in the z direction) is switched on, the energy changes by an amount

$$\Delta E = -\boldsymbol{\mu} \cdot \mathbf{B} = \frac{e}{m_{\text{e}}}S_z B.$$

Since the possible values of S_z are

$$S_z = \pm\tfrac{1}{2}\hbar,$$

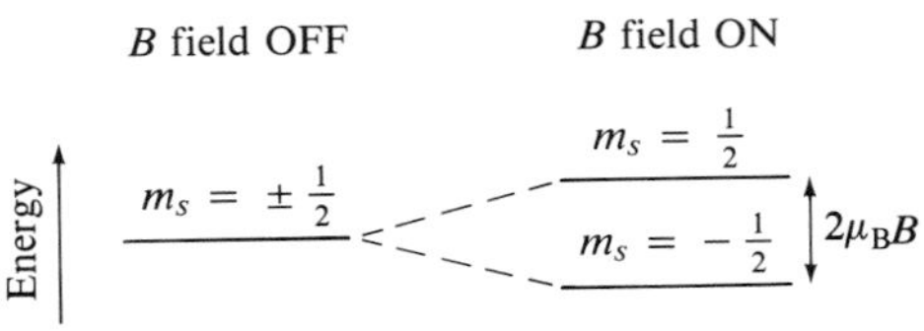

FIGURE 10.5 Any s level in hydrogen is split by a B field into two levels because of the electron's spin magnetic moment. The separation of the levels is $2\mu_B B$.

it follows that

$$\Delta E = \pm \frac{e\hbar}{2m_e} B = \pm \mu_B B. \qquad (10.27)$$

If the electron has spin up, its energy is raised by $\mu_B B$; if it has spin down, its energy is lowered by the same amount. The resulting separation of the two levels is therefore

$$\text{separation of levels} = 2\mu_B B. \qquad (10.28)$$

Notice that this separation is twice the value (10.19) predicted for the normal Zeeman effect; this is because the spin gyromagnetic ratio is twice the orbital ratio.

We conclude that any $l = 0$ level in hydrogen should be split into two neighboring levels by a magnetic field. This splitting is sketched in Fig. 10.5. That these levels are observed to split into two levels is strong evidence that the electron has a spin $s = \frac{1}{2}$ (and hence just two possible orientations). The agreement of the observed level separation with (10.28) confirms the expression (10.26) for μ_{spin}.

The Zeeman effect has been observed in many different levels of dozens of different atoms, and in all cases the results confirm that the electron's total magnetic moment is given by (10.25) as $-(e/2m_e)(\mathbf{L} + 2\mathbf{S})$, that the angular momentum vector $\mathbf{S}$ has magnitude $\sqrt{s(s+1)}\hbar$ with s always equal to $\frac{1}{2}$, and that S_z has the two possible values $\pm\frac{1}{2}\hbar$.

10.7 Fine Structure (optional)

The Zeeman effect is a splitting of atomic energy levels caused by an *externally applied* magnetic field. In most atoms there is a permanent, *internal* magnetic field due to the motion of the charges inside the atom. Even when there is no external magnetic field, this internal field can cause a small splitting of the energy levels and, hence, of the atomic spectrum. These splittings due to the internal magnetic field are called **fine structure.** As an illustration we describe briefly the fine structure of hydrogen, many of whose spectral lines were found to be *doublets* consisting of two closely spaced lines.

Let us consider the states of a hydrogen atom with some definite energy (given by quantum number n) and definite nonzero orbital angular momentum (given by quantum number l). We can understand the fine structure of these states from the following semiclassical argument: In the rest frame of the electron, the proton orbits around the electron, as shown in Fig. 10.6. Therefore, the electron finds itself in the magnetic field produced by the current loop of an orbiting positive charge. This field is proportional to the orbital frequency of the proton (as seen in the electron's rest frame), which in turn is proportional

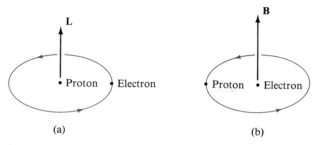

(a)   (b)

FIGURE 10.6 (a) In the proton's rest frame the electron orbits around the proton, with orbital angular momentum **L**. (Because the proton is so much heavier than the electron, this frame is very close to the rest frame of the atom as a whole, and is the frame usually considered.) (b) In the electron's rest frame, the proton orbits around the electron. The sense of the orbit is the same in both pictures. Since the proton's charge is positive, it produces a magnetic field **B** (given by the right-hand rule) in the direction shown. Therefore, **B** is in the same direction as **L**.

to the orbital angular momentum **L** of the electron, as seen in the proton's rest frame. Therefore, the electron sees a field **B** that is proportional to **L**:

$$\mathbf{B} \propto \mathbf{L}. \tag{10.29}$$

As can be seen in Fig. 10.6, the direction of **B** is the same as that of **L**. Therefore, the constant of proportionality in (10.29) is positive.

Since the electron has a magnetic moment* $\boldsymbol{\mu}_{\text{spin}} = -(e/m_e)\mathbf{S}$, the magnetic field **B** gives it an additional energy

$$\Delta E = -\boldsymbol{\mu}_{\text{spin}} \cdot \mathbf{B} = \frac{e}{m_e}\mathbf{S} \cdot \mathbf{B} \propto \mathbf{S} \cdot \mathbf{L}. \tag{10.30}$$

Because it is proportional to $\mathbf{S} \cdot \mathbf{L}$, this energy is often described as the **spin-orbit** energy. We shall see that the spin-orbit energy is usually very small by atomic standards. (Typically, it is a very small fraction of an eV.) This means that the hydrogen energy levels calculated in Chapter 9, which ignored all effects of spin, are still an excellent approximation. Nevertheless, the correction (10.30) does cause a small splitting of the levels, as we now show.

According to (10.30), the electron has a magnetic energy that depends on the orientation of **S** relative to **L**. Now, the spin **S** can have two possible orientations relative to any definite direction. Therefore, the term $\mathbf{S} \cdot \mathbf{L}$ in (10.30) can have two possible different values. This means that the states with any particular values of n and l actually belong to two sightly different energy levels. Those states in which **S** is parallel to **L** have a slightly higher energy; those in which **S** is antiparallel to **L** are slightly lower.

The argument just given applies to any state with $l \neq 0$. If $l = 0$, then since $\mathbf{B} \propto \mathbf{L}$ it follows that the B field seen by the electron is zero and hence that there is no spin-orbit splitting for s states.

The splitting of each level (with $l \neq 0$) into two levels, implies a corresponding splitting of the spectral lines of hydrogen. As an example, let us consider transitions in which a hydrogen atom in one of its $2p$ states drops to the ground state. We have seen that the ground state is not split by the spin-orbit interaction, whereas the $2p$ states are split into two levels. Thus the energy-level

---

* We need consider only the spin magnetic moment because the orbital momentum is always zero in the electron's rest frame.

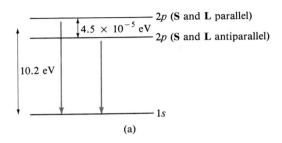

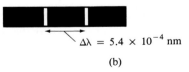

FIGURE 10.7 Fine structure in hydrogen. (a) The $1s$ states have a unique energy, while the $2p$ states belong to two slightly different energy levels, those with **S** parallel to **L** being slightly higher. (The separation of these levels is exaggerated by a factor of 50,000.) Therefore, transitions from $2p$ to $1s$ involve photons with two slightly different energies and produce a doublet of spectral lines, as shown in (b).

diagram for these states is as shown in Fig. 10.7, and there are two different possible photon energies. Therefore, the transitions from $2p$ states to the ground state produce a doublet of spectral lines, as shown in Fig. 10.7(b).

To estimate the separation of these two lines we must first find the magnetic field B of the orbiting proton, as seen by the $2p$ electron. A straightforward calculation (Problem 10.19) shows that

$$B \approx 0.39 \text{ T}. \tag{10.31}$$

According to (10.28), this implies that the $2p$ levels are separated by an energy,

$$
\begin{aligned}
\text{(separation of } 2p \text{ levels)} &= 2\mu_B B \\
&\approx 2 \times \left(5.8 \times 10^{-5} \frac{\text{eV}}{\text{T}} \right) \times (0.39 \text{ T}) \\
&\approx 4.5 \times 10^{-5} \text{ eV}. \tag{10.32}
\end{aligned}
$$

This is extremely small compared to the distance between the $2p$ and $1s$ levels, which is $(13.6 - 3.4) = 10.2$ eV. Therefore, the difference in wavelengths of the emitted photons is approximately

$$
\begin{aligned}
\Delta\lambda \approx \frac{d\lambda}{dE_{\text{ph}}} \Delta E_{\text{ph}} &= \frac{hc}{E_{\text{ph}}^2} \Delta E_{\text{ph}} \\
&= \frac{1240 \text{ eV} \cdot \text{nm}}{(10.2 \text{ eV})^2} \times (4.5 \times 10^{-5} \text{ eV}) \\
&= 5.4 \times 10^{-4} \text{ nm},
\end{aligned}
$$

as indicated in Fig. 10.7.

It should be emphasized that in our discussion of fine structure we have treated the electron as a classical orbiting particle. Also, we have all along treated the hydrogen atom nonrelativistically, and although this is certainly an excellent approximation, there are small corrections required to allow for relativity. It turns out that these relativistic corrections are of the same order of magnitude as the spin-orbit energy discussed here. (See Problems 10.19 to 10.21.) Thus a correct analysis of the fine structure of hydrogen needs to be fully quantum mechanical and to take account of relativity. Under these conditions our calculation of the splittings can only be regarded as an order-of-magnitude estimate. That the answer (10.32) is correct to two significant figures is simply a happy accident. Nevertheless, all of our general conclusions are qualitatively correct.

IDEAS YOU SHOULD NOW UNDERSTAND FROM CHAPTER 10

Orbital and spin angular momenta
Electron spin:
  magnitude—quantum number, $s = \frac{1}{2}$
  $S_z = \pm \frac{1}{2}\hbar$
Magnetic moment μ
Magnetic potential energy $= -\mu \cdot B$

Electron's orbital magnetic moment,
  $\mu_{orb} = -(e/2m_e)L$
Zeeman splitting
The Bohr magneton, μ_B
Electron's spin magnetic moment, $\mu_{spin} = -(e/m_e)S$
[Optional section: fine structure in hydrogen]

PROBLEMS FOR CHAPTER 10

SECTION 10.2 (SPIN ANGULAR MOMENTUM)

10.1 • One can visualize the quantized values of the spin angular momentum **S** with a semiclassical "vector model," as described in Section 9.6 for the orbital angular momentum **L**. In particular, the quantization of S_z requires that the vector **S** must lie on certain cones, like the one sketched in Fig. 9.15. (a) Make a sketch similar to Fig. 9.14 showing the two possible orientations of **S** for an electron. (b) What is the angle between **S** and the z axis for these two states?

10.2 • There exist subatomic particles with spin magnitudes different from that of the electron. However, in all cases they obey the same rules: The magnitude of **S** is $\sqrt{s(s+1)}\hbar$, where s is a fixed number, integer or half-integer; and the possible values of S_z are $m_s\hbar$, where m_s has the values $s, s-1, \ldots, -s$. (a) For a particle with $s = \frac{3}{2}$, how many different values of S_z are there, and what are they? (b) Draw a vector-model diagram similar to Fig. 9.14 showing the possible orientations of **S**. (c) What is the minimum possible angle between **S** and the z axis?

10.3 • Answer the same questions as in Problem 10.2, but for a particle with spin quantum number $s = 1$.

10.4 • Make a table showing the values of the four quan-

tum numbers n, l, m, m_s for each of the 18 states of the hydrogen atom with energy $E = -E_R/9$.

10.5 • Make a table showing the values of the four quantum numbers n, l, m, m_s and the energies for each of the 10 lowest-lying quantum states (not energy levels) of a hydrogen atom.

10.6 •• We have said that a classical picture of the electron as a spinning ball of matter is unsatisfactory. To illustrate this, consider the following: Modern measurements show that the electron's radius is certainly less than 10^{-18} m. Write an expression for the angular momentum of a uniform spinning ball of mass m_e, radius r, and equatorial speed v. By equating this to the observed spin $\sqrt{3}\hbar/2$, find the minimum possible value of v. What is v/c?

SECTION 10.3 (MAGNETIC MOMENTS)

10.7 • A current of 0.4 A flows around a single circular loop of radius 1 cm. (a) What is the resulting magnetic moment, μ? (b) If the loop is placed in a magnetic field $B = 1.5$ T, with μ perpendicular to **B**, what is the torque on the loop? (c) What is the difference in energy between the cases that μ is parallel to **B** and antiparallel?

10.8 • The SI units of magnetic moment μ are ampere·meter². According to (10.13), the magnetic moment of an orbiting electron is

$$\boldsymbol{\mu} = -\frac{e}{2m_e}\mathbf{L}. \qquad (10.33)$$

Verify that this has the correct units.

10.9 • A typical atomic magnetic moment is of order 10^{-23} A·m². Assuming that this is the result of a current i circulating around a single circular loop of radius 0.1 nm (a typical atomic radius), how big is i?

10.10 •• The energy of a magnetic moment $\boldsymbol{\mu}$ in a magnetic field B pointing along the z axis is $-\mu_z B$. For an electron in orbit around a proton, μ_z is given by (10.33) as $-(e/2m_e)L_z$. If $B = 10$ T (a very large field by laboratory standards) and if the electron is in a p state with $L_z = \hbar$, what is the magnetic energy due to the orbital magnetic moment (in joules and in eV)?

SECTION 10.4 (THE ZEEMAN EFFECT)

10.11 • (a) Using the known SI values of e, $\hbar$, and m_e, find the SI value of the Bohr magneton $\mu_B = e\hbar/2m_e$. (b) Given that the units ampere·meter² are the same thing as joule/tesla, find μ_B in eV/tesla.

10.12 •• (a) Verify that the Bohr magneton $\mu_B = e\hbar/2m_e$ has the units of magnetic moment: namely, ampere·meter². (b) Verify that the units ampere·meter² are the same thing as joule/tesla. (Remember that the tesla is the unit of B field, defined by the Lorentz-force equation $\mathbf{F} = q\mathbf{v} \times \mathbf{B}$.)

10.13 •• A helium atom is in an energy level with one electron occupying an s state ($l = 0$) and the other an f state ($l = 3$). The two electron spins are antiparallel so that the spin magnetic moments cancel. The atom is placed in a magnetic field $B = 0.8$ T. (a) Sketch the resulting splitting of the original energy level. (b) What is the energy difference between adjacent levels of the resulting multiplet?

10.14 •• Imagine a hydrogen atom in which the electron has no spin [so that the only magnetic moment is the orbital magnetic moment given by (10.33)]. The atom is placed in a magnetic field $B = 1.5$ T along the z axis. (a) Describe the effect of the B field on the $1s$ and $2p$ states of the hypothetical atom. Sketch the energy levels. (b) When $B = 0$, there is a single spectral line corresponding to the $2p \rightarrow 1s$ transition; how many lines does this become when B is switched on? (c) What is the fractional separation $\Delta f/f_0$ between adjacent lines?

10.15 ••• Consider two levels of the helium atom in both of which the spins are antiparallel and one electron is in an s state ($l = 0$). In the higher level the second electron occupies a d state ($l = 2$) and in the lower level it occupies a p state ($l = 1$). (a) Sketch the splitting of both levels resulting from a magnetic field along the z axis. (b) Imagine a transition from one of the d states, with $L_z = m_i\hbar$, to one of the p states, with $L_z = m_f\hbar$. Since m_i can be 2, 1, 0, -1, or -2 and m_f can be 1, 0, or -1, there are 5×3 or 15 distinct conceivable transitions. How many different photon energies would these 15 transitions produce? (c) Not all of these 15 transitions occur. In fact, it is found that the only transitions observed are those for which

$$(m_f - m_i) = 1, \quad \text{or } 0, \quad \text{or } -1. \qquad (10.34)$$

(A restriction like this on the transitions that take place is called a *selection rule,* as we discuss in Chapter 15.) Prove that because of the restriction (10.34) there are only *three* distinct photon energies produced in all possible transitions. (This means that the normal Zeeman effect always produces just three spectral lines, however large the angular momenta involved.)

SECTION 10.5 (SPIN MAGNETIC MOMENTS)

10.16 • The electron's total magnetic moment $\boldsymbol{\mu}$ is given by (10.25). (a) What are the possible values of μ_z for an electron with $l = 0$? (b) Compare these with the values of μ_z for a hypothetical spinless electron with $l = 1$.

SECTION 10.6 (THE ANOMALOUS ZEEMAN EFFECT)

10.17 • Consider a hydrogen atom in its ground level, placed in a magnetic field of 0.7 T along the z axis. (a) What is the energy difference between the spin-up and spin-down states? (b) An experimenter wishes to excite the atom from the lower to the upper level by sending in photons of the appropriate energy. What energy is this? What is the wavelength? What kind of radiation is this? (Visible? UV? . . .)

10.18 •• Consider a hydrogen atom in the $3d$ state with $L_z = 2\hbar$ and $S_z = \frac{1}{2}\hbar$. How much does its energy change if it is placed in a magnetic field $B = 0.6$ T along the z axis? [*Hint:* The total magnetic moment is given by (10.25) and the energy shift is $\Delta E = -\boldsymbol{\mu}_{tot} \cdot \mathbf{B}$.]

SECTION 10.7 (FINE STRUCTURE)

10.19 •• The fine structure of an atomic spectrum results from the magnetic field "seen" by an orbiting electron. In this question you will make a semiclassical estimate of the B field seen by a $2p$ electron in hydrogen. The B field at the center of a circular current loop, i, of radius r is known to be $B = \mu_0 i/2r$. (a) Treating the electron and proton as classical particles in circular orbits (each as seen by the other), show that the B field seen by the electron is

$$B = \frac{\mu_0}{4\pi} \frac{eL}{m_e r^3}, \qquad (10.35)$$

where L is the electron's orbital angular momentum ($L = m_e vr$ for a circular orbit). Remember that the current produced by the orbiting proton is $i = ev/2\pi r$, where v is the speed of the proton as seen by the electron (or vice versa). (b) For a rough estimate, you can give L and r their values for the $n = 2$ orbit of the Bohr model, $L = 2\hbar$ and $r = 4a_B$. Show that this gives $B \approx 0.39$ T and hence that the separation, $2\mu_B B$, of the two $2p$ levels is about 4.5×10^{-5} eV.

It should be clear that this semi-classical calculation is only a rough estimate. You have used the Bohr values for L and r. If, for example, you had used the quantum value $L = \sqrt{2}\hbar$, this would have changed your answer by a factor of $\sqrt{2}$. There is another very important reason that the argument used here is only roughly correct: The electron's rest frame is noninertial (since it is accelerated) and a careful analysis by the British physicist L. H. Thomas showed that the energy separation calculated here should include an additional factor of $\frac{1}{2}$. That our answer, 4.5×10^{-5} eV, is correct to two significant figures is just a lucky accident.

10.20 •• (a) Use Eq. (10.35) (with the Bohr values $L = 2\hbar$ and $r = 4a_B$) to show that the fine-structure separation $\Delta E_{FS} = 2\mu_B B$, of the two $2p$ levels of hydrogen can be written as

$$\Delta E_{FS} = \frac{m_e (ke^2)^4}{32\hbar^4 c^2}. \qquad (10.36)$$

(*Hint:* Since $\mu_0 \epsilon_0 = 1/c^2$ and $k = 1/4\pi\epsilon_0$, you can replace μ_0 by $\mu_0 = 4\pi k/c^2$.) (b) Show that you can rewrite (10.36) as

$$\Delta E_{FS} = \frac{\alpha^2 E_R}{16}, \qquad (10.37)$$

where α is the dimensionless *fine-structure constant*

$$\alpha = \frac{ke^2}{\hbar c}. \qquad (10.38)$$

(c) Show that $\alpha \approx 1/137$, which, together with (10.37), shows that fine structure is indeed a small effect.

10.21 •• In Problem 6.8 it was shown that the speed of an electron in a Bohr orbit is of order $v \sim \alpha c$, where α is the fine-structure constant (10.38). Using this value, you can estimate the importance of relativistic corrections to the hydrogen energies, as follows: (a) Write down the correct relativistic expression for the electron's kinetic energy and use the binomial series (Appendix B) to show that

$$K \approx \frac{1}{2} mv^2 + \frac{3}{8} \frac{mv^4}{c^2},$$

provided that $v \ll c$ (which is certainly true if $v \sim \alpha c$). Thus the relativistic correction ΔE_{rel} to the energy is about $mv^4/8c^2$. (b) Substitute $v \sim \alpha c$ and show that this gives

$$\Delta E_{rel} \sim \frac{3\alpha^2 E_R}{4}.$$

Comparing this with (10.37), we see that relativistic corrections to the hydrogen energy are of the same order as the spin-orbit correction. Therefore, a correct treatment of fine structure includes both effects.

CHAPTER

11

Multielectron Atoms; the Pauli Principle and the Periodic Table

11.1 Introduction

In Chapter 9 we saw how the Schrödinger equation could be used to predict the properties of the hydrogen atom. Another early triumph for the Schrödinger equation was that, unlike the Bohr model, it could be extended successfully to atoms with more than one electron. Today it appears that the Schrödinger equation can correctly account for the structure of all of the hundred or so multielectron atoms and for the way in which these atoms come together to form molecules. Thus, in principle at least, the Schrödinger equation can explain all of chemistry. It is the basis of this impressive accomplishment that we sketch in this chapter. We complete the story in Chapter 16 when we describe how atoms combine to form molecules.

Our first step is to find a way to handle quantum systems containing several particles. The simplest method — and the method we describe here — is called the independent-particle approximation. When this method was first applied to multielectron atoms, it predicted many atomic levels, some of which exist but many of which do not. This partial failure led to the discovery of a new law that applies to multielectron systems. We describe this law, called the Pauli exclusion principle, in Section 11.4. Armed with this principle, we can successfully describe the general properties of all multielectron atoms from helium through uranium and beyond, as we sketch in Sections 11.5 to 11.8.

11.2 The Independent-Particle Approximation

The Schrödinger equation for one electron moving around a nucleus can be solved exactly, as was described in Chapter 9. With two or more electrons, an exact solution is not possible, and one must resort to various approximations. The starting point for almost all calculations of multielectron atoms is called the **independent-particle approximation** or **IPA.** This approach is familiar from the classical theory of the solar system, in which one starts by treating the motion of each planet independently, taking account of just the dominant force, the attraction of the sun. Once one has found the orbits of all the planets separately, one can, if desired, improve these approximate orbits by correcting for the small attraction of the planets for one another.

In treating the motion of one planet it is a very good approximation to ignore the forces of the other planets in comparison with the force of the sun. The corresponding approximation for an atom is much less satisfactory: If, for example, we consider an atom with 20 electrons (calcium), it is true that the single largest force on any one electron is the force of the nucleus, with charge $20\,e$; but it would be a very poor approximation to consider *only* this force, and to ignore completely the repulsion of the 19 other electrons, with a combined charge of $-19e$. What we do instead is to treat the motion of each electron independently, taking account of the force of the nucleus *plus* the force of the *average, static distribution of the $Z - 1$ other electrons.* We shall refer to the potential energy $U(\mathbf{r})$ of each electron treated in this way as the *IPA potential energy.* The problem is now reduced to finding what the IPA potential energy $U(\mathbf{r})$ should be, and then, given $U(\mathbf{r})$, to solving the Schrödinger equation to find the possible wave functions and energies of each electron in the multielectron atom.

To implement this approach quantitatively requires a whole series of successive approximations. One must first make some reasonable *guess* for the electron wave functions; from these wave functions one can calculate the charge distribution in the atom and hence the IPA potential-energy function $U(\mathbf{r})$ of each electron. Using this potential-energy function one can solve the Schrödinger equation for each electron and obtain an improved set of wave functions. Using these improved wave functions, one can calculate an improved IPA potential-energy function $U(\mathbf{r})$; and so on. This iterative procedure is called the Hartree–Fock method and, with the aid of a large computer, can yield quite accurate atomic wave functions and energy levels.

Fortunately, we do not need to go into any details of the Hartree–Fock procedure here. Using simple known properties of the atomic charge distribution, we can get an excellent qualitative picture of the IPA potential-energy function $U(\mathbf{r})$. Using this knowledge of $U(\mathbf{r})$, we can get a good—sometimes even quantitative—understanding of the electron wave functions and hence of atomic structure.

The essential feature of the independent-particle approximation is that each electron can be considered to move independently in the average field of the $Z - 1$ other electrons plus the nucleus. In most atoms it is a good approximation to assume further that the distribution of the $Z - 1$ other electrons is spherically symmetric around the nucleus. (With this additional assumption the IPA is often called the **central-field approximation**.) We shall therefore assume that the charge distribution "seen" by any one electron is spherically symmetric, which means that the IPA potential-energy $U(\mathbf{r})$ is independent of θ and ϕ, and can be written as $U(r)$. This greatly simplifies our discussion, for as we saw in Chapter 9, when $U(r)$ is spherically symmetric, the angular part of the Schrödinger equation always has the same solutions, characterized by the familiar angular-momentum quantum numbers l and m.

The main features of $U(r)$ are easily understood if we recall two properties of spherical charge distributions, both of which follow from Gauss's law. First, if an electron is *outside* a spherical distribution of total charge Q, the electron experiences exactly the same force as if the entire charge Q were concentrated at its center:

$$F = k \frac{Qe}{r^2}. \tag{11.1}$$

Second, if the electron is *inside* a spherical shell of charge, it experiences no force at all from the shell.

If the electron in which we are interested is close enough to the nucleus, it will be inside all the other electrons and will experience the force of the nuclear charge Ze but no force at all from the other electrons; in other words, it feels the full attractive force of the nucleus:

$$F = \frac{Zke^2}{r^2} \qquad [r \text{ inside all other electrons}]. \tag{11.2}$$

If we now imagine the electron moving outward from the nucleus, it will steadily move outside more and more of the other electrons; thus the force will still be given by (11.1) but with Q equal to the nuclear charge Ze *reduced by* the charge of those electrons inside the radius r. Eventually, when the electron is outside all of the other $Z - 1$ electrons, Q is given by Ze minus $(Z - 1)e$; that is, $Q = e$ and

$$F = \frac{ke^2}{r^2} \qquad [r \text{ outside all other electrons}]. \tag{11.3}$$

Therefore, an atomic electron that is outside all its fellow electrons experiences the same force as the one electron in hydrogen.

The potential energy $U(r)$ of the electron is the integral of the force. It follows from the discussion above that when r is outside all the other electrons

$$U(r) = \frac{-ke^2}{r} \qquad [r \text{ outside all other electrons}]. \tag{11.4}$$

On the other hand, as r approaches zero,

$$U(r) \approx -\frac{Zke^2}{r} \qquad [\text{as } r \to 0], \tag{11.5}$$

since then the electron of interest is inside all the others.* Between these two regions, $U(r)$ connects these two functions smoothly, as shown qualitatively in Fig. 11.1.

We can express the behavior shown in Fig. 11.1 by writing $U(r)$ as

$$U(r) = -Z_{\text{eff}}(r)\frac{ke^2}{r}. \tag{11.6}$$

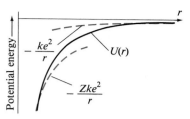

FIGURE 11.1 The IPA potential energy $U(r)$ of an atomic electron in the field of the nucleus plus the average distribution of the $Z - 1$ other electrons. As $r \to \infty$, U approaches $-ke^2/r$; as $r \to 0$, U approaches $-Zke^2/r$, as in Eq. (11.5).

Here $Z_{\text{eff}}(r)$ gives the **effective charge** "felt" by the electron and depends on r. When r is inside the other electrons, Z_{eff} approaches the full nuclear charge:

$$Z_{\text{eff}} \approx Z, \qquad [r \text{ inside all other electrons}]. \tag{11.7}$$

As r increases and the nuclear charge is shielded, or *screened,* by more and more of the other electrons, Z_{eff} decreases steadily, until as r moves outside all other electrons, Z_{eff} approaches 1:

$$Z_{\text{eff}} \approx 1 \qquad [r \text{ outside all other electrons}]. \tag{11.8}$$

11.3 The IPA Energy Levels

Once the potential energy $U(r)$ is known, our next step is to find the energy levels and wave functions of each electron. The potential energy $U(r)$ is sufficiently like the hydrogen potential energy that we can get a good qualitative understanding of the solutions by analogy with what we already know about hydrogen and hydrogen-like ions.

Just as with hydrogen, $U(r)$ depends only on r, and the Schrödinger equation separates. In particular, the two angular equations are exactly the same as for hydrogen. This means that the states of definite energy have angular momentum given by the familiar orbital quantum numbers l and m, and all of the $2l + 1$ different orientations given by $m = l, l - 1, \ldots, -l$ have the same energy. Since $U(r)$ does not involve the spin at all, the energy is also the same for both orientations, $m_s = \pm\frac{1}{2}$, of the spin. Thus each level has a degeneracy of at least $2(2l + 1)$.

As was the case with hydrogen, the solutions of the radial equation are characterized by a principal quantum number n; and a quantum state is completely specified by the four quantum numbers n, l, m, and m_s. The lowest energy level is $1s$ (that is, $n = 1$, $l = 0$) and is twofold degenerate because of the two possible orientations of the spin. Just as in hydrogen, the $1s$ wave function is concentrated closer to the nucleus than any other wave functions. This means that in the region where the $1s$ wave function is large, $U(r)$ is close to the hydrogen-like potential energy with $Z_{\text{eff}} \approx Z$. Therefore, the $1s$ wave function approximates that of a hydrogen-like ion with nuclear charge Ze; the $1s$ energy is close to $-Z^2 E_R$,

$$E_{1s} \approx -Z^2 E_R; \tag{11.9}$$

and the most probable radius is about a_B/Z (as described in Section 9.10).

Just as with hydrogen, the next energy level has $n = 2$. But here an impor-

---

* It follows from (11.2) that $U(r)$ is $-Zke^2/r$ *plus some constant.* However, as $r \to 0$ this constant is negligible compared to $-Zke^2/r$, and we have ignored it in (11.5). The corresponding constant in (11.4) is exactly zero since we are defining U to be zero at $r = \infty$.

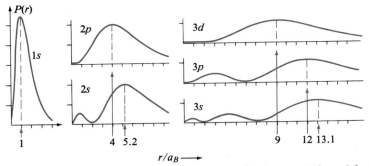

FIGURE 11.2 The radial distributions for the $n = 1$, 2, and 3 states in hydrogen. The numbers shown are the most probable radii in units of a_B.

tant difference emerges. In hydrogen, the $2s$ and $2p$ states are degenerate, whereas in multielectron atoms the $2s$ states are lower in energy. This difference is easy to understand if we look at the $2s$ and $2p$ radial distributions shown in Fig. 9.23, which we reproduce here as Fig. 11.2. These two distributions peak at about the same radius, four or five times further out than the $1s$ distribution. This means that the $2s$ and $2p$ wave functions are concentrated in a region where the nuclear charge Ze is screened by any electrons in the $1s$ states, and the $2s$ and $2p$ electrons see an effective charge Z_{eff} which is less than Z. However, the $2s$ distribution (unlike the $2p$) has a secondary maximum much closer in. That is, a small part of the $2s$ distribution penetrates the region where Z_{eff} is close to the full nuclear value $Z_{\text{eff}} \approx Z$. Therefore, on average, a $2s$ electron is more strongly attracted to the nucleus than is a $2p$ electron. This means that the $2s$ electron is more tightly bound and has lower energy.

With the $n = 3$ states there is a similar separation of energies. In Fig. 11.2 it can be seen that both the $3s$ and $3p$ distributions have secondary peaks near $r = 0$, with one of the $3s$ peaks much closer in. Therefore, the $3s$ state penetrates closest to the nucleus and has the lowest energy, the $3p$ is next, and finally the $3d$. This trend is repeated in all higher levels: For each value of n, states with smaller l penetrate closer to the nucleus and are lower in energy. This systematic lowering of the energy for states with lower l is shown schematically in Fig. 11.3.

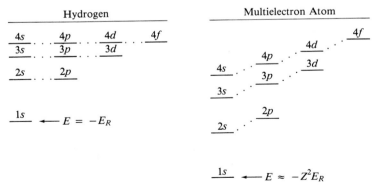

FIGURE 11.3 Schematic energy-level diagrams for a hydrogen atom and for one of the electrons in a multielectron atom. In hydrogen, all states with the same n are degenerate. In multielectron atoms, states with lower l are more tightly bound because they penetrate closer to the nucleus. In many atoms this effect results in the $4s$ level being lower than the $3d$, as shown here.

In many atoms the lowering in energy of the "penetrating orbits" becomes so important that the order of certain levels can be reversed, as compared to hydrogen. This is illustrated in Fig. 11.3, which shows the $4s$ level slightly *below* the $3d$. We shall see that such reversals of the order of energy levels have an important effect on the chemical properties of many elements.

For a hydrogen-like ion we saw that all states with a given n tend to cluster in a spatial shell, with radius roughly equal to the Bohr value $r \approx n^2 a_B/Z$. This same clustering into spatial shells occurs in multielectron atoms and is, in fact, more pronounced. The $n = 1$ states are closest to the nucleus and feel nearly the full nuclear charge Ze; therefore, their most probable radius is close to the Bohr value, a_B/Z, for the $n = 1$ state of a hydrogen-like ion with charge Ze. The states with successively higher n are concentrated at progressively larger radii where they feel an effective charge $Z_{eff}e$ that is progressively *smaller*. Now, the most probable radius for these states is roughly

$$r_{mp} \approx \frac{n^2 a_B}{Z_{eff}}. \qquad (11.10)$$

Since Z_{eff} gets *smaller* as n gets *larger*, the proportionate separation of the spatial shells is even greater in a multielectron atom than it is in hydrogen. Notice that this clustering into spatial shells is according to n, exactly as it is in hydrogen. This contrasts with the energy levels, whose order deviates, as we have seen, from the simple hydrogen ordering.

11.4 The Pauli Exclusion Principle

Knowing the possible states of each electron in a multielectron atom, we are now ready to discuss the possible states of the whole atom. Let us consider first the atomic ground states. For these our problem is to decide how the electrons are to be distributed among their possible states so that the atom as a whole has the minimum possible energy.

One might expect that the ground state of any atom would be found by placing all of its Z electrons into the lowest, $1s$, state; but this is not what is observed to happen. The explanation for what *does* happen was discovered by the Austrian physicist Pauli, who proposed a new law, which has come to be called the **Pauli exclusion principle**. This principle states that:*

> **PAULI EXCLUSION PRINCIPLE**
> No two electrons in a quantum system
> can occupy the same quantum state.

(11.11)

Pauli was led to this law by a study of the states of many atoms, and to this day, the best evidence for the Pauli principle is its success in explaining the diverse properties of all the atoms. There is, however, evidence from many other fields as well. For example, the electrons in a conductor are found to obey the exclusion principle, and many of the observed properties of conductors (conductivity, specific heat, magnetic susceptibility, etc.) depend in a crucial way on the validity of the principle.

* The Pauli principle applies to many other particles besides the electron. For example, it applies to protons and to neutrons, and has important consequences for the energy levels of nuclei, as we shall see in Chapter 12. In this chapter, however, we are concerned only with electrons.

WOLFGANG PAULI (1900–1958, Austrian). At the age of 21, Pauli published a review of relativity that is still regarded as a masterpiece. He made many fundamental contributions to quantum physics, including the exclusion principle (1925), for which he won the 1945 Nobel Prize, the neutrino hypothesis (Chapter 13), and work in relativistic quantum field theory.

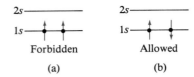

Forbidden   Allowed

(a)   (b)

FIGURE 11.4 The ground state of helium has both electrons in the 1s level. Because of the Pauli principle, their spins have to be in opposite directions.

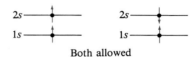

Both allowed

FIGURE 11.5 The lowest excited state of helium has one electron in the 1s and one in the 2s level. The Pauli principle places no restrictions on the spin orientations in this case.

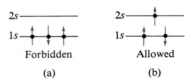

Forbidden   Allowed

(a)   (b)

FIGURE 11.6 (a) It is impossible to put three electrons in the 1s level without violating the Pauli principle. (b) Therefore, the ground state of lithium has two electrons in the 1s level and one in the 2s level. The third electron's spin can point either way.

To illustrate the exclusion principle and some of the evidence for it, let us consider two simple atoms, helium and lithium. First, let us imagine putting together a helium atom ($Z = 2$) from a helium nucleus and two electrons. If we add one electron to the nucleus, its lowest possible state is the 1s state ($n = 1$, $l = m = 0$) with its spin either up or down ($m_s = \pm\frac{1}{2}$). If we next add the second electron, it too can go into the 1s state. But according to the exclusion principle the two electrons cannot occupy exactly the same quantum state. Since they have the same values of n, l, and m, they must have different values of m_s; that is, if both electrons are in the 1s state, their spins must be antiparallel. This situation is sketched in Fig. 11.4. Figure 11.4(a) shows two electrons in the 1s state with spins parallel, a situation that is never observed; Fig. 11.4(b) shows two electrons in the 1s state with spins antiparallel, the situation that *is* observed. The two possibilities shown in Fig. 11.4 would be easily distinguishable, since the first would have a nonzero magnetic moment, while the second has $\mu = 0$. That the helium ground state is always found to have $\mu = 0$ is clear evidence for the exclusion principle.

The situation with the excited states of helium is different. For example, there is an excited state with one electron in the 1s level and the other in the 2s level. In this case the two electrons are certainly in different quantum states, whatever their spin orientations (parallel or antiparallel), as shown in Fig. 11.5. Thus the Pauli principle does not forbid either of these arrangements, and both are observed, the first with $\mu \neq 0$ and the second with $\mu = 0$.

As a second example, let us imagine putting together a lithium atom ($Z = 3$) from a lithium nucleus and three electrons. When we add the first two electrons they can both go into the 1s level, provided that their spins are antiparallel. But since there are only two possible orientations of the spin, there is now no way in which the third electron can go into the 1s level [Fig. 11.6(a)]. The Pauli principle requires that the third electron go into some higher level, the lowest of which is the 2s. Therefore, the ground state of lithium has to be as shown in Fig. 11.6(b), with the third electron in the 2s level and its spin either up or down.

In general, the Pauli principle implies that any s level (1s, 2s, etc.) can accommodate two electrons but no more. Levels with higher angular momentum can accommodate more electrons, because their degeneracy is larger. For example, any level with $l = 1$ contains six distinct quantum states ($6 = 2 \times 3$, since there are two orientations of **S** and three orientations of **L**); therefore, any p level ($l = 1$) can accommodate six electrons, but no more. Similarly, any d level, with $l = 2$, has ten distinct states (2×5) and can accommodate ten electrons, but no more.

11.5 Ground States of the First Few Elements

To determine the ground state of an atom we have only to assign its Z electrons to the lowest individual energy levels consistent with the Pauli principle (that no two electrons occupy the same quantum state). In this section we use this procedure to find the ground states of the lightest few atoms, starting with hydrogen ($Z = 1$) and going as far as sodium ($Z = 11$).

The ground state of hydrogen has its one electron in the 1s level, with its spin pointing either way. The energy is $E = -E_R = -13.6$ eV, which means that the energy needed to remove the electron — the **ionization energy** — is 13.6 eV.

Moving on to helium, we already saw that the ground state has both electrons in the $1s$ level, with their spins antiparallel. Because of the greater nuclear charge ($Z = 2$) the $1s$ level of helium is much lower in energy than that of hydrogen. If we write the energy of either electron as $-Z_{\text{eff}}^2 E_R$, then Z_{eff} will not be equal to the full nuclear charge, 2, since each electron is screened by the other. Nevertheless, Z_{eff} should be appreciably more than 1. Thus we would expect that helium should be significantly more tightly bound than hydrogen. This prediction is well borne out by experiment: The first ionization energy of helium (the energy to remove one electron) is found to be 24.6 eV, nearly twice that of hydrogen.

Another measure of an atom's stability is its first excitation energy, the energy to lift it to its first excited state. In both H and He, this involves lifting one electron from the $1s$ to the $2s$ level. In helium this should require a larger energy by a factor of roughly Z_{eff}^2, the same ratio as for the ionization energies. This, too, is confirmed by experiment: The first excitation energy of He is 19.8 eV, compared to 10.2 eV in H.

The ionization and excitation energies of an atom are important indicators of the atom's stability. On both of these counts helium is about twice as stable as hydrogen. In fact, helium has the largest ionization and excitation energies of *any* atom. Since high stability tends to imply low chemical activity, we might guess that helium should be chemically inactive; and this proves to be the case. Helium is one of the six **noble,** or **inert, gases,** which show almost no chemical activity at all, which form no really stable compounds, and which can bind together into liquid or solid form only at relatively low temperatures.

Another important difference between the hydrogen and helium atoms concerns their sizes. We have seen that the wave functions of a hydrogen-like ion are scaled inward by a factor of $1/Z$, compared to the corresponding wave functions of hydrogen. Therefore, the radius of the $1s$ wave function of helium should be about $1/Z_{\text{eff}}$ times that of hydrogen, and the He atom should therefore be roughly two-thirds the size of the H atom. This prediction also is correct. The precise value of the atomic radius depends on how one chooses to define it, but representative values are 0.08 nm for H and 0.05 nm for He.

The differences in energy and radius between hydrogen and helium reflect the larger nuclear charge ($Z = 2$) of helium. When we consider the lithium atom ($Z = 3$), we encounter a new kind of difference, due to the Pauli exclusion principle. Let us consider first an electron in the $1s$ level of Li. Because of the greater nuclear charge, this $1s$ electron is more tightly bound, and concentrated at a smaller radius, than a corresponding electron in either He or H. However, the Pauli principle allows only two of the lithium's three electrons to occupy the $1s$ level; the third electron must occupy the $2s$ level and is much less tightly bound. In fact, we can easily estimate the binding energy of this outermost electron: Because it is outside the two other electrons, it sees an effective charge of order $Z_{\text{eff}} = 1$, about the same as for the one electron in hydrogen. Therefore, since it is in the $n = 2$ level, it should have about the same ionization energy as an $n = 2$ electron in hydrogen, 3.4 eV. This estimate agrees reasonably with lithium's observed ionization energy of 5.4 eV. (That the actual value, 5.4 eV, is a bit larger than our estimate of 3.4 eV shows that the outer electron is not perfectly shielded by the inner two, and sees an effective charge somewhat

TABLE 11.1

Ionization energies and radii of the first four atoms. Atomic numbers Z are shown as subscripts on the left of chemical symbols. The energy levels are not to scale, since corresponding levels get deeper as Z increases.

	$_1$H	$_2$He	$_3$Li	$_4$Be
Ionization energy (eV):	13.6	24.6	5.4	9.3
Radius (nm):	0.08	0.05	0.20	0.14
Occupancy of energy levels:				

greater than $Z_{eff} = 1$.) This ionization energy is the fifth smallest of any stable atom and means that the lithium atom can easily lose its outermost electron. This is the main reason why lithium is so chemically active, as we describe in Section 16.2.

Because the outer electron of lithium is in the $n = 2$ level, the Li atom should have a much larger radius than either He or H. This prediction is confirmed by the data in Table 11.1, which shows the ionization energies and radii of the first four atoms, $_1$H, $_2$He, $_3$Li, and $_4$Be. (When convenient we indicate the atomic number, Z, by a subscript on the left of the atomic symbol — not to be confused with the mass number, A, which is sometimes shown as a *superscript* on the left.)

In beryllium ($Z = 4$) the fourth electron can join the third electron in the $2s$ level, provided that their spins are antiparallel. Because of the larger nuclear charge, this level is more tightly bound, and its radius smaller, than in $_3$Li. Therefore, the $_4$Be atom should have a larger ionization energy and a smaller radius, as the data in Table 11.1 confirm.

To some small extent the $_4$Be atom with its filled $2s$ level is similar to the $_2$He atom with its filled $1s$ level. But the differences are more important than the similarities. In particular, the $_2$He atom is not only hard to ionize, it is also hard to excite, 19.8 eV being needed to lift one of the electrons from the $1s$ to the $2s$ level. Excitation of the $_4$Be atom requires only that one of the $2s$ electrons be lifted to the nearby $2p$ level, just 2.7 eV higher (see Fig. 11.7). This means that one of the electrons in Be can easily move to the higher level. As we shall see in Chapter 16 (Problem 16.28), this allows Be to bond to other atoms. For this reason Be, unlike He, is chemically active and forms a number of compounds.

After beryllium ($_4$Be) comes boron ($_5$B). The first four electrons of $_5$B go into the $1s$ and $2s$ levels, just as with $_4$Be. But because of the Pauli principle, the last electron of $_5$B must occupy the $2p$ level. Therefore, in moving from $_4$Be to $_5$B we see two competing effects: The increase in Z causes any given level to be somewhat more tightly bound, but the final electron has to occupy a level that is slightly higher and hence somewhat less tightly bound. As far as ionization energy is concerned, the second effect wins: The ionization energy of $_5$B is 8.3 eV, just a little less than the 9.3 eV of $_4$Be. On the other hand, the radius of $_5$B is less than that of $_4$Be and continues the trend of shrinking radii with increasing Z. In neither case is the difference large.

The six elements after $_4$Be are:

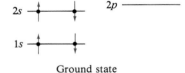

Ground state

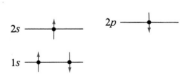

First excited state

FIGURE 11.7 Excitation of beryllium ($Z = 4$) requires only 2.7 eV to lift one of the $2s$ electrons to the nearby $2p$ level. In the excited state the spins of the $2s$ and $2p$ electrons can point either way.

Element:	boron	carbon	nitrogen	oxygen	fluorine	neon
Symbol:	$_5$B	$_6$C	$_7$N	$_8$O	$_9$F	$_{10}$Ne

In all of these atoms, the first four electrons fill the $1s$ and $2s$ levels. Since the $2p$ level can hold six electrons, the remaining electrons all go into the $2p$ level. Thus, as we move from $_5$B to $_{10}$Ne, the additional electron of each succeeding atom goes into the same level, and the main differences should be due to the increasing nuclear charge Z; in particular, the ionization energy should increase and the radius should decrease. These trends show up clearly in Fig. 11.8, in which the ionization energies and atomic radii are plotted as functions of Z for all the first 11 atoms. With one small exception the two graphs change steadily in the expected directions (ionization energy increasing, radius decreasing) as Z increases from 5 to 10. The one exception is the small drop in ionization energy as one moves from $_7$N to $_8$O; we shall return to this anomaly later.

When we reach $_{10}$Ne the $2p$ level has its full allotment of six electrons. Therefore, when we move on to sodium, $_{11}$Na, the last electron must go into the next, and much higher, level—the $3s$ level. This reverses all of the trends set by the last eight atoms: The ionization energy drops abruptly and the radius increases abruptly, as is shown clearly in Fig. 11.8.

Both of the graphs in Fig. 11.8 suggest a parallel between $_3$Li and $_{11}$Na. Both atoms have unusually low ionization energies and unusually large radii. The low ionization energies mean that both atoms can easily lose one electron. As we shall see, this allows lithium and sodium to combine with other atoms to form many different chemical compounds and is the reason why both atoms are chemically so active.

The similarity between $_3$Li and $_{11}$Na is an example of the *periodic* behavior of the elements: As we consider elements with successively higher Z, chemical similarities recur at certain regular, or periodic, intervals. Another example of this periodicity is the pair of elements $_2$He and $_{10}$Ne. Both are very stable (large ionization and excitation energies) and very small in size. In Chapter 9 we mentioned that the word "shell" is often used for a group of energy levels that are close to one another and well separated from any others. From the first

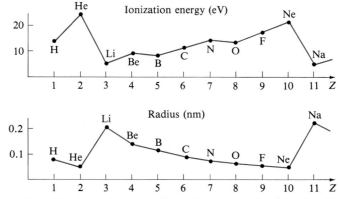

FIGURE 11.8 The ionization energies and atomic radii of the first 11 elements.

graph in Fig. 11.8 it is clear that the $1s$ level should be considered as one shell by itself, and $2s$ and $2p$ together as another. For this reason, helium and neon are called **closed-shell atoms.** We shall see that there are six closed-shell atoms in all and that they are the six noble gases. In the same way lithium and sodium can be described as being **closed-shell-plus-one,** and are the first of six such elements, called the **alkali metals.**

Just before the stable $_{10}$Ne is fluorine, $_9$F. The ionization energy of fluorine is 17.4 eV, which is the third largest ionization energy of any atom. One might, therefore, imagine that fluorine would be chemically inactive, but such is definitely *not* the case. Fluorine is one of the most active of all the elements. The reason for this activity is that the fluorine atom is **closed-shell-minus-one,** since its $2p$ level is one short of the full complement of six electrons. Because of the large nuclear charge, the $2p$ level is very well bound (as the large ionization energy testifies). In fact, the $2p$ level of F is so well bound that it can bind an extra, sixth electron. That is, the negative ion, F$^-$, is stable, with the extra electron just filling the $2p$ level. The tendency of an atom to bind an extra electron is measured by its **electron affinity.** This is defined as the energy released when the atom captures an extra electron and forms a negative ion (or, equivalently, the energy needed to remove one electron from the negative ion). The electron affinity of fluorine is 3.4 eV, the third largest for any element. As we shall discuss later, it is because of its ability to bind an extra electron that fluorine is so active.

11.6   The Remaining Elements

In Section 11.5 we examined the ground states of the first 11 elements. In this section we sketch a similar analysis of some of the remaining 90 or so elements. This will emphasize what was already becoming apparent. Because of the Pauli principle, the properties of atoms do not vary smoothly and uniformly as functions of Z. Rather, as we examine atoms with successively more electrons, their properties vary more-or-less smoothly as long as each extra electron can be accommodated in the same shell; but each time a shell is filled and a new shell comes into play, there is an abrupt change in the properties, reversing the previous smooth trends. As we saw in Section 11.5, this leads to the periodic occurrence of atoms with similar physical and chemical properties.

To find the ground state of an atom we must assign the Z electrons to the lowest levels consistent with the Pauli principle. For $_1$H and $_2$He the electrons go into the $1s$ level. For $_3$Li through $_{10}$Ne the $1s$ level is full, and the outer, or **valence,** electrons go into the $2s$ and then $2p$ levels. Similarly, when we move on to the elements 11 through 18 (sodium through argon) the $1s$, $2s$, and $2p$ levels are all full, and the valence electrons go into the $3s$ and then $3p$ levels.

Perhaps the most descriptive way to show the assignment of electrons to energy levels is with energy-level diagrams like those in Fig. 11.7. Unfortunately, these diagrams become increasingly cumbersome as we discuss atoms with more electrons. A more compact way to show the same information is to give the **electron configuration,** which is just a list of the occupied levels, each with a superscript to indicate the number of electrons in it. For example, the electron configuration of the sodium ground state is

$$_{11}\text{Na:}\quad 1s^2 2s^2 2p^6 3s^1.$$

TABLE 11.2

Electron configurations of the ground states of the first 18 elements.

First shell	Second shell	Third shell
$_1$H : $\quad 1s^1$	$_3$Li : $\quad 1s^2 2s^1$	$_{11}$Na: $\quad 1s^2 2s^2 2p^6 3s^1$
$_2$He: $\quad 1s^2$	$_4$Be : $\quad 1s^2 2s^2$	$_{12}$Mg: $\quad 1s^2 2s^2 2p^6 3s^2$
	$_5$B : $\quad 1s^2 2s^2 2p^1$	$_{13}$Al : $\quad 1s^2 2s^2 2p^6 3s^2 3p^1$
	$_6$C : $\quad 1s^2 2s^2 2p^2$	$_{14}$Si : $\quad 1s^2 2s^2 2p^6 3s^2 3p^2$
	$_7$N : $\quad 1s^2 2s^2 2p^3$	$_{15}$P : $\quad 1s^2 2s^2 2p^6 3s^2 3p^3$
	$_8$O : $\quad 1s^2 2s^2 2p^4$	$_{16}$S : $\quad 1s^2 2s^2 2p^6 3s^2 3p^4$
	$_9$F : $\quad 1s^2 2s^2 2p^5$	$_{17}$Cl : $\quad 1s^2 2s^2 2p^6 3s^2 3p^5$
	$_{10}$Ne: $\quad 1s^2 2s^2 2p^6$	$_{18}$Ar : $\quad 1s^2 2s^2 2p^6 3s^2 3p^6$

With this notation, the ground states of the first 18 elements are as shown in Table 11.2.

The properties of elements 11 to 18 closely parallel those of elements 3 to 10. As we have already noted, $_{11}$Na and $_3$Li are both easily ionized and are chemically very active. As one moves from $Z = 11$ to $Z = 18$ and the $3s$ and $3p$ levels fill, the ionization energies increase (with two small exceptions) and the atomic radii decrease, just as occurred between $Z = 3$ and $Z = 10$. At $Z = 18$ the $3p$ level is completely full, and $_{18}$Ar, like $_{10}$Ne, is very stable and chemically inert. Just before $_{18}$Ar is chlorine ($_{17}$Cl) which, like $_9$F, is able to accept an extra electron into the one vacancy in its outer p level, and is therefore chemically active.

When we move beyond $_{18}$Ar to $_{19}$K (potassium) the story becomes more complicated. One might expect that the next level occupied would be the $3d$ level. In fact, however, the tendency for levels with low angular momentum (the "penetrating orbits") to have lower energy causes the $4s$ level to be slightly lower than the $3d$, as discussed in Section 11.3. Therefore, the configuration of the ground state of $_{19}$K is:

$$_{19}\text{K}: \quad 1s^2 2s^2 2p^6 3s^2 3p^6 4s^1.$$

The order in which the energy levels fill is shown in Fig. 11.9. In this picture we have also shown the grouping of the energy levels into **shells** containing levels that are close to one another but well separated from other levels. (Note that, in this context, an individual level within a shell is sometimes called a **subshell**.) We see, for example, that the $3s$ and $3p$ levels are close together, but that the gap from $3p$ to $4s$ is large. Thus $3s$ and $3p$ form a shell by themselves, just like $2s$ and $2p$. Therefore, $_{18}$Ar is a closed-shell atom, like $_{10}$Ne and $_2$He, and is the third of the noble gases. The next element, potassium ($_{19}$K), is a *closed-shell-plus-one* atom, with low ionization energy, and is one of the alkali metals, similar to its two predecessors $_{11}$Na and $_3$Li.

It can be seen in Fig. 11.9 that, as it happens, the lowest level in each shell is always an s level, and the highest is a p level (except in the first shell, which has only the $1s$ level). Thus all the closed-shell-plus-one atoms have a single s electron outside a closed-shell core; all closed-shell-minus-one atoms (often called **halogens**) have one vacancy, or *hole,* in the p level of an otherwise filled shell.

After $_{19}$K comes calcium ($_{20}$Ca), whose $4s$ level is full. Then with scandium ($_{21}$Sc) the $3d$ level begins to fill. Since any d level can hold 10 electrons, the

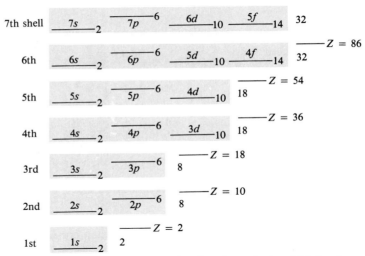

FIGURE 11.9 Schematic diagram showing the order in which levels are occupied as one considers atoms with successively higher Z. This is not the energy-level diagram for any one atom; it just gives the order in which levels are occupied as Z increases. The shaded rectangles indicate the groupings of nearby levels into energy shells. The figure to the right of each level or shell gives the number of electrons that can be accommodated; the figures on the far right are the atomic numbers of the closed-shell atoms. Note that some levels are too close to be ordered unambiguously; for example, $5d$ is occupied before $4f$, but $4f$ is completely filled before $5d$. The exact sequence of occupancy can be found from the electron configurations shown inside the back cover.

filling of the $3d$ level takes us from $_{21}$Sc through $_{30}$Zn (zinc). These 10 elements are called *transition metals* and are alike in several ways. This is because the $3d$ wave functions peak at a much smaller radius than the $4s$ functions (Fig. 11.10). Thus, although it is the $3d$ level that is filling in this sequence of elements, it is the $4s$ electrons that are farthest out. Since it is the outer electrons that can interact with other atoms, all of the transition metals from $_{21}$Sc to $_{30}$Zn have similar chemical properties.

Once the $3d$ level is full at $_{30}$Zn, the $4p$ level starts to fill, and the sequence of six elements from $_{31}$Ga (gallium) to $_{36}$Kr (krypton) is analogous to the sequences from $_{13}$Al to $_{18}$Ar and from $_5$B to $_{10}$Ne, in which the $3p$ and $2p$ levels were filling. In particular, $_{36}$Kr is a closed-shell atom and is the fourth of the noble gases.

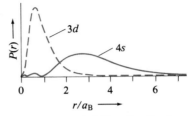

FIGURE 11.10 Typical radial probability distributions for a $3d$ and $4s$ electron in the transition elements $_{21}$Sc through $_{30}$Zn. Note how the $4s$ distribution peaks more than four times farther out than the $3d$. For this reason it is the $4s$ electrons that determine the chemical properties of the transition metals. (The distributions shown are for $_{29}$Cu; courtesy Dr. Steven O'Neil.)

Beyond $_{36}$Kr the fifth, sixth, and finally, seventh shells are occupied, but since there are no really important new features we need not go into details. The periodic recurrence of similar atomic properties can be clearly seen in plots of a variety of properties as functions of Z. In Fig. 11.11 we show ionization energies, atomic radii, and electron affinities for $Z = 1$ through 86. (The first two graphs are extensions of those shown in Fig. 11.8.) Notice how the ionization energies have pronounced maxima at the closed-shell noble gases, and minima at the closed-shell-plus-one alkali metals. The ionization energies have smaller maxima at several other points, and most of these are easily explained. For example, the small peak at $_4$Be is because the $2s$ level (or subshell) is filled; that at $_{30}$Zn is because the $3d$ level is filled.

The small peak in the ionization energy at $_7$N is a little harder to explain and is due to a subtle, but important, change when we move from $_7$N to $_8$O. The configuration of nitrogen is

$$_7\text{N:} \quad 1s^2 2s^2 2p^3,$$

with three electrons in the $2p$ level. The six distinct quantum states of the $2p$

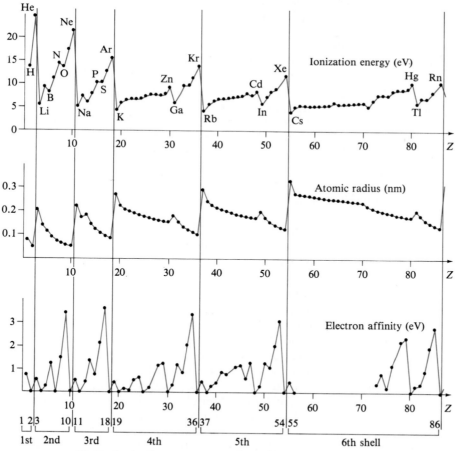

FIGURE 11.11 Ionization energy, atomic radius, and electron affinity as functions of atomic number Z. The vertical lines separate complete shells. The electron affinities for elements 57 through 72 have not been measured. (Data on electron affinities courtesy of Professor Carl Lineberger.)

level arise from the three possible wave functions (the "orbitals" $2p_x$, $2p_y$, and $2p_z$ shown in Fig. 9.21*) each with two possible spin orientations. Therefore, the three outer electrons in nitrogen can occupy three distinct orbitals. This arrangement keeps the electrons well separated, reducing their electrostatic repulsion, and makes the nitrogen atom relatively stable, with a comparatively large ionization energy. In oxygen, the fourth $2p$ electron must occupy the same orbital as one of the others (with opposite spin, of course). Because the distributions of these two electrons overlap strongly, their electrostatic repulsion is relatively large, and the ionization energy of oxygen is a little less. It is for this reason that the plot of ionization energy in Fig. 11.11 has the small drop after $_7$N. For the same reason, there is a small drop in ionization energy after $_{15}$P, also visible in Fig. 11.11.

The graph of atomic radius against Z shows the expected trends, with R falling to a minimum at each of the noble gases and jumping abruptly to a maximum at each closed-shell-plus-one atom. The downward slope from $_{21}$Sc to $_{30}$Zn is very gentle because the last electrons of these transition elements go into the $3d$ level, which is concentrated at a smaller radius than the occupied $4s$ level. Thus the atomic radius is determined by the $4s$ electrons and changes little from $Z = 21$ to 30 (from 0.21 to 0.15 nm). The even gentler slope from $Z = 57$ to 71 is a similar effect: With these *inner transition* elements, the level being filled is the $4f$ level, which has *much* smaller radius than the occupied $6s$ level; thus the atomic radius is almost completely determined by the $6s$ radius and changes very little (from 0.27 to 0.22 nm).

The third graph in Fig. 11.11 shows the electron affinity, the energy released when an atom captures an extra electron and forms a negative ion. As expected, this shows sharp maxima at the closed-shell-minus-one atoms (the halogens, $_9$F, $_{17}$Cl, etc.) and drops abruptly to zero at the closed shells ($_2$He, $_{10}$Ne, etc.). It is also zero at those atoms with closed subshells, such as $_4$Be ($1s^2 2s^2$) and $_{12}$Mg ($1s^2 2s^2 2p^6 3s^2$). Some atoms with *half-filled* subshells also have zero electron affinity. For example, an eighth electron in $_7$N would have to occupy the same orbital, $2p_x$, $2p_y$, or $2p_z$, as one of the other $2p$ electrons, and the resulting Coulomb repulsion would raise its energy so high that it would not be bound.

To conclude this section, we mention one more atomic property that can be predicted from our knowledge of the electron states — the angular momentum of the whole atom. For many atoms, calculation of the total angular momentum is difficult and well beyond our present discussion. However, there are several cases that are quite straightforward. This is because the total angular momentum of all the electrons in any filled level is zero. To understand this useful result, we have only to note that in a filled level there are equal numbers of electrons with spin up and spin down; therefore, the total spin $\Sigma\mathbf{S}$ is certainly zero. Further, for each electron with a given value of L_z there is another with exactly the opposite value (since all the values $m = l, l-1, \ldots, -l$ are occupied); therefore, the total orbital angular momentum $\Sigma\mathbf{L}$ is also zero. It follows that the total angular momentum of any closed-shell, or closed-sub-shell, atom must be zero. For example, the total angular momenta of $_2$He and $_{10}$Ne, and of $_4$Be and $_{30}$Zn, should be and are zero.

---

* One can arrive at the same conclusion using the wave functions with $m = 1, 0, -1$, but the argument is a little more complicated.

The total angular momenta of the closed-shell-plus-one alkali atoms are also easy to predict. These atoms all contain various filled levels, with zero total angular momenta, plus a single electron in an s level ($l = 0$). Therefore, the total angular momentum of any alkali atom ($_3$Li, $_{11}$Na, etc.) is equal to the spin angular momentum ($s = \frac{1}{2}$) of its one valence electron.

One can also predict the angular momentum of the closed-shell-minus-one atoms. These all consist of filled levels, with zero total angular momentum, plus a p level containing five of its six possible electrons. We know that the *addition of one more p* electron (with $l = 1$) would produce a filled level, with $l = 0$. Now, the addition of one vector to another can give zero only if the two vectors have the same magnitude. It follows that the total orbital angular momentum of the original five electrons must also be $l = 1$. A similar argument shows that the total spin of the five electrons must be $s = \frac{1}{2}$. Therefore, the closed-shell-minus-one atoms must have $l = 1$ and $s = \frac{1}{2}$, and this is what is observed. The resulting *total* angular momentum depends on the relative orientation of **L** and **S**; this, also, can be predicted, but would take us too far afield at present.

11.7 The Periodic Table

The periodic recurrence of atoms with similar properties had been noticed long before the discovery of quantum mechanics and the understanding of atomic structure. In 1869, a German chemist, Lothar Meyer, had plotted several different properties of the elements as functions of their atomic masses. In particular, he had noticed that his plot of atomic volume (which was very similar to the plot of atomic radius in Fig. 11.11) was divided by five sharp peaks into six distinct sections, or periods, corresponding (as we now know) to the successive filling of the first six shells.

In the same year, 1869, the Russian chemist Mendeleev had proposed his **periodic table,** an ingenious array in which the elements are arranged so as to highlight their periodic properties. In most modern periodic tables, the elements are arranged in order of increasing atomic number, placed from left to right in rows, with a new row starting each time a shell is closed. Within each row, the elements are placed so that atoms with similar properties are contained in the same vertical column of the table.

One popular arrangement of the entire table is shown in Fig. 11.12. The horizontal rows, or **periods,** contain different numbers of elements, corresponding to the different capacities of the various shells. The first period contains just two elements, H and He; the next two periods contain eight elements each; and the next two contain 18 each. The sixth period contains 32 elements, from $Z = 55$ to 86; to save space, the part of this period from $Z = 57$ to 71 is usually detached and arranged in a separate row below. For the same reason, the corresponding elements (from $Z = 89$ to 103) in the seventh period are placed in a second additional row.

The vertical columns, or **groups,** contain elements with corresponding arrangements of their valence electrons. The leftmost group, often called group I, contains the closed-shell-plus-one atoms, each with a single s electron outside a closed core. The next column, group II, has the closed-shell-plus-two atoms, each with two s electrons outside a closed core. On the right side of the table are

DMITRI MENDELEEV (1834–1907, Russian chemist). Several chemists had recognized that, when arranged by atomic mass, the elements show periodic behavior, but Mendeleev was unique in his recognition that certain elements were clearly missing from this scheme. When three of the missing elements were discovered with exactly the properties he had predicted, his periodic table gained international acceptance.

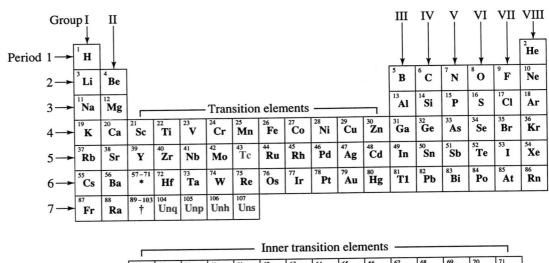

FIGURE 11.12 The periodic table of elements, showing atomic numbers and chemical symbols. Each horizontal row is called a period and each vertical column a group. Our labeling of the groups is one of several schemes in common use. To save space, the inner transition elements have been detached from the main table at the positions indicated by the star and dagger. Elements shown in color are "artificial" in the sense that they do not occur in appreciable amounts on earth.

the six columns, groups III to VIII, whose atoms have a filled or partially filled p level.* At the extreme right is group VIII (sometimes called group 0), which contains the closed-shell noble gases. Ten interior columns contain the **transition elements,** in which a d level is being filled. Since the first d level ($3d$) is in the fourth shell, the transition elements begin only in the fourth period. The sixth and seventh shells contain f levels ($4f$ and $5f$); the **inner transition elements,** in which these f levels are being filled, are the elements which are placed in two separate rows at the bottom of the table. The $4f$ elements are called the **lanthanides** or **rare earths** and the $5f$ elements are known as **actinides.** Each period of the periodic table can be divided into sections according to the levels being filled. These divisions are shown in Fig. 11.13.

Because of the way the periodic table is constructed, one often finds that elements which are close together in the table have many properties in common. An example is the division of the elements into metals and nonmetals as shown in Fig. 11.14. The elements in the upper right part of the table are all nonmetals, including oxygen, nitrogen, and all of the noble gases. Slightly to the left of the nonmetals are a few intermediate elements that are hard to classify; these include the semiconductors germanium and silicon, of such importance in modern electronics. Finally, all elements on the left of the table are metals (except hydrogen, which is generally considered to form a class by itself). We shall discuss the properties of metals further in Chapter 17, but we can already understand why they are all found on the left of the periodic table: The high

* The one exception is the noble gas He, with its closed $1s$ level.

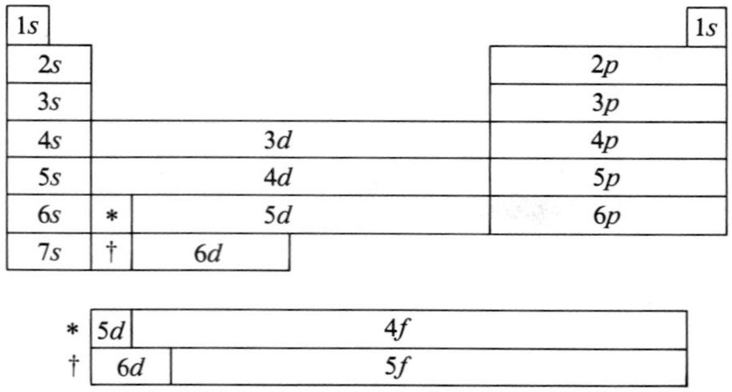

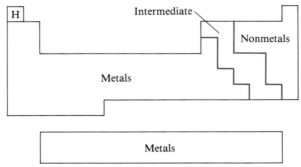

The periodic table figure (11.13):

1s						1s
2s						2p
3s						3p
4s			3d			4p
5s			4d			5p
6s	*		5d			6p
7s	†	6d				

*	5d	4f
†	6d	5f

FIGURE 11.13 The periodic table showing the levels that are filling in different parts of each period.

electrical conductivity of metals requires that there be one or more electrons that are easily detached from their atoms. From our knowledge of atomic ionization energies (Fig. 11.11) we would, therefore, expect to find the metals toward the beginning, and not at the end, of each period. In other words, metals should be on the left of the periodic table, exactly as is observed to be the case.

The periodic table is a convenient way to tabulate many different atomic properties. A large table (turned on its side to fit the page better) can be found inside the back cover and includes the atomic number, atomic mass, name, symbol, and electron configuration of every element. With a little use you will quickly learn to locate elements in the table. If you have any difficulty, you can first find the atomic number, Z, in one of the alphabetical lists of elements in Appendix C and then locate the element itself in the periodic table.

11.8 Excited States of Atoms (optional)

So far in this chapter we have discussed only the ground states of multielectron atoms. In this section we give a brief introduction to the excited states. Recall that, in the atomic ground state, the Z electrons occupy the lowest possible levels consistent with the Pauli principle. To obtain the excited states, we must

FIGURE 11.14 The division of elements into metals and nonmetals. Since the properties of the elements vary more-or-less continuously, the boundaries are only approximate and are sometimes placed a little differently. The elements labeled "intermediate" include the so-called semiconductors, like silicon and germanium, and various elements that are hard to classify, like carbon, which can be a conductor (graphite) or an insulator (diamond).

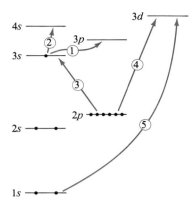

FIGURE **11.15** The energy levels for any one electron in sodium. In the ground state of the atom the 11 electrons are distributed as shown by the dots. The five arrows indicate five ways in which electrons could be excited to higher levels. (Although the ordering of the levels shown here is correct, they are not drawn to scale; in particular, the $n = 1$ and $n = 2$ levels are much deeper than shown.)

simply raise one or more of the electrons to higher levels, as illustrated in Fig. 11.15. This picture shows the energy levels of any one electron in the sodium atom ($Z = 11$). The 11 dots show the 11 electrons as they are distributed in the atom's ground state. We obtain the excited states by moving electrons to higher levels; in the picture we have indicated five possibilities—for example, the arrow labeled 1 represents an excitation in which the outer, valence, electron is raised from the $3s$ to the $3p$ level.

Even for an atom as simple as sodium, there is an enormous number of different excited states, corresponding to the many possible arrangements of the Z electrons among the numerous different levels. Fortunately, not all of them are equally important: States in which several electrons have been raised to higher levels are usually less important than those involving a single electron. For example, when an atom absorbs light, the most probable excitations involve moving just a single electron from its ground-state position (as with any one of the arrows in Fig. 11.15). Here we shall confine attention to these one-electron excitations.

It is clear in Fig. 11.15 that the excitations that require the least energy are those that involve moving a valence electron to a higher level (arrows 1 or 2, for example). Since the energy required to excite a valence electron is almost always a few eV, which is the energy of a visible (or nearly visible) photon, these excitations are often called **optical excitations.** Because these are the lowest excitations, a photon (or other projectile) with just enough energy to produce optical excitations cannot produce any other kinds of excitation. Thus it is a simplifying feature of the optical excitations that they often occur in a context where no other kinds of excitation are possible.

If we increase the energy of the photons (or other projectiles) colliding with an atom, they may be able to excite some of the inner electrons (arrows 3, 4, and 5 in Fig. 11.15). It is clear from Fig. 11.15 that the highest-energy excitations are those that involve the innermost, $1s$, electrons. The most striking thing about these is the very large energies involved. We saw in (11.9) that the energy of the $1s$ level is

$$E_{1s} \approx -Z^2 E_R.$$

For sodium ($Z = 11$) this is more than 1 keV, and for a heavy atom such as uranium ($Z = 92$) it is more than 100 keV. Because of the Pauli principle, it is impossible to excite a $1s$ electron to any level that is already full. (Thus, in Fig. 11.15 one cannot excite a $1s$ electron to either of the $n = 2$ levels.) Therefore a $1s$ electron can only be excited by lifting it all the way to the valence level or higher. Since the binding energy of these upper levels is only a few eV, the energy required to excite one of the innermost electrons (in all but the very lightest atoms) is several keV or more, the energy of X-ray photons. For this reason, excitations of the inner electrons are called **X-ray excitations.**

Once a vacancy has been created in an inner shell, any one of the higher electrons can drop down into the vacancy, emitting an X-ray photon as it does. This is the origin of the X rays studied by Moseley and described in Section 6.9. The discussion there was based on the Bohr model, but depended only on the formula $E_n \approx -Z^2 E_R/n^2$ for the inner energy levels.* Therefore, the results of Section 6.9 carry straight over to a modern quantum treatment and need not be repeated here.

OPTICAL EXCITATIONS IN ALKALI ATOMS

To conclude this section we take a slightly closer look at those excitations involving only the valence electrons—the so-called optical excitations. To simplify the discussion further, we consider only atoms with a single valence electron. These are the closed-shell-plus-one atoms, Li, Na, K, Rb, Cs, Fr, which appear in group I of the periodic table and are called the *alkali atoms* (or alkali metals, since they all form metals).

In each of the alkali atoms, the single valence electron is concentrated outside a "core" consisting of the nucleus (with charge Ze) and $Z - 1$ electrons, all arranged in spherically symmetric closed shells. The total charge of this core is $+e$, the same as that of the single proton in hydrogen. If the valence electron were entirely outside this core, it would have exactly the same allowed energies as the electron in hydrogen. In reality, the wave function of the valence electron penetrates the core, where the positive charge of the nucleus is less shielded and attracts the electron more strongly. Thus all the levels of the valence electron of an alkali atom lie a little lower than the corresponding levels in hydrogen. Since the $l = 0$ states penetrate closest to the nucleus, they are affected the most. For the states of higher l, the wave functions penetrate the core very little and the energies are very close to the corresponding energies in hydrogen.

The close relationship between the optical levels of an alkali atom and the levels of hydrogen is illustrated in Fig. 11.16, which shows the levels of the valence electron in lithium ($Z = 3$). In the ground state the valence electron occupies the $2s$ level, which is appreciably lower (by nearly 2 eV) than the $n = 2$ level in hydrogen, as we would expect. The higher s levels shown are all visibly lower than the corresponding levels of hydrogen, although the difference is smaller for the states of higher n, since these are concentrated at larger radii and penetrate the core less. The difference between corresponding p states is much

---

* Notice that in the discussion of Section 6.9 we took for granted that an electron in one of the levels $n = 2, 3, . . .$ could only drop to the $n = 1$ level after one of the $n = 1$ electrons had been ejected. Thus we were implicitly assuming the Pauli principle in that discussion.

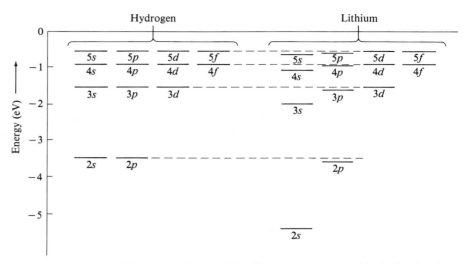

FIGURE **11.16** The energy levels of the H atom compared with the levels of the valence electron in the lithium atom.

smaller, and for the d and f states the difference cannot be seen on the scale of Fig. 11.16.

The optical levels of the other alkali atoms have the same general behavior as those of lithium, although the lowest level of the valence electron is one shell higher in each successive atom, Na, K, and so on. Thus the lowest level of the valence electron in sodium ($Z = 11$) is the $3s$ level. The next level in Na is the $3p$ level, about 2.1 eV above the ground state. As we see in the following example, the $3p \rightarrow 3s$ transition in Na produces the yellow light characteristic of the sodium-vapor lamps used to light many streets.

EXAMPLE 11.1 (*a*) The first excited state of the sodium atom is the $3p$ level, 2.10 eV above the $3s$ level. What is the wavelength of the light emitted in the $3p \rightarrow 3s$ transition? (*b*) Because of the spin-orbit interaction (Section 10.7) the $3p$ level is actually two levels, 2.1×10^{-3} eV apart (whereas the $3s$ is still just a single level). What is the difference $\Delta\lambda$ between the two wavelengths produced by $3p \rightarrow 3s$ transitions in Na?

(*a*) The wavelength of a photon with energy $E_{ph} = 2.10$ eV is

$$\lambda = \frac{hc}{E_{ph}} \tag{11.12}$$

$$= \frac{1240 \text{ eV} \cdot \text{nm}}{2.10 \text{ eV}} = 590 \text{ nm}.$$

Light with this wavelength is yellow, which is why sodium lamps produce their characteristic yellow glow.

(*b*) Differentiating (11.12) with respect to E_{ph}, we find that

$$\Delta\lambda \approx \frac{d\lambda}{dE_{ph}}\Delta E_{ph} = -\frac{hc}{E_{ph}^2}\Delta E_{ph} = -\lambda\frac{\Delta E_{ph}}{E_{ph}}$$

$$= -(590\ \text{nm}) \times \frac{2.1 \times 10^{-3}}{2.1} \approx -0.6\ \text{nm}.$$

The actual wavelengths of the two transitions are indicated in Fig. 11.17. It is because of this small "fine structure" splitting that the yellow line of sodium is actually a doublet of two lines, making the sodium spectrum very easy to identify.

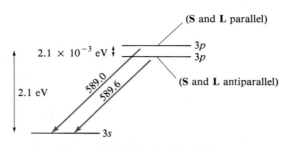

FIGURE 11.17 The spin-orbit interaction splits the $3p$ level in Na into two levels, with the states in which **S** and **L** are parallel, having slightly higher energy. The wavelengths of the two transitions are shown in nanometers.

IDEAS YOU SHOULD NOW UNDERSTAND FROM CHAPTER 11

Independent-particle approximation:
    potential-energy function
    effective charge
    energy levels
Pauli exclusion principle—allowed and
    forbidden states
Electron configurations
Periodic behavior of ionization energy, atomic
    radius, and electron affinity

Shells:
    closed-shell atom = noble gas
    closed-shell-plus-one = alkali metal
    closed-shell-minus-one = halogen
Periodic table
[Optional section: optical and X-ray excitations]

PROBLEMS FOR CHAPTER 11

SECTION 11.2 (THE INDEPENDENT-PARTICLE APPROXIMATION)

11.1 •• Find the electric field $\mathscr{E}$ at $r = a_B$ in the $1s$ state of a hydrogen atom. Compare with the breakdown field of dry air, about 3×10^6 V/m. [*Hint:* Use Gauss's law; treat the atomic electron as a static charge distribution with charge density $\rho(r) = -e|\psi(r)|^2$; and use the result of Problem 9.37.]

11.2 ••• The IPA potential-energy function $U(\mathbf{r})$ is the potential energy "felt" by an atomic electron in the average field of the other $Z - 1$ electrons plus the nucleus. If one knew the average charge distribution $\rho(\mathbf{r})$ of the $Z - 1$ other electrons, it would be a fairly simple matter to find $U(\mathbf{r})$. The calculation of an accurate distribution $\rho(\mathbf{r})$ is very hard, but it is easy to make a fairly realistic guess. For example, one might guess that $\rho(\mathbf{r})$ is spherically symmetric and given by

$$\rho(r) = \rho_0 e^{-r/R}$$

where R is some sort of mean atomic radius. (a) Given that $\rho(r)$ is the average charge distribution of $Z - 1$ electrons, find ρ_0 in terms of Z, e, and R. (b) Use Gauss's law to find the electric field $\mathscr{E}$ at a point r, due to the nucleus and the charge distribution ρ. (c) Verify that as $r \to 0$ and $r \to \infty$, $\mathscr{E}$ behaves as required by (11.2) and (11.3). [Hint: The integrals needed in parts (a) and (b) are in Appendix B.]

SECTION 11.3 (THE IPA ENERGY LEVELS)

11.3 • (a) Estimate the energy of the innermost electron of lead. (b) What is its most probable radius? (Appendix C has a list of atomic numbers.)

11.4 • Answer the same questions as in Problem 11.3, but for silver.

11.5 • • The ground state of sodium ($Z = 11$) has two electrons in the $1s$ level, two in the $2s$, six in the $2p$, and one in the $3s$. Consider an excited state in which the outermost electron has been raised to a $3d$ state (but all of the inner electrons are unchanged). Because the $3d$ wave function is not very penetrating, you can treat the outer electron as if it were completely outside all the other electrons. (a) In this approximation, what is the potential energy function $U(r)$ felt by the outer electron? (b) In the same approximation, what should be the energy of an electron in a $3d$ state? Compare your answer with the observed value of -1.52 eV. Why is the observed value lower than your estimate?

11.6 • • The ground state of lithium ($Z = 3$) has two electrons in the $1s$ level and one in the $2s$. Consider an excited state in which the outermost electron has been raised to the $3p$ level. Since the $3p$ wave functions are not very penetrating, you can estimate the energy of this electron by assuming it is completely outside both the other electrons. (a) In this approximation, what is the potential-energy function felt by the outermost electron? (b) In the same approximation, write the formula for the energy of the outer electron if its principal quantum number is n. (c) Estimate the energy of the $3p$ electron in this way and compare with the observed value of -1.556 eV. (d) Repeat for the case that the outer electron is in the $3d$ level, whose observed energy is -1.513 eV. (e) Explain why the agreement is better for the $3d$ level than for the $3p$. Why is the observed energy for $3p$ lower than that for $3d$?

SECTION 11.4 (THE PAULI EXCLUSION PRINCIPLE)

11.7 • (a) How many electrons can be accommodated in an electron energy level with $l = 2$? (b) How many if $l = 3$? (c) Give a formula (in terms of l) for the number of electrons that can be accommodated in a level with arbitrary l.

11.8 • (a) Imagine an electron (spin $s = \frac{1}{2}$) confined in a one-dimensional rigid box. What are the degeneracies of its energy levels? (b) Make a sketch of the lowest few levels showing their occupancy for the lowest state of six electrons confined in the same box. (Ignore the Coulomb repulsion among the electrons.)

11.9 • • (a) If the electron had spin $s = \frac{3}{2}$ (but was unchanged in every other respect), how many different orientations would its spin have? (b) Sketch energy-level diagrams similar to Fig. 11.6(b) showing how the levels would be occupied in the ground states of helium ($Z = 2$) and lithium ($Z = 3$) if the electron had $s = \frac{3}{2}$.

11.10 • • Repeat Problem 11.8 assuming that the electron had spin $s = \frac{3}{2}$.

11.11 • • Imagine several identical spin-half particles all confined inside the two-dimensional rigid square box discussed in Section 9.3. Assume that the particles do not interact with one another, and hence that the allowed energies for each particle are exactly as shown in Fig. 9.2. (a) What are the lowest four allowed energies for any one particle? How many particles can be accommodated in each of these levels, given that they obey the Pauli exclusion principle? (Hint: In figuring degeneracies do not forget that each particle has two possible orientations of its spin.) (b) Assuming that there are six particles in the box, draw an energy-level diagram similar to Fig. 11.6(b) showing the distribution of particles that gives the state of lowest energy for the system as a whole. (c) Do the same for the case that there are 10 particles in the box.

SECTION 11.5 (GROUND STATES OF THE FIRST FEW ELEMENTS)

11.12 • Draw two energy-level diagrams similar to Fig. 11.7 showing the ground state and first excited state of a boron atom ($_5$B).

11.13 • Draw four energy-level diagrams, similar to those in Fig. 11.7, to illustrate the ground states of the following atoms: $_5$B, $_9$F, $_{10}$Ne, $_{11}$Na.

11.14 • Draw two energy-level diagrams similar to Fig. 11.7 showing the ground state and first excited state of a neon atom.

11.15 • Consider the graph of ionization energy against atomic number Z in Fig. 11.8. It is clear that within each shell ($Z = 1$ to 2 and $Z = 3$ to 10), the ionization energy tends to increase with Z. However, there is a small *drop* as one moves from $_4$Be to $_5$B. Explain this drop. (The second small drop, between $_7$N and $_8$O, is explained in Section 11.6.)

11.16 •• The first ionization energy of an atom is the minimum energy needed to remove one electron. For helium this is 24.6 eV. The second ionization energy is the additional energy required to remove a second electron. (*a*) Calculate the second ionization energy of helium. (*b*) What is the total binding energy of helium (that is, the energy to take it apart into two separated electrons and a nucleus)?

11.17 •• In this question you will estimate the total energy of a helium atom. (*a*) What would be the total energy of a helium atom (in its ground state) in the approximation that you ignore completely the electrostatic force between the two electrons? (*Hint:* In this approximation you can treat each electron separately as if it were in a hydrogen-like ion. The total energy is just the sum of the two separate energies.) (*b*) Your answer in part (*a*) should be negative (indicating that the system is bound) and *too* negative, since you ignored the positive potential energy due to the repulsion between the two electrons. To get a rough estimate of this additional potential energy, imagine the electrons to be in the first Bohr orbit, with radius $a_B/2$ (the appropriate radius for a hydrogen-like ion with $Z = 2$). To minimize their energy, the two electrons would move around the same circular orbit, always on opposite sides of the nucleus, a distance a_B apart. Use this semiclassical model to estimate the potential energy of the two electrons. Combine this with your answer to part (*a*) to estimate the total energy of the He atom. Compare with the observed value of -79.0 eV.

11.18 •• (*a*) If the electron had spin $s = \frac{3}{2}$ (but was unchanged in every other respect), how many different orientations would its spin have? (*b*) Sketch energy-level diagrams showing how the levels would be occupied in the ground states of $_4$Be and $_6$C if the electron had $s = \frac{3}{2}$.

SECTION **11.6** (THE REMAINING ELEMENTS)

11.19 • Use the energy-level diagram of Fig. 11.9 to write down the electron configurations of $_{30}$Zn, $_{35}$Br, $_{54}$Xe, $_{85}$At, $_{87}$Fr.

11.20 • Use the energy-level diagram of Fig. 11.9 to find the electron configurations of $_{20}$Ca, $_{23}$V, $_{32}$Ge, $_{53}$I, $_{88}$Ra.

11.21 • Explain the abrupt drops in the ionization energies shown in Fig. 11.11 between Cd and In and between Hg and Tl.

11.22 • Write down the electron configuration for each of the six alkali-metal atoms.

11.23 •• (*a*) Use the information in Fig. 11.9 to find the ground-state configurations of the following atoms: $_{30}$Zn, $_{80}$Hg, $_{37}$Rb, $_{55}$Cs. (*b*) What is the total angular momentum of each of these atoms?

11.24 •• (*a*) Use the information in Fig. 11.9 to find the ground-state configurations of the following atoms: $_{30}$Zn, $_{31}$Ga, $_{39}$Y, $_{48}$Cd, $_{53}$I, $_{54}$Xe, $_{55}$Cs. (*b*) What is the total spin angular momentum for each of these atoms? (*c*) What are their total orbital angular momenta?

11.25 •• Give the ground-state configurations that the atoms of $_8$O, $_{10}$Ne, and $_{21}$Sc would have if the electron had spin $s = \frac{3}{2}$. (See Problem 11.18.)

11.26 •• The degeneracies of the levels in hydrogen are 2, 8, 18, 32, ... The numbers of electrons in successive shells of multielectron atoms are 2, 8, 8, 18, 18, 32, ... Given the groupings of levels into shells shown in Fig. 11.9, explain the similarities and differences between these two sets of numbers.

SECTION **11.7** (THE PERIODIC TABLE)

11.27 • Use the periodic table inside the back cover to write down the full electron configurations of $_3$Li, $_{10}$Ne, $_{12}$Mg, $_{19}$K, $_{28}$Ni, and $_{48}$Cd. (*Hint:* That periodic table gives the configuration of the outer-shell electrons only; to find the full configuration, you must add the configuration of the preceding closed-shell element. For example, for $_{12}$Mg, add the configuration of $_{10}$Ne.)

11.28 • Use the periodic table inside the back cover to find the full configurations of the following atoms: iron, silver, iodine, polonium. (See the hint for Problem 11.27, and use the alphabetical list in Appendix C, if necessary.)

11.29 • Use the periodic table inside the back cover to find the names, atomic numbers, and full electron configurations of the following atoms: Ga, Xe, W, At, Md, Unh. (See the hint for Problem 11.28.)

11.30 • (*a*) Find the ground-state configurations of nickel and copper from the periodic table inside the back cover. (*b*) Draw energy-level diagrams, similar to those in Fig. 11.7, to illustrate these two ground states. (*Note:* In Fig. 11.9 we showed the $4s$ level below $3d$, since it fills first. Nevertheless, $4s$ and $3d$ are very close together; as Z increases, the $3d$ level becomes almost exactly degenerate with $4s$, and $4s$ can lose one electron to $3d$—as happens in copper.)

11.31 •• Because of the way that atomic properties vary smoothly along the rows and columns of the periodic table, one can often predict the properties of an element from the known properties of its neighbors. (This is how Mendeleev predicted the existence and several properties of the elements now called scandium, gallium, and germanium.) (a) The ionization energies of $_{20}$Ca and $_{38}$Sr are 6.11 eV and 5.70 eV. If one guessed that ionization energies should change linearly as one moves through a group, what would one predict for the ionization energy of $_{56}$Ba? (The observed value is 5.21 eV.) (b) Use the same argument to predict the electron affinity of $_{35}$Br, given that $_{17}$Cl has electron affinity 3.61 eV and $_{53}$I has 3.06 eV. (The observed value is 3.36 eV.) (c) Predict the radius of the $_9$F atom, given that $_7$N and $_8$O have radii of 0.075 and 0.065 nm. (The actual value for $_9$F is 0.057 nm.)

11.32 •• (a) The electron affinities of $_{26}$Fe and $_{27}$Co are 0.163 and 0.661 eV. Using arguments similar to those outlined in Problem 11.31, predict the electron affinity of $_{28}$Ni. (The observed value is 1.156 eV.) (b) In the same way, predict the boiling point of $_{32}$Ge, given that the boiling point of $_{14}$Si is 3540 K and that of $_{50}$Sn is 2876 K. (The observed value is 3107 K.) (c) Similarly, predict the melting point of $_{86}$Rn, given that those of $_{36}$Kr and $_{54}$Xe are 116 K and 161 K. (The observed value is 202 K.)

11.33 •• The simple estimation of atomic properties as described in Problems 11.31 and 11.32 does not always work very well. Here is a well-known example where it is fairly unsuccessful: The electron affinities of $_{17}$Cl and $_{35}$Br are 3.62 and 3.36 eV. Assuming that electron affinity varies linearly within the groups of the periodic table, what would you predict for the electron affinity of $_9$F? The observed value is 3.40 eV.

SECTION **11.8** (EXCITED STATES OF ATOMS)

11.34 • The ground-state configuration of the lithium atom is $1s^2 2s^1$. Give the configurations of the lowest five excited levels. (See Fig. 11.16 and ignore fine structure.)

11.35 • Give the configurations of the lowest four levels of a sodium atom. (See Fig. 11.15 and ignore fine structure.)

11.36 • The spectrum of atomic lithium has a red line at $\lambda = 671$ nm, arising from the transition $2p \rightarrow 2s$. On close inspection, this line is seen to be a doublet of lines separated by 0.0152 nm. What is the fine structure splitting of the $2p$ level in lithium?

11.37 • Give the configurations of the lowest six levels of the He atom.

12

The Structure of Atomic Nuclei

In Parts I and II we have developed the two theories, relativity and quantum mechanics, that revolutionized twentieth-century physics. In Parts III and IV we shall describe some of the many applications of these two theories. In Part III we discuss subatomic systems (nuclei and subnuclear particles), and in Part IV, systems containing more than one atom (mostly molecules and solids). As far as practicable, we have made the chapters of Parts III and IV independent of one another, so that you can select just those topics that interest you. In particular, you can, if you wish, omit Part III entirely and go straight to Part IV.

Part III contains three chapters, 12 through 14. Chapters 12 and 13 describe the properties of atomic nuclei, while Chapter 14 treats the subnuclear, "elementary" particles. The division between the two chapters on nuclear physics is that Chapter 12 is concerned with the properties of an isolated, static nucleus, while Chapter 13 treats nuclear reactions, in which nuclei change their energy and often their identity as well. These two chapters are not completely independent of one another, but if your main interest is in nuclear reactions (Chapter 13), you need to read only Sections 12.1 to 12.3 before skipping to Chapter 13. If your main interest is in particle physics (Chapter 14), you could read just Sections 12.1 to 12.3 and 13.1 to 13.4 before going directly to Chapter 14.

12.1 Introduction

In discussing the properties of atoms, we needed to know surprisingly little about the atomic nucleus. In fact, for most purposes we needed only three facts: The nucleus carries a positive charge Ze; the nucleus is so much heavier than the electrons that we can usually treat it as fixed; and the nucleus is so small that — for the purpose of atomic physics — we can usually treat it as a point particle.

In the next two chapters we turn our attention to the internal structure of atomic nuclei. We shall find that in some ways the motion of the protons and neutrons inside a nucleus is analogous to the motion of the electrons in an atom. Many of the ideas developed for atomic physics (the Schrödinger equation, the Pauli principle, spin, etc.) are equally useful in nuclear physics.

Nevertheless, there are significant differences between atomic and nuclear physics, of which perhaps the most important is the difference between the forces involved. The force that holds electrons in their atomic orbits is the familiar electrostatic force, which was well understood long before the advent of atomic physics. In nuclear physics the situation was quite different, at least historically. As we shall argue in Section 12.3, the force that holds protons and neutrons together in a nucleus is not a force that was known in classical physics, and, even today, we do not have a complete theoretical understanding of nuclear forces. Therefore, the history of nuclear physics during the last 60 years is the story of a simultaneous effort to elucidate the nuclear force and to explain the properties of nuclei. In this sense, the development of nuclear physics has been more difficult than that of atomic physics. Nevertheless, we shall see that nuclear physicists have achieved a remarkable understanding of nuclear properties. We shall find that this understanding is both interesting for its own sake and of great practical importance.

12.2 Nuclear Properties

We begin our study of nuclear physics with a survey of several nuclear properties, some of which we met briefly in Chapter 4. In later sections we shall see that many of these properties can be understood with the help of the same quantum principles developed in Chapters 7 through 11.

CONSTITUENTS

We have already stated that nuclei are made up of protons and neutrons. The number of protons we denote by Z, and the number of neutrons by N. Since each proton has charge $+e$, while neutrons are uncharged, the total charge of the nucleus is $+Ze$.

The mass of a nucleus is

$$m_{\text{nuc}} = Zm_{\text{p}} + Nm_{\text{n}} - \frac{B}{c^2}, \qquad (12.1)$$

where B is the total binding energy, the energy required to pull the nucleus apart into its $Z + N$ separate constituents. As discussed in Chapter 4, the binding-energy term, $-B/c^2$, is usually less than 1% of m_{nuc}. Further, the difference between the masses m_{p} and m_{n} is only 1 part in 1000. Hence we can neglect the last term in (12.1) and set $m_{\text{n}} \approx m_{\text{p}}$ to give the approximation

$$m_{\text{nuc}} \approx (Z + N)m_{\text{p}} = Am_{\text{p}},$$

where, as before, we have defined the **mass number,** or **nucleon number,**

$$A = Z + N,$$

which is the total number of nucleons (protons *and* neutrons) in the nucleus.

It is sometimes convenient to write the numbers A, Z, and N explicitly with the chemical symbol. The agreed placement around the chemical symbol X is now*

$$^A_Z X_N.$$

Some examples are

$$^1_1H_0, \ {}^2_1H_1, \ {}^4_2He_2, \ {}^{12}_6C_6, \ {}^{63}_{29}Cu_{34}, \ {}^{238}_{92}U_{146}. \qquad (12.2)$$

Fortunately, it is usually not necessary to give all of the three numbers on these symbols, and we shall generally include only as many as seem necessary for the discussion at hand. [This may be a good time to remind you that the first three nuclei in (12.2) have their own special names: ^1H is, of course, the proton, denoted p; the nucleus of heavy hydrogen, ^2H, is called the deuteron, and is sometimes denoted D; and the nucleus ^4He is called the alpha particle, denoted α.]

Two different nuclei, such as $^{16}_8O_8$ and $^{17}_8O_9$, which have the same number of protons are called **isotopes.** As discussed in Chapter 4, the corresponding atoms are chemically indistinguishable, or very nearly so. Two nuclei, such as $^{14}_6C_8$ and $^{14}_7N_7$, with the same mass number are called **isobars;** and two nuclei, such as $^{13}_6C_7$ and $^{14}_7N_7$, with the same number of neutrons are called **isotones.**

The naturally occurring elements have atomic numbers in the range $1 \leq Z \leq 92$ and neutron numbers with $0 \leq N \leq 146$. The values of Z and N for all naturally occurring nuclei are plotted in Fig. 12.1, from which one sees that for light nuclei ($A \lesssim 40$) N and Z are nearly equal, while for heavier nuclei N is a little larger than Z, with $N \approx 1.5Z$ in the heaviest nuclei. The tendency to have $N \approx Z$ is called the *symmetry effect.* We shall show in Section 12.6 that the symmetry effect results from the Pauli principle and the so-called charge independence of nuclear forces, and that the tendency for N to be somewhat bigger than Z is due to the electrostatic repulsion of the protons in nuclei with large Z.

Another important tendency among nuclei is the observed preference for Z and N to be even, as illustrated in Table 12.1. The stable nuclei in which Z and N are both even outnumber comfortably those in which either Z or N is odd,

TABLE 12.1
Numbers of stable nuclei

Z	N	Number of stable nuclei
Even	Even	148
Even	Odd	51
Odd	Even	49
Odd	Odd	4

* This arrangement has changed from time to time, and many older books use a different convention.

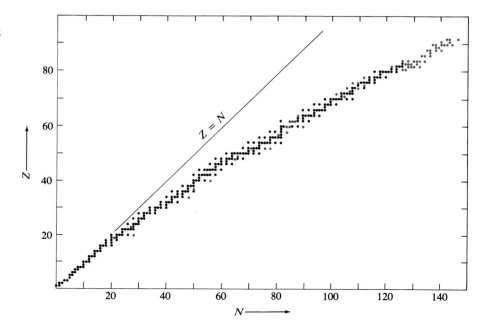

and there are only four stable nuclei in which *both* Z and N are odd. This tendency is the result of the so-called pairing effect, which is described in Section 12.8.

SIZES AND SHAPES OF NUCLEI

We described in Chapter 4 how Rutherford probed the atomic nucleus using energetic α particles. (The main reason that he used α particles was simply that they were readily available, since they are ejected by many naturally occurring radioactive substances.) Rutherford's experiments established that nuclei are much smaller than atoms, with radii of just a few femtometers, or fermis (1 fm = 10^{-15} m). More recent experiments have used artificially accelerated beams of electrons, protons, and α particles to probe the size and shape of atomic nuclei. The results of these experiments are summarized in the next few paragraphs.

The majority of nuclei are spherical. A number are slightly nonspherical, some being prolate (elongated, like an American football) and some oblate (flattened, like most pumpkins). In nonspherical nuclei the longest diameter is typically a few percent greater than the shortest diameter, although in a few cases this difference can be as large as 30%. It is usually a reasonable approximation to treat nonspherical nuclei as spherical, and, for the most part, we shall consider nuclei to be spherical.

The radii of nuclei increase steadily from about 2 fm for light nuclei (helium, for example) to about 7 fm for heavy nuclei (uranium, for example). It is found that the radii of all nuclei are given approximately by the simple formula

$$R = R_0 A^{1/3}, \qquad (12.3)$$

where A is the number of nucleons and R_0 is a constant,

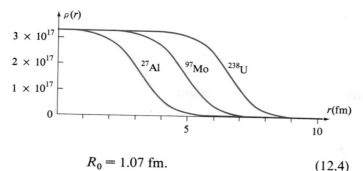

FIGURE 12.2 The density, in kg/m³, as a function of *r* in the nuclei of aluminum, molybdenum, and uranium. These graphs can be accurately fitted by an analytic expression called the Fermi function (see Problems 12.12 to 12.14).

$$R_0 = 1.07 \text{ fm.} \tag{12.4}$$

The significance of (12.3) is clearer if we consider the *volume* of the nucleus,

$$V = \tfrac{4}{3}\pi R^3 = \tfrac{4}{3}\pi R_0^3 A$$
$$= V_0 A, \tag{12.5}$$

where V_0 is the constant $\tfrac{4}{3}\pi R_0^3$. Evidently, (12.3) implies that the volume of a nucleus is proportional to A, the number of constituent nucleons. This means that the volume *per nucleon* V_0 is approximately the same in all nuclei. In other words, the density of nucleons inside the nucleus is the same in all nuclei.

The volume of a single isolated nucleon turns out to be about $V_0/2$. Thus we can say that nucleons in a nucleus take up half of the total volume. This means that the nucleons are quite closely packed inside a nucleus, quite like the molecules in a liquid—and entirely unlike the electrons in an atom.*

The radius of a nucleus is not an exactly defined quantity. This is evident in Fig. 12.2, which shows the distribution of density in three representative nuclei. In no case is there a unique radius at which the density falls exactly to zero. The usual definition of R, and the definition assumed in (12.3) and (12.4), is the radius at which the density $\rho(r)$ falls to half its maximum value $\rho(0)$. Notice that, as expected, the central density is the same in all three pictures. Notice also that all three graphs fall to zero at about the same rate; that is, the thickness of the surface region is about the same in all three nuclei.

The actual density inside a nucleus is seen from Fig. 12.2 to be about 3×10^{17} kg/m³, some 14 orders of magnitude greater than the density of ordinary liquids or solids. (Remember that water has density 10^3 kg/m³.) The enormous density inside nuclei is easy to understand if we recall that almost all the mass of an atom is in the nucleus, whose volume is some 10^{14} times smaller than the atom (Problem 12.7). Therefore, the nuclear density is 10^{14} times bigger than the atomic density, which is itself about the same as that of ordinary liquids or solids.

NUCLEAR ENERGY LEVELS

Like atoms, nuclei are found to have quantized energy levels. However, the spacing of the levels in nuclei is many times larger than that in atoms. In heavy nuclei the level spacing is usually several tens of keV, while in light nuclei it can range up to 10 MeV and more. Because their excitation energies are so much larger than the energies involved in chemical reactions, nuclei remain

---

* Of course, because of their wave functions, the electrons' probability distribution is spread out through the atom. This does not change the fact that the electron itself is very small compared to the atom. In fact, current theory holds that the electron is a point particle—unlike nucleons, which have a measurable size.

locked in their ground states during normal chemical processes. This is why it is often possible in atomic physics to treat the nucleus as a structureless body with no internal motion.

If a nucleus is raised to one of its excited states (when struck by an energetic proton, for instance), it can return to the ground state by emitting a photon. This is analogous to the emission of photons in atomic transitions, the only important difference being that the photons emitted by nuclei are usually far more energetic. Photons emitted in nuclear transitions (with energies ranging from keVs to MeVs) are called γ rays. Measurement of their energy tells us the difference between the two nuclear energy levels involved.

ANGULAR MOMENTA OF NUCLEI

Like the electron, both the neutron and proton have spin $\frac{1}{2}$. Thus the total angular momentum $\mathbf{J}$ of a nucleus is the sum of the orbital angular momenta of all of its nucleons *plus* all of their spins. As with all angular momenta, the magnitude of $\mathbf{J}$ is quantized; the possible values are labeled by a quantum number j, with

$$J = |\mathbf{J}| = \sqrt{j(j+1)}\,\hbar. \tag{12.6}$$

12.3   The Nuclear Force

The force that holds electrons in their atomic orbits is the familiar electrostatic attraction between the positive nucleus and the negative electrons. On the other hand, it should be clear that electrostatic attraction is *not* what holds the nucleons in the nucleus. First, the neutron is uncharged and feels no electrostatic force at all. Second, the protons are all positive and the effect of the electrostatic force is therefore to push the protons apart.

Since the gravitational force acts on all matter, charged or not, and is always attractive, one might wonder if gravity could be the force that holds the nucleus together. Unfortunately, gravity is many orders of magnitude too weak, as the following example illustrates.

EXAMPLE 12.1   Compare the gravitational attraction between two protons with their electrostatic repulsion.

The gravitational attraction is

$$F_{\text{grav}} = \frac{Gm_p^2}{r^2}$$

while the repulsive electrostatic force is

$$F_{\text{elec}} = \frac{ke^2}{r^2}.$$

The ratio of these forces is independent of r and is equal to

$$\frac{F_{\text{grav}}}{F_{\text{elec}}} = \frac{Gm_p^2}{ke^2} = \frac{(6.7 \times 10^{-11}\ \text{N} \cdot \text{m}^2/\text{kg}^2) \times (1.7 \times 10^{-27}\ \text{kg})^2}{(9.0 \times 10^9\ \text{N} \cdot \text{m}^2/\text{C}^2) \times (1.6 \times 10^{-19}\ \text{C})^2} \approx 10^{-36}.$$

Clearly, the gravitational force is far too weak to overcome the electrostatic repulsion of the protons and hold the nucleus together.

Since neither the electrostatic force nor gravity holds the nucleus together, there must be some other force between nucleons, and we call it the **nuclear force.** The nuclear force must be stronger than the electrostatic force, in the sense that it must more than balance the electrostatic repulsion of the protons. For this reason the nuclear force is often called the **strong force.**

A useful way to view the strength of the nuclear force is in terms of the energy of a nucleon in the nucleus. As a crude model, we can imagine any one nucleon moving inside a cubical box whose side a is of order 6 fm (the approximate diameter of a fairly light nucleus such as aluminum, ^{27}Al). We saw in Chapter 9, equation (9.32), that the minimum kinetic energy for such a particle is

$$K_{\min} \approx \frac{3\hbar^2\pi^2}{2ma^2} = \frac{3\pi^2(\hbar c)^2}{2mc^2a^2} \tag{12.7}$$

$$\approx \frac{3\pi^2(200 \text{ MeV}\cdot\text{fm})^2}{2 \times (940 \text{ MeV}) \times (6 \text{ fm})^2} \approx 20 \text{ MeV}.$$

That is, the minimum kinetic energy of any one nucleon in a nucleus is of order 20 MeV. In order that the nucleus hold together, its total energy must be negative. Therefore, the nuclear force on any nucleon due to the other $A - 1$ nucleons must produce a potential-energy well of depth at least several tens of MeV.

While the nuclear potential well must be very deep, it must also be of very short range; that is, the force between nucleons must fall rapidly to zero as their separation increases. Perhaps the simplest evidence for this comes from experiments in which a beam of protons is scattered by nuclei. Those protons that pass by a nucleus at large distance are influenced only by the Coulomb repulsion between the proton's charge and the nuclear charge. Even protons passing within 2 or 3 fm of the nuclear surface scatter in the pattern predicted on the basis of a simple Coulomb repulsion. Only when the protons approach within 1 or 2 fm of the nuclear surface does the pattern of scattering change, indicating the presence of a strong, attractive nuclear force. We conclude that the nuclear force between nucleons is negligibly small when they are more than 2 or 3 fm apart.

At separations of 1 or 2 fm the force between nucleons is strongly attractive, as we have just argued. However, at separations less than about 1 fm the nuclear force becomes strongly repulsive. The simplest argument for this last claim is that, as we saw in Section 12.2, the density inside a nucleus is approximately constant and is the same for all nuclei. If the nuclear force were purely attractive, we would expect nucleons at the center of a large nucleus to be forced inward more tightly than in a small nucleus. The easiest way to explain the observed constancy of nuclear densities is to assume that at very short distances the nucleons are held apart by a repulsive force, and this proves to be the case.*

Today, much of the best evidence for the details of the nuclear force comes from nucleon–nucleon scattering experiments, in which a beam of neutrons or protons is scattered by the protons in a hydrogen target, much as Rutherford's α particles were scattered by the nuclei in a metal foil. The main conclusions of these experiments are summarized in Fig. 12.3, which shows the

---

* In fact, the constancy of the density in nuclei involves several other effects in addition to the short-range repulsion. Nevertheless, the conclusion stated is correct.

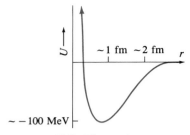

FIGURE 12.3 The nuclear potential energy of two nucleons as a function of their separation r.

potential energy of the nuclear force between two nucleons, plotted as a function of their separation r. It will be seen that the main features of this graph agree with our qualitative conclusions: The nuclear potential energy approaches zero rapidly beyond about 2 fm. Inside 2 fm the force is attractive and the potential energy becomes negative, dropping to a minimum of about -100 MeV at about 1 fm. Inside this minimum, the force is repulsive and the potential energy rapidly becomes positive and large.*

CHARGE INDEPENDENCE OF NUCLEAR FORCES

Two final properties of the nuclear force should be mentioned. First, it is found that for a given quantum state, the nuclear force between two nucleons is the same whether they are both protons, or both neutrons, or one proton and one neutron. This property is called the **charge independence of nuclear forces** — the nuclear force on a nucleon is independent of whether the nucleon is charged (a proton) or uncharged (a neutron). It is charge independence that is responsible for the observed tendency for nuclei to have equal numbers of neutrons and protons as we describe in Section 12.6.

Much detailed evidence for the charge independence of nuclear forces comes from studies of the interactions between two nucleons (p–p, p–n, n–n) in collision experiments. Further evidence comes from the study of energy levels in nuclear *isobars,* that is, two or more nuclei that have the same total number of nucleons and differ only in their numbers of protons and neutrons. Consider, for example, the pair of nuclei $^7_3\text{Li}_4$ and $^7_4\text{Be}_3$. The only difference between these is that one neutron in $^7_3\text{Li}_4$ has been replaced by a proton in $^7_4\text{Be}_3$. Figure 12.4 shows the lowest five energy levels of each of these two nuclei. Because ^7Be has one extra charge, all of its levels are raised by 1.7 MeV due to the additional Coulomb repulsion. Apart from this obvious difference, the levels of the two nuclei are almost identical. The close correspondence of the energy levels (both in excitation energy and angular momentum) is strong evidence that the nuclear force acts equally on neutrons and protons.

FIGURE 12.4 Energy levels of the isobars $^7_3\text{Li}_4$ and $^7_4\text{Be}_3$. The numbers labeling each level are its excitation energy in MeV and its angular-momentum quantum number j. Corresponding levels are connected by a dashed line; they have angular momenta that are exactly equal and excitation energies that are very nearly so. This close agreement is evidence for the charge independence of nuclear forces.

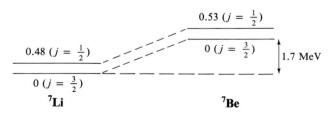

* Strictly speaking, the potential energy sketched in Fig. 12.3 is that for two nucleons with zero angular momentum. The force between nucleons depends somewhat on their angular momentum, but for our present purposes the potential energy shown is the most important.

Although the nuclear force is so crucially important for neutrons and protons, it does not act on electrons at all—in just the same way that the electrostatic force does not act on particles that are electrically neutral. The principal evidence for this claim comes from experiments in which high-energy electrons collide with nuclei; even when the electrons penetrate deep into the nucleus, there is no evidence that they are affected by the strong, nuclear force. This makes the electron a very useful probe of nuclei, since its interaction with the nucleons is just the well-understood electromagnetic interaction.

12.4   Electrons versus Neutrons as Nuclear Constituents

With our general knowledge of nuclear properties and forces we are now ready to examine several questions in more detail. The first question that we address concerns the constituents of nuclei.

Although we now know that nuclei consist of protons and neutrons, this was not always obvious. That nuclei contain protons was established in 1919 when Rutherford observed protons ejected from collisions between α particles and nitrogen nuclei. The neutron was not discovered for another 12 years, and at first, it was natural to suppose that nuclei were made of protons and *electrons*. Recall that the two basic facts about nuclei are that they have charge Ze and mass Am_p (approximately). Today we explain these by saying that the nucleus has Z protons, to give it charge Ze, and $A - Z$ neutrons (with $m_n \approx m_p$), to give it total mass Am_p. However, the same two facts could be equally well explained by supposing that the nucleus has A protons and $A - Z$ electrons. Because m_e is so small, the mass would still be about Am_p; and because the electron has charge $-e$, the $A - Z$ electrons would reduce the total charge to Ze, as observed.

The picture of the nucleus as made of protons and electrons had two very appealing features. First, it was economical: If correct, it meant that all matter was made of just two kinds of particle, protons and electrons. Second, since electrons and protons have opposite charges, it was conceivable that the nucleus was held together by electrostatic attraction.

The question that we address here is this: How do we know that the picture of nuclei as made of protons and neutrons is correct and the proton–electron picture incorrect? We present two arguments.

THE ENERGY OF NUCLEAR CONSTITUENTS

We have seen that the kinetic energy of a proton or neutron inside a nucleus is at least several MeV. This requires that the nuclear force be strong enough to confine nucleons with these energies, and we have seen that the observed nuclear force *is* sufficiently strong. Suppose, however, that the nucleus contained electrons. It is a simple matter to estimate the minimum kinetic energy of an electron in a nucleus, as in the following example.

EXAMPLE 12.2   Assuming that an electron could be confined inside a nucleus of radius about 3 fm (for example, ^{27}Al) estimate the electron's minimum kinetic energy.

To estimate the kinetic energy of nucleons in a nucleus we treated the nucleus as a rigid cubical box. We can do the same thing for an electron, but we need to be careful because the electron mass is only 0.5 MeV/c^2, so we must anticipate that its motion may be relativistic. Therefore, we shall start from the general relation

$$E = \sqrt{(pc)^2 + (mc^2)^2} \tag{12.8}$$

and find the minimum possible value of p^2.

We saw in Section 9.3 that the requirement that the wave function vanish at the walls of the box led to a minimum for each component of momentum:

$$p_x = \hbar k_x = \frac{h}{2a}, \tag{12.9}$$

where a is the length of the cube's edge, and the same minimum applies to p_y and p_z as well.* Therefore, the minimum possible value of p^2 is

$$p^2 = p_x^2 + p_y^2 + p_z^2 = \frac{3h^2}{4a^2}.$$

Setting $a \approx 6$ fm (the diameter of the nucleus) we find that

$$(pc)^2 = \frac{3(hc)^2}{4a^2} \approx \frac{3 \times (1240 \text{ MeV} \cdot \text{fm})^2}{4 \times (6 \text{ fm})^2} \approx 3.2 \times 10^4 \text{ MeV}^2.$$

Substituting into (12.8), we obtain for the minimum energy

$$E = \sqrt{(pc)^2 + (mc^2)^2} \approx \sqrt{(3 \times 10^4) + (0.25)} \text{ MeV} \approx 180 \text{ MeV}.$$

Note that this answer is much larger than $mc^2 = 0.5$ MeV, and hence that the electron would indeed be moving relativistically. Therefore, the kinetic energy, $K = E - mc^2$, is close to the total energy and

$$K_{\text{min}} \approx 180 \text{ MeV}. \tag{12.10}$$

This example shows that an electron confined inside a nucleus would have kinetic energy of order at least 100 MeV. However, there is no known force that could confine an electron with this much kinetic energy. In the 1920s the only known possibility was the electrostatic Coulomb force, and since we now know that the nuclear force does not act on electrons, the Coulomb force is still the only candidate. It is easy to estimate the electrostatic potential energy of an electron inside any given nucleus. For ^{27}Al, for example, one finds (Problem 12.23)

$$U \approx -9 \text{ MeV}.$$

This energy is negative and would tend to bind the electron, but comparison with (12.10) shows that it is hopelessly too small to overcome the minimum kinetic energy of the electron, and we conclude that there can be no electrons bound inside a nucleus.

---

* The minimum, $p_x = h/2a$, in (12.9) reflects that the maximum possible wavelength λ is $\lambda = 2a$.

Our second argument against nuclei made of protons and electrons concerns their magnetic moments. We saw in Chapter 10 that the electron has a magnetic moment whose magnitude is given by the Bohr magneton,

$$\mu_B = \frac{e\hbar}{2m_e}, \qquad (12.11)$$

where m_e is the electron's mass. (The exact magnetic moment depends on the orbital quantum number l but is always equal to μ_B times a number of order 1.) In the same way we would expect the nucleons' magnetic moment to be of order

$$\mu_N = \frac{e\hbar}{2m_N}, \qquad (12.12)$$

where m_N denotes the mass of the nucleon. This proves to be correct: Both the proton and neutron* have magnetic moments whose magnitudes are close to the value (12.12), which is therefore called the **nuclear magneton.**

The important difference between μ_B in (12.11) and μ_N in (12.12) is that, since m_e is about 2000 times *smaller* than m_N, the Bohr magneton μ_B is 2000 times *greater* than μ_N. Thus if there were any electrons in the nucleus, their large magnetic moment would dominate, and the total magnetic moment would be of order μ_B. In fact, however, the measured magnetic moments of all nuclei are far smaller than this and are all of order μ_N — in keeping with the proton–neutron theory.

For these and other reasons (Problems 12.21 and 12.22) it is clear that there are no electrons inside the atomic nucleus, which consists, instead, of protons and neutrons.

12.5   The IPA Potential Energy for Nucleons

In studying atomic structure we relied heavily on the independent-particle approximation, or IPA†. In this approximation each of the Z electrons was assumed to move independently in the field produced by the average distribution of the other $Z-1$ electrons. Working within this approximation, we found the possible states of each separate electron and hence those of the atom as a whole. In this way we were led to the shell structure of atoms and to a remarkably successful account of the periodic properties of the elements.

The question naturally arises whether the corresponding approximation would be useful in nuclear physics. In the early days of nuclear physics it was generally believed that the answer would be no. The main reasons were that the nuclear force is so strong and of such short range. Figure 12.5(a) is a schematic representation of the potential energy of a nucleon moving along the x axis and interacting with just one other nucleon. (This is just the graph of Fig. 12.3 folded over, because x — unlike r — can be positive or negative, corresponding to the nucleon being to the right or left of its fixed companion.) Figure 12.5(b)

---

* Since the neutron is chargeless, one might expect it to have no magnetic moment. However, we now know that although its total charge is zero, there *are* charges inside the neutron, and it is the motion of these charges that gives the neutron its magnetic moment.

† Nuclear physicists often call the IPA the independent-particle *model,* abbreviated IPM.

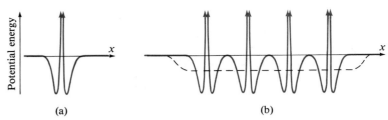

FIGURE 12.5 (a) The nuclear potential energy of one nucleon confined to the *x* axis and interacting with a second nucleon on the *x* axis. (b) The potential energy of one nucleon in the proximity of four others fluctuates rapidly. Nevertheless, its effect is well approximated by the smoothed average shown as a dashed curve.

shows the potential energy of one nucleon in a one-dimensional nucleus containing four other nucleons. The dashed curve shows a smoothed, average potential energy, and at first sight it would seem an extremely bad approximation to replace the rapidly fluctuating solid curve by the smooth dashed curve.

In fact, however, the effect of the rapidly fluctuating nuclear potential energy is approximated surprisingly well by its smoothed IPA average. There are several reasons for this, some of which depend on subtle cancellations between the attractive and repulsive parts of the nuclear force. However, one simple reason is just that the system in question is a quantum system. Figure 12.5(b) represents the potential energy of a nucleon in the field of four classical nucleons, each at a definite position in the nucleus. But quantum mechanics tells us that the four nucleons are themselves spread out in accordance with their wave functions. This quantum spreading of the other particles is an important factor in the success of the IPA in describing the motion of nucleons in nuclei, as it is also for the case of electrons in an atom.

Figure 12.5 is one-dimensional, while the real nucleus is, of course, three-dimensional. We have seen that the nucleons are distributed nearly uniformly through a sphere of radius

$$R(\text{nuclear surface}) = (1.07 \text{ fm})A^{1/3}. \tag{12.13}$$

Therefore, the average potential energy seen by each nucleon is found by averaging the nucleon–nucleon potential energy over this sphere. The resulting IPA potential energy, $U(r)$, is spherically symmetric and is sketched in Fig. 12.6. The range of the potential well is given approximately by

$$R(\text{force range}) = (1.17 \text{ fm})A^{1/3}. \tag{12.14}$$

Comparing this with (12.13) we see that, as one would expect, the force range is a little larger than the nuclear radius, since the force can "reach out" 1 fm or so beyond the distribution of nucleons. The depth of the IPA well is about 50 MeV and is approximately the same for most nuclei.

The potential-energy function in Fig. 12.6 is the averaged nuclear potential energy seen by each nucleon. Because of the charge independence of the nuclear force, this IPA potential energy is the same for neutrons as for protons. In the case of protons, however, we must add the potential energy of the electrostatic Coulomb force. To a good approximation this is the potential energy of a proton in the electric field of a uniform sphere of charge $(Z - 1)e$. Outside the nucleus $(r > R)$ this is

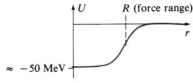

FIGURE 12.6 The nuclear IPA potential well, $U(r)$, seen by one nucleon under the influence of the other $A - 1$ nucleons in a nucleus. The force range R can be defined as the radius at which the potential energy is half its maximum depth and is found to be somewhat larger than the radius of the nucleus.

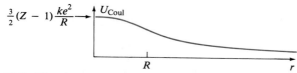

FIGURE 12.7 The electrostatic potential energy of a proton in the electric field of the $Z - 1$ other protons in a nucleus.

$$U_{Coul}(r) = (Z - 1)\frac{ke^2}{r}. \tag{12.15}$$

Inside the nucleus it has the shape of the inverted parabola shown in Fig. 12.7 and is maximum at the center, where it is 1.5 times its value at the nuclear surface (Problem 12.28):

$$U_{Coul}(0) = \frac{3}{2}(Z - 1)\frac{ke^2}{R}. \tag{12.16}$$

The Coulomb potential energy (12.16) is easily calculated for any given nucleus (Problems 12.27 and 12.28). The answer varies from 1 or 2 MeV for the lightest nuclei to about 30 MeV for the heaviest. For protons in light nuclei it is therefore a fair approximation to ignore U_{Coul} compared to the nuclear potential energy of Fig. 12.6. For protons in heavy nuclei this is certainly not a reasonable approximation, and we shall have to use the total potential energy,

$$U(r) = U_{nuc}(r) + U_{Coul}(r).$$

This function is plotted in Fig. 12.8, which shows the IPA potential well for a proton in the middle-mass nucleus ^{120}Sn (tin).

It is often convenient to display the IPA potential-energy functions for protons and neutrons side by side. One convenient way to do this is to draw them "back to back" as in Fig. 12.9. It is important to remember that although we have drawn these two functions separately, they describe the potential energy of a proton and a neutron in one and the same nucleus.

Using the IPA potential-energy functions one can calculate the energy levels of each nucleon in a nucleus. In this way we can construct a shell model of the nucleus, quite analogous to the atomic shell model described in Chapter 11. As we describe in Section 12.8, the nuclear shell model was developed about 1950 and is very successful in explaining many (though certainly not all) nuclear properties. Before taking up the shell model we discuss two other ideas that depend on the existence, though not the details, of the energy levels of the wells shown in Fig. 12.9.

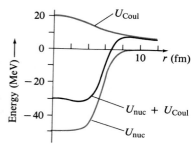

FIGURE 12.8 The IPA potential energy of any one proton in the nucleus ^{120}Sn (tin) is the sum of two parts, U_{nuc} and U_{Coul}.

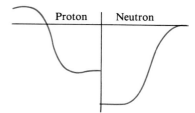

FIGURE 12.9 The IPA potential energy functions for a proton and a neutron in a medium-mass nucleus.

12.6 The Pauli Principle and the Symmetry Effect

We saw earlier, in Fig. 12.1, that there is a tendency for stable nuclei to have nearly equal numbers of protons and neutrons, especially when A is small. This tendency is often called the **symmetry effect.** Using the ideas of Section 12.5, we can now show that the symmetry effect is a simple consequence of the Pauli principle and the charge independence of nuclear forces.

Like electrons, both protons and neutrons obey the Pauli exclusion principle. That is, no two protons can occupy the same quantum state and, similarly, no two neutrons can occupy the same quantum state. It is important to

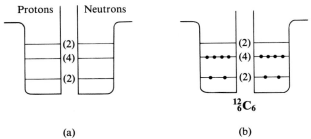

(2)
(4)
(2)

(2)
(4)
(2)

$^{12}_{6}C_6$

(a)   (b)

FIGURE 12.10 (a) Schematic diagram of the IPA wells and levels for protons and neutrons in a light nucleus. The two wells are essentially identical because of charge independence. The numbers in parentheses are the observed degeneracies. (b) The ground state of $^{12}_{6}C_6$ is found by putting its six protons in the lowest available proton levels and its six neutrons in the lowest available neutron levels.

understand that the Pauli principle only restricts the states of two identical particles; it places no restrictions on the states occupied by two particles of different type. Thus it is perfectly possible to have a proton and neutron both occupying exactly the same quantum state.

The ground state of a nucleus is found by assigning its Z protons and N neutrons to their lowest-possible states consistent with the Pauli principle. To see how this works, let us first ignore the Coulomb repulsion of the protons, so that the "wells" in which the protons and neutrons move are identical. (This will be a good approximation for light nuclei, not as good for heavier nuclei.) Figure 12.10(a) shows the two wells and the lowest few levels in each, along with their degeneracies. Since the wells are identical, the same is true of the levels. The Pauli principle requires that we place the Z protons in the lowest proton levels, with the number of protons in a level never exceeding that level's degeneracy; and similarly with the neutrons. Figure 12.10(b) shows the resulting ground state of the nucleus $^{12}_{6}C_6$. (The precise degeneracies of the levels do not matter for the present discussion, but we have shown their actual observed values—two for the lowest level and four for the next.)

The example shown in Fig. 12.10(b), $^{12}_{6}C_6$, is a nucleus with $Z = N$. Let us now consider a nucleus with the same number of nucleons, but $Z \neq N$, that is, one of the isobars of $^{12}_{6}C_6$. For example, consider the nucleus $^{12}_{5}B_7$ with five protons and seven neutrons. The ground state of $_5B_7$ is shown on the left of Fig. 12.11, which also shows the ground states of $_6C_6$ and $_7N_5$ for comparison.

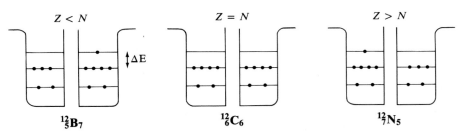

$Z < N$ $Z = N$ $Z > N$

$\uparrow \Delta E$

$^{12}_{5}B_7$ $^{12}_{6}C_6$ $^{12}_{7}N_5$

FIGURE 12.11 The ground states of the three isobars ^{12}B, ^{12}C, and ^{12}N. Because of the Pauli principle the two nuclei with $Z \neq N$ have higher energy by the amount shown as ΔE.

Comparing the ground state of $_5B_7$ with that of $_6C_6$, we see that because of the Pauli principle the seventh neutron in $_5B_7$ has to occupy a higher level than any of the neutrons or protons in $_6C_6$. Therefore, the nucleus $_5B_7$ has higher total energy than $_6C_6$.

If we look on the other side of $^{12}_6C_6$ at $^{12}_7N_5$, we find a similar situation. The seventh proton of $_7N_5$ must occupy a higher level than any of the nucleons in $_6C_6$, and $_7N_5$ has higher total energy than $_6C_6$. If we move still farther away from $Z = N$ (to $_8O_4$, for example), the total energy has to be even higher.

Similar arguments can be applied to any set of isobars. Based on these arguments (in which we have so far ignored the Coulomb energy of the protons) we can state the following general conclusion: Among any set of isobars (nuclei with the same total number of nucleons) the nucleus (or nuclei) with Z closest to N will have the lowest total energy.

To understand what this conclusion implies, we need to anticipate a property of radioactive decay, which we discuss in Chapter 13. If a nucleus has appreciably more energy than a neighboring isobar, the nucleus with greater energy is unstable and will eventually convert itself into the isobar with lower energy. The process by which this occurs is called β decay and is discussed in Section 13.4, but for now the important point is this: Any nucleus with Z much different from N is unstable and will convert itself into a neighboring isobar with Z closer to N and hence lower total energy. Therefore, the stable nuclei, and hence the nuclei normally found to occur naturally, should all have $Z \approx N$.

In discussing the symmetry effect we have so far ignored the electrostatic energy of the protons. This is easily taken into account: As discussed in Section 12.5, the Coulomb energy simply raises the proton well relative to the neutron well. This is shown in Fig. 12.12(a), which shows the two wells and a schematic set of energy levels. If the neutron well is appreciably lower than the proton well, there may be several neutron levels below any of the proton levels. In this case the most stable isobar will have more neutrons than protons, as indicated schematically in the figure.

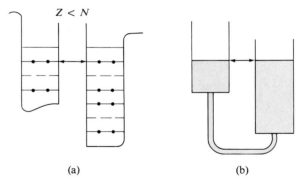

(a)   (b)

FIGURE 12.12   (a) In nuclei with many protons the Coulomb repulsion pushes the proton well appreciably higher than the neutron well. For a given number of nucleons, the lowest energy is obtained when the highest occupied proton and neutron levels have the same energy (as indicated by the arrow). This requires that Z be somewhat less than N. (b) The situation is closely analogous to a pair of buckets of water, whose total energy is lowest when the top surfaces are level, and the higher bucket has less water. A pipe between the buckets lets the water levels adjust to this state of lowest energy, whatever the initial configuration. In nuclei the corresponding adjustment occurs through radioactive decay.

For light nuclei the difference between the proton and neutron wells is small, and our original conclusion that stable nuclei should have $Z \approx N$ is unaltered. However, as we move to heavier nuclei, the Coulomb energy steadily rises and the ratio N/Z should increase slowly, as is confirmed by the data in Fig. 12.1.

It is perhaps worth reiterating the important role of the Pauli principle in the symmetry effect. In Fig. 12.12(a) it is clear that without the Pauli principle, the state of lowest energy would consist in all (or almost all) the nucleons being neutrons, all occupying the lowest neutron level. Under these conditions the majority of elements that make up our world would not exist.

12.7   The Semiempirical Binding-Energy Formula

The binding energy B of a nucleus is the total energy needed to separate completely its A individual nucleons. Binding energies of different nuclei are a useful measure of their relative stabilities as we shall see later. In this section we discuss how the binding energy depends on the nuclear parameters Z, N, and $A = Z + N$. We start with a brief discussion of the measurement of binding energies.

As mentioned in Section 4.5, nuclear binding energies are large enough to cause an observable difference (of order 1%) between the mass of a nucleus and the sum of the separate masses of its Z protons and N neutrons:

$$m_{\mathrm{nuc}} = Zm_{\mathrm{p}} + Nm_{\mathrm{n}} - \frac{B}{c^2}, \qquad (12.17)$$

where m_{nuc} denotes the mass of the nucleus concerned. Given accurate measurements of the masses involved, we can use this relation to find B as

$$B = (Zm_{\mathrm{p}} + Nm_{\mathrm{n}} - m_{\mathrm{nuc}})c^2. \qquad (12.18)$$

Most mass tabulations, including that in Appendix D, give the mass m_{atom} of the *atom,* rather than that of the bare nucleus, since it is m_{atom} that is usually measured. The mass m_{atom} is equal to m_{nuc} *plus* the mass of the Z atomic electrons*

$$m_{\mathrm{atom}} = m_{\mathrm{nuc}} + Zm_{\mathrm{e}}.$$

Now, we can add Zm_{e} to m_{nuc} in (12.18) if we also add Zm_{e} to the term Zm_{p}. This replaces Zm_{p} by

$$Z(m_{\mathrm{p}} + m_{\mathrm{e}}) = Zm_{\mathrm{H}},$$

where m_{H} is the mass of the hydrogen atom (^{1}H). This means that we can rewrite (12.18) as

$$B = (Zm_{\mathrm{H}} + Nm_{\mathrm{n}} - m_{\mathrm{atom}})c^2 \qquad (12.19)$$

and work entirely with atomic masses.

---

* There is, of course, a small correction due to the binding energy of the electrons. Since average electron binding energies are at most several keV, while nuclear binding energies are several MeV, this correction is almost always negligible.

EXAMPLE 12.3 Use the atomic masses in Appendix D to find the binding energy of the nucleus $^{35}_{17}Cl_{18}$.

According to (12.19) the required binding energy is

$$B = [17m_H + 18m_n - m_{atom}(^{35}Cl)]c^2,$$

or, with the data from Appendix D (rounded to five significant figures),

$$B = [17(1.0078 \text{ u}) + 18(1.0087 \text{ u}) - (34.969 \text{ u})]c^2$$

$$= [35.289 \text{ u} - 34.969 \text{ u}]c^2 \qquad (12.20)$$

$$= (0.320 \text{ u})c^2 \times \frac{931.5 \text{ MeV}/c^2}{1 \text{ u}}$$

$$= 298 \text{ MeV}. \qquad (12.21)$$

Notice in (12.20) that the mass of the separate constituents is *greater* than that of ^{35}Cl, confirming that we must *add* energy to pull the nucleus apart. Notice also that (12.20) gives B as the difference of two masses that are very nearly equal. This difference is always of order 1% or less. Thus we must know the masses to at least two significant figures more than are required in our answer. Here we needed five significant figures in the masses to get three in B.

As the example above shows, one can find the binding energy of a nucleus if one knows the corresponding atomic mass accurately. Using a mass spectrometer it is possible to measure atomic masses very accurately (see Section 12.9), and this is therefore one of the best ways to find nuclear binding energies.

The direct measurement of binding energies, by complete separation of a nucleus, is seldom possible. However, it is often possible to find the binding energy of one nucleus in terms of the known binding energy of some other nucleus. For example, one can measure the energy needed to separate one neutron from a nucleus. This is called the *neutron separation energy*,

$$S_n(Z, N) = \text{energy to separate one neutron from nucleus } (Z, N).$$

It is easy to see that the binding energy of the nucleus (Z, N) is

$$B(Z, N) = S_n(Z, N) + B(Z, N - 1); \qquad (12.22)$$

that is, the energy to dismember (Z, N) completely equals the energy to pull off one neutron *plus* the energy to dismember the remaining nucleus $(Z, N - 1)$. Thus if the binding energy of $(Z, N - 1)$ is known, we can use (12.22) to find the binding energy of (Z, N), or vice versa.

EXAMPLE 12.4 The binding energy of $^{16}_8O_8$ is known to be 127.6 MeV. It is found that 4.2 MeV is needed to separate a neutron from $^{17}_8O_9$. What is the binding energy of $^{17}_8O_9$?

According to (12.22) the required binding energy is

$$B(^{17}O) = S_n(^{17}O) + B(^{16}O)$$

$$= (4.2 \text{ MeV}) + (127.6 \text{ MeV}) = 131.8 \text{ MeV}.$$

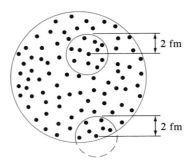

FIGURE 12.13 A nucleon in a nucleus is bound only to those neighbors within a sphere of radius about 2 fm (upper small sphere); thus all interior nucleons are bound to about the same number of neighbors. A nucleon near the surface is bound to fewer neighbors (lower small sphere), and this slightly reduces the total binding energy.

In the remainder of this section we derive a formula that gives the binding energy of nuclei as a function of the numbers Z, N, and $A = Z + N$. All of the terms in this formula have at least partial theoretical justifications, but the values of the coefficients that appear are generally found by fitting the formula to the measured binding energies. For this reason the formula is called the **semiempirical binding-energy formula,** or the **Weizsacker semiempirical formula,** after its inventor. If we insert the formula for B into the equation $m_{nuc} = Zm_p + Nm_n - B/c^2$, the resulting equation is called the *semiempirical mass formula.*

To an excellent approximation the binding energy of a nucleus is found to be proportional to the number, A, of its nucleons. This result is easily understood by the following classical argument: We have seen that the nuclear force has a range of about 2 fm, which is smaller than most nuclear sizes. For most nuclei, then, each nucleon is bound to only a fraction of the other nucleons, namely, those neighbors within a sphere of radius about 2 fm, as shown in Fig. 12.13. We have also seen that the density of nucleons is about the same inside all nuclei. It follows that the number of neighbors to which each nucleon is bound is approximately the same for all nuclei. That is, the average binding energy of *each* nucleon is the same in all nuclei, so that the *total* binding energy is proportional to A. Since A is proportional to the volume of the nucleus, this approximation is called the *volume term* and is written

$$B \approx B_{vol} = a_{vol}A \qquad (12.23)$$

where a_{vol} is a positive constant.

There is an immediate correction to the approximation (12.23). Those nucleons near the nuclear surface have fewer neighbors to attract them (Fig. 12.13) and are less tightly bound than those in the interior. The number of these surface nucleons is proportional to the surface area of the nucleus, $4\pi R^2 = 4\pi R_0^2 A^{2/3}$. Accordingly, (12.23) must be reduced by an amount proportional to $A^{2/3}$, and our next approximation is

$$B \approx B_{vol} + B_{surf},$$

where the *surface correction* B_{surf} is negative and has the form

$$B_{surf} = -a_{surf}A^{2/3}, \qquad (12.24)$$

with a_{surf} a positive constant.

A further correction is needed because of the electrostatic repulsion of the protons. This reduces the binding energy by an amount equal to the potential energy of the total nuclear charge. The potential energy of a uniform sphere of charge Ze is (Problem 12.37)

$$U_{Coul} = \frac{3}{5}\frac{k(Ze)^2}{R}. \qquad (12.25)$$

Since $R = R_0 A^{1/3}$ this implies a correction of the form

$$B_{Coul} = -a_{Coul}\frac{Z^2}{A^{1/3}}, \qquad (12.26)$$

where

$$a_{Coul} \approx \frac{3}{5}\frac{ke^2}{R_0}.$$

Since the nucleus is not precisely a uniform sphere of charge, this is not the exact value of a_{Coul} (Problem 12.37).

The next correction arises from the symmetry effect discussed in Section 12.6, where we saw that for any set of isobars, the nucleus with $Z = N$ has highest binding energy. Since B gets less as $Z - N$ increases in either direction, this suggests a negative term proportional to $(Z - N)^2$. The observed binding energies of all nuclei are best fitted by a correction of the form

$$B_{\text{sym}} = -a_{\text{sym}} \frac{(Z - N)^2}{A}. \tag{12.27}$$

The factor of A in the denominator reflects that for larger nuclei the level spacing is smaller, and the difference of energy between neighboring isobars (shown as ΔE in Fig. 12.11) is therefore smaller.

The final term in our formula for the binding energy reflects a phenomenon that we shall discuss in more detail in the next section. We mentioned in Section 12.2 that the majority of stable nuclei (148, in fact) have both Z and N even; there are 100 nuclei with Z even but N odd, or vice versa; and there are only 4 in which both Z and N are odd. This preference for Z and N to be even is explained by a property of the nuclear force called the **pairing effect.** For now we simply state that this effect gives a correction to the binding energy with the form

$$B_{\text{pair}} = \epsilon \frac{a_{\text{pair}}}{A^{1/2}} \tag{12.28}$$

where a_{pair} is some constant and

$$\epsilon = \begin{cases} 1 & \text{if } Z \text{ and } N \text{ are both even,} \\ 0 & \text{if } Z \text{ or } N \text{ is even and the other is odd,} \\ -1 & \text{if } Z \text{ and } N \text{ are both odd.} \end{cases} \tag{12.29}$$

Putting all five of these contributions together, we obtain the *semiempirical binding-energy formula*

$$
\begin{aligned}
B &= B_{\text{vol}} + B_{\text{surf}} + B_{\text{Coul}} + B_{\text{sym}} + B_{\text{pair}} \\
&= a_{\text{vol}} A - a_{\text{surf}} A^{2/3} - a_{\text{Coul}} \frac{Z^2}{A^{1/3}} - a_{\text{sym}} \frac{(Z - N)^2}{A} + \epsilon \frac{a_{\text{pair}}}{A^{1/2}}.
\end{aligned} \tag{12.30}
$$

where ϵ is defined in (12.29). The five coefficients are generally adjusted to fit all the measured binding energies as well as possible. One set of values that gives a good fit is as follows:

$$
\left. \begin{aligned}
a_{\text{vol}} &= 15.75 \text{ MeV} \qquad a_{\text{surf}} = 17.8 \text{ MeV} \\
a_{\text{Coul}} &= 0.711 \text{ MeV} \qquad a_{\text{sym}} = 23.7 \text{ MeV} \\
a_{\text{pair}} &= 11.2 \text{ MeV}.
\end{aligned} \right\} \tag{12.31}
$$

Because the dominant behavior of B is $B \propto A$, it is usual to discuss B/A, the **binding energy per nucleon.** Figure 12.14 shows the measured value of B/A for a large number of nuclei. The smooth curve is the value of B/A predicted by (12.30) and is seen to give an excellent fit for all but the lightest nuclei (those with A below about 20). It is not surprising that the smooth function (12.30) does not work well for the light nuclei, in which the quantization of energy

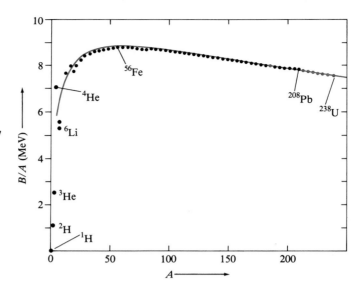

FIGURE 12.14 The binding energy per nucleon, B/A, as a function of A. The points are measured values. (Black dots indicate stable nuclei; colored, unstable.) The curve is the prediction of the binding-energy formula (12.30). [Values of B/A are shown for the most stable isobar of each given A. For $A > 7$ only even–even nuclei are shown, and for clarity, only half the data are included. The curve was calculated from (12.30) assuming that both Z and N are even.]

levels is proportionately more important. Nevertheless, even when A is small, the *general* trend of B/A is correctly given by (12.30).

The most obvious feature of Fig. 12.14 is that B/A is close to 8 MeV for almost all nuclei, with the maximum value of 8.8 MeV for iron (Fe). Beyond iron, B/A slopes gently down to about 7.6 MeV for ^{238}U; this decrease is due mainly to the increasing importance of the Coulomb repulsion of the protons. On the other side of ^{56}Fe, when A decreases below about 20, B/A falls rapidly to zero for ^{1}H (which has no binding energy); this decrease occurs because almost all nucleons in a small nucleus are close to the surface and the negative surface correction is proportionately large.

The lower values of B/A when A is large imply that energy is released when a heavy nucleus splits, or *fissions,* into two lighter nuclei. The lower values of B/A when A is very small imply that energy is also released when two light nuclei *fuse* to form one heavier nucleus. Nuclear fission is the source of energy in nuclear power stations, while fusion provides the energy of the stars, including our own sun. Both fission and fusion are used in nuclear weapons. Clearly, the behavior of B/A, as shown in Fig. 12.14, is of great practical significance.

EXAMPLE 12.5   Use the binding-energy formula (12.30) to predict the binding energy of ^{35}Cl and compare the result with the measured value found in (12.21).

Inserting the values $A = 35$, $Z = 17$, $N = 18$, and the coefficients (12.31) into the formula (12.30) we find that

$$B = a_{vol}A - a_{surf}A^{2/3} - a_{Coul}\frac{Z^2}{A^{1/3}} - a_{sym}\frac{(Z-N)^2}{A} + \epsilon\frac{a_{pair}}{A^{1/2}}$$
$$= (551.2 - 190.5 - 62.8 - 0.7 + 0)\text{ MeV}$$
$$= 297\text{ MeV},$$

which agrees very well with the measured value of 298 MeV. Notice that the pairing term is zero in this example because Z is odd and N even.

12.8 The Shell Model

The semiempirical binding-energy formula gives a remarkably accurate picture of the general trends of nuclear binding energies. Nevertheless, it is a smoothed, average picture and shows none of the fluctuations that one would expect if nucleons, like atomic electrons, occupy energy levels that fall into distinct shells. Recall that atomic binding energies fluctuate periodically with Z, being greatest when the electrons exactly fill a closed shell, and dropping steeply just beyond each closed shell. In this section we discuss the corresponding shell properties of nuclei.

THE MAGIC NUMBERS

The total numbers of electrons in the closed-shell atoms are known to be

atomic closed-shell numbers:
$$2, \quad 10, \quad 18, \quad 36, \quad 54, \quad 86. \qquad (12.32)$$

By the late 1940s, substantial evidence had accumulated that nuclei have analogous closed-shell numbers which lead to unusual stability, and that these **magic numbers,** as they came to be called, were:

nuclear magic numbers:
$$2, \quad 8, \quad 20, \quad 28, \quad 50, \quad 82, \quad 126. \qquad (12.33)$$

A nucleus is especially stable if Z or N is equal to one of these magic numbers, and even more so if both are. The magic numbers are the same for protons as for neutrons, as one would expect from the charge independence of nuclear forces.* On the other hand, the nuclear magic numbers are quite different from the atomic closed-shell numbers (12.32), reflecting the great difference between the dominant forces in nuclei and atoms.

The evidence for the magic numbers (12.33) consists of many observations, few of which are conclusive by themselves. Nuclei in which Z is magic tend to have larger numbers of stable isotopes than average. For example, tin, with Z equal to the magic number 50, has 10 stable isotopes, more than any other element. Similarly, nuclei in which N is magic tend to have more stable isotones than average; $N = 82$ holds the record, with 7 stable isotones. The binding energy per nucleon, B/A, in each of the "doubly magic" nuclei $_2\text{He}_2$, $_8\text{O}_8$, $_{20}\text{Ca}_{20}$, $_{20}\text{Ca}_{28}$, and $_{82}\text{Pb}_{126}$ is markedly higher than the prediction of the binding-energy formula. (This is easily discernible in the case of $_2\text{He}_2$ on the graph of B/A in Fig. 12.14.)

Perhaps the best evidence for the magic numbers (12.33) concerns the separation energies S_p and S_n. These are the energies required to remove one proton or one neutron from a nucleus and are analogous to the ionization energy of an atom. Recall that a plot of atomic ionization energy against number of electrons (Fig. 11.11) shows a maximum at each of the closed shells, with an abrupt drop immediately thereafter. The nuclear separation energies depend on two variables, Z and N, and are harder to plot. One simple procedure is to eliminate one of these variables by an appropriate averaging. If we consider, for example, the proton separation energy $S_p(Z, N)$, then for each value of Z we can compute the average $\bar{S}_p(Z)$ for all stable isotopes with this Z. A plot

---

* Note, however, that nuclei with $Z = 126$ have not been observed. Therefore, there is no experimental evidence for a proton magic number in this region.

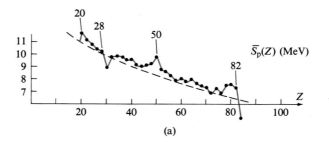

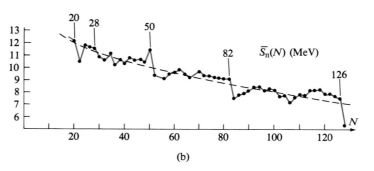

FIGURE 12.15 **(a)** The proton separation energy $\bar{S}_p(Z)$, averaged over all stable isotopes for each Z. **(b)** The neutron separation energy $\bar{S}_n(N)$, averaged over all stable isotones for each N. The dashed curves show the predictions of the binding-energy formula (12.30). Only even values of Z were included in (a) and only even N in (b).

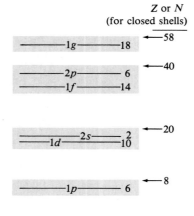

FIGURE 12.16 Filling order of the levels for a proton or neutron, calculated for the nuclear potential well of Fig. 12.6. The number to the right of each level is its degeneracy. The levels fall into distinct energy shells, but the closed-shell numbers do not agree with the observed magic numbers (12.33). It follows that these levels are *not* correct.

of $\bar{S}_p(Z)$ against Z is shown in Fig. 12.15(a) and, as expected, has pronounced drops immediately after each of the magic numbers indicated. (We have not included data for $Z < 20$, since our averaging procedure obscures shell effects for very light nuclei.) The dashed curve in Fig. 12.15(a) shows the proton separation energy predicted by the binding-energy formula. This reproduces the general trend very well but entirely misses the fluctuations associated with the magic numbers.

Figure 12.15(b) shows the average neutron separation energies $\bar{S}_n(N)$, each averaged over all stable isotones of a given N. This shows abrupt drops beyond the same magic numbers, 20, 28, 50, and 82, as in Fig. 12.15(a), and also at $N = 126$.

ORIGIN OF THE MAGIC NUMBERS

Having established the existence of the nuclear magic numbers (12.33), physicists naturally hoped to explain them in terms of nucleon energy levels and the grouping of those levels into energy shells. The IPA potential energy felt by any one proton or neutron was described in Section 12.5 and sketched in Fig. 12.6. The energy levels for this well can be calculated numerically, and the ordering of the levels is shown in Fig. 12.16.

The levels shown in Fig. 12.16 are labeled according to the scheme normally used in nuclear physics. This scheme is nearly, although not quite, the same as that used in atomic physics. Just as in atomic physics, each level is identified by two quantum numbers, n and l. The number l identifies the magnitude of the nucleon's orbital angular momentum, $L = \sqrt{l(l + 1)}\hbar$. The possible values $l = 0, 1, 2, 3, 4, \ldots$ are indicated by the code letters $s, p, d, f, g,$ $\ldots$. For each value of l, there are $2l + 1$ possible values of $L_z = m\hbar$ (with $m = l, l - 1, \ldots, -l$) and two possible values of $S_z = \pm\frac{1}{2}\hbar$. The degeneracy of each level is therefore $2(2l + 1)$. Evidently, all of the parameters associated with angular momentum are the same as in atomic physics. On the other hand, nuclear physicists define the quantum number n so that the levels of a given l are simply numbered $n = 1, 2, 3, \ldots$. Thus the levels with $l = 1$ are labeled $1p$, $2p$, $3p$, . . . in order of increasing energy; the levels with $l = 2$ are $1d$, $2d$, $3d$, . . .; and so on.*

The levels (or subshells) shown in Fig. 12.16 are grouped into well-defined energy shells, and the total numbers of protons or neutrons needed to produce closed shells are indicated on the right. The first three of these numbers agree exactly with the observed magic numbers, but beyond this all of the predicted closed-shell numbers turn out to be wrong.

At first it would seem natural to assume that a different shape for the IPA potential well might produce the correct magic numbers. In fact, however, no reasonable well gives the observed numbers. The correct explanation was discovered independently in 1949 by Mayer and by Jensen and co-workers. It turns out that the nuclear force depends strongly on the relative orientation of each nucleon's spin and orbital angular momentum. Specifically, the IPA potential well of Fig. 12.6 is the correct potential-energy function for a nucleon with $l = 0$, but if $l \neq 0$, then the nucleon has an additional potential energy that depends on the orbital angular momentum $\mathbf{L}$ and its orientation relative to the spin $\mathbf{S}$. This additional energy ΔU is found to have the form

$$\Delta U \propto SL \cos \theta,$$

where θ is the angle between $\mathbf{S}$ and $\mathbf{L}$; that is,

$$\Delta U \propto \mathbf{S} \cdot \mathbf{L}. \tag{12.34}$$

For this reason, the energy ΔU is called the **spin-orbit energy**.

Since the spin-orbit energy of each nucleon is proportional to $\mathbf{S} \cdot \mathbf{L}$, each energy level of Fig. 12.16 (with $l \neq 0$) is split into two levels according to the two possible orientations of $\mathbf{S}$ relative to $\mathbf{L}$. (Recall that $\mathbf{S}$ has just two possible orientations relative to any direction.) A spin-orbit splitting of this kind is familiar in atomic physics, where it produces the so-called fine structure of atomic spectra. As described in Section 10.7, an orbiting electron sees a magnetic field $\mathbf{B} \propto \mathbf{L}$; since the electron has a magnetic moment $\boldsymbol{\mu} \propto \mathbf{S}$, this leads to a magnetic energy $-\boldsymbol{\mu} \cdot \mathbf{B} \propto \mathbf{S} \cdot \mathbf{L}$. However, the nuclear spin-orbit splitting is *not* a magnetic effect. Rather, it is a new property of the nuclear force. The nuclear splitting is much larger than the atomic effect, and is in the opposite direction: For an electron in an atom the state with $\mathbf{S}$ and $\mathbf{L}$ parallel has slightly *higher* energy; for a nucleon in a nucleus the corresponding state has much *lower* energy. The first clear evidence for this nuclear spin-orbit energy came from studies of nuclear shell structure, but scattering experiments with separate spin-up and spin-down beams of nucleons have since confirmed its existence and properties.

The effect of the spin-orbit energy on the levels of Fig. 12.16 is shown in Fig. 12.17, which shows the resulting levels (or subshells) with their degeneracies and their grouping into energy shells. As one would expect, the effect of the spin-orbit energy increases with increasing l: The $1s$ level, like all s levels, is not split at all; the $1p$ level, like all p levels (and also d levels), *is* split, but the resulting two levels are still in the same shell; for the $1f$ level, and all levels

MARIE GOEPPERT–MAYER (1906–1972, German–US). After getting her PhD under Born at Gottingen in 1933, Goeppert–Mayer moved to the US. In 1949, while at Chicago, she proposed the nuclear shell model, for which she shared the 1963 Nobel Prize with Jensen.

JOHANNES JENSEN (1907–1973, German). Jensen was a professor first at Hamburg and then at Heidelberg. In 1949 he and Goeppert–Mayer independently proposed the nuclear shell model. Together they wrote a book on the model in 1955, and they shared the Nobel Prize in 1963.

---

* Recall that in atomic physics the lowest p state is called $2p$, the lowest d state $3d$, and so on. This scheme is very convenient in the hydrogen atom (where it means that all states of given n have the same energy) but has no particular merit in nuclear physics.

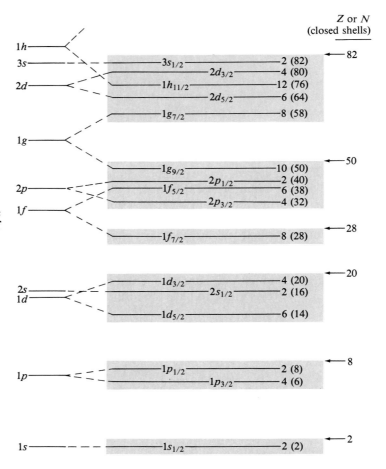

FIGURE 12.17 Filling order of the levels through Z or $N = 82$ for a single nucleon in a nucleus, including the spin-orbit energy proposed by Mayer and Jensen. The levels on the left are the corresponding levels in the absence of the spin-orbit energy. The numbers to the right of each level are the level's degeneracy and (in parentheses) the running total of protons or neutrons needed to fill through that level. On the far right are the closed-shell numbers, which agree perfectly with the observed magic numbers (12.33). The ordering of certain nearby levels is ambiguous (just as it is in atoms) and can be different for protons and neutrons. Beyond 82, where the proton well is strongly distorted by Coulomb repulsion, the level orderings for protons and neutrons are significantly different.

with $l \geq 3$, the splitting is so large that the resulting two levels belong to different shells. It can be seen that as a result of these splittings, the closed-shell numbers agree perfectly with the observed magic numbers.

To understand the labeling and degeneracies of the levels in Fig. 12.17, we must discuss briefly the total angular momentum $\mathbf{J}$ of a nucleon,

$$\mathbf{J} = \mathbf{L} + \mathbf{S}.$$

The magnitude of $\mathbf{L}$ is $L = \sqrt{l(l+1)}\,\hbar$ with $l = 0, 1, 2, \dots$, and that of $\mathbf{S}$ is $S = \sqrt{s(s+1)}\,\hbar$ with s always equal to $\frac{1}{2}$. It is found that the magnitude of $\mathbf{J}$ is given by a similar formula,

$$J = \sqrt{j(j+1)}\,\hbar. \tag{12.35}$$

If its orbital angular momentum is zero ($l = 0$), the nucleon's total angular momentum is just the spin, and hence $j = \frac{1}{2}$. If $l \neq 0$, it is found that there are two possible values of j, depending on the orientation of $\mathbf{S}$ relative to $\mathbf{L}$: If $\mathbf{S}$ is parallel to $\mathbf{L}$, then

$$j = l + \tfrac{1}{2} \qquad (\mathbf{S} \text{ and } \mathbf{L} \text{ parallel}), \tag{12.36}$$

but if $\mathbf{S}$ is antiparallel to $\mathbf{L}$, then

$$j = l - \tfrac{1}{2} \quad \text{(S and L antiparallel).} \tag{12.37}$$

In Fig. 12.17 the value of j in each level is shown by a subscript following the code letter for l. Thus the original $1p$ level, shown on the left, is really two levels with

$$j = l \pm \tfrac{1}{2} = 1 \pm \tfrac{1}{2} = \tfrac{3}{2} \quad \text{or} \quad \tfrac{1}{2}.$$

These are shown on the right of Fig. 12.17 as $1p_{3/2}$ and $1p_{1/2}$, with $1p_{3/2}$ somewhat lower in energy.

For each magnitude of the total angular momentum, there are several possible orientations of **J**, all with the same energy. These correspond to the different possible values of J_z, which are given by

$$J_z = m_j \hbar,$$

where m_j can have any of the $2j + 1$ values

$$m_j = j, j - 1, \ldots, -j.$$

Thus each of the levels of Fig. 12.17 has a degeneracy of $(2j + 1)$, as indicated to the right of each level.

EXAMPLE 12.6 Find the degeneracies of the $1p_{3/2}$ and $1p_{1/2}$ levels. Show that the sum of these is the same as the total degeneracy of the original $1p$ level on the left of Fig. 12.17.

The required degeneracies are

$$1p_{3/2}: \quad \text{degeneracy} = 2j + 1 = (2 \times \tfrac{3}{2}) + 1 = 4 \tag{12.38}$$

and

$$1p_{1/2}: \quad \text{degeneracy} = 2j + 1 = (2 \times \tfrac{1}{2}) + 1 = 2. \tag{12.39}$$

In the absence of the **S·L** energy, there would have been a single $1p$ level whose degeneracy would have been $2(2l + 1) = 6$ (that is, two orientations of the spin times three of **L**). This is equal to the sum of (12.38) and (12.39).

It can be shown quite generally (Problem 12.55) that for any given values of n and l, the sum of the degeneracies of the two corresponding levels (with $j = l \pm \tfrac{1}{2}$) is equal to the total degeneracy of the single level, nl, that would have existed in the absence of the **S·L** energy. Thus the effect of the **S·L** term is simply to redistribute the degeneracies so as to produce the observed magic numbers.

ANGULAR MOMENTA OF NUCLEI

In Section 11.6 we saw that one can predict the angular momentum of many atoms once one knows the order in which the electron energy levels are filled. We can now apply similar reasoning to nuclei. We first note that any filled level of given j contributes zero total angular momentum. (This is because for every nucleon with $J_z = m_j \hbar$ there is a second with $J_z = -m_j \hbar$.) This implies that any nucleus in which all of the occupied levels are closed (no partially filled levels) should have zero total angular momentum. That is,

$$J_{\text{tot}} = \sqrt{j_{\text{tot}}(j_{\text{tot}} + 1)}\, \hbar = 0$$

in this case. For example, in $_6C_6$ the protons and neutrons both completely fill their $1s_{1/2}$ and $1p_{3/2}$ levels. Therefore, $_6C_6$ should have, and does have, zero total angular momentum, $j_{tot} = 0$. Referring to Fig. 12.17, we can predict in the same way that the closed-subshell nuclei $_8O_8$, $_{14}Si_{16}$, and $_{20}Ca_{28}$ should all have $j_{tot} = 0$, and in every case this is found to be correct.

Let us consider next a nucleus with a single nucleon outside otherwise-filled subshells. For example, in $_8O_9$ the eight protons and eight of the nine neutrons fill the levels $1s_{1/2}$, $1p_{3/2}$, and $1p_{1/2}$, but the ninth neutron must occupy the $1d_{5/2}$ level all by itself. Since the closed levels all have zero angular momentum, the angular momentum of the whole nucleus is just that of the final, odd neutron. Thus we predict that $_8O_9$ should have $j_{tot} = \frac{5}{2}$, and this proves to be correct.

Throughout this section we have been discussing the ground states of nuclei, but we can also predict the properties of some excited states. For example, we just saw that the ground state of $_8O_9$ has its last neutron in the $1d_{5/2}$ level. In Fig. 12.17 we see that just above the $1d_{5/2}$ level is the $2s_{1/2}$. Hence we would expect that the first excited state of $_8O_9$ would be obtained by lifting the last neutron to the $2s_{1/2}$ level, and should have $j_{tot} = \frac{1}{2}$. This is observed to be the case.

PAIRING ENERGY

Let us next consider a nucleus with two identical nucleons (two protons or two neutrons) outside otherwise filled subshells. We could, for example, consider $_8O_{10}$ with two neutrons in the $1d_{5/2}$ level. The filled levels all have zero angular momentum, but since each of the last two neutrons has $j = \frac{5}{2}$ there are several possible values of j_{tot}, depending on the relative orientation of the two separate angular momenta. However, it turns out that the state in which the two angular momenta are antiparallel ($j_{tot} = 0$) has significantly lower energy. Therefore, the ground state of $_8O_{10}$ has $j_{tot} = 0$.

It is found quite generally that any pair of identical nucleons in the same level has lowest energy when their total angular momentum is zero. The extra binding energy in the $j_{tot} = 0$ state is called the **pairing energy,** and is typically a few MeV. This energy is the origin of the term B_{pair}, (12.28), in the semiempirical binding-energy formula.

Whenever a level is partially filled, the angular momenta of its nucleons can combine to give various different total angular momenta. However, the pairing effect implies that in the state of lowest energy the nucleons will be arranged, as far as possible, in pairs with zero angular momentum. If the number of nucleons in the level is even, this means that the state of lowest energy has $j_{tot} = 0$; if the number of nucleons is odd, the state of lowest energy has all but one nucleon arranged in $j = 0$ pairs and the total angular momentum is just that of the odd, unpaired nucleon.

These properties lead us to the following useful rules:

> The ground state of any nucleus in which both Z and N are even (an "even–even nucleus") has zero total angular momentum. (12.40)

and

> If a nucleus has A odd (that is, Z is even and N odd, or vice versa), the total angular momentum of its ground state is just that of the last, unpaired nucleon. (12.41)

Nuclei in which both Z and N are odd are harder to analyze,* but there are only four such stable nuclei in any case.

The rule (12.40) is true for all known even–even nuclei. The rule (12.41), together with the level ordering shown in Fig. 12.17, correctly predicts the angular momentum of most odd-A nuclei; the only exceptions involve non-spherical nuclei, for which the ordering of Fig. 12.17 is not always applicable. (See Problem 12.53.) The use of both rules is illustrated in the following example.

EXAMPLE 12.7 What are the quantum numbers j_{tot} for the total angular momenta of the nuclei ^{22}Ne and ^{43}Ca?

Since ^{22}Ne has $Z = 10$ and $N = 12$, it is an even–even nucleus and hence has $j_{tot} = 0$. The nucleus ^{43}Ca has $Z = 20$ and $N = 23$. Since Z is even, the protons have zero total angular momentum. The total angular momentum of the neutrons is just that of the twenty-third neutron, which is in the $1f_{7/2}$ level (Fig. 12.17). Hence the nucleus ^{43}Ca should, and does, have $j_{tot} = \frac{7}{2}$.

12.9 Mass Spectrometers (optional)

In this section we describe mass spectrometers, the devices used to measure nuclear masses. As we saw earlier, measurements of nuclear masses contributed to our understanding of nuclear structure in two important ways. First, early measurements, which could distinguish masses differing by a few percent, revealed the existence of isotopes. Second, more precise measurements were able to determine the small relativistic shifts (about 1%) in nuclear masses due to the binding energy. The variation of binding energy with mass number, A, revealed several properties of the nuclear force, as we discussed in Section 12.7. Today, mass spectrometers can measure masses with an accuracy of order 1 part in 10^8, which allows binding energies of typical nuclei to be calculated within 1 keV or so. Most recently, the mass spectrometer has developed into a powerful tool of chemical analysis, as we describe shortly.

In practice, it is generally the mass of an ionized atom (or molecule) that is measured in a mass spectrometer. The substance to be analyzed is introduced into an electric discharge, which causes some atoms to lose one electron and become positive ions of charge e. In the arrangement shown in Fig. 12.18, the ions are next accelerated by a potential difference V_0. Finally, they are bent into a circular path, of radius R, by a magnetic field B. Measurement of V_0, R, and B allows one to calculate the ions' mass, m, as we shall see directly. Since m is the mass of the ion, the mass of the corresponding neutral atom is calculated by adding one electron mass, m_e; that of the bare nucleus is found by subtracting $(Z - 1)m_e$.

---

* The difficulty is to predict which relative orientation of the angular momenta of the last proton and last neutron gives the lowest energy.

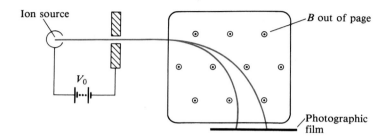

Ion source

V_0

B out of page

Photographic film

FIGURE 12.18 A mass spectrometer. Positive ions are produced in the ion source by an electric discharge and are accelerated by a potential difference V_0. They then enter a magnetic field and follow a curved path. Ions with different masses follow paths of different curvature and strike the photographic film at different places. The entire apparatus operates in a vacuum.

To calculate the mass, we note that when the ions are accelerated by the potential difference V_0, they gain a kinetic energy,

$$K = V_0 e.$$

Their momentum p is then*

$$p = \sqrt{2mK} = \sqrt{2mV_0 e}, \tag{12.42}$$

where m is the mass of the ion. When the ions enter the magnetic field B, they follow a circular path with radius given by (3.48) as

$$R = \frac{p}{eB}. \tag{12.43}$$

Substituting (12.42) and solving for m, we find that

$$m = \frac{eB^2 R^2}{2V_0}. \tag{12.44}$$

If B and V_0 are fixed, different masses will have different radii, and measurement of R lets one find m from (12.44). In most early mass spectrometers, R was measured by locating the spots where the ions struck a photographic film, as in Fig. 12.18. In modern spectrometers a computer monitors the accelerating voltage V_0, the field B, and the location of an electronic ion detector; after initial calibration with ions of known mass, the computer can display the results directly in atomic mass units.

When an element is analyzed in a mass spectrometer, its various isotopes are separated because of their different masses. Figure 12.19 shows the spectrum of masses obtained when natural lead is analyzed in a modern spectrometer, which displays its results directly as a graph of number of particles against mass.

Today, mass spectrometers are widely used in chemical analysis to help determine the composition of molecules. This is illustrated in Fig. 12.20, which shows the spectrum obtained when the simple diatomic compound HCl is introduced into the ion source of a mass spectrometer. Since chlorine has two stable isotopes, ^{35}Cl and ^{37}Cl, there are two large peaks with masses close to 36 u

---

* We are treating the motion as nonrelativistic, which is usually an excellent approximation (see Problem 12.60).

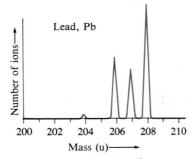

FIGURE 12.19 Results of analyzing natural lead in a mass spectrometer. The vertical axis shows the number of ions detected, and the horizontal axis is calibrated in atomic mass units. The different heights of the four peaks reflect the different natural abundances of the four isotopes of lead. (Courtesy Fred White.)

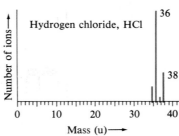

FIGURE 12.20 Mass spectrum obtained from HCl. The peaks labeled 36 and 38 were produced by the molecules $^1H^{35}Cl$ and $^1H^{37}Cl$, containing the two naturally occurring isotopes ^{35}Cl and ^{37}Cl. The smaller peaks at 35 and 37 were produced by atoms of ^{35}Cl and ^{37}Cl that were detached in the ion source.

($^1H^{35}Cl$) and 38 u ($^1H^{37}Cl$). In addition, smaller peaks can be seen at the positions one would expect for single atoms of ^{35}Cl and ^{37}Cl. These occur because the electric discharge in the ion source dissociates some of the molecules into H and Cl atoms. If the HCl analyzed in the spectrometer were an unknown substance, its composition could be deduced from the following observations: (1) the total molecular mass has two values, 36 and 38; (2) the molecule has fragments with masses of 35 and 37; and (3) the ratio of the peaks at 36 and 38 (and those at 35 and 37) is the same as that of naturally occurring chlorine.

The mass spectrum obtained for a more complex molecule is shown in Fig. 12.21, together with a diagram of the molecule's structure. The molecule shown, methionine, has mass 149 u and produces the peak labeled 149 in the spectrum. As was the case with HCl, the molecule can be broken into smaller fragments in the ion source, and it is these fragments that produce the several peaks at lower masses. For example, the peaks labeled 61 and 88 arise because the molecule breaks easily at the point indicated by the arrow in Fig. 12.21(b), producing fragments of masses 61 u and 88 u. The masses of the various segments of this complex organic molecule provide valuable clues to its structure.

Close inspection of the major peaks in Fig. 12.21 shows that each is accompanied by several smaller peaks. These smaller peaks differ from the dominant peaks by one or two mass units and come about in two ways: (1) a molecule, or molecular fragment, can gain or lose one or two hydrogen atoms in the ion source; and (2) about 1% of the carbon atoms in any natural sample are the isotope ^{13}C (as opposed to the dominant ^{12}C). The details of these subsidiary peaks are additional clues to the composition of the molecule.

Current research in molecular structure has extended mass spectroscopy to molecules with masses of thousands of atomic mass units. Mass spectrometers with this capability are commercially available and can quickly and automatically produce spectra like those in Figs. 12.20 and 12.21 for almost any substance that needs to be analyzed.

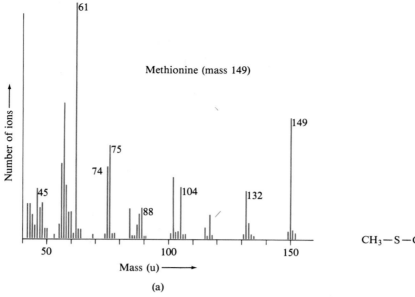

FIGURE 12.21 **(a)** Mass spectrum obtained from methionine $C_5H_{11}O_2NS$ with mass 149 u. The lower mass peaks are produced by fragments of the molecule that were detached in the ion source. **(b)** The molecule's structure. The arrow shows a point where the molecule can easily be broken in the ion source; the resulting fragments have masses of 61 u and 88 u and produce the peaks labeled 61 and 88 in the spectrum. (Courtesy McGraw-Hill.)

IDEAS YOU SHOULD NOW UNDERSTAND FROM CHAPTER 12

Isotopes, isobars, and isotones
Nuclear radius, $R = R_0 A^{1/3}$
The nuclear force:
    range and strength
    charge independence
Symmetry effect
Semiempirical binding-energy formula

Binding energy per nucleon
Shell model:
    magic numbers
    spin-orbit splitting
    total angular momentum; quantum number j
[Optional section: mass spectrometers]

PROBLEMS FOR CHAPTER 12

SECTION 12.2 (NUCLEAR PROPERTIES)

12.1 • Use Appendix D to find the numbers of protons and neutrons in the following nuclei: ^1H, ^3He, ^7Li, ^{20}Ne, ^{40}Ar, ^{63}Cu, ^{206}Pb.

12.2 • Use Appendix D to find the numbers of protons and neutrons in the following nuclei: ^2H, ^4He, ^9Be, ^{29}P, ^{64}Zn, ^{209}Bi, ^{232}Th.

12.3 • (a) Use Appendix D to make a list of all the stable isotopes, isobars, and isotones of ^{70}Ge. (b) Do the same for ^{27}Al and (c) ^{88}Sr.

12.4 • Make lists of all the stable isotopes, isobars, and isotones of each of the following: ^{56}Fe, ^{36}S, ^{208}Pb.

12.5 • Find the mass of the ^{12}C atom in Appendix D.

What fraction of its mass is contained in its atomic electrons?

12.6 • The mass of a lead atom's nucleus is 3.5×10^{-25} kg and its half-density radius R is 6.5 fm. Estimate its density by assuming its volume to be $\frac{4}{3}\pi R^3$. Compare your result to the density of solid lead (11 g/cm³).

12.7 • Atomic radii range from a little less than 0.1 nm to 0.3 nm, while nuclear radii range from about 2 fm to about 7 fm. Taking the representative values $R_{atom} \approx 0.2$ nm and $R_{nuc} \approx 4.5$ fm, find the ratio of the volume of an atom to that of a nucleus. What is the ratio of the corresponding average densities?

12.8 • The nuclear radius R is well approximated by the formula $R = R_0 A^{1/3}$, where $R_0 = 1.07$ fm. Find the approximate radii of the most abundant isotopes of each of the following elements: Al, Cu, and U. (Mass numbers and abundances can be found in Appendix D.)

12.9 • The first excitation energy of the ⁷Li nucleus is 0.48 MeV. What is the wavelength of a photon emitted when an ⁷Li nucleus drops from its first excited state to the ground state? Find the wavelength for the corresponding transition in the Li *atom* (excitation energy, 1.8 eV).

12.10 • The first excitation energy of the helium nucleus, ⁴He, is about 20 MeV. What is the wavelength of a photon that can just excite the helium nucleus from its ground state? Compare this with the wavelength of a photon that can just excite a helium *atom* (excitation energy 19.8 eV).

12.11 •• (a) Treating the nucleus as a uniform sphere of radius $(1.07 \text{ fm}) A^{1/3}$, calculate the density (in kg/m³) of the ⁵⁶Fe nucleus and compare it to that of solid iron (7800 kg/m³). (b) Calculate the electric charge density of this nucleus ($Z = 26$) in coulombs/m³. Compare this charge density to that which can be uniformly distributed in an insulating sphere of 1 m radius in air without sparking. The maximum electric field strength allowable at the surface of the sphere without sparking is 3×10^6 V/m.

12.12 •• The density of mass inside a nucleus is shown in Fig. 12.2. There is no simple theory that predicts the exact shape shown, but it is found that the shape can be approximated by the following mathematical form, known as the Fermi function:

$$\rho(r) = \frac{\rho_0}{1 + e^{(r-R)/t}}. \qquad (12.45)$$

Here ρ_0 is a constant, $\rho_0 = 3.25 \times 10^{17}$ kg/m³, R is the nuclear radius, and t is another constant, $t = 0.55$ fm. Plot the function (12.45) for a nucleus of radius $R = 5$ fm. First find $\rho(r)$ for $r = 0$, 5 fm,

and 10 fm and then choose a suitable grid of intermediate points.

12.13 •• The density $\rho(r)$ in a nucleus is well approximated by the formula (12.45) in Problem 12.12, with $\rho_0 = 3.25 \times 10^{17}$ kg/m³, $t = 0.55$ fm, and $R = R_0 A^{1/3}$, where $R_0 = 1.07$ fm. Plot $\rho(r)$ as a function of r from $r = 0$ to $r = 15$ fm for ²⁰⁸Pb.

12.14 •• The density in a nucleus can be approximated by the function (12.45) in Problem 12.12. (a) Prove that $\rho = \rho_0/2$ at $r = R$. (b) Show that the maximum value of ρ occurs at $r = 0$ and is very close to ρ_0 if R is much greater than t (as it usually is). (c) Sketch ρ as a function of r. (d) Prove that as r increases, the density falls from 90% to 10% of ρ_0 in a distance $\Delta r = 4.4t$. (Thus t characterizes the *thickness* of the surface region.)

SECTION **12.3** (THE NUCLEAR FORCE)

12.15 • If there were no nuclear attraction, what would be the acceleration of either of two protons released at a separation of 4 fm? Compare your answer with g, the acceleration of gravity. (Large as your answer is, it is still small compared to nuclear accelerations. See Problem 12.18.)

12.16 • Use Eq. (12.7) to estimate the minimum kinetic energy of a nucleon in a nucleus of diameter 8 fm.

12.17 •• (a) The minimum kinetic energy, K_{min}, of a nucleon in a rigid cubical box of side a is given by (12.7). By setting $a \approx 2R$, the nuclear diameter, and $R = R_0 A^{1/3}$ (with $R_0 = 1.07$ fm), derive a formula for K_{min} as a function of the mass number A. (b) Find K_{min} for ⁹Be, ²⁷Al, and ²³⁸U.

12.18 •• A representative value for the kinetic energy of a nucleon in a nucleus is 20 MeV. To illustrate the great strength of the nuclear force, do the following classical calculation: Suppose that a nucleon in a nucleus oscillates in simple harmonic motion with amplitude about 4 fm and peak kinetic energy 20 MeV. (a) Find its maximum acceleration. (b) By how many orders of magnitude does this exceed the acceleration of gravity, g? (c) Find the force required to produce this acceleration.

12.19 •• The charge independence of nuclear forces implies that in the absence of electrostatic forces, the energy levels of ⁷Li and ⁷Be would be the same. The main effect of the electrostatic forces is simply to raise all of the levels of $^7_4$Be compared to those of $^7_3$Li. Approximating both nuclei as uniform spheres of charge $Q = Ze$ and the same radius R, estimate the difference in the binding energies for any level of $^7_4$Be and the corresponding level of $^7_3$Li. (The electrostatic energy of a uniform charged sphere is $3kQ^2/5R$—see Problem 12.43. The observed radius R is about

2.5 fm.) Compare your rough estimate with the observed difference of about 1.7 MeV. (Note that the true charge distribution is not a uniform sphere with a well-defined radius R, but is spread out somewhat beyond R. Therefore, the observed electrostatic energies would be expected to be somewhat smaller than your estimates.)

12.20 •• Do the same calculations as in Problem 12.19 but for the nuclei ^{23}Na and ^{23}Mg. The observed difference is 4.8 MeV; use the formula $R = R_0 A^{1/3}$ to get the nuclear radius.

SECTION **12.4** (ELECTRONS VERSUS NEUTRONS AS NUCLEAR CONSTITUENTS)

12.21 • One argument against the proton–electron model of the nucleus concerns the total spins of nuclei. The proton, electron, and neutron all have spin $\frac{1}{2}$, and the total spin of any number of spin-half particles takes the familiar form $\sqrt{s(s+1)}\hbar$. If there is just one particle, then of course, $s = \frac{1}{2}$. With two particles, the spins can be parallel or antiparallel, giving $s = 1$ or 0. It can be shown that the general rule is this: For an *odd* number of particles, the total spin has some half-odd-integer value for s ($s = \frac{1}{2}, \frac{3}{2}, \frac{5}{2}, \ldots$). For an *even* number of particles, the total spin has s equal to some integer ($s = 0, 1, 2, \ldots$). In light of this rule consider the nucleus ^{14}N. (*a*) Assuming that ^{14}N is made of 7 protons and 7 neutrons, predict the character of its total spin (integer or half-odd integer). (*b*) Repeat for the proton–electron model of ^{14}N. (*c*) The observed total spin of ^{14}N is integer; which model does this support?

12.22 • Repeat Problem 12.21 for the deuteron, ^{2}H, whose observed total spin is given by $s = 1$.

12.23 •• Supposing that nuclei could contain electrons, estimate the potential energy of one such electron at the center of a nucleus of ^{27}Al. Compare your answer with the minimum kinetic energy (12.10) of an electron in a nucleus. [*Hint:* The required potential energy is $-eV(0)$, where $V(r)$ is given by Eq. (12.49) below. The radius R is given by (12.3), and the total charge acting on the electron is $Q = (Z + 1)e$.]

12.24 •• (*a*) Following the suggestions in Problem 12.23, find the potential energy that an electron would have at the center of a $^{12}_{6}$C nucleus. (*b*) Repeat for $^{208}_{82}$Pb. (*c*) Compare your answers with the minimum kinetic energy, which is of order 100 MeV.

12.25 •• Because of the proton's magnetic moment, every energy level of the hydrogen atom, as calculated previously, actually consists of two levels, very close together. The proton's magnetic moment creates a magnetic field, which means that the energy of the H atom is slightly different depending on the relative orientations of the electron and proton moments—an effect known as **hyperfine splitting.** You can estimate the magnitude of the hyperfine splitting as follows: (*a*) The magnetic field at a distance r from a magnetic moment $\boldsymbol{\mu}$ is

$$\mathbf{B} = \frac{\mu_0 \boldsymbol{\mu}}{2\pi r^3} \qquad (12.46)$$

where $\mu_0 = 4\pi \times 10^{-7}\ N/A^2$ is the permeability of space. [For simplicity, we have given the field at a point on the axis of $\boldsymbol{\mu}$. At points off the axis the field is somewhat different, but is close enough to (12.46) for the purposes of this estimate.] Given that the magnetic moment of the proton is roughly equal to the nuclear magneton μ_N defined in (12.12), estimate the magnetic field at a distance a_B from a proton. (*b*) Taking your answer in part (*a*) as an estimate for the B field "seen" by an electron in the $1s$ state of a hydrogen atom, show that the atom's energy differs by roughly 10^{-6} eV for the cases that the proton and electron spins are parallel or antiparallel. [The energy of a magnetic moment in a B field is given by (10.10).] (*c*) This means that the $1s$ state of hydrogen is really two very closely spaced levels. Compare your rough estimate with the actual separation of these levels given that the photon emitted when a hydrogen atom makes a transition between them has $\lambda = 21$ cm. (This 21-cm radiation is used by astronomers to identify hydrogen atoms in interstellar space.)

12.26 ••• It is often a useful approximation to treat a nucleus as a uniformly charged sphere of radius R and charge Q. In this problem you will find the electrostatic potential $V(r)$ inside such a sphere, centered at the origin. (*a*) Write down the electric field $E(r)$ at any point a distance r from the origin with $r > R$. (Remember that this is the same as the field of a point charge Q at the origin.) (*b*) Use the definition

$$\Delta V = V(r_2) - V(r_1) = -\int_{r_1}^{r_2} E(r)\, dr \qquad (12.47)$$

to find the potential difference between points at r_1 and r_2 (both greater than R). It is usual to define $V(r)$ so that $V(r)$ approaches 0 as $r \to \infty$. By choosing $r_1 = \infty$ and $r_2 = r$, show that

$$V(r) = \frac{kQ}{r} \qquad (r \geq R). \qquad (12.48)$$

(*c*) Now find $E(r)$ for $r \leq R$. (Remember that Gauss's law tells us this is kQ'/r^2, where Q' is the total charge *inside* the radius r.) Check that your answers for parts (*c*) and (*a*) agree when $r = R$. (*d*) Use Eq. (12.47) to find $V(r_2) - V(r_1)$ for any two points *inside* the sphere. Now choose $r_2 = r$ and

$r_1 = R$, and use the value of $V(R)$ from part (*b*) to prove that

$$V(r) = \frac{kQ}{2R}\left(3 - \frac{r^2}{R^2}\right) \qquad (r \le R). \quad (12.49)$$

In Problems 12.23, 12.24, 12.27, and 12.28 this result is needed to find the potential energy of a charge inside a nucleus.

SECTION 12.5 (THE IPA POTENTIAL ENERGY FOR NUCLEONS)

12.27 • (*a*) Use the result (12.49) of Problem 12.26 to estimate the electrostatic potential energy of a proton at the center of a ^{12}C nucleus. (*b*) Repeat for a nucleus of ^{208}Pb.

12.28 • • The electrostatic potential energy U_{Coul} of a proton in a nucleus can be approximated by assuming that it moves in a uniform sphere of charge $(Z-1)e$ and radius R. (*a*) Use the results (12.48) and (12.49) of Problem 12.26 to write down and sketch $U_{Coul}(r)$. (*b*) Prove that $U_{Coul}(0) = 1.5\, U_{Coul}(R)$. (*c*) Setting $R = (1.07\ \text{fm})\, A^{1/3}$, estimate $U_{Coul}(0)$ for a proton in ^{4}He and for a proton in ^{238}U.

SECTION 12.6 (THE PAULI PRINCIPLE AND THE SYMMETRY EFFECT)

12.29 • Enlarge Fig. 12.11 to include the nuclei ^{12}Be and ^{12}O. By how much (in terms of the quantity ΔE shown in Fig. 12.11) does the energy of each isobar with $Z \ne N$ exceed the energy of ^{12}C?

SECTION 12.7 (THE SEMIEMPIRICAL BINDING-ENERGY FORMULA)

12.30 • Use the data of Appendix D to find the binding energy of ^{12}C.

12.31 • (*a*) Find the mass (in u) of the ^{4}He atom in Appendix D. (*b*) Find the mass of the ^{4}He nucleus to seven figures (but ignore corrections due to the atomic electrons' binding energy). (*c*) Do any of these seven figures change if you take into account the electrons' binding energy (about 80 eV total)?

12.32 • Use the data of Appendix D to find the binding energies of ^{4}He, ^{40}Ca, and ^{120}Sn.

12.33 • Use the data of Appendix D to deduce the energy required to remove one neutron from ^{14}C to produce ^{13}C and a neutron at rest, well separated from the ^{13}C nucleus.

12.34 • • (*a*) Use conservation of energy to write an equation relating the mass of a nucleus (Z, N) and its neutron separation energy $S_n(Z, N)$ to the masses of the nucleus $(Z, N-1)$ and the neutron. (*b*) Use the data in Appendix D to find S_n for ^{13}C, ^{120}Sn, and ^{200}Hg.

12.35 • • (*a*) The proton separation energy S_p (energy to remove one proton) for ^{198}Hg is 7.1 MeV. Given that the total binding energy of ^{197}Au is 1559.4 MeV, find the total binding energy of ^{198}Hg. (*b*) Compare your answer with the answer obtained directly from the mass of ^{198}Hg given in Appendix D.

12.36 • • (*a*) To estimate the importance of the surface term in the binding-energy formula, consider a model of ^{27}Al that consists of 27 identical cubical nucleons packed into a $3 \times 3 \times 3$ cube. Each small cube has six faces. What fraction of all these faces are exposed on the nuclear surface? (*b*) Repeat for ^{125}Te considered as a $5 \times 5 \times 5$ cube. (*c*) Comment on the difference in your answers.

12.37 • • The Coulomb energy of the charges in a nucleus is positive and hence *reduces* the binding energy B. If we approximate the nucleus as a uniform sphere of charge $Q = Ze$, then according to Eq. (12.50) of Problem 12.43, the Coulomb correction to B should be

$$B_{Coul} = -U_{Coul} = -\frac{3}{5}\frac{k(Ze)^2}{R}.$$

Putting $R = R_0 A^{1/3}$, show that this agrees with the form (12.26) of the Coulomb correction given in Section 12.7. Use your answer here to calculate a_{Coul} in MeV and compare with the empirical value $a_{Coul} = 0.711$ MeV. (You should not expect precise agreement since the nucleus is certainly not a perfectly uniform sphere, but you should get the right order of magnitude.)

12.38 • • (*a*) Use the binding-energy formula (12.30) to predict the binding energy of ^{40}Ca. (*b*) Compare your answer with the actual binding energy found from the masses listed in Appendix D.

12.39 • • Use the semiempirical binding-energy formula to find B/A for ^{20}Ne, ^{60}Ni, ^{259}No. Make a table showing all five terms for each nucleus and comment on their relative importance. Do your final answers follow the general trend of Fig. 12.14?

12.40 • • Use the binding-energy formula (12.30) to find the total binding energy of $^{56}_{28}$Ni, $^{56}_{27}$Co, $^{56}_{26}$Fe, and $^{56}_{25}$Mn. Which terms are different in the four cases, and why?

12.41 • • If ^{235}U captures a neutron to form ^{236}U in its ground state, the energy released is $B(^{236}\text{U}) - B(^{235}\text{U})$. (*a*) Prove this statement. (*b*) Use the binding-energy formula (12.30) to estimate the energy released, and compare with the observed value of 6.5 MeV. (*Note:* We have assumed here that ^{236}U is formed in its ground state, and the 6.5 MeV is carried away, by a photon, for example. An important alternative is that ^{236}U can be formed in an excited state, 6.5 MeV above the ground state. This excitation energy can lead to oscillations that cause the nucleus to fission.)

12.42 •• Compare the surface area of a sphere to that of a cube of the same volume. How much energy, in MeV, would be needed to distort the normally spherical ^{60}Ni nucleus into a cubical shape if the volume and Coulomb energies changed by negligible amounts?

12.43 ••• In this problem you will calculate the electrostatic energy of a uniform sphere of charge Q with radius R. The electric potential $V(r)$ at any radius $r < R$ is given by (12.49) in Problem 12.26, and the potential energy of a charge q at radius r is $qV(r)$. (a) Write down the total charge contained between radius r and $r + dr$, and hence find the potential energy of that charge. (b) If you integrate your answer to (a) from $r = 0$ to $r = R$, you will get *twice* the total potential energy of the whole sphere, since you will have counted the energy of any two charge elements twice. Show that the total Coulomb energy of the sphere is

$$U_{Coul} = \frac{3}{5}\frac{kQ^2}{R}. \qquad (12.50)$$

12.44 ••• (a) Substitute the binding-energy formula (12.30) into the relation

$$m_{nuc} = Zm_p + Nm_n - \frac{B}{c^2}$$

to obtain the semiempirical *mass* formula. [To simplify matters, take A to be odd, so that the pairing term in (12.30) is zero.] (b) Among any set of isobars, the nucleus with lowest mass is the most stable. To identify this nucleus, first write your mass formula in terms of the variables A and Z (that is, replace N, wherever it appears, by $A - Z$). Now, with A fixed, differentiate m with respect to Z. The minimum mass is determined by the condition $\partial m/\partial Z = 0$. Show that this leads to a relation of the form

$$Z = \frac{A}{2}\frac{1 + \alpha}{1 + \beta A^{2/3}} \qquad (12.51)$$

where α and β are related to the coefficients (12.31) as follows:

$$\alpha = \frac{(m_n - m_p)c^2}{4a_{sym}} \quad \text{and} \quad \beta = \frac{a_{Coul}}{4a_{sym}}.$$

(c) For $A = 37$, 115, 185 find the values of Z that give the most stable nuclei and compare with the observed values from Appendix D.

12.45 ••• (a) Use the formula (12.51) of Problem 12.44 to make a table showing the most stable values of Z and N for $A = 25, 45, 65, \ldots, 245$. (b) Plot a curve of Z versus N for the most stable nuclei found in this way, and compare your picture with the observed curve in Fig. 12.1.

SECTION 12.8 (THE SHELL MODEL)

12.46 • Which of the following nuclei has a closed subshell for Z? Which for N? ^{10}B, ^{11}B, ^{17}O, ^{27}Al, ^{28}Si, ^{42}Ca, ^{48}Ca.

12.47 • Use the levels shown in Fig. 12.17 to predict j_{tot} for the ground states of the nine nuclei ^{40}Ca, ^{41}Ca, . . . , ^{48}Ca. Compare with the data in Appendix D.

12.48 • Use the levels shown in Fig. 12.17 to predict the total angular momenta of the following nuclei: ^{29}Si, ^{33}S, ^{37}Cl, ^{59}Co, and ^{71}Ga.

12.49 • Use the shell model to predict j_{tot} for the ground states of $^{39}_{19}$K$_{20}$, $^{40}_{20}$Ca$_{20}$, and $^{41}_{21}$Sc$_{20}$.

12.50 • The ground state of ^{7}Li has $j_{tot} = \frac{3}{2}$ and its first excited state has $j = \frac{1}{2}$. Explain these observations in terms of the shell model.

12.51 • (a) Use the levels shown in Fig. 12.17 to draw an occupancy diagram like Fig. 12.10(b) for the ground state of ^{7}Be. (b) What is j_{tot}? (c) What would you predict for the first excited state? (Draw an occupancy diagram and give j_{tot}.)

12.52 • Answer the same questions as in Problem 12.51 for ^{17}F.

12.53 • In applying the shell model we usually use the level ordering of Fig. 12.17, which is based on an assumed spherical potential well. For certain nonspherical nuclei this assumption is incorrect and the ordering of the levels is different. Examples are ^{19}F and ^{121}Sb; check this by using the rule (12.41) and Fig. 12.17 to predict the angular momenta of these two nuclei and comparing with the observed values in Appendix D.

12.54 •• (a) Use the energy levels of Fig. 12.17 to draw an occupancy diagram like Fig. 12.10(b) for ^{13}C. (b) What is j_{tot} for the ground state? (c) The first excited state is formed by lifting one neutron from $1p_{3/2}$ to $1p_{1/2}$. Draw an occupancy diagram for this state and explain what you would predict for j_{tot} of this state. (Does your answer agree with the observed value, $j_{tot} = \frac{1}{2}$?) (d) Now predict j_{tot} for the ground and first excited states of ^{13}N. (In this case the excitation involves a proton.)

12.55 •• For each choice of n and l (with $l \neq 0$) a nucleon has two different possible energy levels with $j = l \pm \frac{1}{2}$. Prove that the sum of the degeneracies of these two levels is equal to the total degeneracy that the level nl would have had in the absence of any $\mathbf{S} \cdot \mathbf{L}$ splitting.

12.56 ••• The first magic number for both nuclei and atoms is 2. The second is 8 for nuclei and 10 for atoms. The difference occurs because what we call the $1p$ and $2s$ levels in nuclei are widely separated, whereas the corresponding levels in atoms (called $2s$ and $2p$ by atomic physicists) are very close together.

To explain this difference, compare the shape of an atomic potential energy ($\propto 1/r$), to that of the nuclear potential well. Noting that states with higher l have probability distributions that are pushed out to larger radii, qualitatively explain the difference in the positions of these two levels.

12.57 • • • The spin-orbit energy (12.34) is proportional to $\mathbf{S} \cdot \mathbf{L}$. To evaluate $\mathbf{S} \cdot \mathbf{L}$ note that since $\mathbf{J} = \mathbf{S} + \mathbf{L}$,

$$J^2 = (\mathbf{S} + \mathbf{L}) \cdot (\mathbf{S} + \mathbf{L}) = S^2 + L^2 + 2\mathbf{S} \cdot \mathbf{L}.$$

You can solve this for $\mathbf{S} \cdot \mathbf{L}$ and then put in the values for J^2, S^2, and L^2. [For instance, $J^2 = j(j+1)\hbar^2$, where $j = l \pm \frac{1}{2}$.] Show that if $j = l + \frac{1}{2}$ then $\mathbf{S} \cdot \mathbf{L} = l\hbar^2/2$, and if $j = l - \frac{1}{2}$, then $\mathbf{S} \cdot \mathbf{L} = -(l+1)\hbar^2/2$. Use these answers to prove that the spin-orbit splitting increases with increasing l, as indicated in Fig. 12.17.

SECTION 12.9 (MASS SPECTROMETERS)

12.58 • Where does the methionine molecule in Fig. 12.21 break to account for the two peaks at 74 u and 75 u in its mass spectrum?

12.59 • Repeat Problem 12.58 for the peaks at 45 u and 104 u.

12.60 • • (a) Find the radius R of the curved path for a once-ionized nitrogen atom in a mass spectrometer with $B = 0.05$ T and $V_0 = 10$ kV, using (12.44).

(b) Equation (12.44) was derived nonrelativistically. What would R be if calculated relativistically? How many significant figures must you include to see the difference?

12.61 • • Conditions in the ion source of a mass spectrometer can be adjusted to allow formation of multiply ionized atoms. (a) Rewrite Eq. (12.44) for the case of an n times ionized atom and solve for R. (b) Taking the masses of ^{4}He, ^{12}C, and ^{16}O to be 4 u, 12 u, and 16 u, respectively, write down expressions for the radii, R, for He$^+$, C^{3+}, and O^{4+}. (c) Using the actual masses of the atoms, find the fractional differences in these three radii (He versus C and C versus O). The tiny differences in these radii can be measured very accurately and allow precise comparisons of the three masses involved (one of which, ^{12}C, defines the atomic mass scale).

12.62 • • The chemical composition of a molecule can often be deduced from a very accurate measurement of its mass. As a simple example consider the following. Gas from a car's exhaust is analyzed in a mass spectrometer. Two of the resulting peaks are very close to 28 u. (a) Can you suggest what gases produced these two peaks? (b) Careful measurements indicate that the masses of the neutral molecules concerned are 28.006 u and 27.995 u. What are the two molecules?

CHAPTER

13

Radioactivity and Nuclear Reactions

13.1 Introduction

In Chapter 12 we discussed the structure of isolated, static nuclei — their constituents, masses, energy levels, and angular momenta. In this chapter we discuss various nuclear processes, in which nuclei change their state — and often their identity as well — by emitting and absorbing energy in the form of photons or other particles. The first of these processes to be studied was the decay of naturally occurring unstable nuclei. Such nuclei were called *radioactive,* and we discuss radioactivity in Sections 13.2 through 13.5.

In Sections 13.6 through 13.8 we discuss nuclear reactions, the processes in which two nuclei collide and induce a variety of nuclear transformations. The two best-known examples of nuclear reactions, fission and fusion, are of great practical importance as sources of enormous energy, both peaceful and destructive, as we discuss in Sections 13.7 and 13.8.

In the optional Section 13.9 we describe the quantum theory of the α decay of certain radioactive nuclei. This was an early application of the Schrödinger equation to nuclear physics. It is an example of the process called quantum tunneling, or barrier penetration, which (in the case of electrons) makes possible some of the electronic devices mentioned in Chapter 17.

In the optional Sections 13.10 and 13.11 we describe some of the devices used to accelerate and detect nuclear particles. In studying microscopic bodies such as small organisms, one shines light on them and examines the deflected light with a microscope. Nuclei are, of course, far too small to be seen with ordinary light. Instead, nuclei are probed by bombarding them with energetic particles (electrons, protons, α particles, etc.) and observing the deflected particles with suitable detectors. The accelerators that boost the particles to the required energies and the devices that detect them can be regarded as the "light" sources and "microscopes" of the nuclear physicist. Finally, in Section 13.12 we describe the various units used to measure amounts of radiation.

If your only interest in this chapter is to prepare yourself for Chapter 14 on elementary particles, you need read only Sections 13.2 through 13.4 (and perhaps 13.10 and 13.11).

13.2 Radioactivity

In 1896, Becquerel discovered that uranium salts emit radiation that could darken a photographic film, even if the film was wrapped in black, lightproof paper. Evidently, the radiation, just like ordinary light, could ionize the atoms in photographic emulsion and hence "expose" the film. Unlike ordinary light, however, it could pass through black paper. The emission of this new radiation came to be called **radioactivity,** and within two years Marie and Pierre Curie had identified three more radioactive elements: thorium, polonium, and radium. (The Curies and Becquerel shared the 1903 Nobel prize for physics for their work on radioactivity.) We now know that radioactivity originates in the atomic nucleus. Thus although the existence of nuclei was not established for another 15 years, we can date the beginning of nuclear physics from Becquerel's discovery in 1896.

The radiation from radioactive materials was found to ionize air, and the measurement of electric currents passing through the ionized air became a common method of detecting and measuring the radiation. It was soon found that the radiation could be divided into three types, which Rutherford called α, β, and γ rays. Alpha rays ionize air the most and are the most easily stopped as they pass through matter. (They can be stopped by a single sheet of paper, for example.) Beta rays are intermediate in the ionization they cause and in their ability to penetrate matter; gamma rays ionize the least and penetrate farthest. (Beta rays are stopped by a few millimeters of metal; gamma rays penetrate an order of magnitude farther.*) Electric and magnetic deflection show that α rays carry a positive charge and β rays a negative charge, while γ rays are neutral.

Using electric and magnetic deflections, much as Thomson had with electrons, experimenters measured the mass-to-charge ratio, m/q, for α and β rays. For α rays it was found that $m/q = 2m_H/e$. This ratio could be explained if

MARIE CURIE (1867–1934, Polish–French). Marie Curie was the first person to win two Nobel Prizes. She shared the 1903 prize in physics with her husband, Pierre, and Becquerel for work on radioactivity, and then won the 1911 prize in chemistry for the discovery of the elements polonium and radium. Her daughter and son-in-law, the Joliot–Curies, also won the Nobel Prize (in 1935 for discovering man-made radioactivity).

---

* These differences are useful in identifying the different radiations. They originate in the different masses, speeds, and charges of the particles involved. The details are quite complicated and are not worth discussing here.

α rays consisted of once-ionized hydrogen molecules, H_2^+, or twice-ionized helium atoms, He^{2+}, or other possibilities, such as six-times-ionized carbon 12. In a beautifully direct experiment, Rutherford established that He^{2+} was the correct choice. He had a glass vessel fitted with two electrodes and part of its wall made thin enough that α rays could enter its evacuated interior. When a source of α rays was placed nearby for several days, a detectable quantity of helium accumulated inside the vessel, as demonstrated by an electric discharge that produced the spectrum characteristic of helium. Thus α particles were shown to be He^{2+} ions or, as we now know, He nuclei.

It was later established that the radiation emitted by radioactive materials originates in the atomic nucleus. All radioactive nuclei have excess energy, which makes them unstable. In α-radioactive nuclei the excess energy is released by the emission of an α particle. If the original nucleus has Z protons and N neutrons, we can write this process of *alpha decay* as

$$(Z, N) \rightarrow \alpha + (Z - 2, N - 2). \tag{13.1}$$

Notice that the remaining, "offspring," nucleus has two fewer protons and two fewer neutrons than the original "parent" and is therefore the nucleus of a different element, with an atomic number two less than that of the parent. It was for establishing that radioactivity involved this transformation of elements that Rutherford won the Nobel prize for chemistry in 1908.

In the case of β rays it was found that $m/q = -m_e/e$, suggesting that β rays were just electrons, or "cathode rays" as they were then called. In due course it was established that β rays are indeed electrons, although the energy of β electrons is much larger than that of typical cathode rays: The former can be as much as 10 MeV, whereas the latter seldom exceed 100 keV. The energy of β particles is also much larger than their rest energy (0.5 MeV). Thus the electrons emitted by radioactive nuclei were the first particles observed moving at relativistic speeds.

The radioactive nuclei that emit β particles do so because they have too many neutrons; that is, the ratio, N/Z, of neutrons to protons is larger than the value that gives maximum stability. As mentioned in Chapter 12, one of the excess neutrons can change into a proton by creating and ejecting an electron. The resulting process of *beta decay* can be written as *

$$(Z, N) \rightarrow (Z + 1, N - 1) + e. \tag{13.2}$$

Like α decay, β decay causes a transformation of the element involved — in this case an increase of one in the atomic number Z.

The γ rays emitted by radioactive nuclei were found to have many properties in common with X rays, and it was eventually proved that like X rays, γ rays are electromagnetic radiation, the only difference being their energies: X-ray photons (that is, photons produced in X-ray tubes) seldom exceed 100 keV or so, because of the difficulty of maintaining higher voltages; γ-ray photons have energies up to about 10 MeV, depending on the radioactive nucleus involved. As we mentioned in Chapter 12, the origin of γ rays is exactly analogous to that of the photons given off by atoms: When a nucleus is in an excited state, it can emit a photon and drop to a lower level.

---

* As we discuss in Section 13.4, the equation (13.2) is missing one term on the right: In addition to the electron, a neutral particle called a neutrino is produced. This particle is so hard to detect that it was not observed directly until 1956.

13.3 The Exponential Decay Law

The radiation from a radioactive sample consists of many individual particles, each of which came from its own separate parent nucleus. In this section we discuss the laws governing the rate at which these radioactive decays occur, and the variation of that rate as time goes by. Our discussion applies equally to all three kinds of radioactivity (α, β, and γ) and, in fact, to the decay of any unstable quantum system. Thus the same ideas apply to the decay of unstable elementary particles (such as the pions discussed in Section 2.4) and to the decay of excited atomic states (as we discuss in Chapter 15).

If the unstable nuclei of a given species were identical clock-like mechanisms obeying the laws of classical physics, we would expect all of them to decay at the same definite time after their formation. Instead, they are found to decay after a wide range of different times. The explanation of this behavior lies in the probabilistic nature of quantum mechanics. We have seen, for example, that quantum mechanics does not give the precise location of an electron in an atom; instead, it gives only the probabilities of the various possible locations. In the same way, quantum mechanics cannot predict the precise time at which a radioactive nucleus will decay; instead, it gives only the probabilities of the various times at which the decay may occur.

We shall denote by r the probability that a particular nucleus will decay in any unit interval of time. This parameter is called the **decay constant** of the nucleus in question. If, for example, $r = 0.02$ per second (s^{-1}), the probability is 2% that the nucleus will decay in any 1-second interval. It is an experimental fact that for any given kind of nucleus, the probability r is a constant, independent of how long ago the nucleus was formed.* Given this fact, we can set up and solve the equation that determines the rate at which the nuclei in a sample will decay.

Suppose that at time t our sample contains a large number, $N = N(t)$, of radioactive nuclei, all of the same species (and hence with the same value of r). The decay constant r is the probability that any one nucleus will decay in unit time. Therefore, the total number of nuclei expected to decay in unit time is the probability r multiplied by the number of nuclei $N(t)$; that is, the total rate $R(t)$ at which nuclei will decay is

$$R(t) = rN(t). \tag{13.3}$$

Now, each time a nucleus decays, the number $N(t)$ decreases by 1. Thus R is just the rate of decrease of N:

$$R = -\frac{dN}{dt}. \tag{13.4}$$

Combining (13.3) and (13.4), we find that

$$\frac{dN}{dt} = -rN. \tag{13.5}$$

---

* We can reexpress this loosely by saying that a nucleus is unaware of its own past history: At the beginning of each 1-second interval it has the same probability of decay, whether it has been in existence for 1 millisecond or one century. There is a partial analogy here with the probabilities in throwing a die: Provided that the die is true, the probability of throwing an ace is the same 1/6, whether the last 10 throws were all aces or none of them were.

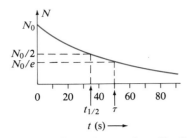

FIGURE 13.1 The number, N, of nuclei in a radioactive sample decreases exponentially, as in Eq. (13.6). At time $t = 0$, the number is $N = N_0$. At time $t_{1/2}$ it has dropped to $N_0/2$; at time τ, it has dropped to N_0/e. The decay constant is $r = 0.02 \text{ s}^{-1}$ in this example.

This is a first-order differential equation for $N(t)$ and asserts that the derivative of N is proportional to N itself. The only function that satisfies this equation is the exponential function (Problem 13.10)

$$N(t) = N_0 e^{-rt}, \tag{13.6}$$

where N_0 is a constant. To identify N_0, we set $t = 0$ and find that N_0 is simply the number of nuclei at time $t = 0$. Therefore, we have the following important conclusion: If a sample consists of N_0 radioactive nuclei at time $t = 0$, and if $N(t)$ denotes the number of the nuclei remaining at any later time t, then the function $N(t)$ decreases exponentially with time, as given by (13.6). This behavior is sketched for the case $r = 0.02 \text{ s}^{-1}$ in Fig. 13.1.

There are several ways to characterize the rate at which a radioactive nucleus decays. One is to give the decay constant r, the probability that any one nucleus will decay in unit time. An alternative is to give the reciprocal, $1/r$, which is denoted by τ:

$$\tau = \frac{1}{r}. \tag{13.7}$$

Since r has the dimensions (time)$^{-1}$, the parameter τ has the dimensions of time, and τ is called the **lifetime** of the nucleus. Its significance follows easily from (13.6): With $t = \tau = 1/r$,

$$N(\tau) = N_0 e^{-1} = \frac{N_0}{e};$$

that is, τ is *the time in which N drops to the fraction $1/e$ of its original value,* as indicated in Fig. 13.1. For this reason, τ is sometimes called the $1/e$ time of the nucleus. One can also show (Problem 13.11) that τ is the *average* time for which the nuclei in a sample survive, and τ is therefore also called the **mean life** of the nucleus. Note that in terms of τ the decay law (13.6) can be rewritten

$$N(t) = N_0 e^{-t/\tau}. \tag{13.8}$$

Another way to characterize the decay rate of a nucleus is to give its **half-life** $t_{1/2}$, the time in which the number of nuclei drops to half its original value:

$$N(t_{1/2}) = \tfrac{1}{2} N_0. \tag{13.9}$$

According to (13.6), $t_{1/2}$ is determined by the equation

$$N_0 e^{-r t_{1/2}} = \tfrac{1}{2} N_0.$$

Canceling the N_0 and taking the natural logarithm of this equation, we find that

$$-r t_{1/2} = \ln(\tfrac{1}{2}) = -\ln 2$$

or

$$t_{1/2} = \frac{\ln 2}{r} = \frac{0.693}{r} \tag{13.10}$$

or since $\tau = 1/r$,

$$t_{1/2} = \tau \ln 2 = 0.693\tau. \qquad (13.11)$$

The half-life is also indicated on Fig. 13.1.

It is a simple exercise (Problem 13.9) to check that the exponential decay law (13.6) can be rewritten as

$$N(t) = \frac{N_0}{2^{t/t_{1/2}}}. \qquad (13.12)$$

In this expression, the quantity $t/t_{1/2}$ is simply the number of half-lives that have gone by in time t:

$$n = (\text{number of half-lives in time } t) = \frac{t}{t_{1/2}}. \qquad (13.13)$$

Thus (13.12) can be rewritten as

$$N(t) = \frac{N_0}{2^n}; \qquad (13.14)$$

that is, in n half-lives, the population halves itself n times over. This is the property of half-lives that was introduced in Section 2.4. Note that (13.12) can also be rewritten, in close parallel with (13.8), as

$$N(t) = N_0 2^{-t/t_{1/2}}.$$

It is important to be aware that the parameters $t_{1/2}$ and τ are both in common use, and one needs to feel at home with either. In particular, nuclear physicists tend to use the half-life $t_{1/2}$, while particle physicists tend to prefer the mean life τ.

EXAMPLE 13.1 The nucleus of thorium 231 is β radioactive, with half-life $t_{1/2} = 25.6$ hours. At $t = 0$ a sample of ^{231}Th contains $N_0 = 56{,}000$ nuclei. How many ^{231}Th nuclei will remain after 3 days? What is the total rate at which electrons are emitted at $t = 0$? What will the rate be after 3 days?

To get a rough estimate for the number of nuclei after 3 days, we note that $t_{1/2}$ is about a day. Therefore, 3 days is about 3 half-lives, and N will have halved itself about three times:

$$N(3 \text{ days}) \approx \frac{N_0}{2^3} = \frac{56{,}000}{8} = 7000.$$

If we want a more accurate answer, we have only to calculate the number of half-lives more accurately:

$$n = \frac{t}{t_{1/2}} = \frac{72 \text{ hours}}{25.6 \text{ hours}} = 2.81.$$

Therefore,

$$N(3 \text{ days}) = \frac{N_0}{2^{2.81}} = \frac{56{,}000}{7.01} = 8000, \qquad (13.15)$$

to two significant figures.

It is perhaps worth emphasizing that we could do the same calculation in terms of the exponential decay constant r, as given by (13.10),

$$r = \frac{\ln 2}{t_{1/2}} = 0.0271 \text{ hour}^{-1}.$$

With $t = 3$ days $= 72$ hours, the exponential decay law (13.6) gives

$$N(3 \text{ days}) = N_0 e^{-rt} = (56{,}000)e^{-(0.0271 \times 72)} \approx 8000,$$

as before.

To find the rate at which electrons are emitted, we note that each decay gives one electron, so that the rate of electron emission is the same as the rate R of decays. According to (13.3), this is

$$R(t) = rN(t), \qquad (13.16)$$

which, with $t = 0$, gives

$$R_0 = rN_0 = (0.0271/\text{hour}) \times 56{,}000$$

$$= 1520 \text{ electrons/hour} \approx 25 \text{ electrons/minute}.$$

According to (13.16), the rate $R(t)$ is proportional to the number of nuclei, $N(t)$. Thus $R(t)$ shows the same exponential decay as $N(t)$, and in any time t the rate R decreases by the same factor as does N. In particular, we saw in (13.15) that in 3 days N drops to $1/7.01$ of its original value; therefore, the same is true of R:

$$R(3 \text{ days}) = \frac{R_0}{7.01} = 3.6 \text{ electrons/minute}.$$

RADIOACTIVE DATING

To conclude this section we mention an application of radioactivity that makes explicit use of the exponential decay law: the dating of archeological and geological specimens. If a specimen contains a measurable amount of a known radioactive material (N atoms, say) and if one can reliably estimate the number, N_0, of atoms that were present when the specimen was formed, then the ratio N/N_0 should be the exponential decay factor $\exp(-rt)$. Knowing N, N_0, and r, one can calculate the age of the specimen.

Perhaps the best known example is the method of radiocarbon dating. The isotope ^{14}C is β radioactive with a half-life of 5730 years. It occurs naturally because it is produced continually in collisions between cosmic rays and nitrogen nuclei in the atmosphere. The number of ^{14}C atoms in the atmosphere represents an equilibrium between the rate of production by cosmic rays and the rate of loss by decay. This number holds nearly constant throughout the atmosphere at about 1.3 atoms of ^{14}C in every 10^{12} atoms of all isotopes of carbon. Since ^{14}C is chemically indistinguishable from its stable isotopes, this proportion persists through all chemical reactions. Plants obtain their carbon from the atmosphere, in the form of CO_2, and fix it by photosynthesis. Animals get their carbon when they eat plants or plant-eating animals. Thus all living organisms contain ^{14}C in the same proportion as the atmosphere. However, as soon as an organism dies its intake of fresh carbon ceases, and the proportion of ^{14}C atoms starts to diminish exponentially with a known decay constant. By measuring the proportion of ^{14}C atoms in a specimen of wood, bone, or cloth, one can calculate how long it is since the specimen was part of a living organism.

Other radioactive nuclei are used similarly to date geological samples. In this way the oldest known rocks have been shown to have solidified about 4 billion years ago, and the oldest fossils about 3 billion.

EXAMPLE 13.2 In radiocarbon dating the amount of ^{14}C in a specimen is usually found by measuring the rate at which it ejects electrons (using a Geiger counter for example). If 30 grams of carbon from the wood from a prehistoric dwelling eject 200 electrons per minute, about how old is the dwelling?

When the 30 g of carbon was part of a branch that had just been cut from a living tree, we know that the proportion of ^{14}C atoms was 1.3×10^{-12}. At that time (which we call $t = 0$) the number of ^{14}C atoms in the 30 g of carbon was

$$N_0 = (1.3 \times 10^{-12}) \times \frac{30 \text{ grams}}{12 \text{ gram/mole}} \times \left(6.02 \times 10^{23} \frac{\text{atoms}}{\text{mole}}\right)$$

$$= 1.96 \times 10^{12} \text{ atoms.}$$

Since the half life of ^{14}C is $t_{1/2} = 5730$ years, the decay constant is

$$r = \frac{\ln 2}{t_{1/2}} = \frac{0.693}{5730 \times 365 \times 24 \times 60 \text{ min}} = 2.30 \times 10^{-10} \text{ min}^{-1}.$$

Therefore, the total decay rate from our sample at $t = 0$ was

$$R_0 = rN_0 = (2.30 \times 10^{-10} \text{ min}^{-1}) \times (1.96 \times 10^{12}) = 451 \text{ decays/min.}$$

Today (time t) the rate of decays is measured to be

$$R = 200 \text{ decays/min.}$$

This rate is roughly half of R_0. Thus the time that has elapsed since the wood was part of a living tree is about a half-life, or roughly 6000 years. More precisely, the number, n, of half-lives is found by solving the equation

$$\frac{R}{R_0} = \frac{1}{2^n}.$$

Taking the logarithm of this equation, we find that

$$\log \frac{R_0}{R} = n \log 2$$

or

$$n = \frac{\log(R_0/R)}{\log 2} = \frac{\log(451/200)}{\log 2} = 1.17.$$

That is, the age of the wood (and hence presumably the dwelling) is 1.17 half-lives, or

$$t = 1.17 t_{1/2} = 1.17 \times (5730 \text{ years}) = 6700 \text{ years.}$$

13.4 Beta Decay and the Neutrino

In several ways β decay is the most interesting of the radioactive decay processes. In particular, for 20 years there was a serious puzzle over the number of particles ejected in β decay. It is this puzzle and its resolution — with the discovery of the neutrino — that we describe in this section.

We saw in Section 12.6 that because of the Pauli principle, nuclei with $Z \approx N$ tend to have lower rest energy than their isobars with Z much different from N. This claim is easily verified using the experimentally measured masses of any series of isobars. For example, the rest masses of five isobars with $A = 11$ are plotted in Fig. 13.2. The masses are shown in atomic mass units and are all very close to 11 u, as one would expect. Nevertheless, it is clear that ^{11}B, with $Z = 5$ and $N = 6$, has the lowest mass and hence lowest rest energy. As one moves away from $Z \approx N$, the rest energies become progressively greater.

In heavier nuclei the Coulomb repulsion of the protons implies that the minimum occurs with Z appreciably less than N. This is illustrated in Fig. 13.3, which shows the masses of the $A = 231$ isobars, with the minimum at ^{231}Pa, for which $Z = 91$ and $N = 140$. In any case, the important point is that for any given set of isobars there is a definite value of Z, usually somewhat less than N, for which the rest energy is a minimum. Any nucleus with too many neutrons, such as ^{231}Th, or with too many protons, such as ^{231}U, has a rest energy larger than this minimum. (Sometimes there can be two minima; see Problem 13.25.)

Let us now consider a nucleus, such as ^{231}Th in Fig. 13.3, that has too many neutrons. That ^{231}Th has higher energy than ^{231}Pa would imply that ^{231}Th is unstable, *provided* that there were some mechanism to let ^{231}Th transform itself into the lower-energy ^{231}Pa; and indeed, experiment shows that there *is* such a mechanism. The excess energy in ^{231}Th creates and ejects an electron, and one neutron is simultaneously converted into a proton. The resulting transformation,

$$^{231}_{90}\text{Th}_{141} \rightarrow {}^{231}_{91}\text{Pa}_{140} + e^-, \tag{13.17}$$

is the process that Rutherford called β decay. Two such decays are indicated on

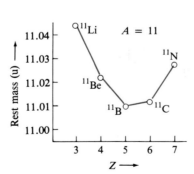

FIGURE 13.2 The rest masses, in atomic mass units, of the five most stable isobars with $A = 11$. Although all are close to 11 u, boron (with $Z = 5$ and $N = 6$) has the lowest mass and hence lowest rest energy.

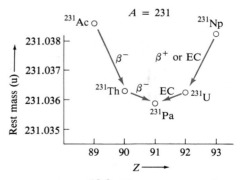

FIGURE 13.3 The rest masses of the isobars with $A = 231$ have their minimum at protactinium with $Z = 91$ and $N = 140$. The possible decays of the other species are indicated as β^-, β^+, and EC (for electron capture), as described in the text.

Fig. 13.3, where they are labeled β^- to emphasize that they produce negative electrons.

A nucleus such as ^{231}Np, with too many protons, is also unstable. In fact, for most such nuclei there are two possible modes of decay. The excess energy in ^{231}Np can create and eject a *positron* (the positively charged version of the electron, mentioned briefly in Section 3.8) changing a proton into a neutron. The resulting transformation,

$$^{231}_{93}\text{Np}_{138} \rightarrow ^{231}_{92}\text{U}_{139} + e^+, \qquad (13.18)$$

is called β^+ decay. Alternatively, the ^{231}Np nucleus can convert a proton into a neutron by capturing one of the atomic electrons:

$$^{231}_{93}\text{Np}_{138} + e^- \rightarrow ^{231}_{92}\text{U}_{139}, \qquad (13.19)$$

a process called **electron capture.** Figure 13.3 indicates both of these processes for ^{231}Np. (Note in the same picture that ^{231}U decays by electron capture but not by β^+ emission. The impossibility of β^+ decay in ^{231}U is related to the slightly different energy requirements for these two processes. See Problem 13.23.)

The puzzling aspect of β decay, mentioned earlier, emerges when we consider the energy of the ejected electron or positron in β decay. Let us consider the case of β^- decay, for which the observed process can be written as

$$(Z, N) \rightarrow (Z + 1, N - 1) + e^-. \qquad (13.20)$$

If the parent nucleus (Z, N) is at rest, and if the process really involves only the two final bodies indicated in (13.20), then the energies of those two bodies are uniquely determined by conservation of energy and momentum. With the parent at rest, the total kinetic energy of the final particles is

$$K = \Delta Mc^2 = \{m_{\text{nuc}}(Z, N) - [m_{\text{nuc}}(Z + 1, N - 1) + m_e]\}c^2. \qquad (13.21)$$

We can simplify this expression for ΔM by adding and subtracting Zm_e to give

$$K = \Delta Mc^2 = \{m_{\text{atom}}(Z, N) - m_{\text{atom}}(Z + 1, N - 1)\}c^2; \qquad (13.22)$$

that is, ΔM can be found by taking the difference of the masses of the two corresponding *atoms.*

Since the electron is so much lighter than the recoiling nucleus, the latter takes very little kinetic energy. Therefore, to an excellent approximation (about 1 part in 10^5 for a heavy nucleus—see Problem 13.34) all the kinetic energy (13.22) must go to the electron. We conclude that if the only particle ejected in β^- decay were the electron, as in (13.20), then all the electrons in a sample of β^- radioactive material (of one type) would emerge with exactly the same kinetic energy ΔMc^2, as given by (13.22).

EXAMPLE 13.3 Assuming that the electron's energy is given correctly by (13.22), find the kinetic energy of the electrons ejected in the β^- decay of bismuth 210 to polonium 210.

The atomic masses concerned are given in Appendix D as

$$m(^{210}\text{Bi}) = 209.984096 \text{ u}$$
$$m(^{210}\text{Po}) = 209.982848 \text{ u}$$

and the mass difference is therefore $\Delta M = 0.001248$ u. Thus, according to (13.22), the electron's kinetic energy should be

$$K = \Delta Mc^2 = (0.001248 \text{ u}) \times \left(\frac{931.5 \text{ MeV}/c^2}{1 \text{ u}} \right) c^2 = 1.16 \text{ MeV}.$$

Experiment shows that our prediction that all electrons should be ejected with the same energy is completely wrong. Figure 13.4 shows the distribution of energies of the electrons emitted by ^{210}Bi. Evidently, the energies range continuously from $K = 0$ up to the predicted value of 1.16 MeV. While a few electrons have nearly the predicted energy, the great majority have much less. This apparent failure of conservation of energy was beginning to be recognized in the late 1920s, and efforts were made to find the missing energy. For example, it was conceivable that some energy was carried away by undetected radiation. To check this possibility, β-radioactive materials were enclosed in thick-walled calorimeters, but here too the apparent release of energy—that is, the heat produced in the calorimeter—was appreciably less than required by conservation of energy.

The suggestion was even made that conservation of energy might not apply to β decay. However, a less radical suggestion due to Pauli proved to be the explanation. In the early 1930s Pauli suggested that there might be another particle emitted in β decay, which interacted only very weakly with matter and hence could carry away the missing energy undetected. This suggestion would immediately explain the spread in the measured energies of the electrons: With two ejected particles (in addition to the residual nucleus) the released energy could be shared between the two particles in many different ways. For example, the electron could be left at rest while the hypothetical particle took all the available energy, or vice versa. Thus the electron's kinetic energy could range anywhere from zero to some K_{max}, as observed.

From the observed facts about β decay we can immediately infer the main properties of Pauli's proposed particle—its charge, its mass, and its interaction with matter. First, its charge must be zero: The process (13.20) already conserves charge, since the right-hand side has one extra charge of each sign (one proton and one electron). If there really is an additional particle on the right-hand side, conservation of charge requires that this additional particle be neutral.

Concerning the particle's mass, we saw in connection with Fig. 13.4 that the observed kinetic energies of the ejected electrons range up to the value ΔMc^2 predicted on the assumption that there was no additional particle. In other words, the total energy taken by the proposed additional particle varies *all the way down to zero*. Since the minimum possible energy of a particle is its rest energy mc^2, this requires that the new particle have zero rest mass. Since the new particle has zero charge and zero mass, it came to be called the **neutrino** (little neutral one), denoted by ν, the Greek lowercase letter nu.*

Concerning its interactions, the great difficulty of detecting the neutrino

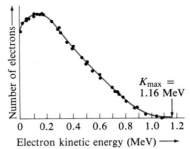

FIGURE 13.4 Distribution of kinetic energies of the electrons ejected in the β^- decay of ^{210}Bi. The observed values range from $K = 0$ up to a maximum denoted K_{max}. It is found that K_{max} is exactly equal to the value 1.16 MeV predicted by (13.22) on the assumption that the electron is the only particle ejected.

* Because of the shape of the curve in Fig. 13.4, it is hard to measure K_{max} precisely. Current measurements show that K_{max} could be a few eV less than ΔMc^2, in which case the neutrino would have a mass of a few eV/c^2. Although some recent experiments seem to suggest a nonzero rest mass, the indicated mass is so small that for our purposes it is negligible.

implies that it interacts very weakly with all matter. In fact, we now know that, like electrons, neutrinos are completely immune to the strong nuclear force; in addition, they are immune to all electromagnetic forces. The only force to which neutrinos are subject is the so-called weak force, which we discuss in Chapter 14 (and also gravity, but this is usually negligible in nuclear processes). For this reason, the neutrino was not directly detected until 1956. Nevertheless, the faith in conservation of energy was so strong that well before then most physicists confidently believed that the neutrino would eventually be found. This belief was reinforced by the realization that if there were no neutrino, then β decay would also violate the conservation laws of momentum (Problem 13.24) and angular momentum.

It eventually became clear that there are two closely related particles, the neutrino ν and the antineutrino $\bar{\nu}$, and it is actually the antineutrino that is emitted in β^- decay:

$$(Z, N) \rightarrow (Z + 1, N - 1) + e^- + \bar{\nu}. \tag{13.23}$$

On the other hand, it is the neutrino that emerges in β^+ decay,

$$(Z, N) \rightarrow (Z - 1, N + 1) + e^+ + \nu, \tag{13.24}$$

and also in electron capture (see Problem 13.23). We shall discuss the difference between the ν and $\bar{\nu}$ in Chapter 14. Even though the two particles are different, it is common practice (which we follow) to use the one name "neutrino" for both ν and $\bar{\nu}$ in any context where the distinction is unimportant.

INSTABILITY OF THE NEUTRON

Before we describe the discovery of the neutrino, we must mention briefly one particularly striking example of β decay: the decay of the neutron. A glance at the masses inside the front cover shows that the rest mass of the neutron is greater than that of the proton by enough to create the rest energy of an electron. This means that the neutron is unstable and can decay via the β^- decay

$$n \rightarrow p + e^- + \bar{\nu}. \tag{13.25}$$

This rather surprising prediction is borne out by experiment: An isolated neutron is indeed unstable and decays as in (13.25) with a half life of about 10 minutes.

This immediately raises the question: If the neutron is unstable, how is it that normal stable matter contains large numbers of neutrons? To answer this, consider again the $A = 11$ isobars shown in Fig. 13.5. A nucleus like ^{11}Be, with too many neutrons, is indeed unstable, because the total energy of ^{11}Be is greater than that of ^{11}B, and it is energetically favorable for one of the neutrons in ^{11}Be to undergo β^- decay, changing ^{11}Be to ^{11}B. If we look instead at ^{11}B, we find a different story: Because of the properties of the nuclear force and the Pauli principle, the next nucleus over, ^{11}C, has *higher* total energy. Therefore, it is *not* energetically favorable for a neutron in ^{11}B to convert to a proton, and ^{11}B is stable. The situation is clearly this: Even though the isolated neutron is β^- unstable, a nucleus (Z, N) may or may not be β^- unstable, depending on whether its neighboring isobar $(Z + 1, N - 1)$ has less or more rest energy.

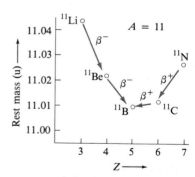

FIGURE 13.5 The neutrons in ^{11}Be and ^{11}Li are unstable; that is, it is energetically favorable for a neutron to change into a proton via β^- decay. In ^{11}C and ^{11}N it is the protons which are unstable. In ^{11}B both neutrons and protons are *stable*. (Both ^{11}C and ^{11}N can also decay by electron capture.)

If we go farther to the right in Fig. 13.5 and look at ^{11}C, we find that here the balance has tipped against the protons: In this case it is energetically favorable for a *proton* to change into a neutron. More generally, in any atom (Z, N) with more rest energy than its neighbor $(Z - 1, N + 1)$ it is energetically favorable for a proton to convert to a neutron, and such atoms are unstable against β^+ decay or electron capture.

OBSERVATION OF THE NEUTRINO

The difficulty of observing the neutrino is that it interacts so weakly with matter. Theoretical arguments indicate that if the β decay (13.25) is possible, so also is the so-called *inverse β decay,*

$$\bar{\nu} + p \rightarrow e^+ + n \tag{13.26}$$

in which a $\bar{\nu}$ collides with a proton and converts it to a neutron plus positron. If an antineutrino from a β decay were to pass through a material containing protons (such as water), it could, in principle, cause the reaction (13.26) and hence be detected. Unfortunately, this reaction, although possible, is exceedingly improbable. It has been calculated, for example, that on average an antineutrino would travel several hundred light years through water before producing a single inverse β decay.

Nevertheless, the reaction (13.26) was observed by the American physicists Cowan and Reines in 1956. Their first requirement was obviously to have a source of an enormous number of antineutrinos. For this they chose a nuclear fission reactor. As we discuss in Section 13.7, reactors produce large quantities of material that is β^- radioactive and hence generate many antineutrinos. Their detector, a large tank of water, was placed close to a reactor where, they calculated, some 10^{20} antineutrinos would pass through it each hour. The product of this enormous number of antineutrinos and the tiny probability for each $\bar{\nu}$ to induce the reaction (13.26) gave their expected number of events — about 3 per hour.

To register the occurrence of the reaction (13.26), they needed some way to detect the newly created positron and neutron. The positron rapidly slows down in the water and, within about a nanosecond, annihilates with an atomic electron to give two photons, each of about 0.5 MeV:

$$e^+ + e^- \rightarrow \gamma + \gamma$$

(see Section 3.8). The neutron moves off more slowly. To detect its presence Cowan and Reines added cadmium chloride to the water, since the cadmium nucleus has a high probability of capturing neutrons. Within about a microsecond, the neutron is captured, and the cadmium gives off a photon (or several photons) of total energy about 9 MeV:

$$n + {}^{113}Cd \rightarrow {}^{114}Cd + \gamma.$$

Thus both the positron and neutron eventually signal their presence by producing γ-ray photons of known characteristic energies, as shown in Fig. 13.6(a).

The γ rays were detected by sandwiching the water tank between large tanks of scintillator, a fluid that gives off visible light when traversed by γ rays, as shown in Fig. 13.6(b). The visible light was detected by photocells surrounding the scintillator tanks, and in this way both the timing and energy of the γ rays could be recorded. Finally, the entire apparatus was surrounded by thick

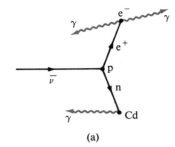

(a)

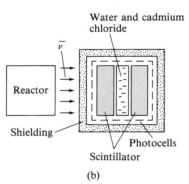

(b)

FIGURE 13.6 (a) Sequence of events for detecting the inverse β decay, $\bar{\nu} + p \rightarrow e^+ + n$. The positron annihilates with an electron to give two photons; the neutron is captured by a cadmium nucleus, which emits one or more photons. All of these events take place in the water tank; when the photons pass through the neighboring tanks of scintillator they produce visible light, which is detected by photocells. (b) Arrangement of the reactor (which produces the neutrinos) and the tanks of water and scintillator. The shielding excludes all particles except the neutrinos, but is completely "transparent" to neutrinos.

shielding (mostly lead and paraffin) to keep out any particles other than neutrinos.

After more than 1000 hours of observation and numerous consistency checks, Cowan and Reines were able to announce the positive identification of the antineutrino by verifying the inverse β decay (13.26) with the characteristic time sequence of photons that it produced.

13.5   The Natural Radioactive Series

Most of the naturally occurring radioactive nuclei are heavy nuclei, with $A \geq$ 210, that can be grouped into three radioactive series. These three series owe their existence to the occurrence of three parent nuclei with exceptionally long half-lives:

$$
\begin{array}{lll}
\textbf{Parent} & \textbf{Half-Life} & \\
{}^{232}\text{Th} & \text{14 billion years} & \\
{}^{235}\text{U} & \text{0.70 billion years} & (13.27) \\
{}^{238}\text{U} & \text{4.5 billion years} &
\end{array}
$$

(where 1 billion = 10^9). These nuclei are believed to have been formed, along with many other radioactive nuclei, in the nuclear processes of nearby supernova explosions shortly before the birth of the solar system. When the solar system condensed, it contained some of this radioactive material, most of which has decayed in the 5 billion or so years since then. However, because the three half-lives (13.27) are of order 1 billion years, the original supply of these three nuclei is still not exhausted.*

When nuclei of the three species (13.27) decay, they give birth to radioactive "children," which in turn produce radioactive grandchildren, and so on. This process continues, producing three radioactive series, or chains, which stop only when a stable descendant is reached. For example, the series that begins with ^{238}U ends at ^{206}Pb. Some of the stages in between (with half-lives shown in years, days, or hours) are as follows:

$$
\underset{(4.5 \times 10^9 \text{ y})}{{}^{238}\text{U}} \xrightarrow{\alpha} \underset{(24 \text{ d})}{{}^{234}\text{Th}} \xrightarrow{\beta^-} \underset{(6.7 \text{ h})}{{}^{234}\text{Pa}} \xrightarrow{\beta^-} \cdots \xrightarrow{\beta^-} \underset{(140 \text{ d})}{{}^{210}\text{Po}} \xrightarrow{\alpha} \underset{(\text{stable})}{{}^{206}\text{Pb}}
$$

Notice that the half-lives of the descendants can be, and are, much shorter than that of the original parent. The natural occurrence of ^{234}Th, for example, does not depend on its having survived for 5 billion years; rather, it is being produced continually by its long-lived parent ^{238}U.

Notice also that both α and β decays occur in the same radioactive series. In addition, the offspring nuclei are often produced in excited states that deexcite by γ decay. Thus all three types of decay will be observed in any natural sample. Further, a natural sample of any of the three original parents is bound to contain a mixture of all the descendants, each decaying at its own characteristic decay rate. Under the circumstances it is remarkable that the early workers — the Curies, Rutherford, and others — were able to unravel what was happening as quickly as they did.

---

* Notice, however, that the half-life of ^{235}U is much shorter than that of ^{238}U. This is why the isotope ^{235}U is much less abundant today. See Problem 13.28.

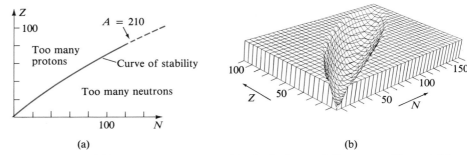

(a)   (b)

FIGURE 13.7 **(a)** The curve of stability is the locus of the most stable nuclei in a plot of Z against N. For $A \geq 210$ even those nuclei on the curve are unstable. **(b)** The valley of stability. The variables plotted horizontally are Z and N, exactly as in (a). The variable plotted upward was chosen to be the *mass excess*, $m - Au$, since this excludes the automatic, and hence uninteresting, increase of m as A increases. (The plot was truncated at a mass excess of 100 MeV/c^2 to show the valley more clearly.) The curve in (a) can be seen as an aerial view of the valley floor in (b). (Courtesy Dale Prull.)

To understand better the sequence of decays in the three radioactive series, let us consider a plot of Z and N for the stable nuclei. This *curve of stability* was shown in Fig. 12.1 and is reproduced more schematically in Fig. 13.7(a). If we imagine rest masses plotted upward, out of the paper, stable nuclei would be lower than neighboring unstable nuclei, and the curve in Fig. 13.7(a) would become the floor of a **valley of stability,** as sketched in Fig. 13.7(b). Nuclei on the sloping sides of this valley, to the right or left of the curve in Fig. 13.7(a), have too many neutrons or too many protons and decay by β^- or β^+ emission.

For heavy nuclei, the increasing Coulomb repulsion of the protons raises the valley floor as A increases, and even those nuclei on the valley floor are unstable for $A \geq 210$. It turns out that the most favorable decay for such nuclei is α decay, as the following example illustrates.

EXAMPLE 13.4   Use the measured masses in Appendix D to find the kinetic energy released in the α decay of ^{232}Th at rest

$$^{232}\text{Th} \rightarrow {}^{228}\text{Ra} + \alpha. \tag{13.28}$$

Compare this with the kinetic energy released in the hypothetical neutron decay

$$^{232}\text{Th} \rightarrow {}^{231}\text{Th} + \text{n}. \tag{13.29}$$

The atomic masses involved in the α decay are (Appendix D)

Initial	Final
$m(^{232}\text{Th}) = 232.0381$ u	$m(^{228}\text{Ra}) = 228.0311$ u
	$m(^4\text{He}) \quad = \quad 4.0026$ u
Total $M_i = 232.0381$ u	Total $M_f = 232.0337$ u

Therefore, the kinetic energy released is

$$K = (M_i - M_f)c^2 = (0.0044 \text{ u}) \times \frac{931.5 \text{ MeV}}{1 \text{ u}} = 4.1 \text{ MeV}.$$

Strictly speaking, the decay (13.28) is a nuclear process, and we should have used the masses of the three *nuclei,* rather than the three atoms. However, we get the same answer using the atomic masses because there are equal total numbers of atomic electrons on both sides of (13.28), and their masses cancel when we take the difference $M_i - M_f$.

The corresponding masses in the hypothetical neutron decay (13.29) are:

Initial	Final
$m(^{232}\text{Th}) = 232.0381$ u	$m(^{231}\text{Th}) = 231.0363$ u
	$m_n \quad = \quad 1.0087$ u
Total $M_i = 232.0381$ u	Total $M_f = 232.0450$ u

We see that the combined rest mass of the final bodies is greater than the initial mass, and the proposed neutron decay (13.29) cannot occur.

The reason that α decay is so highly favored is the exceptionally large binding energy of the α particle. In each of the decays (13.28) and (13.29), the binding energy per nucleon (B/A) of the offspring nucleus is greater than that of the parent. However, in (13.29) the ejected neutron has no binding energy at all, and the net effect of the decay proves to be a loss of binding energy—or a gain of rest energy—and the process is forbidden. In (13.28), on the other hand, the large binding energy of the α particle results in a net gain of binding energy, and the process is energetically favored. The same considerations apply to all unstable nuclei near the floor of the valley of stability: Alpha decay is the favored, and often the only, mode of decay.

When a nucleus emits an α particle, it loses two protons and two neutrons. This moves it in a 45°, "southwest" direction on the plot of Z against N, as shown in Fig. 13.8(a). If the parent is on the valley floor, this means that the offspring will be on the neutron-rich side of the valley. If the offspring is far enough up the valley side, it will undergo β^- decay to move it northwest back to the valley floor, as in Fig. 13.8(b). Otherwise, it will undergo further α decays, each of which carries it further up the neutron-rich side of the valley, until finally one or more β^- decays take it back to the valley floor, as in Fig. 13.8(c).

We see that a radioactive series will consist of a succession of α and β^- decays (plus γ decays when an offspring is formed in an excited state), which zigzag down the valley of stability. Because of the orientation of the valley floor, the path never strays appreciably onto the proton-rich side of the valley. Therefore, β^+ decays (and electron capture) have no part in the natural radioactive series. The whole of the series that begins with ^{232}Th is shown in Fig. 13.9.

At the start of the series shown in Fig. 13.9 an α decay is followed immediately by two β^- decays. The effect of these three decays is that N decreases by 4 while Z returns to its original value; that is, nucleus number 4 (^{228}Th) is an isotope of, and hence chemically indistinguishable from, number 1 (^{232}Th). The presence of different isotopes of several elements in the radioactive series was a source of confusion to the early nuclear physicists, but eventually led to the first clear understanding that there can be several different nuclei all belonging to the same chemical element.

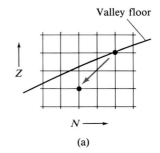

(a)

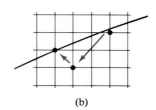

(b)

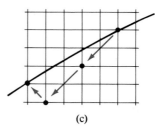

(c)

FIGURE 13.8 (a) In α decay both Z and N decrease by 2, and the offspring nucleus lies exactly southwest of the parent. If the parent is on the valley floor, this puts the offspring on the neutron-rich side of the valley. (b) If the offspring is high enough on the valley side, it will undergo β^- decay, producing a grandchild nearer the valley floor. (c) Depending on the position and orientation of the valley floor, a series may make two or more α decays before returning to the valley floor by β^- decay.

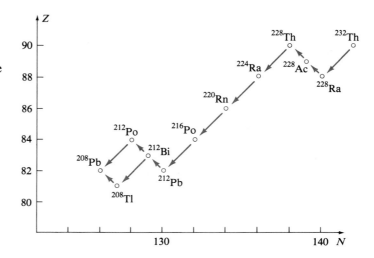

FIGURE 13.9 The radioactive series that originates in ^{232}Th follows a zigzag path, staying as close as possible to the floor of the valley of stability. Notice the branch at ^{212}Bi.

Notice also in Fig. 13.9 the *branch* at ^{212}Bi, which can undergo both α and β^- decay. When this happens we cannot predict which decay will actually occur; rather, we can state only the probabilities of the two modes: 64% for β^- and 36% for α, in the case of ^{212}Bi.

To conclude this section, it is perhaps worth emphasizing the great importance of the natural radioactive series, both in the history of nuclear physics and in many other fields today. The discovery of natural radioactivity in 1896 was the beginning of nuclear physics. Rutherford's identification of the atomic nucleus in 1911 used α particles from the radioactive series, and the predicted energy dependence of Rutherford scattering was verified by using α particles of different energies from different sources. As we have just seen, the important concept of isotopes emerged from the study of radioactive series. As we describe in the next section, α particles from the radioactive series produced the nuclear reactions that established the presence of protons in nuclei and the existence of the neutron. The invention in the early 1930s of machines that could accelerate charged particles reduced sharply the importance of natural α particles as nuclear probes. Nevertheless, α particles from radioactive materials are still used to make convenient sources of neutrons, and such sources were used to produce artificial elements and to induce nuclear fissions.

Today natural radioactivity is used in several ways to study the earth's crust. The measured ratios of parent to offspring atoms in a rock can date its solidification. An important member of the radioactive series is radon, a noble gas that moves easily through porous rock because it is chemically inactive and not easily absorbed.* Because radon is radioactive, even minute quantities seeping out of the earth's surface can be detected quite easily. Sudden increases in radon levels have been associated with the rock fractures preceding earthquakes. Above all, the principal cause of the slow movement of the earth's crust appears to be convection currents that are driven by heat from the natural radioactivity distributed throughout the earth's volume.

---

* For this reason radon can seep into buildings, where it is a potential health hazard, since it can be inhaled into the lungs. Those radons that decay while in the lungs produce the chemically active polonium, which sticks to the lung walls, so that all its subsequent decays irradiate the nearby tissue.

13.6   Nuclear Reactions

We have so far discussed nuclear transformations that result from the spontaneous decay of a single nucleus. While there is much to learn from such processes, there is even more to learn from **nuclear reactions,** the transformations that occur when two nuclei collide. The first such reaction was observed by Rutherford in 1919 when he bombarded nitrogen nuclei with α particles (from a natural radioactive source) and produced the reaction

$$\alpha + {}^{14}N \rightarrow p + {}^{17}O. \tag{13.30}$$

There is an obvious parallel between a nuclear reaction such as (13.30) and a chemical reaction, for example

$$H_2 + CO_2 \rightarrow H_2O + CO, \tag{13.31}$$

in which molecules (or atoms and molecules) collide and rearrange their constituent atoms. However, there is an important practical difference: Under normal conditions on earth, chemical reactions occur naturally and abundantly, whereas nuclear reactions occur hardly at all. The reason for this difference is easy to see. For two atoms or molecules to react chemically they only need to approach one another closely enough for their outer electrons to overlap, and this condition is easily met at normal temperatures and densities. But even in the most violent chemical reactions the nuclei, buried deep inside their protective shells of electrons, remain very far apart compared to the nuclear force range. Even if two nuclei were stripped of their electrons, a large energy would still be needed to bring them close together, because of the Coulomb repulsion between their positive charges. For example, the energy needed to bring a proton to the surface of an α particle is about 1 MeV (Problem 13.38). For this reason, nuclear reactions occur naturally only very rarely on earth.* On the other hand, they *do* occur—and are the main source of energy—in stars, where the temperature is so high that the kinetic energy of thermal motion is sufficient to overcome the Coulomb repulsion.

To produce a nuclear reaction in the laboratory one must first produce nuclei with energies of order 1 MeV and then direct them at other nuclei (contained in a suitable solid, liquid, or gas). The energy of the projectiles lets them sweep through the electrons of a target atom, with some of them approaching the nucleus close enough to induce a nuclear reaction. Until the early 1930s the only available projectiles were the α particles from natural radioactive sources. Then, two developments resulted in several new possibilities. First, the development of particle accelerators made it possible to accelerate many different nuclei to an energy sufficient to induce nuclear reactions. Second, soon after the discovery of the neutron, it was realized that because they are electrically neutral, even low-energy neutrons can approach and penetrate a nucleus. The study of nuclear reactions induced by low-energy neutrons quickly became a popular and fruitful area of research.

In this section we describe just a few nuclear reactions of historical importance, starting with the reaction (13.30). In Sections 13.7 and 13.8 we discuss the special cases of fission and fusion.

---

* There are a few examples, mostly induced by high-energy cosmic-ray particles, for instance the reaction $n + {}^{14}N \rightarrow p + {}^{14}C$, which produces carbon 14 in the atmosphere.

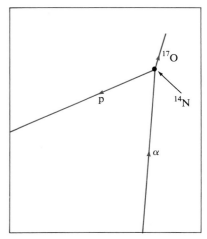

FIGURE 13.10 An early cloud chamber photograph of the reaction $\alpha + {}^{14}\text{N} \rightarrow \text{p} + {}^{17}\text{O}$. Numerous α particles, from ${}^{212}\text{Bi}$, enter the picture from below. One of them strikes a nitrogen nucleus (stationary and hence invisible) and produces a long-range proton and an oxygen nucleus. Note that the source produced α particles of two different energies; those with low energy all stop less than halfway up the picture; those with high energy almost all reach the top.

IDENTIFICATION OF THE PROTON AS A NUCLEAR CONSTITUENT

The first nuclear reaction to be observed was the reaction (13.30), which we can rewrite in more detail as

$$\text{}^4_2\text{He} + {}^{14}_7\text{N} \rightarrow {}^1_1\text{H} + {}^{17}_8\text{O}. \tag{13.32}$$

This reaction was induced by the α particles from ${}^{212}\text{Bi}$ as they passed through air. Rutherford noticed that particles were being produced with a range in air much longer than that of the α particles, but similar to that of hydrogen nuclei of comparable energy. He immediately suspected (and eventually proved) that these particles *were* hydrogen nuclei ejected from nitrogen nuclei in the air. He then argued that hydrogen nuclei are presumably contained in *all* nuclei and provide the positive charge and much of the nuclear mass. It was for this reason that he proposed a special name for the hydrogen nucleus: namely, *proton* (Greek for "first one").

At about this time the British physicist Wilson developed the cloud chamber, in which moving charged particles leave tracks of condensed water droplets. In this way it became possible to photograph the tracks of particles involved in nuclear reactions, as in Fig. 13.10, which shows a cloud chamber photograph of reaction (13.32).

We can use reaction (13.32) to illustrate two conservation laws that apply to all nuclear reactions. Consider first the mass number, or nucleon number, A, which appears as a superscript on each symbol. If you add up the superscripts on either side of (13.32), you will see that the total number of nucleons is the same before and after the reaction. Thus

$$\Sigma\, A_{\text{initial}} = 4 + 14 = 18$$

and

$$\Sigma\, A_{\text{final}} = 1 + 17 = 18$$

and hence

$$\Sigma A_{\text{initial}} = \Sigma A_{\text{final}}. \tag{13.33}$$

It is found that this equality holds in all nuclear reactions; the total number of nucleons does not change. We refer to this observation as the **law of conservation of nucleon number.*** It is a useful aid in checking whether or not a proposed reaction is possible: Any nuclear reaction that does not satisfy (13.33) is impossible.

A second conservation law that applies to all reactions is the familiar law of conservation of charge; the total charge cannot change:

$$\Sigma q_{\text{initial}} = \Sigma q_{\text{final}} \tag{13.34}$$

In (13.32) the only charged particles are protons, and conservation of charge implies simply that the total number of protons (shown as a subscript for each nucleus) cannot change:

$$\Sigma Z_{\text{initial}} = 9 = \Sigma Z_{\text{final}}$$

In general, however, the number of protons *can* change in processes that involve other charged particles as well. For example, in the β decay

$$(Z, N) \rightarrow (Z + 1, N - 1) + e^- + \bar{\nu}$$

the number of protons increases by one, but this increase of charge is balanced by the production of the negative electron. The law of conservation of charge is another useful aid in checking whether a given process can occur.

DISCOVERY OF THE NEUTRON

The neutron was discovered by Rutherford's colleague Chadwick in the reaction

$$_2^4\text{He} + {}_4^9\text{Be} \rightarrow {}_0^1\text{n} + {}_6^{12}\text{C}. \tag{13.35}$$

(You should check quickly that this satisfies the two conservation laws discussed in the last two paragraphs.) This reaction had actually been studied by the French physicists Irène and Frédéric Joliot-Curie, who had found that when α particles strike beryllium a neutral "radiation" is produced, which can eject protons from any material containing hydrogen, such as paraffin wax (Fig. 13.11). Since the neutron was not yet known, the Joliot-Curies made the natural suggestion that the neutral radiation consisted of photons. However, as Chadwick emphasized, protons were ejected from the wax with momenta up to about 100 MeV/c. It is easy to show (Problem 13.45) that if a photon strikes a proton and gives it 100 MeV/c of recoil momentum, the photon itself must have had a momentum of at least 50 MeV/c, and hence an energy of 50 MeV.

---

* As we shall see in Chapter 14 this law has to be generalized. Nucleons are just two members of a whole class of particles called *baryons,* and it is actually the number of baryons that cannot change. However, in most nuclear reactions the only baryons involved are neutrons and protons, and (13.33) therefore holds.

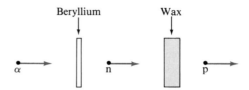

FIGURE 13.11 Discovery of the neutron. An unknown neutral particle, now known to be the neutron, was produced when α particles collided with beryllium nuclei. The particle's presence was evident, since it could knock protons out of a block of wax; that it was neutral was clear because it produced so little ionization.

(Recall that $E = pc$ for a photon.) But it is hard to see how photons with this much energy could possibly have been produced when the beryllium was struck by the original α particles, whose energy was only about 5 MeV.

Chadwick proposed instead that the "radiation" consisted of neutral particles, which he called neutrons,* whose mass was close to that of the proton. This suggestion immediately solved the problem of the photon energy: In the collision of two *equal* masses, the target particle (the proton) can take *all* the kinetic energy of the incident projectile (the neutron). Thus the neutrons needed to have only the kinetic energy of the recoil protons, which, for $p = 100$ MeV/c, was only about 5 MeV ($K = p^2/2m \approx 5$ MeV), and this was entirely reasonable.

In a series of experiments Chadwick let neutrons from the same source strike various different nuclei. By measuring the recoil energy of the different targets he was able to estimate the neutron's mass as about 1.15 u. For a more accurate measurement he used the mass-energy relation and the known masses of the nuclei involved in the neutron-producing reaction (Problem 13.39). He concluded that the neutron mass "probably lies between 1.005 and 1.008," in remarkable agreement with the current value of 1.009 u.

REACTIONS PRODUCED BY MAN-MADE ACCELERATORS

As we describe in Section 13.10, the early 1930s saw the development of several *accelerators*—machines that accelerate charged particles, at first to energies of order 10^5 eV and, today, up to 10^{12} eV and more. The first nuclear reaction produced by an accelerator was the reaction

$$^1H + {}^7Li \rightarrow {}^4He + {}^4He, \tag{13.36}$$

which was observed in 1932 by Cockcroft and Walton (both associates of Rutherford) using 600-keV protons from an accelerator of their own design.

We can use the reaction (13.36) to illustrate another of the conservation laws that apply to all reactions—the conservation of relativistic energy. Since the energy of each body in (13.36) has the form $K + mc^2$, conservation of energy implies that

$$K_i + M_i c^2 = K_f + M_f c^2,$$

where K_i and K_f denote the initial and final total kinetic energies in (13.36) and

---

* The idea of the neutron, and the name, had been suggested several years earlier by Rutherford.

M_i and M_f are the corresponding total rest masses. Therefore, the total gain in kinetic energy is just the total loss of rest energy:

$$\Delta K = K_f - K_i = (M_i - M_f)c^2. \qquad (13.37)$$

(The gain ΔK of kinetic energy in a reaction is often called the **Q value** and denoted by Q.) Cockcroft and Walton were able to measure ΔK and got the value 17.2 MeV, in satisfactory agreement with the modern value 17.35 MeV for $(M_i - M_f)c^2$. In this way they gave the first experimental verification of Einstein's mass-energy relation.

ENRICO FERMI (1901–1954, Italian–US). Fermi was a great theorist, an innovative experimentalist, and an inspiring teacher. While a professor at Rome, he made important contributions to the theory of beta decay and carried out an experimental study of neutron-induced reactions (which led ultimately to the discovery of fission). After receiving the Nobel Prize in 1938 (for the neutron work), he fled from Fascist Italy to the US, where he soon began work on neutron-induced fission. This resulted in the first controlled nuclear chain reaction, which was a key step in the development of nuclear weapons.

NEUTRON-INDUCED REACTIONS

Because the neutron is electrically neutral it can easily approach and penetrate nuclei, even at very low energies. One of the first to exploit the neutron as a probe of nuclei was the Italian physicist Fermi. As a source of neutrons he used a mixture of a radioactive α emitter and powdered beryllium, in which the α particles produced neutrons by Chadwick's reaction (13.35). (This arrangement is still used today as a convenient portable source of neutrons.) Fermi systematically fired the neutrons at about 70 of the known elements and produced a variety of nuclear reactions and radioactive products. Most of the reactions that he observed were among the following three:

$$\text{n} + (Z, N) \rightarrow (Z, N + 1) + \gamma \qquad \text{[neutron capture]}, \qquad (13.38)$$
$$\rightarrow (Z - 1, N + 1) + \text{p} \qquad \text{[(n, p) reaction]}, \qquad (13.39)$$
$$\rightarrow (Z - 2, N - 1) + \alpha \qquad \text{[(n, α) reaction]}. \qquad (13.40)$$

In all three cases the neutron-to-proton ratio N/Z increased, and most of the resulting nuclei were β^- radioactive. As we describe in the next section, another reaction that Fermi produced was the neutron-induced fission of uranium, but this reaction went unrecognized for a few more years.

CROSS SECTIONS

It is important to be able to characterize the probabilities (either measured or calculated) of different nuclear reactions. This is traditionally done with a parameter called the *reaction cross section*, which is defined as follows: We imagine first a single target particle [for example, the nucleus (Z, N) in the reaction (13.38)]. We now send in a beam of projectiles [like the neutron in (13.38)] such that a number n_{inc} of projectiles are incident per unit area perpendicular to the incident direction:

$$n_{inc} = \text{flux of projectiles} \qquad \text{(particles/area)}.$$

Now let N denote the number of reactions that occur (of the particular kind in question). As one might expect, N is proportional to the flux, n_{inc}, of projectiles, and we define the **reaction cross section** σ as the ratio of these two quantities:

$$\sigma = \frac{N}{n_{inc}} = \frac{\text{number of reactions}}{\text{flux of projectiles}}. \qquad (13.41)$$

The cross section σ is large for reactions that are very probable, and small for those that are improbable. Since σ has the dimensions of an area we can think of it as the size of the target as "seen" by the projectile. For example, the cross section for the inverse β decay $\bar{\nu} + p \rightarrow e^+ + n$ (a very improbable reaction) is found to be about 6×10^{-48} m². We can therefore say that, as seen by a neutrino, the proton is exceedingly small, in the sense that almost all neutrinos will pass it unaffected.

For most nuclear reactions, the cross section is found to be of the order of the geometrical size of nuclei. Since typical nuclear dimensions are of order 10^{-14} m, a natural unit of area for cross sections is 10^{-28} m², and this unit has come to be called the *barn*:

$$1 \text{ barn} = 10^{-28} \text{ m}^2. \tag{13.42}$$

For example, the cross section for 10-keV neutrons to be captured by ^{238}U is about 0.7 barn. On the other hand, the cross section for 0.1-eV neutrons to be captured by ^{113}Cd is unusually large, nearly 10^5 barns. As "seen" by low-energy neutrons a cadmium nucleus is extremely big, in the sense that the probability of their capture is very large. This is why cadmium was used by Cowan and Reines to catch neutrons, as described in Section 13.4.

Once a reaction cross section σ has been measured or calculated, one can use it to predict the number of reactions that will occur in a given situation. We can solve (13.41) for the number of reactions produced by a single target nucleus,

$$N(\text{for one target nucleus}) = \sigma n_{\text{inc}}. \tag{13.43}$$

In practice all targets contain many nuclei, and we must multiply (13.43) by the total number, N_{tar}, of target nuclei to give the total number of reactions

$$N = \sigma n_{\text{inc}} N_{\text{tar}}. \tag{13.44}$$

EXAMPLE 13.5 In the Cowan and Reines experiment, their water tank contained about 3×10^{28} protons ($N_{\text{tar}} = 3 \times 10^{28}$). In each hour the flux of neutrinos from the reactor was about $n_{\text{inc}} = 1.7 \times 10^{19}$ m^{-2}. Given that the cross section for inverse β decay is 6×10^{-20} barn, how many events could they expect in each hour?

According to (13.44), the number of events should have been

$$N = \sigma n_{\text{inc}} N_{\text{tar}}$$
$$= (6 \times 10^{-20} \times 10^{-28} \text{ m}^2) \times (1.7 \times 10^{19} \text{ m}^{-2}) \times (3 \times 10^{28})$$
$$= 3,$$

which is what they observed.

13.7 Nuclear Fission

In many nuclear processes there is a large release of kinetic energy; for example, in typical α decays, 4 to 9 MeV is released. Nuclear fission and fusion are two processes in which the large energy produced can be put to practical use on a macroscopic scale. If the energy is released rapidly, it can cause a violent explosion, and if it is released more slowly, it can be used to produce steam and hence to generate electricity.

To understand what fission and fusion are, and why they produce so

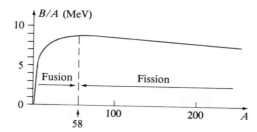

FIGURE 13.12 The binding energy per nucleon, B/A, as a function of A. Both the fusion of light nuclei and the fission of heavy nuclei lead to a higher binding energy and hence to a release of kinetic energy.

much energy, we consider again the graph of the nuclear binding energy per nucleon, B/A, as a function of the mass number A. This graph was shown in Fig. 12.14, which we reproduce here as Fig. 13.12. It shows that B/A increases with A, up to a maximum at $A = 58$, and then decreases steadily all the way to the heaviest nuclei. This means that if we combine two light nuclei, the resulting single nucleus will have higher total binding energy. Therefore, the fusion of two light nuclei, for example,

$$^2H + {}^2H \rightarrow {}^4He + \text{energy}, \tag{13.45}$$

would release energy, as we discuss in Section 13.8. On the other hand, when a heavy nucleus splits into two pieces, for example,

$$^{236}U \rightarrow {}^{92}Kr + {}^{144}Ba + \text{energy}, \tag{13.46}$$

the total binding energy of the two fragments will be greater than that of the single parent, and again there will be a release of kinetic energy. It is this process of **nuclear fission** * that we discuss first.

We can easily estimate the energy released in a nuclear fission such as (13.46). From Fig. 13.12 we see that B/A for the two fragments (with $A \sim 100$) is nearly 1 MeV higher than that of the parent nucleus ($A \sim 200$). Therefore, the binding energy *per nucleon* increases by about 1 MeV, so the kinetic energy released from a single fission is roughly 200 MeV. The magnitude of this release is apparent if one recalls that the energy released in a typical chemical process, such as the burning of coal, is only a few eV for each separate reaction.

Given that the fission of heavy nuclei is so energetically favorable, we must ask why heavy nuclei do not immediately fission into two smaller fragments. To answer this, we must consider the potential energy of the two fragments during the fission process. It is found that this process is quite analogous to the breakup of a charged liquid drop, as is suggested by the sketches in Fig. 13.13. In the first picture the original parent nucleus is shown in its normal, approximately spherical shape. Internal oscillations can cause the nucleus to elongate (second picture) and develop a distinct neck (third picture). At about this stage the Coulomb repulsion of the two pieces begins to dominate the weakening nuclear attraction and flings the pieces apart as two separate nuclei.

In Fig. 13.14 we show the potential energy of the two fragments as a function of their separation distance x. To understand this curve it helps to imagine the fission process taking place in reverse, starting with the two frag-

* The name *fission* is generally reserved for the case that the two fragments have more or less comparable size. Although α decay is qualitatively similar, it is not generally regarded as fission.

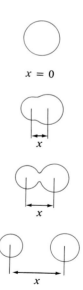

FIGURE 13.13 Excitation of a nucleus can cause it to change shape, develop a neck, and fission into two separate pieces. The separation can be characterized by the center-to-center distance x.

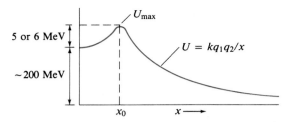

FIGURE 13.14 The fission barrier (not to scale). The potential energy U of two fission fragments in a typical fissionable nucleus such as ^{236}U is shown as a function of the separation distance x. For x large, the repulsive Coulomb force dominates, and $U = kq_1q_2/x$; for small separations, the attractive nuclear force dominates, and U forms a potential well. The maximum of U at separation x_0 is called the fission barrier, and the ground state of the parent nucleus is trapped inside this barrier with energy $E < U_{max}$.

ments far apart. When x is very large ($x \to \infty$) the two fragments exert no force on one another, and we can choose U to be zero. If we now wish to move the two fragments together, we must do work against their Coulomb repulsion, and they gain a potential energy $U = kq_1q_2/x$. When the fragments come within 1 or 2 fm (surface to surface), the attractive nuclear force comes into play and eventually exceeds the Coulomb repulsion. Therefore, the potential energy reaches a maximum at some separation x_0 and then drops again for $x < x_0$, as shown in Fig. 13.14. If the two fragments are closer than x_0, and if their energy is less than the maximum U_{max}, they are trapped and fission is impossible. For this reason, the peak in U near $x = x_0$ is called the **fission barrier;** this barrier is the reason that all naturally occurring nuclei are stable against prompt fission.*

NEUTRON-INDUCED FISSION

We see in Fig. 13.14 that 5 or 6 MeV of additional energy is needed to bring a typical heavy nucleus to the top of the fission barrier. Therefore, if one could produce a nucleus in an *excited state* 5 or 6 MeV above the ground state, it would have enough energy to pass over the barrier, and fission could occur, releasing some 200 MeV of kinetic energy. The simplest way to accomplish this uses the capture of slow neutrons, in a process such as

$$n + {}^{235}U \to {}^{236}U. \tag{13.47}$$

Using the ground-state masses listed in Appendix D, one can easily calculate that the total rest energy of the neutron plus ^{235}U is 6.5 MeV greater than that of the ground state of ^{236}U (Problem 12.41). Therefore, if the nucleus ^{236}U is formed by the process (13.47), it must be formed in an excited state at least 6.5 MeV above its ground state, and in this state the nucleus has enough energy to fission. We can write this sequence of events as

$$n + {}^{235}U \to {}^{236}U^* \to \text{fission}, \tag{13.48}$$

where we have introduced the star to indicate that the ^{236}U nucleus is in an excited state.

* If the nucleus were a classical system, then with $E < U_{max}$ it would be completely trapped inside the fission barrier. As we shall see in Section 13.9, quantum mechanics allows the fragments to "tunnel" through the barrier, and this explains the occurrence of spontaneous fission. However, for all the nuclei with which we are concerned here, the probability of spontaneous fission is exceedingly small.

Since 99% of all natural uranium is the isotope ^{238}U, it is natural to consider the possibility of neutron-induced fission of ^{238}U:

$$n + {}^{238}U \rightarrow {}^{239}U^* \rightarrow \text{fission}. \tag{13.49}$$

This process is closely analogous to (13.48). Nevertheless, there is an important difference: Because of the pairing effect, the ground state of ^{236}U (which is even–even) is better bound than that of ^{239}U (which is even–odd). The result is that for given neutron energy, the ^{236}U* in (13.48) is produced with a higher excitation energy than the ^{239}U* in (13.49). In particular, even with very slow neutrons, the ^{236}U* in (13.48) has enough excitation energy to fission, whereas the ^{239}U* in (13.49) does not.

This difference between ^{235}U and ^{238}U has important consequences, since it is found that the probability for neutron-induced fission is large enough for most applications only with very slow neutrons. Therefore, ^{238}U is usually not suitable as a fuel for nuclear fission. It was for this reason that the production of a uranium bomb in World War II required the separation of the rare isotope ^{235}U. For the same reason, many nuclear reactors that generate electricity require, for their fuel, uranium in which the proportion of ^{235}U is at least partially enriched.

The probability for the neutron-induced fission (13.48) is sufficiently large that Fermi undoubtedly produced fission when he bombarded uranium with neutrons in 1934, but he failed to identify the process. In fact, the possibility of fission was not recognized for another five years, even though neutron bombardment of uranium was studied intensively. This surprising delay had two causes: First, all known nuclear reactions produced changes of Z by one or two units at most; thus neutron bombardment of uranium (with $Z = 92$) was expected to produce elements in the neighborhood of $Z = 92$, but certainly *not* in the range $Z = 35$ to 60. Second, it turns out that many of the common fission fragments are chemically similar to the neighbors of uranium. For example, two typical fragments are strontium ($Z = 38$) and barium ($Z = 56$); a glance at the periodic table shows that these are in the same group as, and hence easily mistaken for, radium ($Z = 88$). Given the small quantities involved, it is easy to see why these fission fragments were wrongly identified as being neighbors of uranium.

The credit for discovering fission is generally given to Hahn and Strassman, who established (in 1938) the presence of the fission product barium after bombarding uranium with neutrons. However, the correct interpretation of their discovery (and the name fission) came from Meitner and Frisch in 1939. It was then quickly recognized that fission could possibly be induced on a large enough scale to produce macroscopic amounts of energy, either for explosives or for the generation of electricity.

CHAIN REACTIONS

The possibility of large-scale fission depends on two circumstances: The first is that fission can be induced by neutrons, as in (13.48); the second is that the fission reaction itself produces additional neutrons. Once the reaction begins, it can therefore sustain itself in what has come to be called a **chain reaction,** the neutrons produced in each fission going on to induce further fissions. This possibility was recognized almost as soon as fission was discovered, and by 1942 a group led by Fermi in Chicago had constructed the first nuclear chain reactor, using nearly 50 tons of uranium and uranium oxides.

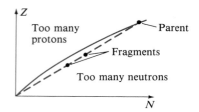

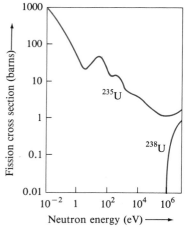

FIGURE 13.15 If a nucleus fissions into two comparable pieces, the fragments will lie close to the line $Z/N =$ constant through the parent (dashed line). This puts them on the neutron-rich side of the valley of stability (solid curve).

FIGURE 13.16 The cross sections for neutron-induced fission of ^{235}U and ^{238}U. (Recall that the cross section for nuclear reactions is just a convenient measure of their probability.) The cross section for ^{235}U rises to very large values at low neutron energies; that for ^{238}U is zero below about 1 MeV. Note that both scales are logarithmic. Both curves have many rapid fluctuations that do not show on this scale (and are unimportant in most applications).

To see why fission tends to produce neutrons, we consider, yet again, the valley of stability in a plot of Z against N (Fig. 13.15). Let us suppose for a moment that the parent divides into just two fragments, for example,

$$n + {}^{235}U \rightarrow {}^{236}U^* \rightarrow {}^{92}Kr + {}^{144}Ba. \qquad (13.50)$$

Since the protons and neutrons are uniformly distributed in a heavy nucleus, the proton-to-neutron ratio will be approximately the same for either offspring as for the original parent:

$$(Z/N)_{\text{offspring}} = (Z/N)_{\text{parent}}.$$

This means that the two offspring lie close to the line $Z/N =$ constant shown in Fig. 13.15. Because of the way the valley floor curves, both offspring are some distance away from the valley floor on the neutron-rich side.

Because the fragments have so many excess neutrons, they promply eject several separate neutrons to bring them closer to the valley floor.† Thus, in place of (13.50) one might observe

$$n + {}^{235}U \rightarrow {}^{236}U^* \rightarrow {}^{90}Kr + {}^{143}Ba + 3n. \qquad (13.51)$$

The actual number of neutrons produced in a fission can vary. On average the neutron-induced fission of ^{235}U produces 2.5 neutrons, and it is these that make possible further fissions in a chain reaction.

That each fission produces 2.5 neutrons is not by itself enough to guarantee that the reaction will be sustained. The important parameter is the number, k, of neutrons from each fission that do in fact induce another fission. If this number, called the **reproduction factor,** is less than 1 ($k < 1$), the reaction will die out; if k is greater than or equal to 1 ($k \geq 1$), the reaction will continue; and if k is much greater than 1, the reaction will accelerate explosively.

The value of the reproduction factor k depends on several things. For example, the probability of a neutron-induced fission like (13.51) is much larger for slow neutrons ($K \sim 0.01$ eV) than for the fast neutrons ($K \sim 1$ MeV) that usually emerge from the fission, as can be seen in Fig. 13.16. Therefore, the fuel in reactors is usually embedded in a **moderator,** a material (such as carbon or water) containing light nuclei that carry away a significant fraction of a neutron's energy as they recoil from elastic collisions (Problem 13.51). Repeated collisions slow down the neutron and improve its chance of inducing a fission before it escapes or gets captured. In this way, the moderator increases the reproduction factor k.

Perhaps the most obvious factor on which k depends is the size of the fissionable sample. In a small sample, most neutrons will escape before they have the opportunity to induce a fission; the larger the sample, the better the chance for a neutron to induce a fission before it can escape. Under given conditions (density and shape of the material, for example) there is therefore a **critical mass,** below which k is less than 1 and a chain reaction dies out, but above which $k \geq 1$ and the chain reaction is sustained. For example, the critical mass for an isolated sphere of pure ^{235}U is on the order of 50 kg.

A fission bomb depends on the sudden creation of a supercritical mass. The two bombs used in World War II worked in quite different ways. In the bomb dropped on Hiroshima two separate subcritical masses of ^{235}U were

---

† Even so, the offspring still have an excess of neutrons and will undergo several β^- decays. This is why the waste products of fission reactors are extremely β^- radioactive.

forced together into one supercritical mass by conventional explosives. In the bomb dropped on Nagasaki a single hollow sphere of ^{239}Pu was compressed, also by conventional explosives, until it became critical.

In reactors that are used for peaceful purposes it is obviously important to control k so that it remains very close to 1. This is usually done with *control rods,* made of a material such as cadmium that absorbs neutrons strongly. When the rods are fully inserted in the reactor, they absorb many neutrons and k is less than 1. The rods can be withdrawn until $k = 1$ and can be continually adjusted to keep k close to 1. As the reaction proceeds, the kinetic energy of the fission fragments is shared among the atoms of the fuel rods, whose temperature rises. The heat is removed by a coolant, such as water (which also acts as a moderator to slow down the neutrons). The coolant is usually sealed inside the reactor system, but transfers the heat to an external steam boiler which drives the turbines of an electrical generating plant.

Most commercial reactors today use ^{235}U as their fuel (either in its normal mix with ^{238}U, or somewhat enriched). Because of the limited supply of this isotope, such reactors cannot possibly meet society's energy needs indefinitely. One alternative is to use plutonium 239. This fissions even more easily than ^{235}U and, although not naturally available, can be produced from the relatively plentiful ^{238}U by the neutron capture

$$n + {}^{238}U \rightarrow {}^{239}U + \gamma, \tag{13.52}$$

followed by two β^- decays,

$$^{239}U \rightarrow {}^{239}Np + e^- + \bar{\nu} \tag{13.53}$$

and

$$^{239}Np \rightarrow {}^{239}Pu + e^- + \bar{\nu}. \tag{13.54}$$

A single reactor can be designed in which ^{238}U is converted to ^{239}Pu *and* ^{239}Pu fissions to produce heat. Some of the neutrons from the fission of ^{239}Pu induce further fissions, while others convert ^{238}U into more ^{239}Pu. Reactors of this type are called *breeder reactors,* and a few are in operation (although not in the United States).

13.8 Nuclear Fusion

We saw at the beginning of Section 13.7 that energy is released when two light nuclei fuse to form a heavier nucleus. An example of such **fusion** is the reaction

$$^1H + {}^2H \rightarrow {}^3He + \gamma + 5.5 \text{ MeV}, \tag{13.55}$$

in which two hydrogen nuclei (one the ordinary ^{1}H, and one the heavy isotope ^{2}H, called deuterium) fuse to form helium 3. Reactions of this kind produce the energy radiated by most stars, including our own sun. Some 40 years ago the same kinds of reaction were first used in the hydrogen bomb, and since then extensive research has been devoted to harnessing fusion reactions for peaceful uses. The principal attraction of fusion as a source of power is that the fuels — light elements like hydrogen — are abundant and cheap. Another important advantage is that it would produce far less radioactivity than fission reactors.

Unfortunately, there is an enormous obstacle to the large-scale use of nuclear fusion. The two initial nuclei are positively charged, and large initial energies (many keV at least) are needed to overcome the Coulomb repulsion and allow fusion. This is easily accomplished on a small scale using an accelerator; but the number of reactions produced in this way is hopelessly insufficient, even to replace the energy needed to run the accelerator.

It seems that the only way to produce macroscopic amounts of energy by nuclear fusion is to heat the fuel to high temperatures, where the kinetic energy of thermal motion is sufficient to overcome the Coulomb barrier. (This is the origin of the name **thermonuclear reaction**—a nuclear reaction induced by high temperature.) Unfortunately, the temperatures required are extremely high, as we can easily see: For two hydrogen nuclei (ordinary or heavy) to react, they must certainly come within 10 fm of one another (and probably closer). At this separation their potential energy is

$$U = \frac{ke^2}{r} = \frac{1.44 \text{ MeV} \cdot \text{fm}}{10 \text{ fm}} = 144 \text{ keV}.$$

Thus to bring them close enough in a head-on collision, we must give each nucleus about 70 keV of kinetic energy. Now, it is known from kinetic theory that the average energy of a particle at absolute temperature T is

$$K_{av} = \tfrac{3}{2} k_B T, \tag{13.56}$$

where k_B is Boltzmann's constant,*

$$k_B = 8.62 \times 10^{-5} \text{ eV/K}$$

(where K denotes the kelvin, the SI unit of thermodynamic temperature). Thus if the average thermal energy is to be about 70 keV, the required temperature is

$$T = \frac{2}{3} \frac{K_{av}}{k_B} = \frac{2}{3} \times \frac{7 \times 10^4 \text{ eV}}{8.6 \times 10^{-5} \text{ eV/K}} \approx 5 \times 10^8 \text{ K}. \tag{13.57}$$

The temperature (13.57) is actually an overestimate of the temperature needed to produce thermonuclear reactions. In the first place (13.56) is the *average* kinetic energy at temperature T; even at lower temperatures, *some* particles will still have this much energy. Second, because of a quantum effect called tunneling (which we describe in Section 13.9), even particles with insufficient energy to surmount the Coulomb barrier *can* undergo fusion. For both these reasons, fusion can occur at temperatures an order of magnitude below (13.57) at about†

$$T \approx 5 \times 10^7 \text{ K}, \tag{13.58}$$

still a very high temperature!

The problems of producing temperatures of order 10^7 K on earth are obvious. The only successes so far are, unfortunately, in nuclear explosions. A fission bomb can produce temperatures of this order and act as the trigger for a fusion bomb. The principal fuel for the fusion bomb is deuterium or heavy hydrogen, ^{2}H (hence the name "hydrogen bomb"). One of the fusion reactions is

$$^2\text{H} + ^2\text{H} \rightarrow ^3\text{H} + ^1\text{H} + 4.0 \text{ MeV}, \tag{13.59}$$

---

* In many introductory texts, Boltzmann's constant is introduced in the form $k_B = R/N_A$, where R is the universal gas constant ($R = 8.31$ J/mol·K) and N_A is Avogadro's number.

† The precise temperature needed naturally depends on the circumstances; for example, the temperature in the sun (where thermonuclear reactions occur rather slowly) is somewhat cooler.

which produces the isotope ^{3}H, called tritium. This in turn can fuse with deuterium in the deuterium–tritium (DT) reaction:

$$^2H + {}^3H \rightarrow {}^4He + n + 17.6 \text{ MeV.} \qquad (13.60)$$

Notice that although the energy release in these fusion reactions is smaller than the 200 MeV released in fission, the energy released *per nucleon* (and hence per unit mass) is much larger in fusion than in fission.

The production of controlled thermonuclear reactions for peaceful purposes is much more difficult than producing an explosion. The difficulty is to contain the fuels while they are heated to many millions of degrees. Obviously, any conventional container would vaporize long before such temperatures were reached. Two methods of confinement are being investigated: a confinement by magnetic fields, and a brief compression by bursts from several high-power lasers. Most experts seem to agree that the enormous problems of controlled fusion will eventually be solved. However, even after 40 years of intense research in several countries, the method is still a long way from practical realization.

FUSION IN STARS

The temperatures needed to produce thermonuclear fusion are available naturally inside stars, and nuclear fusion is indeed the principal source of stellar energy.* A relatively young and small star like our own sun consists mainly of hydrogen, and the main source of energy is a sequence of three reactions, called the **proton–proton cycle**. The first reaction produces deuterons (^{2}H nuclei):

$$^1H + {}^1H \rightarrow {}^2H + e^+ + \nu + 0.4 \text{ MeV.} \qquad (13.61)$$

This reaction involves the conversion of a proton to a neutron by β^+ decay, during the fleeting collision of the two protons, and the reaction (13.61) is therefore extremely unlikely (Problem 13.54). Nevertheless, proton–proton collisions are so frequent in the star that this reaction proceeds at a sufficient rate.

A deuteron produced in the reaction (13.61) can next collide with another proton to produce helium 3:

$$^2H + {}^1H \rightarrow {}^3He + \gamma + 5.5 \text{ MeV.} \qquad (13.62)$$

Finally, two of the ^{3}He nuclei produced in this way can collide to give the very stable ^{4}He:

$$^3He + {}^3He \rightarrow {}^4He + {}^1H + {}^1H + 12.9 \text{ MeV.} \qquad (13.63)$$

Notice that the net effect of the whole proton–proton cycle is to convert four protons into a ^{4}He nucleus (plus assorted electrons and neutrinos)

$$4{}^1H \rightarrow {}^4He + 2e^+ + 2\nu + 24.7 \text{ MeV.} \qquad (13.64)$$

(The two positrons quickly annihilate with two electrons to produce a further 2.0 MeV, so the total yield is actually more like 27 MeV.) In this way, the sun

---

* Before nuclear reactions were understood, the source of stellar energy was one of the outstanding mysteries of astronomy. For example, it is easy to show that if stars depended on gravity for their energy, they would collapse in just a few million years, whereas we know that our own sun is several billion years old.

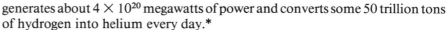

HANS BETHE (born 1906, German–US). After postdoctoral work with Rutherford in Cambridge and Fermi in Rome, Bethe taught in Germany for a few years before coming to the US in 1935. Among many contributions to atomic and nuclear physics, he is best known for finding the two nuclear cycles by which most stars get their energy. For this discovery he won the 1967 Nobel Prize.

generates about 4×10^{20} megawatts of power and converts some 50 trillion tons of hydrogen into helium every day.*

When a star has "burned" most of its hydrogen to helium, its source of energy is temporarily exhausted, and the star can no longer support itself against its own gravitational attraction. The star therefore contracts, and its central temperature rises until new fusion processes, forming carbon from helium, can begin. In large stars, when the helium is exhausted, there is a second transition to processes in which carbon is fused to form heavy elements all the way up to iron. Since iron has the highest value of B/A, it is the heaviest element that can be formed by fusion with the release of energy. Heavier elements, from iron to uranium and beyond, are formed by a succession of neutron captures followed by β decays. This process can occur slowly in an old, but stable star or very rapidly in a supernova explosion. This type of explosion occurs when a heavy star, having exhausted its supply of fusion fuel, undergoes a catastrophic collapse and then explodes. The explosion ejects an enormous outer shell of matter and leaves behind a tiny but massive neutron star or black hole. The matter that is ejected, including some of the recently formed heavier elements, is dispersed through interstellar space and, in due course, is formed into new stars like our own sun. The presence of the heavy elements in the sun and planets is evidence that they, and we, are descendants of many long-dead stars.

13.9   The Theory of Alpha Decay (optional)

We saw in Chapter 8 that a quantum particle can be found in regions that are classically forbidden. We shall see now that this allows a quantum particle to escape from a region in which a classical particle would remain forever trapped. This phenomenon is called **barrier penetration** or **tunneling** and is the basis of many modern devices, such as the tunneling microscope, mentioned in Chapter 17. Barrier penetration was discovered in 1928 by Gamow, and independently by Condon and Gurney, as the explanation of alpha decay.

The facts about alpha decay that needed to be explained can be quickly summarized: Many unstable heavy nuclei emit alpha particles. The kinetic energy released in these decays varies rather little among the various nuclei — from about 4 to 9 MeV. On the other hand, the half-lives vary over an astonishing range, from less than a microsecond to more than 10 billion years. These facts are illustrated in Table 13.1, which lists five alpha-radioactive nuclei, with the energy release and half-lives of their decays. These five examples illustrate what is found to be generally true — that the half-lives are correlated with the energy released: the shorter the half-life, the higher the energy released, and vice versa.

Given these facts, the questions we must answer are these: How do nuclei emit alpha particles? Given that they can emit alpha particles, why do some nuclei wait more than 10 billion years before doing so? And why is the half-life so extraordinarily sensitive to the energy release K, with a factor of 2 difference in K corresponding to nearly 25 orders of magnitude in $t_{1/2}$?

The theory of alpha decay starts from the plausible assumption that two of

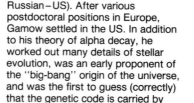

GEORGE GAMOW (1904–1968, Russian–US). After various postdoctoral positions in Europe, Gamow settled in the US. In addition to his theory of alpha decay, he worked out many details of stellar evolution, was an early proponent of the "big-bang" origin of the universe, and was the first to guess (correctly) that the genetic code is carried by triplets of nucleotides. Gamow was also the author of many entertaining books on science.

---

* In heavier stars the conversion of hydrogen to helium is accomplished primarily by a different sequence of reactions, called the carbon cycle (Problem 13.55).

the protons inside a heavy nucleus can occasionally cluster together with two of the neutrons to form an alpha particle. Experiments* show that this is correct: Alpha particles are continually forming and dissolving inside a nucleus. Averaged over time, it is found that there is of order one alpha particle inside any heavy nucleus.

To find how quickly alpha particles escape from the nucleus, we must first estimate how frequently they appear at the nuclear surface. An alpha particle is formed from four nucleons, each with potential energy about -50 MeV and kinetic energy about 25 MeV. Therefore, the alpha particle should have a kinetic energy of roughly 100 MeV, corresponding to a speed $v \sim c/4$. For a nuclear diameter D of about 15 fm, the frequency with which an alpha particle appears at the surface of the nucleus would be roughly

$$f \sim \frac{v}{D} \sim \frac{c/4}{15 \text{ fm}} \sim 5 \times 10^{21} \text{ s}^{-1}. \tag{13.65}$$

We see that alpha particles appear at the nuclear surface more than 10^{21} times per second. We would not necessarily expect these particles to escape, since the strong nuclear force pulls them back into the nucleus. Nevertheless, the observed fact is that an alpha particle does occasionally escape. Our task is to see how this occurs and to find the probability of its doing so.

Let us denote by P the probability that a single alpha particle striking the nuclear surface will escape. Our ultimate goal is to find the decay constant, r, of the nucleus. Since r is the probability per unit time of an alpha particle escaping, it is the product of f (the frequency with which alpha particles collide with the surface) and P (the probability of escape in any one collision):

$$r = fP. \tag{13.66}$$

We have already estimated f in (13.65), and we must now find the probability P.

In quantum mechanics the probability of finding a particle in some region is given by the square of its wave function. The wave function in turn is determined by the Schrödinger equation, whose solution requires knowledge of the potential energy function. The potential energy of the alpha particle is sketched in Fig. 13.17 and is qualitatively similar to the potential energy of two fission fragments (shown in Fig. 13.14). If the alpha particle is outside the nucleus, then U is just the Coulomb potential energy $U(x) = 2Zke^2/x$, where x is the separation of the alpha from the residual nucleus, and Ze is the charge of the residual nucleus. (Remember the charge of the alpha is $2e$.) For x less than some x_0, about equal to the nuclear radius R, the attractive nuclear force dominates, and the potential energy drops abruptly. There is therefore a *barrier,* exactly analogous to the fission barrier, and it is this barrier that prevents the alpha particle from escaping instantly.

It is important to establish the approximate parameters of the barrier confining the alpha particle. Its left-hand boundary is at the point shown as x_0, which is about equal to the nuclear radius (or, more precisely, the sum of the radii of the residual nucleus and the alpha):

$$x_0 \approx 8 \text{ fm}.$$

---

* For example, reactions involving ejection or transfer of alpha particles from nuclei.

TABLE 13.1

Five alpha-emitting nuclei in order of increasing half-life. The second column shows the kinetic energy released in the decay and the third the half-life.

Nucleus	K (MeV)	$t_{1/2}$
^{216}Ra	9.5	0.18 μs
^{194}Po	7.0	0.7 s
^{240}Cm	6.4	27 days
^{226}Ra	4.9	1600 years
^{232}Th	4.1	14 billion years

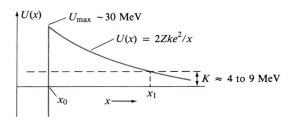

FIGURE 13.17 The barrier that keeps an alpha particle from escaping promptly from a nucleus. The curve shown is the potential energy $U(x)$ of the alpha particle at a distance x from the center of the offspring nucleus. For $x > x_0$ it is just the Coulomb potential energy, while for $x < x_0$ it is dominated by the attractive nuclear force and is strongly negative. (The actual curve would be somewhat rounded at x_0.) The observed kinetic energy when the alpha particle escapes is shown as K.

As can be seen in Fig. 13.17, the height U_{max} of the barrier is simply the electrostatic potential energy at $x = x_0$:

$$U_{max} = 2Z \frac{ke^2}{x_0} \sim 2 \times 90 \times \frac{1.44 \text{ MeV} \cdot \text{fm}}{8 \text{ fm}} \approx 30 \text{ MeV}.$$

We know that when the alpha particle emerges from the nucleus and moves far away it has kinetic energy, and hence total energy, between 4 and 9 MeV.* By conservation of energy, this must be the same as its energy before the decay occurred. Therefore, its energy inside the nucleus was also between 4 and 9 MeV (as shown by the dashed horizontal line in Fig. 13.17). This is far below the top of the barrier ($U_{max} \sim 30$ MeV), and if the alpha were a classical particle inside the nucleus, it would be permanently trapped by the barrier. Our main purpose in this section is to see how a quantum particle can tunnel through the barrier and appear on the outside.

One more important parameter shown in Fig. 13.17 is the distance x_1. This is the separation at which an alpha particle coming *inward* with initial kinetic energy K would be stopped by the barrier. The value of x_1 is determined by the condition

$$2Z \frac{ke^2}{x_1} = K$$

or

$$x_1 = 2Z \frac{ke^2}{K} \approx 2 \times 90 \times \frac{1.44 \text{ MeV} \cdot \text{fm}}{4 \text{ to } 9 \text{ MeV}} \approx 30 \text{ to } 65 \text{ fm}.$$

The interval from x_0 to x_1 is the classically forbidden region. Its length, $L = x_1 - x_0$, is the thickness of the barrier through which the alpha particle must tunnel. By nuclear standards L is rather long and varies appreciably (from roughly 20 to 55 fm as K ranges from 9 down to 4 MeV). We shall see that these

---

* We are treating the process nonrelativistically, as is certainly an excellent approximation since $mc^2 \approx 4000$ MeV for an alpha particle. Thus "total energy" here means the classical energy, $E = K + U$, and if $U = 0$, then $E = K$.

are the main reasons why the escape probability P is so low and why it varies so much with the energy of the emerging alpha particle.

BARRIER PENETRATION

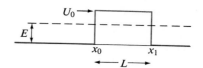

FIGURE 13.18 A rectangular barrier of height U_0 and width L.

To see how a quantum particle that is trapped inside a barrier like that of Fig. 13.17 can tunnel through and escape, we consider first the simpler rectangular barrier shown in Fig. 13.18. We imagine a particle moving on the left of the barrier, with an energy E that is less than the barrier height U_0. Classically, such a particle is trapped in the region $x < x_0$; when it reaches x_0 it rebounds with 100% probability. To see what happens to a quantum particle when it hits this barrier, one must solve the Schrödinger equation with this potential-energy function $U(x)$. Although the solution is not especially difficult, we do not need to go through it, since we can already understand its main features from our discussion of the finite square well in Section 8.8.

Consider first a barrier whose length (shown as L in Fig. 13.18) is *infinite*. This extreme case is shown in Fig. 13.19 and is precisely the same as the right wall of the finite square well of Section 8.8 (Fig. 8.19). In Fig. 13.19 we have also shown the wave function $\psi(x)$ for an energy level with $E < U_0$. Inside the well ($x < x_0$), the kinetic energy, $K = E - U$, is positive and $\psi(x)$ is an oscillating sinusoidal wave, with the form $A \sin kx$. In the barrier ($x > x_0$), $E - U$ is negative and $\psi(x)$ is a decreasing exponential with the form

$$\psi(x) = Be^{-\alpha x}, \tag{13.67}$$

where, you may recall [see (8.78) if you don't],

$$\alpha = \sqrt{\frac{2m(U_0 - E)}{\hbar^2}}. \tag{13.68}$$

As we emphasized in Section 8.8, the quantum particle has a nonzero probability of being found in the classically forbidden region where $E - U$ is negative.

In the infinitely long barrier of Fig. 13.19, $\psi(x)$ goes steadily to zero as x increases. But if the barrier has finite length, the situation is as shown in Fig. 13.20. Just as in Fig. 13.19, $\psi(x)$ is sinusoidal (with amplitude A_{in}) inside the well ($x < x_0$); and just as in Fig. 13.19, it decreases exponentially within the barrier ($x_0 < x < x_1$). But when we reach x_1 the barrier stops and, once again, $E - U$ is positive. Therefore, before $\psi(x)$ has decreased to zero, it starts oscillating again with amplitude A_{out}. The probability that the particle is inside the well, approaching the barrier, is proportional to A_{in}^2; the probability that it is outside the well and beyond the barrier is proportional to A_{out}^2. Therefore, there

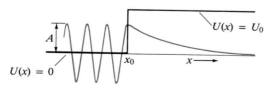

FIGURE 13.19 A rectangular barrier of infinite length, with $U(x) = U_0$ for $x_0 < x < \infty$. The wave function is sinusoidal with amplitude A when $x < x_0$, and decreases exponentially when $x > x_0$. (Courtesy Eloise Trabka.)

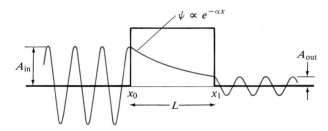

FIGURE 13.20 Wave function for a barrier of finite length. Inside the well ($x < x_0$) $\psi(x)$ is sinusoidal, with amplitude A_{in}; in the barrier it decreases exponentially; then for $x > x_1$ it is sinusoidal again, with amplitude A_{out}. (Courtesy Eloise Trabka.)

is a nonzero probability, P, that a particle striking the barrier from the left will escape to the right:*

$$P = \left(\frac{A_{out}}{A_{in}}\right)^2. \tag{13.69}$$

From Fig. 13.20 we see that A_{out} is less than A_{in} because of the exponential decrease of $\psi(x)$ within the barrier. Specifically,†

$$\frac{A_{out}}{A_{in}} \approx \frac{e^{-\alpha x_1}}{e^{-\alpha x_0}} = e^{-\alpha L}. \tag{13.70}$$

Therefore, the probability that a particle which strikes the rectangular barrier of Fig. 13.20 with energy $E < U_0$ will tunnel through and emerge on the outside is

$$P \approx e^{-2\alpha L}, \tag{13.71}$$

where

$$\alpha = \sqrt{\frac{2m(U_0 - E)}{\hbar^2}} \tag{13.72}$$

and m is the mass of the particle concerned.

The escape probability (13.71) depends on two variables: α, as given by (13.72), and L, the barrier thickness. If the barrier is high and wide (as is the case in alpha decay), then both α and L are large and the escape probability is very small. (Remember that if x is large, then e^{-x} is extremely small; for example, if $x = 10$, then $e^{-x} = 0.00005$.) Furthermore, if α is large, then $P = e^{-2\alpha L}$ is very sensitive to the barrier thickness L. This is the main reason that the observed half-lives vary over such an enormous range.

Equation (13.71) gives the escape probability for any particle confined by a rectangular barrier. Our interest is in an alpha particle confined by the nonrectangular barrier of Fig. 13.17. To find the corresponding escape probability we can approximate the actual barrier by a succession of n rectangular barriers as in Fig. 13.21. The total escape probability is the product of the n individual probabilities

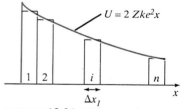

FIGURE 13.21 The actual curved barrier can be approximated as n rectangular barriers, each with thickness Δx_i.

$$P = P_1 P_2 \cdots P_n$$
$$= e^{-2\alpha_1 \Delta x_1} e^{-2\alpha_2 \Delta x_2} \cdots e^{-2\alpha_n \Delta x_n} = \exp\left(-2 \sum \alpha_i \Delta x_i\right), \tag{13.73}$$

---

* Our discussion here is a little oversimplified. Strictly speaking, we should be using wave functions e^{ikx} that travel in a definite direction, rather than sinusoidal functions. Nevertheless, the result (13.69) is close enough for our purposes.

† Here too we are oversimplifying. In Fig. 13.20 you can see that the value of the exponential wave function at x_0 is not quite the same as the amplitude A_{in}. (Similarly, the value of x_1 is not quite equal to A_{out}.) Nevertheless, under the conditions of interest, (13.70) is a satisfactory approximation.

where each α_i is given by (13.72) with the appropriate barrier height U_i. In the limit that all $\Delta x_i \to 0$, the sum in (13.73) becomes an integral, and we obtain the desired probability,

$$P = \exp\left[-2 \int_{x_0}^{x_1} \alpha(x)\, dx\right], \qquad (13.74)$$

where $\alpha(x)$ is found from (13.72),

$$\alpha(x) = \sqrt{\frac{2m[U(x) - K]}{\hbar^2}}, \qquad (13.75)$$

by replacing U_0 with $U(x)$ and noting that E is just K, the kinetic energy when the alpha particle is far away from the nucleus and U is zero.

For the case of alpha decay we know that $U(x) = 2Zke^2/x$ and the integral in (13.74) can be evaluated (Problem 13.60) to give

$$P = \exp\left[-\frac{2\pi ke^2\sqrt{2m}}{\hbar} ZK^{-1/2} + \frac{8\sqrt{mke^2}}{\hbar}(ZR)^{1/2}\right], \qquad (13.76)$$

where Z and R denote the charge and radius of the residual nucleus. It is convenient to rewrite this result as

$$P = \exp[-aZK^{-1/2} + b(ZR)^{1/2}] \qquad (13.77)$$

where the two constants a and b are easily evaluated (Problem 13.56) to give

$$a = \frac{2\pi ke^2\sqrt{2m}}{\hbar} = 3.97\ (\text{MeV})^{1/2} \qquad (13.78)$$

and

$$b = \frac{8\sqrt{mke^2}}{\hbar} = 2.98\ (\text{fm})^{-1/2}. \qquad (13.79)$$

If we substitute the typical values $Z = 90$, $K = 6$ MeV, and $R = 8$ fm into (13.77), we find that

$$P \approx \exp(-146 + 80) \approx 2 \times 10^{-29}.$$

As anticipated, the probability P that an alpha particle striking the surface of a nucleus will escape is extremely small.

HALF-LIVES

Since experimental data are generally given in terms of half-lives, let us assemble our results to give an expression for the half-life of an alpha-emitting nucleus. The half-life is $(\ln 2)/r$ and the decay constant r is $r = fP$, where f is the frequency with which alpha particles strike the nuclear surface. Therefore,

$$t_{1/2} = \frac{\ln 2}{r} = \frac{0.693}{fP}.$$

Since P is given by (13.77) as an exponential, it is convenient to take natural logarithms to give

$$\ln t_{1/2} = -\ln P + \ln \frac{0.693}{f}$$

or, from (13.77),

$$\boxed{\ln t_{1/2} = aZK^{-1/2} - b(ZR)^{1/2} + c.} \qquad (13.80)$$

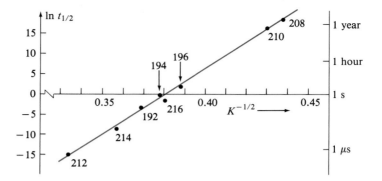

FIGURE 13.22 Plot of $\ln t_{1/2}$ against $K^{-1/2}$ for eight alpha-emitting isotopes of polonium. (Half-lives, $t_{1/2}$, in seconds and energy release, K, in MeV.) The number beside each point is the mass number of the isotope. The line is the least-squares fit to the data, and the axis on the right shows $t_{1/2}$ itself.

In this formula Z and R are the charge and radius of the residual nucleus. The constants a and b are given by (13.78) and (13.79). The constant c is $\ln(0.693/f)$, where f was estimated as roughly 5×10^{21} s^{-1} [see (13.65)]; therefore,

$$c = \ln \frac{0.693}{f} \approx -50. \qquad (13.81)$$

The simplest way to test our theory is to consider a set of alpha-emitting nuclei which all have the same value of Z. For example, we could consider the isotopes ^{192}Po, ^{194}Po, ^{196}Po, ^{208}Po, ^{210}Po, ^{212}Po, ^{214}Po, ^{216}Po of polonium, all of which decay exclusively by alpha emission.* Since these nuclei all have the same value of Z and very nearly the same radius R, the second term in (13.80) is approximately constant, and (13.80) reduces to

$$\ln t_{1/2} = aZK^{-1/2} + \text{constant}. \qquad (13.82)$$

That is, our theory of alpha decay predicts that a plot of $\ln t_{1/2}$ against $K^{-1/2}$ for a set of isotopes should be a straight line. The remarkable success of this prediction can be seen in Fig. 13.22, which shows the data for the eight isotopes of polonium mentioned before. The line shown is the least-squares fit to the data. Its slope is 328, in good agreement with the value predicted by (13.82):

$$\text{slope} = aZ = 3.97 \times 82 = 326.$$

The line's intercept with the vertical axis (not shown in the picture) is at -125, while our predicted value from (13.82) is

$$\text{``constant''} = -b(ZR)^{1/2} + c \approx -120 \qquad (13.83)$$

(if we take $R \approx 7$ fm). Given the rough approximations that went into this value, this agreement is entirely satisfactory. The agreement between theory and experiment is even better if we allow for the small variations of R in the "constant" of (13.83).

If we wish to use the result (13.80) to compute half-lives of various different elements, we must recognize that the three variables K, Z, and R can all vary. However, neither Z nor R varies very much. (For example, $81 \le Z \le 90$ for all residual nuclei in the natural radioactive series.) Thus the single most important effect is still the energy dependence of the term $aZK^{-1/2}$ in $\ln t_{1/2}$.

---

* We must exclude any isotope with other decay modes in addition to alpha emission, since these reduce $t_{1/2}$. We have also excluded isotopes with A odd since these all have nonzero angular momentum, which (it turns out) complicates the analysis and alters $t_{1/2}$ appreciably.

Small differences in K correspond to appreciable differences in this term and hence enormous differences in $t_{1/2}$. In particular, as K ranges from 4 to 9 MeV, this term is the main reason that $t_{1/2}$ drops by some 24 orders of magnitude, as noted at the beginning of this section.

OTHER EXAMPLES OF TUNNELING

We have already noted the similarity of alpha decay and nuclear fission. In particular, for both processes, there is a barrier that prevents the decay from occurring instantaneously. In Section 13.7 we discussed *induced fission,* in which neutrons produce a nucleus with sufficient excitation energy to pass immediately over the top of the fission barrier. It should now be clear that even without any excitation energy, nuclei should undergo **spontaneous fission** when the fission fragments tunnel through the fission barrier.

The probability of spontaneous fission can be calculated using the theory of barrier penetration. If we suppose, for simplicity, that the barriers are rectangular, the escape probability is given by (13.71) as

$$P = e^{-2\alpha L}, \tag{13.84}$$

where

$$\alpha = \sqrt{\frac{2m(U_0 - E)}{\hbar^2}}. \tag{13.85}$$

Comparing the fission barrier of Fig. 13.14 and the alpha barrier of Fig. 13.17, we see that the fission barrier is somewhat lower ($U_0 - E$ is less) and shorter (L is less). Both of these observations would suggest that spontaneous fission would occur more easily than alpha decay. However, there is a third factor working the other way. In (13.85) we see that the parameter α depends on the mass of the emitted particle. Since fission fragments are much heavier than alpha particles, the parameter α is appreciably *larger* for fission than for alpha decay. In all naturally occurring nuclei this effect is dominant, and spontaneous fission is much less probable than alpha decay. Nevertheless, spontaneous fission has been observed in several nuclei, and is actually the dominant decay mode in some artificial nuclei, such as ^{250}Cm and ^{258}No.

Another example of barrier penetration occurs in some electrical wiring. When two wires are joined by simply twisting them together, there is usually a thin layer of oxide separating the two wires. Since many oxides are insulators, this layer forms a barrier, and the flow of current through the junction is only possible because the electrons can tunnel through. In this case the barrier is sufficiently small that the tunneling probability is close to 100% and the barrier has almost no effect on the circuit. Barrier penetration plays an important role in several electronic devices, where a thin layer of insulator is inserted between two conductors to control the flow of current in various useful ways.

13.10   Accelerators (optional)

By far the most fruitful source of information about nuclei is the *scattering,* or *collision, experiment,* in which a beam of energetic projectiles, such as α particles or protons, is directed at a target containing the nuclei to be studied. The projectiles may simply scatter off the target, or they may induce various reactions; but in either case one can infer many properties of the nucleus by studying the results of the collisions. Almost all the great discoveries described in this

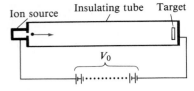

FIGURE 13.23 In a dc accelerator a large potential difference V_0 applied between the ends of an evacuated tube accelerates positive ions from the ion source to the target. If the ion has charge q, this gives it an energy $V_0 q$.

chapter — of the nucleus, of the neutron, of fission, and many more — were collision experiments of this kind. In this section we describe **accelerators,** the machines that accelerate the projectiles to the energies required for interesting collisions. In Section 13.11 we describe the detectors that allow us to identify the particles produced in collisions and to measure their properties.

The earliest collision experiments all used for their projectiles α particles produced by naturally occurring radioactive nuclei. These α particles had energies restricted to a few values in the range from 4 to 9 MeV. By the 1920s it was clear that it would be useful to do collision experiments using particles whose energy could be varied continuously over a larger range, and with other projectiles besides α particles. It was for this reason that several physicists set out to design "accelerators" to accelerate a variety of particles to various energies. Since then many different accelerators have been developed, of which we describe a few. All these machines use electric fields to accelerate the projectiles, which must therefore be electrically charged. The projectiles used include protons, electrons, and the positive and negative ions of many different atoms. The general principles are the same whether the particle's charge is positive or negative; to be definite, we shall usually suppose in what follows that the particle is positive; that is, is a proton or other positive ion.

DC ACCELERATORS

The most obvious way to accelerate a charged particle is to use an electric field applied along the length of a single evacuated tube, as sketched in Fig. 13.23. The field is produced by applying a potential difference between the ends of the tube, and since this potential difference is of a fixed sign, we refer to accelerators of this kind as *dc accelerators.*

While dc accelerators are very simple in principle, there was nevertheless a large practical difficulty in their initial development: To find out anything about nuclear structure the projectile must approach the target nucleus within 1 or 2 fm (and preferably closer). Because of the Coulomb repulsion between the projectile and the nucleus, this requires a large energy; for example, to bring two protons within 1 fm requires nearly 1.5 MeV. Thus the accelerator must give the projectile a kinetic energy of order 1 MeV — or at least several hundred keV — to be of any use.* Now, if the projectile has charge e (as is usually the case) and the accelerating potential is V_0, the kinetic energy given to the projectile is $K = V_0 e$. This means that V_0 needs to be at least several hundred kilovolts, and preferably a megavolt or more. In the 1920s there was no known way to produce such large dc voltages. Thus before a useful accelerator could be built, a way had to be found to generate these huge voltages.

In the quest for high voltages, one group went so far as to stretch a cable between two mountaintops to exploit the high potentials found in thunderstorms. A spherical terminal was suspended from the cable, and sparks several hundred feet long jumped from the terminal to the valley floor during thunderstorms. Plans were made to install an accelerating tube, but were abandoned when one of the investigators was killed by lightning.

* We are here assuming that the projectile is positive and is therefore *repelled* by the target nucleus. One might think that one could get around this large energy requirement by using a negative projectile, but this is not so: The only possibility is an electron, and, because of its low mass, a low-energy electron has a wavelength far bigger than a nucleus. This makes low-energy electrons almost useless as probes of nuclei. In fact, an electron needs an energy of hundreds of MeV before its wavelength is comparable to the nuclear size.

The first successful high-voltage accelerator was produced by the English physicists Cockroft and Walton, who used ac transformers and rectifiers to put a steady voltage across a bank of capacitors. In this way they accelerated protons to 500 keV and, in 1932, observed the first purely man-made nuclear reaction, described in Section 13.6. Their accelerator is shown in Fig. 13.24. Modern versions of the Cockroft–Walton accelerator can achieve about 1 million volts and are widely used to inject protons into higher-energy accelerators such as the linear accelerators described later.

Another successful source of high voltages was the electrostatic generator designed by the American physicist van de Graaff in the early 1930s. In this machine, the high-voltage terminal is a large, hollow conductor, usually spherical or oval in shape. A motor-driven belt carries electrostatic charges to the inside of this terminal, where they are deposited and immediately move to the outside surface. In this way a large total charge can be placed on the terminal and produces a correspondingly high voltage. An early version of an accelerator using a van de Graaff generator is shown in Fig. 13.25. Modern versions enclose the entire device in a tank of high-pressure gas to inhibit loss of charge by sparking; in this way they can achieve voltages of about 10 MV and are used to accelerate many different particles, ranging from electrons to heavy ions such as uranium.

FIGURE 13.24 The accelerating tube of Cockroft and Walton's proton accelerator can be seen above the small enclosure where the α particles from the reaction $p + {}^7Li \rightarrow \alpha + \alpha$ were observed on a scintillating screen. The observer in this photograph is Cockroft himself. (Courtesy Cavendish Lab.)

AC ACCELERATORS

A dc accelerator accelerates particles to high energy in a single large boost, produced by a correspondingly large voltage. An obvious alternative is to use a succession of smaller boosts. This removes the need for such enormous voltages and, if one uses *many* small boosts, lets one achieve a higher final energy than is possible with any dc accelerator. Most accelerators in use today are of this multiple-boost type. Since they all use an alternating current to produce the accelerating fields, we refer to them as *ac accelerators.* (Since the frequency of the alternating voltage is usually in the range 10 to 200 MHz, it is a radio frequency and is often abbreviated RF.) Here we describe just two important types of ac accelerator, the linear accelerator and the cyclotron. We confine ourselves mostly to machines in which relativistic effects are not too large — proton energies up to a few hundred MeV, for example. A few of the more recent accelerators that produce particles with thousands and even millions of MeV are described in Section 14.11.

THE LINEAR ACCELERATOR

In the **linear accelerator,** or **linac,** each particle travels through a succession of many conducting cylinders, mounted in a single straight line (whence the name *linear* accelerator). Inside each cylinder, or *drift tube,* there is no electric field and the particle feels no force. However, the potentials of the tubes oscillate in such a way that whenever the particle reaches a gap between two tubes there is an accelerating electric field in the gap. This field boosts the particle from one tube into the next, giving it an additional energy of $V_0 q$, where V_0 is the potential difference across the gap. If the linac has N gaps in all, the particle gains an energy of $NV_0 q$ in its passage through the whole accelerator. In most linacs the particle is given a preliminary boost (by a Cockroft–Walton accelerator, for example), so that it enters the first tube with kinetic energy K_0 and leaves the last with $K = K_0 + NV_0 q$.

There are various ways to arrange for the required accelerating voltages in

FIGURE 13.25 An early van de Graaff accelerator. The belt that carried charges into the egg-shaped terminal is visible at the upper right; the accelerating tube is directly below the terminal. The rings around the accelerating tube, connected by high resistances, help to ensure a uniform field, which reduces sparking. (Courtesy Carnegie Institution.)

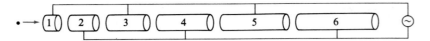

FIGURE 13.26 An early type of linear accelerator, or linac, for protons or other positive ions. The ions are injected from the left, usually after being given a preliminary boost in a smaller accelerator. They then receive a boost, of energy $V_0 q$, when they pass each of the gaps between successive tubes. The tubes get progressively longer to ensure that the accelerating ions always arrive at a gap at the moment of maximum potential drop.

a linac. One simple method, used in some early machines, is sketched in Fig. 13.26, which shows a linac with just six cylindrical electrodes, and hence five accelerating gaps. An ac supply is connected to the six tubes, one side to tubes 1, 3, and 5, the other to tubes 2, 4, and 6. Thus the potential difference between successive tubes oscillates between two values $\pm V_0$, tubes 1, 3, 5 being at maximum potential when tubes 2, 4, 6 are at minimum, and vice versa. Consider now a group of positive ions injected into tube 1 so that they arrive at the first gap just when tube 1 is at maximum potential (and, hence, tube 2 at minimum). These ions are accelerated from tube 1 to tube 2 by the potential difference V_0, and each gains an energy $V_0 q$.

Next, the length of tube 2 is chosen so that the time for the ions to traverse tube 2 is exactly one half the period of the ac supply.* Thus, by the time the ions reach the gap between tubes 2 and 3, tube 2 is at maximum potential, and tube 3 at minimum. Therefore, the ions receive a second boost of energy $V_0 q$.

Each successive tube is designed in the same way, so that the ions take a half period to traverse the tube and hence arrive at the next gap at the moment of maximum potential drop. Since the ions are accelerating, this requires each tube to be somewhat longer than its predecessor. A typical frequency for the ac supply is of order 100 MHz, for which a half period is 5×10^{-9} s. In this time a proton with about 100 MeV, and hence $v \approx 0.5c$, travels nearly 1 m. Therefore, linear accelerators, sometimes with hundreds of tubes, can be very large machines. The 800-MeV proton linac at Los Alamos, shown in Fig. 13.27, is about half a mile long; the 50-GeV electron linac at Stanford is 2 miles long.

THE CYCLOTRON

To reach useful energies a linac must have many accelerating gaps. In 1929 it occurred to the American physicist Lawrence to use a magnetic field to make the accelerating particles move in circles. In this way the particles could be brought back repeatedly to the *same* accelerating gap.

Lawrence's accelerator, which came to be called the **cyclotron,** is shown schematically in Fig. 13.28. The magnetic field is produced by a magnet with two large circular pole faces. This can be seen in the side view of Fig. 13.28(a), with its uniform magnetic field pointing straight up between the poles. Also visible in the side view are the grounded vacuum chamber within which the ions are accelerated, and an electrode, called a "dee," which accelerates the ions.†

The dee is more clearly visible (and the origin of its name very apparent)

---

* Modern linacs work somewhat differently: In some, the accelerating field is produced by a standing wave in the tank containing the drift tubes, which are all grounded. The ions are inside the tubes, and hence shielded from the field, during the half-cycle when the field points the wrong way; they are in the gaps, and exposed to the field, during the half-cycle when it points the right way. Many high-energy linacs use a traveling electromagnetic wave (Section 14.11).

† Some cyclotrons have two or more electrodes to give a more rapid energy gain. For simplicity, we describe a single-electrode machine.

in the overhead view of Fig. 13.28(b); it is a hollow semicircular conductor, inside which the ions spend half the period of their acceleration. The ions are produced in an electric discharge, close to the center of the dee; the discharge is fed by a gas whose nuclei are to be accelerated (hydrogen for protons, deuterium for deuterons, helium for α particles, and so on). The dee is connected to an ac voltage of amplitude V_0, arranged so that the dee is at $-V_0$ and attracts the ions as they enter the dee, but is at $+V_0$ and repels them as they leave it. Thus the ions gain an energy $V_0 q$ each time they enter the dee and again each time they leave it. Meanwhile the magnetic field forces the ions to follow a circular path with radius [Eq. (3.48)]

$$R = \frac{p}{qB} = \frac{mv}{qB}. \qquad (13.86)$$

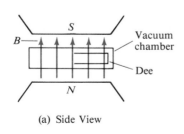

(a) Side View

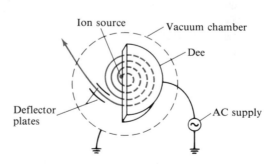

(b) Overhead View

FIGURE **13.28** The cyclotron. **(a)** Side view, showing the poles of the large magnet, the vacuum chamber in which the ions move, and the accelerating electrode, called the "dee." **(b)** Overhead view. The ions are produced at the center and gradually spiral outward. They gain an energy $V_0 q$ each time they enter the dee and again each time they leave it.

ERNEST LAWRENCE (1901–1958, US). Lawrence, who spent almost all of his career at Berkeley, invented the cyclotron, one of the most successful machines for accelerating subatomic particles to the energies needed to probe the atomic nucleus. This invention, which earned him the 1939 Nobel Prize, was the precursor of the modern synchrotron, which can accelerate protons to energies of a million MeV.

(For the moment we assume the ions move nonrelativistically, so that $p = mv$.) Each time the ions speed up, this radius increases, so the ions spiral steadily outward from the ion source, as shown in Fig. 13.28. Eventually, when they approach the outer edge of the B field, a pair of charged plates deflects them outward to extract them from the machine. The photograph in Fig. 13.29 shows the dee of a cyclotron at the University of Colorado.

The frequency f with which the ions orbit is easily found from (13.86):

$$f = \frac{v}{2\pi R} = \frac{qB}{2\pi m}. \tag{13.87}$$

The beautiful feature of this result — first appreciated by Lawrence — is that it is independent of v: With each boost v increases, but R also increases in direct proportion, so that the time for each orbit stays the same. Therefore, if we fix the frequency of the ac supply at the value of the *cyclotron frequency* (13.87), the ions will continue to enter and leave the dee in time with the maximum accelerating field, throughout the acceleration process. That a cyclotron can be driven by an ac supply of fixed frequency was a major simplification and made it a practical accelerator used in many nuclear laboratories around the world.

EXAMPLE 13.6 A cyclotron accelerates protons to 10 MeV (and hence $v = 4.38 \times 10^7$ m/s) with a maximum orbital radius of 40 cm. (a) Find the strength B of the magnetic field. (b) What must be the frequency of the ac supply? (c) If the amplitude of the accelerating voltage is 50 kV, how many complete orbits are required to reach full energy?

(a) The maximum speed ($v = 4.38 \times 10^7$ m/s) is reached at the maximum radius ($R = 40$ cm). Thus, from (13.86),

$$B = \frac{mv}{qR} = \frac{(1.67 \times 10^{-27}\text{ kg}) \times (4.38 \times 10^7\text{ m/s})}{(1.60 \times 10^{-19}\text{ C}) \times (0.40\text{ m})} = 1.14\text{ T}.$$

(b) The frequency of the ac supply must equal the cyclotron frequency (13.87):

$$f = \frac{qB}{2\pi m} = \frac{(1.6 \times 10^{-19}\text{ C}) \times (1.14\text{ T})}{2\pi \times (1.67 \times 10^{-27}\text{ kg})} = 17.4\text{ MHz}.$$

(c) A proton gains 50 keV each time it enters and each time it leaves the dee. Thus, in each complete orbit it gains 100 keV, and the total number of orbits is

$$N = \frac{10\text{ MeV}}{100\text{ keV}} = 100\text{ orbits}.$$

In the simple form described here, the cyclotron depends crucially on the result (13.87) that the orbital frequency remains constant as the ions speed up. Unfortunately this remains true only as long as the ions move nonrelativistically. Once they become relativistic, we must include a factor of γ in the expression (13.86) for R (since $p = \gamma mv$), and the frequency (13.87) becomes $f = qB/(2\pi m\gamma)$, which decreases like $1/\gamma$. As long as γ remains close to 1, this is a

FIGURE **13.29** The dee of the Colorado cyclotron was photographed while the machine was opened for maintainance. When in operation, the ion source was about 1 cm from the small metal block at the center of the dee, and the ions began their first orbit through the hole in this block.

small effect that can be accommodated (for example, by arranging that B increases gradually with R), but for energies above several hundred MeV, new and more elaborate methods are needed, as we describe in Section 14.11.

13.11 Particle Detectors (optional)

Most nuclear measurements involve the detection of particles—particles ejected from radioactive nuclei, particles produced from accelerators to probe nuclei, and particles created in nuclear reactions. In addition to detecting these particles, one must usually measure some of their properties—their mass, charge, energy, momentum and so on. In the business of detecting and measuring particles there is a sharp distinction between charged and neutral particles. When a charged particle passes through matter (solid, liquid, or gas) it can ionize or excite many of the atoms which it passes; that is, it can eject electrons or raise them to excited states. This ionization or excitation is easily detected and is the basis for most detectors of charged particles. Neutral particles, such as the photon and neutron, are usually not as easily detected, and most detectors of neutral particles work by having the neutral produce a charged particle (as when a photon produces an electron in the photoelectric effect) and then detecting the charged particle.

PHOTOGRAPHIC FILM

The discovery of radioactivity by Becquerel involved the darkening of photographic film by nuclear particles. Of course, the original purpose of photographic film was to detect visible photons and hence to record pictures. The essential component of photographic film is a silver halide* in the form of many small grains. When a photon passes through the film, it can knock an electron

* Silver bromide, silver chloride, or silver iodide.

FIGURE 13.30 A film badge used to monitor exposure to radiation.

out of an atom in its path. This free electron initiates a sequence of events leading to the precipitation of silver atoms when the film is developed. This leaves the film darkened at those points where many photons passed through and produces a negative image of the original picture.

In much the same way, photographic film is sensitive to other kinds of photons, and it is, for example, widely used to detect X-ray photons for medical purposes. In the same way, also, film is sensitive to charged particles. Any charged particle passing through a film produces many free electrons, which cause the deposit of silver atoms and darken the film. This was the basis of Becquerel's discovery in 1896, and since that time photographic film has been widely used to detect nuclear and subnuclear particles. In Chapter 12 (Fig. 12.18) we saw that it was used in early mass spectrometers to detect the arrival of ions. In Chapter 14 (Fig. 14.4) we show the paths of the nuclear particles called pions and muons recorded in photographic films. Photographic film can even be used to detect neutrons, since the neutrons can knock protons out of water molecules in the film, and the protons then darken the film in the usual way.

For all its versatility as a particle detector, photographic film has largely been replaced by various other kinds of detectors, several of which we describe below. One important application of photographic film that *is* still in wide use today is in the *film badge* worn by workers who may be exposed to nuclear radiation (Fig. 13.30). Film in these badges is removed periodically and developed to determine whether workers have been exposed to radiation. The darkness of the film is measured to monitor any exposure and to ensure that safety regulations are being followed properly.

CLOUD AND BUBBLE CHAMBERS

Photographic film can produce a visible track of a nuclear particle, even though the particle itself is totally invisible. There are various other detectors that can do the same thing and are, in some ways, more convenient than photographic film. The first such was the **cloud chamber,** invented by the Scottish physicist C. T. R. Wilson in about 1910. This is a chamber containing air that is supersaturated with water vapor. When a charged particle passes through the chamber, water droplets condense around the ions produced by the particle. These droplets form a visible trail, which can be photographed for later study. If the particle happens to collide with a nucleus in the chamber, any charged particles produced in the resulting nuclear reaction will also produce visible trails. Thus many details of the reaction are revealed in a form that is easy to study. This was illustrated in Fig. 13.10, which is a photograph of the reaction $\alpha + {}^{14}\text{N} \rightarrow \text{p} + {}^{17}\text{O}$ in a cloud chamber.

A modern variant of the cloud chamber is the **bubble chamber,** which contains superheated liquid hydrogen. The ions formed by any passing charged particle trigger boiling in the nearby liquid, producing tiny bubbles that give a visible record of the particle's path. One of the advantages of bubble chambers is that the density of nuclei in the liquid of the bubble chamber is some 1000 times that in the gas of a cloud chamber. Thus the probability that a passing particle will produce an interesting reaction in a bubble chamber is much higher than in a cloud chamber. Photographs of reactions in bubble chambers are reproduced in Figs. 14.5, 14.6, and 14.11.

If a magnetic field is applied to a cloud or bubble chamber, the path of any charged particle will be curved, and the particle's momentum can be found by measuring the radius of curvature [see Eq. (3.48)]. Further, the thickness of the trail left by a particle depends on its charge and mass, and under favorable conditions, the particle can be completely identified by its track.

ION CHAMBERS

The ion chamber is a detector that makes direct use of the ionization produced by any charged particle. Figure 13.31 shows a simple ion chamber that contains two plates maintained at different potentials by a battery. Any charged particle passing through the air, or other gas, between the plates leaves a trail of electrons and positive ions. The electrons are drawn to the positive plate (or anode), and the positive ions are drawn to the negative plate (or cathode). For each electron that arrives at the anode, one negative charge flows around the circuit and neutralizes one positive charge at the cathode.

Although the current produced by the passage of a single charged particle is very small, it can be measured if amplified electronically. Thus the ion chamber can register the passage of individual charged particles, each of which produces a separate measurable pulse of current.

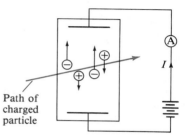

FIGURE 13.31 Schematic sketch of an ion chamber. A charged particle passing through the gas in the chamber produces a trail of electrons and positive ions, which are drawn to the plates and cause a measurable current I in the circuit.

One convenient variant of the ion chamber is the *Geiger counter,* in which the two plates in Fig. 13.31 are replaced by a conducting cylinder and a wire running down its axis. The potential difference produces a large electric field near the narrow wire. When a charged particle ionizes an atom, some of the resulting charges are drawn near to the wire, where they accelerate rapidly and ionize more atoms. This chain reaction quickly produces a current that is large enough to detect without expensive amplifiers and can, for example, be made to cause an audible click in a small loudspeaker. Thus the Geiger counter is an economical, and often portable, detector of nuclear radiation.

In addition to detecting particles, an ion chamber can measure their energies. When a charged particle ionizes an atom, it loses an energy that has, on average, the same value whatever the particle's speed. Therefore, the total energy that the particle loses in the chamber is proportional to the number of atoms ionized, and hence to the total charge that flows in the circuit. In particular, if the chamber is large enough, the particle will come to rest, and, by measuring the charge flow, one can find the total kinetic energy with which the particle entered the chamber.

EXAMPLE 13.7 Alpha particles ejected in the radioactive decay of ^{235}U are detected in an ion chamber. Each atom ionized costs an alpha particle about 30 eV of energy. Given that the alphas were ejected with about 4.4 MeV and assuming that they come to rest in the chamber, find the total charge that flows in the circuit when one particle is detected.

The total number of atoms ionized is just the incident energy of the α particle divided by the energy lost in each ionization:

$$N = \frac{4.4 \times 10^6 \text{ eV}}{30 \text{ eV}} = 1.5 \times 10^5.$$

The total charge flowing around the circuit is therefore

$$Q = Ne = 2.4 \times 10^{-14} \text{ C}.$$

With modern electronic amplifiers one can measure such charges with an accuracy of about 10^{-17} C. Thus one can measure Q, and hence the particle's original energy, with an accuracy of about 1 in 1000; that is, the kinetic energy can be measured within a few keV.

To measure a particle's energy with an ion chamber, one must be sure that the particle stops before it can leave the chamber. The higher a particle's energy, the farther it can travel before coming to rest, and for particles with many MeV this requires the gas-filled chamber to be inconveniently large. One way around this difficulty is to replace the gas with a solid: Because the densities of solids are some 1000 times greater than those of gases, the particle meets 1000 times more atoms (in a given distance) and therefore loses all its energy in one thousandth of the distance. We next describe two detectors that use a solid medium: solid-state and scintillation detectors.

SOLID-STATE DETECTORS

One cannot simply replace the gas of an ion chamber by any solid. If the solid is an insulator, the charges produced by ionization cannot flow to the collecting plates; if the solid is a conductor, a current will flow all the time, making it difficult to detect the small extra current caused by a passing particle. There are, however, certain materials called semiconductors that can be arranged to act as insulators *except* when a charged particle passes through them. (We describe semiconductors in Section 17.6.) By placing a suitable semiconductor between two collecting plates, one can make a *solid-state detector* that acts much like a gas-filled ion chamber, but can stop a high-energy particle — and hence measure its energy — in a much smaller volume.

SCINTILLATION DETECTORS

When charged particles pass through matter they not only ionize atoms; they also elevate atoms to excited states. These excited atoms then give off light as they fall back to the ground state, and this light is exploited in the *scintillation detector*. One of the earliest such detectors was the zinc sulfide screen used by Rutherford in many of his experiments with α particles. Each time an α hit the screen, the tiny flash of light that it produced was observed and recorded by the experimenter — a tedious and tiring job, which could only be done in a totally dark room.

Today, the light from a scintillation detector is monitored automatically. A photoelectric cell converts the light into an electric pulse, which, amplified by a *photomultiplier* if necessary, can be fed directly into a computer for processing and recording.

Most modern scintillation detectors use materials, such as NaI crystals and certain plastics, that are transparent to the light which they produce. This lets one use a block of material that is thick enough to stop the particles and hence to measure their energy. Some detectors use a liquid scintillator, which allows a radioactive sample to be placed right inside the scintillator, to improve the chance of detecting weak or low-energy signals.

WIRE CHAMBERS

Ion chambers, solid-state detectors, and scintillation detectors all produce electric signals that can be fed directly to a computer. In addition to detecting a particle, they can also measure its energy, *provided* that the particle comes to rest in the detector. However, many modern experiments use particles with thousands and even millions of MeV, and such particles cannot be stopped in any detector of reasonable size. Instead, their energy is usually found by measuring the curvature of their path in a magnetic field, and this requires a detector that records the path of the particle. We have already described three such detectors — photographic films, cloud chambers, and bubble chambers — but all of them record their information in the form of a picture, which has to be interpreted and measured by a human experimenter. The *wire chamber* is a detector that came into use in the 1970s and records a particle's trajectory directly in the form of electrical pulses.

In a wire chamber, many closely spaced, parallel wires are positively charged to attract the electrons produced by a fast ionizing particle. The wire closest to the path of the particle captures most of the electrons and produces a signal to record this point on the path. A second grid with its wires perpendicular to the first allows both x and y coordinates of the path to be measured. Several such pairs of wire grids determine the particle's entire trajectory through a magnetic field, and from this its momentum can be calculated.

NEUTRAL PARTICLES

When a charged particle such as a proton passes through matter, its electric field pulls on the electrons in the atoms along its path and leaves a trail of electrons, ions, and excited atoms. This is the basis for all the detectors described in this section. Neutral particles do not produce trails of ionization and excitation, but they can occasionally collide with atoms and produce moving charged particles. These charged particles can be detected in the usual way. To detect neutrons, for example, hydrogen (in one form or another) is introduced into a detector; some of the neutrons collide with the hydrogen nuclei (protons) and the recoiling protons are detected as usual. Photons are detected by three main mechanisms, all of which produce moving electrons: (1) the Compton effect, in which the photon scatters off an atomic electron, causing it to recoil; (2) the photoelectric effect, in which a photon is absorbed and gives its energy to an electron; and (3) pair production, in which a photon's energy creates an electron–positron pair. A cloud-chamber photograph of the latter process is reproduced in Fig. 14.1, which shows clearly the absence of any visible track for the photon, but two clear tracks for the electron and positron.

13.12 Units of Radiation

There are various units used to measure amounts of radiation. These units are especially important in radiology, the study of the effects of radiation on biological systems. Radiation has been used successfully to treat several diseases, particularly cancer, and radioactive "tracers" have proved invaluable in following the paths of biological processes. However, excessive exposure to radiation can cause serious illness and even death. Therefore, safety regulations are needed for radiologists and other people exposed to radiation. To formulate these regulations, one must have units with which to measure the radiation and its effects.

The various units of radiation fall into two main classes: (1) units of *source activity,* which characterize the radiation *produced* by a radioactive source, and (2) units of *absorbed radiation,* which measure the *effects* of radiation on substances exposed to it. As in most areas of physics, the units defined by early workers in the field have been replaced by internationally accepted, SI, units. However, because the older units are still widely used, we describe both the SI units and their older counterparts.

UNITS OF SOURCE ACTIVITY

The SI unit that measures the amount of radioactivity in a source is the **becquerel,** or Bq:

> 1 Bq of a radioactive substance is that amount which produces 1 decay per second.

As an example of the use of this unit, consider the naturally occurring radioactive gas radon 222, whose decay constant is $r = 2.1 \times 10^{-6}$ s^{-1}. To get one decay per second, we would need

$$N = \frac{1}{2.1 \times 10^{-6}} = 4.8 \times 10^5$$

atoms of radon. Therefore, 1 Bq of radon consists of 4.8×10^5 atoms and has a mass

$$\text{(mass of 1 Bq of radon)} = (4.8 \times 10^5) \times (222 \text{ u}) = 1.8 \times 10^{-19} \text{ kg.}$$

Recent studies of the radon that is released by natural radioactivity in the ground and seeps into the basements of houses suggest that in the average U.S. home, the level of radon is about 50 Bq/m^3. This is disturbingly close to the level of 150 Bq/m^3 above which the U.S. Environmental Protection Agency recommends remedial action because of increased risk of cancer.

The earlier unit of activity, still in wide use today, was the *curie,* or Ci, which is now defined in terms of the Bq as

$$1 \text{ Ci} = 3.7 \times 10^{10} \text{ Bq} = 3.7 \times 10^{10} \text{ decays/second.}$$

The origin of this definition is that 1 Ci is, at least approximately, the activity of 1 gram of radium 226.

UNITS OF ABSORBED RADIATION

The becquerel measures the strength of a radioactive source, but the biological effect of radiation depends more on the amount, or *dose,* that is absorbed by a living tissue. The **absorbed dose** resulting from any exposure to radiation is defined as the energy absorbed per unit mass of the exposed tissue. The SI unit for absorbed dose is the **gray,**

$$1 \text{ Gy} = 1 \text{ gray} = 1 \text{ J/kg;}$$

that is, if the tissue absorbs energy at a rate of 1 joule per kilogram, we say that it

has absorbed a dose of 1 gray. The older unit for absorbed dose is the *rad**
(radiation absorbed dose), defined as

$$1 \text{ rad} = 0.01 \text{ Gy} = 0.01 \text{ J/kg}.$$

The damage caused to living tissue by a given dose of radiation depends
on the nature and energy of the radiation. For example, exposure to neutrons
with energies of a few hundred keV can cause about 10 times the damage
produced by the same dose of X rays. The **relative biological effectiveness** or
RBE of any radiation is defined by comparing its effects with those of 200-keV
X rays. Specifically, the RBE of a given type of radiation is defined as the dose of
200-keV X rays that would produce the same biological damage as a unit dose
of the given radiation. Table 13.2 shows approximate values for the RBE (also
called the quality factor, QF) for some types of radiation.

The biological damage caused by an exposure to radiation is determined
by the product of the absorbed dose and the RBE of the radiation concerned.
This product is called the **effective dose:**

$$\text{effective dose} = (\text{absorbed dose}) \times \text{RBE}.$$

The SI unit for effective dose is called the **sievert** or Sv; thus

$$(\text{effective dose in Sv}) = (\text{absorbed dose in Gy}) \times \text{RBE}.$$

With this definition, 1 sievert of one kind of radiation causes roughly the same
damage as 1 sievert of any other kind. The corresponding older unit is the **rem**
(rad equivalent in man), so that

$$(\text{effective dose in rem}) = (\text{absorbed dose in rad}) \times \text{RBE}.$$

Most normal exposures to radiation result in effective doses that are
conveniently measured in millisieverts (mSv). The effective dose resulting from
a single medical X ray is of order 0.1 mSv. The annual effective dose from
cosmic rays and other natural sources of radiation is of order 1 mSv for the
average person living at sea level. (In the Rocky Mountain states, this number is
nearer to 2 mSv because of the higher intensity of cosmic rays and the greater
concentration of natural radioactivity in the ground.) These figures for natural
exposure do not include the effects of breathing natural radon gas into the lungs;
it has recently been realized that this may well contribute an annual dose of
another 2 or 3 mSv.

Catastrophic exposures are typically thousands of mSv. An effective dose
of 5000 mSv to the whole body at one time is fatal in about 50% of cases. A
single exposure above 1000 mSv is known to increase a person's chance of fatal
cancer by about 1.5×10^{-5} per mSv of exposure. In the setting of safety standards, it is generally assumed that this same rate applies at low levels of radiation, although there is substantial evidence that at lower doses the rate is actually less.†

The main units of radiation are summarized in Table 13.3.

TABLE 13.2
The relative biological
effectiveness, or RBE, of some
types of radiation.

Type of Radiation	RBE
Photons (200-keV X ray)	exactly 1
Photons (4-MeV γ ray)	0.6
Electrons	about 1
Neutrons	2 to 10
Protons	about 10
Alpha particles	up to 20

---

* An earlier unit was the *roentgen,* which was about 0.87 rad. This was the dose that
produced 1 electrostatic unit of ions (3.3×10^{-10} C) in 1 cm³ of air.

† Report of the Committee on the Biological Effects of Ionizing Radiation, National Acad-
emy of Sciences, Washington, DC (1988).

TABLE 13.3
Principal units of radiation.

Quantity	SI unit	Traditional unit
Source activity	Bq (becquerel) = 1 decay/second	Ci (curie) = 3.7×10^{10} Bq
Absorbed dose	Gy (gray) = 1 J/kg	Rad (radiation absorbed dose) = 10^{-2} Gy
Relative biological effectiveness	RBE	RBE
Effective dose [(eff dose) = (abs dose) × RBE]	Sv (sievert) [(eff dose in Sv) = (abs dose in Gy) × RBE]	Rem (rad equivalent in man) = 10^{-2} Sv [(eff dose in rem) = (abs dose on rad) × RBE]

IDEAS YOU SHOULD NOW UNDERSTAND FROM CHAPTER 13

Radioactivity; α, β, γ
Exponential decay:
    decay constant r
    mean life $\tau = 1/r$
    half-life $t_{1/2} = \tau \ln 2$
Radioactive dating
$\beta^{\pm}$ decay and electron capture
Neutrinos
The natural radioactive series
The valley of stability

Nuclear reactions:
    conservation of nucleon number
    conservation of charge
    cross sections
Neutron-induced fission
Chain reaction
Nuclear fusion
Fusion in stars
[Optional sections: barrier penetration or tunneling;
    accelerators (dc, linear, cyclotron); particle de-
    tectors (photo film, cloud and bubble chambers,
    ionization chambers, scintillators, wire cham-
    bers); units of radiation]

PROBLEMS FOR CHAPTER 13

SECTION 13.3 (THE EXPONENTIAL DECAY LAW)

13.1 • The decay constant of ^{210}Rn is $r = 0.29$ hour^{-1}. If a sample of ^{210}Rn contains 10^{12} atoms at $t = 0$, how many will remain: (a) after 1 hour, (b) after 1 day, (c) after 2 days?

13.2 • The decay constant for a certain nucleus is $r = 0.01$ min^{-1}. At $t = 0$ a sample contains 10^6 of these nuclei. (a) How many of these nuclei will remain after an hour? (b) What is the total rate of decays, $R(t)$, at $t = 0$? (c) What is $R(t)$ after an hour?

13.3 • (a) Use the data in Appendix D to find the decay constant and mean life of uranium 232. (b) Do the same for polonium 214.

13.4 • The isotope cobalt 60 has a decay constant $r = 4.2 \times 10^{-9}$ s^{-1}. Consider a pure sample of 1 μg of ^{60}Co. (a) How many atoms are in this sample? (b) How many ^{60}Co atoms will remain after one year? (c) What is the half-life of ^{60}Co?

13.5 • (a) Given that ^{238}U has a half-life of 4.47×10^9 years, find its decay constant r. (b) How many atoms are there in 1 g of pure ^{238}U? (c) How many decays are there per second in 1 g of ^{238}U?

13.6 • The nucleus of polonium 208 has a decay constant $r = 0.239$ year^{-1}. (a) What is the half-life of ^{208}Po? (b) How many half-lives are there in 9 years? (c) If a sample of ^{208}Po contains 8000 atoms today, how many will remain in 9 years' time?

13.7 • From the exponential decay law (13.6) prove that

$$\frac{N(t + \Delta t)}{N(t)} = e^{-r\Delta t};$$

that is, the fractional decrease in N during any time interval from t to $t + \Delta t$ depends on Δt, but not on t itself.

13.8 • • The decay constant of a certain nucleus is 0.05 s^{-1}. At $t = 0$ a sample contains 6×10^{15} nuclei. (a) How many nuclei remain at $t = 10$ s? (b) How many remain at $t = 60$ s and at 70 s? (c) Evaluate the ratios N_{10}/N_0 and N_{70}/N_{60} and explain why they are equal.

13.9 • • Prove that the exponential decay law $N(t) = N_0 e^{-rt}$ can be rewritten as $N(t) = N_0 2^{-t/t_{1/2}}$, where $t_{1/2} = (\ln 2)/r$. (Hint: $e^{\ln 2} = 2$.)

13.10 • • In Section 13.3 we proved that the number $N(t)$ of radioactive nuclei surviving after a time t satisfies the equation $dN/dt = -rN$, where r is the decay constant of the nuclear species in question. Rewrite this equation as $dN/N = -r\, dt$ and by integrating prove that $N = N_0 \exp(-rt)$, where N_0 is any constant.

13.11 • • Half of the nuclei in a sample decay before the time $t = t_{1/2}$, and the other half after; that is, $t_{1/2}$ is the *median* life. On the other hand, $t_{1/2}$ is *not* the *mean* (or average) life. To find the mean life, note that the number of nuclei that decay between times t and $t + dt$ is $R(t)\, dt$. Thus the fraction of the original N_0 nuclei that live for a time between t and $t + dt$ is $R(t)\, dt/N_0$. If you multiply this fraction by t and integrate over all possible values of t, you will obtain (by definition of the average) the average life t_{av}:

$$t_{av} = \int_0^\infty \frac{tR(t)}{N_0}\, dt.$$

Evaluate t_{av} in terms of the decay constant r. How is t_{av} related to the lifetime τ defined in Eq. (13.7)?

13.12 • • At $t = 0$ a sample contains $N_0 = 64,000$ radioactive nuclei all of the same species. If the decay constant is $r = 0.01$ s^{-1}, how many of the nuclei remain after (a) 2 min and (b) 20 min? (c) Is your answer in part (b) the actual number of nuclei you would expect to find after 20 min? Explain. (Remember that a nucleus is a quantum system, and quantum mechanics deals in probabilities.)

13.13 • • A radioactive nucleus A has decay constant r and decays to a stable offspring nucleus B. At $t = 0$, a pure sample of A contains N_0 atoms. Derive an expression for the number $N_B(t)$ of offspring nuclei in the sample at time $t \geq 0$. Sketch $N_B(t)$ as a function of t.

13.14 • • A radioactive nucleus A has decay constant r_A. It decays to an offspring nucleus B, which decays much more rapidly, with decay constant $r_B \gg r_A$. The population of A decreases exponentially as usual [$N_A(t) = N_0 \exp(-r_A t)$]. Because B decays so rapidly, each B decays almost as soon as it is produced. Therefore, the rate of production of the offspring B (by decay of the parents A) is approximately equal to the rate of loss (by decay of the B nuclei themselves). Use this condition to find an expression for $N_B(t)$ in terms of N_0, r_A, r_B, and t. Show in particular that N_B remains proportional to N_A.

13.15 • • The carbon in living organisms contains about 1.3 atoms of ^{14}C for every 10^{12} atoms of all kinds of carbon, and the half-life of ^{14}C is 5730 years. (a) How many decays would one expect per minute from 7 g of carbon taken from a living tree? (b) If 7 g of carbon is extracted from a wooden beam of a prehistoric dwelling, and if they make 10 decays per minute, about how old is the dwelling?

13.16 • • (a) At what rate would 2 g of carbon from a living organism eject electrons? (b) If archeologists extract 2 g of carbon from a dog's skull and count 21 electrons ejected per minute, how long ago did the dog die?

13.17 • • • The following data were obtained from a Geiger counter placed beside a sample of a single radioactive species. (a) Make a plot of the counting rate $R(t)$ against t, including error bars. (*Note:* The data are the actual number of counts made in intervals of 1 minute. The uncertainty in numbers of this type is approximately the square root of the number.) Sketch an appropriate curve to fit the data and use it to estimate the mean life τ. (b) Do all the same exercises using a graph of $\ln R(t)$ against t. Why is the procedure of part (b) superior?

Time (hours)	Counts in 1 min	Time (hours)	Counts in 1 min
0	99	6	15
1	78	7	15
2	64	8	7
3	36	9	3
4	21	10	6
5	22	11	4

SECTION **13.4** (BETA DECAY AND THE NEUTRINO)

13.18 • Equation (13.22), which gives the kinetic energy released in β decay, was derived on the assumption that there are only two final particles (the electron and the residual nucleus, but no neutrino). Prove, nevertheless, that (13.22) correctly predicts the total kinetic energy available to all three final particles, given that the neutrino is massless.

13.19 • Use the data in Appendix D to find the total kinetic energy released in the β^- decay of ^{14}C.

13.20 • Use the data in Appendix D to find the total kinetic energy released in the β^- decay of ^{208}Tl.

13.21 • Calculate the total kinetic energy released in the β decay of a single free neutron.

13.22 • • Write down an equation [similar to Eq. (13.25)] to describe the β decay of each of the following nuclei: (a) ^{60}Co, (b) ^{3}H (tritium), (c) ^{15}O, and (d) the decay by electron capture of ^{231}U.

13.23 • • The total kinetic energy, K, released in the β^- decay $(Z, N) \rightarrow (Z + 1, N - 1) + e^- + \bar{\nu}$ is given by (13.22). In particular the criterion that β^- decay is possible is just that K must be positive, or

$$m_{atom}(Z, N) > m_{atom}(Z + 1, N - 1) \qquad \text{(for } \beta^- \text{ decay).}$$

In other words, one has only to compare the atomic masses involved to see if β^- decay is possible. (a) Find an expression for the total kinetic energy released in the β^+ decay

$$(Z, N) \rightarrow (Z - 1, N + 1) + e^+ + \nu,$$

and show that the condition for this process to be possible is

$$m_{atom}(Z, N) > m_{atom}(Z - 1, N + 1) + 2m_e$$
$$\text{(for } \beta^+ \text{ decay).} \qquad (13.88)$$

(b) Do the same for the electron capture

$$(Z, N) + e^- \rightarrow (Z - 1, N + 1) + \nu, \qquad (13.89)$$

assuming that the original electron is at rest, and show that the condition for electron capture is the same as (13.88) *except* that the term $2m_e$ is missing. (c) The mass of ^{231}U is 231.03626 u, while that of ^{231}Pa is 231.03588 u. Show that ^{231}U can decay by electron capture but not by β^+ emission.

13.24 • • • Observations show that if there were no neutrino, β decay would violate conservation of momentum, as well as energy. For example, in a decay of ^{6}He at rest, the two final charged particles were observed with momenta as shown in Fig. 13.32. (a) Show that this decay would violate conservation of momentum if there were no neutrino involved. (b) Assuming that a neutrino *is* ejected and that

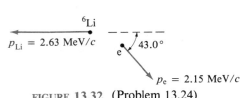

^{6}Li

$p_{Li} = 2.63$ MeV/c 43.0°

e

$p_e = 2.15$ MeV/c

FIGURE 13.32 (Problem 13.24)

total momentum is conserved, calculate the momentum (magnitude and direction) of the unobserved neutrino. (c) Assuming conservation of energy, find the neutrino's energy. (d) Show that your answers in parts (b) and (c) are consistent with the neutrino's having zero rest mass (within the accuracy of the given numbers).

13.25 • • • (a) Use the semiempirical binding-energy formula (12.30) to write an expression for the mass of a nucleus in terms of Z and A. (b) Now consider a set of isobars with A odd. Use your mass formula to show that a plot of mass against Z, as in Fig. 13.2, should fit a parabola ($m = \alpha Z^2 + \beta Z + \gamma$). (c) Now consider a set of isobars with A even. Show that because of the pairing energy, the masses will alternate between two parabolas, one for the nuclei with Z even and the other for Z odd. (d) Sketch the plot of m against Z and explain why one can have two stable isobars if A is even. (e) Find two examples of this phenomenon in Appendix D.

13.26 • • • In an observed β decay $X \rightarrow Y + e^- + \bar{\nu}$ of a nucleus X that is exactly at rest, it happens that the electron is also exactly at rest. For this special case, derive an exact (relativistic) expression for the energy of the $\bar{\nu}$ in terms of m_X, m_Y, and m_e. Show that to an excellent approximation, the neutrino takes all of the available kinetic energy, ΔMc^2, when the electron is at rest. In a typical case [$m_X \approx 200$ u and $(m_X - m_Y) \approx 0.001$ u] how good is this approximation? (Give your answer as a fractional error.)

SECTION **13.5** (THE NATURAL RADIOACTIVE SERIES)

13.27 • Use the data in Appendix D to find the kinetic energy released in each of the following conceivable decays:

$$^{228}\text{Th} \rightarrow {}^{227}\text{Th} + n$$
$$\rightarrow {}^{227}\text{Ac} + p$$
$$\rightarrow {}^{228}\text{Pa} + e^- + \bar{\nu}$$
$$\rightarrow {}^{224}\text{Ra} + \alpha$$
$$\rightarrow {}^{216}\text{Po} + {}^{12}\text{C}.$$

Which of these processes is possible? (*Note:* You will find that the last one *is* possible. However, its probability is so small, compared to α decay, that it has not been observed.)

13.28 • • The supernova explosion that is believed to have preceded the formation of our solar system created large quantities of radioactive nuclei. There are reasons to think that the isotopes ^{235}U and ^{238}U would have been created in nearly equal numbers. That is, the ratio N_{238}/N_{235} was close to 1 at $t = 0$ (the time of the explosion). (a) Show that at any later time

$$\frac{N_{238}}{N_{235}} = \exp[(r_{235} - r_{238})t],$$

where r_{235} and r_{238} denote the decay constants of ^{235}U and ^{238}U. (b) Use the data in Appendix D to find r_{235} and r_{238}. (c) Use the observed abundances of ^{235}U and ^{238}U, on earth today, to estimate how long ago the supernova explosion occurred.

13.29 •• (a) Use the data in Appendix D to calculate the total kinetic energy released in the α decay of ^{220}Rn. (b) Find the total kinetic energy released in the β^- decay of ^{220}Rn. Is this decay possible? (c) Answer the same questions for ^{212}Bi. (d) What is the relevance of your answers in consideration of Fig. 13.9?

13.30 •• Use the data in Appendix D to write down the whole of the radioactive series that begins with ^{238}U. Draw the series on a plot of Z against N as in Fig. 13.9.

13.31 •• Use the data in Appendix D to write down the whole of the radioactive series that begins with ^{235}U. Draw the series on a plot of Z against N as in Fig. 13.9.

13.32 •• The mass number A decreases by 4 in α decay, but is unchanged in β or γ decay. This means that throughout a radioactive series A changes only in multiples of 4. Thus if A happens to be a multiple of 4 ($A = 4n$) in the original parent, it will remain a multiple of 4 in all members of the series. This is what happens in the ^{232}Th series, which is therefore called the $4n$ series. In the same way, there are three other possible series: $(4n + 1)$, $(4n + 2)$, and $(4n + 3)$. (a) What are the original parents of the $(4n + 2)$ and $(4n + 3)$ series? (b) Use the data in Appendix D to explain why the $(4n + 1)$ series is not found naturally.

13.33 •• (a) Use the data in Appendix D to predict the total kinetic energy released in the α decay of ^{227}Th. (b) Figure 13.33 shows the observed energy spectrum in the α decay of ^{227}Th nuclei (all in their ground state). Can you explain the three spikes labeled B, C, and D? (c) Use Fig. 13.33 to predict the excitation energy of three excited states of ^{223}Ra.

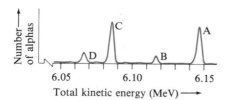

FIGURE 13.33 (Problem 13.33)

13.34 •• Consider the α decay of a nucleus X at rest, $X \rightarrow Y + \alpha$, where Y denotes the offspring nucleus. (a) Prove that because the α is much lighter than Y, almost all the available kinetic energy goes to the α; specifically, prove that

$$K_\alpha = \frac{m_Y}{m_\alpha} K_Y. \qquad (13.90)$$

(*Hint:* Use conservation of momentum and treat all particles nonrelativistically.) (b) In the α decay of a typical naturally radioactive nucleus, what percentage of the available kinetic energy does the α particle take? (c) Assuming that β decay has the form $X \rightarrow Y + e$, use the relation corresponding to (13.90) to find what percentage of the available kinetic energy goes to the electron. (Your answer here is really an upper bound, since some energy can go to the neutrino. It is also only approximate, since the electron can be fairly relativistic.)

SECTION **13.6** (NUCLEAR REACTIONS)

13.35 • In most nuclear reactions the total nucleon number is conserved; that is, $\Sigma A_{\text{initial}} = \Sigma A_{\text{final}}$, where $A = Z + N$ for each nucleus, as usual. Show that, except in reactions involving β decay, both ΣZ and ΣN are separately conserved. (*Hint:* Total charge is always conserved, and except in β decay, all the charge is carried by protons.)

13.36 • Use Eq. (13.37) and the data of Appendix D to predict ΔK, the kinetic energy released in the reaction $p + {}^7\text{Li} \rightarrow \alpha + \alpha$.

13.37 • When Rutherford first observed the reaction (13.30) he assumed that the α particle simply knocked a proton out of the ^{14}N in the process $\alpha + {}^{14}\text{N} \rightarrow \alpha + p + {}^{13}\text{C}$. Given that the incident α particles had about 5.6 MeV, was this reaction possible?

13.38 • Before two nuclei can react they must come within 1 fm or so of one another, and because of their Coulomb repulsion this requires a lot of energy. (a) Calculate the energy needed to bring a proton from infinity to a separation of 1 fm from an α particle. (1 fm = distance from surface to surface.) (b) Do the same for an α particle and a uranium nucleus. [Use Eq. (12.3) to estimate the radii involved.]

13.39 • To find the mass of the neutron, Chadwick used the reaction

$$\alpha + {}^{11}\text{B} \rightarrow n + {}^{14}\text{N}.$$

The best available values for the masses of the three nuclei were $m_\alpha = 4.00106$ u, $m_B = 11.00825$ u, and $m_N = 14.0042$ u. The kinetic energies Chadwick measured as $K_\alpha = 5.26$ MeV, $K_B = 0$, $K_n =$

3.26 MeV, and $K_N = 0.57$ MeV. Using these data, find Chadwick's value for the mass of the neutron, and compare with the current value.

13.40 • The cross section for 0.01-MeV neutrons to be captured by uranium 238 is 0.7 barn. Two grams of ^{238}U is exposed to a flux of 3×10^9 neutrons per square meter (with energy 0.01 MeV). How many atoms of ^{239}U will be produced?

13.41 • A hunter observes 100 crows settling randomly among the leaves of a large oak tree, where they are no longer visible. He fires 500 shots at random into the tree and hits 25 of the crows. If the area of the tree (as seen by the hunter) is 1200 ft², what is σ, the cross-sectional area of a single crow? (Notice how, in a classical experiment of this type, σ is the actual cross-sectional area of the target.)

13.42 • Which of the following reactions violate the laws of conservation of charge or nucleon number (and are therefore impossible)?
(a) $p + {}^{63}Cu \rightarrow n + {}^{63}Ni$
(b) $n + {}^{64}Zn \rightarrow {}^2H + {}^{63}Cu$
(c) $n + {}^{56}Fe \rightarrow \alpha + {}^{53}Mn$
(d) $n + {}^{14}N \rightarrow p + {}^{13}C$

13.43 • • Use the laws of conservation of energy, charge, and nucleon number to decide which of the following processes are impossible. Give your reasons.
(a) ${}^{207}Po \rightarrow {}^{207}At + e^- + \bar{v}$
(b) ${}^{213}Rn \rightarrow {}^{208}Po + \alpha$
(c) ${}^{212}Po \rightarrow {}^{208}Pb + \alpha$
(d) $p + {}^{14}N \rightarrow \alpha + {}^{12}C$
(e) $n + {}^{22}Na \rightarrow p + {}^{22}Ne$

13.44 • • Reactions in which kinetic energy is lost ($\Delta K < 0$) are called **endothermic** and cannot occur below a minimum (threshold) initial energy. (Reactions in which kinetic energy is released are called **exothermic**.) Which of the following reactions are endothermic, and what are their threshold kinetic energies?
(a) $n + {}^{208}Pb \rightarrow p + {}^{208}Tl$
(b) $n + {}^{207}Pb \rightarrow \alpha + {}^{204}Hg$
(c) $n + {}^{187}Os \rightarrow p + {}^{187}Re$
(d) $n + {}^{197}Au \rightarrow {}^2H + {}^{196}Pt$

13.45 • • When the Joliot-Curies observed the reaction $\alpha + {}^9Be \rightarrow n + {}^{12}C$, they suggested (erroneously) that the neutral particle was a photon. Chadwick's main argument against this concerned the large momentum (up to 100 MeV/c) of the protons that these supposed photons ejected from wax. Assuming that it *was* photons that ejected the protons, and given that a proton gets the largest momentum in a head-on collision, do the following: (a) To a fair approxi-

mation the photon would bounce back with its energy unchanged; in this approximation, what is the incident photon energy required to eject a proton with momentum 100 MeV/c? (b) Using conservation of energy and momentum, redo this calculation without the approximation of part (a).

13.46 • • • One of the ways that Chadwick estimated the neutron mass was to direct neutrons (all of the same velocity) at two different targets (A and B, say). The neutrons struck the target nuclei and caused them to recoil. (a) If we denote the maximum recoil speeds (which occur in head-on collisions) by v_A and v_B, prove that

$$m_n = \frac{m_B v_B - m_A v_A}{v_A - v_B}$$

(use nonrelativistic mechanics). (b) Chadwick used for his targets hydrogen and nitrogen, and his measured speeds were $v_H = 3.3 \times 10^7$ m/s and $v_N = 4.7 \times 10^6$ m/s. What was his value for m_n based on these data?

SECTION **13.7** (NUCLEAR FISSION)

13.47 • Use the masses given in Appendix D to find the energy released in the fusion reaction (13.45) and the fission (13.46).

13.48 • One can get a rough estimate of the energy released in fission by considering Fig. 13.14, which shows the potential energy of the two fragments. The energy released is roughly the potential energy at $x = x_0$, which, in turn, is roughly $kq_1 q_2/x_0$. (Both of these approximations will actually be overestimates.) Taking $x_0 \sim 12$ fm, estimate the energy released in the fission ${}^{236}U \rightarrow {}^{92}Kr + {}^{144}Ba$.

13.49 • Consider the fission ${}^{236}U \rightarrow {}^{92}Kr + {}^{144}Ba$. (a) Using the data in Appendix D, find the number of neutrons by which the fragment ${}^{92}Kr$ exceeds its nearest stable isotope. (b) Do the same for ${}^{144}Ba$. (c) If each fragment threw off two neutrons during the process of fission, would it then be stable? If not, what kind of radioactivity would you expect it to show? (Explain.)

13.50 • • A large electrical generating plant produces about 1 GW of power. (1 GW = 10^9 W.) (a) What mass of uranium would a 1-GW nuclear plant consume in one year? (Assume that each fission of ${}^{235}U$ produces about 200 MeV, that the conversion of this energy to electrical energy is about 30% efficient, and that the fuel is natural uranium—of which 99.3% is ${}^{238}U$, which produces no energy.) (b) What mass of coal would a 1-GW coal-burning plant consume in a year? (1 kg of coal produces roughly 33 MJ of energy, and, since coal-burning plants can operate at higher temperatures, their efficiency is about 40%.) (c) A modern railroad car can carry about 100 tonnes. (1 tonne = 1000 kg). How many car loads of coal does the coal plant use in a year? (*d*) How many car loads of uranium does the nuclear plant use?

13.51 •• To improve the chance that the neutrons produced in fission will themselves induce further fissions, it is important to slow them down. A *moderator* is a substance with which neutrons can collide and lose energy. It must be chosen so that the neutrons lose as much energy as possible (but are not actually captured). In practice the best process for reducing the neutrons' energy is an elastic collision. Consider a head-on elastic collision between a neutron (mass m) and a nucleus (mass M). (*a*) Use conservation of energy and momentum (nonrelativistic) to derive an expression for the neutron's fractional loss of energy, $\Delta K/K$, as a function of the mass ratio $\mu = m/M$. (*b*) Show that the loss is maximum when the masses are equal ($\mu = 1$). (*c*) This result suggests that hydrogen—in water, for example—would be the best moderator. Unluckily, ordinary hydrogen, ^{1}H, has the disadvantage that it can capture neutrons readily to form deuterium. Two practical alternatives, with masses at least fairly close to the neutron mass, are deuterium—in "heavy" water—and carbon—in the form of graphite. What is $\Delta K/K$ for these two moderators?

13.52 ••• To get a rough feeling for critical sizes, consider the following: A stream of 1-MeV neutrons impinges normally on a slab of ^{235}U, 1 cm thick. (1 MeV is the average energy of neutrons ejected in fission.) (*a*) Given that the fission cross section at 1 MeV is 1.25 barns, find the fraction, N/N_{inc}, of incident neutrons that induce a fission as they pass through the slab. [*Hint:* In Eq. (13.44) we wrote the number of reactions as $N = \sigma n_{inc} N_{tar}$ where n_{inc} is the incident flux of projectiles (number/area) and N_{tar} is the total number of target particles. Here it is more convenient to use the equivalent expression

$$N = \sigma N_{inc} n_{tar},$$

where N_{inc} is the total number of neutrons and n_{tar} is the number of ^{235}U atoms per unit area of the slab. To find n_{tar} you need to know that the density of uranium is about 19 g/cm^3.] (*b*) On average, the number of neutrons produced in a fission is 5/2; therefore, in a self-sustaining reaction at least 2/5 of the neutrons produced must induce further fission. Use your answer to estimate how many centimeters of slab are needed to give $N/N_{inc} = 2/5$. This answer gives the order of magnitude of the critical radius of a solid sphere of ^{235}U. (*c*) Using this estimate of the critical radius, find the approximate value of the critical mass of a sphere of ^{235}U.

13.53 • The energy released by nuclear weapons is often expressed in megatons of TNT. (1 megaton of TNT is 4.3×10^{15} J, the energy released by the explosion of 1 million tons of TNT.) (*a*) Approximately what mass of ^{235}U would release 1 megaton by fissioning? (*b*) In a fusion bomb the two reactions (13.59) and (13.60) consume altogether three deuterons (^{2}H nuclei) and produce 21.6 MeV. What mass of deuterium is needed to produce 1 megaton by these combined processes?

13.54 •• The sun's energy comes from the proton–proton cycle, which begins with the reaction (13.61)

$$^1\text{H} + {}^1\text{H} \rightarrow {}^2\text{H} + e^+ + \nu.$$

This reaction is exceedingly improbable, since it requires conversion of a proton to a neutron by β^+ decay during the brief collision of the two protons. Estimate the probability that a proton–proton collision will lead to this reaction, as follows: Estimate the time that the first proton (with about 70 keV) is within range of the second (time to traverse about 10 fm, say). Take the mean life for a β decay of this kind to be of order 1 minute, and find the probability that the decay will occur while the protons are within range. [*Hint:* Recall Eq. (13.7) and the definition of r.]

13.55 •• The principal source of energy in stars like the sun is the proton–proton cycle, the net effect of which is to fuse four protons into ^{4}He (plus two positrons and two neutrinos). In hotter stars the same effect is produced by a different cycle, called the **carbon,** or **CNO, cycle.** This cycle involves six reactions, parts of which are as follows:
1. ^{1}H + ^{12}C $\rightarrow$ _____ + γ
2. _____ $\rightarrow$ ^{13}C + e^+ + ν
3. ^{1}H + ^{13}C $\rightarrow$ _____ + γ
4. ^{1}H + _____ $\rightarrow$ ^{15}O + γ
5. ^{15}O $\rightarrow$ _____ + e^+ + ν
6. ^{1}H + _____ $\rightarrow$ ^{12}C + ^{4}He

(*a*) Fill in the blanks in these reactions. (*b*) Verify that the net effect of one complete cycle is to replace four protons by a ^{4}He nucleus plus two positrons, two neutrinos, and some photons. (*c*) Obviously, the possibility of the cycle depends on the star's having some ^{12}C to act as a "catalyst." In addition, it requires higher temperatures than are needed for the proton–proton cycle. Explain why. Very roughly, how much higher would you expect the required temperature to be? (Remember that the high temperatures are needed to overcome the Coulomb repulsion.)

SECTION 13.9 (THE THEORY OF ALPHA DECAY)

13.56 • Verify the values (13.78) and (13.79) for the two constants a and b that appear in the theory of alpha decay.

13.57 •• One of the longest-lived alpha emitters is ^{232}Th, with a half-life $t_{1/2}(^{232}\text{Th}) = 14$ billion years, and one of the shortest is ^{216}Ra with $t_{1/2}(^{216}\text{Ra}) = 0.18\ \mu s$. (a) Compute the ratio of these two measured half-lives and its natural log. (b) Use the formula (13.80) to predict this same natural log. (The energies released are given in Table 13.1 at the beginning of Section 13.9.)

13.58 •• Use the data in Appendix D to find the half-lives of the isotopes ^{232}U, ^{234}U, ^{236}U, and ^{238}U and the kinetic energy released in their alpha decays. Make a plot of $\ln t_{1/2}$ against $K^{-1/2}$ to see how well these fit a straight line as predicted by (13.82).

13.59 •• The half lives of certain alpha emitting isotopes of radium, and the corresponding kinetic energies released, are as follows:

A	K(MeV)	$t_{1/2}$ (s)
216	9.53	1.82×10^{-7}
218	8.55	1.4×10^{-5}
220	7.59	2.3×10^{-2}
222	6.68	38.0
224	5.79	3.16×10^{5}
226	4.87	5.05×10^{10}

(a) Make a plot of $\ln t_{1/2}$ against $K^{-1/2}$. (b) How well does the slope of your graph agree with the value predicted by Eq. (13.82)?

13.60 ••• Evaluate the integral in Eq. (13.74) and verify the expression (13.76) for the escape probability of an α particle hitting the surface of a nucleus. [*Hints:* Remember that $U(x) = 2Zke^2/x$ and $K = 2Zke^2/x_1$. These substitutions lead to an integral of the form

$$\int_{x_0}^{x_1} \sqrt{\frac{x_1}{x} - 1}\ dx;$$

with the change of variables $x = x_1 \cos^2\theta$, you can evaluate this integral as $x_1(\theta_0 - \sin\theta_0 \cos\theta_0)$, where $\theta_0 = \cos^{-1}\sqrt{x_0/x_1}$. Because $x_0 \ll x_1$, you can approximate θ_0 as $\pi/2 - \sqrt{x_0/x_1}$ and $\sin\theta_0$ as 1. Finally, you can use the relation $K = 2Zke^2/x_1$ to eliminate x_1.]

13.61 ••• Occasionally, some radioactive nuclei eject a particle other than an α or β particle. In particular, it has been found that several nuclei, including ^{223}Ra, can emit a ^{14}C nucleus, although this mode is very rare (of order one of these decays for every 10^9 alpha decays.) (a) What is the offspring nucleus when ^{223}Ra emits a ^{14}C nucleus, and what is the energy released? (b) Assuming that ^{14}C emission is analogous to alpha decay, one can calculate its probability by the methods of Section 13.9. Explain why the probability $P(^{14}\text{C})$ for escape of a ^{14}C nucleus can be obtained from Eq. (13.76) simply by replacing $2Z$ by $6Z$ and taking m to be the mass of ^{14}C. (c) Calculate $P(^{14}\text{C})$ and find the ratio $P(^{14}\text{C})/P(\alpha)$ for the parent ^{223}Ra. [*Caution:* Z is the charge of the offspring. The answers are quite sensitive to the value of the nuclear radius R; to be definite, take $R = 7.5$ fm in both calculations.]

SECTION 13.10 (ACCELERATORS)

13.62 • A proton is accelerated from rest by a constant potential difference of 0.8 MV applied to an accelerating tube of length 1.5 m. How long does this take?

13.63 •• Protons are injected into a linear accelerator so that they have $K = 250$ KeV as they coast through the first drift tube. The frequency of the ac voltage applied to the tubes is 50 MHz and the peak potential difference across each gap is 200 kV. Consider a proton that passes from the first to the second tube properly synchronized at the peak potential drop. (a) How long must the second tube be if the proton is to arrive at the next gap in proper synchronization with the accelerating voltage? (b) How long must the tenth tube be?

13.64 •• A cyclotron with a magnetic field $B = 0.9$ T and a maximum orbital radius of 50 cm is used to accelerate deuterons (^{2}H nuclei). (a) What is the maximum energy to which the deuterons can be accelerated? (b) What should be the frequency of the ac accelerating voltage? (c) If the accelerating voltage has amplitude 40 kV, how many revolutions are needed to bring the deuterons to full energy?

SECTION 13.11 (PARTICLE DETECTORS)

13.65 • An alpha particle from a radioactive source is directed into a cloud chamber in a uniform magnetic field of 0.80 T. If its path is curved with radius 0.40 m, what are its momentum and energy?

13.66 •• A thin foil of uranium is placed inside an ion chamber of the sort described in Example 13.7, and the foil is irradiated with neutrons to induce fissions. When a fission occurs the ionization produced by both fission fragments can be collected as a single pulse of charge flowing around the circuit. If the total charge of one of these pulses is 9.9×10^{-13} C, what was the combined kinetic energy of the two fission fragments? (This is the method by which the large energy released in nuclear fission was discovered.) Assume that the energy loss per ionization is the same as given in Example 13.7 for α particles.

13.67 • (a) What is the mass of 1 Ci of ^{224}Ra? (b) What is it for 1 Ci of ^{226}Ra? [Part (b) illustrates the original definition of the curie.]

13.68 • Suppose that a 60-kg person absorbs an effective dose of 5 Sv of X rays (which, in a single exposure, would probably be fatal). (a) What is the total energy absorbed by the person? (b) Taking the person's specific heat to be about that of water, find their resulting temperature rise. (Your answer will show that heating does not play an important role in the damaging effect of radiation.)

13.69 • Medical technicians place an ion chamber 20 cm from a radioactive source (which is in its shielded container) and measure an intensity that would produce an effective dose of 3 mSv per hour in human tissue. If safety regulations limit a worker's exposure to 50 mSv per year, at what minimum distance from the source should people normally work? (Assume that they work a 40-hour week for 50 weeks per year, and that absorption of the radiation in air is negligible.)

13.70 •• A 25-kg child swallows 2×10^9 Bq (about 50 mCi) of a β emitter, which ejects electrons with average energy 1.2 MeV. Assuming that all the electrons are absorbed in the body, what is the child's average absorbed dose per hour? Given that 5 Gy is a lethal dose, roughly how quickly must doctors flush out the radioactivity to save the child?

14

Elementary Particles

Now the smallest Particles of Matter may cohere by the strongest Attractions and compose bigger Particles of weaker Virtue; and many of these may cohere and compose bigger Particles whose Virtue is still weaker, and so on for diverse successions, until the Progression ends in the biggest Particles on which the Operations in Chymistry and the Colours of natural Bodies depend, and which by cohering compose Bodies of a sensible Magnitude.

There are therefore Agents in Nature able to make the Particles of Bodies stick together by very strong Attractions. And it is the Business of Experimental Philosophy to find them out.

Issac Newton, *Opticks,* 1704

14.1 Elementary Particles: The Story So Far*

"Elementary particle" is the modern name for the fundamental units of matter. As we described in Chapter 4, the concept of an ultimate, indivisible unit of matter goes back more than two thousand years to the Greek philosophers, who also gave us the name "atom," meaning indivisible. Experimental evidence for the existence of atoms emerged gradually, beginning about 1800, and for a brief period before 1900, it was reasonable to suppose that the 90 or so different atoms were indeed the ultimate constituents of matter.

We now know that atoms are themselves divisible and so are not elementary particles in the sense that we are using the term. The first evidence that atoms can in fact be subdivided was the discovery of the electron by Thomson in 1895. This was followed by Rutherford's identification of the atomic nucleus in 1910 and the discovery that nuclei are made of protons (Rutherford in 1918) and neutrons (Chadwick in 1932). With the discovery of the neutron, it appeared that there were just three different elementary particles: the electron, proton, and neutron.

Even as the pieces of this beautifully simple picture were falling into place, evidence was accumulating that the true story was more complicated. In particular, it became clear that there were several additional particles with just as much right to be considered elementary as the electron, proton, and neutron. We have already met several of these particles in earlier chapters. For example, the positron, or antielectron, was discovered by Anderson in 1932. As we have seen, this particle has the same mass and spin as the electron, but has charge $+e$. The positron has no part in the construction of normal matter, but is generally agreed to be just as elementary as its more familiar partner, the electron.

A second "extra" particle that we have already met is the neutrino, the neutral particle produced in nuclear beta decay, as proposed by Pauli in the early 1930s. As we have argued, the neutrino is not normally present in nuclei. Nevertheless, all of space is filled with numerous neutrinos coming from nuclear reactions in stars, and it would be difficult to exclude the neutrino from any list of elementary particles.

One other familiar particle that was already known in the 1930s is the photon. This is certainly not a *constituent* of matter in the way that the electron, proton, and neutron are; but it has a fundamental role in the structure of matter, since it is the carrier of electromagnetic forces (in a sense that we shall discuss shortly). Thus, whether or not we describe the photon as an elementary particle, it is undeniably a particle with an important function in matter.

By the mid-1930s it was clear that the simple picture of a world containing just three kinds of elementary particle, the e, p, and n, was not the end of the story. In this chapter we describe the continuing quest for the truly elementary particles, whatever they may be. We shall see that from the 1930s to the 1960s the picture steadily became more complicated, with the discovery of more and more particles, most of them unstable but all of them apparently elementary. Then, in the mid-1960s a new order began to appear, as evidence emerged that most of the particles that had seemed to be elementary were in fact composites, all made up from a small number of more fundamental particles called quarks (a whimsical name borrowed by the American physicist Gell-Mann from James

---

* The material of this chapter depends on some of the material in Chapters 12 and 13. Specifically, you should have read Sections 12.1 through 12.3 and 13.1 through 13.4 before beginning this chapter.

Joyce's *Finnegan's Wake*). Of the particles that were formerly considered to be elementary, only a few are still regarded as such; among these are the electron, neutrino, and photon, none of which have yet shown any evidence of an internal structure. Most of the rest, including the proton and neutron, now appear to be composites, each made up of two or three quarks.

Before we continue the story of particle physics, we should mention that this field of study is often called **high-energy physics.** The reason for this second name is easy to see: Elementary particles have sizes of 10^{-15} m and less, and certainly cannot be observed with ordinary visible light, whose wavelength is about 10^{-6} m. Instead, as we saw in our discussion of nuclear physics, these subatomic systems are probed with beams of particles, such as electrons, protons, and high-energy photons. The wavelength of these particles is given by the de Broglie relation,

$$\lambda = \frac{h}{p},$$

and to investigate sizes of order 10^{-15} m one needs wavelengths of this same order or less. A simple calculation using the de Broglie relation shows that this requires a kinetic energy of order 1 GeV = 10^9 eV (Problem 14.1). Some experiments today probe distances down to 10^{-17} m or less, and this requires energies of hundreds of GeV. By atomic standards this is an extremely high energy and amply justifies the name "high-energy physics." To a large extent the story of particle physics has been the story of the development of larger and larger machines to accelerate particles to these high energies. Today (1991) the largest of these accelerators is the Tevatron near Chicago, which accelerates protons to 1 TeV = 1000 GeV = 10^{12} eV and is some 4 miles in circumference. We describe some of these machines in Section 14.11.

14.2 Antiparticles

In this section we review and extend the story of the positron, which proved to be the first example of what is now called an antiparticle.

THE PREDICTION AND DISCOVERY OF THE POSITRON

The existence of the positron was predicted theoretically several years before it was verified experimentally. In 1928, the English physicist Dirac had proposed a relativistic generalization of the Schrödinger equation. Although Dirac's original goal was simply to find a form of the Schrödinger equation that was consistent with relativity, he accomplished much more. He found that his equation required the existence of spin, with magnitude given by the quantum number $s = \frac{1}{2}$. Therefore, if one assumed (correctly, as it turned out) that Dirac's equation described electrons, one could say that Dirac had predicted the existence of the electron's spin. Better still, the equation also predicted a magnetic moment, with the observed value (10.24).

An unexpected feature of Dirac's equation, as applied to electrons, was that it required the existence of a second particle with charge $+e$. Since the only known particle with positive charge was the proton, Dirac thought for a while that his equation for electrons was predicting the existence of protons. However, it soon became clear that the newly predicted particle would have to have the same mass as the electron.

PAUL DIRAC (1902–1984, English). Dirac's relativistic version of the Schrödinger equation predicted the existence of the positron (discovered a couple of years later by Anderson) and earned him the 1933 Nobel Prize, which he shared with Schrödinger. He was the author of a seminal book on quantum mechanics, first published in 1930.

At the time the situation was quite confused, but in retrospect it can be easily summarized. If one accepted the existence of negatively charged electrons obeying Dirac's equation, the equation predicted the existence of a second kind of particle with the same mass and spin as the electron, but with charge $+e$. This particle is, of course, what we now call the positron. Furthermore, the theory predicted that an energy greater than $2m_ec^2$ should be able to create an electron–positron pair; for example, if a photon with energy greater than $2m_ec^2$ collides with a nucleus, the photon can disappear in the process of *pair production*:

$$\gamma + \text{nucleus} \rightarrow e^- + e^+ + \text{nucleus}, \tag{14.1}$$

where e^- and e^+ denote the electron and positron, respectively. (The need for the nucleus in this reaction is explained in Problem 14.6.) Conversely, if a positron meets an electron they can annihilate each other with all of their rest energy (plus any kinetic energy) going into photons:

$$e^- + e^+ \rightarrow \text{photons}. \tag{14.2}$$

These predictions were unexpectedly verified in 1932 by the American experimentalist Anderson. Anderson was studying cosmic rays, the particles that bombard the earth's atmosphere from space. The primary cosmic rays are high-energy particles, mostly protons and alpha particles, with a few heavier nuclei. When these particles collide with atoms in the earth's atmosphere they produce large numbers of secondary particles, including photons, electrons, and positrons. Anderson was studying cosmic rays using a cloud chamber, in which any moving charged particle leaves a narrow track of condensed liquid drops. These tracks can be photographed and then studied in detail. If the chamber is placed in a magnetic field, the resulting curvature of any track gives information about the particle's charge and momentum. Further, it is found that the track's thickness depends on the particle's mass, with heavier particles tending to leave thicker tracks. Thus, with care and experience, an experimenter can find most of the properties of any charged particle that passes through his cloud chamber.

Anderson had observed many tracks of electrons and protons in cosmic rays, but in 1932 he reported seeing some tracks that seemed to be made by a particle with the mass of an electron but the charge of a proton. In particular, he obtained a picture like that shown in Fig. 14.1, which is an early photograph of the pair-production process (14.1). In the picture the incoming photon and the nucleus are invisible—the photon because it is chargeless and the nucleus because it is stationary—but the tracks of the resulting electron and positron are clearly visible. The opposite curvatures of the two tracks show that the two particles have opposite charges. That the two tracks have about equal thickness suggests (to the experienced observer) that the two particles have roughly equal masses. In the following years Anderson published conclusive evidence for the existence of a positive particle with the same mass as the electron, and there is now no doubt that his original paper was indeed the first report of what we now call the positron (named by Anderson) and that Dirac's daring predictions were correct.

THE ANTIPROTON

If the electron has an "antiparticle," the question naturally arises: Do *all* particles have corresponding **antiparticles?** The answer appears to be that they do. For every particle, there is a corresponding antiparticle with the same mass

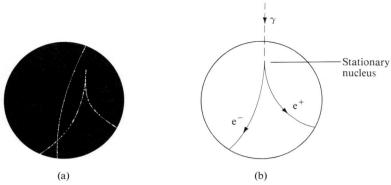

FIGURE 14.1 (a) An early photograph of pair production in a cloud chamber. (b) A tracing of the same event showing the incident photon (which is invisible in the photograph). A magnetic field directed into the page caused the electron and positron to curve in opposite directions.

and spin, and with charge of the same magnitude but opposite sign. (Notice that for a neutral particle this means that the antiparticle is also neutral.) Further, when a particle meets its antiparticle the two can annihilate one another, converting all of their rest energy into electromagnetic radiation in the form of photons. For example, a proton and antiproton can disappear in the reaction*

$$p + \bar{p} \rightarrow \text{photons} \qquad (14.3)$$

where $\bar{p}$ denotes the antiproton.

Since charge is conserved and since the photon is uncharged, the possibility of annihilation as in (14.3) makes clear why every antiparticle must have charge opposite to that of its particle. For example, the proton has charge $+e$ and, since the charge on both sides of (14.3) must balance, the antiproton must have charge $-e$.

Similar considerations apply to several other conservation laws. For example, we saw in Chapter 13 that the total number of nucleons does not change in nuclear reactions. We can reword this observation by introducing a quantity — somewhat analogous to charge — called **nucleon number.** We assign both the proton and neutron a nucleon number of $+1$, and other particles, like the electron and photon, a nucleon number of zero. Then we can say simply that the total nucleon number of a system is conserved in all the nuclear reactions discussed in Chapter 13. With some small modifications that we discuss later, the conservation of nucleon number appears to be universally true.† In the light of this conservation law, let us reexamine the annihilation (14.3): The proton has nucleon number $+1$, while the photon has nucleon number 0. It then follows from (14.3) that the antiproton must have nucleon number -1. Evi-

---

* Other reactions are also possible; for example, the p and $\bar{p}$ can also annihilate into pions (discussed in Section 14.3). As usual in quantum physics, we cannot predict with certainty which reaction will occur — just their probabilities. In the case of the p and $\bar{p}$, annihilation into photons as in (14.3) is actually less probable. Nonetheless, it does occur, and this possibility can be regarded as characteristic of any particle–antiparticle pair.

† The main modification is that the nucleons, n and p, turn out to be just two of a large class of particles called *baryons,* and it is the total baryon number (not just nucleon number) that is conserved.

dently, any particle and its antiparticle, in addition to having opposite charges, must also have opposite nucleon numbers.

Since the neutron has zero charge, so does the antineutron; that is, with regard to charge, there is no difference between n and $\bar{n}$. Nevertheless, the neutron and antineutron *are* different. An antineutron can annihilate a neutron:

$$n + \bar{n} \rightarrow \text{photons}, \tag{14.4}$$

whereas a neutron cannot:

$$n + n \xrightarrow{\times} \text{photons}. \tag{14.5}$$

If the latter process were possible, matter as we know it would not be stable, since the neutrons in nuclei would quickly disappear. Notice that in the process (14.4) the total nucleon number is conserved $[1 + (-1) = 0]$, whereas in (14.5) it would not be $[1 + 1 \neq 0]$; therefore, we can say that the impossibility of (14.5) is a consequence of the law of conservation of nucleon number.

The existence of antiprotons and antineutrons was not confirmed experimentally until 1955. Like the positron, antinucleons do not occur naturally in ordinary matter since they would be promptly annihilated. To create an antinucleon requires an enormous energy. For example, a proton–antiproton pair can be produced when a photon collides with a nucleus,

$$\gamma + \text{nucleus} \rightarrow p + \bar{p} + \text{nucleus},$$

but the photon needs to have an energy of at least $2m_p c^2$, or roughly 2 GeV. Cosmic rays do include particles with these energies, and one might hope to discover antiprotons in cosmic rays, just as Anderson had discovered positrons. Unfortunately, 2-GeV photons are extremely rare compared to the 1-MeV photons needed to produce positrons. Thus no antiprotons had been definitely identified in cosmic rays prior to 1955.

If antiprotons could not be found in a naturally occurring process, the only alternative was to produce them using an energetic particle that had been accelerated artificially. In 1955, Chamberlain, Segrè, and co-workers at Berkeley completed an accelerator that could accelerate protons to the energy required to produce antiprotons. These protons were then fired at a target containing more protons (actually, a piece of copper, the protons being those in the copper nuclei) and produced proton–antiproton pairs in the reaction

$$p + p \rightarrow p + p + p + \bar{p}. \tag{14.6}$$

To identify the antiprotons required painstaking care, since about 40,000 other particles were produced for every antiproton generated by the reaction (14.6). (In this way the experiment was typical of many modern experiments in high-energy physics, where the events of interest must be sifted out from a background of thousands of uninteresting events.) First, a magnetic field was used to select particles with negative charge and with the momentum expected for the antiprotons. The velocity of each selected particle was then calculated from its measured time of flight between two detectors; and from the particle's momentum and velocity, one could find its mass. In this way, Chamberlain et al. identified several hundred particles with the mass and charge expected for the antiproton.

Some of these presumed antiprotons were sent through a stack of photographic films. When a charged particle passes through photographic emulsion it leaves a trail of ionized atoms, which reveal the particle's path as a string of dark spots when the films are developed. In this way Chamberlain et al. confirmed their discovery by finding the tracks of an antiproton and the several particles produced when it annihilated with a proton.

THRESHOLD ENERGY

In designing the accelerator to produce antiprotons, it was obviously essential to know how much energy was needed to allow the reaction (14.6):

$$p + p \rightarrow p + p + p + \bar{p}. \tag{14.6}$$

At first glance one might guess that a kinetic energy $K_{min} = 2mc^2$ would be sufficient, and this would be correct *if* the four final particles could be produced at rest, with total energy $4mc^2$. Unfortunately, this is not possible if the target proton is at rest: The incident proton has a large momentum. Therefore, by conservation of momentum, the four final particles must have nonzero momentum, and hence nonzero kinetic energy; that is, their total energy has to be greater than $4mc^2$. Thus the required incident kinetic energy is actually more than $2mc^2$.

Using conservation of energy and momentum, one can show (Problem 14.8) that the minimum, or *threshold,* energy to produce the reaction (14.6) is actually

$$K_{min} = 6mc^2 \approx 5.6 \text{ GeV}. \tag{14.7}$$

This result explains why the Berkeley accelerator was designed to accelerate protons to about 6 GeV, just a little more than the threshold for production of antiprotons by the reaction (14.6). An important feature of the result is how very much energy ($6mc^2$) is needed compared to the $2mc^2$ one would at first expect. If one could arrange that the two original protons approached one another with equal and opposite momenta, the total momentum would be zero. In this case, the final four particles *could* all be produced at rest, with a total energy of just $4mc^2$, and the threshold kinetic energy would be just $2mc^2$. This is why many modern experiments searching for extremely heavy particles are arranged with two colliding beams of particles rather than one beam incident on a stationary target. Despite the difficulties of working with two colliding beams, the saving in energy makes this arrangement worthwhile (see Problem 14.40).

OTHER ANTIPARTICLES

Chamberlain and Segrè were awarded the 1959 Nobel prize for their discovery of the antiproton. In the same series of experiments they also found antineutrons, produced in processes such as

$$p + p \rightarrow p + p + n + \bar{n}. \tag{14.8}$$

It is now well established that every particle has a corresponding antiparticle. For example, the π^+, a particle with mass 139.6 MeV/c^2, spin 0, and charge $+e$, has an antiparticle (denoted π^-) with exactly the same mass and spin, but with charge $-e$. Similarly, the Δ^{++}, a particle with mass 1232 MeV/c^2, spin $\frac{3}{2}$, and charge $+2e$, has an antiparticle with the same mass and spin but with charge $-2e$.

A few particles are the same as their antiparticle. For example, the photon

TABLE 14.1

Some of the particles discussed so far, with their antiparticles. Masses are given in MeV/c^2. In every case the mass and spin of particle and corresponding antiparticle are the same.

Name	Particle			Antiparticle			Both	
	Symbol	Charge	Nucleon number	Symbol	Charge	Nucleon number	Mass	Spin
Photon	γ	0	0	—same as particle—			0	1
Neutrino	ν	0	0	$\bar{\nu}$	0	0	0	$\frac{1}{2}$
Electron	e^-	$-e$	0	e^+	$+e$	0	0.5	$\frac{1}{2}$
Pion	$\begin{cases} \pi^\circ \\ \pi^+ \end{cases}$	$\begin{matrix} 0 \\ +e \end{matrix}$	$\begin{matrix} 0 \\ 0 \end{matrix}$	—same as particle— π^-	$-e$	0	$\begin{matrix} 135 \\ 140 \end{matrix}$	$\begin{matrix} 0 \\ 0 \end{matrix}$
Proton	p	$+e$	$+1$	$\bar{p}$	$-e$	-1	938	$\frac{1}{2}$
Neutron	n	0	$+1$	$\bar{n}$	0	-1	940	$\frac{1}{2}$

is its own antiparticle and the same is true of the neutral pion, π°, which we discuss in Section 14.3. It is clear that a particle can be the same as its antiparticle only if it is electrically neutral, since particle and antiparticle must have opposite charges. Similarly, it must have nucleon number zero. For the particles that satisfy these conditions it is often hard to be sure whether or not the particle and antiparticle are different. For example, it is still not completely certain that neutrinos are different from antineutrinos, although the conventional view is that they are (that is, that $\bar{\nu} \neq \nu$).

In Table 14.1 we list some of the particles discussed so far, together with their antiparticles. The list includes some examples in which particle and antiparticle are identical and several in which they are not. Notice that as you read across each line the charge and nucleon number of each antiparticle are always equal in magnitude, but opposite in sign, to those of the corresponding particle.

ANTIMATTER

The existence of an antiparticle for every particle implies the possibility of *antimatter*. Antimatter would be made of positrons and antinucleons, just as normal matter is made of electrons and nucleons. Indeed, several antinuclei, such as antihelium, comprising two antiprotons and two antineutrons, have been observed; and experiments are under way in which it is hoped to produce antihydrogen atoms, comprising a positron and an antiproton. Antimatter would, of course, be extremely unstable in the presence of ordinary matter, but would be perfectly stable in isolation. It is, therefore, possible that some distant galaxies, well separated from any normal galaxies, are made entirely of antimatter, but there is, at present, no evidence that such galaxies exist.

14.3 Pions and Muons

In 1935 the Japanese physicist Hideki Yukawa predicted the existence of a particle with mass of order $100 \; MeV/c^2$. Because this mass lay between that of the electron ($0.5 \; MeV/c^2$) and that of the nucleons (about $940 \; MeV/c^2$), Yukawa's proposed particle came to be called the **meson,** meaning particle of intermediate mass. When several different particles of medium mass were

HIDEKI YUKAWA (1907–1981, Japanese). In 1935 Yukawa predicted the existence of the particle now called the pion, arguing that the strong nuclear force must be "carried" by a pion just as the electromagnetic force is "carried" by the photon. In 1947 the pion was discovered, with almost exactly the mass predicted by Yukawa. For this prediction he was awarded the 1949 Nobel Prize.

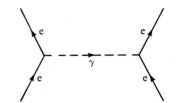

FIGURE 14.2 A Feynman diagram representing a process in which the electron on the left emits a photon, which is then absorbed by the electron on the right.

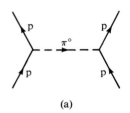

(a)

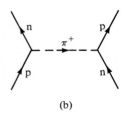

(b)

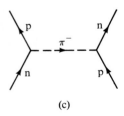

(c)

FIGURE 14.3 Three Feynman diagrams representing processes in which two nucleons exchange a pion. Two more possibilities, not shown, are that a proton and neutron can exchange a $\pi°$, and two neutrons can exchange a $\pi°$.

found, they were all referred to as mesons, and the particle that proved to be Yukawa's was named the π meson, or pion.

Yukawa's prediction was based on the idea that every force is, in some sense, "carried" by a corresponding particle, and that the pion is required as the "carrier" of the strong nuclear force. To understand this idea, let us consider the relation of the familiar photon to electromagnetic forces. In the nineteenth century, Maxwell had explained electromagnetic forces in terms of the electromagnetic fields $\mathscr{E}$ and $\mathbf{B}$. As we saw in Chapter 5, it became clear at the beginning of the twentieth century that the energy in electromagnetic fields is carried by photons. Thus, in some sense at least, we can say that electromagnetic forces are carried by photons.

The exact sense in which photons carry the electromagnetic force emerges from a complete quantum theory of the electromagnetic field. This theory, called quantum field theory, was developed in 1927 by Dirac but is beyond the scope of this book. The rough picture that emerges from quantum field theory is that the electromagnetic forces between two charged particles result from one particle emitting a photon which is then absorbed by the other. This exchange of a photon between two electrons, or any other charged particles, can be illustrated by a **Feynman diagram** as in Fig. 14.2. These diagrams are named for the American physicist Richard Feynman, who won the 1965 Nobel prize for his contributions to quantum field theory. As suggested in Fig. 14.2, the emission of a photon by the left-hand electron changes the electron's momentum; the subsequent absorption of the photon by the right-hand electron changes its momentum by an equal and opposite amount. This exchange of momentum has exactly the effect of a classical force between the two particles.*

Yukawa argued that if electromagnetic forces are carried by a particle (the photon), the same could be true of the strong nuclear forces. That is, the strong force between two nucleons might be the result of the nucleons exchanging some other particle, much as the photon is exchanged in Fig. 14.2. To explain the known properties of the nuclear force between all possible pairs of nucleons (p–p, p–n, n–n) it was found that there must be three kinds of exchanged particle. We now call all three of these particles **pions** and denote them π^+, $\pi°$, and π^-, indicating their charges of $+e$, 0, and $-e$. This means that there are several different processes involving exchange of a pion between two nucleons. Figure 14.3 shows three possibilities. In (a) two protons exchange a $\pi°$; in (b) a proton and neutron exchange a π^+; and in (c) a neutron and proton exchange a π^-. Notice that in the last two processes the emission or absorption of a charged pion changes a proton into a neutron, and vice versa. These two *charge exchange* processes make important contributions to the resultant force between a neutron and proton.

THE MASS OF THE PION

There is a relation between the mass of any force-carrying particle and the range of the corresponding force. In the case of the pion and the strong nuclear force, this relation is confirmed experimentally and lends strong support to

---

* However, it should be emphasized that Feynman diagrams are classical representations of an essentially quantum process. They ignore many quantum effects (such as the uncertainty principle) and can lead to contradictions if taken too literally. For example, it is hard to see how the exchange of Fig. 14.2 can lead to an attractive force, as between opposite charges. Nevertheless, a full quantum treatment *does* predict attraction between unlike charges.

Yukawa's theory. The connection between mass and range originates in the time-energy uncertainty relation and illustrates the essentially quantum nature of the exchange processes of Fig. 14.3. We first note that a process in which an isolated nucleon emits a pion,

$$N \rightarrow N + \pi, \tag{14.9}$$

violates conservation of energy. To see this we have only to consider the rest frame of the original nucleon. In this frame the energy of the initial nucleon is just

$$E_i = m_N c^2,$$

whereas the energy of the final nucleon plus pion is $m_N c^2 + m_\pi c^2 + K$. Therefore,

$$E_f \geq m_N c^2 + m_\pi c^2 > E_i$$

and the process (14.9) would violate conservation of energy by at least $m_\pi c^2$.

How, then, do exchange processes like those shown in Fig. 14.3 occur? The answer lies in the uncertainty relation $\Delta t \, \Delta E \geq \hbar/2$, which implies that when a system undergoes a process lasting for a time Δt, its energy has an intrinsic uncertainty

$$\Delta E \geq \frac{\hbar}{2\Delta t}.$$

RICHARD FEYNMAN (1918–1988, US). Feyman was a colorful personality, who contributed to many areas of theoretical physics. He won the 1965 Nobel Prize for his part in developing the quantum theory of electromagnetism (particularly the Feynman diagram), but he was equally known for his beautiful expositions of physics. His several books are ''goldmines'' of physical insight.

Thus a violation of energy conservation by an amount $m_\pi c^2$ would be meaningless, since it could never be detected (even in principle), provided that

$$m_\pi c^2 \leq \frac{\hbar}{2\Delta t}.$$

It follows that emission of a pion *is* possible, as long as it is reabsorbed within a time

$$\Delta t \leq \frac{\hbar}{2m_\pi c^2}. \tag{14.10}$$

The maximum speed with which the pion can travel is, of course, the speed of light c. Therefore, the maximum distance the pion can travel is $c \, \Delta t = \hbar/2m_\pi c$. If we identify this distance with the range R of the strong nuclear force, we conclude that

$$R \approx \frac{\hbar}{2m_\pi c}. \tag{14.11}$$

Knowing that the nuclear force has range $R \approx 1$ fm, we solve (14.11) for m_π and find that

$$m_\pi \approx \frac{\hbar c}{2Rc^2} \approx \frac{200 \text{ MeV} \cdot \text{fm}}{2 \times (1 \text{ fm})c^2} = 100 \text{ MeV}/c^2. \tag{14.12}$$

This is essentially Yukawa's original estimate for m_π. It is only a rough estimate, since different assumptions about the pion's speed or about R could change the estimate by a factor of 2 or so. Thus the observed mass, $m_\pi \approx 140$ MeV/c^2, is perfectly consistent with our estimate of 100 MeV/c^2. The relation (14.11) between range and mass can be applied successfully to other forces and their corresponding force-carrying particles, as discussed in Problems 14.9 and 14.10.

It is important to recognize that the exchange of a π meson between two nucleons is possible only because the pion is reabsorbed almost immediately after it is emitted. According to (14.10), the maximum time the pion can exist is of order 10^{-23} s, and there is absolutely no possibility that the exchanged pion could be observed directly. For this reason the exchanged pions in Fig. 14.3 (and similarly the exchanged photon in Fig. 14.2) are sometimes called *virtual particles*. As we shall see shortly, however, pions *can* be observed directly in collisions where the incident kinetic energy is sufficient to supply the rest energy, $m_\pi c^2$, of the pion.

DISCOVERY OF THE MUON

In 1937, Anderson (the discoverer of the positron) and his student Neddermeyer discovered, among some cloud chamber tracks of cosmic rays, a charged particle with mass around 100 MeV/c^2. This mass matched Yukawa's prediction almost perfectly, and the discovery had the important effect of drawing attention to Yukawa's work. Nevertheless, it became clear during the next 10 years that the particle discovered by Anderson and Neddermeyer was not Yukawa's meson. For this reason Anderson and Neddermeyer's particles were eventually called **muons** or μ particles. (Actually, there were two such particles μ^+ and μ^-, with charges $+e$ and $-e$.) As already noted, Yukawa's particles came to be called pions, denoted π^+, π°, and π^-.

In retrospect, at least, the most important difference between the pions and muons is this: As carrier of the strong nuclear force, Yukawa's pion must necessarily interact strongly with nuclei, but Anderson and Neddermeyer's muon was found to interact very weakly with nuclei. For example, it was observed that some negative muons were captured into "atomic" orbits around an atomic nucleus. If the muon were Yukawa's particle, then, once in an atomic orbit, it would be absorbed promptly into the nucleus by the strong nuclear force (Problem 14.15) and would release enough energy to disintegrate the nucleus. The orbiting muon, however, frequently remained in orbit until it decayed of its own accord; that is, it decayed in about the same time that it takes in isolation.

For this reason (and several more) it is quite clear that the muons discovered by Anderson and Neddermeyer were not the particles predicted by Yukawa as the carriers of the strong nuclear force. In fact, the muons seem to have no connection at all with the strong nuclear force. Rather, the μ^- and μ^+, which are particle and antiparticle to one another, appear in almost every way to be overgrown siblings of the electron and positron, e^- and e^+. Like the electron and positron, the muons have spin $\frac{1}{2}$ and have magnetic moments in excellent agreement with the predictions of Dirac's theory. Like the e^- and e^+, muons are completely immune to the strong force.

DISCOVERY OF THE PION

By 1947, convinced that Anderson and Neddermeyer's muon was not the particle predicted by Yukawa, physicists were searching cosmic rays for signs of a second middle-mass particle that could be Yukawa's pion. If nucleons could emit and absorb virtual pions in the exchange processes of Fig. 14.3, high-energy collisions between two nucleons should be able to produce pions in processes of the form

$$N + N \rightarrow N + N + \pi \tag{14.13}$$

or, given sufficient energy,

$$N + N \rightarrow N + N + \pi + \pi + \cdots + \pi. \tag{14.14}$$

Thus one would expect many pions to be created in the upper atmosphere when high-energy cosmic rays collide with atmospheric nuclei. On the other hand, because pions were expected to interact strongly with atomic nuclei, the probability that they could traverse the several miles of atmosphere without being stopped by a nucleus is extremely small (Problem 14.14). Therefore, it seemed unlikely that many of the pions formed in cosmic-ray collisions would find their way down to sea level. Accordingly, a British group led by Powell began observations on the tops of mountains in the Pyrenees and the Andes. Photographic plates were left for a week or more on the mountain tops and, when developed, showed tracks of charged particles that had passed through them. After examining many hundreds of these photographs, Powell and his colleagues established that there were indeed two different kinds of medium-mass particles: the familiar muons with mass 105 MeV/c^2, and the long-expected pions with mass 140 MeV/c^2.

The first pions to be discovered were the charged π^+ and π^-. Powell et al. obtained pictures in which the pions decayed into muons in the processes

$$\pi^+ \rightarrow \mu^+ + \nu \tag{14.15}$$

and

$$\pi^- \rightarrow \mu^- + \bar{\nu}. \tag{14.16}$$

(The neutrinos, being neutral, cause no ionization and leave no visible tracks; nonetheless, their presence was evident from conservation of momentum.) It has since been shown that the processes (14.15) and (14.16) are the most probable decay modes of the charged pions and are the principal source of the many muons observed in cosmic rays. The lifetimes of both π^+ and π^- were found to be of order 10^{-8} s.

Powell et al. found a few pictures in which a pion decayed into a muon, as in (14.15) or (14.16), and the muon then decayed by its principal decay mode, into an electron and two neutrinos:*

$$\mu^\pm \rightarrow e^\pm + \nu + \bar{\nu}. \tag{14.17}$$

An early picture of this sequence of events is shown in Fig. 14.4(a).

It was quickly verified that pions interact strongly with nuclei. For example, the negative π^- can be captured into an "atomic" orbit around a nucleus, but is then rapidly drawn into the nucleus and absorbed. The energy released breaks the nucleus apart into several fragments, as illustrated in Fig. 14.4(b).

Soon after their discovery in cosmic rays, it became possible to produce pions artificially in processes like (14.14), with protons from accelerators such as the 6-GeV machine at Berkeley. This was obviously much more convenient than waiting on mountaintops for the occasional arrival of a cosmic-ray pion,

---

* Two neutrinos are required in this decay because all the particles involved have spin $\frac{1}{2}$. If there were only one neutrino on the right, the final total angular momentum would have an integer quantum number, whereas the lone μ has half-integer angular momentum.

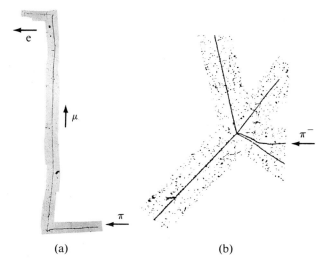

(a)   (b)

FIGURE 14.4 (a) Tracks of a cosmic ray pion and its decay products in photographic emulsion. The pion decays into a muon and an invisible neutrino; the muon then decays into an electron and two invisible neutrinos. (b) A "star" formed when a π^- is caught in an atomic orbit and absorbed in the nucleus; the energy released blows the nucleus apart into several charged fragments. (Courtesy Science Photo Library and Professor Fowler.)

and most of the accurate measurements of pion properties were made with pions produced by accelerators.

As expected from Yukawa's theory, it was found that in addition to the charged π^+ and π^-, there is a neutral π°. While the two charged pions, $\pi^\pm$, have exactly the same mass (being particle and antiparticle) the π° is slightly lighter:

$$m(\pi^\pm) = 139.6 \text{ MeV}/c^2,$$

whereas

$$m(\pi^\circ) = 135.0 \text{ MeV}/c^2.$$

A much more dramatic difference between the charged and neutral pions concerns their decay. The π^+ and π^- decay mainly by the processes (14.15) and (14.16), with equal mean lives of 2.6×10^{-8} s. (The mean life of any unstable particle is equal to that of its antiparticle.) The π° decays mainly to two photons,

$$\pi^\circ \rightarrow \gamma + \gamma \tag{14.18}$$

with the much shorter mean life

$$\tau(\pi^\circ) = 8.3 \times 10^{-17} \text{ s.} \tag{14.19}$$

The reason for the enormous difference between the lifetimes of the charged and neutral pions is that the decay (14.18) of the π° is induced by electromagnetic forces, while the decay of the charged pions depends on the much feebler weak force, which we discuss in the next section.

14.4 The Four Fundamental Forces

After the discovery of the muon and pion, many more particles were found, and it slowly became clear that the growing family of elementary particles could be classified into smaller groups according to their relationship with the fundamental forces of nature. In this section we discuss these fundamental forces.

By the mid-1930s it was evident that in addition to the familiar electromagnetic and gravitational forces, there had to be two other fundamental forces. First, there was the strong nuclear force that binds nucleons together in nuclei. Second, it was found that the phenomenon of β decay could not be explained in terms of the strong, electromagnetic, or gravitational forces. Therefore, there had to be a fourth force that caused β decays, and this was called the **weak force** because β decays occur so slowly compared to processes involving the strong force.

The ranges and strengths of the four forces are widely different. Both gravity and electromagnetism obey inverse-square laws and can, in principle, be felt at any distance. In this sense we say that these forces have infinite range. By contrast, the range of the strong force is only about 10^{-15} m, and that of the weak force is even smaller, being of order 10^{-18} m. At macroscopic distances, therefore, the strong and weak forces are absolutely negligible.

Even when two particles are very close, it is not necessarily the case that all four forces act between them. For example, no matter how close we bring two electrons, there is no strong force between them. However, when all four forces *do* act, the strong force is the most important, followed, in order of decreasing strength, by the electromagnetic, weak, and gravitational forces. In particular, the effects of gravity on individual particles are so small that they can usually be ignored entirely.

In discussing the fundamental forces, physicists tend to use the word **"interaction"** rather than "force." The reason for this preference is that "force" has the connotation of a simple push or pull. This connotation is entirely appropriate when we speak of the electromagnetic force that holds an electron in an atomic orbit. However, the same basic phenomenon — electromagnetism — is responsible for processes like the creation of a photon from the energy released in an atomic transition; and in this context, it seems inappropriate to speak of the electromagnetic *force*. Today, particle physicists use "electromagnetic interaction" to describe all manifestations of electromagnetism, from the simple force between two charges to the production of photons in atomic and nuclear transitions. In the same way, "strong interaction" is used to describe the force between two nucleons, but also processes like the production of pions in energetic collisions between protons. Similarly, "weak interaction" describes the mechanism that produces the electron and neutrino in nuclear β decays.

We have already seen that a few particles, including the electron, muon, and neutrino, are not affected by the strong interactions. These particles are called **leptons** (from the Greek for "light," since the first leptons discovered were the lightest known particles). As we describe in Section 14.5, we know of just six leptons and, as far as we can tell, all six are truly elementary particles. That is, there is no evidence that the leptons are composites made up of smaller particles.

Those particles that do feel the strong interactions are called **hadrons** (from a Greek word for "strong"). The obvious examples of hadrons are the

TABLE 14.2

Classification of the several hundred observed subatomic particles into three main groups.

Name of group	Characteristic	Examples	Current status
Gauge particles	Force carriers	photon	Elementary
Leptons	Immune to strong force	e, μ, ν	Elementary
Hadrons	Subject to strong force	p, n, π	Composite (made of quarks)

proton, neutron, and pion, but we now know of several hundred others. As we describe in Section 14.7, it now appears that none of these particles is truly elementary. Rather, it seems that they are all built up from just six different particles called quarks, and that these six quarks are structureless elementary particles like the six leptons.

Finally, in addition to the leptons and hadrons, there are the particles that carry the forces (or, more generally, interactions). These force carriers are called **gauge particles,** * and we now believe that every force is mediated by corresponding gauge particles. For example, the gauge particles for the electromagnetic interaction are photons. Those for the strong force between quarks are called gluons, and those for the weak interaction are called W and Z particles. We describe gluons in Section 14.8 and the W and Z in Section 14.9.

Our classification of the several hundred known subatomic particles into three main groups is summarized in Table 14.2. The last column in that table indicates that according to current evidence, the gauge particles and leptons are truly elementary, whereas the hadrons are composites, each made up of two or three elementary quarks. It is, of course, possible that future experiments will show that some of the particles which now appear to be elementary are themselves composites of some even more basic entities.

14.5 Leptons

The lepton family consists of spin-$\frac{1}{2}$ particles that are immune to the strong interaction, including the electron, muon, and neutrino. Apart from the large difference in their masses, the first two of these, e^- and μ^-, are strikingly alike. Both carry charge $-e$, and both have an antiparticle with charge $+e$. For almost every process that involves one, there is a corresponding process involving the other. For example, the usual decay mode of the π^- is to a μ^- and antineutrino,

$$\pi^- \rightarrow \mu^- + \bar{\nu}, \tag{14.20}$$

but the π^- can also decay into an e^- and antineutrino:†

$$\pi^- \rightarrow e^- + \bar{\nu}. \tag{14.21}$$

The other lepton that we have already met is the neutrino ν (plus its antiparticle $\bar{\nu}$, of course). In 1962 it was found that there are at least two

---

* This use of the word "gauge" originates in classical electromagnetic theory, where it is used in connection with the measurability of the electromagnetic potentials.

† The decay (14.21) is very rare compared to (14.20). However, this difference can be explained by the large difference in the e and μ masses (see Problem 14.19).

different kinds of neutrino: The neutrino produced in association with the μ, as in (14.20), is different from that produced with the e, as in (14.21). These two kinds of neutrino are now called the μ-neutrino, v_μ (with antiparticle $\bar{v}_\mu$), and the e-neutrino, v_e (with antiparticle $\bar{v}_e$). Thus the two reactions (14.20) and (14.21) should be rewritten as

$$\pi^- \rightarrow \mu^- + \bar{v}_\mu \tag{14.22}$$

and

$$\pi^- \rightarrow e^- + \bar{v}_e. \tag{14.23}$$

That these two neutrinos really are different was proved by showing that the μ-neutrinos produced in (14.22) could induce certain reactions involving muons, but could not induce the corresponding electron reactions (which *can* be induced by e-neutrinos).

In 1975 it was found that there are two more leptons: a charged τ^- and a corresponding τ-neutrino v_τ. The τ^- and its antiparticle τ^+ were produced by colliding electron and positron beams, in the reaction

$$e^- + e^+ \rightarrow \tau^- + \tau^+ \tag{14.24}$$

and both decayed in a variety of modes all of which involved τ-neutrinos. Like the muon, the τ particle is best described as an overgrown electron, but in this case the mass difference is even more enormous — m_τ is more than 3000 times greater than m_e:

$$m_\tau = 1784 \text{ MeV}/c^2.$$

Like all of the other four leptons, the τ and its neutrino both have spin $s = \frac{1}{2}$.

The τ particle and its τ-neutrino complete the list of known leptons. There are, as far as we know, just six leptons (each with a corresponding antiparticle). These six particles fall naturally into three pairs, or **generations,** each with one charged particle and a corresponding neutrino. This situation is summarized in Table 14.3.

Of the six particles listed in Table 14.3 only the electron has an obvious role in the structure of ordinary matter. To some extent one can understand why this is. The electron is apparently the lightest charged particle. Therefore, there is no particle to which the electron could decay, and it *has* to be stable. By contrast, the μ and τ are both very unstable and obviously cannot be a part of normal matter. Unfortunately, we have no theory to explain why the μ and τ exist in the first place, nor why their masses are so enormous compared to m_e. It is tempting to suggest that the electron could be a bound state of some smaller objects, and that the μ and τ are excited states of this same system. However, there is, at present, no evidence to support this suggestion. In fact, current

TABLE 14.3

The six known leptons fall into three "generations." Masses are given in MeV/c^2, and those for the neutrinos are upper limits. It is possible — and, many believe, probable — that all neutrino masses are exactly zero. For each particle there is a corresponding antiparticle.

Generation	Charged lepton (mass)	Neutrino (mass)
First	electron, e^- (0.5110)	e-neutrino, v_e ($<2 \times 10^{-5}$)
Second	muon, μ^- (105.7)	μ-neutrino, v_μ (<0.25)
Third	tau, τ^- (1784)	τ-neutrino, v_τ (<35)

evidence shows that, down to a scale of about 10^{-18} m, the e, μ, and τ are all structureless, elementary particles. Similarly, there is no evidence that the three neutrinos have any internal structure; and finally, no one really understands why each of the three generations includes one of these massless (or nearly massless) members.

14.6 Discovery of More Hadrons

The hadron family consists of all particles that feel the strong interaction and includes the familiar proton, neutron, and pion. The advent in the early 1950s of machines that could accelerate particles to several GeV began a period of discovery of many new hadrons. In this section we describe how these particles are detected and mention some of their properties.

THE LONGER-LIVED NEW HADRONS

A few of the new hadrons have mean lives of order 10^{-8} to 10^{-10} s. We shall see that, by the standards of particle physics, this is a long time. In particular, a particle traveling near the speed of light can easily travel several centimeters in this time. If the particle is charged, one may be able to see the trail that it leaves in a suitable detector. This is illustrated in Fig. 14.5, which is a photograph made in a bubble chamber containing liquid hydrogen near its boiling point. When a charged particle traverses the chamber, tiny bubbles of hydrogen gas form near the particle's path, making a visible trail analogous to the trail of condensation left in a cloud chamber. The advantage of the bubble chamber is that the density of nuclei in the liquid is some 1000 times greater than that in the vapor of a cloud chamber; this increases the frequency of collisions by the same factor and gives the experimenter much more data in a reasonable time. In Fig.

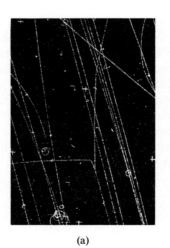
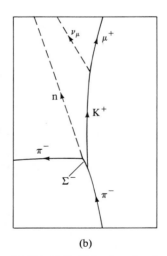

(a)   (b)

FIGURE 14.5 (a) Photograph made in a hydrogen bubble chamber traversed by several negative pions. One π^- has hit a stationary hydrogen nucleus and induced the reaction $\pi^- + p \rightarrow K^+ + \Sigma^-$. (b) Tracing of the interesting tracks, including the paths of the neutral particles, which are invisible in the photograph. The Σ^- has decayed into a neutron and π^- after traveling a short distance; the K^+ has decayed into a neutrino and μ^+ shortly before leaving the picture. (Courtesy Lynn Stevenson.)

14.5 the tracks of several negative pions are visible entering the picture from below. One of these pions has hit a stationary proton (a hydrogen nucleus) and produced two new charged hadrons, called K^+ and Σ^-, in the reaction

$$\pi^- + p \rightarrow K^+ + \Sigma^-. \qquad (14.25)$$

Both of the new particles produced in the reaction (14.25) have left visible trails in Fig. 14.5, and their subsequent decays can also be seen, via the processes

$$K^+ \rightarrow \mu^+ + \nu_\mu \quad \text{and} \quad \Sigma^- \rightarrow \pi^- + n. \qquad (14.26)$$

The track of each charged particle is curved because of an applied magnetic field, and from the measured curvature one can calculate the particle's momentum. Under favorable conditions (Problem 14.27) one can also find the energy and hence speed. Knowing the particle's speed and how far it traveled, one can calculate how long it lived. Then, if one has several examples to average over, one can find the mean life.

If a particle is neutral, it leaves no track in a bubble chamber or similar device. Nevertheless, its path and properties can be inferred if its decay products are charged. This is illustrated in Fig. 14.6, which shows a bubble-chamber photograph of a reaction that produces two neutral hadrons:

$$\pi^- + p \rightarrow K^\circ + \Lambda^\circ. \qquad (14.27)$$

Neither of these neutral particles is visible, but each decays into two charged particles, producing a characteristic "vee" on the picture.

RESONANCE PARTICLES

Many of the new particles decay far too quickly to leave an identifiable track like those in Figs. 14.5 and 14.6. To illustrate how these so-called resonance particles can be identified, let us consider the case of the hadron called

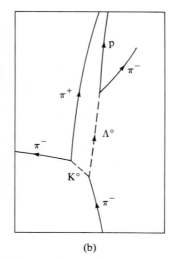

(a)   (b)

FIGURE 14.6 (a) Bubble-chamber photograph of the reaction $\pi^- + p \rightarrow K^\circ + \Lambda^\circ$. Neither of the neutral particles is visible, but the subsequent decay into two charged particles is easily seen. (b) Tracing of the same sequence, including the paths of the two neutral particles. (Courtesy Lawrence Berkeley Lab.)

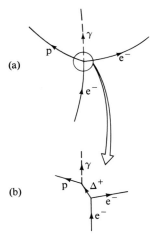

(a)

(b)

FIGURE 14.7 (a) The reaction $e + p \rightarrow e + \Delta^+$ followed by the decay $\Delta^+ \rightarrow \gamma + p$ as seen, for example, in a bubble chamber. The Δ^+ decays too quickly to leave any track, and the observed process is indistinguishable from the single reaction $e + p \rightarrow e + \gamma + p$. (The photon would, of course, leave no visible track, but could be detected in several other ways.) (b) An imaginary enlargement by some 13 orders of magnitude shows the actual sequence of events (14.28).

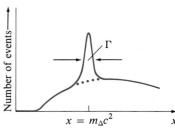

FIGURE 14.8 Schematic plot of the number of events like that in Fig. 14.7(a) against the variable x defined in Eq. (14.33). Those events in which the outgoing γ and proton came from decay of a Δ^+ all have $x = m_\Delta c^2$ and cause the spike at that value. The parameter Γ is called the width of the Δ^+ and is traditionally defined as the width of the spike at half its maximum height.

Δ^+. This can be produced in the reaction

$$e + p \rightarrow e + \Delta^+$$
$$\quad\quad\quad\quad \rightarrow \gamma + p, \quad\quad (14.28)$$

where we have also indicated the subsequent decay of the Δ^+ into $\gamma + p$.* The Δ^+ lifetime is so short that it could only travel a distance of order 10^{-15} m, which is less than a nuclear diameter and is far too short to show as a visible track in any detector. Thus a photograph of the sequence of reactions (14.28) would *look* just like the single direct reaction

$$e + p \rightarrow e + \gamma + p \quad\quad (14.29)$$

as shown in Fig. 14.7(a). Nevertheless, it *is* possible to identify the Δ^+ by examining many events with the overall appearance of (14.29) [some of which would have proceeded via (14.28) and some directly]. To understand this, we must consider the energy and momentum of the particles concerned.

The energy and momentum of the Δ^+ in (14.28) must satisfy the Pythagorean relation $E^2 = (pc)^2 + (mc^2)^2$, which implies that

$$m_\Delta c^2 = \sqrt{E_\Delta^2 - (\mathbf{p}_\Delta c)^2}, \quad\quad (14.30)$$

where m_Δ is the rest mass of the Δ^+. Because the Δ^+ leaves no visible track, one cannot measure E_Δ or $\mathbf{p}_\Delta$ directly. However, by conservation of energy and momentum,

$$E_\Delta = E_\gamma + E_p \quad\quad \text{and} \quad\quad \mathbf{p}_\Delta = \mathbf{p}_\gamma + \mathbf{p}_p, \quad\quad (14.31)$$

where E_γ, E_p, and $\mathbf{p}_\gamma$, $\mathbf{p}_p$ are the energies and momenta of the outgoing γ and proton, all of which *can* be measured. Substituting (14.31) into (14.30), we see that

$$m_\Delta c^2 = \sqrt{(E_\gamma + E_p)^2 - (\mathbf{p}_\gamma + \mathbf{p}_p)^2 c^2}. \quad\quad (14.32)$$

The relation (14.32) applies to all events involving a Δ^+ as in (14.28). However, the process (14.29) can also occur directly without formation of a Δ^+, and such direct processes are not subject to (14.32). Given a single picture like Fig. 14.7(a), we cannot tell whether it involved a Δ^+ or not, so we cannot apply (14.32) to find m_Δ. Nevertheless, it *is* possible to identify the existence of the Δ^+ particle, and to find its mass, provided that one has many similar pictures.

For each picture like Fig. 14.7(a), we measure the energies and momenta of the outgoing γ and proton and calculate the quantity

$$x = \sqrt{(E_\gamma + E_p)^2 - (\mathbf{p}_\gamma + \mathbf{p}_p)^2 c^2}. \quad\quad (14.33)$$

From (14.32) it is clear that x must equal $m_\Delta c^2$ for all those events in which the γ and p came from decay of a Δ^+. On the other hand, if one calculates x for those events (14.29) that did *not* involve a Δ^+, one obtains a continuous range of values. Therefore, if we draw a histogram, plotting number of events against the observed values of x, we should get a curve like that shown in Fig. 14.8. The smooth curve (including the dotted curve under the spike) corresponds to the direct events in which a Δ^+ was not produced; the sharp spike at $x = m_\Delta c^2$

---

* We choose this comparatively rare decay mode for convenience in the following discussion. The commonest decay modes are actually $\Delta^+ \rightarrow \pi^\circ + p$ and $\Delta^+ \rightarrow \pi^+ + n$.

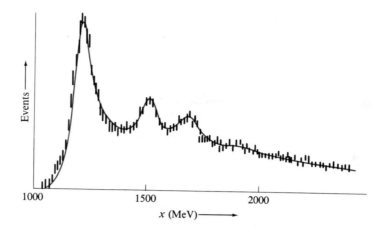

FIGURE 14.9 A plot similar to Fig. 14.8, but based on actual data for electron–proton collisions. The first spike corresponds to the Δ^+ particle with mass 1232 MeV/c^2; the other two spikes correspond to two other unstable particles, called N(1520) and N(1680).

corresponds to those events (14.28) in which the outcoming γ and proton came from decay of a Δ^+, and for all of which $x = m_\Delta c^2$. Turning this argument around, a plot like Fig. 14.8 is evidence for the existence of an unstable particle, and the value of x at which the spike occurs tells us the particle's rest energy mc^2. Figure 14.9 shows real data from e–p collisions. The spike at 1232 MeV corresponds to the Δ^+ particle (with $m_\Delta c^2 = 1232$ MeV) and the other two spikes correspond to two other unstable particles.

A particle whose presence has to be inferred as in Fig. 14.8 or 14.9 is called a **resonance particle.**[*] In practice, resonance particles are those particles whose lifetimes are too short to be measured directly. Nevertheless, their lifetimes can be measured using the time-energy uncertainty relation. As we saw in Section 7.9, an unstable quantum system has a minimum uncertainty in its energy:

$$\Delta E = \frac{\hbar}{2\Delta t} \tag{14.34}$$

where Δt is the mean life of the system — in this case the lifetime τ. For an unstable particle, ΔE is the uncertainty in the rest energy, and this uncertainty is the reason that the spike in Fig. 14.8 cannot be perfectly sharp. Particle physicists usually characterize the width of this kind of spike by the parameter Γ that is shown in Fig. 14.8 and is called the **full width at half maximum.** Since we have always defined ΔE as the half-width, it follows that $\Gamma = 2\Delta E$, and (14.34) can be rewritten as $\Gamma = \hbar/\Delta t = \hbar/\tau$ or

$$\tau = \frac{\hbar}{\Gamma}. \tag{14.35}$$

By measuring the width Γ from Fig. 14.8 we find the mean life τ.

---

[*] The name is based on an analogy with resonances in LC circuits. If an ac voltage of frequency f is applied to an LC circuit, a large current I results if f is close to the resonant frequency, but I is small otherwise. Thus a graph of I against f has the same general shape as Fig. 14.8, with a sharp spike at the resonant frequency.

EXAMPLE 14.1   The rest energy of the Δ^+ particle is uncertain, with a measured width $\Gamma = 115$ MeV. What is the mean life of the Δ^+?

According to (14.35), the mean life τ is given by

$$\tau = \frac{\hbar}{\Gamma} = \frac{6.6 \times 10^{-16} \text{ eV} \cdot \text{s}}{115 \times 10^6 \text{ eV}} = 5.7 \times 10^{-24} \text{ s}.$$

This value justifies our earlier claim that a Δ^+ particle — even if traveling at a substantial fraction of c — can only cover a distance of order 10^{-15} m.

THE STRANGE PARTICLES

One of the early puzzles about the many new hadrons found in the 1950s was the wide range in their lifetimes. Based on the known strength of the strong interaction, calculations show that an unstable hadron should have a mean life of order 10^{-23} s (like the Δ^+). Thus the real puzzle was the enormously long lives of particles like the Σ and Λ with lifetimes of order 10^{-10} s (some 13 orders of magnitude longer than expected). For this reason these particles were called **strange particles.** The theoretical efforts to understand their long lives contributed to the development of the quark model of hadrons, which we describe in Section 14.7.

MULTIPLETS

A striking feature of many of the new hadrons is that they occur in families, or **multiplets,** whose members have similar masses but different charges. For example, the Δ^+ is one of a quartet of particles, $\Delta^{++}, \Delta^+, \Delta^\circ, \Delta^-$, all of whose masses are close to 1232 MeV/c^2 but whose charges are $+2e, +e, 0$, and $-e$. A more familiar example is the doublet consisting of the proton and neutron, and another is the triplet of pions, π^+, π°, and π^-. Several examples of multiplets, ranging from a singlet to a quartet, are shown in Fig. 14.10. Apart from the difference in their charges, the particles of a multiplet seem to be almost identical; in particular, they respond to the strong interaction in the same way. The existence of these multiplets was another stimulus for the development of the quark model of hadrons (Section 14.7).

BARYON NUMBER

We are going to see in the next section that many of the new hadrons prove to be excited states of the neutron and proton. For example, the Δ^+ is now recognized to be an excited state of the proton. This conclusion is suggested by the process already discussed,

$$e + p \rightarrow e + \Delta^+$$
$$ \hookrightarrow \gamma + p. \qquad (14.36)$$

It is possible to interpret the first event as the excitation of the proton to a state of higher energy (called Δ^+); in the subsequent decay, the excess energy creates a photon and the excited state drops back to its ground state, the proton. (Historically, the excited state was not recognized as such, and was given the separate name Δ^+.) In this view the process (14.36) is closely analogous to the Franck–Hertz experiment (Section 6.10)

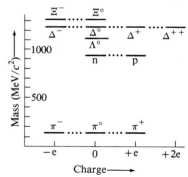

FIGURE 14.10   A plot of mass against charge for five examples of the many hadron multiplets. Except with the nucleon doublet (n and p), the same letter is used for all members of each multiplet, with the different charges shown as superscripts. On this scale, the mass differences within each multiplet cannot be seen.

$$e + Hg \rightarrow e + Hg^*$$
$$\hookrightarrow \gamma + Hg, \qquad\qquad (14.37)$$

in which an electron excited a mercury atom, which subsequently emitted a photon and dropped back to its ground state. In either case, the energy of the excited state (Δ^+ or Hg*) can be deduced by observing the outgoing photon or, equivalently, by measuring the energy lost by the incident electron.

In Chapter 13 we noted that in all the processes of nuclear physics the total number of nucleons is conserved. In the light of reactions of the type $e + p \rightarrow e + \Delta^+$, this law must obviously be modified to include the excited states of nucleons. The word *nucleon* has been reserved for the neutron and proton, and the word **baryon** (from the Greek "heavy") has been introduced to mean *either* one of the two nucleons *or* any of a family of similar particles, including excited states like the Δ^+. (We shall see in Section 14.7 that a baryon can be succinctly defined as any bound state of three quarks.) No process has ever been observed in which the total number of baryons changes; that is, as far as we know, the baryon number, denoted B, of an isolated system is an absolutely conserved quantity.

All of the baryons, except for the neutron and proton, are very unstable. Therefore, the only baryons present in normal matter are the two nucleons. In the reactions discussed in Chapter 13 there was insufficient energy to produce any of the heavier baryons; in that case, therefore, the conservation of baryon number reduced to conservation of nucleon number.

14.7 The Quark Model of Hadrons

We have said that many of the new hadrons seem to be excited states of the neutron or proton. This requires that the neutron and proton must themselves be composite systems, constructed from other, more truly elementary objects. In the early 1960s the American physicists Gell-Mann and (independently) Zweig showed that if one assumed that there are three different kinds of particles, which Gell-Mann called **quarks,** then every known hadron could be explained as a composite of two or three quarks.

The three quarks that Gell-Mann and Zweig proposed have come to be called "up" or u, "down" or d (both with mass around 350 MeV/c^2) and "strange" or s (with mass around 500 MeV/c^2). Each is a spin-$\frac{1}{2}$ particle, and each has a corresponding antiparticle ($\bar{u}$, $\bar{d}$, and $\bar{s}$). In Table 14.4 we list the principal assumed characteristics of these three quarks. (As we describe later, there are now thought to be six quarks in all.) The most startling of the properties listed in Table 14.4 is the charge of the quarks. Gell-Mann and Zweig found that they had to assume the quarks carry *fractional charge*—$2e/3$ for the u and $-e/3$ for the d and s. Since no one had ever observed fractional charges, this was

MURRAY GELL–MANN (born 1929, US). Gell–Mann got his PhD from MIT at 22 and was a full professor at Cal Tech by 26. In 1961 he predicted the existence of the omega-minus particle, which was found in 1964 (see Fig. 14.11), and in 1964 he proposed the idea that all hadrons are made up of quarks. He won the Nobel Prize in 1969.

TABLE 14.4
The original three quarks proposed by Gell-Mann and Zweig.

Name	Symbol	Charge	Baryon number	Mass (MeV/c^2)	Spin
Up	u	$2e/3$	$\frac{1}{3}$	≈ 350	$\frac{1}{2}$
Down	d	$-e/3$	$\frac{1}{3}$	≈ 350	$\frac{1}{2}$
Strange	s	$-e/3$	$\frac{1}{3}$	≈ 500	$\frac{1}{2}$

a very bold assumption. Nevertheless, it has now been confirmed in several ways.

Using the properties shown in Table 14.4, it is easy to understand the quark model of hadrons. All baryons (the proton, neutron, and their heavier relatives) are explained as bound states of three quarks. Since each quark has baryon number $B = \frac{1}{3}$, this guarantees that all baryons have $B = 1$, as one would expect. The proton is the lowest bound state of two up quarks and one down, and the neutron is the lowest state of one up and two downs:

$$\text{proton} = (\text{uud}), \qquad \text{neutron} = (\text{udd}). \tag{14.38}$$

Notice how this gives both p and n their correct charges:

$$q_{\mathbf{p}} = (\tfrac{2}{3} + \tfrac{2}{3} - \tfrac{1}{3})e = e, \qquad q_{\mathbf{n}} = (\tfrac{2}{3} - \tfrac{1}{3} - \tfrac{1}{3})e = 0.$$

Since the u and d quarks have nearly the same mass, it immediately follows that the same is true of the proton and neutron. Because both the nucleons are known to have spin $\frac{1}{2}$, it is clear from (14.38) that two of the three quarks must have their spins antiparallel to give the observed total spin $s = \frac{1}{2}$.*

The Δ^+ is an excited state of the same three quarks (uud) that make up the proton. In fact, the four particles $\Delta^{++}, \Delta^+, \Delta^\circ, \Delta^-$ are simply the corresponding states of the four combinations

$$\Delta^{++} = (\text{uuu}), \qquad \Delta^+ = (\text{uud}), \qquad \Delta^\circ = (\text{udd}), \qquad \Delta^- = (\text{ddd}). \tag{14.39}$$

Since the observed spin of the Δ particles is $s = \frac{3}{2}$, the three quark spins must all be parallel in this case.†

The strange baryons, like the Σ and Λ described in Section 14.6, are simply those baryons that include a strange quark s. For example, the particle called Σ^+ consists of two u quarks and one s:

$$\Sigma^+ = (\text{uus}).$$

The Σ^+ is the lowest state of these three quarks, but because the s quark is heavier than the d, the Σ^+ is heavier than the proton (uud). In fact, the Σ^+ is sufficiently heavy to decay into a proton and π°. However, this requires that the s quark convert into a d quark, and the only interaction that can convert one type of quark into another is the weak interaction. Thus the process $\Sigma^+ \rightarrow p + \pi^\circ$ can occur only via the weak interaction, and this explains the "strangely" long life of the Σ^+.

The π mesons are explained as quark–antiquark pairs; for example,

$$\pi^+ = (\text{u}\bar{\text{d}}), \qquad \pi^- = (\bar{\text{u}}\text{d}).$$

In fact, the name *meson* is now used to denote any bound state of one quark and one antiquark. For example, the particles K^+ and K° mentioned in Section 14.6 are strange mesons, containing one strange quark (actually, an $\bar{\text{s}}$):

$$K^+ = (\text{u}\bar{\text{s}}), \qquad K^\circ = (\text{d}\bar{\text{s}}).$$

---

* Actually, the "spin" of the nucleon is the total angular momentum of the three quarks, and this could include orbital angular momentum as well as spin. However, in the lowest states the orbital angular momentum is zero, and the conclusion stated is correct. As one would expect, there are excited states with nonzero orbital angular momentum as well.

† You may reasonably ask why in (14.38) the two nucleons do not have two more companions consisting of the combinations (uuu) and (ddd), as in (14.39). The answer (whose details are too complicated to give here) is that the Pauli principle forbids these two combinations for the $s = \frac{1}{2}$ states (but allows them for $s = \frac{3}{2}$).

Notice that since all quarks have baryon number $B = \frac{1}{3}$, while antiquarks have $B = -\frac{1}{3}$, it follows that all mesons have baryon number zero.

It is found that *all* hadrons are either baryons (three quarks) or mesons (a quark and an antiquark). One could imagine various other combinations — two quarks or two antiquarks, for example — but these seem never to occur. We shall return to this point in Section 14.8.

The quark model is much more than a mere description of how hadrons are built up from quarks. For example, the scheme predicts the existence of multiplets of hadrons, such as the Δ quartet, which, we have seen, consists of corresponding spin-$\frac{3}{2}$ states of the four combinations

$$\text{(uuu)},\quad \text{(uud)},\quad \text{(udd)},\quad \text{(ddd)} \;.\;.\;.\;.\;.\; \Delta(1232).$$

(The parentheses after the symbol Δ show the mass in MeV/c^2.) Because the u and d quarks have very nearly the same mass, the same is true of these four Δ particles. Because the s quark is a little heavier than the u and d, the corresponding states containing s quarks are heavier than their nonstrange counterparts. Thus the quark model requires three baryons, somewhat heavier than the Δ particles, each with one s quark:

$$\text{(uus)},\quad \text{(uds)},\quad \text{(dds)} \;.\;.\;.\;.\;.\; \Sigma(1385).$$

These have been observed and are called Σ particles. In addition, there should be two corresponding particles each with two s quarks:

$$\text{(uss)},\quad \text{(dss)} \;.\;.\;.\;.\;.\; \Xi(1530),$$

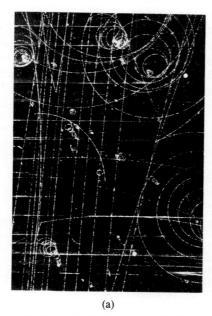

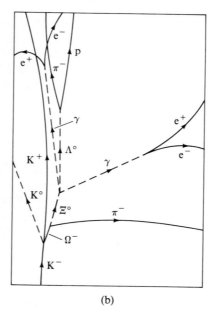

(a)
(b)

FIGURE **14.11** Photograph of the production of an Ω^- baryon in a liquid-hydrogen bubble chamber. Many K^- mesons entered the picture from below, and one produced the Ω^- in the reaction $K^- + p \rightarrow \Omega^- + K^+ + K^0$. The Ω^- traveled about an inch before decaying into a π^- and Ξ^0. The neutral Ξ^0 is invisible, as are its decay products, γ, Λ^0, and γ; but all of these neutrals eventually reveal themselves, the photons by producing e^+e^- pairs and the Λ^0 in the decay $\Lambda^0 \rightarrow \pi^- + p$. (Courtesy Brookhaven National Lab.)

TABLE 14.5

The lowest-mass baryon supermultiplet. The eight members of this octet are arranged in four multiplets. Each multiplet is labeled by its symbol, followed by its average mass in MeV/c^2; each particle symbol is followed by its actual mass. All the members of a given supermultiplet have the same spin, in this case $s = \frac{1}{2}$.

Multiplet	Particle	Quark content	Mean life (s)
N(939)	p(938.3)	uud	∞
	n(939.6)	udd	896
Λ(1116)	Λ(1115.6)	uds	2.63×10^{-10}
Σ(1193)	Σ^+(1189.4)	uus	7.99×10^{-11}
	Σ°(1192.5)	uds	7.4×10^{-20}
	Σ^-(1197.4)	dds	1.48×10^{-10}
Ξ(1318)	Ξ°(1314.9)	uss	2.90×10^{-10}
	Ξ^-(1321.3)	dss	1.64×10^{-10}

called Ξ (Greek capital letter xi), and finally, a singlet with three s quarks:

$$(\text{sss}) \ . \ . \ . \ . \ . \ \Omega(1672),$$

called the Ω^-. Altogether the quark model predicts a **supermultiplet** of 10 particles, all with the same wave functions and with masses that increase with the number of s quarks. The prediction of this supermultiplet was made before the Ω particle had been observed, and the subsequent discovery of the Ω, with the predicted mass and decay modes, was an important step in the development and acceptance of the quark model. A photograph made in 1964 of the production and decay of the first Ω^- to be discovered is shown in Fig. 14.11.

In Tables 14.5 to 14.7 we have listed the members of three low-mass supermultiplets, showing how all of these particles are made up from just the three quarks, u, d, and s. Table 14.5 shows the spin-$\frac{1}{2}$ baryon octet, which includes the proton and neutron; 14.6 shows the spin-$\frac{3}{2}$ decuplet just described; and 14.7 shows the spin-0 meson octet, which includes the three pions. These

TABLE 14.6

The second lowest baryon supermultiplet is the spin-$\frac{3}{2}$ decuplet. Although the masses of the particles within each multiplet vary slightly, these differences are not shown since they are all much smaller than the widths. Similarly, the small differences in lifetimes are not included.

Multiplet	Particle	Quark content	Mean life (s)
Δ(1232)	Δ^{++}(1232)	uuu	
	Δ^+(1232)	uud	6×10^{-24}
	Δ°(1232)	udd	
	Δ^-(1232)	ddd	
Σ(1385)	Σ^+(1385)	uus	
	Σ°(1385)	uds	2×10^{-23}
	Σ^-(1385)	dds	
Ξ(1530)	Ξ°(1530)	uss	7×10^{-23}
	Ξ^-(1530)	dss	
Ω(1672)	Ω^-(1672)	sss	8×10^{-11}

TABLE 14.7

The lowest meson octet, all of whose members have spin $s = 0$. The $\pi°$ and η are mixtures of $u\bar{u}$, $d\bar{d}$, and $s\bar{s}$. The $K°$ and $\overline{K}°$ each have two components with different mean lives.

Multiplet	Particle	Quark content	Mean life (s)
$\pi(138)$	$\pi^+(139.6)$	$u\bar{d}$	2.6×10^{-8}
	$\pi°(135.0)$	$u\bar{u}$ and $d\bar{d}$	8.4×10^{-17}
	$\pi^-(139.6)$	$d\bar{u}$	2.6×10^{-8}
$K(496)$	$K^+(493.6)$	$u\bar{s}$	1.2×10^{-8}
	$K°(497.7)$	$d\bar{s}$	5.2×10^{-8} and 8.9×10^{-11}
$\overline{K}(496)$	$\overline{K}°(497.7)$	$s\bar{d}$	5.2×10^{-8} and 8.9×10^{-11}
	$K^-(493.6)$	$s\bar{u}$	1.2×10^{-8}
$\eta(549)$	$\eta°(548.8)$	$u\bar{u}$, $d\bar{d}$, and $s\bar{s}$	6.1×10^{-19}

tables illustrate several general properties of the quark model. For example, the theory predicts that the number of particles in a baryon supermultiplet can only be 1 (a singlet), 8 (an octet as in Table 14.5), or 10 (a decuplet as in Table 14.6); similarly, the number of particles in a meson supermultiplet can only be 1 or 8 (as in Table 14.7). All of the several observed supermultiplets obey these predictions. The number of hadrons is far too large for each particle to be identified by a separate letter. Instead, as illustrated in the tables, a single letter is used to label several similar particles, and the identification is completed by giving the particle's charge as a superscript, and its mass (in MeV/c^2) in parentheses [for example, $\Sigma^+(1189.4)$ in Table 14.5 and its excited state $\Sigma^+(1385)$ in Table 14.6].

Each of the supermultiplets in Tables 14.5 to 14.7 is made up of several multiplets whose members are very close in mass. All of the particles in a single supermultiplet have many properties in common, and this is even more so for the particles within a single multiplet. For example, all members of a given supermultiplet have the same spin; the masses in a given supermultiplet are at least roughly equal, while those within a multiplet are very close. On the other hand, the lifetimes of particles within a single supermultiplet are often very different. This is because relatively small differences in mass can radically affect the possibilities for decay. For example, we see in Table 14.5 that the lifetime of the neutron is 896 s while that of the proton is infinite (that is, the proton is stable); this is because m_n, although very close to m_p, is nevertheless greater than $m_p + m_e$, which allows the neutron to decay, whereas the proton cannot. Similar circumstances explain the anomalously long life of the Ω in Table 14.6 and the huge difference between the lives of the π^+ and $\pi°$ in Table 14.7.

The 26 particles listed in Tables 14.5 to 14.7 are only a sampling of the many hadrons that have been observed. However, the majority of the known hadrons are simply excited states of those listed here, and we do not need to list them all now that we understand how they come about (any more than we needed to list the excited states of the hydrogen atom once we understood their origin).*

---

* A complete listing of the known particles, called "Particle Properties Data Booklet," is published every two years and can be obtained from the Technical Information Department, MS 90-2125, Lawrence Berkeley Laboratory, Berkeley, CA 94720.

DIRECT EVIDENCE FOR QUARKS

If hadrons are constructed from quarks, one would naturally hope to be able to break a hadron apart and observe its individual constituents. Since quarks carry fractional charges ($2e/3$ and $-e/3$) they should be easily identifiable once they are separated. However, extensive searches have produced no verifiable observations of any fractional charges. One explanation could be that the energy needed to split a hadron apart is more than the energy available in any experiment currently possible. However, this seems unlikely, since cosmic-ray particles have been observed with energies of 10^{22} eV; although energies this high are admittedly rare, one would nevertheless expect them to produce *some* free quarks, a few of which should be observed. The alternative is that although hadrons are constructed of quarks, it is actually *impossible* to break the quarks apart. This is the explanation that is currently favored. As we describe in Section 14.8, the present theory of the strong force between quarks suggests that an infinite energy is needed to pull apart either the three quarks in a baryon or the quark–antiquark pair in a meson (a phenomenon called **quark confinement**).

Although no one has observed an isolated quark, and many physicists believe no one ever will, there is compelling evidence that there are quarks inside hadrons. The first such evidence came from Stanford in California, where a 2-mile-long linear accelerator produced electrons with many GeV of energy. These electrons were fired at protons and neutrons, and the resulting reactions observed. Because the electron is a point particle and interacts mainly via the well-understood electromagnetic interaction, it is a particularly suitable probe of the structure of nucleons. The first clear conclusion of these experiments was simply that nucleons are not point particles; rather, they are extended objects with charge* spread over a distance of order 10^{-15} m.

The original experiments at Stanford have been repeated at ever-increasing energies, and hence with improving spatial resolution. The results of these experiments are conclusive: The electrons scatter off the nucleon at much larger angles than would be expected if the nucleon were a continuous distribution of matter. Rather, the evidence implies that the nucleon contains several much smaller charged objects, whose individual size is 10^{-18} m or less. This conclusion is closely analogous to Rutherford's discovery of the atomic nucleus (Section 4.3), where the scattering of α particles at unexpectedly large angles indicated a localized concentration of charge and mass inside the atom. The distribution of scattered electrons observed in the Stanford experiments implies that there are three particles inside the nucleon, that they have spin $\frac{1}{2}$, and that they carry charges of $2e/3$ or $-e/3$. These conclusions are reinforced by experiments using other probes, such as muons. Taken together these scattering experiments are accepted by most physicists as proof of the existence of quarks, just as the Rutherford scattering experiments were accepted as proof of the existence of the atomic nucleus.

MORE QUARKS

We have so far discussed three types of quark, the u, d, and s. The possibility of a fourth type arose in connection with the decay of the meson called K$^\circ$. It was expected that the K$^\circ$ should be able to decay into two muons,

---

* Of course, the neutron has zero total charge. Nevertheless, it is found that there are charges (equal quantities of positive and negative) distributed inside the neutron.

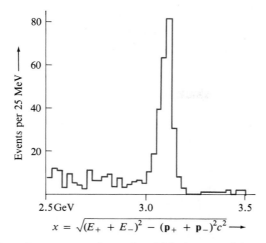

FIGURE 14.12 Data from an experiment in which the ψ particle was created in the process $p + p \rightarrow \psi +$ other particles, and promptly decayed into an e^+e^- pair: $\psi \rightarrow e^+ + e^-$. This is a beautiful example of a resonance plot of the type discussed in Section 14.6 and illustrated in Fig. 14.8. The variable x is calculated from the e^+ and e^- energies and momenta; it is equal to $m_\psi c^2$ for all events that involved a ψ, but is continuously distributed for those that did not. One sees clearly that $m_\psi c^2 = 3.1$ GeV. (Courtesy Professor Ting.)

$$K^\circ \rightarrow \mu^+ + \mu^-, \tag{14.40}$$

but extensive searches had found absolutely no sign of this decay. In 1970 it was shown (by an argument that is beyond our scope here) that if one postulated a fourth quark, with mass two to four times that of the s quark, the decay (14.40) would be unobservably improbable — thereby explaining the failure to observe it. Whether such a quark, called c for "charm," actually existed was entirely unknown, but in 1974 the c quark was discovered in the form of a new meson that could only be explained as a bound state of a new quark (the c) and its antiparticle ($\bar{c}$). This new meson is called the ψ particle (or J/ψ in tribute to the two groups that discovered it, one naming it J and the other ψ):

$$\psi = (c\bar{c}).$$

The mass of the ψ was found to be 3.1 GeV/c^2 and that of the c quark about* 1.5 GeV/c^2, comfortably inside the range predicted. A reasonance plot from one of the experiments that discovered the ψ particle is shown in Fig. 14.12.

A few years later yet a fifth quark, the b or "bottom," was found in a similar way: A new meson, the Y (Greek capital letter upsilon), was discovered and was shown to be a bound state of the fifth quark b and its antiparticle $\bar{b}$:

$$Y = (b\bar{b}).$$

---

* Because no quarks have been observed in isolation, their masses must be inferred from the masses of their observed bound states (such as the ψ). The mass of a bound state depends on the masses of its constituents *and* their kinetic and potential energies. Since the kinetic and potential energies are not known accurately, the same is true of the quark masses. For this reason all quark masses are very approximate.

TABLE 14.8

The six leptons and six quarks, showing their groupings into three generations. Notice that all the particles in each column have the same charge, as indicated on the bottom line.

Generation	Leptons		Quarks	
First	e^-	ν_e	u	d
Second	μ^-	ν_μ	c	s
Third	τ^-	ν_τ	t	b
Charge	$-e$	0	$\dfrac{2e}{3}$	$\dfrac{-e}{3}$

The b quark is much heavier than its four predecessors, with $m_b \approx 5.0 \text{ GeV}/c^2$. Today it is widely believed that there is a sixth quark, t or "top," with $m_t \geq 20 \text{ GeV}/c^2$, although the evidence is not conclusive.

Surprisingly enough, the discovery of six different kinds of quark* actually simplifies our picture of the elementary particles. Since there are six leptons, there is now a pleasing parallel between quarks and leptons. This parallel is sharpened when we realize that much like the six leptons, the six quarks fall naturally into three "generations." The justification for this grouping, which is shown in Table 14.8, is that when a quark of one type changes to a second type (which can only occur via the weak interaction), this change occurs predominantly within one generation. That is, a u changes to a d, or vice versa, much more readily than a u changes to an s, and so on. Further, just as with the leptons, it is only the first generation of quarks that plays an obvious role in normal stable matter. Although nobody knows whether this parallel between leptons and quarks is anything more than a coincidence, it does introduce a gratifying order into the roster of elementary particles.

Because the new quarks (c, b, t) are so much more massive than the first three (u, d, s), it turns out to be much easier to calculate the energies of their bound states. A light quark (u, d, or s) inside a hadron has kinetic energy larger than its rest energy and moves relativistically. For this reason it has, to date, proved impossible to calculate the allowed energies of hadrons containing light quarks. By contrast, the kinetic energy of a heavy quark inside a hadron is much less than the rest energy, and the motion is nonrelativistic. This lets particle physicists use the familiar (nonrelativistic) Schrödinger equation to calculate the energies of those hadrons containing only heavy quarks. For example, using the theory of strong forces sketched in Section 14.8, one can write down the potential energy of the c and $\bar{c}$ that comprise the ψ meson and its various excited states. Using this potential-energy function, one can solve the Schrödinger

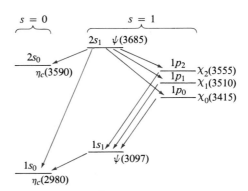

FIGURE 14.13 Energy levels of the ($c\bar{c}$) system can be calculated using the Schrödinger equation and the potential-energy function (14.41) (next section). The levels are labeled according to the scheme nl_j (for example, $2s_1$ means $n = 2$, $l = 0$, $j = 1$) and s denotes the total spin ($s = 0$ or 1). All levels shown have been observed and their experimental names are indicated, the number in parentheses being the mass in MeV/c^2. The transitions shown are photon transitions and have all been seen.

---

* Particle physicists use the word "flavor" for "kind of quark"; thus we speak of the six flavors u, d, s, c, b, t.

equation to give the allowed energies, and hence masses, of the (c$\bar{\text{c}}$) system. The resulting masses agree remarkably with the observed masses shown in Fig. 14.13. The states are labeled by the same scheme as used in nuclear physics, with three quantum numbers n, l, and j. Here $n = 1, 2, \ldots$ for the levels of given orbital angular momentum, $l = 0, 1, \ldots$ (coded as $s, p, d, \ldots$), and j gives the total angular momentum. The three quantum numbers are given in the order nl_j; for example, $1p_2$ means $n = 1$, $l = 1$, $j = 2$. In addition, since both quarks have $s = \frac{1}{2}$, the total spin can be $s = 0$ or $s = 1$, although the $s = 0$ states are harder to observe and very few are known.

14.8 The Strong Force and QCD

The strong force was originally introduced as the force between any two nucleons in a nucleus. However, since we now know that nucleons are bound states of quarks, it is clear that the strong force must ultimately be a force between quarks. It is natural to suppose that the strong force originates in a *strong charge* carried by quarks, in the same way that the electromagnetic force originates in the electric charge carried by charged particles. Although there is just one kind of electric charge, there appear to be three kinds of strong charge, which have come to be called **colors**—red, green, and blue. These whimsical names do not, of course, imply that quarks are colored in the ordinary sense of the word. Nonetheless, the name "color" is particularly apt, since an equal combination of all three strong charges produces a neutral strong charge, just as a superposition of the ordinary colors red, green, and blue can produce the neutral color white. It appears that each of the six kinds of quark (u, d, s, c, b, t) can carry any of the three different color charges. Thus any combination of three quarks can occur in an arrangement that has a total color charge of zero. In fact, all observed hadrons are color neutral in this way, and as we shall see directly, this may explain why the observed hadrons can be separated from one another whereas individual quarks cannot.

The quantum theory of the force between electric charges is called quantum electrodynamics, or QED. By analogy, the theory of the strong force between color charges is called **quantum chromodynamics,** or QCD. In QED the force-carrying particle (or gauge particle) is of course the massless photon. In QCD we assume that there is a corresponding massless force carrier called the **gluon.** Although there are close parallels between QED and QCD, the differences are perhaps even more important. In particular, the photon is electrically neutral, whereas the gluon can carry color charge, and this leads to a striking difference between the electromagnetic and strong forces. The electric force between two electric charges falls off with distance, as indicated by the spreading lines of force in Fig. 14.14(a); by contrast, the color force between two quarks retains its strength as the quarks move apart because the color-charged gluons attract one another and produce the lines of force shown in Fig. 14.14(b). The spreading electric lines of force describe a force $F \propto 1/r^2$ and hence a potential energy $U \propto 1/r$. The color lines of force do spread out for a short distance but then become essentially parallel; this gives a constant force and hence a potential energy that grows linearly. As a result, the potential energy of two quarks a distance r apart is thought to have the general form

$$U(r) = \alpha r - \frac{\beta}{r},$$

$\qquad(14.41)$

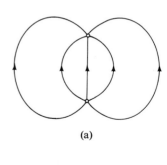

(a)

(b)

FIGURE 14.14 (a) The electric lines of force between two opposite electric charges. (b) Corresponding lines of force between two quarks; because the force-carrying gluons are themselves color charged, the color lines of force are drawn together and produce a constant force as the quarks move apart.

where α and β are constants. This is the potential that was used to calculate the levels for the $c\bar{c}$ mesons shown in Fig. 14.13. When the two quarks are close together, the potential (14.41) approximates $-\beta/r$ and resembles that of two electric charges. However, when r increases, the potential (14.41) increases without limit, so that an infinite energy would be needed to separate the quarks completely.

If these ideas are correct, no two systems with overall color charge can be separated (which explains why no isolated quark has ever been observed). On the other hand, if we consider two systems that are both color neutral, then at large separations the color forces sum to zero, and the two systems *can* separate. The obvious examples of color-neutral systems are those comprising three differently colored quarks, for which the three colors are overall neutral, and a quark–antiquark pair, in which the color of the quark is canceled by the anticolor of the antiquark. These two examples are, of course, the observed baryons and mesons.

It is interesting to consider what would happen if one tried to pull apart two colored charges. For instance, Fig. 14.15(a) shows a q and $\bar{q}$ flying apart after being formed in a high-energy collision. As the quarks move apart the potential energy of the color field grows and eventually becomes so large that quark–antiquark pairs are created as in part (b) of the figure. Further separa-

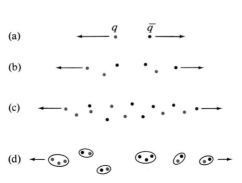

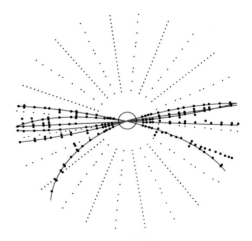

FIGURE 14.15 (a) A $q\bar{q}$ pair is formed in a high-energy collision and begins to separate. (b) and (c) As the q and $\bar{q}$ move apart, energy is stored in the color field and then creates more $q\bar{q}$ pairs. (d) The many resulting quarks cluster together into color-neutral baryons and mesons, which can separate indefinitely as two "jets" of hadrons.

FIGURE 14.16 Two "jets" of hadrons produced in the wakes of a $q\bar{q}$ pair, which was formed in the high-energy collision of an e^+ and e^- at the center of the picture. The e^+ and e^- are not shown since they came in perpendicular to the picture, one from above and one from below the page. A magnetic field has curved the outgoing trajectories, allowing measurement of the momenta. The small dots show the positions of individual wires that detect the particles. The large dots show the positions at which particles were detected; this information was fed directly to a computer, which reconstructed and drew the particles' trajectories. (Courtesy DESY.)

tion of the original quarks produces still more pairs, as in (c), until all available energy is exhausted. At this stage the quarks can cluster together, as in part (d), to form color-neutral combinations that can separate indefinitely. Thus, what began as two outgoing quarks becomes two streams of outgoing mesons and baryons. This phenomenon is illustrated by the data from an electron–positron collision shown in Fig. 14.16, where we see two "jets" moving away from the collision region. These jets comprise the hadrons which materialized in the wakes of the q and $\bar{q}$ that were formed when the e^- and e^+ collided. The tracks in this picture were drawn by a computer. In this respect Fig. 14.16 is typical of modern particle experiments, in which data on the positions and momenta of the particles are recorded directly in computers; the particles' paths can then be reconstructed and drawn by the computer to facilitate analysis by experimenters.

It is important to note that although the force between two color-neutral hadrons is negligible at large separations, when the hadrons are close together the individual quark–quark forces do not cancel exactly, and there is a residual net force. There is a close parallel here with the electric attraction between two atoms: When two neutral atoms are far apart the electric force between them is completely negligible, but at distances of order the atomic size, the charges inside the atoms can distribute themselves such that the attractive forces between opposite charges are greater than the repulsive forces between like charges. (As we shall see in Chapter 16, this is exactly the mechanism by which neutral atoms bind together to form molecules.) In an analogous way, the different colors inside two color-neutral nucleons produce no attraction at large separations but can produce overall attraction at separations on the order of the nucleon size, and this is precisely the attractive force that binds nucleons together in a nucleus.

The attraction between two nucleons was explained by Yukawa as arising from the exchange of virtual pions, as illustrated in Fig. 14.17(a). According to QCD, this is not the ultimate explanation of the nucleon–nucleon force. Nevertheless, the success of Yukawa's theory can be understood within the framework of QCD, as illustrated in Fig. 14.17(b), where each of the nucleons is shown as a bound state of three quarks. The three quarks in each nucleon are continually emitting and absorbing virtual gluons. (This is indicated by the shading in the picture.) When the two nucleons approach within 1 fm or so, the virtual emissions and absorptions within either nucleon can interact with those in the other. In the event shown in Fig. 14.17(b), one of the u quarks in the incoming proton has emitted a virtual gluon, which subsequently created a d$\bar{d}$ pair. The recoiling u and the newly created $\bar{d}$ both move across to the neutron, where the $\bar{d}$ is annihilated by one of the incoming d quarks and the u is captured. The net effect is that the proton changes into a neutron, and vice versa, and two quarks are exchanged, one u and one $\bar{d}$. Since the u and $\bar{d}$ are precisely the constituents of a π^+ meson, we recognize that Fig. 14.17(b) is simply the quark-model version of the π^+ exchange shown in Fig. 14.17(a). The quark model of the nucleon–nucleon interaction allows many other processes besides that of Fig. 14.17(b). Nonetheless, the success of Yukawa's theory shows that those processes [like that of Fig. 14.17(b)] that are equivalent to exchange of a pion are especially important.

Although QCD successfully explains the strong interaction in terms of gluons, direct evidence for the existence of gluons is even harder to find than that for quarks. Nevertheless, the general success of QCD has convinced most

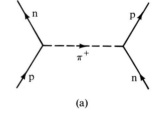

(a)

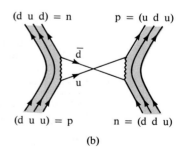

(b)

FIGURE 14.17 (a) Yukawa explained the force between two nucleons in terms of the exchange of virtual pions. (b) The same process as explained in QCD. The two wavy lines represent virtual gluons.

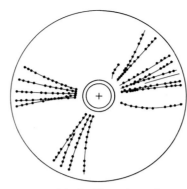

FIGURE 14.18 Three jets of hadrons produced in the head-on collision of an e⁺ and e⁻ (which came in from directly above and below the plane of the picture). This event is similar to the two-jet event of Fig. 14.16, but here the e⁺ and e⁻ created a q and a q̄, one of which radiated a high-energy gluon. The three jets were created in the wakes of the q, the q̄, and the gluon. (Courtesy DESY.)

particle physicists that gluons really do exist. Furthermore, there *is* some evidence, of which we mention two examples: The electron–nucleon scattering experiments that have shown the presence of three quarks inside the nucleon indicate that the total momentum of the three quarks is *less* than the momentum of the nucleon. It is widely believed that the missing momentum is carried by the virtual gluons that are continually bouncing back and forth among the constituent quarks.

Further indirect evidence for gluons comes from e⁺e⁻ collisions, which sometimes produce three jets of hadrons such as those shown in Fig. 14.18. This picture is interpreted as resulting when the e⁺e⁻ collision produces a qq̄ pair (as in the two-jet event of Fig. 14.16) but one of the quarks promptly ejects a high-energy gluon. Like the quarks, the gluon carries color charge and cannot escape by itself. Therefore, like the two quarks, it creates a stream of qq̄ pairs in its wake, and these emerge as a jet of hadrons. The third jet in such three-jet events is widely accepted as proof of the existence of gluons.

14.9   Electroweak Interactions: The W and Z Particles

Perhaps the greatest triumph for particle physics in the last 20 years has been the discovery that the electromagnetic and weak interactions are different manifestations of a single phenomenon called the **electroweak interaction.** This unification has been compared to the discovery in the nineteenth century that electric and magnetic forces are two aspects of the single phenomenon that we now call electromagnetism. At first it appeared that electric and magnetic forces were independent phenomena, and their unification by Faraday and Maxwell was one of the high points of nineteenth-century physics. In a similar way, the electromagnetic and weak interactions appeared to be entirely different, and their unification was a major achievement, for which its principal authors — the Americans Glashow and Weinberg and the Pakistani Salam — were awarded the 1979 Nobel prize for physics.

The theory of the electroweak interaction begins by postulating that the electromagnetic and weak interactions are carried by four massless gauge particles, two electrically charged and two neutral. With this assumption, there would be an almost complete symmetry between the electromagnetic and weak forces. In particular, with massless force carriers, the weak force would be a long-range force. [Remember the relation (14.11), $R \approx \hbar/(2mc)$, between the range of a force and the mass of its carrier; in particular, if $m = 0$, the force range is infinite — as is the case in electromagnetism.] Since the observed range of the weak force is very short, it is clear that this theory is not yet complete. Rather, it is postulated that the observed four force carriers must be mixtures of the original massless gauge particles, one of them still massless but three of them with mass. (The mechanism by which the originally massless particles acquire mass is called spontaneous symmetry breaking.) The surviving massless particle is the familiar photon. The three massive particles, called W⁺, W⁻, and Z°, are the carriers of the weak interaction. The theory predicts their masses as

$$m(\text{W}^\pm) = 82.4 \text{ GeV}/c^2 \quad \text{and} \quad m(\text{Z}°) = 93.3 \text{ GeV}/c^2, \quad (14.42)$$

both with uncertainties of about 1 GeV/c^2. These enormous masses account for the observed short range of the weak interaction. In fact, if we substitute the masses (14.42) in the relation $R \approx \hbar/(2mc)$, we find a range of order 10^{-18} m as observed (Problem 14.35).

Because of the large masses of the weak force carriers, the observed weak interactions are very different from the electromagnetic. Nevertheless, these differences arise in a well-understood way from a theory that is basically symmetric. At extremely high energies (many hundreds of GeV) the difference between the masses (14.42) and the photon mass (zero) would be negligible and the underlying electroweak symmetry should become apparent. It is to test this idea (among others) that physicists are planning to build accelerators that can produce particles with many TeV of energy (1 TeV = 1000 GeV = 10^{12} eV).

The electroweak symmetry is important to astrophysicists interested in the birth of the universe. The present outward motion of the galaxies indicates that the universe originated in an explosion at a single point. Immediately after this "big bang" the temperatures, and hence energies, must have been high enough for the electroweak symmetry to be important and to have influenced the development of the universe we see today. Even at more mundane temperatures and energies, where the electroweak symmetry is not readily apparent, the electroweak theory gives us a consistent theory of electromagnetic and weak interactions and an accurate method to calculate the probabilities of electroweak processes. For example, recent experiments on cesium atoms have dramatically confirmed the tiny contribution of the weak interaction to atomic transitions and verified that the electroweak theory applies even at energies of only a few eV.

DISCOVERY OF THE W AND Z PARTICLES

Although the success of the electroweak theory gave strong indirect evidence for the W and Z, physicists naturally hoped to observe these particles directly. Since their masses are 80 to 90 GeV/c^2, a minimum energy of order 100 GeV was obviously needed to produce them. In 1976 an accelerator called the super proton synchroton or SPS was completed at the CERN laboratory in Europe.* This machine accelerates protons as they circulate round an evacuated circular pipe some 4 miles in circumference. The protons can be accelerated to 400 GeV, which might seem more than enough to produce the W and Z particles. Unfortunately, it is actually hopelessly inadequate, because of the problem mentioned in Section 14.2: When a high-energy particle collides with a stationary particle, very little of the incident energy is available to create new particles. The reason is that conservation of momentum requires the final particles to have a large total momentum (equal to that of the original projectile). Thus the final particles have a certain minimum kinetic energy, and only the remainder of the original kinetic energy is available to create new particles. In particular, if a 400-GeV proton collides with a stationary proton, only 30 GeV of the original energy is available to create new particles (see Problem 14.40). This is totally inadequate to produce a W or Z particle.

The solution to this difficulty is obvious in theory but extremely hard to realize in practice. If one could make two high-energy particles collide head-on, in such a way that their total momentum was zero, the final particles could have zero momentum and hence zero kinetic energy, and all of the incident kinetic energy would be available to create new particles. To create two high-energy beams and arrange for them to collide head-on is obviously much harder than creating a single beam that collides with a fixed target (see Problem 14.36).

---

* CERN—Conseil Européen pour la Recherche Nucléaire—is a laboratory run jointly by 13 European countries and situated outside Geneva, straddling the Swiss–French border.

FIGURE **14.19** Part of the 2000-ton UA1 detector used to find the W$^\pm$ and Z particles at CERN. The paths and energies of the outgoing particles were fed directly to computers for storage and later analysis. (Courtesy CERN.)

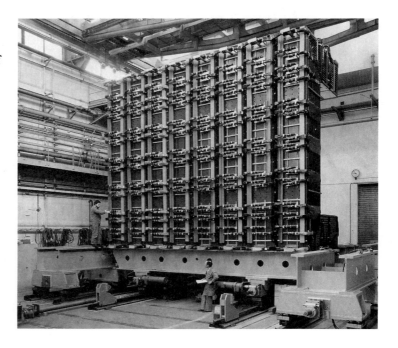

Nevertheless, the SPS accelerator at CERN was modified successfully to produce two colliding beams.

One of the colliding beams was the proton beam, already available at the SPS. The second beam consisted of antiprotons. These were created in collisions of 26-GeV protons with a metal target. Such collisions produce many different particles, but suitable electric and magnetic fields could select the antiprotons and steer them to an an *antiproton accumulator*—an evacuated pipe, forming a circle of diameter about 50 m, in which the antiprotons could circulate held in orbit by a ring of magnets. Antiprotons were steadily fed into the accumulator for about a day, until there were enough for the experiment.* A burst of antiprotons was then guided out of the accumulator ring and boosted into the same SPS that accelerates the protons. Since the p and p̄ have the same mass but opposite charges, they could be accelerated simultaneously in the same machine, provided that they traveled in opposite directions. In this way the p and p̄ beams were brought up to 270 GeV each.

The two beams were arranged so that their paths crossed at six locations around the 4-mile circumference of the SPS tunnel. A large assembly of detectors placed all around each collision area could monitor the particles produced in the resulting pp̄ collisions. Each detector could produce an electrical signal in response to the passage of a particle in its vicinity, and all detectors were connected directly to computers that recorded the energies of the outgoing particles and their positions and times at several points along their paths. A part of one detector assembly, which had a total mass of 2000 tons, is shown in Fig. 14.19.

The mean lives of the W and Z were expected to be of order 10^{-24} s.

---

* Antiprotons were stored successfully for as long as a month—dramatically confirming the belief that when isolated from matter, antimatter is perfectly stable.

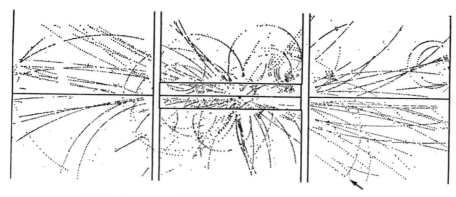

FIGURE 14.20 Computer reconstruction of the process $p + \bar{p} \rightarrow$ $W^- + $ hadrons, followed by the decay $W^- \rightarrow e^- + \bar{\nu}_e$. The p and $\bar{p}$ enter from left and right along the horizontal solid lines. The path of the outgoing e^- is indicated with an arrow. The path of the $\bar{\nu}_e$ had to be inferred by measurement of all the other outgoing particles. (Courtesy CERN.)

Therefore, they had to be identified by detecting their decay products: for example,

$$Z^\circ \rightarrow e^+ + e^- \qquad \text{and} \qquad W^- \rightarrow e^- + \bar{\nu}_e. \qquad (14.43)$$

The problems of identifying these decay products were formidable indeed. First, it was calculated that only 1 collision in 10 million would produce an identifiable W or Z. Second, every collision (including those few that produce a W or Z) creates a large number of hadrons. Therefore, the CERN experiments had to observe a prodigous number of events, and then, for each event, sort through an enormous clutter of hadrons to identify the hoped-for decay products of the W or Z. In addition, in the case of the W, one of the decay products was always a neutrino, as in (14.43), and the presence of this neutrino could only be inferred by measuring the energy and momentum of all other outgoing particles and comparing with the total incident energy and momentum.

Data from the collisions were taken and recorded by a large number of computers. Each event was examined briefly, and the vast majority immediately rejected as uninteresting. On average about one event per second was selected as promising enough to be worth recording on magnetic tape — with the paths and energies of all the outgoing particles that had been detected. These events could then be analyzed fully in the coming months. A few events were immediately selected as especially promising, and were sent to an "express lane" for prompt processing.

After seven years of planning by nearly 200 scientists from some 20 universities and laboratories in Europe and the United States, the first full experimental run began in September 1982. After two months of almost continuous operation, the first candidate for a W particle was reported, and by the end of 1983 there were some 90 definite observations of the W and 13 of the Z. A computer reconstruction of an early example of a W^- decaying into an electron and an antineutrino is shown in Fig. 14.20. The data from these observations were enough to determine the masses of the W and Z rather accurately, and as shown in Table 14.9, the agreement with theory was spectacular. These results were an outstanding triumph for the electroweak theory and were recognized by the award of the 1984 Nobel prize for physics to two of the leaders of the experiment, Rubbia from Italy and Van der Meer from Holland.

TABLE 14.9

Measured and predicted masses of the $W^\pm$ and Z° in GeV/c^2.

	$W^\pm$	Z°
Measured	82.1 ± 1.7	93.0 ± 1.8
Predicted	82.4 ± 1.1	93.3 ± 0.9

14.10 Summary and Outlook

Let us first summarize the roster of elementary particles as sketched so far.

THE STANDARD MODEL

There are 12 elementary spin-$\frac{1}{2}$ particles — six leptons and six quarks — which fall into three generations, each with two leptons and two quarks. (See Table 14.8 in Section 14.7.) The leptons, which include the familiar electron, can exist in isolation; the quarks exist only inside baryons and mesons. All 12 leptons and quarks are subject to the electroweak interactions, which are carried by four gauge particles, the massless photon and the massive $W^\pm$ and $Z°$, all of which have been observed directly. In addition, the six quarks are subject to the strong force which is carried by the massless — and so far not directly observed — gluon.

This picture has come to be called the **standard model,** a name which reflects the confidence of most physicists that the basic facts are well established and unlikely to change. On the other hand, we should emphasize that many questions remain unanswered: The model does not explain why the leptons and quarks have the masses observed. Why is the electron so light compared to all the other massive particles? Why are the second and third generations of particles so much heavier than the first? And indeed, why do the second and third generations occur at all?

GRAND UNIFIED THEORIES

The electroweak theory has unified the electromagnetic and weak interactions. Many physicists believe that eventually the strong interactions, also, will be incorporated in a single *grand unified theory* (GUT). Considerable progress has been made in this direction and some startling predictions have been made. For example, most proposed GUTs predict that the proton should be unstable with a mean life τ_p of order 10^{31} years. This life is prodigously long, being some 21 orders of magnitude greater than the age of the universe, and for all practical purposes the proton would appear to be stable (in agreement with observation). Nevertheless, experiments have been set up that should be able to detect proton decays. So far no verifiable decays have been observed, and the experimental lower bound on τ_p is already somewhat larger than the value predicted by the simpler grand unified theories. That is, the simpler GUTs seem to be incompatible with the observed facts. Notice that because the proton is the lighest baryon, its decay — for example, $p \rightarrow \pi° + e^+$ — would violate conservation of baryon number. In fact, it is an essential feature of most grand unified theories that conservation of baryon number is violated, very weakly at least.

Another prediction of most GUTs is the existence of a magnetic monopole — a particle with a north pole but no south pole, or vice versa. Such particles have never been observed, and recent searches have still not found any. Obviously, it is important to current grand unified theories that we establish whether the proton is unstable and whether there are magnetic monopoles. (Indeed, both of these questions are extremely interesting, irrespective of their relevance to GUTs.) Meanwhile, it should be emphasized that grand unified theories have scored some impressive successes. For example, certain parameters of the standard model* are predicted by grand unified theories, and the predicted values agree well with observation.

Some physicists are already seeking a theory, even more ambitious than the grand unified theories, in which *all four* interactions—strong, electromagnetic, weak, and gravitational—are unified. We have emphasized that on the level of elementary particles gravity is usually completely negligible. Nevertheless, as a matter of principle, it is desirable to have a theory that could explain *all* of the known forces. The attempts to unify gravity with the other forces go back to Einstein, who spent many years trying unsuccessfully to unify gravity and electromagnetism. Until recently the problem of incorporating gravity in a quantum theory had seemed completely intractable; but it now seems that a new approach, called the *theory of superstrings,*† may produce a satisfactory quantum theory of all four interactions. Nevertheless, this theory is still very speculative, and no one would claim that any answers are final.

Another speculation that deserves mention is that the leptons and quarks may themselves be composite, constructed from objects that are more truly elementary. As already noted, the occurrence of the three generations of leptons and quarks suggests this idea, and several such theories have appeared. However, there is so far no direct evidence for such a substructure, and these theories have yet to prove their worth.

THE COSMIC CONNECTION

Particle physics, by its very nature, is concerned with the smallest scales known to man—currently about 10^{-18} m. At the opposite extreme is astrophysics and, especially, cosmology, the study of the structure and origin of the whole universe—whose size is of order 10^{26} m. Paradoxically, these two extreme branches have been growing increasingly dependent on one another for many years. Astronomers studying stellar spectra have needed to understand—and have contributed to—atomic physics for more than a century. A proper understanding of the energy source and evolution of stars came only after the development of nuclear physics in the 1930s. Most recently, it has become apparent that during the first moments after the big bang, the whole universe was a natural laboratory for elementary-particle physics. For the first 3 minutes the temperature was so high that no atoms or nuclei could exist, and the universe was a fiery gas of leptons and hadrons. During the first microsecond or so, the temperature may have been high enough that hadrons could not exist and there were only leptons and quarks. One of the motivations for building particle accelerators with higher and higher energies is to allow physicists to observe collisions that approximate conditions in these earliest moments of the universe.

Because the first moments of the universe presumably determined its subsequent evolution, an understanding of elementary particles has become very important to astrophysics, and recent years have witnessed several papers in astrophysics published by particle theorists, and vice versa. To conclude, we mention two examples of this kind of interplay. First, a question that has

---

* For example, the so-called Weinberg angle, which gives the degree of mixing among the massless gauge particles of the electroweak theory.

† In superstring theory the elementary particles are treated as short line segments rather than points.

puzzled many physicists is why the universe apparently consists mainly of what we call matter (electrons and nucleons) rather than antimatter (positrons and antinucleons). It now seems conceivable that the universe was originally symmetric, with equal amounts of matter and antimatter; then, at a later time, a chance fluctuation in favor of matter could have precipitated a steady disappearance of almost all antimatter to produce the present imbalance. This scenario would require that baryon number is not conserved, as is the case in certain grand unified theories; thus an improved understanding of these theories is needed before we can decide this interesting question.

Second, there is evidence that the mass in the universe is considerably more than the mass of the visible stars and interstellar gas. Several suggestions, involving various new kinds of particles, have been made to account for this "dark" mass. One simple suggestion is that neutrinos might have a small, but nonzero mass. Enormous numbers of neutrinos are produced by the nuclear reactions in stars; even if the neutrino mass were less than 20 eV/c^2 (the present experimental limit on the mass of the ν_e), this could well account for the large invisible mass in the universe. It is for this reason (and others) that considerable efforts are being made to measure even more accurately the masses of the neutrinos.

CONCLUSION

Obviously, extraordinary progress has been made in the last 50 years toward understanding the fundamental constituents of matter. Nevertheless, there are still plenty of questions—both theoretical and experimental—to be answered, and it is almost certain that many interesting discoveries will be made in the coming decades.

14.11 High-Energy Accelerators (optional)

In Section 13.10 we described some of the accelerators that give charged particles kinetic energies of tens or even hundreds of MeV. These energies suffice for the study of atomic nuclei, but the study of elementary particles requires energies from several GeV to several thousand GeV. The two types of accelerator that have succeeded in producing these immense energies are the *linear accelerator,* or *linac,* which accelerates the particles along a straight path, and the *synchrotron,* in which the particles follow a circular path of fixed radius.

LINACS

We described a low-energy linac in Section 13.10. Most high-energy linacs work in a quite different way: Instead of giving the particle discrete boosts in the gaps in a series of drift tubes, these machines accelerate the particle continuously with the electric field of an electromagnetic wave traveling along the evacuated pipe. The charged particles ride along with the wave, much as a surfer rides a water wave.

To achieve high energies, linacs must be very long. For example, the world's highest-energy linac, the Stanford linear accelerator, which accelerates electrons and positrons to 50 GeV, is 2 miles long. [For comparison, the circular proton synchrotron at the Fermi National Accelerator Laboratory, which accelerates protons to 1000 GeV (or 1 TeV), has a diameter of 1.3 miles.]

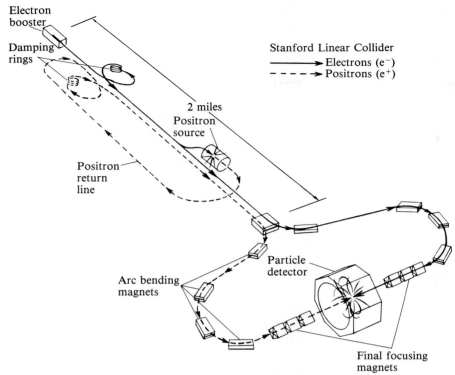

Electron
booster

Damping
rings

Stanford Linear Collider
———————▶ Electrons (e⁻)
— — — — ▶ Positrons (e⁺)

2 miles
Positron
source

Positron
return
line

Particle
detector

Arc bending
magnets

Final focusing
magnets

FIGURE **14.21** The Stanford linear collider. Electrons and positrons are simultaneously accelerated to 50 GeV in the 2-mile linac. (The positrons are created by diverting some of the electrons about two-thirds of the way along the linac; these diverted electrons collide with a metal plate and create copious positrons, which are then guided back to the start of the linac. The damping rings near the start of the linac are used to focus the electrons and positrons into dense bunches of suitable shape.) At the end of the linac, magnetic fields separate the electrons and positrons and guide them to a head-on collision at the center of the detector.

However, in accelerating electrons and positrons, the linac has a crucial advantage over circular machines: Because of their small mass, electrons and positrons emit intense electromagnetic radiation when forced to follow a circular path (see Problems 15.5 to 15.7). At extremely high energies (many hundreds of GeV), the resulting loss of energy becomes so great that it is impractical to accelerate electrons or positrons in circular machines, and the linac is the only practical alternative.

The Stanford linear accelerator, shown in Fig. 14.21, can accelerate electrons and positrons simultaneously — the oppositely charged particles riding on opposite sides of a wave crest as it travels along the tube. As they leave the accelerator, the two kinds of particle are separated by a magnetic field, which bends them in opposite directions. The separated beams are steered around two large arcs, shaped like a pair of giant tongs, and then meet in a head-on collision with total energy up to 100 GeV.

SYNCHROTRONS

In Section 13.10 we described the cyclotron, which uses a uniform magnetic field to hold charged particles in an approximately circular orbit. This lets the particles pass repeatedly through the same accelerating electrodes and

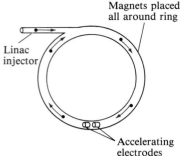

Magnets placed all around ring

Linac injector

Accelerating electrodes

FIGURE 14.22 A synchrotron. Protons (or other charged particles) are injected from a preliminary accelerator, such as a linac, into the ring-shaped evacuated tube. They are held in the circular orbit by magnets placed all around the ring (oriented to make B vertically out of the page in this picture). Each time they go round the ring they are given a boost by the electric field produced by one or more accelerating electrodes. As the particles speed up, the frequency of the accelerating fields must increase to remain synchronized with the orbit, and the magnetic field must increase to hold them in the orbit. The same synchrotron can accelerate protons and antiprotons simultaneously, with the antiprotons circulating in the opposite direction, since the magnetic force $\mathbf{F} = q\mathbf{v} \times \mathbf{B}$ is unchanged if we reverse both q and $\mathbf{v}$.

makes the cyclotron a relatively compact machine. As the particles gain energy, they move faster, and the radius of their orbit grows in proportion:

$$R = \frac{p}{qB} = \frac{mv}{qB} \qquad \text{[for nonrelativistic motion].} \qquad (14.44)$$

Thus the particles spiral steadily outward. However, because R is proportional to v, the period of their orbit ($\tau = 2\pi R/v$) remains constant, and the accelerating fields of a cyclotron can therefore be supplied by ac of fixed frequency.

Unfortunately, the simple proportionality of R and v breaks down once the motion becomes relativistic. Since $p = \gamma mv$, (14.44) must be replaced by

$$R = \frac{p}{qB} = \frac{\gamma mv}{qB} \qquad \text{[relativistic].} \qquad (14.45)$$

Thus the orbital period, $\tau = 2\pi R/v$, begins to change once γ becomes appreciably different from 1. If we continued to drive the machine at constant frequency, the particles would get out of step with the accelerating fields until they ceased to gain energy. There are several ways around this difficulty. One is to vary the frequency of the ac supply to keep it synchronized with the changing orbital frequency. (A cyclotron modified in this way is called a *synchro-cyclotron.*) However, a second difficulty develops as we push for higher and higher energies: From (14.45) it follows that $p = qBR$. Thus the higher the final momentum, the larger must be the outside radius of the circular magnet, and the cost of building large enough magnets quickly becomes prohibitive.

In the **synchrotron,** the spiral path of the cyclotron is replaced by a circular path of fixed radius, as sketched in Fig. 14.22. The particles move around a ring-shaped, evacuated pipe of radius R. They are repeatedly accelerated by one or more electrodes located around the ring and are held in their circular orbit by magnets placed all around the ring. In this way a magnetic field is required only in the immediate vicinity of the radius R (not for *all* $r \leq R$), and an enormous saving in cost results. There is, of course, a price for this saving. Since $p = qBR$, and since R is fixed, the B field has to be increased as p increases. In addition, the orbital period, $\tau = 2\pi R/v$, gets shorter as v increases; thus the frequency with which the accelerating fields are applied must also increase as the particles speed up. Thus, in a synchrotron, both B and f must be varied in step with the acceleration of the particles. One example of a synchrotron is the Super Proton Synchrotron (SPS) at CERN near Geneva, where the W and Z particles were discovered, as described in Section 14.9; another is the Tevatron in Illinois, shown in Fig. 14.23.

Since $p = qBR$, the maximum momentum of the particles is limited by the maximum attainable magnetic field and the radius of the ring: $p_{max} = qB_{max}R$. In the last decade or so, it has become possible to achieve very large B fields by using superconducting magnets (that is, electromagnets whose current flows in superconducting wires). This was first successfully accomplished with the Tevatron (Fig. 14.23), which was originally built with conventional magnets (B_{max} around 2 T) and could accelerate protons to around 400 GeV. With the introduction of superconducting magnets (B_{max} around 4 T), this machine can now reach nearly 1000 GeV using the same underground tunnel that houses the original ring. (It is because 1000 GeV = 1 TeV that the machine is now called the Tevatron.) A second important advantage of superconducting magnets is that they waste far less energy in Joule heating.

FIGURE 14.23   Aerial view of the Tevatron at the Fermi National Accelerator Laboratory in Illinois, which accelerates protons and antiprotons to about 1 TeV (that is, 1000 GeV). The large circle is the main synchrotron ring and is more than a mile in diameter. (The building on the left is 16 stories high.) (Courtesy Fermilab.)

EXAMPLE 14.2   If the maximum B field in a synchrotron is 3 T, what must be the machine's circumference if it is to accelerate protons to 1 TeV? How long will the protons take to make one orbit once they reach full energy?

At 1 TeV, the rest energy is such a small fraction of the total energy that we can take $E = pc$. Therefore,

$$R = \frac{p}{eB} = \frac{E}{ceB} = \frac{(10^{12} \text{ eV}) \times (1.6 \times 10^{-19} \text{ J/eV})}{(3 \times 10^8 \text{ m/s}) \times (1.6 \times 10^{-19} \text{ C}) \times (3 \text{ T})} = 1.1 \times 10^3 \text{ m}$$
$$\approx 1 \text{ km},$$

and the circumference is about 6 km (which is in fact the circumference of the Tevatron in Fig. 14.23).

At full energy (1 TeV), v is very close to c, and

$$\tau = \frac{2\pi R}{c} = \frac{2\pi \times (1.1 \times 10^3 \text{ m})}{3 \times 10^8 \text{ m/s}} = 2.3 \times 10^{-5} \text{ s}.$$

At this rate the protons make some 40,000 trips around the 6-km circumference in each second. In fact, the protons are injected into the synchrotron ring with an energy of 150 GeV, and they are already traveling close to c, so the orbital frequency changes very little during this final acceleration process.

A new synchrotron, called the *superconducting supercollider* or SSC, is now being planned to accelerate protons to about 20 TeV. As its name implies, it will use superconducting magnets, placed around a ring of circumference about 50 miles. It will be built as a collider, simultaneously accelerating two beams of protons traveling in opposite directions in two separate tubes, to produce collisions with a total energy of 40 TeV. The particles will be injected into the main ring by a booster synchrotron, itself about as big as the Tevatron at the Fermi Laboratory.

IDEAS YOU SHOULD NOW UNDERSTAND FROM CHAPTER 14

The positron and other antiparticles
Force carriers:
    the photon and pion
    the mass-range relation
The four fundamental interactions
The lepton family; generations
The hadron family
Resonance particles
Conservation of baryon number

The quark model:
    fractional charge of quarks
    multiplets and supermultiplets
    generations
QCD and gluons:
    color charge
    confinement
Electroweak theory; the W and Z particles
[Optional section: linacs; synchrotrons]

PROBLEMS FOR CHAPTER 14

SECTION **14.1** (ELEMENTARY PARTICLES; THE STORY SO FAR)

14.1 • To investigate distances of order 10^{-15} m requires a probe whose wavelength λ is of this same order. (a) Supposing that the probe particle has mass m, derive an expression for its total (relativistic) energy in terms of λ. (b) Show that the minimum energy that gives $\lambda \approx 10^{-15}$ m is of order 1 GeV = 10^9 eV.

14.2 •• (a) Given particles of total relativistic energy E and mass m, what is the minimum distance, d_{min}, that can be probed with these particles? (b) For given E, what mass m would give the best resolution? (c) If $E = 1$ TeV = 10^{12} eV, what is the best possible resolution?

SECTION **14.2** (ANTIPARTICLES)

14.3 • Consider a positron colliding with a stationary electron. Use conservation of energy and momentum to prove that the e$^+$ and e$^-$ cannot annihilate into a single photon. [This explains why Eq. (14.2) was written as e$^+$ + e$^-$ → photons, with "photons" in the plural.]

14.4 • The first antiprotons to be observed were produced in the reaction

$$p + p \rightarrow p + p + p + \bar{p}, \qquad (14.46)$$

which required a minimum kinetic energy of 6 GeV. Below are listed three reactions that would require less energy, if they were possible. Explain why none of them is possible.
(a) p + p → p + $\bar{p}$
(b) p + p → p + p + $\bar{p}$
(c) p + p → p + e$^+$ + e$^+$ + $\bar{p}$

14.5 • Suppose that n particles all have the same rest mass m and all move with the same velocity **u**. Prove that their total energy and total momentum satisfy the Pythagorean relation $E_{tot}^2 = (p_{tot}c)^2 + (Mc^2)^2$, appropriate to a single particle of rest mass $M = nm$. (This result is needed in Problem 14.8.)

14.6 •• Prove that a single photon (isolated from any other particles) cannot convert spontaneously into an e$^+$e$^-$ pair:

$$\gamma \not\rightarrow e^+ + e^-.$$

Hint: Use the Pythagorean relation (3.23) to relate the energy and momentum on each side of this reaction, and then show that the reaction cannot conserve both energy and momentum. [This result explains the need for the nucleus in Eq. (14.1); the recoil of the nucleus allows energy and momentum to be conserved.]

14.7 •• To identify the antiproton its discoverers found its mass by separately measuring its momentum and speed. The many particles produced in the original collisions were sent through a magnetic field chosen to select only particles with momentum 1.2 GeV/c and charge $-e$. These particles were then timed in their flight between two scintillation counters some 12 m apart. (*a*) How long would an antiproton with $p = 1.2$ GeV/c take to travel this distance? (*b*) To get an idea of the difficulties of the experiment, calculate how long a π^- ($m = 140$ MeV/c^2) would take to travel the same distance. (*c*) If the experimenters could measure flight times to about a nanosecond, could they tell the difference between a $\bar{p}$ and a π^- in this way?

14.8 •• The first antiprotons to be observed were produced in the reaction (14.46) (Problem 14.4), with one of the initial protons at rest. The minimum value of the incident kinetic energy K_i needed to induce the reaction is called the threshold energy for the reaction and is surprisingly large compared to $2mc^2$. To calculate this threshold energy, note that the minimum of K_i occurs when the final kinetic energy is as small as possible, to maximize the fraction of K_i available to create particles. This requires that all final particles have the same velocity, so that there is no kinetic energy "wasted" in relative motion. Write down the Pythagorean relation for the total energy and momentum of the final particles. (Use the result of Problem 14.5.) Then use conservation of energy and momentum to rewrite this relation in terms of the energy and momentum, E_i and p_i, of the incident projectile. Finally, use the Pythagorean relation for E_i and p_i to eliminate p_i. Show that $E_i = 7mc^2$ and hence $K_i = 6mc^2$.

SECTION 14.3 (PIONS AND MUONS)

14.9 • Equation (14.11), $R \approx \hbar/(2mc)$, relates the range R of a force to the mass m of the particle that carries the force. (*a*) Apply this relation to the electromagnetic force, whose carrier is the massless photon. (Note that because electromagnetic forces obey an inverse-square law they can, in principle, be felt at any distance, however large.) (*b*) It is now known that the so-called weak force is carried by a particle called the W, and that the range of the weak force is of order 10^{-18} m. Estimate the mass of the W particle.

14.10 • The force between two nucleons has a short-range repulsive component that is believed to be carried by

a particle called the ω meson, whose mass is 783 MeV/c^2. Estimate the range of this force.

14.11 • Find the maximum time Δt for which a nucleon can emit a virtual pion, $N \rightarrow N + \pi$.

14.12 • Yukawa's theory of the strong nuclear force requires that a nucleon be able to emit a virtual pion, $N \rightarrow N + \pi$. Explain why the pion spin, s_π, must be an integer. (*Hint:* The time-energy uncertainty relation allows the nonconservation of energy in this virtual process, but the same is not true of angular momentum; that is, total angular momentum is always conserved, even in virtual processes.)

14.13 •• Explain why one would never observe a process in which a massive particle x emits a particle y, while x itself is completely unchanged, $x \rightarrow x + y$.

14.14 •• Numerous pions are created by cosmic rays in the upper atmosphere, but very few reach sea level. One reason is that most of the pions decay before they reach sea level, and another is that they are quickly absorbed by the nuclei in the atmosphere. To estimate this second effect, answer the following: (*a*) About how many nuclei (of nitrogen and oxygen) are contained in a column of atmosphere of area 1 m^2, extending from sea level to the top of the atmosphere? (Atmospheric pressure at sea level is about 10^5 N/m^2.) (*b*) About how many of these nuclei are needed to cover completely 1 m^2 in a layer one nucleus thick? [Recall Eq. (12.3) for the nuclear radius.] (*c*) Use these two results to estimate the number of nuclei through which a pion would have to pass traveling from the upper atmosphere down to sea level.

14.15 •• Consider a negative pion caught in a $1s$ orbit around an oxygen nucleus. Because the pion interacts strongly with nucleons, it will be absorbed promptly if there is any probability of its penetrating the nucleus. Show that the probability of finding the pion inside the nucleus is about 0.25% (which is ample to ensure that the pion is rapidly absorbed). *Hints:* The radial probability density is given by Eq. (9.83), with the constant A given by Eq. (9.86), except that $a_B = \hbar^2/(m_e ke^2)$ must be replaced by $a = \hbar^2/(m_\pi Zke^2)$. The required probability is that of finding the pion in the range $0 \leq r \leq R \approx 3$ fm. Notice that $R \ll a$, which lets you approximate the factor $e^{-2r/a}$ by 1 inside the integral.

SECTION 14.4 (THE FOUR FUNDAMENTAL FORCES)

14.16 • In Section 14.4 it is stated that the gravitational force between two nuclear particles is negligible compared to the electrostatic force. Verify this claim by computing the ratio F_{grav}/F_{elec} for the forces between two stationary protons.

14.17 • To contrast the strengths of the strong and electromagnetic interactions between two protons do the following: Find the nuclear potential energy of two protons 1 fm apart (see Fig. 12.3). Calculate the electrostatic potential energy of the same two protons and compare.

14.18 • Of the four fundamental interactions, which is the strongest that can cause each of the following processes?
(a) $^{12}\text{N} \rightarrow {}^{12}\text{C} + e^+ + \nu$
(b) $\pi^+ + p \rightarrow \pi^+ + p$
(c) $e + p \rightarrow e + p$
(d) $\nu + e \rightarrow \nu + e$
(e) $e^- + e^+ \rightarrow e^- + e^+$
(f) $e^- + e^+ \rightarrow$ photons
(g) $n \rightarrow p + e^- + \bar{\nu}$
(h) $p + n \rightarrow p + n + \pi^\circ$
(i) (excited atom) $\rightarrow$ (ground state) $+ \gamma$

SECTION **14.5** (LEPTONS)

14.19 • The charged pion π^- usually decays into a muon and a neutrino,

$$\pi^- \rightarrow \mu^- + \bar{\nu}_\mu$$

but occasionally into an electron and a neutrino,

$$\pi^- \rightarrow e^- + \bar{\nu}_e.$$

The relative frequency of the electron decay (compared to the muon decay) is on the order of 1 in 10^4. The large difference between the probabilities for these two decays can be explained in terms of the $\mu - e$ mass difference as follows: The theory of weak interactions predicts that the probability for either decay is proportional to $(1 - v/c)$, where v is the speed of the outgoing μ^- or e^-. Calculate the quantity $(1 - v/c)$ for each decay, and compute their ratio. Note that this ratio has the same order of magnitude as the observed relative frequency. (Use the result of Problem 3.36; $m_\pi \approx 140$, $m_\mu \approx 106$, and $m_e \approx 0.51$ MeV/c^2.)

14.20 • The electron is the lightest known charged particle. Explain why this means that an isolated electron is perfectly stable.

14.21 •• It appears that there are three conserved quantities, N_e, N_μ, and N_τ, called **lepton numbers.** These are analogous to nucleon number and are defined as follows: The e^- and ν_e have $N_e = 1$ (each), the e^+ and $\bar{\nu}_e$ have $N_e = -1$, and all other particles have $N_e = 0$. If we define N_μ and N_τ correspondingly, then, as far as is known, the total value of each of these quantities is always conserved. (a) Verify this claim for the processes $n \rightarrow p + e^- + \bar{\nu}_e$, $\mu^- \rightarrow e^- + \bar{\nu}_e + \nu_\mu$, and $\pi^- \rightarrow \mu^- + \bar{\nu}_\mu$. (b) Use the laws to fill in the blanks

in the following decays (where each blank is some kind of neutrino or antineutrino):

$$\tau^+ \rightarrow e^+ + \underline{\quad} + \underline{\quad}$$
$$\tau^- \rightarrow \mu^- + \underline{\quad} + \underline{\quad}.$$

SECTION **14.6** (DISCOVERY OF MORE HADRONS)

14.22 • A Σ^+ particle is formed in a bubble chamber in a magnetic field of 0.4 T and leaves a track whose radius of curvature is measured to be $R = 0.5$ m. The charge of the Σ^+ is known to be $+e$. What is its momentum, in kg·m/s and in MeV/c? [Remember the relation (3.48) between R and the momentum.]

14.23 • (a) The measured width in the rest energy of the $\psi(3097)$ meson is $\Gamma = 0.068$ MeV. What is this meson's mean lifetime? (b) What is the mean life of the $\Sigma(1385)$ baryon, for which $\Gamma = 37$ MeV?

14.24 • The Σ^+ baryon has a mean life $\tau = 8.0 \times 10^{-11}$ s. What is the full width Γ of the intrinsic uncertainty in its rest energy? The present *experimental* uncertainty in the Σ^+ mass is 0.06 MeV/c^2; does the intrinsic uncertainty affect the measurement of the Σ^+ mass significantly?

14.25 • Consider the reaction

$$\pi^- + p \rightarrow K^\circ + \Lambda^\circ$$
$$\hookrightarrow \pi^- + p.$$

We know that the proton has baryon number $B = +1$ and that pions have $B = 0$. Given that baryon number is always conserved, deduce the baryon number of the Λ° and the K°.

14.26 •• (a) Consider the reaction

$$\pi^- + p \rightarrow K^+ + \Sigma^-$$
$$\hookrightarrow \pi^+ + \pi^\circ.$$

Knowing that the proton is a baryon and the pions are not, deduce the baryon numbers B of the K^+ and Σ^-. (b) Given the reaction

$$\pi^+ + p \rightarrow K^+ + \Sigma^+,$$

deduce the baryon number of the Σ^+. (c) Is the Σ^- the antiparticle of the Σ^+?

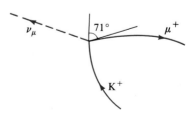

FIGURE **14.24** (Problem 14.27)

14.27 ••• Figure 14.24 shows the track of a K^+ meson in a bubble chamber. The K^+ has decayed into a μ^+ and an invisible neutrino: $K^+ \rightarrow \mu^+ + \nu_\mu$. The paths of the charged particles were curved by a magnetic field $B = 0.50$ T directed out of the page. The measured radii of the curved paths are $R_K = 1.0$ m for the K^+ and $R_\mu = 1.7$ m for the μ^+. (a) Use this information to find the momenta of the K^+ and μ^+. [Recall Eq. (3.48), relating R to p.] (b) The angle between $\mathbf{p}_K$ and $\mathbf{p}_\mu$ at the moment of decay is 71° as shown. Deduce the momentum of the invisible neutrino. (c) Assuming the known masses of the μ^+ and ν_μ ($m_\mu = 106$ MeV/c^2 and $m_\nu = 0$), calculate their energies. (d) Deduce the energy of the K^+ and hence find its mass and speed.

SECTION **14.7** (THE QUARK MODEL OF HADRONS)

14.28 • According to the quark model, the reaction $\pi^+ + p \rightarrow K^+ + \Sigma^+$ should be understood as follows:

$$(u\bar{d}) + (uud) \rightarrow (u\bar{s}) + (uus).$$

Rewrite each of the following reactions in the same way.
(a) $\pi^+ + n \rightarrow K^\circ + \Sigma^+$
(b) $K^+ + n \rightarrow K^\circ + p$
(c) $\Delta^{++} \rightarrow p + \pi^+$
(d) $e^- + p \rightarrow e^- + \Delta^{++} + \pi^-$
(The quark constituents of all the hadrons concerned can be found in Tables 14.5 to 14.7.)

14.29 • Analyze the following reactions in terms of the quark constituents of the hadrons involved, as in Problem 14.28:
(a) $\pi^- + p \rightarrow K^+ + \Sigma^-$
(b) $\Sigma^- \rightarrow n + \pi^-$
(c) $e^- + n \rightarrow e^- + \Delta^\circ$
(d) $\Delta^\circ \rightarrow \pi^- + p$

14.30 • The Ξ° (mass 1315 MeV/c^2) is a "doubly strange" baryon with composition (uss) and is a member of a doublet of two baryons with almost exactly the same mass. Based on this information, what are the quark composition of the other member and its charge, and what should be its symbol?

14.31 • The Σ^+ is a strange baryon containing one s quark, $\Sigma^+ = (uus)$, and is one of a triplet of particles with different charges but almost exactly the same masses. Given this information, what would you suggest for the quark constituents of the two companions of the Σ^+? What should they be called, and what are their charges?

14.32 • Use Fig. 14.13 to find the energy of the photon emitted in each of the decays of the excited states of the $(c\bar{c})$ meson. What sort of photons are these? (Visible? UV? etc.)

SECTION **14.8** (THE STRONG FORCE AND QCD)

14.33 • Figure 14.25 shows a proton and Σ^+ intereacting by exchange of a K° meson. Redraw this picture in

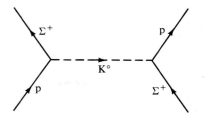

FIGURE **14.25** (Problems 14.33 and 14.34)

detail [as in Fig. 14.17(b)] to show the quark constituents of all the particles involved.

14.34 • Figure 14.25 (Problem 14.33) shows a proton and Σ^+ interacting by exchange of a K° meson. (a) Draw the corresponding diagram in which a proton and Σ° exchange a K^+. (b) Redraw your diagram to show the quark constituents of all the particles involved.

SECTION **14.9** (ELECTROWEAK INTERACTIONS: THE W AND Z PARTICLES)

14.35 • Given that the weak force is carried by the W particle with mass about 80 GeV/c^2, estimate the range of the weak force.

14.36 • In a fixed-target experiment the target is usually a liquid or solid with a high density of individual target particles. In a colliding-beam experiment the "target" is one of the two beams, and the corresponding density is much lower. This means that colliding-beam experiments yield a much smaller number of events and are hence much harder to perform. Estimate the magnitude of this difference as follows: (a) Find the density of protons (number per unit volume) in a liquid hydrogen target. (The mass density of liquid hydrogen ≈ 0.07 g/cm^3.) (b) Estimate the density of protons in a beam with the following characteristics: beam diameter ≈ 1 mm, beam current ≈ 1 A, speed of particles $\approx c$. (c) Compare the "target" densities in the two types of experiment. (The rate of events depends on several other factors as well; nevertheless, low target density is one of the main problems to be overcome in colliding-beam experiments.)

14.37 • When a neutron (udd) undergoes β decay what actually happens is that one of the d quarks emits a virtual W^-, which decays into an e^- and $\bar{\nu}_e$. Draw a Feynman diagram to represent the β decay of a neutron, showing all quark constituents.

14.38 •• The π° decays electromagnetically with $\tau(\pi^\circ) = 8.7 \times 10^{-17}$ s, but the π^+ decays only weakly with $\tau(\pi^+) = 2.6 \times 10^{-8}$ s. You can explain the large difference between these lifetimes in terms of the different ranges of the electromagnetic and weak interactions: First, recall that any meson is composed of two quarks. Next, note that, according to the electroweak theory, the intrinsic strengths of the electromagnetic and weak interactions are the same; this means that the probabilities for either kind of interaction are about the same, *provided* that two quarks are within the relevant range. Now, the two quarks in a meson are always within the (infinite) range of the electromagnetic force; on the other hand, they are very seldom within the range of the weak force. Given that the two quarks move more or less randomly inside the volume of the meson (radius of order 1 fm) and that the range of the weak force is of order 10^{-3} fm, estimate what fraction of the time the two quarks are within the range required for weak interaction. Use these considerations to estimate the ratio $\tau(\pi^\circ)/\tau(\pi^+)$, and show that your answer is of the same order as the observed ratio.

14.39 • • • Lifetimes of mesons that decay via the strong interaction are of order 10^{-23} s. Lifetimes for weak decays vary considerably, depending on the energy released, but a representative value is 10^{-10} s. Explain this difference using an argument parallel to that sketched in Problem 14.38. (*Hint:* In this problem you are comparing the weak and *strong* interactions. The difference in the intrinsic strengths of these interactions would, by itself, make the strong decays some 10^4 times faster. The two quarks inside the meson are always within the range of the strong color force.)

14.40 • • • Suppose that a particle z is expected to be produced in a reaction $x + y \rightarrow z$. (For example, the Z° particle could be produced in the reaction $p + \bar{p} \rightarrow Z^\circ$.) (*a*) By far the simplest experimental arrangement is to fire the x particle at a stationary y target (or vice versa). Prove that in this case the required incident total energy $(E = E_x + m_y c^2)$ is equal to

$$E = \frac{m_z^2 - m_x^2 + m_y^2}{2m_y} c^2 \qquad (y \text{ stationary}).$$

Hint: Use conservation of energy and momentum, and apply the Pythagorean relation (3.23) to particles x and z. (*b*) A much more efficient arrangement (although much harder experimentally) is to use colliding beams, in which the x and y collide head-on with equal but opposite momenta. Prove that in this case the required total energy $(E = E_x + E_y)$ is

$$E = m_z c^2 \qquad (\text{colliding beams}).$$

(*c*) Suppose that $m_x = m_y$ (as is the case in pp or p$\bar{p}$ collisions). Show that the particle z that can be produced has

$$m_z c^2 = \sqrt{2m_y c^2 E} \qquad (\text{stationary target})$$

for a stationary target, but

$$m_z c^2 = E \qquad (\text{colliding beams})$$

for colliding beams. If $E = 400$ GeV and m_x and m_y equal the proton mass, what is the mass m_z that can be produced in each case? (*d*) If $m_z = 90$ GeV/c^2 (about the Z° mass) and m_x and m_y equal the proton mass, what is the energy, E, needed to produce the z particle in each case?

SECTION **14.10** (SUMMARY AND OUTLOOK)

14.41 • The proton is the lightest known baryon. Explain why if we ever observe the decay of a proton (as predicted by certain "grand unified theories"), baryon number cannot be conserved.

14.42 • Certain "grand unified theories" predict that the proton should be unstable with a lifetime of order 10^{31} years. Estimate how long one would have to wait for one of the protons in a gallon (about 4 liters) of water to decay.

14.43 • Experiments are under way to detect the supposed decay of the proton, by monitoring an underground cave full of water. Assuming that the proton's mean life is about 10^{31} years, estimate how much water is needed to give one decay per hour.

14.44 • The mean kinetic energy of particles at temperature T is $\overline{K} = 3k_B T/2$, where k_B is Boltzmann's constant, $k_B = 8.62 \times 10^{-5}$ eV/K. (*a*) Given that mean energies of order 10 keV are sufficient to strip the electrons off most atoms, find the temperature T_{atom} above which one would find few atoms. (For its first million years or so, the universe was above this temperature and so contained just electrons and nuclei.) (*b*) Given that mean energies of order 1 MeV will knock most nuclei apart, find the temperature T_{nuc} above which one would find few nuclei. (For its first 3 minutes or so, the universe was above T_{nuc} and contained just electrons, protons, and neutrons.)

SECTION **14.11** (HIGH-ENERGY ACCELERATORS)

14.45 • In the Tevatron, protons are injected from a ring that uses conventional magnets into the final, superconducting ring with about 150 GeV. They then gain roughly 1 MeV per orbit as they move around the final ring. (*a*) Roughly how many orbits does a proton have to make in the final ring to gain the full energy of 1 TeV? (*b*) Given that the ring has radius 1 km, how far does the proton travel inside the final ring, and how long does this final acceleration take?

14.46 • One might hope that the next generation of accelerators after the proposed SSC will produce particles with energies of at least 200 TeV (10 times more than the SSC). Assuming that the machine is a proton synchrotron and that the maximum practical B field is 5 T (the same as for the SSC), what would be the required diameter of such a machine? (Your answer should suggest that some completely new technology will probably be needed before a machine of this energy can be undertaken.)

14.47 • In the proposed superconducting supercollider, protons will accelerate to 20 TeV, held in an approxi-mately circular orbit by a magnetic field of roughly 5 T. (a) Estimate the required diameter of the ring. (b) How long will a proton take to go once around the ring (at full energy)?

14.48 •• We have said that as the particles in a synchro-tron speed up, the frequency f of the accelerating field must be increased. Given that protons are in-jected into the Tevatron ring with 150 GeV and are then accelerated to 1 TeV, what is the percent varia-tion in f required in this acceleration?

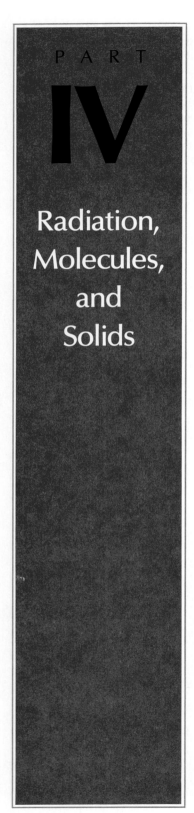

Emission and Absorption
of Radiation

In Part IV (Chapters 15 to 17) we treat applications of quantum mechanics to systems that are larger than a single atom. The obvious examples of such systems are molecules and solids, but before we treat these, we take up a branch of quantum theory that is a little harder to classify — the emission and absorption of radiation by quantum systems, like atoms and molecules. We have repeatedly seen the importance of such radiation as a source of information on a system's energy levels, but we have not yet described the mechanism by which the radiation is emitted and absorbed. Although a complete theory of these processes is beyond the scope of this book, we can give *some* understanding of how photons are emitted and absorbed, and this forms the subject of Chapter 15.

In Chapter 16 we discuss how atoms combine to form molecules. Traditionally, this topic is one of the principal concerns of chemistry, but the boundary between physics and chemistry is now quite blurred. Whether we regard the study of molecules as part of physics or chemistry, it is certainly an important application of quantum mechanics. Finally, in Chapter 17 we describe how atoms or molecules combine to form solids. The solid is our first example of a *macroscopic* system for whose understanding we require quantum mechanics. Solid-state physics has spawned a dazzling array of electronic "solid-state" devices — transistors, LEDs, photovoltaic cells, integrated circuits — that have made possible the electronic revolution of the last few decades.

The three chapters of Part IV are nearly independent of one another and can be read in almost any order. The only exception is that before starting Chapter 17 you should have read at least the first four sections of Chapter 16.

15.1   Introduction

In this chapter we discuss the emission and absorption of radiation by atoms. Our general observations will also apply to a wide variety of other quantum systems, including ions, molecules, and nuclei. We begin with a classical description of radiation and apply it to a classical model of the atom with electrons in circular orbits. In doing so, we discover some useful concepts and see how the classical view of an atom fails to agree with observation. Next we shall see how radiation striking a quantum atom can cause transitions between quantum states. This process will be shown to account for both the emission and absorption of radiation by atoms. Armed with this understanding, we conclude Chapter 15 by describing one of the most exciting applications of quantum theory, the laser.

15.2   Electromagnetic Radiation

A moving charge produces electric and magnetic fields that change with time. Faraday's law of induction tells us that a changing magnetic field induces an electric field, and, as first proposed by Maxwell, a changing electric field induces a magnetic field. This suggests that a moving charge can bring about an intertwined process in which changing electric and magnetic fields continually induce one another, with the fluctuations in these fields moving away from the charge as electromagnetic radiation. This suggestion proves to be correct, with the single exception that a charge whose velocity is constant does not radiate.* In other words, it is found that *any accelerated charge radiates.*

In Maxwell's theory of electromagnetic radiation these ideas are made precise by showing that the laws of magnetic and electric induction can be combined to give a wave equation for both the magnetic and electric fields. The wave speed u for disturbances of either field is given by the combination of electromagnetic constants

$$u = \frac{1}{\sqrt{\epsilon_0 \mu_0}} = 3 \times 10^8 \text{ m/s},$$

---

* To see why no radiation occurs when $\mathbf{v}$ is constant, consider the inertial frame in which $\mathbf{v} = 0$. In this frame the magnetic field is zero and the electric field is constant, so no radiation occurs. It follows from the principle of relativity that no radiation occurs in any frame.

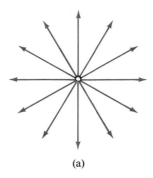

(a)

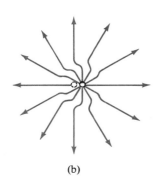

(b)

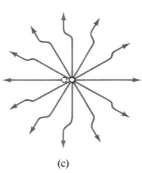

(c)

FIGURE 15.1 (a) Electric field lines from a static charge are radial. (b) When the charge is displaced abruptly, changes in its electric field propagate outward at speed c; distant portions of the field still point outward from the original (dashed) position. (c) The transverse disturbance linking near and far fields continues to move outward.

which equals the observed speed of light. This suggested to Maxwell what has since been amply verified, that light consists of electromagnetic waves in which electric and magnetic fields travel together.

To get some picture of how a moving charge produces radiation, let us examine the electric field of a single charge that is moved abruptly. In Fig. 15.1(a) we show the initially static charge and its radial field. When the charge is moved the news of its changed position moves outward at the speed of light, so distant portions of the field lines are offset from near portions. The transition zone between the near and far fields necessarily contains a transverse component, as shown in Fig. 15.1(b) and (c). While the radial component of the electric field falls like $1/r^2$, it can be shown that the transverse component falls like $1/r$. Consequently, at large distances it is the transverse component that dominates and carries radiated energy away from the charge.

The total power P radiated by an accelerated charge q (moving nonrelativistically*) can be shown to be

$$P = \frac{2kq^2a^2}{3c^3},\qquad(15.1)$$

where a is the acceleration. This formula accurately describes the power radiated by any macroscopic system of moving charges. For example, in TV or radio broadcasting, electric charges are made to oscillate inside the rods of an antenna, and the resulting radiated power is given by (15.1) (see Problem 15.2).

The formula (15.1), and the classical theory from which it derives, can sometimes be applied to microscopic systems as well. For example, when an electron in an X-ray tube collides with the anode, it undergoes rapid deceleration. The classical theory correctly predicts that this deceleration will produce X-ray radiation, known as bremsstrahlung or braking radiation (Section 5.4). Since the classical theory takes no account of the quantization of radiation, it cannot, of course, be correct in every detail. For example, classical theory predicts that some bremsstrahlung X rays will be produced at all frequencies. In reality, each photon carries energy hf, and no photons can be produced for which hf exceeds the kinetic energy K of the incident electron. Therefore, the spectrum of emitted X rays is cut off abruptly at $f = K/h$—a result called the Duane–Hunt law and described in Section 5.5.

When formula (15.1) was applied to a classical atom, difficulties arose almost immediately, and it was this problem that suggested to Bohr that classical physics needed modification, as described in Section 6.5. The problem in question is that an electron orbiting in a classical atom is necessarily accelerating. Therefore, according to (15.1), it must be radiating electromagnetic energy. In the following examples, we see that the rate of radiation predicted by (15.1) is so large that a classical atom would collapse completely in a time on the order of 10^{-11} s.

EXAMPLE 15.1 Find the power that would be radiated by a classical electron in the $n = 1$ Bohr orbit of a hydrogen atom.

* To see how (15.1) is modified when the motion is relativistic, see Problem 15.5.

The orbiting electron has a centripetal acceleration $a = v^2/r$ and must therefore radiate. Perhaps the simplest way to find the acceleration is to note that it is given by Newton's second law as $a = F/m$, with $F = ke^2/r^2$. Substituting into (15.1), we find that

$$P = \frac{2ke^2}{3c^3}\left(\frac{ke^2}{mr^2}\right)^2 = \frac{2(ke^2)^3 c}{3(mc^2)^2 r^4}, \qquad (15.2)$$

or, putting $r = a_B = 5.3 \times 10^{-2}$ nm,

$$P = \frac{2 \times (1.44 \text{ eV}\cdot\text{nm})^3 \times (3 \times 10^{17} \text{ nm/s})}{3 \times (5.1 \times 10^5 \text{ eV})^2 \times (5.3 \times 10^{-2} \text{ nm})^4}$$
$$= 2.9 \times 10^{11} \text{ eV/s}, \qquad (15.3)$$

where we have used, again, the useful combination $ke^2 = 1.44$ eV·nm. Since typical energies of electrons in atoms are a few eV, the rate (15.3) is a very rapid energy loss.

As the orbiting electron gives off energy, its orbital radius decreases, and when the radius reaches that of the proton, the atom has collapsed completely. We can use the rate of energy loss (15.3) to estimate the time required for this collapse.

EXAMPLE 15.2 Use the relation (6.10) between r and E for an electron in a circular orbit to find dr/dE. Then use the result of Example 15.1 to find the rate at which r shrinks:

$$\frac{dr}{dt} = \frac{dr}{dE}\frac{dE}{dt} = \frac{dr}{dE}(-P), \qquad (15.4)$$

where, since P is the rate of energy *loss*, we have replaced dE/dt with $-P$. Evaluate (15.4) at $r = a_B$ and, assuming that dr/dt remains constant at this value, make a rough estimate of the time required for the complete collapse of the classical atom.

The relation (6.10) between E and r for a circular orbit is

$$E = -\frac{ke^2}{2r}.$$

Differentiating, we find that

$$\frac{dE}{dr} = \frac{ke^2}{2r^2} \qquad \text{or} \qquad \frac{dr}{dE} = \frac{2r^2}{ke^2}. \qquad (15.5)$$

The rate of decrease of orbital radius is then given by (15.4) as

$$\frac{dr}{dt} = \frac{dr}{dE}(-P) = \frac{2r^2(-P)}{ke^2}.$$

Taking P from (15.3) in Example 15.1, we find that

$$\left|\frac{dr}{dt}\right| = \frac{-2r^2P}{ke^2} = \frac{-2 \times (0.053 \text{ nm})^2 \times (2.9 \times 10^{11} \text{ eV/s})}{1.44 \text{ eV} \cdot \text{nm}} = -1.1 \times 10^9 \text{ nm/s}.$$

On the scale of an atom ($r \sim 0.1$ nm) this is an enormous rate, leading to a very rapid collapse. If this rate remained constant, the time Δt for complete collapse would be simply

$$\Delta t = \frac{\Delta r}{dr/dt} = \frac{-0.053 \text{ nm}}{-1.1 \times 10^9 \text{ nm/s}} \approx 5 \times 10^{-11} \text{ s},$$

where we have neglected the radius of the proton because it is so small compared to a_B.

An exact calculation of the time required for collapse of a hydrogen atom gives $\Delta t = 1.6 \times 10^{-11}$ s (see Problem 15.13). This dramatic instability of the classical atom was what led Bohr to postulate the existence of quantized orbits to which the classical laws of radiation did not apply. As we saw in Chapter 8 (and review in Section 15.3), the Schrödinger theory avoids this problem since it predicts that atoms have *stationary* charge distributions.

15.3 Stationary States

In Section 15.2 we saw that radiation from a classical atom would lead to a rapid collapse. We now turn to a proper quantum treatment of electrons in atoms and will see that this problem is avoided. As we discussed in Section 8.3, the wave function of an electron generally depends on time as well as on spatial coordinates. We saw that for a state with definite energy E, the complete wave function $\Psi(\mathbf{r}, t)$ is a product of a spatial part $\psi(\mathbf{r})$ and an oscillating factor $e^{-i\omega t}$:

$$\Psi(\mathbf{r}, t) = \psi(\mathbf{r})e^{-i\omega t}, \tag{15.6}$$

where ω is given by the de Broglie relation $\omega = E/\hbar$. In the problems considered so far, we have only had to consider the spatial part $\psi(\mathbf{r})$. However, since radiation is a time-dependent phenomenon, we must now use the full time-dependent wave function $\Psi(\mathbf{r}, t)$.

As we discussed in Chapter 7, the probability density P of an electron is given by the square of the absolute value of its wave function

$$P(\mathbf{r}, t) = |\Psi(\mathbf{r}, t)|^2, \tag{15.7}$$

which for the wave function (15.6) becomes

$$P(\mathbf{r}, t) = |\psi(\mathbf{r})|^2 \times |e^{-i\omega t}|^2 = |\psi(\mathbf{r})|^2 \tag{15.8}$$

since $|e^{-i\omega t}| = 1$ (Section 8.3). We see in (15.8) that the probability density (and hence the charge density) of an electron in the state (15.6) does not depend on time. It is for this reason that these states are called *stationary states*. Their charge distributions do not accelerate and hence do not radiate.

The simple Schrödinger theory outlined above successfully explains the existence of stable atomic states. Unfortunately, it goes too far, since it predicts that all energy levels should be perfectly stable. Thus our task now is to explain why, with the single exception of the ground state, all energy levels are found to be unstable; that is, all excited states eventually drop to some lower level by radiating a photon. To understand this, it helps to begin by considering atomic transitions caused by externally applied radiation. This is what we take up in Section 15.4. Then, in Section 15.5, we return to the case of an isolated atom and explain the instability of its excited states.

15.4 Absorption and Stimulated Emission

We consider now an atom at which we direct radiation from an external source. We shall find that this external radiation can cause transitions between the atom's energy levels. For radiation of a definite angular frequency ω_r (where the subscript r stands for "radiation") the electric field $\mathscr{E}$ oscillates so that

$$\mathscr{E} = \mathscr{E}_0 \cos \omega_r t, \tag{15.9}$$

where $\mathscr{E}_0$ is the amplitude of the incident wave.

To simplify our discussion we will consider an electron in a one-dimensional infinite square well (the rigid box). The conclusions we reach, however, are quite general and apply to real three-dimensional atoms. The energy levels and wave functions for the ground state and first excited state of the rigid box are shown in Fig. 15.2(a). We imagine first that a constant uniform electric field $\mathscr{E}$ is applied in the positive x direction. This field attracts the electron to the left side of the box. The potential energy is then lowest at the left side of the box and highest at the right, as in Fig. 15.2(b), where the bottom of the potential energy well is shown tilted by the electric field.

We can find the effect of the tilting of the potential well by examining the Schrödinger equation

$$\frac{d^2\psi(x)}{dx^2} = \frac{2m}{\hbar^2}[U(x) - E]\psi(x). \tag{15.10}$$

On the left side of the well the quantity in brackets in (15.10) has a larger magnitude than it does on the right. Thus $d^2\psi/dx^2$, which gives the curvature of ψ, also has a greater magnitude on the left of the box. This means that ψ is more sharply curved on the left than on the right, so both ψ_1 and ψ_2 take on the asymmetric shapes shown in Fig. 15.2(b).

Let us suppose first that the electron is initially in the ground state, ψ_1. In Fig. 15.3 we have shown the original wave function ψ_1 and the modified function* ψ_1' that results from applying the field $\mathscr{E}$. In addition, we have shown the change, $\Delta\psi_1 = \psi_1' - \psi_1$, in the wave function. A striking property of $\Delta\psi_1$ is that it closely resembles the wave function ψ_2 of the first excited state. That is, applying the electric field causes the wave function ψ_1 to pick up a small component proportional to ψ_2:

$$\psi_1' = \psi_1 + a\psi_2. \tag{15.11}$$

where a is a small coefficient. (Although it is not obvious from Fig. 15.3, ψ_1 actually picks up smaller components proportional to several of the other excited states ψ_2, ψ_3, ...)

So far we have considered a constant applied field $\mathscr{E}$. However, our real interest is in the field of the applied radiation, which oscillates with frequency ω_r

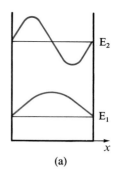

(a)

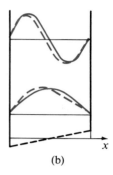

(b)

FIGURE 15.2 **(a)** The first two levels for an electron in a rigid box. **(b)** An electric field directed to the right reduces the potential energy on the left and raises it on the right. The new wave functions (dashed lines) are slightly altered from the original wave functions (solid lines).

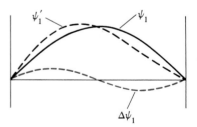

FIGURE 15.3 The difference between the original wave function ψ_1 of Fig. 15.2 (solid curve) and the new one ψ_1' (dashed curve) is $\Delta\psi_1$ (colored curve). Note how $\Delta\psi_1$ has the same shape as ψ_2 in Fig. 15.2.

* Here, the prime is used to indicate the modified wave function resulting from the changed potential; do not confuse it with the use of the prime to indicate differentiation.

as in (15.9). The oscillating field causes the potential well of Fig. 15.2 to rock back and forth with the applied frequency ω_r, and this can produce two very different effects. For most values of ω_r, the wave function cannot respond appreciably to the oscillations of the well, and it remains essentially unchanged. However, for certain values of the applied frequency, called *resonant frequencies,* the wave function *can* respond. For example, if ω_r is close to the value

$$\omega_r = \omega_2 - \omega_1, \tag{15.12}$$

the wave function begins to pick up a component proportional to ψ_2 as in (15.11). As the radiation continues, the coefficient a grows, and after the radiation has passed by, the wave function is left with the form (15.11)

$$\psi_1' = \psi_1 + a\psi_2, \tag{15.13}$$

with the coefficient a nonzero. This function is a superposition of the states ψ_1 and ψ_2, which means that the electron now has a nonzero probability (proportional to $|a|^2$, in fact) of being found in level 2. Similarly, if ω_r is close to the value $\omega_r = \omega_n - \omega_1$, for any n, the wave function picks up a component proportional to ψ_n, and when the radiation has passed there is a definite probability of finding the electron in level n.

There is a partial analogy between the effect of radiation on an atom, as just described, and the effect of a sound impinging on a stretched string. We know that a string can vibrate at a number of different frequencies, called the harmonics, or resonant frequencies, of the string. If the sound striking the string has frequency equal to one of these, the string resonates, and the harmonic is excited with an appreciable amplitude. If the applied frequency is *not* equal to a resonant frequency, the amplitude of any excitation is negligible.

We have found that when radiation of frequency $\omega_r = \omega_2 - \omega_1$ strikes an atom in the ground state, the resulting final wave function has the form* $\psi_1 + a\psi_2$, where $|a|^2$ gives the probability of finding the atom in level 2, that is, the probability that the atom will be found to have made a transition to level 2. The significance of this statement is clearer if we consider a large number, N, of atoms, each with an electron in its ground state and all exposed to the same radiation with $\omega_r = \omega_2 - \omega_1$. After the radiation has passed by, the number of electrons that have been excited to level 2 is $|a|^2N$. This, of course, corresponds to what is observed when a group of atoms is struck by radiation with the appropriate frequency; some of them are excited, others are not. The number actually excited is determined probabilistically, as is always the case in quantum mechanics. That is, we cannot say *which* atom will be excited, only how many.

ABSORPTION OF RADIATION

In a complete quantum treatment of the transitions caused by radiation, we recognize that the radiation is quantized. Thus, when a transition from state 1 to state 2 occurs, a photon of radiation disappears, so that total energy is conserved. The frequency condition that we have found, $\omega_r = \omega_2 - \omega_1$, is

---

* Strictly speaking, there should be a coefficient on the term ψ_1 in order that the total wave function remain normalized. However, as long as the coefficient a remains small (as it usually does), the coefficient of ψ_1 stays very close to 1.

exactly the condition for conservation of energy, since the photon energy equals the energy gained by the atom:

$$E_{ph} = \hbar \omega_r = \hbar (\omega_2 - \omega_1) = E_2 - E_1 = \Delta E_{atom}. \qquad (15.14)$$

In this discussion we have considered excitation of an atom from its ground state. However, the same considerations apply to any level. If the atom were initially in an excited state, it could absorb radiation and make a transition to a higher excited state. For example, if it were initially in the $n = 2$ state and if $\omega_r = \omega_3 - \omega_2$, the atom could absorb radiation and make a transition to the $n = 3$ state.

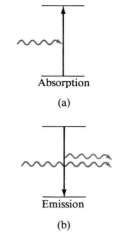

Absorption

(a)

Emission

(b)

FIGURE 15.4 Schematic diagram of absorption and emission. **(a)** When an atom absorbs radiation, it makes an upward transition. **(b)** In stimulated emission an incident photon stimulates the atom to give off a second photon.

STIMULATED EMISSION

In our discussion so far, we have seen that radiation applied to an atom in some initial state ψ_i can cause a transition to a higher level. If the atom is initially in an excited state, there is another important possibility. Suppose the atom is initially in the $n = 2$ state; then, as we have just seen, an *upward* transition to $n = 3$ can occur if $\omega_r = \omega_3 - \omega_2$. However, if $\omega_r = \omega_2 - \omega_1$, it turns out that a *downward* transition to $n = 1$ can occur. In the latter case, conservation of energy requires that the atom release the energy $E_2 - E_1$. It does this by emitting a photon of frequency $\omega_r = \omega_2 - \omega_1$, and the radiation field *gains* this photon. Since this emission of a photon is caused, or *stimulated,* by the applied radiation, it is called the **stimulated emission of radiation.** Both of these possibilities are illustrated schematically in Fig. 15.4. The possibility of stimulated emission, depicted in part (b), is the basis of the laser, as we discuss in Sections 15.7 and 15.8.

15.5 Spontaneous Emission

We have seen that radiation directed at an excited atom can cause it to make a downward transition. However, we also know that an excited atom can make a downward transition even when it is isolated from all external radiation. This process is called **spontaneous emission.** As we discussed in Section 15.3, an isolated atom in a truly stationary state, excited or not, would remain forever in that state. How, then, do excited atoms give off light without the aid of stimulating radiation? The answer is that the radiation required to stimulate downward transitions is *always* present, even when an atom is perfectly isolated from all apparent sources. The origin of this ubiquitous radiation has to do with the quantization of the electromagnetic field.

We know that the classical description of the electromagnetic field fails to explain the quantization of radiation. The complete quantum treatment of electromagnetic radiation is called **quantum electrodynamics** (QED) and is beyond the scope of this book. However, we can easily describe its predictions relevant to our present discussion: At any given frequency, ω_r, the energy of radiation is quantized and can only change by integer multiples of $\hbar \omega_r$, corresponding to the now familiar concept of photons. Furthermore, QED predicts that the lowest possible energy of the radiation field is *not* zero. That is, when one treats the electromagnetic field as a quantum system, its lowest energy level is found to be greater than zero.

We have seen similar behavior in other quantum systems. For instance, the minimum energy of a particle in a box is nonzero. As we saw in Section 7.8,

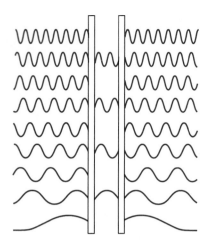

FIGURE 15.5 In the Casimir effect zero-point radiation exerts forces on two uncharged parallel metal plates. Outside the plates, all wavelengths of radiation are allowed; between the plates, only certain discrete wavelengths are allowed (just as only certain discrete wavelengths are allowed on a stretched string). If the plates are close enough together, this difference produces a measurable inward force.

the reason for this is the Heisenberg uncertainty relation between position and momentum. A similar uncertainty relation applies to the electromagnetic fields $\mathscr{E}$ and B and implies that they cannot both be exactly zero, even in empty space. As a result, the electromagnetic field has a nonzero amplitude at every frequency. This minimum possible field is called the *zero-point field*.

In many situations the effects of the zero-point fields are negligible. Nevertheless, the presence of these fields can be demonstrated in an experiment first proposed by the Dutch physicist Casimir. As illustrated in Fig. 15.5, this experiment measures a tiny force pushing two uncharged parallel metal plates together. This force would not exist if space were not filled by the zero-point fields.

For our present purposes, a far more important consequence of the zero-point fields is their effect on an isolated atom. If the isolated atom is in an excited state, the zero-point fields can stimulate it to emit a photon and drop to a lower state. Since this process occurs without externally applied radiation, it is called spontaneous emission. However, we should recognize that stimulated and spontaneous emission are essentially the same phenomenon. Stimulated emission is caused by an externally applied field. Spontaneous emission is just stimulated emission at its lowest possible rate, which occurs when the electromagnetic fields are at their minimum possible (zero-point) level.

Although the zero-point fields cause spontaneous downward transitions, conservation of energy does not allow them to cause upward transitions. For an upward transition to occur, the energy gained by the atom must be balanced by a loss of energy from the radiation field. This is clearly impossible when the energy of the radiation field is already at its lowest possible value. In the special case of the ground state, this means that there can be no spontaneous transitions at all. Upward transitions cannot occur because energy could not be conserved, and downward transitions do not occur because there are no lower states. Thus the ground state of an isolated atom is truly stable.

15.6 Lifetimes and Selection Rules

An excited atomic state can be formed in a variety of ways—absorption of radiation and collision with another atom are examples. Whatever the mechanism of excitation, an excited atom, once formed, will make downward transitions until it reaches the ground state. In this section we discuss the lifetimes of excited states that make transitions by spontaneous emission. Of course, it is possible that some other process, such as collision with another atom, could intervene before spontaneous emission occurs. However, we will consider just spontaneous emission—the only mechanism possible for isolated atoms.

In Section 15.4 we considered radiation incident on an atom in an initial state $\psi_i(\mathbf{r})$, whose full time-dependent wave function has the form (15.6)

$$\Psi_i(\mathbf{r}, t) = \psi_i(\mathbf{r})e^{-i\omega_i t}. \tag{15.15}$$

We saw that radiation of the proper frequency can introduce into the wave function ψ_i a component of a different state ψ_f, as in (15.13), to give $\psi' = \psi_i + a\psi_f$. For the full, time-dependent wave function, this implies that

$$\Psi'(\mathbf{r}, t) = \psi_i(\mathbf{r})e^{-i\omega_i t} + a\psi_f(\mathbf{r})e^{-i\omega_f t}. \tag{15.16}$$

The probability density of an electron in any state Ψ is

$$|\Psi|^2 = \Psi^*\Psi, \tag{15.17}$$

where Ψ^* denotes the complex conjugate of Ψ (see Problem 15.17). Thus, substituting (15.16) into (15.17), we find that*

$$|\Psi|^2 = (\psi_i e^{i\omega_i t} + a\psi_f e^{i\omega_f t})(\psi_i e^{-i\omega_i t} + a\psi_f e^{-i\omega_f t})$$
$$= \psi_i^2 + a^2\psi_f^2 + a\psi_i\psi_f e^{i(\omega_i - \omega_f)t} + a\psi_f\psi_i e^{-i(\omega_i - \omega_f)t}. \quad (15.18)$$

The first two terms are independent of time, but the last two oscillate at the difference frequency $\omega_i - \omega_f$ (see Problem 15.18). Thus part of the probability density of an atomic electron undergoing a transition oscillates at the frequency $\omega_i - \omega_f$. This frequency is the frequency of the electromagnetic radiation given off in the transition, and we can think of this radiation as being generated by the oscillations in the electron's charge density.†

In Section 15.2 we stated that the power radiated by a charge is proportional to the square of its acceleration as given by (15.1). For an electron oscillating with amplitude x_0 and frequency ω_r, the average power P_{av} found from (15.1) is given by

$$P_{av} = \frac{ke^2 x_0^2 \omega_r^4}{3c^3} \quad (15.19)$$

(see Problem 15.1). This radiation is quantized into photons each carrying an energy $\hbar\omega_r$. Thus the rate of emission r of photons is equal to the radiated power divided by the energy per photon:

$$r = \frac{P_{av}}{\hbar\omega_r} = \frac{ke^2 x_0^2 \omega_r^3}{3\hbar c^3}. \quad (15.20)$$

Since we have not calculated the amplitude x_0 for the oscillating charge in an atomic transition, we can only state that

$$r = C\omega_r^3, \quad (15.21)$$

where C is a constant, which depends upon details of the initial and final wave functions, ψ_i and ψ_f. The factor of ω_r^3 in (15.21) means that, other things being equal, transitions with larger energy differences (and hence larger ω_r) tend to occur more rapidly.

The significance of the rate r in (15.21) is that it gives the number of photons emitted per second. This means that the average time interval τ between the emission of successive photons is $1/r$:

$$\tau = \frac{1}{r}. \quad (15.22)$$

Our discussion here has been classical, and the quantities r and τ both require proper probabilistic interpretations in quantum mechanics. A single excited atom cannot, of course, emit a stream of photons at a steady rate r, since with the emission of the first photon the atom is no longer in the original excited state. Instead of being the number of photons emitted per second, r is the *probability,* per second, for emission of a photon. Correspondingly, $\tau = 1/r$ is not the average time interval between the emission of successive photons;

---

* In general, the wave functions ψ_i and ψ_f and their coefficients can be complex. Since this does not affect our conclusion, we have treated them as real.

† Be aware that this discussion is semiclassical. A completely correct treatment requires the use of quantum electrodynamics. Nevertheless, our semiclassical discussion does help us understand the origin of the radiation, and can even yield quantitative predictions.

rather, τ is the average time between the formation of ψ_i and a transition to the state ψ_f. For this reason τ is called the **mean life,*** or **lifetime,** of the excited state ψ_i. If a group of N atoms are all in the excited state, each with transition probability r, the total rate of transitions will be

$$R = Nr = \frac{N}{\tau}. \tag{15.23}$$

EXAMPLE 15.3 A 1-cm³ sample of argon gas is at a pressure of 10^{-4} atm and a temperature of $0°C$. All of its atoms are in the ground state, when a flash of radiation elevates 1% of them to an excited state with lifetime 1.4×10^{-8} s. Find the initial rate of emission of photons from this gas sample.

First, we must find the number of atoms within the 1-cm³ volume of the sample. We use the ideal gas law

$$pV = nRT,$$

where R, the universal gas constant, has the value 0.0821 liter·atm/mol·K. Thus the number of moles of gas, n, in this sample is

$$n = \frac{pV}{RT} = \frac{(10^{-4} \text{ atm}) \times (10^{-3} \text{ liter})}{(0.0821 \text{ liter·atm/mol·K}) \times (273 \text{ K})}$$
$$= 4.46 \times 10^{-9} \text{ mol}.$$

Each mole contains $N_A = 6.02 \times 10^{23}$ atoms, so

$$N = (4.46 \times 10^{-9} \text{ mol}) \times (6.02 \times 10^{23} \text{ atoms/mol}) = 2.7 \times 10^{15} \text{ atoms}.$$

The initial number of excited atoms, N^*, is 1% of N, so we have

$$N^* = 2.7 \times 10^{13} \text{ excited atoms}.$$

The rate of emission of photons is given by (15.23)

$$R = \frac{N^*}{\tau} = \frac{2.7 \times 10^{13}}{1.4 \times 10^{-8} \text{ s}} = 1.9 \times 10^{21} \text{ s}^{-1}.$$

This rate can be compared to the number of photons striking an area of 1 m² in bright sunlight, about 4×10^{21} s⁻¹. Clearly, the flash from the excited argon is very bright; remember, however, that it lasts only for a time of order 10^{-8} s.

SELECTION RULES

Transitions whose rate r is large are called **allowed transitions.** Because these occur rapidly, a large flux of light is produced by even a small sample, so allowed transitions are easily detected. Transitions whose rate r is very small are called **forbidden transitions.** These are not completely impossible, but have such a low rate that they are difficult to observe in a sample of reasonable size and were not detected by early spectroscopists.

In Figure 15.6 the energy levels of the hydrogen atom are shown arranged in columns according to the quantum number l of the orbital angular momentum. Some of the transitions that are observed are shown in the figure, and it

* For a more complete discussion, see Section 13.3.

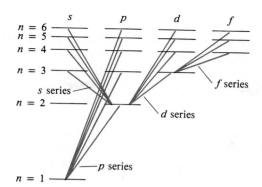

FIGURE 15.6 Some of the allowed transitions observed in the hydrogen atom. Note that each involves a change of l by one unit, as is found to be the case for *all* allowed transitions. Note also that the traditional labels s (sharp), p (principal), d (diffuse), and f (fundamental) were originally applied to transitions, not levels.

can be seen that these allowed transitions always involve a change in l by one unit, up or down. That is,

$$\Delta l = \pm 1 \qquad \text{for allowed transitions,} \qquad (15.24)$$

where Δl denotes the difference between initial and final values of l. The rule (15.24) is an example of a **selection rule** — a simple rule that identifies allowed transitions. Among the wide variety of transitions that one could imagine, only those that satisfy this rule are allowed. For example, any transition from an s state to another s state is forbidden. Similarly, $p \to p$, $s \to d$, $d \to s$, $\cdots$, are all forbidden.

The occurrence of selection rules is explained by quantum mechanics. When the time-dependent Schrödinger equation is used to calculate the transition rate r between any pair of states ψ_i and ψ_f, it is found that r is proportional to the square of an integral involving the product of the wave functions ψ_i and ψ_f. When this integral is evaluated, one finds that almost all transitions fall into two distinct groups. For certain transitions, the rate r is appreciable, and these are, of course, the allowed transitions. For most transitions, the predicted rate is much smaller — usually by many orders of magnitude — and these are the so-called forbidden transitions. The allowed transitions always involve initial and final states whose quantum numbers satisfy various *selection rules*, like that in (15.24). Other examples of selection rules appear in Problems 15.21 to 15.25.

METASTABLE STATES

Looking back at Fig. 15.6, we see that the $2s$ level of hydrogen has no allowed downward transitions, because there is no $l = 1$ level below the $2s$ level. This would seem to imply that the $2s$ state is perfectly stable. In fact, there exist other processes that deexcite the $2s$ state, such as collision with other atoms.* However, these processes all occur very slowly and the lifetime of the $2s$ state is exceptionally long. There are many other atoms that have similar excited states with no allowed downward transitions. Whenever this occurs the long-lived excited state is called a **metastable state**. These states are important in the operation of many lasers, as we see in the next section.

---

* Another such process, which occurs even in isolated atoms, is the simultaneous emission of two or more photons, but this occurs extremely slowly.

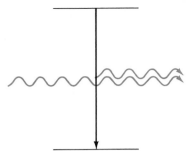

FIGURE 15.7 Incident radiation stimulates a transition from an excited state to a lower state. The emitted radiation is exactly in phase with the incident radiation.

CHARLES TOWNES (born 1915, US). The principle of the laser was first developed for microwaves, the corresponding device being the *maser* (an acronym for Microwave Amplification by Stimulated Emission of Radiation). Townes had a large part — both theoretical and practical — in development of the maser. He also contributed substantially to the laser (which amplifies light). Both masers and lasers have revolutionized almost every branch of science; for example, the maser is the basis of the atomic clock. For his role in all this, Townes won the 1964 Nobel Prize.

15.7 Lasers

In stimulated emission a beam of radiation incident on excited atoms causes them to emit more radiation of the same frequency and therefore to amplify the original beam. The laser is a device that exploits this effect to amplify light of a definite frequency. The name *laser* stands for "light amplification by the stimulated emission of radiation."

In stimulated emission the time dependence of the oscillation induced in each atom is in lock-step with the stimulating radiation. This means that the emitted radiation is exactly in phase with the radiation causing the transition, as suggested in Fig. 15.7. This **coherence** of the emitted photons means that the resulting light wave is an almost perfect sinusoidal wave. This contrasts with the **incoherent** light from ordinary sources where the phase of each atom's radiation is random with respect to all the others. It can also be shown that the photons produced in stimulated emission are ejected in the same direction as the stimulating radiation. This *spatial coherence,* as it is called, means that the laser beam can have a very well defined direction in space.

We have described the amplification of light in a laser as caused by stimulated emission from excited atoms. Normally, however, most atoms are in their ground state, and radiation that strikes atoms in their ground state will be absorbed, not amplified. To achieve amplification, we must arrange that a majority of the atoms are in an excited state. This reversal of the normal population of the levels is called a **population inversion.** Different lasers use different means to achieve population inversion, as we describe in the following accounts of some important types of lasers. Some lasers produce short pulses of light, while others produce a continuous beam. Since the details of operation of these two types are somewhat different, we describe them in turn, starting with the pulsed laser.

PULSED LASERS

The first successful laser for visible light* was a pulsed laser, developed by the American physicist Maiman in 1960. The essential element of this laser was a ruby rod containing chromium ions, which have a metastable state 1.79 eV above the ground state. Transitions between these two states produce photons with a wavelength $\lambda = 694$ nm, in the deep red portion of the visible spectrum.

The general design of the ruby laser is shown in Fig. 15.8. In the first step of operation, a brilliant burst of light from a flashlamp causes most of the Cr ions in the ruby rod to be excited out of their ground state. The majority of these excitations populate a short-lived state which makes a prompt transition to the metastable state at 1.79 eV, as shown in Fig. 15.9. (The process of moving the ions from their ground state to the metastable level is called **pumping.**) Because the metastable level is relatively long-lived (about 4 ms) a majority of the Cr ions are caught, briefly, in this level. Once the number of ions in the metastable state is greater than the number in the ground state ($N_2 > N_1$), any light produced by a spontaneous $2 \rightarrow 1$ transition is amplified by stimulated emission from the excited ions it encounters. As this amplified light sweeps through the rod, it rapidly deexcites many excited ions and forms an intense, short pulse of

---

* We should mention that the first use of stimulated emission to amplify any kind of radiation was with microwaves. Devices that amplify microwaves in this way are called *masers.*

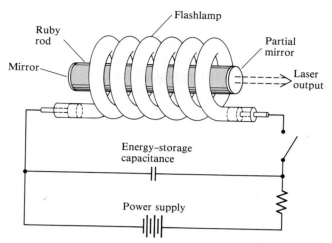

FIGURE 15.8 Schematic drawing of a pulsed ruby laser. The flashlamp excites Cr ions in the ruby rod. The mirrors cause light to reflect back and forth in the rod to increase the probability of stimulating further emission. One mirror is not 100% reflective, so some light escapes to form the external beam. (Courtesy University Science Books.)

light. When the majority of the ions have returned to their ground state, the entire process can be repeated.

To give the light ample opportunity to stimulate emission, it is reflected back and forth along the rod, whose ends are polished and silvered to form two mirrors.* One of these mirrors is made partially transparent to allow light to escape and form the external beam of laser light. In the original laser, the xenon flashlamp was powered by a capacitor bank of several hundred microfarads charged to several kilovolts. The principal parts of this laser are shown in Fig. 15.10 on page 450.

There are three major differences between the pumping light from the flashlamp and the light produced by the laser action itself. First, the flashlamp emits a broad range of wavelengths (white light), while the laser light is concentrated in a narrow spectral line at $\lambda = 694$ nm. Second, the flashlamp light is incoherent, while the laser light is coherent. Third, the flashlamp light radiates in all directions, whereas the laser beam has a well-defined direction in space. This spatial coherence is one of the most striking characteristics of laser light.

Since the development of the ruby laser, a variety of other materials have been found that can be pumped to a metastable level and used to make pulsed lasers. These lasers vary widely as to total energy and pulse length. For the ruby laser, a representative value for the energy of a single pulse is on the order of 10 J. Since each pulse has a typical duration of order† 100 μs, this gives an instantaneous power of order 100 kW. This high instantaneous power is the basis of many laser applications: Small pulsed lasers are used in medicine to cause coagulation and to suture tissues by forming tiny scars. Larger versions are widely used in industry for welding, perforating, and machining. Even larger versions are being investigated for military use.

---

* The distance L between the mirrors must satisfy the condition $L = n\lambda/2$, to ensure constructive interference of the multiply reflected waves.

† As we discuss in Section 15.8, this pulse is really a cluster of much shorter pulses.

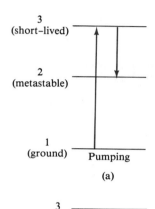

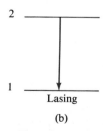

FIGURE 15.9 The relevant levels of the chromium ions in a ruby laser are: (1) the ground state, (2) the metastable state at 1.79 eV, and (3) a short-lived state. (a) Absorption of flashlamp light elevates many Cr ions to the short-lived state, which promptly decays to the metastable state, where a large population builds up. (b) The process of stimulated emission then causes the laser transition to occur in many Cr ions.

FIGURE 15.10 The original laser, built by Maiman. The ruby rod, about 1 cm in diameter, can be seen inside the coiled flashlamp. (Courtesy Hughes Research Lab.)

CONTINUOUS LASERS

For many purposes it is convenient to have a continuous wave of laser light. Most **continuous-wave** (or **CW**) **lasers** use *four* levels, instead of the three levels used in the ruby laser. This avoids an important problem with any three-level laser: In a three-level laser, more than half the atoms must be excited out of their ground state to achieve the necessary population inversion, with $N_2 > N_1$. Since nearly all atoms are normally in the ground state, a great deal of energy must be supplied by the pumping flash to accomplish this. A **four-level laser** greatly reduces this requirement by making the lower level of the laser transition an excited state which is normally empty. As shown in Fig. 15.11, the laser transition from level 3 to level 2 avoids the heavily populated ground state, so that the condition of population inversion ($N_3 > N_2$ in this case) is easily achieved. Because of their lower power requirements, such lasers can operate continuously.

A popular type of continuous, four-level laser is the helium–neon laser. This uses a mixture of helium and neon gases and produces red light with $\lambda = 633$ nm. The uppermost level of this laser is in the helium atoms, and the other three are in the neon, as shown in Fig. 15.12. The He atoms are pumped by high-speed electrons rather than by a flashlamp. This is achieved by a steady electric discharge (like that in a neon sign) within the He–Ne mixture. The energetic electrons in the discharge strongly excite a metastable level in He whose energy is very close to one of the excited levels of Ne. In collisions between the excited He atoms and unexcited Ne atoms, there is a high probability for transfer of the excitation energy, with the He atom dropping to its ground state while the Ne atom is excited. (The upper level in Ne is not so readily excited directly by fast electrons, hence the need for He and the transfer process.) The laser transition occurs in the Ne, producing the characteristic red light at 633 nm.

Not many atoms are required to be in the upper Ne level because the level below it rapidly empties by a fast transition to the ground state. The condition $N_3 > N_2$ is therefore achieved with only modest power requirements for the electric discharge.

Figure 15.13 is a schematic sketch of an He–Ne laser. The gas mixture fills a glass tube fitted with electrodes to produce a continuous electric discharge. As in the ruby laser, mirrors reflect light back and forth to increase the opportunity for stimulated emission. The external beam passes out through one of the mirrors, which is partially transmitting. Typical small models produce a light beam with a power of 10^{-3} W and consume a few watts of electric power.

FIGURE 15.11 In a four-level laser, the laser transition is between levels 3 and 2. Since few atoms are in level 2, the condition $N_3 > N_2$ is readily achieved.

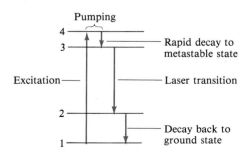

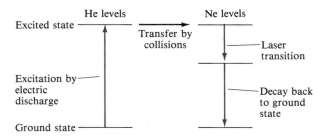

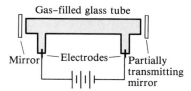

FIGURE 15.13 Key components of an He–Ne laser are the gas-filled glass tube, mirrors, and power supply.

FIGURE 15.12 The level initially pumped in the He–Ne laser is in the He atoms. Collisions transfer this energy to a level in the Ne atoms, which then produce stimulated emission, terminating in a nearly empty excited state.

SCIENTIFIC APPLICATIONS

The advent of the laser has totally transformed research in many branches of science. The sharply defined frequency of laser light has made possible measurements of quantized energy levels in atoms, molecules, and solids, with unprecedented accuracy. Because laser light is so intense, measurements can be made with very small samples—transitions in *single* atoms have even been observed. Using lasers with pulses that last just a few femtoseconds (1 fs = 10^{-15} s), chemists can follow the detailed evolution of chemical reactions.

Lasers also allow extremely accurate measurements of distances. For example, in the lunar ranging experiment, the distance from the earth to the moon is measured by timing the round trip of a laser pulse that is fired at the moon and reflected back by a mirror placed there by Apollo astronauts; in this way, changes in the earth–moon distance of a few centimeters can be detected. Similarly, geophysicists can monitor the extremely slow motions of the tectonic plates that comprise the earth's crust. Laser interferometers (described in Section 15.8) allow length measurements with accuracies of order one hundreth of the wavelength of light (that is, accuracies of a few nanometers). With the help of lasers the speed of light can be measured so accurately that it has now become a defined constant, in terms of which the SI meter is specified.

We have already mentioned some commercial applications of lasers, and it is easy to list more—in communications using optical fibers, in sound recording using compact discs, in computing, and so on. To conclude, we mention just one more application, the laser gyroscope, which has replaced the conventional gyroscope in many commercial aircraft and is shown in Fig. 15.14.

15.8 Further Properties of Lasers (optional)

ANGULAR DIVERGENCE

The directional properties of laser beams depend on the paths followed by light within the laser cavity (the space between the two mirrors). Figure 15.15(a) shows light reverberating back and forth in a straight line between the mirrors as its amplitude is increased by stimulated emission from excited atoms. A laser operating in this manner is called a single-mode laser. If the laser medium were perfectly homogeneous and the mirrors perfectly flat, the external beam from single-mode operation would be nondiverging if light were not a wave. However, because of the wave nature of light, diffraction occurs and causes the light to diverge slightly. The minimum angle of divergence, $\delta\theta$ (defined as the angle

FIGURE 15.14 A laser gyroscope, consisting of three perpendicular, ring-shaped lasers, all bored in a single 5-inch cube of glass. Each ring contains two laser beams, rotating in opposite directions (only parts of which are visible in this picture). The interference between the two beams is extremely sensitive to rotations of the ring, and with three such rings, the device can be used like an ordinary gyroscope to keep track of an aircraft's orientation. (Courtesy Litton Guidance and Control Systems.)

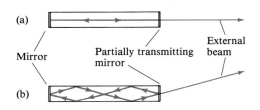

(a)

Mirror   Partially transmitting   External
          mirror   beam

(b)

FIGURE 15.15 **(a)** A laser operating on only one mode, in which light reflects back and forth along a straight path between the mirrors. **(b)** A more complicated mode which produces light traveling at an (exaggerated) angle.

from center to first minimum), depends on the diameter d of the beam and is given by the diffraction relation*

$$\delta\theta \approx \frac{\lambda}{d}, \qquad (15.25)$$

where λ is the wavelength of light (and $\delta\theta$ is measured in radians). With typical laser dimensions, the actual angle of divergence $\delta\theta$ is of order 10^{-4} rad, which is extremely small compared to the spreading from most other light sources. Nevertheless, over large distances, even laser beams suffer appreciable spreading, which reduces their intensity, as we see in the following example.

EXAMPLE 15.4   A single-mode He–Ne laser ($\lambda = 633$ nm) with an initial beam diameter of 3 mm has an angular divergence limited only by diffraction. Find the approximate diameter of the beam at a distance of 300 m. By what factor is the intensity reduced at this distance?

According to (15.25), $\delta\theta \approx \lambda/d$; therefore, the radius R of the beam at a distance L is (Fig. 15.16)

$$R \approx L\delta\theta \approx \frac{L\lambda}{d}.$$

With $L = 300$ m this gives

$$R = \frac{(300\text{m}) \times (6.3 \times 10^{-7} \text{ m})}{3 \times 10^{-3} \text{ m}} \approx 60 \text{ mm}.$$

Therefore, the beam diameter is about 120 mm.

Since intensity is power per unit area, it is inversely proportional to the cross-sectional area of the beam. Thus the ratio of final to initial intensities is

$$\frac{I_f}{I_i} = \frac{\pi d_i^2/4}{\pi d_f^2/4} = \left(\frac{d_i}{d_f}\right)^2 \approx \frac{1}{1600}.$$

This loss of intensity is sufficiently large that it must obviously be taken into account when planning to use lasers over large distances.

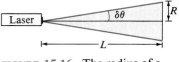

Laser $\delta\theta$ R

L

FIGURE 15.16   The radius of a laser beam at a distance L is approximately $R \approx L\delta\theta$. (Because R is much larger than the original beam size, we can neglect the latter.)

More complex modes, like that in Fig. 15.15(b), can also occur. These modes produce nonaxial light that increases the angular divergence of the laser beam; typically, $\delta\theta \approx 10^{-3}$ rad. A small aperture can be inserted in the laser

---

* For a circular aperture, this formula is generally quoted as $\delta\theta = 1.22\lambda/d$. For our purposes of estimation, we have dropped the factor of 1.22.

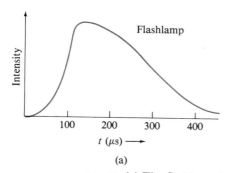

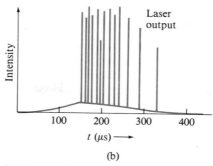

FIGURE 15.17 (a) The flashlamp intensity in a pulsed laser typically lasts several hundred μs. (b) The output of the laser contains a weak background of incoherent spontaneous emission and a series of intense spikes of coherent light.

cavity to limit laser action to the axial mode and hence reduce the angular divergence, but this usually results in a loss of total power.

TIME DEPENDENCE OF PULSED LASER BEAMS

In a pulsed laser the flashlamp that initiates the pulse usually produces light over a period of several hundred microseconds, as shown in Fig. 15.17(a). The output of the laser generally consists of a series of spikes, as in Fig. 15.17(b). The occurrence of these brief spikes is easily understood. Initially, in the first 100 μs or so, $N_2 < N_1$, so laser action cannot occur. Only spontaneous emission, which is incoherent, occurs during this time. When N_2 finally exceeds N_1, laser amplification begins and the light level within the cavity builds rapidly. In fact a "runaway" occurs because the *rate* of stimulated emission grows as the amplitude of the radiation increases. Very quickly, then, N_2 decreases until $N_2 < N_1$ and laser amplification ceases, typically in a fraction of a microsecond. However, the pumping radiation from the flashlamp is still present and again causes N_2 to increase until $N_2 > N_1$, producing another spike. This process continues until the level of pumping radiation falls below that which can make N_2 greater than N_1.

Q-SWITCHING

The size of the pulses produced by a ruby laser is limited because laser amplification begins as soon as $N_2 > N_1$ and ends quickly when N_2 drops back to $N_2 \approx N_1$. Larger pulses could be produced if N_2 could be greatly increased before laser amplification begins. This can be accomplished by preventing amplification temporarily until N_2 is large and then permitting laser action to occur. This method of producing single, very intense pulses is called **Q-switching.** The name originates in the term **quality factor,** generally abbreviated as Q. This parameter is a measure of the time light can reverberate within the laser cavity before it dies away.* If Q is large, the light loses energy slowly and hence is easily able to stimulate laser amplification. If Q is small, the light loses energy quickly, and if Q is sufficiently small, laser action does not occur.

---

* More precisely, Q is the ratio of the total energy in the cavity to the energy lost per cycle.

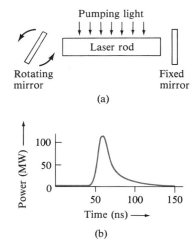

Pumping light

Laser rod

Rotating
mirror

Fixed
mirror

(a)

(b)

FIGURE 15.18 (a) One type of
Q-switching uses a rotating
mirror, which prevents laser
action until the instant the two
mirrors are parallel. (b)
Oscilloscope trace showing
output power against time for a
laser of this type.

In the lasers described so far the cavity is designed with high Q by using low-loss mirrors and a clear laser medium so that laser amplification is easily achieved. This causes laser action to begin very soon after $N_2 > N_1$ and limits the size of the pulse. However, if the Q of the cavity is reduced to a low level, laser action cannot occur even with all the atoms in the excited state. A very large pulse can then be produced by abruptly increasing Q (hence the name "Q-switching").

A variety of schemes are used for the temporary reduction of Q. The simplest method to understand uses a rotating mirror for one of the cavity mirrors as sketched in Fig. 15.18. The laser cannot operate until the mirrors are parallel. The flashlamp is fired a few hundred microseconds before the mirrors become parallel so that the metastable level becomes highly populated. The pulse that occurs when the mirrors become parallel rapidly drains the population of level 2, producing a single extremely intense pulse. Figure 15.18(b) shows the pulse produced by a laser of this type with a peak power of 100 MW (1 megawatt = 10^6 watts). (The peak power of such lasers can be as large as thousands of megawatts.)

Another Q-switching technique uses a dye contained in a transparent cell that is placed between the ruby rod and one of the mirrors. The color of a dye is due to its absorption of light at wavelengths corresponding to the energy difference between its ground state and a group or *band* of excited states. By use of a dye that absorbs strongly at the laser wavelength, Q is reduced so much that laser action cannot occur. However, after the flashlamp has operated for some time, virtually all of the dye molecules have been excited out of their ground state so that absorption ceases.* At this point the dye becomes transparent, increasing Q, and a very large laser pulse occurs.

COHERENCE LENGTH

To conclude this section, we return to the property of coherence of laser light as compared to light from other sources. In particular, we introduce the notion of **coherence length** as a measure of this property. To understand better the notion of coherence, let us consider first a perfectly sinusoidal wave. Such a wave has the property that its oscillations at any one point A are perfectly correlated, or coherent, with the oscillations at any other point B, no matter how far A is from B. This means, for example, that if the distance $AB = n\lambda$ (with n an integer, small or large), the oscillations at A and B are exactly in step; and if we could somehow combine the waves from A and B we would obtain perfectly constructive interference. Similarly, if $AB = (n + \frac{1}{2})\lambda$, the oscillations at A and B are exactly out of step and, if combined, would interfere perfectly destructively.

No real wave is perfectly sinusoidal. Among other things, it is found that light waves from most sources are produced with continual random shifts in their phase. (For instance, in an ordinary light filament, photons emitted in different regions are totally unrelated and have random relative phase.) This means that the relative phase of the oscillations at any two points A and B is continually and randomly changing. As one might expect, these random fluctuations are small if A and B are close together, but become steadily larger as the distance AB increases. Thus if $AB = n\lambda$ with n equal to a *small* integer, the

* A dye with this property is called a saturable dye and is said to have been bleached by absorption.

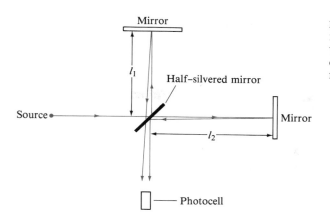

Mirror

l_1

Half-silvered mirror

Source

Mirror

l_2

Photocell

FIGURE 15.19 A Michelson interferometer. Light is split into two beams, which travel different distances, $2l_1$ and $2l_2$, before recombining.

oscillations at A and B remain very nearly in step and, if combined, would interfere almost completely constructively. On the other hand, if AB is sufficiently *large,* the relative phase of the oscillations at A and B fluctuates by 180° or more. In this case, even if AB is exactly $n\lambda$, the oscillations at A and B will sometimes be in step and sometimes completely out of step. If combined, these waves would interfere, sometimes constructively and sometimes destructively. Since the fluctuations are rapid, the observed intensity would be the average of the maximum and minimum intensities (namely, half the maximum). Evidently, once the distance AB is large enough we would get this same result whether or not $AB = n\lambda$. Under these conditions we say that the oscillations at A and B are no longer coherent. Roughly speaking, we are going to define the coherence length of a wave as the largest distance AB for which the oscillations at A and B are still coherent.

We can test these ideas using a Michelson interferometer, as sketched in Fig. 15.19. The half-silvered mirror splits the beam into two parts, which travel out and back along arms of lengths l_1 and l_2, and are then recombined and measured by a photocell. Since the two waves started as a single wave and traveled distances that differ by $\Delta = 2(l_1 - l_2)$, the photocell is in effect measuring the superposition of the oscillations at two points, A and B, in the same wave, the distance between A and B being the path difference Δ. By varying the length of either arm, we can study the interference as a function of the distance $AB = \Delta$.

If the original wave were perfectly sinusoidal, then as Δ increases the observed intensity would alternate between maxima, all of the same height I_{max}, and minima, $I_{min} = 0$, as in Fig. 15.20(a). These alternations would continue no matter how large we made Δ. In reality, the wave is not perfectly sinusoidal and the situation is as shown in Fig. 15.20(b). When Δ is small, the intensity alternates between I_{max} and 0, much as in Fig. 15.20(a). But as Δ increases, the two waves begin to lose their coherence, and the maxima begin to shrink and the minima to grow. Finally, as $\Delta \to \infty$, the contrast between the maxima and minima disappears entirely, and I approaches the constant value $I_{max}/2$.

It is clear from Fig. 15.20(b) that one cannot define a unique coherence length Δ_c beyond which the two waves abruptly lose their coherence. Rather, the coherence fades out continuously, and one can, for example, define Δ_c as the distance at which the difference between successive maxima and minima has dropped to the value I_{max}/e. This is the definition indicated in Fig. 15.20(b).

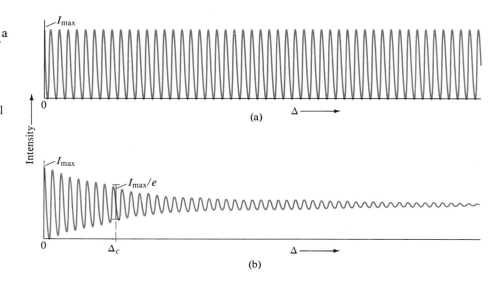

FIGURE 15.20 Intensity I as a function of path difference Δ in a Michelson interferometer. (a) If the light source were perfectly coherent, I would alternate between I_{max} and 0 and back to I_{max} each time Δ increased by one wavelength. (b) For any real source one finds that as Δ increases the coherence diminishes, and the difference between the maxima and minima slowly decreases. The coherence length Δ_c can be defined as the value of Δ for which this difference has decreased to I_{max}/e.

Before the advent of lasers, the coherence length of typical light sources was less than a millimeter or so, although lengths of order 10 cm could be achieved with difficulty. Since the coherence length of lasers can easily exceed 1 km, they have enormously increased the usefulness of interferometers for measuring distances. In a typical application, one arm of the interferometer is kept fixed while the mirror on the other is moved through the distance that is to be measured. In this way distances can be measured to a fraction of a wavelength. However, since interference cannot be observed once Δ is much more than Δ_c, the distances that can be measured cannot be much larger than Δ_c.

FIGURE 15.21 In this laser interferometer one of the paths is inside the device while the other extends to a mirror attached to the point whose displacement is to be measured. This photo shows a reading of 1 microinch, the displacement of the I-beam caused by the weight of a penny. (Courtesy *Hewlett-Packard Journal.*)

Therefore, the laser has increased by many orders of magnitude the distances that can be measured using interferometers. Perhaps equally important, laser interferometers can measure displacements of objects that are far removed from the interferometer (and hence have $l_2 \gg l_1$) as with the arrangement shown in Fig. 15.21.

IDEAS YOU SHOULD NOW UNDERSTAND FROM CHAPTER 15

Radiation from accelerated charges
Inevitable collapse of the classical atom
Absence of radiation from stationary states
Transitions induced by external radiation:
  absorption
  stimulated emission
Spontaneous emission
Zero-point fields
Lifetime of excited states
Selection rules

Metastable states
Lasers:
  coherence
  population inversion
  pumping
  pulsed and continuous lasers
  three-level and four-level lasers
[Optional section: angular divergence of laser beams, time dependence of laser pulses, Q-switching, coherence length]

PROBLEMS FOR CHAPTER 15

SECTION 15.2 (ELECTROMAGNETIC RADIATION)

15.1 • A charge q executes simple harmonic motion with position $x = x_0 \sin \omega t$. (a) Find P, the total power of the radiation emitted by this oscillating charge. (b) Show that the average power over one complete cycle is

$$P_{av} = \frac{kq^2\omega^4 x_0^2}{3c^3}.$$

15.2 • In the antenna of a TV or radio station, charges oscillate at some frequency f and radiate electromagnetic waves of the same frequency. As a simple model of such an antenna, imagine that a single charge $q = 250$ nC is executing simple harmonic motion at 100 MHz with amplitude 0.3 m. (1 nC = 10^{-9} coulomb.) Use the result of Problem 15.1 to calculate the average total power radiated by this antenna.

15.3 • A CB transmitter in a car radiates 2 W of power at about 30 MHz. (a) Find the rate of emission of photons by the transmitter. (b) In this case, is there an appreciable difference between the correct quantum view and the classical picture in which the radiation is emitted continuously?

15.4 • Carry out the calculation of Example 15.1 using SI units throughout. [Note: If you use a calculator and exceed its maximum exponent range (usually $10^{\pm 99}$), you should calculate the mantissa and exponent separately.]

15.5 •• Many particle accelerators, including the cyclotron and the synchrotron (Sections 13.10 and 14.11), hold charged particles in a circular orbit using a suitable magnetic field. The centripetal acceleration, $a = v^2/r$, can be very large and can lead to serious energy loss by radiation, in accordance with Eq. (15.1). (a) Consider a 10-MeV proton in a cyclotron of radius 0.5 m. Use the formula (15.1) to calculate the rate of energy loss in eV/s due to radiation. (b) Suppose that we tried to produce electrons with the same kinetic energy in a circular machine of the same radius. In this case the motion would be relativistic and formula (15.1) is modified by an extra factor* of γ^4:

$$P = \frac{2kq^2 a^2 \gamma^4}{3c^3}. \tag{15.26}$$

* Note that this is for the case of circular motion. For linear motion the factor is γ^6.

Find the rate of energy loss of the electron and compare with that for a proton. (Your answer for the electron should be enormously larger than for the proton. This explains why most electron accelerators are linear, not circular, since the acceleration in a linear accelerator—once $v \sim c$—is far smaller than the centripetal acceleration considered here.)

15.6 •• Answer the same questions as in Problem 15.5 but assume that both the proton and electron have kinetic energy 10 GeV and move in a circle of radius 20 m. [In this case both particles are relativistic and you must use the relativistic formula (15.26).]

15.7 •• The device called PEP at Stanford in California stores electrons and positrons orbiting in opposite directions around a circle* of radius 170 m. Because of the centripetal acceleration, $a = v^2/r$, the particles lose energy in accordance with Eq. (15.26) [which is the appropriate relativistic form of Eq. (15.1)]. (a) Find the rate of energy loss of a single 15-GeV electron. (b) If a total of 2×10^{12} particles are radiating at this rate, what is the total power, in watts, needed to keep them at 15 GeV? (For comparison, the power used by a typical household appliance is on the order of 100 W.)

15.8 •• Figure 15.22 shows an electric field line from a charge q that was moved abruptly through a small displacement **d**. The kink in the field line occurs between r and $r + \Delta r$, where $r \approx ct$ and t is the time since q was moved. The radiation from q is carried by the transverse component of $\mathcal{E}$, shown as $\mathcal{E}_{tr}$ in the picture. Prove that as a function of θ, $\mathcal{E}_{tr}$ is proportional to $\sin \theta$. (This means that the maximum radiation is at $\theta = 90°$ and there is no radiation at $\theta = 0$. For this reason broadcasting antennas are oriented at 90° to the direction in which they need to transmit.)

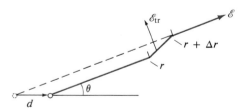

FIGURE **15.22** (Problem 15.8)

15.9 •• (a) A classical point charge q of mass m is in a circular orbit of radius r around a fixed charge Q (with q and Q of opposite sign, of course). Starting from Eq. (15.1) derive a formula for the radiated

*The actual device contains both curved and straight sections but is reasonably described as a single circle for the purposes of this problem.

power P in terms of q, m, r, and Q. (b) By what factor is P changed if we double q (leaving m, r, and Q unchanged)? (c) What if we double Q (with m, r, q unchanged)?

15.10 •• One of the difficulties with classical models of the atom was that they failed to predict the correct frequency for the radiation emitted. According to classical electromagnetic theory, the frequency of emitted radiation should equal the frequency of the orbiting electron. (a) Calculate the orbital frequencies, $f_{orb}(1)$ and $f_{orb}(2)$, of a classical electron in the $n = 1$ and $n = 2$ Bohr orbits of a hydrogen atom. (b) Now find the frequency $f_{ph}(2 \rightarrow 1)$ of the actual photon emitted in the $2 \rightarrow 1$ transition. Show that $f_{ph}(2 \rightarrow 1)$ is *not* equal to either $f_{orb}(2)$ or $f_{orb}(1)$ (or their average or their difference). (c) It turns out, however, that as $n \rightarrow \infty$ the orbital frequency $f_{orb}(n)$ of the nth orbit does approximate the frequency $f_{ph}(n \rightarrow n - 1)$ of a photon emitted in the transition $n \rightarrow n - 1$. Prove that $f_{orb}(n) = E_R/(\pi \hbar n^3)$. Derive an expression for $f_{ph}(n \rightarrow n - 1)$ and show that it approaches $f_{orb}(n)$ as $n \rightarrow \infty$. (This result—called the **correspondence principle**—played an important role in the development of the Bohr model.)

15.11 •• Consider the electron in a classical He$^+$ ion. Using the method of Example 15.1, find the radiated power predicted by the classical radiation formula (15.1), for a radius equal to that of the first Bohr orbit of He$^+$. Compare your result with that predicted for hydrogen.

15.12 ••• Using the method of Example 15.2, estimate the lifetime of a classical He$^+$ ion starting in the $n = 1$ Bohr orbit for He$^+$. Compare with the answer for hydrogen.

15.13 ••• In Example 15.2 we estimated the time Δt for collapse of a classical hydrogen atom, making the approximation that it shrinks at a constant rate. Recalculate Δt without making that approximation, as follows: (a) Use Eq. (6.10) to evaluate dE/dr, where E is the electron's energy in an orbit of radius r. (b) Using Eq. (15.2), you can immediately write down dE/dt, the rate at which the electron would lose energy by radiation, as a function of r. (c) Combine your results to give dr/dt, the rate at which the orbit shrinks, and find the time for this classical atom to collapse entirely, by evaluating

$$\Delta t = \int_{a_B}^{0} \frac{dt}{dr} \, dr.$$

SECTION **15.4** (ABSORPTION AND STIMULATED EMISSION)

15.14 • (a) The first excited state of the sodium atom is 2.11 eV above the ground state. What wavelength radiation can cause transitions between these two levels? What sort of radiation is this? (Visible, UV, etc.?) (b) Answer the same questions for the $2p_{3/2}$ and $2p_{1/2}$ levels in hydrogen, which are 4.5×10^{-5} eV apart. (This is the fine-structure splitting discussed in Section 10.7.) (c) Answer the same questions for the lowest two levels of the ^{7}Li *nucleus,* which are 0.48 MeV apart.

15.15 • The atoms of a certain monatomic gas have five energy levels: $E_1 = 0$, $E_2 = 5.4$, $E_3 = 8.2$, $E_4 = 8.6$, and $E_5 = 12.4$, all measured up from E_1 in eV. (a) If infrared light of 3100 nm is shone through the gas, what transitions could it cause? If the gas were so cool that all atoms were in the ground state, would you expect to observe these transitions? (b) Answer the same questions for ultraviolet light of wavelength 100 nm. (c) What wavelength light could deexcite the atoms from level 3 to level 2?

15.16 •• (a) Sketch the $n = 3$ wave function ψ_3 for an electron in an infinitely deep square well. On the same picture sketch the corresponding wave function ψ_3' for a well whose bottom is tilted as in Fig. 15.2. (b) Sketch the difference $\Delta\psi_3 = \psi_3' - \psi_3$.

SECTION 15.6 (LIFETIMES OF EXCITED STATES)

15.17 • When a quantum wave function Ψ is complex (with both real and imaginary parts) its probability density is $|\Psi|^2$, where $|\Psi|$ is the absolute value of Ψ, defined by Eq. (7.12) as

$$|\Psi| = \sqrt{\Psi_{\text{real}}^2 + \Psi_{\text{imag}}^2}.$$

Prove that $|\Psi|^2 = \Psi^*\Psi$, where Ψ^* is the complex conjugate of Ψ. (The complex conjugate z^* of any complex number $z = x + iy$, where x and y are real, is defined as $z^* = x - iy$.)

15.18 • The probability density $|\Psi|^2$ for an electron undergoing a transition is given by Eq. (15.18), in which the last two terms depend on time. Show that the sum of these two terms has the form $A \cos(\omega_i - \omega_f)t$ and find the constant A in terms of a, ψ_i, and ψ_f. (The relations between complex exponentials and trigonometric functions are given in Appendix B.)

15.19 • An isolated hydrogen atom in a $3p$ state can drop spontaneously to the $1s$ state or the $2s$ state. (Transitions to the $2p$ states are "forbidden"—that is, extremely unlikely—and can be ignored.) The probabilities of these two possible transitions are $r(3p \rightarrow 1s) = 1.64 \times 10^8$ s^{-1} and $r(3p \rightarrow 2s) = 0.22 \times 10^8$ s^{-1}. What is the total probability per second, r, that a hydrogen atom in a $3p$ state will drop to a lower state? What is its mean life $\tau = 1/r$?

15.20 • The outermost (valence) electron of sodium is in a $3s$ state when the atom is in its ground state (Table 11.2 in Section 11.6). The valence electron can be

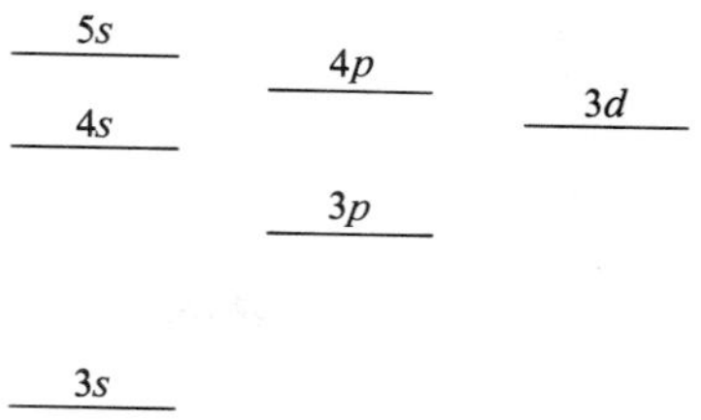

FIGURE 15.23 Some low-lying levels in sodium (Problem 15.20).

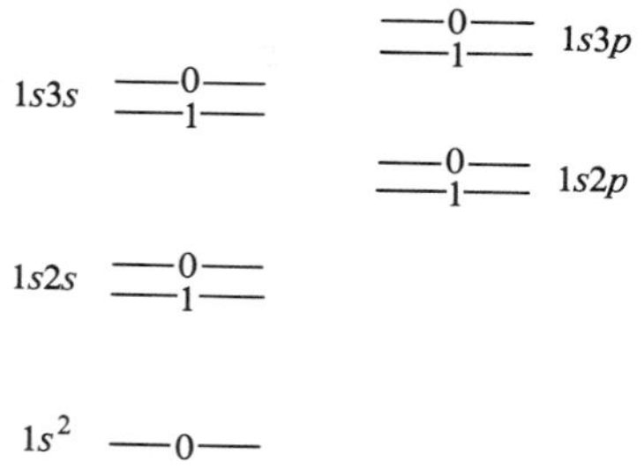

excited to higher levels, the first few of which are shown in Fig. 15.23. Given the selection rule that only those transitions for which $\Delta l = \pm 1$ are allowed, indicate on this energy-level diagram all allowed transitions among the levels shown.

FIGURE 15.24 Some low energy levels of helium. The numbers 0 and 1 indicate the quantum number s_{tot} of the total spin (Problem 15.21).

15.21 • Figure 15.24 shows some of the lowest energy levels of the He atom. They are labeled by their configuration (for example, $1s2p$ means that the atom has one electron in the $1s$ level and one in the $2p$ level). The energy depends somewhat on the orientation of the two electrons' spins: If the spins are antiparallel, the total spin is zero (quantum number $s_{\text{tot}} = 0$); if the spins are parallel, the total spin has $s_{\text{tot}} = 1$. For a given configuration the state with $s_{\text{tot}} = 1$ is slightly lower. (a) Explain why the $1s^2$ configuration has only $s_{\text{tot}} = 0$. (b) There is a selection rule $\Delta s_{\text{tot}} = 0$, that is, transitions in which s_{tot} changes are forbidden. Indicate all allowed transitions on the energy-level diagram. (Don't forget the selection rule $\Delta l = \pm 1$.) (c) Which excited levels would you expect to be metastable?

15.22 •• A discharge tube contains a microgram of H_2 gas. A brief electric discharge dissociates some of the H_2 molecules into H atoms and leaves altogether 0.1% of the original H atoms as separate atoms in the $2p$ level. (a) How many atoms does this produce in the $2p$ level? (b) The lifetime of an atom in this level is 1.6×10^{-9} s. What is the initial rate of radiation from these atoms? Give your answer in photons per second and in watts.

15.23 •• There is a selection rule on the total angular momentum of an atom or nucleus. The total angular momentum is given by a quantum number j_{tot}, and it is found that transitions which do not satisfy $\Delta j_{tot} = 0$ or ± 1 are "forbidden" (that is, very improbable). Further, among the forbidden transitions, the larger the value of Δj_{tot}, the more improbable a transition is. This trend applies to β decay in nuclei as well as to radiative transitions. Use these facts to explain why an excited state of the ^{180}Ta nucleus occurs naturally in measurable amounts. (See Appendix D.)

15.24 •• Consider two energy levels of the helium atom, both with the two electrons' spins antiparallel (so that the total spin is zero and the spins can be ignored). In the upper level, one electron has $l = 0$ and the other $l = 2$; in the lower level, one electron has $l = 0$ and the other $l = 1$. The atom is placed in a magnetic field, and (as described in Section 10.4) the upper level splits into five equally spaced sublevels and the lower into three sublevels (with the same spacing). (*a*) Sketch the resulting levels. (*b*) There are, in principle, 15 different possible transitions from the upper ($l = 2$) level to the lower ($l = 1$) level. Show that because the sublevels all have the same spacing, there are actually only seven distinct energy differences. (*c*) The selection rules for these transitions are $\Delta l = \pm 1$ and $\Delta m = 0$ or ± 1; that is, only transitions that satisfy these rules are allowed. Indicate all of the allowed transitions on your energy-level diagram. (*d*) How many distinct photon frequencies will result from allowed transitions between the two levels? This is the normal Zeeman effect described in Section 10.4.

15.25 •• Carry out the same tasks as in Problem 15.24 (with the numbers modified where necessary) using the same upper level ($l = 2$) but with a lower level in which one electron is in an s state and the other in an f state ($l = 3$).

SECTION **15.8** (FURTHER PROPERTIES OF LASERS)

15.26 • A single-mode He–Ne laser has a beam of diameter 3 mm. As the beam propagates away from the laser, its diameter increases because of diffraction. (*a*) Given that this effect simply adds to the initial beam size, find the distance at which the beam diameter is doubled. (*b*) At what distance will the beam have a diameter of 1 m?

15.27 •• In three of the Apollo lunar experiments, astronauts left reflector panels on the moon, so that laser beams from the earth could be reflected off the panels and back to earth. Lasers with $\lambda = 532$ nm send pulses of 0.3 J to the moon, and the round-trip time is measured within $\delta t \approx 0.4$ ns. In this way the one-way distance to the reflector on the moon is determined regularly within $\delta l = c\ \delta t/2 \approx 6$ cm. The beams from the lasers have a diameter of 5 km on the moon. (This is the result of spreading due to diffraction and to atmospheric turbulence.) The reflector panels contain 300 mirrors, each of diameter 4 cm. (*a*) Find the number of photons sent from the earth in a single pulse. (*b*) Find the fraction of these photons that strike any one of the small mirrors. (*c*) Find the angular divergence $\delta\theta \approx \lambda/d$ of the reflected beam from this mirror (where d is the mirror's diameter). (*d*) What is the diameter of the reflected beam when it returns to earth? (*e*) The return light is measured by a photomultiplier at the focus of a telescope. What fraction of the return light is captured by the telescope, whose diameter is 1 m? (*f*) For a single pulse, find the total number of photons captured by the telescope from all 300 mirrors. (The actual number is smaller because of losses in the atmosphere and in the telescope.) This experiment monitors the moon's orbit within a few centimeters (compared to the earth–moon distance of 4×10^8 m), allowing stringent tests of competing theories of gravity.

16

Molecules

16.1 Introduction

A molecule is a stable, or nearly stable, bound state of two or more atoms. An oxygen molecule, O_2, is a bound state of two oxygen atoms; a water molecule, H_2O, is a bound state of two hydrogens and one oxygen. In this chapter we discuss the nature of the forces that bind atoms to one another and some of the properties of the resulting molecules.

The number of atoms in a molecule ranges from two, in a *diatomic*

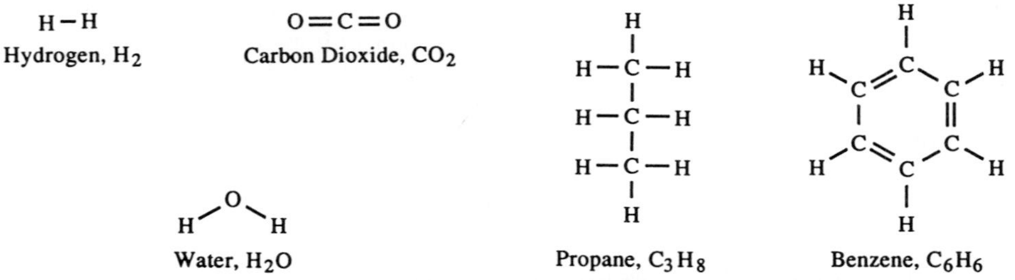

H—H
Hydrogen, H_2

O=C=O
Carbon Dioxide, CO_2

Propane, C_3H_8

Benzene, C_6H_6

Water, H_2O

FIGURE 16.1 The atoms within molecules are arranged in many different ways: straight lines, triangles, chains, rings, and many others. The lines connecting atoms represent molecular bonds, and double lines indicate double bonds, as described in Section 16.4.

461

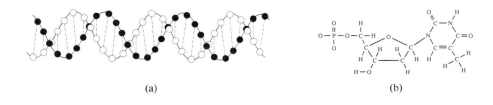

(a)

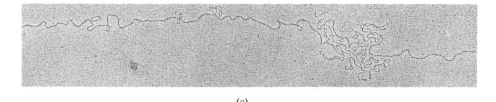

(b)

(c)

FIGURE 16.2 (a) A segment, about 10 nm long, of the double helix of the DNA molecule. (b) Each circle in (a) represents a *base,* comprising about 30 atoms, and there are altogether just four different kinds of base. This picture shows the atomic composition of the base called thymine. (c) Electron micrograph of a segment of DNA about 10 μm long (coated in platinum to make it show up). Although this segment is about a thousand times longer than the segment shown in (a), it is still only a tiny fraction of an entire molecule. The average DNA molecule in a human chromosome is about 4 cm long and contains some 10 billion atoms. (Courtesy David Prescott.)

molecule such as O_2 or CO, to several billions in large biological molecules. In Fig. 16.1 we have illustrated the arrangement of the atoms in five different, relatively simple molecules, and Fig. 16.2 shows one of the largest known molecules, DNA, which carries genetic information in organisms. Even larger numbers of atoms can form stable bound states. For example, if we took any large number of sodium atoms and the same number of chlorine atoms, then at room temperature these could bind together to form a stable salt crystal; but we would usually describe such a crystal as a *solid,* not a *molecule.** If, however, we heated the salt until it vaporized, we would find that the atoms would move apart in bound pairs, one Na atom bound to each Cl, and these pairs are exactly what we do describe as molecules of NaCl. In this chapter we discuss only molecules, such as the NaCl molecule, containing a definite, and reasonably small, number of atoms. Further, we shall consider only individual, isolated molecules. Since the molecules in a solid or liquid can never really be considered to be isolated, this means that this chapter is actually about molecules in a gas. The properties of solids and (very briefly) liquids are discussed in Chapter 17.

A common feature of all molecules is that the force which holds their atoms together is ultimately the electrostatic attraction between opposite charges. At first one might think there could be no electrostatic force between

---

* There is no clear, unambiguous distinction between molecules and solids. In fact, one could argue that a salt crystal is just an enormous molecule. Nonetheless, this is not normal usage. In particular, it is generally considered that any given molecular species should have a definite number of constituent atoms, whereas one can add any number of Na–Cl pairs to a salt crystal and still have a salt crystal, which is just a little larger. By this criterion, then, a salt crystal is not a molecule.

two neutral atoms. However, when two or more atoms are sufficiently close, their charges can redistribute themselves with pairs of opposite charges close together, so that attraction predominates over repulsion, bonding the atoms together to form a molecule.

Although all interatomic bonds are basically electrostatic, we can distinguish five different ways in which the necessary redistribution of charge occurs. The resulting kinds of bond are called:

the ionic bond
the covalent bond
the hydrogen bond
the van der Waals bond
the metallic bond

Ionic and covalent bonds are important in simple molecules and are the main topics of this chapter. The hydrogen bond involves the sharing of a proton between two negative ions. Since its main role is in large organic molecules, which are beyond the scope of this book, we shall not discuss the hydrogen bond here. The van der Waals bond is very weak and is usually important only when other types of bond do not act; it is the bond that holds some liquids and solids together and is discussed briefly in Chapter 17. As its name implies, the metallic bond is what binds metals together and is described in Chapter 17.

In this chapter we focus mostly on diatomic molecules, since they are relatively simple but still exemplify many of the principles of molecular binding in general. We begin with a brief overview of molecular properties in Section 16.2. In Sections 16.3 and 16.4 we give more detailed descriptions of the ionic and covalent bonds. In Section 16.5 we discuss how some molecular bonds have pronounced directional properties. This gives many molecules distinct shapes, which can be predicted using the properties of atomic wave functions described in Chapters 9 and 11. In Section 16.6 we discuss the excited states of molecules. We shall see that the distribution of the energy levels in a molecule is quite different from that in an atom. For this reason molecular spectra are quite different from typical atomic spectra, as we describe in Section 16.7.

GERHARD HERZBERG (born 1904, German–Canadian chemist). Herzberg fled Nazi Germany in 1935 and moved to Canada. His painstaking analyses of molecular spectra have made possible numerous applications, such as the identification of many molecules in interstellar space. He won the Nobel Prize for chemistry in 1970.

16.2 Overview of Molecular Properties

The attraction that binds atoms together to form molecules is the electrostatic attraction between the positive nucleus in either atom and the negative electrons in the other. Because ordinary matter is electrically neutral, we are normally unaware of the enormous magnitude of the electric charges within matter. It is only when the positive and negative charges are rearranged into separate regions that electric forces become evident. This is what happens within molecules on a microscopic scale. To illustrate the magnitude of the charges present in matter, we first consider a thought experiment on a macroscopic scale.

EXAMPLE 16.1   Imagine the charges in 1 gram of hydrogen separated so that all of the electrons are at one place and all of the protons at another, separated from the electrons by a distance equal to the earth's diameter. What would be the force of attraction between the electrons and the protons?

Each atom of hydrogen contains one electron and one proton, and 1 gram of hydrogen contains Avogadro's number, N_A, of atoms. Thus the magnitude of the charge of either sign is

$$Q = N_A e = (6.02 \times 10^{23}) \times (1.60 \times 10^{-19}\text{C}) = 9.63 \times 10^4 \text{ C}.$$

The enormity of these charges becomes apparent when we calculate the attraction between them when separated by an earth diameter:

$$F = k\frac{Q^2}{d^2} = \left(8.99 \times 10^9 \; \frac{\text{N} \cdot \text{m}^2}{\text{C}^2}\right) \times \frac{(9.63 \times 10^4 \text{ C})^2}{(1.27 \times 10^7 \text{ m})^2} = 5.2 \times 10^5 \text{ N}.$$

This is a force of roughly 60 tons!

It is the strong electrostatic interaction between the many charges within ordinary matter that is responsible for the strength of solids, the contact force when one solid presses against another, and the energy released in chemical reactions such as combustion. Most important, for our present purposes, it is the electrostatic interaction that binds atoms together in molecules.

Before we try to calculate the binding strengths of molecules, let us consider some of the observed parameters for a few simple molecules. In Table 16.1 we give a sampling of eight diatomic molecules. The first column shows the chemical symbol and the second the **bond length** R_0, defined as the distance between the two nuclei. The third column gives the **dissociation energy,** or **binding energy,** B, which is the energy needed to separate the molecule into two neutral atoms, and the last column gives the type of bond.

We see from Table 16.1 that typical bond lengths in diatomic molecules are one or two tenths of a nanometer, with ionic molecules tending to be a little larger than covalent. The binding energies are all of order a few eV. To understand the values of these parameters, we must examine the distribution of charge inside the molecule. This distribution depends on the behavior of the atomic electrons as the atoms approach one another, and this behavior is determined mainly by the atoms' positions in the periodic table. We now sketch three types of behavior: first, the almost complete failure of the noble gases to

TABLE 16.1

Sampling of eight diatomic molecules, four of which are bound ionically and four covalently. (Many molecules are bound by a mixture of both kinds of bond.) The second column gives the bond length R_0, and the third gives the binding energy (or dissociation energy) B.

Molecule	R_0 (nm)	B (eV)	Bond
KCl	0.27	4.3	
LiF	0.16	5.9	Ionic
NaBr	0.25	3.7	
NaCl	0.24	4.2	
H_2	0.074	4.5	
HCl	0.13	4.4	Covalent
N_2	0.11	9.8	
O_2	0.12	5.1	

form bonds; second, the formation of ionic bonds between atoms at opposite sides of the periodic table; and finally, the formation of covalent bonds, often between identical or similar atoms.

THE NOBLE GASES

Until recently it was believed that the elements He, Ne, Ar, . . . in group VIII of the periodic table formed no stable molecules, and for this reason they were named "inert" or "noble." Although we now know that the noble atoms do form some molecules, all such molecules are very weakly bound, and the names are still perfectly appropriate.

To see why the noble atoms interact so weakly with other atoms, recall from Section 11.5 that they are closed-shell atoms and are therefore spherically symmetric. Gauss's law for the electric field tells us that a spherically symmetric charge distribution of zero total charge produces no external electric field, and experiences no net force when placed in an applied field.* Equally important, the electron distribution in a noble atom cannot easily be distorted because the excited states are high above the ground state (about 20 eV in He). Thus a noble atom, even when close to another atom, remains nearly spherical, and there is very little force between the two atoms.

This argument shows correctly why there is little force between a noble atom and any other atom, provided that they do not overlap. One could imagine that if the atoms were so close that they overlapped, there might be an attractive force; but even this is not the case: Once the atoms overlap, we need to take account of the quantized energy levels of the electrons, and we shall see later that because the noble atom has all closed shells, there can be no appreciable attraction, even when the atoms overlap.

IONIC BONDING

The simplest example of a pair of atoms that do form a strong bond is an alkali–halide pair, consisting of one alkali and one halogen atom. As we discussed in Section 11.5, the alkali metals (Li, Na, K, . . .) have one loosely bound valence electron outside a closed shell. The halogens (F, Cl, Br, . . .) have one vacancy in their outer shell and hence have large electron affinity. If, for example, the alkali Li approaches the halogen F, the F atom can capture the valence electron of the Li atom, as indicated schematically in Fig. 16.3. After the transfer of an electron from Li to F, the resulting ions are stable and spherical, since each now has a closed outer shell. The strong electrostatic attraction between these two oppositely charged stable spheres then binds them together in a molecule of LiF.

Since both ions are spherically symmetric, they behave like point charges of opposite sign and form an *electric dipole*. The **dipole moment** p of such a pair of equal but opposite charges is defined to be

$$p = qd, \tag{16.1}$$

where q is the magnitude of either charge and d is their separation. We have seen

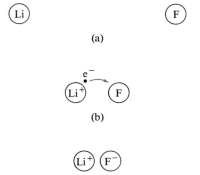

FIGURE 16.3 **(a)** When widely separated, the lithium and fluorine atoms are neutral. **(b)** When they approach one another an electron can transfer from the lithium to the fluorine. **(c)** The resulting charged ions are strongly attracted and form a stable LiF molecule.

---

* This second claim follows from Newton's third law: The spherical distribution exerts no force on any external charge; therefore, the external charge exerts no force on the distribution.

that a typical separation distance is $d \sim 0.2$ nm, so an ionically bonded molecule should have a dipole moment of order

$$p = ed \sim (1.6 \times 10^{-19} \text{ C}) \times (2 \times 10^{-10} \text{ m}) = 3.2 \times 10^{-29} \text{ C} \cdot \text{m}. \quad (16.2)$$

Measurements show that the dipole moments of alkali halide molecules are of this order, confirming that their bonding does indeed involve the transfer of an electron, of the kind illustrated in Fig. 16.3 (see Problem 16.3).

COVALENT BONDING

When two identical atoms approach one another, there is no difference in electron affinities to cause the transfer of an electron as required for an ionic bond. Nonetheless, stable molecules such as H_2, N_2, and O_2 are known to exist. Furthermore, many other diatomic molelcules, such as CO, have dipole moments that are much smaller than that predicted in (16.2) and thus cannot be ionically bonded. The bond that forms in all these cases involves a *sharing,* or *coownership,* of the valence electrons from both atoms, and is called a *covalent* bond.

We shall discuss the covalent bond in detail in Section 16.4, but the main result of that discussion is easily summarized with the help of Fig. 16.4, which is a schematic sketch of the probability density for the valence electrons around two atoms joined by a covalent bond. Part (a) shows the electron distributions around the two well-separated atoms. In (b) the atoms are close together and a large proportion of the electron distribution is concentrated in a region between the two nuclei. The electrostatic attraction of the positive nuclei toward this concentration of negative charge is what bonds the atoms together.

MIXED BONDS

In a purely ionic bond an electron is completely transferred from one atom to another, and the dipole moment is given by (16.2). In a purely covalent bond the shared electrons are equally divided between the atoms, and the dipole moment is zero. While a few molecules (such as H_2 and O_2) are purely covalent, there are no molecules that are 100% ionic, and in most molecules the bonding is a combination of ionic and covalent. That is, there is a partial transfer of charge from one atom to the other and a partial sharing. This is signaled by a dipole moment that is not as large as expected for an ionic bond, but is still greater than zero.

ESTIMATING BOND STRENGTHS

In the next two sections we discuss in some detail the strengths of ionic and covalent bonds. Here we show that it is easy to get a rough estimate of their strengths. In particular, we can see why both kinds of bond lead to binding energies that are of order a few eV, as we saw in Table 16.1.

To estimate the binding energy of an ionic molecule, let us suppose that one electron is completely transferred from one atom to the other. The potential energy of the two resulting ions, at separation R_0, is $U = -ke^2/R_0$, and if we substitute the observed value $R_0 \approx 0.2$ nm, we find that

$$U = -\frac{ke^2}{R_0} \approx -\frac{1.44 \text{ eV} \cdot \text{nm}}{0.2 \text{ nm}} \approx -7 \text{ eV}. \quad (16.3)$$

FIGURE 16.4 Schematic plot of the distribution of the outer electrons in two atoms that bond covalently. **(a)** The two separate atoms. **(b)** When the atoms form a covalent molecule, the wave functions for the outer electrons interfere constructively and produce a concentration of charge in the region between the two nuclei. The two dots show the positions of the two nuclei, and for clarity the distribution of inner electrons is omitted entirely.

We shall see in the next section that this potential energy is the dominant contribution to the molecule's total energy. If we simply ignore all other contributions, we get the rough estimate $E \approx -7$ eV. That this is negative confirms that the molecule is a stable bound state, and the binding energy of the ionic molecule is (in this approximation)

$$B = -E \approx \frac{ke^2}{R_0} \approx 7 \text{ eV}. \tag{16.4}$$

Comparing this value with the measured values in Table 16.1, we see that our estimate is a bit large. As we shall see in Section 16.3, the terms that we have neglected in (16.4) reduce our value of B and give a result that agrees well with experiment. Nevertheless, the rough estimate (16.4) is definitely of the right order of magnitude.

One can get a similar estimate of the binding energy of a covalent molecule by making reasonable assumptions about the distribution of charge in this type of molecule (see Problem 16.11). Here, however, we shall only point out that as is clear from Table 16.1, the binding energy of covalent molecules is of the same order of magnitude as that of ionic molecules, and this similarity can be explained. The potential energy of the covalent molecule is the energy that results from concentrating one or two electrons between two positive charges, separated by a distance R_0 (as shown in Fig. 16.4). This potential energy is naturally of order (Problem 16.11)

$$U \approx -\frac{ke^2}{R_0},$$

which is the same expression (16.3) that we obtained for ionic molecules,* and we conclude—just what is observed—that covalent and ionic molecules would have binding energies of comparable magnitudes.

ENERGY RELEASED IN CHEMICAL REACTIONS

The binding energy of molecules manifests itself on a macroscopic scale in chemical reactions. A chemical reaction is simply a regrouping of the atoms in some initial set of molecules to form different molecules; for example,

$$HCl + NaOH \rightarrow H_2O + NaCl.$$

Since the initial and final molecules generally have different binding energies, this regrouping results in the release or absorption of energy (usually in the form of heat). Binding energies are of order a few eV, so the energy released in an individual regrouping can itself be of this same order. On a macroscopic scale, a few eV is a very small energy, but macroscopic amounts of matter contain enormous numbers of molecules, and the total energy released when a substantial amount of matter reacts can be very large, as the following example illustrates.

---

* We should emphasize that this is only an order-of-magnitude argument. It may help to draw a parallel with the hydrogen atom, whose energy could be estimated as the potential energy of an electron and proton a distance a_B apart. This gives $E \approx -ke^2/a_B$, which is twice the correct answer but is certainly the right order of magnitude.

EXAMPLE 16.2 Gasoline engines use the heat produced in the combustion of the carbon and hydrogen in gasoline. One of the important sources of energy in this combustion is the oxidation of carbon to form carbon dioxide:

$$C + O_2 \rightarrow CO_2 + 11.4 \text{ eV}, \tag{16.5}$$

where the 11.4 eV released comes from the increased binding energy of the CO_2 molecule as compared to that of the separate C and O_2. Find the total energy released when 1 kg of carbon is oxidized. If this energy were used with 10% efficiency (a reasonable practical value) to drive a 1500-kg car up a hill, what elevation gain would it produce?

Because a mole of carbon is 12 grams, the number of carbon atoms in 1 kg is

$$N = \frac{1000 \text{ grams}}{12 \text{ grams/mole}} \times \left(6.0 \times 10^{23} \frac{\text{atoms}}{\text{mole}} \right) = 5.0 \times 10^{25} \text{ atoms}.$$

The oxidation of each carbon atom yields 11.4 eV, and the total energy released is

$$E = N \times (11.4 \text{ eV}) = (5.0 \times 10^{25}) \times (11.4 \text{ eV})$$
$$= 5.7 \times 10^{26} \text{ eV} = 9.1 \times 10^7 \text{ J}. \tag{16.6}$$

This energy results from burning 1 kg of carbon in the reaction (16.5). Since this reaction is the main source of energy in the combustion of gasoline, and carbon is the main constituent of gasoline, we can take (16.6) as a rough estimate of the energy released in burning 1 kg of gasoline. (For a more careful estimate, see Problem 16.9.)

If 10% of the energy (16.6) were used to raise a 1500-kg car through a vertical height h, then, since the gain in potential energy would be mgh, we can calculate h as

$$h = \frac{\text{gain in PE}}{mg} = \frac{9.1 \times 10^6 \text{ J}}{(1500 \text{ kg}) \times (9.8 \text{ m/s}^2)} \approx 620 \text{ m}.$$

This is the elevation gain produced by 1 kg of gasoline. Since a car's gas tank holds at least 40 kg, we see that a tankful of gasoline can lift a car through a vertical gain of some 25,000 m (or 16 miles) — considerably more than the total height of Mount Everest! This startling conclusion emphasizes that gasoline is an extremely concentrated source of energy.*

16.3 The Ionic Bond

We have seen that an **ionic bond** is formed when an electron is transferred from one atom to another, producing a pair of oppositely charged ions that attract one another strongly. This occurs when the first atom has a low ionization energy and can lose an electron easily, while the second has high electron affinity and can bind an extra electron relatively well. We start this section by

---

* We have assumed here that the car drives slowly up a steep incline, so that the work against rolling and air resistance is small compared to the work, mgh, against gravity. In the more usual context of a level road, the mgh term is zero and only rolling and air resistance remain; under these conditions a tankful of gasoline takes one several hundred miles (see Problem 16.5).

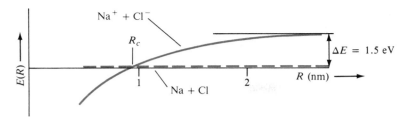

FIGURE 16.5 Energy of sodium and chlorine as a function of their internuclear separation R. The dashed, horizontal line is the energy of the two neutral atoms; the solid curve is that of the two ions Na^+ and Cl^-. When R is less than R_c, the two ions have lower energy than the two netural atoms.

considering the specific case of NaCl, and examine in detail the energy balance in the formation of an NaCl molecule.

To begin, let us imagine that our Na and Cl atoms are far apart and electrically neutral (in which case they exert no force on one another). If we were to transfer an electron from the Na to the Cl while the atoms were still far apart, this transfer would actually *cost* us energy. The energy needed to remove an electron from Na is its ionization energy, $IE(Na) = 5.1$ eV; the energy gained when Cl captures an electron is its electron affinity, $EA(Cl) = 3.6$ eV. Since the former is greater than the latter, there is a net cost of energy,

$$\Delta E = IE(Na) - EA(Cl) = 5.1 \text{ eV} - 3.6 \text{ eV} = 1.5 \text{ eV}. \tag{16.7}$$

Clearly, the spontaneous transfer of an electron between the well-separated atoms is something that will not occur. However, this situation changes as the atoms approach one another, as we now show.

To understand what happens as the two atoms come closer together, consider the graphs in Fig. 16.5. These show the energy, as a function of the distance R between the two nuclei, both for the neutral pair Na + Cl, and for the ions $Na^+ + Cl^-$. We have chosen the zero of energy as the energy of the two neutral atoms when far apart (and at rest). Since the neutral atoms exert no force on one another, their energy does not change as we move R inward, and the corresponding graph (shown as the dashed line) is a constant:

$$E(R) = \text{constant} = 0 \qquad \text{[neutral atoms]}.$$

Suppose instead that while the atoms are far apart ($R = \infty$) we transfer an electron from one to the other, and then slowly move them together. While they are still far apart, the energy of the ions is higher than that of the neutral atoms by the amount $\Delta E = 1.5$ eV calculated in (16.7). However, the ions are oppositely charged, and as they move closer their potential energy, $-ke^2/R$, comes into play, so their total energy is

$$E(R) = \Delta E - \frac{ke^2}{R} \qquad \text{[Na}^+ + \text{Cl}^-\text{]}. \tag{16.8}$$

This energy is shown as the solid curve in Fig. 16.5 and decreases steadily as the ions get closer.

At the critical separation shown as R_c in Fig. 16.5, the energy of the two ions Na^+ and Cl^- has dropped to that of the neutral pair, Na + Cl. This separation is easily calculated by setting $E(R)$ in (16.8) equal to zero, to give

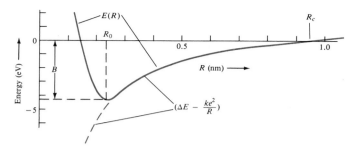

FIGURE 16.6 The general behavior of the energy $E(R)$ of the two ions Na^+ and Cl^- is shown by the solid curve. As long as the two ions do not overlap ($R \gtrsim 0.3$ nm) the energy is just $E(R) = \Delta E - ke^2/R$. Once the ions overlap, repulsive forces quickly dominate and $E(R)$ climbs steeply up again. The minimum of $E(R)$ occurs at R_0, and $E(R_0) = -B$, where B is the molecule's binding energy.

$$R_c = \frac{ke^2}{\Delta E}. \qquad (16.9)$$

For the case of NaCl, with $\Delta E = 1.5$ eV, this gives

$$R_c = \frac{ke^2}{\Delta E} = \frac{1.44 \text{ eV} \cdot \text{nm}}{1.5 \text{ eV}} = 0.96 \text{ nm}.$$

Once R is less than R_c the ions have less energy than the neutral atoms; that is, the potential energy $-ke^2/R$ of the ions more than offsets the cost ΔE of transferring the electron, and the transfer is energetically favored. To predict the probability that the transfer actually occurs would require a detailed quantum calculation, but for our purposes it is sufficient to note that the common occurrence of ionic molecules proves that the transfer does have significant probability.

Once the electron is transferred, the two ions are strongly attracted and can bind together to form a molecule. Their energy does not continue to decrease indefinitely as R approaches zero. Rather, once the two ions overlap appreciably, the force between them becomes repulsive, and the energy $E(R)$ rises as shown in Fig. 16.6. One obvious source of this repulsion is the increasing force between the two positive nuclei as they get closer. Another important contribution is due to the Pauli exclusion principle: Electrons in the same region cannot occupy the same quantum state, and once the atoms overlap, some of the electrons must move to higher levels, increasing $E(R)$ further.

The energy $E(R)$ of the two ions has a minimum at the separation shown as R_0 in Fig. 16.6, and the two ions are in stable equilibrium at this separation. Therefore, R_0 is the center-to-center distance of the two atoms in the lowest state of the stable molecule, and the corresponding value of the energy, $E(R_0)$, is the ground-state energy of the molecule.* That is, $E(R_0) = -B$, where B is the binding energy of the molecule. The observed values of these two parameters for NaCl are

$$R_0 = 0.24 \text{ nm} \qquad \text{and} \qquad B = 4.2 \text{ eV}.$$

* We are still ignoring any kinetic energy of the two atoms. As we discuss in Section 16.6, there is a small, zero-point, kinetic energy, and the ground state is actually slightly above $E(R_0)$.

Without a detailed calculation of the repulsive contributions to $E(R)$, we cannot actually *predict* these values of R_0 and B. However, one can see in Fig. 16.6 that when R is close to R_0 the repulsive contribution to $E(R)$ is still quite small; that is, near $R = R_0$ the exact $E(R)$, shown by the solid curve, is close to the value $(\Delta E - ke^2/R)$ shown by the dashed curve. This suggests that we would get good approximations for $E(R_0)$ and B by ignoring the difference between these two curves; that is, we can approximate

$$B = -E(R_0) \approx \frac{ke^2}{R_0} - \Delta E. \qquad (16.10)$$

This estimate for B is easily understood: The term ke^2/R_0 is the electrostatic binding energy of the two ions, and is reduced by the energy ΔE that was needed to transfer the electron and form the ions.

In Section 16.2 we made the rough estimate

$$B \approx \frac{ke^2}{R_0}. \qquad (16.11)$$

Comparing this with the new estimate (16.10), we see that this earlier approximation simply ignored the transfer energy ΔE, and was therefore an overestimate, as we noted in Section 16.2. In fact, the improved approximation (16.10) is still a slight overestimate, since it ignores the repulsive contribution to $E(R)$ and this contribution necessarily reduces B.

EXAMPLE 16.3 Given the known values of R_0 and ΔE for NaCl, estimate the dissociation energy B of the NaCl molecule, using both of the approximations (16.11) and (16.10).

Since $R_0 = 0.24$ nm, the rough estimate (16.11) gives

$$B \approx \frac{ke^2}{R_0} = \frac{1.44 \text{ eV} \cdot \text{nm}}{0.24 \text{ nm}} = 6.0 \text{ eV},$$

while the improved estimate (16.10) gives

$$B \approx \frac{ke^2}{R_0} - \Delta E = (6.0 - 1.5) \text{ eV} = 4.5 \text{ eV}.$$

As expected, both approximations overestimate the observed value of 4.2 eV. The first is within 40%, and the second a respectable 7%.

VALENCE

The NaCl molecule is formed by the transfer of one electron from an Na atom to a Cl atom, to give the closed-shell ions Na^+ and Cl^-. Similar ionic molecules, involving transfer of one electron, can be formed by combining any alkali atom (closed-shell-plus-one, or group I) with any halogen (closed-shell-minus-one, or group VII).

Ionic molecules that involve the transfer of more than one electron are also possible. For example, an atom like Mg, from group II, has two electrons outside closed shells and can lose these two electrons to become the closed-shell ion Mg^{2+}; meanwhile an atom like O from group VI is two electrons short of a

TABLE 16.2
The nine possible molecules formed by combining the positive ions Na^+, Mg^{2+}, or Al^{3+} with the negative ions F^-, O^{2-}, or N^{3-}. Numbers in parentheses show the valence of each element concerned.

	F^- (1)	O^{2-} (2)	N^{3-} (3)
Na^+ (1)	NaF	Na_2O	Na_3N
Mg^{2+} (2)	MgF_2	MgO	Mg_3N_2
Al^{3+} (3)	AlF_3	Al_2O_3	AlN

closed shell, so can bind two extra electrons to form the closed-shell O^{2-}. Thus Mg and O can combine to form the ionic molecule MgO by transfer of *two* electrons. Similarly, an atom such as Al from group III can lose three electrons, and an atom such as N from group V can gain three; thus one can form the ionic molecule AlN by transfer of three electrons from Al to N.

There are also many possible ionic molecules involving more than two atoms. For example, one can form the ionic molecule MgF_2, in which the Mg atom loses two electrons (to become Mg^{2+}) while each of the two F atoms gains one electron (to become F^-). In Table 16.2 we show the nine possibilities for the ionic molecules formed from Na, Mg, or Al, each in combination with one of the elements F, O, or N.

When an atom can form ionic molecules, we define its **valence** as the number of electrons it gains or loses in forming such a molecule. Elements from groups I and VII (closed-shell ± one) have valence 1; elements from groups II and VI (closed-shell ± two) have valence 2; and so on. Knowing the valences of two elements that form an ionic molecule, one can immediately predict the proportions in which the atoms combine. For example, Mg has valence 2 (closed-shell ion Mg^{2+}) and N has valence 3 (closed-shell ion N^{3-}). To retain overall neutrality we must combine three Mg^{2+} ions with every two N^{3-} ions, and the resulting compound is Mg_3N_2, as shown in Table 16.2. In general, the proportion of constituent atoms in any ionic molecule is just the reciprocal of the ratio of their valences.

16.4   The Covalent Bond

Most diatomic molecules that are not bound ionically are bound instead by the **covalent bond.** As already described, the covalent bond involves a concentration of one or more electrons from each atom in the region between the two nuclei. To understand how this concentration occurs, we begin by considering the simplest of all molecules, the H_2^+ molecular ion, which consists of two protons and a single electron.* Just as an understanding of the one-electron atom (hydrogen) helped us to build a theory of multielectron atoms in Chapter 11, so we shall see here that an understanding of the one-electron molecule (H_2^+) will help us to understand the covalent bonding of multielectron molecules such as H_2, O_2, and H_2O.

LINUS PAULING (born 1901, US chemist). Pauling applied the principles of quantum mechanics to the bonding of atoms in molecules, and his book, *The Nature of the Chemical Bond* (1939), is one of the most influential texts of the period. He was one of the first to suggest that protein molecules are arranged in a spiral. He won the Nobel Prize for chemistry in 1954 and the Nobel Peace Prize, for his fight against nuclear weapons, in 1962.

THE H_2^+ MOLECULAR ION

In H_2^+, the single electron moves in the field of the two protons. Because the protons are so heavy, they move very little compared to the electron, and it is a good approximation to treat them as stationary. (This is the same approximation used in Chapters 9 and 11 to discuss atoms.) Within this approximation, we must solve the Schrödinger equation for the single electron in the field of the two fixed protons and find its lowest allowed energy E. Just as with ionic molecules, this energy will depend on the internuclear distance R; that is, $E = E(R)$. If $E(R)$ has a minimum at some separation R_0, a stable molecule exists, with bond length R_0 and energy $E(R_0)$.

* The H_2^+ ion is readily formed in hydrogen gas by an electric discharge, which can strip an electron from a neutral H_2 molecule. In isolation H_2^+ is perfectly stable. In practice, it eventually picks up a free electron and reverts to neutral H_2. Nevertheless, it survives quite long enough for an accurate determination of its properties.

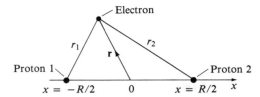

FIGURE 16.7 The H_2^+ molecule consists of two protons and a single electron. We treat the protons as fixed at $x = \pm R/2$ on the x axis. The electron's position relative to the origin is denoted by $\mathbf{r}$. Its electrostatic potential energy is determined by the distances to the two protons, shown as r_1 and r_2.

Before trying to solve the Schrödinger equation we must choose a system of coordinates. A convenient choice is to put the two protons on the x axis at positions $x = \pm R/2$, as shown in Fig. 16.7. The electron's position we denote by $\mathbf{r}$, and we must find the wave function $\psi(\mathbf{r})$ that gives the minimum total energy for the molecule with a given separation R.

A direct analytic solution of the Schrödinger equation for the electron in H_2^+ is possible, but is rather complicated and not especially illuminating. For our purposes it is more convenient to seek an approximate solution, and to this end we begin by supposing that the two protons are far apart — specifically, that R is much larger than the Bohr radius a_B. Under these conditions it is easy to see what are the solutions of the Schrödinger equation. If the electron is close to proton 1, the effect of the distant proton 2 is small, and the lowest possible state is just the ground state for an electron bound to proton 1. The corresponding wave function, which we denote by $\psi_1(\mathbf{r})$, is the familiar $1s$ wave function for a hydrogen atom centered on proton 1:

$$\psi_1(\mathbf{r}) = Ae^{-r_1/a_B} \tag{16.12}$$

where r_1 is the distance from proton 1 to the electron. There is a second state with exactly the same energy, in which the electron is bound to proton 2, with wave function

$$\psi_2(\mathbf{r}) = Ae^{-r_2/a_B}. \tag{16.13}$$

We have sketched these two wave functions in Fig. 16.8, which shows the values of $\psi_1(\mathbf{r})$ and $\psi_2(\mathbf{r})$ along the x axis (that is, the line joining the two protons).

The two wave functions ψ_1 and ψ_2 satisfy the Schrödinger equation with the same energy; that is, they are degenerate. Therefore, any linear combination,

$$\psi = B\psi_1 + C\psi_2$$

(for any two constants B and C) also solves the Schrödinger equation with this same energy. This wave function is a superposition of a state where the electron is bound to proton 1 and another where it is bound to proton 2, and — in principle at least — any such state is a possible state of our system. Two particular such states that we shall find to be especially important are*

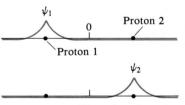

FIGURE 16.8 Wave functions for the one electron in the H_2^+ molecule with the two protons far apart ($R \gg a_B$). The plots show the values of ψ on the internuclear axis. The function ψ_1 describes a state in which the electron is bound in the $1s$ state around proton 1, and is therefore affected very little by the distant proton 2; ψ_2 is the corresponding state with the electron bound to proton 2.

---

* If we wish these functions to be normalized, we should include an overall normalization factor; since this factor does not affect our discussion here, we omit it (but see Problem 16.29).

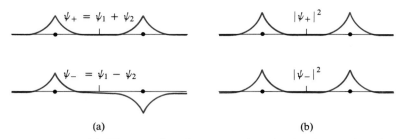

(a)   (b)

FIGURE 16.9 (a) The wave functions ψ_+ and ψ_- for the electron in H_2^+, when the two protons are far apart. The plots show values of $\psi_\pm$ along the internuclear axis. (b) Corresponding plots of the electron's probability density $|\psi_\pm|^2$ (which are identical as long as the protons are far apart).

$$\psi_+ = \psi_1 + \psi_2 \qquad (16.14)$$

and

$$\psi_- = \psi_1 - \psi_2. \qquad (16.15)$$

These two wave functions are sketched in Fig. 16.9(a). Their characteristic property is that they represent states where the electron is distributed equally around both protons. To see this, note that, as is clear from Fig. 16.9(a),*

$$\psi_+(x) = \psi_+(-x) \qquad \text{and} \qquad \psi_-(x) = -\psi_-(-x); \qquad (16.16)$$

that is, $\psi_+(x)$ is an even function of x and $\psi_-(x)$ is an odd function. Therefore, for either wave function, the probability density $|\psi|^2$ is the same at any point x as at the point $-x$:

$$|\psi(x)|^2 = |\psi(-x)|^2 \qquad (16.17)$$

with ψ equal to either ψ_+ or ψ_-. This is illustrated in Fig. 16.9(b) and means that the electron distribution at any point near proton 1 is the same as at the corresponding point near proton 2.

The individual wave functions ψ_1 and ψ_2 are often called **atomic orbitals,** since they describe states where the electron forms an atom with one of the nuclei and is unaffected by the other. [Recall that "orbital" is simply an alternative name for the wave function $\psi(\mathbf{r})$ for any one electron; that is, an orbital specifies the orbital motion—but not the spin state—of one electron. In an atom, an orbital is specified by the quantum numbers n, l, m.] The combinations ψ_+ and ψ_- are called **molecular orbitals** since they describe states where the electron is associated with *both* nuclei and belongs to the molecule as a whole. We shall see that it is the molecular orbitals ψ_+ and ψ_- that become the stationary states of H_2^+ when the two protons come close enough to form a stable molecule.

When the two protons approach one another, the wave function $\psi_1(\mathbf{r})$ overlaps proton 2, and vice versa, and the electron's motion is influenced by *both* protons. The simple atomic wave functions ψ_1 and ψ_2 of Eqs. (16.12) and (16.13) are no longer solutions of the complete Schrödinger equation (which includes the potential energy due to both protons). Nevertheless, we can show that the combinations $\psi_\pm = \psi_1 \pm \psi_2$ remain reasonable approximations. Any

---

* For simplicity we focus here on the wave functions on the x axis [and for brevity write them as $\psi_\pm(x)$]. The corresponding symmetry for arbitrary points $\mathbf{r} = (x, y, z)$ is this: $\psi_+(\mathbf{r}) = \psi_+(-\mathbf{r})$ and $\psi_-(\mathbf{r}) = -\psi_-(-\mathbf{r})$.

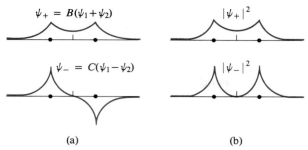

$$\psi_+ = B(\psi_1 + \psi_2) \qquad |\psi_+|^2$$

$$\psi_- = C(\psi_1 - \psi_2) \qquad |\psi_-|^2$$

(a)   (b)

FIGURE 16.10 (a) Sketch of the wave functions ψ_+ and ψ_- for the electron in the H_2^+ molecule, once the distance R between the two protons is comparable to the size of an H atom. At the origin, ψ_+ is larger than either ψ_1 or ψ_2, whereas ψ_- is exactly zero. (The factors B and C are normalization constants; C is a little larger than B—see Problem 16.29—and this is why the peaks of ψ_- are a little taller than those of ψ_+.) (b) The corresponding probability densities. (As in Figs. 16.8 and 16.9, all plots show values along the internuclear axis.)

correct solution should reflect the symmetry of the system, which has two identical force centers at $x = \pm R/2$; that is, the wave functions for the stationary states of H_2^+ should have the symmetry property (16.17). Now, as we have just seen, the wave functions $\psi_\pm = \psi_1 \pm \psi_2$ *do* have this symmetry. Furthermore, these functions, being combinations of ψ_1 and ψ_2 reflect the presence of both protons. This suggests that the functions ψ_+ and ψ_- could be reasonable approximations to correct solutions of the Schrödinger equation, even though ψ_1 and ψ_2 separately are not. Using a more advanced argument (called the variational method) one can prove that this suggestion is correct. Here we shall simply accept that ψ_+ and ψ_- *are* reasonable approximate solutions, and use them to deduce the general behavior of the energy of the H_2^+ molecule for the two states concerned.

Once the two protons are within one or two Bohr radii, the behavior of ψ_+ is as sketched in Fig. 16.10(a). The corresponding probability densities $|\psi_+|^2$ and $|\psi_-|^2$ are shown in Fig. 16.10(b), and the important difference between the two states is immediately clear: The probability density $|\psi_+|^2$ is nonzero at, and near to, the origin; in fact, it is larger in this region than either $|\psi_1|^2$ or $|\psi_2|^2$ separately (see Problem 16.29). This means that for the state ψ_+, the electron distribution is concentrated in the region between the two protons—precisely where it is strongly attracted by both protons. This lowers the molecule's energy (as compared to the energy when R is large) and causes the H_2^+ molecule to bind. By contrast, the wave function ψ_- is exactly zero at the origin. Therefore, $|\psi_-|^2$ is small in the region between the two protons, and the energy is not low enough to bind in this state.

In Fig. 16.11 we show a schematic two-dimensional picture of the states $\psi_\pm = \psi_1 \pm \psi_2$. The left side of each picture represents the state while the atoms are well separated and the distribution consists of two distinct spherical clouds.* The right side of each picture represents the distributions once the wave func-

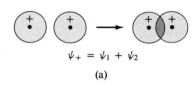

$$\psi_+ = \psi_1 + \psi_2$$

(a)

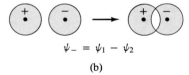

$$\psi_- = \psi_1 - \psi_2$$

(b)

FIGURE 16.11 Schematic representation of the behavior of the molecular orbitals $\psi_\pm = \psi_1 \pm \psi_2$ of H_2^+ as the nuclei approach one another. The plus and minus signs show the signs of the wave functions. When the nuclei are far apart both distributions consist of two separate spherical clouds. Once the clouds overlap, the wave functions ψ_1 and ψ_2 interfere constructively in the state ψ_+ and produce an enhanced density between the two nuclei; in the state ψ_- they interfere destructively and give a reduced density.

---

    * We should emphasize that these pictures are highly schematic. The distribution of ψ_1 or ψ_2 is not a clearly defined sphere but, rather, drops continuously from a maximum at the nucleus and reaches zero only at infinity. Thus the region of overlap is not really as sharply defined as our schematic pictures suggest.

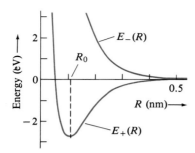

FIGURE 16.12 The energy of the H_2^+ molecule as a function of the distance R between the two protons. The curve $E_+(R)$ is the energy of the "bonding state" ψ_+; and $E_-(R)$ is that of the "antibonding state" ψ_-.

tions ψ_1 and ψ_2 begin to overlap. In the state ψ_+ the two functions ψ_1 and ψ_2 interfere constructively in the region of overlap and concentrate the probability density there. In the state ψ_-, they interfere destructively and reduce the corresponding density.

In Fig. 16.12 we have plotted the energy $E(R)$ of the H_2^+ molecule, as a function of the distance R between the two protons, for both of the states ψ_+ and ψ_-. The zero of energy has been chosen as the energy when the electron is bound to one of the protons and the other proton is far away ($R \rightarrow \infty$). As we have just argued, the energy $E_+(R)$ of the **bonding orbital** ψ_+ decreases as the protons get closer, until it reaches a minimum at a separation R_0. Once R is less than R_0 the repulsion of the two protons is dominant, and $E_+(R)$ increases rapidly as R approaches zero. The important point is that, just as with the ionic molecules discussed in Section 16.3, the energy has a minimum at a separation R_0, and a stable molecule can form, with $R = R_0$ and binding energy $B = -E_+(R_0)$. The calculated values,

$$R_0 = 0.11 \text{ nm} \qquad \text{and} \qquad B = 2.7 \text{ eV},$$

agree perfectly with experiment.

The curve $E_-(R)$ in Fig. 16.12 increases steadily as R decreases and has no minimum; so the H_2^+ molecule has no stable bound state corresponding to the **antibonding orbital** ψ_-.

THE NEUTRAL H_2 MOLECULE

Armed with a theory of the one-electron molecule H_2^+, we can now understand many simple multielectron molecules. We consider first the neutral H_2 molecule with its two electrons. Just as in our discussion of multielectron atoms in Chapter 11, we start by ignoring the repulsion between the electrons. In this approximation each electron moves independently in the field of two protons, and its lowest level is just that of the bonding orbital ψ_+ of the H_2^+ molecule, with energy -2.7 eV. The ground state of H_2, with its two electrons, is obtained by putting both electrons into this lowest level, to give a total energy of -5.4 eV, and hence binding energy $B = 5.4$ eV. This value is an overestimate since it ignores the electrostatic repulsion of the two electrons. Nevertheless, it is quite close to the observed value $B = 4.5$ eV.

We found the ground state of H_2 by giving both electrons the same wave function ψ_+. According to the Pauli exclusion principle, this is possible only if the two electrons occupy different spin states; that is, the ground state of the H_2 molecule must have the two electrons' spins antiparallel and hence have $s_{tot} = 0$ — a prediction that is fully confirmed by experiment. If we try to assemble an H_2 molecule from two H atoms whose electron spins are *parallel,* the exclusion principle requires that one of the electrons move into the antibonding orbital ψ_- or some other higher level. In the former case it is found that the total energy $E(R)$ has no minimum, and there is no corresponding molecule; in the latter, a molecule may form, but with much higher energy — that is, it forms an excited state (as we discuss in Section 16.6).

The ground state of the H_2 molecule typifies the main features of all covalent bonds. Each atom contributes one electron to the bond. The two electrons have their spins antiparallel, which allows them both to occupy the same bonding orbital with a large density between the two nuclei. The resulting "electron pair" produces a strong bond between the two atoms, but we shall see that, because of the exclusion principle, they cannot be joined by a third electron. It is the strong bond produced by a shared pair of electrons with antiparallel spins, and their inability to accept a third electron, that are the essential characteristics of the covalent bond.

SATURATION OF THE COVALENT BOND

One of the most striking features of many molecular bonds is the phenomenon of **saturation**: Molecules containing given elements usually contain some definite number of atoms and cannot bind any additional atoms. For example, hydrogen forms the stable molecule H_2, as we have just seen, but a stable H_3 molecule does not exist.

To understand why this is, we have only to recall part of our discussion of multielectron atoms in Chapter 11. In the two-electron atom, He, both electrons can occupy the lowest ($1s$) orbital and are very tightly bound. In the three-electron atom, Li, two of the electrons can occupy the lowest orbital, but the exclusion principle requires the third electron to occupy the next ($2s$) orbital, which is much less tightly bound.

Almost exactly the same argument applies to the molecules H_2 and H_3: In H_2, both electrons can occupy the lowest molecular orbital (the bonding orbital ψ_+) and this molecule is tightly bound. In H_3, there is again an orbital of lowest energy, with the probability density concentrated inside the triangle formed by the three protons. However, as with the Li atom, only two of the three electrons can occupy this lowest orbital, and the third must occupy a higher orbital.

This qualitative argument shows clearly that an H_3 molecule, if it is bound at all, should be less tightly bound than H_2. To decide whether or not the H_3 molecule (or indeed the Li atom) actually is bound requires a detailed calculation of the energy levels concerned. In the case of the Li atom we know that it *is* bound, although with much lower ionization energy than He. In the case of the H_3 molecule it turns out that the higher levels are so much higher that H_3 is not bound. The big difference between the Li atom and the H_3 molecule results from several effects, of which one of the most important concerns the three protons in either system: In the Li atom, the three protons are tightly bound inside the nucleus by the strong nuclear force; as far as the atom is concerned, the proton–proton repulsion is effectively switched off. In the hypothetical H_3 molecule, the proton–proton repulsion raises the energy by so much (several tens of eV) that the molecule is unstable.

If we consider possible larger hydrogen molecules, such as H_4, this situation only becomes worse. Therefore, there are no hydrogen molecules beyond the familiar, diatomic H_2, and the bonding between n hydrogen atoms "saturates" at $n = 2$.

COVALENT BONDING OF MULTIELECTRON ATOMS

The formation of molecules from multielectron atoms is not as complicated as one might at first fear. This is because it is only the outer, or valence, electrons of each atom that overlap enough to have an important role in the molecular bond. A closed-shell-plus-one atom such as Li (with configuration $1s^2 2s^1$) or Na ($1s^2 2s^2 2p^6 3s^1$) can, for our purposes, be regarded as a single electron in an s state outside an inert positive "core." This suggests that these atoms should behave very like hydrogen, the main difference being that the one valence electron is in a state concentrated at much larger radius (being $n = 2, 3$,

TABLE 16.3

Bond lengths (R_0) and binding energies (B) of the covalent molecules H_2, LiH, and Li_2.

Molecule	R_0 (nm)	B (eV)
H_2	0.074	4.5
LiH	0.16	2.4
Li_2	0.27	1.0

. . .). Thus we would anticipate that a lithium atom could bond covalently with hydrogen to form LiH, or with a second lithium to form Li_2. The molecule Li_2 should be bigger than LiH, which should be bigger than H_2, and because of the greater separations the binding energies should be correspondingly smaller. All of these predictions are confirmed by the experimental data shown in Table 16.3.

Atoms with more than one valence electron are naturally more complicated, but here again the situation is not so bad as one might fear. Consider, for example, the fluorine atom, with its seven valence electrons and configuration

$$\text{F:} \quad 1s^2 2s^2 2p^5. \tag{16.18}$$

Recall that there are three independent $2p$ wave functions, or orbitals, which we can take to be the functions introduced in Section 9.8, labeled $2p_x$, $2p_y$, and $2p_z$, and we can rewrite the electron configuration (16.18) in more detail as

$$\text{F:} \quad 1s^2 2s^2 2p_x^2 2p_y^2 2p_z^1.$$

(It does not matter which of the three $2p$ orbitals gets the single electron; to be definite we suppose that it is the $2p_z$.) The important point is that the orbitals $2s$, $2p_x$, and $2p_y$ are occupied by two electrons each. The electrons in each orbital must therefore have their spins antiparallel; that is, they form an electron pair that cannot participate in a covalent bond. Thus the fluorine atom has just one electron available to share in a covalent bond with another atom. In other words, fluorine is, in this respect, similar to hydrogen. We would expect it to bond covalently with H to form HF, or with another F atom to form F_2 — and both of these molecules are indeed observed.

The oxygen atom has one less electron than fluorine, with the configuration

$$\text{O:} \quad 1s^2 2s^2 2p^4.$$

Two of the electrons in the $2p$ level must occupy the same orbital ($2p_x$, say) but as we argued in Section 11.6, the other two will go, one each, into each of the remaining orbitals,* $2p_y$ and $2p_z$. That is, the configuration of the oxygen atom is

$$\text{O:} \quad 1s^2 2s^2 2p_x^2 2p_y^1 2p_z^1.$$

This leaves *two* unpaired electrons that are free to be shared in covalent bonds with other atoms. Thus oxygen can form a molecule like H_2O, in which the O atom forms one covalent bond with each of the two H atoms. It can also form two covalent bonds, with a second O atom to make the doubly bonded molecule O_2. Because O_2 is doubly bonded, we would expect it to be more strongly bound than the singly bonded F_2, and this prediction is confirmed by the observed binding energies given in Table 16.4.

The nitrogen atom has one less electron than oxygen and has the configuration

$$\text{N:} \quad 1s^2 2s^2 2p_x^1 2p_y^1 2p_z^1.$$

Since it has *three* unpaired electrons that are available for sharing with other atoms, it can, for example, form one bond with each of three H atoms to make

TABLE 16.4

Binding, or dissociation, energies of the singly, doubly, and triply bonded molecules F_2, O_2, and N_2.

Molecule	B (eV)
F_2	1.6
O_2	5.1
N_2	9.8

---

* Recall that with this arrangement the two electrons are farther apart and have slightly lower energy because their Coulomb repulsion is reduced.

an ammonia molecule NH_3. It can also form three bonds with a second N atom to make the triply bonded N_2. Since N_2 has three bonds it is very well bound, as seen in Table 16.4, and for this reason N_2 gas is chemically very inactive.

VALENCE

In the context of covalent bonding we define the **valence** of an atom as the number of electrons that it can share with other atoms in covalent bonds; H, Li, and F all have valence 1, while O has valence 2, and N has valence 3. In the case of ionic molecules such as LiF, we defined the valence of an atom as the number of electrons the atom can gain or lose to form the ionic molecule. (F has valence 1 because it *gains* one electron to become F^-, whereas Li has valence 1 because it *loses* one electron to become Li^+.) Since many atoms can bond both covalently and ionically, it is important to note that in all the simple examples discussed here, our two definitions give the same value for the valence. For instance, a closed-shell-plus-one atom such as Li or Na can *lose* its outer electron to form an ionic bond, and it can *share* the outer electron to form a covalent bond. Either way its valence is 1. Similarly, a closed-shell-minus-one atom such as F or Cl can *gain* an electron (which fills the one vacancy in its outer shell) to form an ionic bond; but as we have just seen, it can *share* the one unpaired electron in its outer shell to form a covalent bond. Either way its valence is also 1.

On the basis of what has been said here, it should be clear that a molecule like LiF could bond ionically *or* covalently and that (at the level of our qualitative treatment) we have no way to predict which mechanism will prevail. By measuring the dipole moment, as discussed in Section 16.2, one can find the extent to which a given molecule is ionically or covalently bonded. In the case of the alkali halide molecules (such as LiF) it is found that the bond is predominantly ionic, but in many other molecules the bond is found to be a combination of both types.

We should emphasize that we have focused our discussion on the simplest molecules and that many molecules are much more complicated. Perhaps the most important example of the complications that can occur, even with fairly simple atoms, concerns carbon. The ground state of carbon has the configuration

$$C: \quad 1s^2 2s^2 2p^2,$$

which suggests that carbon could contribute two electrons to covalent bonds and should have valence 2, as it does in CO. However, carbon usually has valence 4, as is testified by the existence of such molecules as CH_4 (methane) and CO_2. The explanation of this apparent contradiction is actually quite simple: The $2s$ and $2p$ levels are close together, and only a small energy is needed to raise one electron to give the configuration

$$C: \quad 1s^2 2s^1 2p^3. \tag{16.19}$$

In this configuration carbon has four electrons available for sharing in covalent bonds, and the energy gained by forming four bonds amply repays the small energy needed to achieve the configuration (16.19).

Carbon can use its four valence electrons to form bonds in many different ways. For example, it can form the molecule CH_4 as shown schematically in Fig. 16.13(a). However, it can also form the molecule C_2H_4 (ethylene), in which

H ∖ ∕ H
  C
H ∕ ∖ H

Methane, CH₄

(a)

H ∖   ∕ H
   C = C
H ∕   ∖ H

Ethylene, C₂H₄

(b)

          H
          |
H ∖   C ≡ C   ∕ H
    C   C
    | ||
H ∕   C   C   ∖ H
          |
          H

Benzene, C₆H₆

(c)

FIGURE 16.13 Schematic views of three hydrocarbon molecules: CH₄ (methane), C₂H₄ (ethylene), and C₆H₆ (benzene). Each short line represents a covalent bond formed by a shared pair of electrons. The actual molecules are, of course, three-dimensional; for example, the four H atoms in CH₄ are arranged symmetrically at the four corners of a tetrahedron with the carbon at its center.

each carbon is joined to two hydrogens but is also doubly bonded to the other carbon, as shown in Fig. 16.13(b). Another example of the numerous *hydrocarbon* molecules is C_6H_6 (benzene), in which the six carbons are joined in a ring of alternating single and double bonds; this leaves room for one hydrogen to attach to each carbon, as shown in Fig. 16.13(c). Because carbon can bond in so many ways to other carbon atoms, and to other elements, such as H, N, O, and many more, there is a vast array of different carbon compounds. Since many of them play an important role in organisms, all carbon-containing molecules (with a few exceptions such as CO and CO_2) have come to be called *organic molecules*.

16.5 Directional Properties of Covalent Bonds (optional)

In this section we describe how the properties of atomic wave functions determine the shapes of molecules. To begin, we review briefly the formation of a simple diatomic molecule like H_2.

As we saw in Section 16.4, the two electrons in H_2 both occupy the bonding orbital, which corresponds to the combination $\psi_+ = \psi_1 + \psi_2$ of the $1s$ wave functions for each separate atom,

$$\psi_1 = e^{-r_1/a_B} \quad \text{and} \quad \psi_2 = e^{-r_2/a_B}.$$

Since these functions are both positive, they interfere constructively in their region of overlap (for the orbital ψ_+) as was indicated schematically in Fig. 16.11(a). This enhances the charge density between the two nuclei and lowers the energy of the molecule.

Similar considerations apply to any diatomic molecule in which the valence electrons are all in s states, and Fig. 16.11(a) could equally have represented such molecules as Li_2 and LiH. If we consider instead such atoms as N, O, or F, whose valence electrons are in p states, the situation is more interesting. Recall that p states have three possible wave functions, which we can take to be the functions $p_x, p_y,$ and p_z introduced in Section 9.8 and illustrated in Fig. 9.21 (which we have redrawn here as Fig. 16.14). The important point about each of these p states is that its probability density is concentrated in two lobes along one of the three coordinate axes. For example, the $2p_z$ wave function of hydrogen is [Eq. (9.92)]

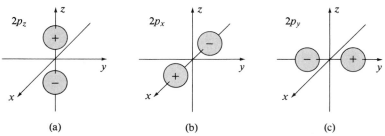

(a)   (b)   (c)

FIGURE 16.14 Each of the three wave functions p_x, p_y, p_z has two lobes where the probability density is largest. (These pictures show the contours where the density is 75% of its maximum value, for the $2p$ wave functions of hydrogen.) The plus and minus signs indicate that the wave function is positive in one lobe and negative in the other.

$$\psi_{2p_z} = Aze^{-r/2a_B}. \qquad (16.20)$$

This gives maximum probability on the z axis and zero probability in the xy plane (where $z = 0$). For our present purposes it is important to note that the function (16.20) is positive in its upper lobe (where $z > 0$) and negative in the lower one (where $z < 0$), and we have indicated these signs by plus and minus signs in Fig. 16.14.

Let us consider first an atom with just one unpaired p electron; for example, fluorine, with configuration

$$F: \quad 1s^2 2s^2 2p_x^2 2p_y^2 2p_z^1.$$

The first eight electrons in F are all paired and can, for our present purposes, be ignored. The single unpaired electron, which is distributed as in Fig. 16.14(a), can pair with the one electron in a hydrogen atom to form the molecule HF. To arrange this, however, we must bring up the H atom in such a way that the wave functions of the shared electrons interfere constructively in their region of overlap; this requires that the H atom approach the positive lobe of the F wave function as in Fig. 16.15(a). If the H atom approaches from the opposite direction, as shown in Fig. 16.15(b), we obtain an antibonding orbital, and no molecule forms.

We can summarize our findings so far: When valence electrons are in s states, their distribution is spherical, with no preferred direction for bonding. When the valence electrons are in p states, their distribution is concentrated along a definite axis and this means that they form directed bonds.

We are now ready to discuss some molecules that are more interesting, such as the water molecule, H_2O. The oxygen atom, with configuration

$$O: \quad 1s^2 2s^2 2p_x^2 2p_y^1 2p_z^1$$

has two unpaired electrons, whose distributions are sketched in Fig. 16.16(a). Each of the unpaired electrons can form a bond with one H atom, but only if the H atoms approach along the y and z axes, as shown in Fig. 16.16(b). This implies that the water molecule should be shaped like a letter "L," as shown schematically in Fig. 16.16(c). It is found that the water molecule does have an "L" shape, but the measured angle between the two arms is about 105°, some 15° larger than predicted. However, the reason for this discrepancy is not hard to find: The two H atoms, having shifted their electrons toward the oxygen, are

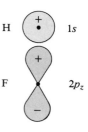

(a) Bonding

(b) Antibonding

FIGURE 16.15 Wave functions for the HF molecule. **(a)** We obtain the bonding orbital by adding the wave functions for the $1s$ electron in H and the $2p_z$ electron in F, *provided that* the H atom approaches the positive lobe of the $2p_z$ function. (The $2p_z$ wave function in F is shown here as a figure of eight to emphasize that both lobes belong to a single atomic wave function.) **(b)** If the H atom approaches from the opposite direction, then adding the two wave functions gives an antibonding orbital. For clarity the two atoms are shown with their wave functions not yet overlapping.

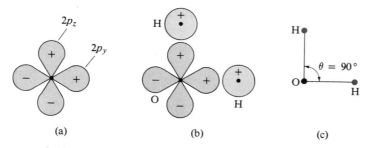

FIGURE 16.16 **(a)** The two unpaired electrons in an O atom occupy the $2p_y$ and $2p_z$ orbitals. **(b)** A water molecule can form if the two H atoms approach the positive lobes of these two orbitals. **(c)** The centers of the atoms in the resulting molecule form an "L" with angle $\theta = 90°$ between the arms. (The experimental value is $\theta \approx 105°$, as explained in the text.)

FIGURE **16.17** Schematic view of the charge distribution in a water molecule. The plus and minus signs here indicate positive and negative charges whose observed magitude is about $e/3$ (Problem 16.30).

left somewhat positive and repel one another, pushing the two arms a little farther apart than the 90° predicted by our simple model.

If the bonds in H_2O were purely covalent, the shared electrons would concentrate at the midpoint between the O and each H. It is found, however, that they concentrate somewhat closer to the oxygen. In other words, the bonds are partly ionic and there is a small net negative charge near the O, leaving a small excess of positive charge near the two hydrogens, as shown schematically in Fig. 16.17. For this reason the H_2O molecule has an electric dipole moment, which is responsible for several familiar properties of water: A water molecule can easily pick up a charge of either sign, as shown in Fig. 16.18(a) and (b), and can carry away charges from a body that is electrostatically charged. This explains why walking across a nylon carpet on a moist day does not cause the annoying electric shocks that we experience when the humidity is low.

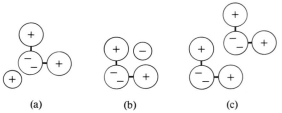

FIGURE **16.18** **(a)** and **(b)** A water molecule can attach itself to a positive or negative charge. **(c)** Water molecules tend to align and attract one another.

For much the same reason, water molecules can dissolve ionic molecules such as NaCl by attaching themselves to the separate Na^+ and Cl^- ions. Finally, when properly oriented, two water molecules attract each other, as in Fig. 16.18(c). This *dipole–dipole* attraction is what holds water together in snow and ice crystals.

Just as with water, H_2O, we can predict the shape of the ammonia molecule, NH_3. The nitrogen atom, with configuration

$$\text{N:} \quad 1s^2 2s^2 2p_x^1 2p_y^1 2p_z^1,$$

has three unpaired p electrons, each concentrated near one of three perpendicular axes. Thus N can bond with three atoms to form NH_3, with one H atom on each of the three axes, as shown in Fig. 16.19. According to our simple model, the angle θ between any two of the N—H bonds should be $\theta = 90°$, but, just as with water, the repulsion between the positive protons increases θ somewhat ($\theta = 107°$ in this case).

Similar considerations apply to many other molecules. For example, it can be shown that the four valence electrons in carbon have lobes that point symmetrically outwards toward the corners of a regular tetrahedron centered on the nucleus.* This has important implications for the shapes of the many hydrocarbon molecules mentioned at the end of the preceding section. For example, the methane molecule, CH_4, is a regular tetrahedron with the four H

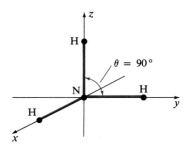

FIGURE **16.19** The ammonia molecule, NH_3, is a pyramid with the N atom at one corner. The H atoms at the other three corners form an equilateral triangle. The angle θ between any two N—H bonds is actually a little more than 90° because of the Coulomb repulsion between the H nuclei.

---

* The valence electrons occupy four orbitals called *s-p* hybrids. Each of these is a combination of the usual s and p wave functions, and each concentrates the electron in a single lobe. The electrostatic repulsion of the electrons pushes these as far apart as possible, so that they point symmetrically outward as described.

atoms at its corners and the C atom at its center. Since the basic ideas remain the same, while the details get much more complicated, we shall not pursue this topic farther here.

16.6 Excited States of Molecules (optional)

We have so far discussed only the ground states of molecules. However, most information about molecules comes from the study of molecular spectra; and since spectra result from transitions, which necessarily involve at least one excited state, we now extend our discussion to include excited states as well.

ELECTRONIC LEVELS

Our analysis of molecular ground states started by considering two atoms, at large separation R, both in their ground states with total energy $E(R)$. We found that as the atoms move closer the curve $E(R)$ may decrease and go through a minimum. If this happens (as it does with H_2, for example), the two atoms can bond to form a stable molecule, whose ground-state energy is simply the minimum value of $E(R)$.

Suppose now that we start instead with one or more of the atom's valence electrons* in excited states, so that the total energy of the two atoms is $E'(R) > E(R)$. As the two atoms approach one another, it may happen that the curve $E'(R)$ goes through a minimum, and if this does occur, we conclude that this molecule has an excited state with energy equal to the minimum value of $E'(R)$. We can describe this excited state as "stable" in the sense that the atoms will not fly apart spontaneously; however, it is not perfectly stable, since it can drop to the ground state (or any other lower level) by emitting radiation. In Fig. 16.20 we have sketched three molecular energy levels (labeled 0, 1, 2) that come about in this way.

The three curves in Fig. 16.20 give the possible energies of the molecule as functions of the separation R. When R is large the energies $E_0(R)$, $E_1(R)$,

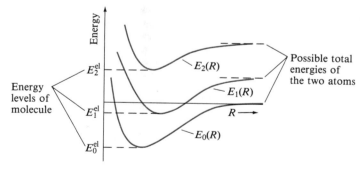

FIGURE 16.20 When two atoms are far apart their total energy can have various different values $E_0(R)$, $E_1(R)$, $E_2(R)$, ... As R decreases each of the curves $E_i(R)$ ($i = 0, 1, 2, . . .$) may have a minimum, and if this happens, there is a corresponding bound state of the molecule. The resulting energy levels of the molecule are denoted by E_i^{el}, as explained in the text.

* Our only concern here is the valence electrons, since inner-shell electrons play little part in molecular bonding.

. . . are the allowed total energies of the two separated atoms. The minimum values of the functions $E_0(R), E_1(R), \ldots$ are the corresponding energy levels of the molecule in question. These levels are called **electronic levels** because they are the energies calculated using the Schrödinger equation for the electrons while ignoring possible motions of the two nuclei; for this reason they are denoted by $E_0^{el}, E_1^{el}, \ldots$, as shown on the left of Fig. 16.20. As the picture suggests, one would expect the spacing of these electronic levels of the molecule to be of the same order as the spacing of the atomic levels from which they evolved. Therefore, since the spacing of valence levels in atoms is usually a few eV, the same is generally true for the electronic levels in molecules. This means that transitions between electronic levels in a molecule involve photons of a few eV — that is, photons in or near the visible range.

MOTION OF THE NUCLEI

If the electronic levels $E_0^{el}, E_1^{el}, \ldots$ just described were the only levels of molecules, molecular spectra would look very like atomic spectra. There is, however, a very important complication that we have so far ignored. Up to now we have assumed that the two nuclei of our diatomic molecule are at rest; that is, we have considered only the motion of the electrons and have completely ignored the nuclear motion. (Note that by "nuclear motion" we do *not* mean the internal motion of the nuclear constituents, but rather the motion of each nucleus as a whole.) We now consider the energy E^{nuc} associated with this motion and shall find that the values of E^{nuc} are quantized, with allowed values $E_0^{nuc}, E_1^{nuc}, \ldots$. Thus the allowed energies of the entire molecule, $E = E^{el} + E^{nuc}$, are

$$E = E_i^{el} + E_j^{nuc} \qquad (i = 0, 1, 2, \ldots; \quad j = 0, 1, 2, \ldots).$$

When the molecule makes a transition, both the states of the electrons (labeled by i) *and* the nuclei (labeled by j) can change. If we consider a transition (up or down) between a level E and a higher level E',

$$E = E_i^{el} + E_j^{nuc} \leftrightarrow E' = E_{i'}^{el} + E_{j'}^{nuc},$$

then the photon absorbed or emitted in the transition has energy

$$\begin{aligned} E_{ph} = E' - E &= (E_{i'}^{el} - E_i^{el}) + (E_{j'}^{nuc} - E_j^{nuc}) \\ &= \Delta E^{el} + \Delta E^{nuc}. \end{aligned} \qquad (16.21)$$

The energies ΔE^{el} are, as we just argued, typically a few eV. On the other hand, the energies ΔE^{nuc} are generally small fractions of an eV and are small corrections to ΔE^{el}. However, there are many possible different values for ΔE^{nuc}. Therefore, (16.21) implies that the spectrum of photons absorbed or emitted, instead of being a single line for each value of ΔE^{el}, will consist of many different lines, all close to ΔE^{el}, corresponding to the many possible small values of ΔE^{nuc}. Unless a spectrometer has high resolution, it may fail to resolve some of these lines, which will then appear instead as a wide *band*. Several such bands in the emission spectrum of N_2 gas are illustrated in Fig. 16.21. In the low-resolution picture one sees several broad lines or bands; in the high-resolution picture three of these bands are seen clearly to be made of many narrowly spaced lines. The occurrence of such **band spectra** at visible, or nearly visible, frequencies is one of the chief characteristics of molecular spectra, and the

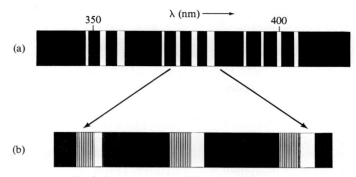

(a)

(b)

FIGURE 16.21 Emission spectrum of N_2 gas in the near ultraviolet. **(a)** Several broad lines or bands between 340 and 410 nm. **(b)** High-resolution enlargement of the three bands near 380 nm shows that each consists of many separate lines.

measurement of their many constituent frequencies is a rich source of information about molecules.

VIBRATIONAL ENERGY

For a diatomic molecule there are two kinds of motion that contribute to E^{nuc}. First, the two nuclei can move in and out along the internuclear axis, with a **vibrational energy** E^{vib}; and second, they can rotate about their center of mass with a **rotational energy** E^{rot}. Thus

$$E^{nuc} = E^{vib} + E^{rot}.$$

We shall consider these two contributions in turn.

The simplest calculation of the vibrational energy of the nuclei uses an approach first proposed by Born and the American physicist Oppenheimer. Because the electrons are so much lighter, they move much faster than the nuclei. Thus the electrons will complete many orbits before the nuclei move appreciably. This means that the electron orbits can be calculated as if for fixed nuclei, and that the electrons will occupy these "fixed-nuclei" orbits, adjusting themselves continuously to follow the slow changing of the nuclear positions. (Although this argument is phrased classically, the same conclusion holds in quantum mechanics as well.) This result has two important consequences: First, our calculation of the electronic energies $E_i(R)$, based on stationary nuclei, is not wasted. As the nuclear separation R varies slowly, the electron orbits adjust themselves and, for each value of R, the energy is still given correctly by $E_i(R)$.

The second important consequence of the Born–Oppenheimer approach is that it shows us how to find the vibrational energies. As the nuclei move slowly in and out, they carry the electrons with them, and we can just as well think of the vibrations as vibrations of the two whole atoms, as illustrated in Fig. 16.22. When the separation of the atoms is R, their total energy is the energy $E_i(R)$ already discussed *plus* the kinetic energy of these vibrations:

$$E_{tot} = K + E_i(R). \tag{16.22}$$

Here K denotes the kinetic energy of the two atoms each moving as a whole, and this is the term that we have hitherto ignored. Now, let us imagine for a moment

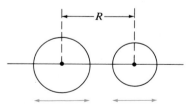

FIGURE 16.22 The vibrational motion of a diatomic molecule is an in-and-out motion of the two atoms along the internuclear axis. (In reality the two atoms overlap one another, but for clarity we have drawn them separately.)

that our two atoms are classical particles with potential energy $U(R)$ and total energy

$$E_{\text{tot}} = K + U(R). \qquad (16.23)$$

Since total energy is conserved, the variations of K must exactly mirror those of $U(R)$ — when $U(R)$ increases, K must decrease equally, and vice versa. That is, $U(R)$ governs the motions of the two atoms. Returning to (16.22), we see that, since E_{tot} is conserved, the variations of K must similarly mirror those of $E_i(R)$. That is, $E_i(R)$ in (16.22) plays the role of the potential energy that governs the motion of the two atoms. Although this discussion is classical, the corresponding result for a quantum system is that $E_i(R)$ in (16.22) is the effective potential-energy function in the Schrödinger equation for the two atoms. To find the allowed vibrational energies of our molecule, we must solve the Schrödinger equation for two bodies (the two atoms) with the potential-energy function $U(R) = E_i(R)$.

In Figure 16.20 we sketched three of the functions $E_i(R)$; what we have now established is that each of these curves defines a potential well, which controls the vibrational motion of the two atoms. The different curves $E_0(R)$, $E_1(R)$, . . . correspond to different states of the electrons, and we see that for each electronic state, $i = 0, 1, \ldots$, there is a different potential-energy function, $U(R) = E_i(R)$, to govern the vibrational motion.

Let us focus attention first on the electronic ground state, $i = 0$, with potential-energy function $U(R) = E_0(R)$. To simplify our discussion we consider a molecule, such as HCl, in which one atom is much lighter than the other. In this case we can assume, to a good approximation, that the heavy atom (mass m') is stationary, and we have only to consider the motion of the light atom (mass $m \ll m'$).* As a further simplification we note that the vibrational

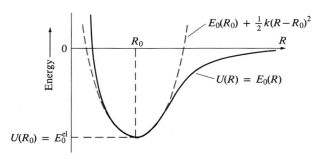

FIGURE 16.23 If the electrons in a molecule occupy their ground state with energy $E_0(R)$, the potential-energy function that controls the slow vibrations of the two atoms is $U(R) = E_0(R)$. Provided that the nuclei remain close to R_0, the equilibrium separation, $U(R)$ can be approximated by a parabola as shown by the dashed curve.

* Actually, the case that m and m' are of the same order is not much more difficult. As described briefly in Section 6.8 and Problem 6.17, the two-body problem with both masses moving is equivalent to the problem with one of the bodies fixed, except that one must use the reduced mass $\mu = mm'/(m + m')$ in place of m. Thus all of the formulas of this section are correct, whether or not $m \ll m'$, provided that one replaces m by μ. Note that if m is much less than m', then $\mu \approx m$ (see Problems 16.37 and 16.38).

motion occurs along the line joining the two nuclei. Thus our problem is reduced to solving the Schrödinger equation for a single body (the lighter atom) moving in one dimension, with potential energy $U(R) = E_0(R)$, as sketched in Fig. 16.23.

As long as the atoms remain reasonably close to their equilibrium separation R_0, the well of Fig. 16.23 can be closely approximated by a parabola of the form

$$E_0(R) \approx E_0(R_0) + \tfrac{1}{2}k(R - R_0)^2, \tag{16.24}$$

for some suitably chosen constant k, as shown by the dashed curve in the picture (see Problems 16.35 and 16.41). This potential energy is the simple harmonic oscillator (or SHO) potential, discussed in Section 8.9, where k was called the force constant. We found there that the allowed energies are [see Eq. (8.96)]*

$$\boxed{E_n^{\text{vib}} = (n + \tfrac{1}{2})\hbar\omega_c \qquad (n = 0, 1, 2, \ldots),} \tag{16.25}$$

where ω_c is the angular frequency of the corresponding classical oscillator (same mass m and force constant k):

$$\omega_c = \sqrt{\frac{k}{m}}.$$

The lowest vibrational level has the "zero-point" energy $\tfrac{1}{2}\hbar\omega_c$, and all the levels given by (16.25) are spaced at equal intervals $\hbar\omega_c$. However, at high energies the approximation (16.24) breaks down and, as seen in Fig. 16.23, the true potential well rapidly becomes wider than its parabolic approximation. This means that the higher levels are somewhat closer together, and there are no levels above $E = 0$. This is illustrated in Fig. 16.24.

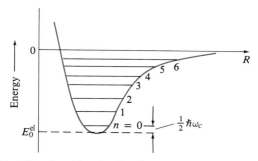

FIGURE 16.24 Vibrational levels of a diatomic molecule in its lowest electronic state. The numbers on the right identify the vibrational quantum number, n. For clarity we have shown a molecule with only seven levels; in most real molecules there are more vibrational levels, more closely spaced.

We see in Fig. 16.24 that since the lowest possible energy is $E_0^{\text{el}} + \tfrac{1}{2}\hbar\omega_c$, the ground state of the molecule is higher by $\tfrac{1}{2}\hbar\omega_c$ than the value E_0^{el} we used in Sections 16.3 and 16.4 (where we ignored motion of the nuclei). However, this zero-point energy is usually quite small compared to electronic energies. For

---

* Equation (16.25) gives the allowed energies measured up from the bottom of the well (16.24), and the total energy is therefore $E_0(R_0) + (n + \tfrac{1}{2})\hbar\omega_c$. The first term $E_0(R_0)$ is the electronic energy E_0^{el}, and the term $(n + \tfrac{1}{2})\hbar\omega_c$ is the energy of vibration, which we denote by E_n^{vib}.

example, in LiBr it is about 0.035 eV, as we see in the following example. Compared to typical electronic energies of order 1 eV, this is indeed fairly small.

EXAMPLE 16.4 The parabola that best approximates the potential well $E_0(R)$ for LiBr has a force constant $k \approx 800$ eV/nm² (see Problem 16.46). Use this to find the zero-point energy $\frac{1}{2}\hbar\omega_c$ for LiBr.

The mass of Br (80 u) is much greater than that of Li (7 u), so we can consider the Br to be fixed, and we get

$$\frac{1}{2}\hbar\omega_c = \frac{\hbar}{2}\sqrt{\frac{k}{m}} = \frac{\hbar c}{2}\sqrt{\frac{k}{mc^2}}$$

$$\approx \frac{200 \text{ eV}\cdot\text{nm}}{2}\sqrt{\frac{800 \text{ eV/nm}^2}{7 \times 931.5 \times 10^6 \text{ eV}}} \approx 0.035 \text{ eV}.$$

Note that the spacing between adjacent vibrational levels is just twice this, $\hbar\omega_c \approx 0.07$ eV, which is also small compared to typical electronic energies.

We can apply similar considerations to any other electronic state. For each electronic state, labeled i, the curve $E_i(R)$ defines the oscillator well that controls the vibrational motion of the atoms, and for each such well there will be a ladder of vibrational levels $E_n^{\text{vib}} = (n + \frac{1}{2})\hbar\omega_c$. The levels of the molecule, including electronic and vibrational motion, are therefore as sketched in Fig. 16.25.

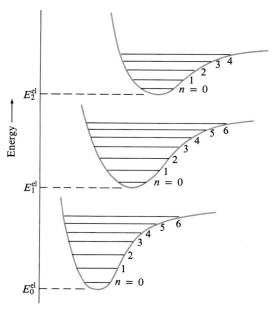

FIGURE 16.25 Energy levels of a typical diatomic molecule, including electronic and vibrational energies. The three electronic levels are shown as E_0^{el}, E_1^{el}, E_2^{el}. For each electronic level, there is a ladder of vibrational energies, $E_n^{\text{vib}} \approx (n + \frac{1}{2})\hbar\omega_c$, with $n = 0, 1, \ldots$; these are drawn inside the appropriate well, with the quantum number n shown at the right.

ROTATIONAL ENERGY

The rotational energies are easily calculated. We can think of the molecule as a dumbell, which can rotate about its center of mass as shown in Fig. 16.26. The classical energy of rotation is $E^{rot} = \frac{1}{2}I\omega^2$, where I is the moment of inertia and ω the angular velocity. Since the angular momentum L is $I\omega$, it follows that $\omega = L/I$ and the rotational energy is

$$E^{rot} = \frac{L^2}{2I}.$$

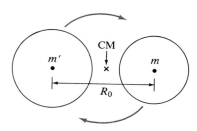

FIGURE 16.26 Rotational motion of a diatomic molecule.

This formula is also valid in quantum mechanics, provided that we use the correct quantized values $l(l+1)\hbar^2$ for L^2. Hence the allowed values of the rotational energy of our diatomic molecule are

$$E_l^{rot} = l(l+1)\frac{\hbar^2}{2I} \qquad (l = 0, 1, 2, \ldots). \qquad (16.26)$$

These values of E^{rot} are shown in Fig. 16.27, where we see — what is also clear from Eq. (16.26) — that the energies are all positive and that the spacing between the levels increases steadily with increasing l.

To estimate the magnitude of the rotational energies (16.26) we again consider a molecule like HCl or LiBr in which one atom is much lighter than the other. In this case, the rotational motion is — to a good approximation — just an orbiting of the light atom (mass m) around the fixed heavy atom, and the moment of inertia is simply

$$I = mR_0^2.$$

In the case of HCl, the observed value of R_0 is 0.13 nm and the light atom (H) has mass $m = 938$ MeV/c^2. Thus, for example, the $l = 1$ level has energy

$$\frac{\hbar^2}{I} = \frac{\hbar^2}{mR_0^2} = \frac{(\hbar c)^2}{(mc^2)R_0^2}$$

$$\approx \frac{(200 \text{ eV} \cdot \text{nm})^2}{(938 \times 10^6 \text{ eV}) \times (0.13 \text{ nm})^2} \approx 2 \times 10^{-3} \text{ eV}. \qquad (16.27)$$

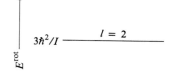

FIGURE 16.27 Quantized energies of rotation, $l(l+1)\hbar^2/2I$, for a diatomic molecule.

Since the electronic energies are of order 1 eV, while the energies of vibration are of order 0.1 eV, we see that rotational energies are much smaller than both electronic and vibrational energies.

TOTAL ENERGY OF THE MOLECULE

We can now put our various results together. The total energy of a diatomic molecule is

$$E = E^{el} + E^{vib} + E^{rot}.$$

All three of these energies are quantized. The allowed values of E^{el} we have denoted by E_i^{el} ($i = 0, 1, \ldots$); those for E^{vib} are denoted E_n^{vib} ($n = 0, 1, \ldots$) and, at least for the lower levels, have the approximate form $(n + \frac{1}{2})\hbar\omega_c$;

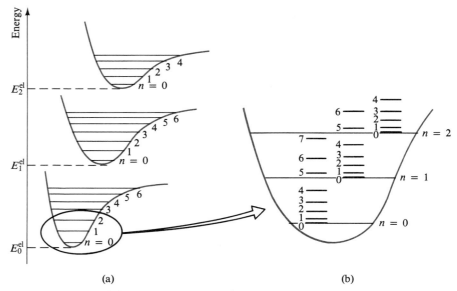

FIGURE 16.28 Energy levels of a diatomic molecule have the form $E_i^{el} + E_n^{vib} + E_l^{rot}$. **(a)** For each electronic level, there is a ladder of vibrational levels, labeled $n = 0, 1, \ldots$. (On the scale of this picture the rotational levels are too closely spaced to be shown.) **(b)** Above each vibrational level is a "subladder" of closely spaced rotational levels, as illustrated by this enlargement of the three lowest vibrational levels. The rotational levels are labeled by the quantum number $l = 0, 1, 2, \ldots$, shown on their left. Note that in most molecules the rotational levels are much more closely spaced than shown here.

finally, the allowed values of E^{rot} are $l(l + 1)\hbar^2/2I$. Representative values for the spacings of the various levels are

$$\Delta E^{el} \sim 1 \text{ eV}$$
$$\Delta E^{vib} \sim 0.1 \text{ eV}$$
$$\Delta E^{rot} \sim 0.001 \text{ eV}.$$

If we ignore the small rotational energies entirely, the energy levels of the molecule appear as shown in Fig. 16.25 [which we have reproduced as Fig. 16.28(a)]. Each electronic state, $i = 0, 1, \ldots$, defines a potential well that determines the vibrational motion of the atoms, and inside each well we have drawn the ladder whose rungs represent the different vibrational energies $E_n^{vib} \approx (n + \frac{1}{2})\hbar\omega_c$ with $n = 0, 1, \ldots$.

If we wish to take into account the rotational energies, then above each rung of every ladder in Fig. 16.28(a) we must draw a "subladder" whose rungs represent the rotational energies $l(l + 1)\hbar^2/2I$ ($l = 0, 1, \ldots$). On the scale of Fig. 16.28(a) the rungs of these subladders are too closely spaced to be distinguished, but Fig. 16.28(b) shows an enlargement of the lowest three levels of part (a), each with its subladder of rotational levels. It is the level structure in Fig. 16.28 — with rotational levels built on vibrational levels built on electronic levels — that determines the main features of molecular spectra, as we describe in the next section.

16.7 Molecular Spectra (optional)

Knowing the structure of molecular energy levels — with their hierarchy of electronic, vibrational, and rotational energies — we can now discuss the spectrum of radiation emitted or absorbed as a molecule changes its energy. The many possible different spectral lines can be classified according to which of the terms in the total energy,

$$E = E^{el} + E^{vib} + E^{rot},$$

changes in the corresponding transitions.

ROTATIONAL TRANSITIONS

We begin by discussing the **rotational transitions,** in which only the rotational state changes. Since the separation of rotational levels is of order 10^{-3} eV, the same is true of the photons emitted or absorbed in rotational transitions. Photons with energy 10^{-3} eV have wavelength of order 1 mm and are classified as *microwave* photons. Thus the rotational spectrum of most molecules lies in the microwave region, although for lighter molecules (in which the rotational energies are larger) it includes the far infrared.

At room temperature and below, the average kinetic energy of the molecules in a gas is of order 0.04 eV or less.* This is so low that collisions between molecules will seldom excite either the electronic or vibrational excited levels. On the other hand, it is quite enough to excite several rotational levels. Thus the molecules of a gas at room temperature remain locked in their lowest electronic and vibrational levels, but are continually being excited to several different rotational levels. In many molecules, transitions between different rotational levels can occur rapidly,† and the gas will emit microwave radiation as these excited molecules drop back to lower levels. Since this radiation is rather weak, the most convenient way to study the rotational spectrum is usually to monitor the absorption of a microwave beam that is directed through the gas and whose frequency can be adjusted. The different transitions then reveal themselves as sharp reductions in the transmitted beam whenever the microwave photons have the right energy to excite the molecule from one level to another.

Not all transitions that are energetically possible occur with appreciable probability. As we discussed in Section 15.6, those transitions that have appreciable probability are called *allowed* and are determined by *selection rules.* The selection rule for transitions between different rotational levels is

$$\Delta l = \pm 1. \tag{16.28}$$

That is, only those transitions in which l changes by one unit occur rapidly enough to be easily observed.‡ Therefore, the only allowed rotational transitions, up or down, have the form

---

* We know from the kinetic theory of gases that the average translational kinetic energy is $3k_B T/2$, where k_B denotes Boltzmann's constant, $k_B = 8.62 \times 10^{-5}$ eV/K. At room temperature ($T \approx 300$ K) this energy is about 0.04 eV.

† The main exception is that in homonuclear molecules (molecules such as H_2 with two identical nuclei) these transitions are forbidden and there is no rotational spectrum.

‡ We have not *proved* the selection rule (16.28). The proof can be found in any book on molecular spectra, but is beyond our scope here.

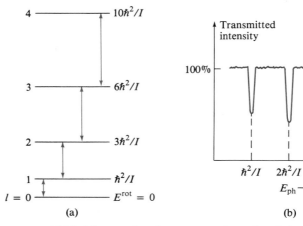

FIGURE 16.29 The rotational spectrum of a molecule is produced by transitions between different rotational levels, while the vibrational and electronic levels remain unchanged. (a) The allowed transitions satisfy $\Delta l = \pm 1$. (b) General appearance of the rotational spectrum in absorption. A microwave beam of adjustable frequency is directed through the gas and is absorbed only when its photons have energy ($E_{ph} = hf$) equal to one of the transition energies $\hbar^2/I$, $2\hbar^2/I$, ...

$$l \leftrightarrow l+1 \qquad (l = 0, 1, 2, 3, . . .), \tag{16.29}$$

as shown in Fig. 16.29(a). Since the energy E_l^{rot} of the rotational levels is $l(l+1)\hbar^2/2I$, the energies of the photons emitted or absorbed in these transitions have the form

$$E_{ph}(l \leftrightarrow l+1) = E_{l+1}^{rot} - E_l^{rot} = [(l+1)(l+2) - l(l+1)]\frac{\hbar^2}{2I}$$

$$= \frac{(l+1)\hbar^2}{I} \qquad (l = 0, 1, 2, 3, . . .). \tag{16.30}$$

This gives a series of equally spaced lines separated by the energy $\hbar^2/I$, as illustrated in Fig. 16.29(b). By measuring the separation of any two adjacent lines in the rotational spectrum of a molecule, one can determine the molecule's moment of inertia, I, and its bond length R_0.

EXAMPLE 16.5 The four absorption lines of longest wavelength in lithium iodide gas are found at $\lambda = 1.13, 0.564, 0.376$, and 0.282 cm. Deduce the bond length of the LiI molecule.

Because the lines with longest wavelength correspond to photons with lowest energy, the four given lines should correspond to the lowest four transitions of Fig. 16.29. The four energies $E_{ph} = hc/\lambda$ are easily calculated:

λ (cm)	E_{ph} (eV)	Transition
1.13	1.10×10^{-4}	$0 \to 1$
0.564	2.20×10^{-4}	$1 \to 2$
0.376	3.30×10^{-4}	$2 \to 3$
0.282	4.40×10^{-4}	$3 \to 4$

These are seen to be in the ratios $1:2:3:4$, as required by (16.30), with

$$\frac{\hbar^2}{I} = \frac{\hbar^2}{mR_0^2} = 1.10 \times 10^{-4} \text{ eV}.$$

Since lithium has $m = 7 \text{ u} = 7 \times 931.5 \text{ MeV}/c^2$, this gives

$$R_0 = \frac{\hbar}{\sqrt{m(1.10 \times 10^{-4} \text{ eV})}} = \frac{\hbar c}{\sqrt{mc^2(1.10 \times 10^{-4} \text{ eV})}}$$

$$= \frac{197 \text{ eV} \cdot \text{nm}}{\sqrt{(7 \times 931.5 \times 10^6 \text{ eV}) \times (1.10 \times 10^{-4} \text{ eV})}} = 0.233 \text{ nm}.$$

Note that in this calculation we treated the iodine nucleus (mass = 127 u) as fixed, so that $I = mR_0^2$ with $m = 7$ u, the mass of the lithium nucleus. If one takes account of the small motion of the iodine, one gets $R_0 = 0.239$ nm (see Problem 16.48).

Rotational spectra of molecules are useful in astronomy, where they make possible the identification of many molecules in the large "molecular clouds" of our galaxy. These clouds are often so cool (a few tens of degrees Kelvin) that intermolecular collisions almost never raise the molecules to excited vibrational or electronic levels; nevertheless, they can excite several rotational levels, which then emit radiation that can be detected by radio telescopes on earth. For example, carbon monoxide is frequently detected in molecular clouds by the microwaves emitted in the transitions $l = 1 \rightarrow 0$ and $l = 2 \rightarrow 1$.

VIBRATIONAL–ROTATIONAL TRANSITIONS

In many molecules one can observe transitions in which the vibrational level changes but the electronic level does not. However, in any such transition the selection rule (16.28) still applies and the rotational level must also change. For this reason, these transitions are called **vibrational–rotational transitions.** We have seen that the spacing of vibrational levels is of order 0.1 eV, while that of rotational levels is much smaller. Therefore, the photons emitted or absorbed in vibrational–rotational transitions have energies of order 0.1 eV and are in the infrared.

In addition to (16.28) there is also a selection rule for the vibrational quantum number:

$$\Delta n = \pm 1. \tag{16.31}$$

That is, among the many conceivable vibrational–rotational transitions, only those in which both n and l change by one unit are observed with appreciable probability. A typical such transition is illustrated in Fig. 16.30. The energies involved have the form

$$E_n^{\text{vib}} = (n + \tfrac{1}{2})\hbar\omega_c \quad \text{and} \quad E_l^{\text{rot}} = \frac{l(l+1)\hbar^2}{2I}, \tag{16.32}$$

with the vibrational spacing ($\hbar\omega_c$) much greater than the rotational. Therefore, if n decreases by one ($n \rightarrow n - 1$), the total energy decreases whichever way l changes ($l \leftrightarrow l + 1$), and a photon is emitted with energy

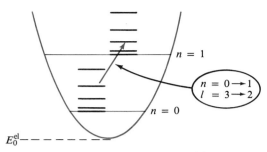

FIGURE 16.30 In a vibrational–rotational transition the electronic state is unchanged, but both n and l change by one unit. In the transition shown, n increases ($n = 0 \rightarrow 1$) and l decreases ($l = 3 \rightarrow 2$). Since the total energy increases, this transition entails the absorption of a photon. (Spacing of the rotational levels is greatly exaggerated.)

$$E_{ph} = (E_n^{vib} - E_{n-1}^{vib}) \pm (E_{l+1}^{rot} - E_l^{rot})$$
$$= \hbar\omega_c \pm \frac{(l+1)\hbar^2}{I} \qquad (l = 0, 1, 2, \ldots). \qquad (16.33)$$

If n increases by one, the molecule's energy increases and a photon has to be absorbed, with energy again given by (16.33). Either way, (16.33) implies that the vibrational–rotational spectrum consists of many lines, equally spaced on either side of $\hbar\omega_c$ with spacing $\hbar^2/I$. Since there are no transitions with $\Delta l = 0$, there is no line with E_{ph} actually equal to $\hbar\omega_c$.

A typical vibrational–rotational absorption spectrum, for HBr, is shown in Fig. 16.31. One sees clearly the equally spaced lines as predicted by (16.33), with a gap at the center corresponding to the absence of transitions with $E_{ph} = \hbar\omega_c$. From the position of the central gap one can measure the vibrational parameter $\hbar\omega_c$, and from the spacing of the lines one can read off the rotational parameter $\hbar^2/I$, as the following example shows.

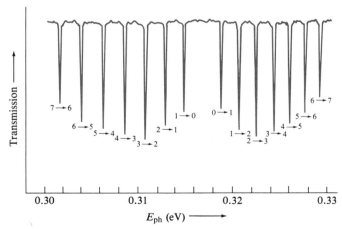

FIGURE 16.31 Vibrational–rotational absorption spectrum of HBr. All lines correspond to the vibrational transition $n = 0 \rightarrow 1$, with various different rotational transitions $l \leftrightarrow l + 1$, as indicated. (Courtesy McGraw-Hill.)

EXAMPLE 16.6 Use the graph in Fig. 16.31 to determine the parameters $\hbar\omega_c$ and $\hbar^2/I$ for the HBr molecule. Hence find the force constant k and bond length R_0 for HBr.

The central gap in a vibrational–rotational spectrum comes at $E_{ph} = \hbar\omega_c$, and from Fig. 16.31 we can read off $\hbar\omega_c$ as

$$\hbar\omega_c = 0.317 \text{ eV}. \qquad (16.34)$$

Since $\omega_c = \sqrt{k/m}$, where m is the mass of the hydrogen atom (as usual we ignore the motion of the heavier atom), a simple calculation gives

$$k = m\omega_c^2 = \frac{mc^2(\hbar\omega_c)^2}{(\hbar c)^2} = \frac{(938 \times 10^6 \text{ eV}) \times (0.317 \text{ eV})^2}{(197 \text{ eV}\cdot\text{nm})^2}$$

$$= 2.4 \times 10^3 \text{ eV/nm}^2.$$

The spacing between adjacent lines in a vibrational–rotational spectrum is $\hbar^2/I$ and from Fig. 16.31, we find

$$\frac{\hbar^2}{I} = 0.0020 \text{ eV}. \qquad (16.35)$$

Comparing (16.34) and (16.35), we see that the rotational energies are smaller than the vibrational by two orders of magnitude, as stated earlier. Given that $I = mR_0^2$ (where m is the mass of hydrogen), we find R_0 to be

$$R_0 = \sqrt{\frac{I}{m}} = \sqrt{\frac{\hbar^2}{m(\hbar^2/I)}} = \frac{\hbar c}{\sqrt{mc^2(\hbar^2/I)}}$$

$$= \frac{197 \text{ eV}\cdot\text{nm}}{\sqrt{(938 \times 10^6 \text{ eV}) \times (0.0020 \text{ eV})}} = 0.14 \text{ nm}. \qquad (16.36)$$

The relative strengths of the absorption or emission lines in any spectrum depend on how many of the molecules occupy the initial levels concerned, and this occupancy depends on the temperature of the gas. (The higher the temperature, the greater the number of levels occupied.) Measurement of the relative intensities of the lines in a spectrum like those of Fig. 16.31 or 16.29(b) allows one to find the temperature of the gas. This is especially useful in astronomy, where the gases of interest are inaccessible to direct measurement.

In fact, rotational and vibrational–rotational spectra are useful to astronomers for several other reasons as well. They can be used to identify the molecules in interstellar clouds that are too cool to emit at higher frequencies. Furthermore, microwave and IR radiations can pass through dust clouds that are completely opaque to visible light. Thus microwaves and IR can give information about regions of our galaxy that are inaccessible to conventional optical astronomy. For example, they allow direct observations of the core of the galaxy, where measurements of Doppler shifted spectra let us determine the speed of rotation of the galaxy's inner spiral arms.

ELECTRONIC TRANSITIONS

Any transition in which the electronic state of a molecule changes is called **an electronic transition.** In electronic transitions both the vibrational and rotational states can also change. A typical electronic transition has the form

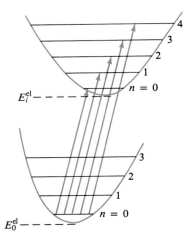

FIGURE 16.32 Typical electronic transitions starting from the lowest electronic and vibrational levels. Since there is no selection rule on n in electronic transitions, many different transitions, $0 \to n$, ($n = 0, 1, 2, 3, \ldots$), are possible for any one final electronic level E_i^{el}.

$$E = E_i^{el} + E_n^{vib} + E_l^{rot} \leftrightarrow E' = E_{i'}^{el} + E_{n'}^{vib} + E_{l'}^{rot}$$

and the energy of the photon emitted or absorbed is

$$E_{ph} = E' - E = \Delta E^{el} + \Delta E^{vib} + \Delta E^{rot}.$$

Since the largest of these three terms, ΔE^{el}, is of order a few eV, the same is true of E_{ph}, and electronic transitions involve photons in the visible, or nearly visible, region.

Just like an atom, a molecule can make both upward and downward transitions, producing absorption and emission spectra. If a gas sample is at room temperature, however, all of the molecules will be in the lowest electronic and vibrational states. In this case they can make no downward electronic transitions and will produce no electronic *emission* spectrum. On the other hand, they can make upward transitions—if exposed to light of the right frequency—and their electronic spectrum can be studied in absorption. Any observed transition must start in the lowest electronic and vibrational levels, since only these are occupied. If we ignore for a moment the small rotational energies, the observed transitions will have the form

$$E_0 = E_0^{el} + E_0^{vib} \to E = E_i^{el} + E_n^{vib}. \tag{16.37}$$

The selection rule $\Delta n \pm 1$, (16.31), which applies to vibrational transitions in a single SHO well, does not apply to transitions between the levels of two different wells. Therefore, the electronic transitions (16.37) can occur with many different values of n:

$$0 \to n \qquad (n = 0, 1, 2, \ldots)$$

as illustrated in Fig. 16.32. For any one final electronic level, these transitions will produce a series of equally spaced absorption lines, corresponding to the several possible final vibrational levels $n = 0, 1, 2, \ldots$ as shown in Fig. 16.33(a).

If we take account of the rotational energies, every one of the levels shown in Fig. 16.32 has a ladder of many, very closely spaced, rotational levels above it. Thus for each of the transitions (16.37), there are in reality many transitions, in which the rotational energy changes by small but different amounts. Thus each of the lines in Fig. 16.33(a) is actually a band and is found, at high enough resolution, to be a series of closely spaced lines, as illustrated in Fig. 16.33(b).

In summary, molecular spectra range from rotational spectra in the microwave region, through vibrational–rotational spectra in the IR, to electronic

FIGURE 16.33 **(a)** The electronic absorption spectrum corresponding to the transitions shown in Fig. 16.32. **(b)** At high resolution, each of the bands in part (a) is seen to be a series of lines, corresponding to the many possible changes of rotational energy.

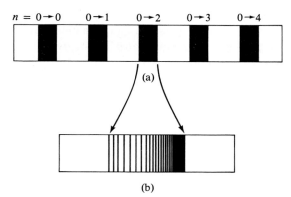

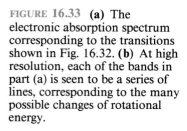

spectra in the visible region. These provide three different windows, often requiring quite different experimental techniques, through which we study the properties of molecules. They are also the basis of many practical applications, including lasers and masers, as discussed in Chapter 15.

IDEAS YOU SHOULD NOW UNDERSTAND FROM CHAPTER 16

Bond length
Binding, or dissociation, energy
Ionic bond
Covalent bond
Dipole moment
Energy release in chemical reactions
The H_2^+ molecular ion
    bonding and antibonding orbitals

Valence in ionic and covalent molecules
[Optional sections: directional character of p orbitals, H_2O, NH_3, CH_4; electronic, vibrational, and rotational energies; $E^{vib} \approx (n + \frac{1}{2})\hbar\omega_c$; $E^{rot} = l(l+1)\hbar^2/2I$; rotational, vibrational–rotational, and electronic transitions]

PROBLEMS FOR CHAPTER 16

SECTION 16.2 (OVERVIEW OF MOLECULAR PROPERTIES)

16.1 • (*a*) If all the electrons in 1 gram of hydrogen could be detached from all the protons and separated from them by 1 cm, what would be the total force between the electrons and protons? (*b*) What would be the electrons' acceleration in g's (assuming that they are somehow held together as a single body)?

16.2 • If all the electrons in the earth could be removed and attached to the sun (leaving all the protons on the earth), what would be the resulting electrostatic force of the sun on the earth, and what acceleration would it produce? Compare this force with the corresponding gravitational force. (Make reasonable estimates. The mass of the earth is about 6×10^{24} kg and that of the sun is 2×10^{30} kg. The distance from earth to sun is 1.5×10^{11} m.)

16.3 • If an ionic molecule results from the transfer of exactly one electron from one atom to the other, it should have a dipole moment $p = eR_0$, where R_0 is the bond length. Use the data in Table 16.1 (Section 16.2) to predict the dipole moments of KCl, LiF, NaBr, and NaCl in coulomb·meters. The observed values are, respectively, 3.42×10^{-29}, 2.11×10^{-29}, 3.04×10^{-29}, and 3.00×10^{-29} C·m; express these as percentages of your predicted values. (These percentages measure the extent to which the molecules concerned *are* ionic.)

16.4 • If a diatomic molecule is ionically bonded by the complete transfer of one electron, its dipole moment should be $p = eR_0$. Given the following data:

Molecule:	NaF	HF	CO
Bond length, R_0 (nm):	0.193	0.0917	0.113
Dipole moment, p (C·m):	2.72×10^{-29}	6.07×10^{-30}	3.66×10^{-31}

discuss the extent to which the molecules concerned are ionically bonded.

16.5 • A 1984 Ford Escort traveling at 50 mi/h experiences a drag force, due to rolling friction and air resistance, of about 360 N (roughly, 80 lb).* The energy released by 1 liter of gasoline is about 3.2×10^7 J and is used with about 18% efficiency. How far can the car travel on 1 liter of gasoline at 50 mi/h? Convert your answer to miles per gallon. (1 mi = 1.6 km, 1 gal = 3.8 liters.)

16.6 • In Example 16.2 (Section 16.2) we found that the energy released in the burning of 1 kg of carbon to give CO_2 is 9.1×10^7 J. In that calculation we as-

---

* E. J. Horton and W. D. Compton, *Science,* vol. 225, p. 587 (1984).

sumed that the original carbon atoms were separate free atoms. [Specifically, the energy in Eq. (16.5) is for the case that the C atom is free and unattached.] If the C atoms are bound in a liquid or solid, some energy is needed to detach them, and the energy yield is smaller. Given that graphite (one of the solid forms of pure carbon) has a binding energy of 7.4 eV per atom, find the energy released when 1 kg of graphite burns in O_2 gas to form CO_2. Compare your answer with the result in Example 16.2.

16.7 • Calculate the energy released if 1 kg of graphite (one of the solid forms of pure carbon) burns in oxygen gas to give carbon monoxide, in the reaction

$$2C + O_2 \rightarrow 2CO.$$

Don't forget that energy is needed to dissociate the carbon atoms from the graphite and the oxygen atoms from the O_2 molecules. The binding energy of graphite is 7.4 eV per atom; the dissociation energy of the O_2 molecule is 5.1 eV, and that of CO is 11.1 eV.

16.8 • (a) The energy needed to dissociate the H_2O molecule into three separate atoms is 9.50 eV. Given that the dissociation energies of H_2 and O_2 are 4.48 and 5.12 eV, calculate the energy released in the reaction

$$2H_2 + O_2 \rightarrow 2H_2O.$$

(b) The airship Hindenburg was held up by about 200 million liters of hydrogen gas. Assuming that this was at STP, estimate the energy released when the Hindenburg exploded. Give your answer in "tons of TNT." (One "ton of TNT" is 4.3×10^9 J, the energy released by 1 ton of the explosive TNT.)

16.9 • • Gasoline is a variable mixture of various hydrocarbons, such as octane (C_8H_{18}). To be definite consider pure octane, and calculate the energy released by burning 1 kg of octane in the reaction

$$2C_8H_{18} + 25O_2 \rightarrow 16CO_2 + 18H_2O,$$

using the following dissociation energies:

Molecule:	C_8H_{18}	O_2	CO_2	H_2O
Dissociation energy (eV):	101.6	5.1	16.5	9.5

Note that these dissociation energies are the energies needed to separate each molecule completely into its individual constituent atoms. (Your answer here should be appreciably less than the value found in Example 16.2, which treated the oxidation of carbon atoms that were already separated.)

16.10 • • Consider a point charge q outside a spherically symmetric, rigid charge distribution of zero total charge. (a) Use Gauss's law to prove that the distribution exerts no force on q. (b) Next show that no

assembly of external charges can exert a force on a spherically symmetric, rigid distribution of zero total charge.

16.11 • • As a simple classical model of the covalent bond, suppose that an H_2 molecule is arranged symmetrically as shown in Fig. 16.34. Write down the total potential energy U of the four charges and, treating the protons as fixed, find the value of the electrons' separation s for which U is a minimum. Show that the minimum value is $U_{min} \approx -4.2 \ ke^2/R_0$. (Note that you have here treated the two protons as fixed at separation R_0. Classically, they could not remain fixed in this way; it is one of the triumphs of quantum mechanics that it explains how the protons have a stable equilibrium separation, as we describe in Section 16.4.)

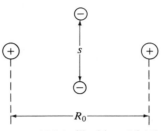

FIGURE 16.34 (Problem 16.11)

SECTION 16.3 (THE IONIC BOND)

16.12 • (a) Use the graphs in Fig. 11.11 to find the energy ΔE needed to transfer an electron between the following pairs of atoms: from Li to Br, from C to O, from Rb to F, from O to O. (b) Two of these pairs bond ionically and two do not; which are which?

16.13 • (a) Use the graphs in Fig. 11.11 to find the energy ΔE needed to transfer an electron between the following pairs of atoms: from Cs to I, from N to O, from N to N, from Na to Br. (b) Two of these pairs bond ionically and two do not; which are which?

16.14 • The ionization energy of K is 4.34 eV and the electron affinity of Cl is 3.62 eV. What is the critical radius R_c inside which the transfer of an electron from a K atom to a Cl is energetically favored?

16.15 • Use the ionization energies and electron affinities listed in Problem 16.18 to find the critical radius R_c inside which the transfer of an electron is energetically favored, for each of the ionic molecules KI, NaBr, LiF.

16.16 • The ionization energy of potassium is 4.34 eV while the electron affinity of chlorine is 3.62 eV. Given that the equilibrium separation of KCl is 0.267 nm, estimate the dissociation energy of KCl using Eq. (16.10). Compare your answer with the observed value $B = 4.34$ eV.

16.17 • The ionization energy of Li is 5.4 eV and the electron affinity of chlorine is 3.6 eV. Given that LiCl is ionically bonded with dissociation energy 4.8 eV, estimate the equilibrium separation R_0 of LiCl. Will your answer be an overestimate or underestimate?

16.18 • Use the following data to estimate the dissociation energies of the three ionic molecules KI, NaBr, and LiF. (IE denotes ionization energy and EA electron affinity.) Compare your answers with the observed values given in the last column. Your answers should all be overestimates; explain why.

IE (eV)	EA (eV)	R_0 (nm)	B (ev)
K: 4.34	I: 3.06	KI: 0.305	3.31
Na: 5.14	Br: 3.36	NaBr: 0.250	3.74
Li: 5.39	F: 3.40	LiF: 0.156	5.91

16.19 • Assuming that the following pairs of elements combine ionically, write down their chemical formulas: Ca and S, Ca and Cl, Li and O, Na and P. (*Hint:* Use the periodic table.)

16.20 • Assuming that the following pairs of elements combine ionically, write down their chemical formulas: Ba and S, Ca and I, Sr and Br, K and N. (*Hint:* Use the periodic table.)

16.21 •• (*a*) Given that Ba and S bond ionically, what would you expect their valence to be? (*b*) Assuming that the bond is purely ionic, predict the electric dipole moment of BaS. (The measured bond length is $R_0 = 0.251$ nm.) (*c*) Express the observed dipole moment, 3.62×10^{-29} C·m, as a percentage of your answer in part (*b*).

SECTION 16.4 (THE COVALENT BOND)

16.22 • Assuming that they bond covalently as described in Section 16.4, predict the chemical formulas of oxygen fluoride and nitrogen fluoride.

16.23 • Hydrogen can bond covalently with phosphorus, tellurium, and bromine. Predict the chemical formulas of the three resulting molecules.

16.24 • (*a*) Octane, C_8H_{18}, is called a *straight-chain* hydrocarbon, because its carbon atoms are arranged in a straight line. Draw a picture of the octane molecule showing all bonds, as in Fig. 16.13. (*b*) Do the same for the straight-chain propane, C_3H_8. (In both cases all carbon atoms have valence 4.)

16.25 • Make a sketch similar to those in Fig. 16.13 of acetylene C_2H_2 (in which both carbons have valence 4). (*Hint:* Remember that more than one bond can join the two carbon atoms.)

16.26 • The observed electric dipole moments of CO, ClF, and NaF are 3.7×10^{-31}, 3.0×10^{-30}, and

2.7×10^{-29} C·m, respectively. The corresponding bond lengths are 0.11, 0.16, and 0.19 nm. Express the observed dipole moments as percentages of the values you would predict if the molecules were purely ionic. Which of the molecules are predominantly ionic and which predominantly covalent?

16.27 •• (*a*) Assuming that the following pairs of elements combine covalently predict the formulas of the resulting molecules.

	p (C·m)	R_0 (nm)
Cl and F	3.0×10^{-30}	0.16
Br and Cl	1.9×10^{-30}	0.21
I and Cl	4.1×10^{-30}	0.23

(*b*) Use the observed dipole moments and bond lengths to confirm that these molecules are predominantly covalent.

16.28 •• (*a*) What should be the valence of Be in its ground state ($1s^2 2s^2$)? (*b*) As discussed in Section 11.5, the $2s$ and $2p$ levels are close together and, when another atom is nearby, it is often energetically favorable to promote one of the $2s$ electrons to a $2p$ state, so that a bond can form. What is the valence of Be in the configuration $1s^2 2s^1 2p^1$? (*c*) Predict the chemical formulas for the compounds of Be with fluorine, with oxygen, and with nitrogen.

16.29 ••• Consider the H_2^+ wave functions $\psi_\pm$ discussed in Section 16.4. In that discussion we did not worry about normalization, but $\psi_\pm$ should strictly have been defined as $\psi_+ = B(\psi_1 + \psi_2)$ and $\psi_- = C(\psi_1 - \psi_2)$, where B and C are normalization constants needed to ensure that $\int |\psi|^2 \, dV = 1$. (*a*) If ψ_1 and ψ_2 do not overlap (or, more precisely, if their overlap is negligible), show that $B = C = 1/\sqrt{2}$. (Assume that ψ_1 and ψ_2 are themselves normalized.) (*b*) If ψ_1 and ψ_2 overlap *a little,* argue that B is a little less than $1/\sqrt{2}$ and hence that at the midpoint between the two protons, $|\psi_+|^2$ is just a little less than $2|\psi_1|^2$. This proves our claim that ψ_+ concentrates the probability density between the two protons. (*c*) Argue similarly that C must be a little larger than $1/\sqrt{2}$.

SECTION 16.5 (DIRECTIONAL PROPERTIES OF COVALENT BONDS)

16.30 •• The water molecule is partially ionic, in that an electron is partially transferred from each hydrogen to the oxygen, as indicated in Fig. 16.35, where q denotes the magnitude of each of the two charges transferred. (*a*) Write down the electric dipole moment p of the H_2O molecule in terms of the charge q, the H-O bond length d, and the angle θ. (*Note:* The dipole moment of two charges q and $-q$ a distance d apart is a vector of magnitude qd pointing from the

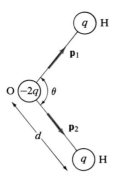

FIGURE 16.35 (Problem 16.30)

negative to the positive charge. Thus H_2O has two dipoles $\mathbf{p}_1$ and $\mathbf{p}_2$ as shown, and the total moment is $\mathbf{p} = \mathbf{p}_1 + \mathbf{p}_2$.) (b) The measured values are $p = 6.46 \times 10^{-30}\,C \cdot m$, $d = 0.0956$ nm, and $\theta = 104.5°$. Find the magnitude q of the charge transferred and express it as a fraction of the electron charge e.

16.31 • • • We mentioned in Section 16.5 that the four bonds of a carbon atom (in a molecule like CH_4) point symmetrically to the four corners of a regular tetrahedron centered on the carbon nucleus. Prove that the angle between any two bonds is 109.5°.

SECTION **16.6** (EXCITED STATES OF MOLECULES)

16.32 • The force constant for HI is $k = 2.0 \times 10^3$ eV/nm^2. Calculate the vibrational energy of the lowest three vibrational levels of HI. (Treat the iodine atom as fixed.)

16.33 • The separation of the lowest two vibrational levels in HBr is 0.33 eV. Find the force constant k of HBr. (Treat the bromine atom as fixed.)

16.34 • Calculate the lowest four rotational energies of the HI molecule in eV and sketch them in an energy-level diagram. (The bond length of HI is 0.16 nm and you can treat the iodine atom as stationary.)

16.35 • • Fig. 16.6 shows the energy $E(R)$ for the Na^+ and Cl^- ions that make up the NaCl molecule. Near its minimum at R_0, $E(R)$ can be approximated by a parabola: $E(R) \approx E(R_0) + \frac{1}{2}k(R - R_0)^2$. Use the information on the graph to make a rough estimate of the force constant k. (*Hint:* To the left of R_0, the graph retains a reasonably parabolic shape as far as the point where it crosses the horizontal axis.)

16.36 • • Consider an O_2 molecule rotating about its center of mass (as in Fig. 16.26). (a) Derive an expression for its moment of inertia I in terms of m, the mass of an O atom, and R_0, the bond length. (b) Use Eq. (16.26) to find the lowest three rotational energy levels in eV. (The bond length of O_2 is 0.121 nm.)

16.37 • • Consider two atoms of masses m and m', bound a distance R_0 apart and rotating about their center of mass (as in Fig. 16.26). (a) Calculate their moment of inertia, I, and prove that it can be written as $I = \mu R_0^2$, where μ is the *reduced mass*

$$\mu = \frac{mm'}{m + m'}. \qquad (16.38)$$

This shows that one can treat the rotational motion of any diatomic molecule as if only one of the atoms were moving, provided that one takes its mass to be μ. (b) Prove that if $m \ll m'$, then $\mu \approx m$.

16.38 • • Consider two classical atoms of mass m and m', separated by a distance R, with their center of mass fixed at the origin. (a) Write down expressions for the atoms' distances, r and r', from the origin, in terms of m, m', and R. (b) Suppose that the two atoms vibrate in and out along the internuclear axis. Show that their total kinetic energy can be expressed as $K = \frac{1}{2}\mu(dR/dt)^2$, where μ is the *reduced mass* defined in Eq. (16.38) (Problem 16.37). This result proves for a classical molecule (what is true in quantum mechanics as well) that the vibrational motion of both atoms can be treated as if only one atom were moving, provided that we take its mass to be μ.

16.39 • • Consider the molecules HCl and DCl, where D denotes the deuterium, or heavy hydrogen, atom, $D = {}^2H$. (a) Explain why the force constant k and bond length R_0 should be about the same for DCl as for HCl. (b) The spacing of adjacent vibrational levels in HCl is $\hbar\omega_c \approx 0.37$ eV. What is the corresponding spacing in DCl? (You may treat the Cl atom as fixed.)

16.40 • • The potential energy of two atoms that form a stable molecule has a minimum at a separation R_0, as shown in Fig. 16.23. Consider two atoms with total kinetic energy K_0 approaching one another from far apart. (a) What will be their KE when they reach the equilibrium separation $R = R_0$? (b) At what point in the picture will the atoms come to rest? Will they remain there? (c) Assuming that the total energy remains entirely in the form KE + PE of the two atoms, can the two atoms come to rest at R_0? What value must R eventually approach? (d) We see that if the atoms are to form a stable molecule, some of their energy must be dissipated while they are near R_0. Suggest a mechanism to do this.

16.41 • • • (a) Use the curve in Fig. 16.12 to get a rough estimate of the force constant k for the H_2^+ molecule. (See the hint for Problem 16.35.) (b) Use your value of k to estimate the zero-point energy of vibration in H_2^+. (Since the two nuclei have equal mass, you will need to use the reduced mass as in Problem 16.38.)

16.42 ••• A simple analytic function that can be used to approximate the potential energy of the two atoms in a diatomic molecule is the Morse potential

$$U(R) = A[(e^{(R_0 - R)/S} - 1)^2 - 1] \quad (16.39)$$

where A, R_0, and S are positive constants, with $R_0 \gg S$. (a) Sketch this function for $0 \le R < \infty$. (b) In terms of A, R_0, and S, write down the bond length and binding energy of a molecule whose potential energy is given by (16.39). (c) Write down the first three terms of the Taylor series for $U(R)$ about R_0 (see Appendix B) and show that they give the parabolic approximation (16.24). (d) Show that the force constant k is $2A/S^2$.

16.43 ••• The potential energy of the two atoms in HCl can be approximated by the Morse potential (16.39) (Problem 16.42) with

$$A = 4.57 \text{ eV}, \quad R_0 = 0.127 \text{ nm}, \quad S = 0.0549 \text{ nm}.$$

(a) Calculate $U(R)$ at intervals of 0.04 nm for $0 \le R \le 0.28$ nm and plot the curve. (Restrict the vertical axis to the range $-5 \le U \le 5$ eV.) (b) What are the bond length and binding energy for HCl? (You may ignore the zero-point vibration.) (c) What is the force constant k for HCl? [See Problem 16.42(d).] (d) Plot the SHO well (16.24) that approximates $U(R)$ near R_0.

SECTION 16.7 (MOLECULAR SPECTRA)

16.44 • The microwave spectrum of HCl consists of a series of equally spaced lines, 6.4×10^{11} Hz apart. Find the bond length of HCl (treating the Cl atom as fixed).

16.45 • The microwave spectrum of HBr consists of a series of equally spaced lines, 5.1×10^{11} Hz apart. Find the bond length of HBr (treating the Br atom as fixed).

16.46 • The vibrational–rotational spectrum (16.33) of a diatomic molecule is in the infrared region. At low resolution it is seen as a single band centered on $E_{ph} = \hbar\omega_c$. The molecule LiBr shows a bright band in the IR centered at $\lambda = 17.8\ \mu m$. Use this to find the force constant k for LiBr (treating the Br atom as fixed).

16.47 •• A spectroscopist finds three lines in the far-infrared spectrum of HI at 259, 195, and 156 μm (and no other lines in between). (a) Find the photon energies and identify the rotational transitions involved. (b) Hence find the bond length of HI.

16.48 •• Repeat the calculation in Example 16.5 (Section 16.7) of the bond length of LiI, but do not make the approximation that the I atom is fixed. (Use the reduced mass introduced in Problem 16.37.) Compare your answer with that of Example 16.5, where we ignored the motion of the iodine atom.

16.49 •• A spectroscopist finds three lines in the microwave spectrum of CO at $\lambda = 2.59$, 1.29, and 0.86 mm (and no other lines in between). (a) Find the corresponding photon energies and identify the transitions involved. (b) Using the reduced mass of Problem 16.37, find the bond length of CO.

16.50 •• The first four lines in the rotational spectrum of CaCl are shown schematically in Fig. 16.36. (The relative strengths of the four lines depend on the temperature and need not concern you here.) Use this information to find the bond length of CaCl. (Since the two masses are comparable, you will need to use the reduced mass introduced in Problem 16.37. Assume that the Cl is chlorine 35.)

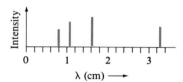

FIGURE 16.36 (Problem 16.50)

16.51 •• The vibrational–rotational spectrum of HI is sketched in Fig. 16.37. Use the information in that figure to find the force constant and bond length of HI.

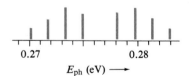

FIGURE 16.37 (Problem 16.51)

16.52 ••• (a) Consider an HCl molecule in which the Cl atom is ^{35}Cl. Allowing for the motion of both atoms, find the wavelengths of the first two lines in the rotational spectrum. (Use the reduced mass introduced in Problem 16.37. Take $R_0 = 0.1270$ nm and keep four significant figures.) (b) Do the same for the case that the Cl atom is ^{37}Cl. (Use the same value for R_0.) (c) Make a sketch similar to Fig. 16.36 of these two lines in the emission spectrum of natural HCl, which contains 3 parts ^{35}Cl to 1 part ^{37}Cl.

C H A P T E R

17

Solids

17.1 Introduction

In Chapter 16 we saw how atoms can bond together to form molecules. In this chapter we study the aggregation of atoms on a larger scale, to form solids. There are many close parallels between solid-state and molecular physics. Several of the bonding mechanisms in solids are exactly the same as in molecules; for example, many solids are covalently or ionically bonded. Also, our treatment of solids closely parallels that of molecules. First, we consider just the motion of the electrons and from this can explain many properties, such as electrical conductivities and cohesive strengths. Then an analysis of the vibrational motion of the atomic cores can explain various further properties such as specific heats.

There are, however, many differences between solid-state and molecular physics. For example, the bonding of metallic solids has no direct parallel in molecules. Also, certain solids, instead of being simple aggregates of atoms, are better viewed as being built in two stages, with atoms bound into molecules and then molecules bound into a solid. (A good example is ice, which is best understood as an aggregate of water molecules.) Another new feature of the solid state is the possibility that atoms (or molecules) can bond together into a symmetric, repetitive crystalline array; the study of the possible crystal structures is an important part of solid-state physics with no exact counterpart in molecular or submolecular physics.

Perhaps the most striking thing about solids is that they were the first example of *macroscopic* systems that were understood with the help of quantum mechanics.* Numerous commonplace properties of solids — specific heat, conductivity, transparency, magnetic susceptibility, and many more — had defied classical explanation, but could be explained and calculated using quantum mechanics. It is this ability to explain — and hence control — the properties of solids that has led to the numerous technical and commercial applications of solid-state physics. These applications are so widespread that the adjective "solid-state" has been adopted into the everyday language to describe any of the high-speed electronic devices that have resulted from the work of solid-state physicists.

In Section 17.2 we describe the principal mechanisms by which atoms or molecules bond to one another to form solids. In Section 17.3 we describe briefly some of the regular crystalline structures in which many solids can form, and we contrast these with the irregular, noncrystalline structure of the amorphous solids and the still more irregular composition of liquids and gases.

We next take up what is perhaps the single most important achievement of solid-state physics, the theory of electrical conductivity. This theory depends on an understanding of the electron energy levels in a solid. In Section 17.4 we discuss these levels and describe how they fall into *bands* of allowed energies. Within these bands the energy levels are distributed almost continuously, but between them are *gaps* where there are no allowed levels. In Section 17.5 we show how the different possible arrangements of these bands explain why some solids are conductors, some are insulators, and some — the so-called semiconductors — are in between. In Section 17.6 we discuss semiconductors and describe a few of their many practical applications in the electronics industry.

In Section 17.7 we make a brief diversion from the themes of conductivity and electron motion to describe the vibrational motions of the atomic cores that form the lattice of a solid. These vibrations are quantized, and their quanta are called phonons (in analogy with photons). They have an important effect on several properties, including specific heats. Finally, in Section 17.8 we describe the phenomenon of superconductivity, in which the resistance of some solids drops to zero when the solid is cooled below a certain critical temperature.

LEV LANDAU (1908–1968, Russian). One of the most influential theorists and teachers of the twentieth century, Landau worked in many areas of physics. His contributions to solid state physics included theories of magnetism, superconductivity, and phase transitions. He won the Nobel Prize in 1962.

17.2 Bonding of Solids

Solids, like molecules, are held together by several different types of bond. Just as with molecules, the distinctions between the various bonds are not always clear cut, and many solids are bonded by a combination of several mechanisms. Nevertheless, we can distinguish four main types of bond — covalent, metallic, ionic, and dipole–dipole — and we describe each of these in turn.

---

* The applications of quantum theory to solids go back to the 1907 paper of Einstein, in which he used quantization to explain the specific heats of solids.

We saw in Chapter 16 that many molecules are held together by covalent bonds, each comprising two valence electrons shared by two of the atoms. In H_2, for example, the two hydrogen atoms share their two electrons to form a single bond; in CH_4 (methane), each of the four valence electrons of the carbon joins with an electron from one of the four hydrogens to form a covalent bond. This same mechanism holds the atoms together in many solids. A beautiful example of such a covalent solid is diamond, which is one of the forms of solid carbon. In diamond each carbon atom is covalently bonded to four neighboring carbon atoms. The four bonds from any one carbon atom are directed symmetrically outward, and each atom in diamond is at the center of a regular tetrahedron formed by its four nearest neighbors. Both silicon and germanium — in the same group as carbon in the periodic table — are bonded covalently in a similar tetrahedral structure. Other examples of covalent solids are silicon carbide (widely used for its hardness and resistance to high temperatures) and zinc sulfide (used by Rutherford and other pioneers in nuclear physics as a scintillation detector).

An important characteristic of a solid is its strength. One simple measure of this strength is the **atomic cohesive energy,** defined as the energy needed, per atom, to separate the solid completely into individual atoms. In Table 17.1 we have listed a number of solids with their cohesive energies. The first three entries are covalent solids, and we see — what is generally true for covalent solids — that their cohesive energies vary from around 3 to about 7 eV per atom. As one would expect, those solids with high cohesive energies, like diamond and silicon carbide, are especially strong and have high melting points.

TABLE 17.1

The cohesive energies of various solids. With two exceptions, the energies listed are atomic cohesive energies (energy per atom to separate the solid into individual atoms).* The exceptions are the molecular solids ice and methane. Since these are properly considered assemblies of molecules, we list their *molecular* cohesive energies (the energy per *molecule* to separate the solid into individual *molecules*). The melting points shown in the last column can be seen to correlate well with the cohesive energies, as one would expect.

Bond	Solid	Cohesive energy	Melting point
Covalent	C (diamond)	7.37 eV/atom	>3770 K
	Ge	3.85 eV/atom	1231 K
	SiC	6.15 eV/atom	2870 K
Metallic	Fe	4.32 eV/atom	2082 K
	Cu	3.52 eV/atom	1631 K
	Pb	2.04 eV/atom	874 K
Ionic	NaCl	3.19 eV/atom	1074 K
	LiF	4.16 eV/atom	1143 K
	CsI	2.68 eV/atom	621 K
Dipole–dipole	H₂O	0.52 eV/molecule	273 K
	CH₄	0.10 eV/molecule	89 K
	Ar	0.08 eV/atom	84 K

* For solids containing two elements, like SiC and NaCl, the cohesive energy is often expressed in eV per *atomic pair*. With this convention SiC would be shown as 12.3 eV per SiC pair, but this number is not directly comparable to the numbers listed for solids containing a single element, like diamond.

Since their electrons are usually well bound and cannot move easily, most covalent solids are poor conductors of electricity. In many covalent solids, the excitation energy of the electrons is higher than the energy of visible photons (2 or 3 eV). For this reason many covalent solids (like diamond) cannot absorb visible photons and are therefore transparent to light.

METALLIC BONDS

More than half of all elements are metals, and all solid metals are held together by the **metallic bond.** In some ways, this bond can be seen as a relative of the familiar covalent bond. In a covalent bond, electrons are shared between pairs of atoms. In the metallic bond, electrons are shared among *all* the atoms: One or two valence electrons are completely detached from each parent atom and can wander freely through the whole metal. Thus a metal can be thought of as a lattice of positive atomic cores, immersed in a "sea" of mobile electrons. It is the attraction between the positive atomic cores and this negative sea of electrons that holds the metal together. As illustrated in Table 17.1, the cohesive energy of typical metals is less than that of many covalent solids, but is, nevertheless, high enough to make most metals rather strong.

All metals are good conductors of electricity because their valence electrons move through the metal almost as if they were free electrons. (For this reason, they are sometimes described as a "gas" of electrons.) Since the electrons can also carry energy, metals are also good thermal conductors. When a photon strikes a metal, it can easily lose energy and momentum to the "free" electrons, and metals are therefore not transparent to light.

IONIC BONDS

We saw in Chapter 16 that certain pairs of elements can bond ionically to form molecules. For exactly the same reasons, these same pairs of elements can bond ionically to form solids. For example, in solid NaCl (common salt) each Na atom has lost one electron to a Cl atom; the resulting closed-shell ions arrange themselves so that each Na^+ ion is surrounded by several Cl^- ions, and vice versa.* The resulting solid is therefore quite tightly bound. As illustrated by the examples in Table 17.1, typical atomic cohesive energies (energy to separate the solid into neutral atoms) are 3 or 4 eV per atom for ionic solids.

Since the electrons of an ionic solid are all bound tightly in closed-shell ions, ionic solids are poor conductors of electricity and heat. For the same reason visible photons cannot be absorbed easily, and ionic solids are transparent to light.

DIPOLE–DIPOLE BONDS

The ionic bond results from the simple Coulomb force between oppositely charged ions. If the basic units (atoms or molecules) are electrically neutral, this force is not present. There is, nevertheless, a related, though usually weaker attraction that *can* act, even between neutral atoms or molecules. A neutral atom or molecule can have its charges distributed so that it has a *dipole moment;* that is, the preponderance of positive charge is centered at one point

---

* Because each Na^+ "belongs" to several Cl^- ions, and vice versa, we should not think of solid NaCl as made up of NaCl molecules; rather, it is a bonding of many Na atoms with many Cl atoms.

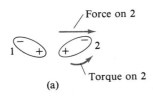

Force on 2

1 — + + — 2

Torque on 2

(a)

(b)

(c)

FIGURE 17.1 (a) If two dipoles are oriented with like charges closest, they repel one another. (b) If unlike charges are closest, the dipoles attract. (c) The attraction is greatest when the dipoles are exactly aligned. In both (a) and (b) the dipoles also exert torques on one another, and these torques tend to rotate the dipoles into the attractive alignment of (c).

and that of negative at another. Two neutral dipoles can exert an electrostatic force on one another, as we now describe.

Let us consider first two **permanent dipoles,** that is, two objects each of which is a dipole even when in complete isolation. (An example of a permanent dipole is the water molecule, as we saw in Chapter 16.) If we bring these dipoles together so that two like charges are closest, as in Fig. 17.1(a), the repulsive forces are slightly greater than the attractive, and the net force is repulsive (Problems 17.3 and 17.4). On the other hand, if the positive charge in one dipole is closest to the negative in the other, as in Fig. 17.1(b), the net force is attractive. In addition to these forces, each dipole exerts a torque on the other, and this torque tends to rotate them into the alignment of maximum attraction (c). Therefore, the net force between two permanent dipoles (that are free to rotate) is automatically attractive. Because it involves substantial cancellations between attractive and repulsive forces, this dipole–dipole attraction is much weaker than the Coulomb forces between the individual charges. Nevertheless, it is the attraction that holds many solids together.

Because of their basic spherical symmetry, no atoms are permanent dipoles. On the other hand, many molecules (the so-called *polar molecules,* like H_2O) *are,* and these molecules can be bound into solids by the dipole–dipole attraction just described. Because the bond holding the molecules to one another is much weaker than the bond holding the atoms together inside each molecule, these solids are properly seen as composed of *molecules* and are sometimes called *molecular solids.* An example of a molecular solid is ice; we see from Table 17.1 that the energy needed to separate ice into H_2O molecules is about 0.5 eV per molecule. As expected, this is appreciably less than the cohesive energies of typical covalent, metallic, and ionic solids (usually several eV per atom). It is also much less than the 5 eV needed to remove any one atom from the H_2O molecule—which confirms that ice is properly regarded as an assembly of H_2O molecules.

The bond between two water molecules actually involves another effect as well. We saw in Section 16.5 that the oxygen atom in water appropriates at least part of the electron from each hydrogen, leaving nearly bare protons at the two ends of the molecule. When water molecules bond to form ice, these protons are shared with nearby oxygen atoms in somewhat the same way that electrons are shared in a covalent bond. This bonding by shared protons is called **hydrogen bonding** and plays an important role in biology: The DNA molecule comprises two intertwined helical strands of atoms. The atoms within each strand are bound together by covalent bonds, while the binding *between* the two strands is by hydrogen bonds, which are much weaker. This is what allows the two strands to unravel from one another, without disturbing their internal structure, in the process of replication.

Although all atoms and many molecules have zero permanent dipole moment, they can acquire an **induced moment** when put in the field of other charges. This is because the field pushes all positive charges one way and pulls all negatives the other. These induced dipoles are responsible for the very weak **van der Waals bond*** between some atoms and molecules, as we describe

---

* The terminology here is a little confused. A few authors use "van der Waals" to describe *both* the attraction between two permanent dipoles *and* the weaker attraction involving induced dipoles. We follow the majority and reserve "van der Waals" to describe only the weaker induced-dipole attraction.

shortly. Because this bond is so weak, it is important only in solids where all other types of bond are absent. For example, the noble gas atoms are subject to none of the bonding mechanisms described earlier, but can form solids because of the van der Waals attraction. Because the bond is so feeble, the noble gases solidify only at rather low temperatures. For instance, as shown in Table 17.1, solid argon has a cohesive energy of only 0.08 eV/atom, and its melting point is 84 K. The van der Waals force is also responsible for the solids of certain *molecules* that have no permanent dipole moment, such as O_2, N_2, and CH_4 (methane).

The origin of the van der Waals force between two atoms (or molecules) that have no permanent dipole moments can be explained by the following classical argument. A classical atom consists of Z electrons orbiting around a fixed nucleus. The statement that it has no permanent dipole moment means (classically) that, *when averaged over time,* the dipole moment is zero. Nevertheless, as the electrons move around, they can produce an instantaneous nonzero dipole moment. Now, consider two such atoms close together. Suppose that at a certain instant atom 1 has a nonzero dipole moment. This produces an electric field, which induces a dipole moment in atom 2. This induced dipole is always aligned such that it is *attracted* by atom 1. (This is illustrated, for one possible orientation of the atoms, in Fig. 17.2. For other orientations, see Problem 17.5.) Therefore, as the two dipoles fluctuate, each induces in the other a dipole that it attracts. This is the origin of the van der Waals attraction between any two atoms or molecules.

FIGURE 17.2 The fluctuating dipole moment of atom 1 induces a dipole moment in atom 2 (and vice versa). The direction of the induced moment is such that the two dipoles attract one another.

17.3 Crystals and Noncrystals

All of the mechanisms that bond atoms (or molecules) together into solids have the same general behavior: At large separations, r, the force between two bonding atoms is attractive, at a certain separation, $r = R$, it is zero, and for r less than R it is repulsive. This behavior is illustrated in Fig. 17.3, which shows the potential energy of two atoms as a function of their separation r.

When just two atoms interact as in Fig. 17.3, they have minimum possible energy, and hence form a stable bound state, at separation R. Three such atoms would be in equilibrium at the corners of an equilateral triangle, with the distance between each pair equal to R. Similarly, four would be in equilibrium at the corners of a regular tetrahedron. However, with five or more atoms, it is impossible to arrange them so that the distances between all pairs are the same. Instead, the atoms must arrange themselves so that any one atom is closest to a few others and farther away from all the rest. It is one of the tasks of solid-state physics to predict, or at least explain, the arrangement of the many atoms in a solid that minimizes their total energy and hence makes the solid stable.

As one might expect, the stable configuration of many solids has the atoms arranged in a regular pattern, in which the grouping of atoms is repeated over and over again. This regular pattern is called a **lattice,** and the resulting solid is called a **crystal.** Experimentally, the structure of crystals is found using diffraction of X rays, neutrons, and electrons, and by the use of various kinds of electron microscope (Fig. 17.4). The mathematical study of the possible crystal structures is an important part of theoretical solid-state physics. Here we shall just describe a few such structures.

The simplest crystal structure is the **cubic lattice,** in which the atoms are

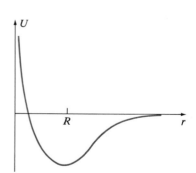

FIGURE 17.3 The potential energy of two atoms (or molecules) that can bond to form a solid is shown as a function of their separation r. Their equilibrium separation is labeled R.

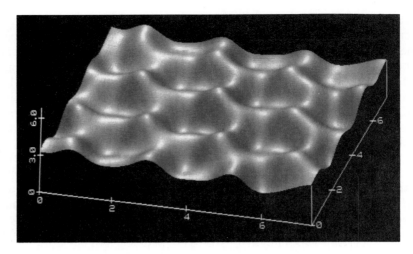

FIGURE 17.4 An image of the surface of graphite made with a scanning tunneling electron microscope, or STM. Each peak repesents a single atom, and the regular crystalline pattern, with the atoms arranged in hexagons in this case, is clearly visible. The STM measures the current of electrons that "tunnel" between the sample and a sharp needle that is scanned across the sample's surface. Since the current varies strongly with the distance between the surface and the needle, this produces a map of the surface, with a resolution that is much less than an atomic diameter. The scale markings are in angstroms (1 Å = 0.1 nm). (Courtesy Digital Instruments Inc.)

arranged at the corners of many adjacent cubes, as shown in Fig. 17.5. In a perfect cubic lattice, all these cubes are identical, and each is placed in the same relation to its neighbors. Therefore, all the information about the structure of the lattice is contained in the single cube, or **unit cell,** shown in Fig. 17.5(c).

A slightly more complicated crystal lattice is the **face-centered cubic** (or **FCC**) lattice, with atoms on each face of a cube as well as at the corners, as in Fig. 17.6(a). This arrangement actually gives a slightly higher density of atoms (more atoms per unit volume) than the simple cubic, and is therefore favored in many solids. For example, common salt (NaCl) is an FCC crystal. As shown in

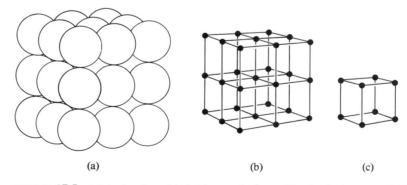

(a) (b) (c)

FIGURE 17.5 **(a)** A simple cubic lattice made from 27 spherical atoms. **(b)** The structure of the lattice is much clearer if one shows just the centers of the atoms, using black dots, and connects the dots with lines. Although the actual appearance of the atoms in a crystal is more like (a) than (b), we shall use the dot-and-line arrangement of (b) because of its greater clarity. **(c)** The unit cell of the same lattice is a single cube. The whole crystal can be constructed from many duplicates of this unit cell.

508 CHAP. 17 SOLIDS

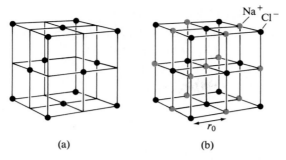

(a)   (b)

FIGURE 17.6   (a) A unit cell of the face-centered cubic (or FCC) lattice. There
is an atom at each corner of the cube and another at the center of each face.
(b) The NaCl structure. The Cl^- ions form one FCC lattice, and the Na^+ ions
another, offset by half the height of the unit cube. Each Na^+ is nearest to six
Cl^- ions, and vice versa. (This is most easily seen for the Na^+ at the center of
the whole cube.) The distance r_0 is the *nearest-neighbor distance* between any
ion and its six neighboring ions of opposite sign.

Fig. 17.6(b), the Cl^- ions are arranged on one FCC lattice. The Na^+ ions lie on a
second FCC lattice, which is offset from the Cl^- lattice. In this way, each Na^+ is
nearest to six Cl^- ions, and vice versa.

EXAMPLE 17.1   Given that the nearest-neighbor separation in NaCl is $r_0 =$
0.28 nm, estimate the atomic cohesive energy of NaCl in eV per atom.

Let us consider first the potential energy of any one ion, for instance, the
Na^+ ion at the center of the cube in Fig. 17.6(b). This ion has six nearest-
neighbor Cl^- ions, which give it a negative potential energy of $-6ke^2/r_0$.
However, the Na^+ ion also has 12 next-nearest-neighbor Na^+ ions at a distance
of $r_0 \sqrt{2}$. (These are all of the remaining 12 Na^+ ions in the picture.) These
contribute a positive potential energy of $+12ke^2/(r_0 \sqrt{2})$. Next, there are eight
Cl^- ions (at the eight corners of the cube in the picture) at a distance of $r_0 \sqrt{3}$,
and these contribute a negative potential energy of $-8ke^2/(r_0 \sqrt{3})$. Continuing
in this way we would find a total potential energy:

$$\text{total PE of any one ion} = -\frac{ke^2}{r_0}\left(6 - \frac{12}{\sqrt{2}} + \frac{8}{\sqrt{3}} - \cdots \right) = -\alpha \frac{ke^2}{r_0}. \quad (17.1)$$

The sum of the infinite series, α, is called the **Madelung constant.** Since the
series is difficult to evaluate, we shall simply state the result, that its sum is
$\alpha = 1.75$ (to three significant figures). Therefore, the potential energy of any
one ion in solid NaCl is

$$\text{total PE of any one ion} = -\alpha \frac{ke^2}{r_0} = -1.75 \times \frac{1.44 \text{ eV} \cdot \text{nm}}{0.28 \text{ nm}}$$
$$= -9.0 \text{ eV}. \quad (17.2)$$

In this calculation we have ignored the short-range repulsive forces due to the
overlap of the electron distributions when the ions get very close. The repulsive contribution to the energy is actually rather small (for exactly the same
reason as it was in ionic molecules; see Section 16.3). Therefore, our estimate
(17.2) is just a little below the observed value

$$\text{total PE of any one ion} = -7.9 \text{ eV.} \qquad (17.3)$$

For the remainder of this calculation, we shall use this observed value.

To find the energy of the whole NaCl crystal, we cannot simply multiply the answer (17.3) by the total number (N) of ions, since this would count each interaction twice over.* Instead, we must multiply by $N/2$, to give a total potential energy of $(-7.9 \text{ eV})N/2$. Thus the energy to pull the solid NaCl apart into separated Na$^+$ and Cl$^-$ ions is $(7.9 \text{ eV})N/2$. Since $N/2$ is the number of Na$^+$–Cl$^-$ *pairs,* we can say that

energy to separate NaCl into Na$^+$ and Cl$^-$ ions = 7.9 eV per Na$^+$–Cl$^-$ *pair.*

If we wish to have separate *neutral atoms,* we must next remove an electron from each Cl$^-$ ion (which requires 3.6 eV), and then attach it to the Na$^+$ ion (which gives us back 5.1 eV). Thus the total work done is

energy to separate NaCl into Na and Cl atoms = $(7.9 + 3.6 - 5.1)$ eV per pair

$$= 6.4 \text{ eV per Na–Cl pair.}$$

Dividing by 2, we find finally that

atomic cohesive energy of NaCl = 3.2 eV/atom.

Many other solids have the FCC lattice, for example, diamond† and several metals, including copper, silver, and gold. Two other important structures are the body-centered cubic (BCC) and the hexagonal close packed (HCP), both sketched in Fig. 17.7. Many metals, such as sodium, potassium, and iron, are BCC; examples of HCP are the solid forms of hydrogen and helium, and several metals, such as magnesium and zinc.

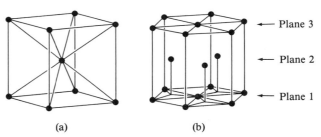

FIGURE 17.7 (a) The body-centered cubic (BCC) lattice. The unit cell has one atom at each corner of a cube and one at the center. (b) The hexagonal close packed structure, which consists of equally spaced planes, each of which is closely packed in a hexagonal pattern. Planes 1, 3, . . . are aligned vertically, as are planes 2, 4, . . . , but the odd planes are all displaced sideways from the even planes. Plane 2 has exactly the same arrangement as 1 and 3, but only three of its atoms fall inside the unit cell shown.

---

* For every pair of ions, i and j, there is a single potential energy of interaction that went into (17.3). However, if we simply multiply (17.3) by N, we will have counted the potential energy of i due to j *and* of j due to i. In other words, we will have counted every interaction twice over.

† This claim does not contradict our earlier statement that diamond has a tetrahedral structure. The carbon atoms are placed on two FCC lattices, one offset from the other. Each atom on either lattice is covalently bonded to its four nearest-neighbor atoms, all of which are on the other lattice at the corners of a regular tetrahedron.

FIGURE 17.8 A crystal of fluorite (CaF$_2$). Note the well-defined faces and edges.

FIGURE 17.9 When magnified 2500 times, stainless steel is seen to be a disordered assembly of many microcrystals, each a few μm across. (Courtesy U.S.X.)

MACROSCOPIC APPEARANCE OF CRYSTALS

When a large enough number of atoms form into a single crystalline lattice, the resulting solid is a macroscopic crystal. As the crystal forms, its macroscopic surfaces tend to coincide with the planes of its microscopic lattice; if a grown crystal is struck, it tends to cleave along these same atomic planes. Either way, the crystal acquires smooth flat surfaces that reflect light like a mirror. If the material is transparent, these same surfaces can cause strong internal reflections, which add to the glittering appearance that we associate with crystals. There are many examples of such macroscopic crystals: the Kohinoor diamond, the fluorite crystal in Fig. 17.8, the humble salt crystal, and many more.

Many solids, including most metals, have a microscopic crystalline structure but do not, nevertheless, *appear* crystalline. This is because, instead of forming as a single large crystal, these solids form in a jumbled array of many tiny microcrystals, as illustrated in the micrograph of stainless steel in Fig. 17.9. Since a typical microcrystal is about 1 μm across, the crystalline nature of such solids is completely invisible to the naked eye. Nevertheless, each microcrystal contains some 10^{10} atoms, and, from a microscopic point of view, the dominant feature of the material is its crystalline structure.

EXAMPLE 17.2 An important task for solid-state physics is to predict the strengths of materials. Given the estimate in Problem 17.22 for the force needed to separate two bonded atoms, make a rough estimate of the tensile strength of a simple cubic material with lattice spacing $r_0 \approx 0.4$ nm.

The tensile strength of a material is the force per unit area needed to rupture it. From Problem 17.22 we find that the force needed to separate a single pair of bonded atoms is of order $F \approx 3 \times 10^{-9}$ newton. The density of atoms in any one crystal plane is $1/r_0^2$. Therefore (if we ignore all but nearest-neighbor interactions), the force to tear apart a unit area of two adjacent planes is F/r_0^2; that is

$$\text{tensile strength} = \frac{F}{r_0^2} \approx \frac{3 \times 10^{-9} \text{ N}}{(4 \times 10^{-10} \text{ m})^2} \approx 2 \times 10^{10} \text{ N/m}^2. \quad (17.4)$$

FIGURE 17.10 These nearly perfect crystalline whiskers of iron (shown magnified about six times) have tensile strengths of order 10^{10} N/m^2, in agreement with (17.4). (Courtesy U.S.X.)

This calculation assumed a cubic lattice and included only nearest-neighbor interactions. Furthermore, although the force that we used is realistic for strong bonds (covalent, ionic, or metallic), it would be a gross overestimate for weaker bonds such as the van der Waals bond. Nevertheless, the answer (17.4) should—and does—give the right order of magnitude for any single crystal of strongly bonded atoms. Single-crystal whiskers of iron (as in Fig. 17.10) and of quartz have tensile strengths of this order. Macroscopic samples of many crystalline materials have tensile strengths much smaller than (17.4) (that for cast iron is some 100 times smaller), but this is because such macroscopic samples are *not* perfect crystals. It is the many imperfections in these materials that break easily and explain their much lower tensile strengths.

AMORPHOUS SOLIDS

Not all solids have a crystalline structure. Instead, the atoms of many solids, such as glass, wax, and rubber, arrange themselves in an irregular pattern. Such solids are described as *amorphous,* from a Greek word meaning "without form." In fact, many materials can solidify in either a crystalline or an amorphous form, depending on their preparation. For example, a material may form in a crystalline state if cooled slowly from a liquid. (Cooling it slowly gives the atoms time to arrange themselves in the lowest-energy, crystalline, state.) On the other hand, the same material may form in an amorphous state if cooled very rapidly.

The difference between crystalline and amorphous solids is shown schematically in Fig. 17.11. Part (a) illustrates a perfect crystal, with its atoms in a rigid and regular lattice. Part (b) shows an amorphous solid, with its atoms bonded in a more-or-less rigid structure, but not in a perfectly regular pattern. For comparison, parts (c) and (d) illustrate a liquid and a gas. When one heats a solid sufficiently, it first melts [part (c)]: Although its atoms remain within the range of interatomic forces, they are no longer held in a rigid pattern and are, instead, free to move around. When heated still further, the liquid becomes a gas [part (d)]: The atoms separate much further, they move more rapidly, and they interact only during their brief collisions.

We can characterize the differences between crystals, amorphous solids, liquids, and gases in terms of their **long-range order.** A perfect crystal has perfect

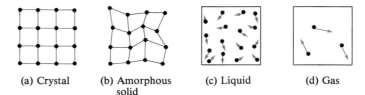

(a) Crystal   (b) Amorphous solid   (c) Liquid   (d) Gas

FIGURE 17.11 Schematic representation of atoms (or molecules) of a given type in (**a**) a crystal, (**b**) an amorphous solid, (**c**) a liquid, and (**d**) a gas. The atoms of the solids (a) and (b) are bound in a rigid lattice, as indicated by the lines, which represent the permanent interatomic bonds. The atoms of the liquid and gas, (c) and (d), are free to move around, as indicated by the arrows; for this reason, a liquid or gas requires a container to confine it. (The atoms of a solid can also move, but only in small-amplitude oscillations about their equilibrium positions.) Although the atoms in the liquid (c) are not rigidly bonded, they remain within the range of the interatomic forces.

long-range order: If we know its crystal structure and the location of one atom, we can predict with certainty the location of all other atoms. This is illustrated in Fig. 17.12(a), which shows the probability density (the probability of finding an atom, plotted against position) for an ideal one-dimensional crystal. If we know that there is an atom at the origin, we can predict with certainty that there are atoms at r_0, $2r_0$, $3r_0$, . . . , for as far as the solid extends.*

The distribution function for an amorphous solid is like Fig. 17.12(b). If we know that there is an atom at the origin, the probability of finding another atom very close to $r = 0$ is small, since two atoms repel one another strongly at short range. The probability increases to a peak at about the same r_0 at which the next atom of a crystal would be located. It then drops again because the repulsion between overlapping atoms prevents another atom from locating too close to the atom of the first peak. Because of the amorphous solid's irregular structure, we can predict the atoms' positions with less and less certainty as we move on, and the density function rapidly approaches a constant, reflecting our complete ignorance of the positions of the more distant atoms. We say that the amorphous solid shows short-range order, but no long-range order.

The probability density for a liquid is almost exactly like that for an amorphous solid, shown in Fig. 17.12(b). That for a gas is shown in part (c). It is much lower than those for solids or liquids, since the density is much less; it shows even less structure than that for the amorphous solid because a gas has even less order.

17.4   Energy Levels of Electrons in a Solid; Bands

We have stated that the reason metals conduct electricity is that some of their electrons can move essentially freely within the boundaries of the metal, but this does not explain *why* the electrons in a metal can move so freely, nor why the electrons in an insulator cannot. To explain this difference we must examine the energy levels for the electrons in a solid. Fortunately, we can do this quite easily, leaning on our discussion of the energy levels of electrons in molecules (Section 16.4). As one might expect, once we understand the energy levels of the electrons, we can explain many properties of solids in addition to their conductivity.

In Section 16.4 we described the energy levels of an electron belonging to either of two atoms as the atoms come together to form a molecule. While the atoms are far apart, we can focus on any one atomic orbital ($1s$, $2s$, . . .). Since the electron can occupy this orbital in either atom, this gives two degenerate orbitals (that is, two possible wave functions with the same energy). We saw in Section 16.4 that as the two atoms come closer together and begin to overlap, this degeneracy splits, and we have instead two orbitals of different energy, as indicated in Fig. 17.13(a).

If we start instead with *three* atoms far apart, then for each atomic orbital there are *three* independent, degenerate wave functions for an electron (corresponding to attaching the electron to any of the three atoms). As we move the atoms together this threefold degeneracy is split, and we get three distinct

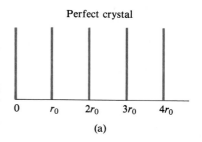

(a)

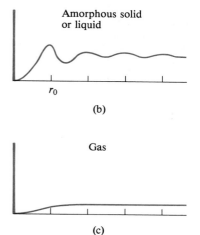

(b)

Gas

(c)

FIGURE 17.12   The probability densities for (a) a perfect crystal, (b) an amorphous solid or a liquid, and (c) a gas, all in one dimension. In each case the origin is placed at the known location of one atom. The probability of finding an atom is plotted up, and the distance away from the original atom is plotted to the right.

---

* Strictly speaking, Fig. 17.12(a) represents a classical crystal at absolute zero. In quantum mechanics the uncertainty principle means that we cannot know the locations with complete certainty, and at nonzero temperatures thermal motions introduce a further uncertainty. For both reasons the ideal spikes of Fig. 17.12 (a) should be a little spread out.

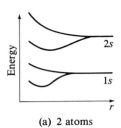

(a) 2 atoms

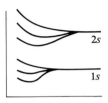

(b) 3 atoms

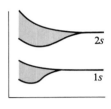

(c) *N* atoms

FIGURE 17.13 Energy levels of an electron in a system of several atoms, shown as functions of the atomic separation *r*. (**a**) For two atoms each atomic level splits into two levels as *r* is reduced. (**b**) With three atoms each level splits into three levels. (**c**) With many atoms (*N*, say) each atomic level fans out into *N* levels, which are so closely spaced that we can regard them as a continuum.

possible energies, as indicated in Fig. 17.13(b). More generally, if we consider *N* atoms far apart, then for each atomic orbital there are *N* degenerate wave functions, and as we move the atoms together these fan out into *N* more-or-less evenly spaced levels. We have shown this in Fig. 17.13(c), where we have indicated the many closely spaced levels by shading.

We can imagine putting together a whole solid in this way, starting with *N* well-separated atoms. While they are still far apart, we can arrange the atoms in the same geometrical pattern as the final solid, and we can then imagine the whole structure slowly contracting to its equilibrium size. As the nearest-neighbor separation *r* decreases, the energy levels of the electrons will vary, their general behavior being that shown in Fig. 17.13(c), which we have redrawn in more detail as Fig. 17.14. For each atomic orbital there are *N* independent wave functions, all of which have the same energy when *r* is large. As we reduce *r*, this degenerate level fans out into *N* distinct levels. We refer to these *N* closely spaced levels, which all evolved from a single atomic level, as a **band.** The width and spacing of these bands varies with *r*, but the bands of the actual solid are found by setting *r* equal to its equilibrium value r_0, as shown on the left of Fig. 17.14.

Figure 17.14 includes some of the terminology used in connection with the energy levels of electrons in solids. The band of levels that evolved from the 1*s* atomic levels is called the 1*s* band, and similarly with 2*s*, 2*p*, and so on. The range of energies in a single band — the *width* of the band — is typically a few eV. If there are *N* levels within the band,* the average spacing δE between adjacent levels is

$$\delta E \approx \frac{\text{a few eV}}{N}.$$

In practice, *N* is very large; even in a microcrystal 1 μm across, there are some 10^{10} atoms (Problem 17.25). Thus the level spacing within a band is very small; in a microcrystal with $N \approx 10^{10}$, δE is of order 10^{-10} eV, and in a larger crystal it is even smaller. Therefore, for almost all purposes, we can think of the electron energy levels in a solid as *continuously distributed within each band.* This is what the shading of the bands in Fig. 17.14 was intended to suggest.

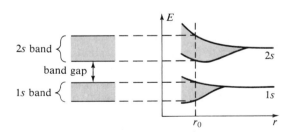

FIGURE 17.14 As we assemble a solid from *N* well-separated atoms, the original atomic energy levels fan out into *bands* of closely spaced levels. The bands of the actual solid are shown on the left and are found by taking the nearest-neighbor distance *r* equal to its equilibrium value r_0.

---

* As we shall see shortly, there are generally *more* than *N* states in a band, but this only strengthens our conclusion that the average spacing is very small.

It can happen that the highest level in one band is above the lowest level in the next-higher band. In this case we say that the two bands overlap. On the other hand, it often happens (as in Fig. 17.14) that two neighboring bands do *not* overlap, and there is instead a **band gap**—a range of energies that are not allowed. We shall see that the occurrence of band gaps, and their widths, are two of the main factors that determine the conductive properties of a solid.

It is important to know whether all the states in a given band of a solid are occupied (just as it was important in our discussion of the periodic table to know whether all the states of a given level in an atom were occupied). To this end, we need to know how many states there are in a given band. For *s* bands (the 1*s* and 2*s* bands of Fig. 17.14, for example) there are *N* independent wave functions and two possible spin orientations, and hence 2*N* states in all. If we consider a *p* band (a band that evolved from an $l = 1$ atomic level), then when the *N* atoms were well separated there were 3*N* independent wave functions (three different values of l_z for each atom) and two spin orientations. Therefore, there are altogether 6*N* states in a *p* band. Since we shall only be concerned with *s* and *p* bands, we shall not give a general formula here (but see Problem 17.28).

17.5 Conductors and Insulators

We are now ready to explain the difference between conductors and insulators. We illustrate our discussion with three examples, starting with lithium, which we know to be a metal and hence a conductor.

LITHIUM: A CONDUCTOR

The lithium atom has the configuration $1s^2 2s^1$, with its 1*s* level full and its 2*s* valence level half full. As we assemble *N* lithium atoms into solid Li, the 1*s* and 2*s* levels fan out into the 1*s* and 2*s* *bands,* which we have shown in Fig. 17.15. The *N* lithium atoms contribute altogether 3*N* electrons, while each of

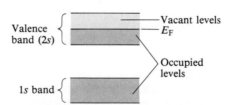

FIGURE 17.15 The bands of allowed electron energies in solid lithium. All levels in the 1*s* band are fully occupied (dark shading); the 2*s* valence band is only half full, with empty levels shown lightly shaded.

the two bands can hold 2*N* electrons. Thus the 1*s* band is completely filled, and the 2*s* *valence band* is half filled. We have shown this occupancy in Fig. 17.15, where dark shading represents occupied, and light shading unoccupied levels. The energy shown as E_F is the *Fermi energy,* which is defined as the energy of the highest occupied level.*

---

* Strictly speaking, this is correct only at absolute zero. At higher temperatures, thermal motions can excite a few electrons to levels somewhat above E_F.

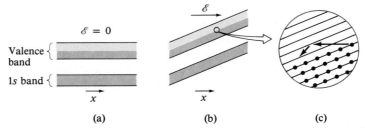

Valence band {

1s band {

$\mathscr{E} = 0$

$\mathscr{E}$

x

x

(a)   (b)   (c)

FIGURE **17.16** Schematic representation of the electron energy levels along a solid Li wire. **(a)** In the absence of an electric field, the levels are the same at all points along the wire. **(b)** An electric field in the x direction raises the potential energy (and hence the total energy) on the right and tilts the whole diagram as shown. **(c)** A magnified portion of the valence band of (b). In response to the electric field, those electrons in the highest occupied levels can now move into *unoccupied* levels on their left, and a current flows. The electrons eventually collide with atoms of the lattice and lose energy, but the process of accelerations followed by collisions repeats itself for as long as there is an electric field. [Although we have drawn the levels of part (c) as discrete levels, for all practical purposes they are a continuum.]

We are now going to argue that the electrons in the partially full valence band can move around in response to an applied electric field — and can contribute to conduction — whereas the electrons in the full $1s$ band cannot. To see this, let us consider a lithium wire oriented along the x axis. In the absence of an electric field, the energy levels are the same at all points along the wire, as indicated in Fig. 17.16(a).

Suppose now that we connect a battery to the ends of the wire to maintain an electric field to the right. This field raises the potential energy of the electrons on the right, and the bands now look like Fig. 17.16(b) or, in a magnified version, Fig. 17.16(c). For any electron at the top of the occupied levels in the valence band, there are now *unoccupied* levels of the same energy (and lower) just to the left. Therefore, each of these electrons begins to move to the left in response to the electric field, as indicated (for one electron) by the horizontal arrow in (c). This motion of electrons to the left is, of course, the expected electric current. In due course, the electrons moving to the left collide with the atomic cores of the lattice and lose some kinetic energy, as indicated by the downward arrow in Fig. 17.16(c). (These collisions are the main cause of electrical resistance in most metals.) The sequence of accelerations and collisions is continually repeated as the electrons move along the wire. The energy lost by the electrons in collisions increases the vibrations of the lattice, and the wire's temperature rises. This temperature rise is just the familiar Joule heating of a wire by a current.

The argument just given does *not* apply to an electron in the full $1s$ band, since there are no vacant states to its immediate left. Thus the electrons in a filled band do not contribute to conduction. For the same reason, electrons deep down in the valence band cannot move either. Only electrons near the top of the occupied levels of the partially filled band contribute to conduction.

This example already illustrates our two most important conclusions: Solids with a partially filled band are conductors; those in which all occupied bands are 100% full are insulators. Our next two examples show that these simple rules have to be applied with care.

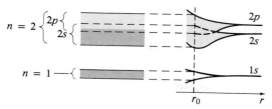

$$n = 2 \begin{cases} 2p \\ 2s \end{cases}$$

$$n = 1 \longrightarrow \{$$

FIGURE 17.17 The band structure of solid beryllium. As we assemble the solid from well-separated atoms, the 2s and 2p levels fan out so far that they overlap and form a single $n = 2$, valence band. Since this band could accommodate up to $8N$ electrons, it is less than half full, and the solid is a conductor.

BERYLLIUM: A CONDUCTOR

We turn now from lithium to the next element in the periodic table, beryllium. Its atomic configuration is $1s^2 2s^2$, and both of its occupied levels are filled. This might suggest that solid Be should have filled 1s and 2s *bands* and should therefore be an insulator, whereas Be is, in fact, a good conductor. The explanation of this apparent contradiction is not hard to find: As we assemble solid Be and its atomic levels fan out into bands, the 2s and 2p bands spread so far that they *overlap* one another, as shown in Fig. 17.17. The 2s and 2p bands thus form a single valence band, which can accommodate altogether $8N$ electrons. (Remember an s band holds $2N$ electrons and a p band $6N$.) Thus, of the $4N$ electrons in the solid, $2N$ fill the 1s band, and the remaining $2N$ can only *partially* fill this valence band. Therefore, Be is a conductor.

DIAMOND: AN INSULATOR

Atomic carbon has the configuration $1s^2 2s^2 2p^2$, with its $n = 2$, valence shell half filled. This suggests that solid carbon should be a conductor, which, indeed, it is in the form of graphite. However, solid carbon in the form of diamond is an outstanding *insulator*. To explain this puzzle, we must look carefully at the behavior of the energy levels as we assemble a diamond. In diamond the carbon atoms are covalently bonded. As we bring the atoms together and the covalent bonds form, the 2s and 2p levels fan out into bands, as we would expect. However, each level splits into *two separate* bands, as shown in Fig. 17.18. (This is closely analogous to the development of the bonding and antibonding orbitals described in Section 16.4.) At the equilibrium separation r_0, the lower two bands overlap each other, as do the upper two. Thus the band structure of diamond is as shown on the left of Fig. 17.18: There is an $n = 1$ band that can hold $2N$ electrons, and then two well-separated $n = 2$ bands, each of which can hold $4N$ electrons. Since carbon has $6N$ electrons in all, the lowest two bands are completely full, and the upper $n = 2$ band is completely empty. Because the lower $n = 2$ band holds all the valence electrons, it is called the valence band, and because electrons in the upper $n = 2$ band (if there were any) would contribute to conduction, this band is called the *conduction band*. With its valence band full and its conduction band empty, diamond is an insulator. It turns out that as indicated in Fig. 17.18, the gap between the valence and conduction bands of diamond is about 7 eV. Under normal conditions, this means that no electrons can be excited into the empty conduction band. (At room temperature, for example, thermal motions *could* excite an electron, but the probability is so low that the expected number of excited electrons in a large

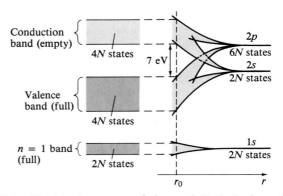

FIGURE 17.18 The band structure of diamond. Both the $2s$ and $2p$ levels split, each forming two separate bands. Each of the $2s$ subbands contains N states, and each of the $2p$ subbands contains $3N$. The lower $2s$ and the lower $2p$ bands then join to form a single band with $4N$ states in all, and likewise the upper $2s$ and upper $2p$.

diamond is far less than 1.) This makes diamond an extremely good insulator. This same large band gap also explains why diamond is transparent. Visible photons have energies of 2 to 3 eV and cannot excite any electrons across the 7 eV gap; therefore, visible light cannot be absorbed in diamond.

We mentioned in Section 17.2 that solid silicon (just below carbon in the periodic table) has the same crystal structure as diamond. The $3s$ and $3p$ levels in Si behave almost exactly like the $2s$ and $2p$ levels of diamond, and silicon, like diamond, is a nonconductor. There is, however, a subtle but crucial difference. The equilibrium separation r_0 in Si is larger than in diamond (0.24 nm compared to 0.15 nm). It is easy to see in Fig. 17.18 that a larger value of r_0 suggests a *smaller* band gap, and, indeed, the band gap in silicon is only about 1 eV (compared with 7 eV in diamond). Since visible photons can easily excite electrons across a 1-eV gap, light is quickly absorbed in silicon, which is therefore *not* transparent. The narrow band gap in silicon also has a dramatic effect on its conductivity. At low temperatures, all the valence electrons in Si are locked in the full valence band, and Si is an insulator. But as the temperature rises, a substantial number of electrons are excited across the narrow gap into the conduction band, and the conductivity increases rapidly. For this reason, silicon and other solids with a similar narrow gap between a full valence band and an empty conduction band are called **semiconductors.** Semiconductors include silicon and germanium, both from group IV of the periodic table, and several compounds of group III with group V, such as gallium arsinide. Semiconductors are so important that we have made them the subject of a separate section.

17.6   Semiconductors

A semiconductor in its ground state has a filled valence band separated by a narrow gap, of order 1 eV or less, from an empty conduction band. A pure semiconductor, such as pure silicon or pure germanium, is sometimes called an **intrinsic semiconductor.** Because the band gap is so narrow, one can easily excite electrons from the valence band into the conduction band, where they can carry a current. Thus the conductivity of an intrinsic semiconductor can be increased by exciting some of its electrons (by heating it, for example). An **impurity semiconductor** is a semiconductor that has been "doped" with impurities. We shall see that this doping lets one alter the conductivity of the semiconductor, and that, by combining two or more semiconductors that are differently doped, one can make an amazing variety of electronic devices. Although it is the impurity semiconductors that are commercially important, we need to say a little more about intrinsic semiconductors first.

ELECTRONS AND HOLES

Let us imagine an electron excited from the valence band of a pure semiconductor into the conduction band. Once in the conduction band, the electron is free to move around and carry a current. However, there is a second type of current carrier that can be just as important: The excitation of the electron leaves a *vacancy* in the valence band, and this allows the remaining valence electrons to move as well. Thus once some electrons have been excited, conduction actually occurs in both the conduction *and* valence bands. We have indicated this schematically in Fig. 17.19, which shows the effect of an electric field directed to the right. In the conduction band, the field causes the excited electron to move to the left (producing a conventional current to the right). In the valence band, the electron just to the right of the vacancy can move into the vacancy; this then allows the next electron to move, and then the next, and so on. Here, too, electrons move to the left, and the conventional current is to the right.

There is a different, and often better, way to view the conduction in the valence band. The removal of an electron leaves the unbalanced charge of a positive ion. Thus the vacancy, or **hole,** in the valence band is associated with a *positive* charge, and we can say simply that the field to the right causes this positive hole to move to the right.* In these terms, we can restate our conclusion: A current flowing to the right in a pure semiconductor is made up of two components: negative electrons moving to the left in the conduction band, and positive holes moving to the right in the valence band. This notion of positive holes moving in the valence band will play a central role in our discussion of impurity semiconductors.

If a photon excites an electron from the valence to the conduction band, as in Fig. 17.20(a), we can say that the photon has created a hole in the valence band and an electron in the conduction band. Conversely, when an electron drops from the conduction band into a hole in the valence band and emits a photon, as in Fig. 17.20(b), we can say that the electron and hole have annihilated one another and created a photon.

IMPURITY SEMICONDUCTORS

Most commercial applications of semiconductors involve *impurity semiconductors,* into which carefully chosen impurities have been introduced in a process known as **doping.** There are basically two types of impurity semiconductors: In the so-called **n-type,** the impurities contribute electrons to the conduction band, making possible conduction by negative (*n*) carriers; in the

---

* It may help to draw an analogy with a bubble floating up from the bottom of a pool of water. What really happens is that gravity pulls the surrounding water *down,* but it is more natural to say that buoyancy causes the bubble (or "hole") to move *up.*

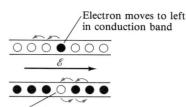

FIGURE **17.19** Once a valence electron has been excited to the conduction band of a pure semiconductor, both the excited electron and the remaining valence electrons can move in response to an electric field. A field to the right makes the electrons in both bands move to the left, as shown, but within the valence band it is often better to think in terms of a positive hole moving to the right.

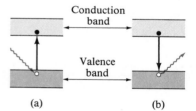

FIGURE **17.20** **(a)** When a photon excites an electron from the valence to the conduction band, we say the photon has created an electron–hole pair. **(b)** If the electron drops back into a hole and emits a photon, we say that the electron and hole have annihilated one another and produced a photon.

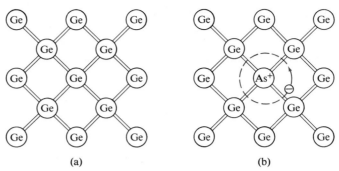

(a)   (b)

FIGURE 17.21 Schematic view of solid germanium. Each atom is covalently bonded to its four nearest neighbors, and each of its four valence electrons participates in one of these bonds. **(b)** If an As atom replaces one of the Ge atoms, it can fit into the Ge lattice using just four of its five valence electrons; that is, it is the As^+ ion that fits into the lattice. The fifth valence electron is weakly bound to the As^+ ion by the Coulomb attraction, but is easily detached to move freely through the crystal. (The "Bohr radius" for the electron is actually much larger than shown here.)

p-type, the impurities introduce holes into the valence band, allowing conduction by positive (p) carriers.

As an example of an n-type semiconductor imagine a sample of germanium, of valence 4, into which have been introduced a few atoms of arsenic, of valence 5. (Concentrations of one As atom per million atoms of Ge are typical.) Figure 17.21(a) is a schematic view of the structure of pure Ge, with each atom joined by covalent bonds to its four neighbors. (Each bond is represented by two lines to symbolize that it involves two electrons.) Suppose now that we replace one atom of Ge (valence 4) by an atom of As (valence 5). The As atom can fit into the crystal structure of the Ge, with four of its five valence electrons forming covalent bonds, as in Fig. 17.21(b), but this leaves one electron unaccounted for. Therefore, the object that is bound into the lattice is actually the As^+ ion, and the last electron is bound to this ion by their Coulomb attraction. It is possible to calculate the "Bohr orbits" for this hydrogen-like atom, and it turns out that the binding energy is of order only* 0.01 eV. That is, the electron is very weakly bound to its parent As atom. At room temperature, the thermal vibrations of the atoms in the solid have energies significantly larger than 0.01 eV, and there is a high probability that the weakly bound electron will be knocked loose, and will then be free to move through the solid. That is, many of the impurity atoms will contribute their outermost electron to the conduction band of the solid. More generally, if we dope either silicon or germanium (both from group IV) with an element from group V (like P, As, Sb), the impurities will contribute electrons to the conduction band, and the resulting n-type material will conduct by motion of negative electrons. Since the impurities in an n-type semiconductor *donate* electrons, they are often called **donors.**

Imagine, instead, that we dope germanium with an element from group III, gallium say. In this case, the Ga atoms can fit into the crystal structure of Ge, but since Ga has only three valence electrons, one of the bonds that it forms will be incomplete, as indicated in Fig. 17.22. This incomplete bond can readily

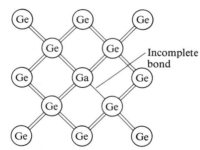

FIGURE 17.22 Solid germanium doped with gallium. Because Ga has valence 3, one of the bonds that it forms is incomplete. This incomplete bond can readily steal a valence electron from a nearby Ge atom and create a hole in the valence band.

* This binding energy is strikingly small compared to that of an electron in a hydrogen atom. The main reason for this difference is that the electron orbiting around the As^+ ion is not moving in a vacuum. Rather, it is moving in solid Ge, which has a large dielectric constant. This reduces the Coulomb force and greatly reduces the binding energy (Problem 17.41).

accept another electron, and only about 0.01 eV is needed to remove one of the valence electrons from a nearby Ge atom and bind it to the **acceptor** impurity. In this way an acceptor impurity creates holes in the valence band, and the resulting *p*-type material conducts by the motion of these positive holes.

THE *n-p* DIODE

The simplest of the semiconductor devices is the *n-p* diode, which consists of a single semiconductor that has been doped to make part of it *n*-type, and the remainder, *p*-type. As we shall see, this has the remarkable property that it can conduct current in one direction but not the other. An *n-p* diode is therefore a one-way valve, which can be used, for example, to rectify an ac voltage to dc, as illustrated in Fig. 17.23.

The working of an *n-p* diode is illustrated in Fig. 17.24. In part (a) of that figure we have shown the diode before it is connected to an EMF. In the *n* region we have shown the negative carriers (colored minus signs) moving freely among the fixed positive charges on the lattice (black plus signs). In the *p* region the situation is the other way around: positive holes moving among the fixed negative charges on the lattice.

In Figure 17.24(b) we have connected an EMF in such a way as to drive the mobile electrons in the *n* region *toward* the *n-p* interface. Since the same EMF also drives the holes in the *p* region toward the interface, we can say simply that it draws the carriers on either side *toward one another*. (An EMF in this direction is said to produce a **forward bias**.) Within the diode, electrons arriving from below at the interface meet and fall into holes arriving from above. Thus

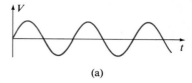

(a)

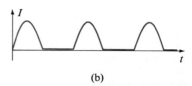

(b)

FIGURE 17.23 An *n-p* diode can pass current in one direction only, and can therefore be used as a rectifier. **(a)** An ac voltage is applied across the rectifier. **(b)** A current flows only when the voltage is in the forward direction.

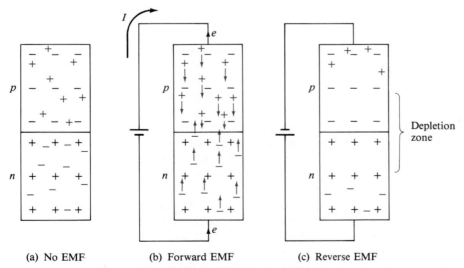

(a) No EMF   (b) Forward EMF   (c) Reverse EMF

FIGURE 17.24 **(a)** An *n-p* diode. In the *n* region, electrons (colored minus signs) move freely among the positive charges of the lattice (black plus signs); in the *p* region, holes (colored plus signs) move among the negative charges of the lattice. Note that in both regions color denotes mobile carriers. **(b)** A forward EMF drives *n* and *p* carriers toward the interface, where they meet and annihilate. The battery replenishes the supply of carriers (*n* at the bottom and *p* at the top), and a steady current, *I*, flows. The small arrows labeled e indicate the flow of electrons in the connecting wires (into the *n* region and out of the *p* region). **(c)** A reverse EMF draws the carriers *away* from the interface leaving a depletion zone, through which no current can flow.

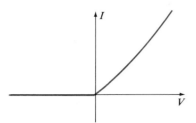

FIGURE 17.25 The current–voltage curve of an n–p diode. A forward voltage produces a current, but a backward voltage does not. (In truth, there *is* a small backward current, but it is normally too small to show on this scale.)

the diode is continually losing carriers, both n and p, at the interface. However, the battery drives additional electrons into the diode at the bottom and hence replenishes the supply of electrons; the battery also pulls electrons out at the top, creating new holes. Thus the diode's supply of carriers, both n and p, is continually replenished, and a steady current flows.

In Figure 17.24(c) we have connected an EMF so as to pull the electrons and holes in the diode *away* from the interface. (An EMF in this direction is described as a **reverse bias.**) As the electrons and holes move away from the interface, they leave a **depletion zone,** in which there are no carriers and hence no current can flow. The fixed lattice charges left behind in the depletion zone produce a field that opposes the outward motion of the carriers. As the depletion zone grows, this field eventually balances the applied field, and no further motion occurs. We conclude that the n-p diode can carry current in one direction but not the other, and has a current–voltage curve with the general shape of Fig. 17.25.*

LIGHT-EMITTING DIODES

As a current flows in the forward direction through an n-p diode, electrons are continually meeting holes at the n-p interface. When an electron drops into a hole, the excess energy is usually emitted in the form of a photon. If this photon is in the visible range, and if the diode is transparent to light, then the diode will emit visible light. Such diodes, which emit light when a current flows, are called **light-emitting diodes** or LEDs and are widely used in the displays of stereos, clocks, and other electrical equipment.

If the current in an LED is large enough, then many electrons flow into the p region and, before they have a chance to drop into holes, they form an appreciable *population inversion* of excited electrons (Section 15.7). The photon emitted when one electron drops spontaneously into a hole can then stimulate many more electrons to drop and emit further photons. This is the basis of the *solid-state laser,* whose dimensions can be very small compared to other types of laser, and is finding application in fiber-optic communication systems.

In the LED and solid-state laser, a current is driven through the device and photons are emitted. It is also possible to reverse this procedure: If photons shine onto an n-p diode, they can create electron–hole pairs, which, in turn, cause a current to flow. The energy to drive this current comes, of course, from the incident photons. This is the basis of the photovoltaic cell (or solar cell), in which light (often sunlight) drives an electric current.

An n-p diode can also be used as a detector of the particles ejected from radioactive nuclei or other nuclear processes. The diode is maintained with a reverse bias, so that normally no current flows. However, when a nuclear particle passes through the depletion zone it creates some electrons and holes, which allow a pulse of current to flow through the diode and around the circuit. Thus each nuclear particle traversing the diode is registered as a pulse of current.

JOHN BARDEEN (1908–1991, US). Bardeen is the first person to have won two Nobel Prizes in physics. The first was in 1956, for the invention of the transistor, and was shared with Brattain and Shockley. The second was in 1972, for the "BCS" theory of superconductivity, and was shared with Cooper and Schrieffer.

* We should emphasize that our simple account of the n-p diode has glossed over some subtle but interesting points. For example, even with no EMF there is a small depletion zone, since some electrons from the n region can diffuse into the p region and drop into holes. Also, even with a reverse EMF there is some current, but it is many orders of magnitude less than the corresponding forward current.

TRANSISTORS

The semiconductor device that has contributed most to the electronic revolution of the last three decades is the **transistor.** This device has literally thousands of applications, many of which depend on its ability to amplify an electric current. For example, transistors can amplify the weak current from a microphone enough to drive a loudspeaker.

The transistor is a semiconductor doped with a narrow p region, called the *base,* separating two n regions, called the *emitter* and *collector,* as shown in Fig. 17.26. (This arrangement is an *n-p-n* transistor; one can also have a *p-n-p* transistor.) We can think of the transistor as two *n-p* diodes placed back to back. The emitter–base junction is given a forward bias (by the battery V_b in the picture), while the base–collector junction is given a reverse bias (by the larger battery V_c in the picture). The forward bias across the emitter–base junction pushes electrons from the emitter to the base. Some of these electrons leave the transistor and form the current shown as I_b. However, because the base is very narrow, most of the electrons from the emitter coast through the base and into the collector. Thus, even though the collector–base junction is reverse biased, a current I_c can now flow through the collector, as shown. It turns out that (over some range of values) the current I_c is proportional to I_b:

$$I_c = \beta I_b,$$

where the constant β may be as large as 100 or more. Small variations in I_b produce large proportional variations in I_c. Thus, by connecting weak input signals into the base circuit, we can produce amplified output signals in the collector circuit.

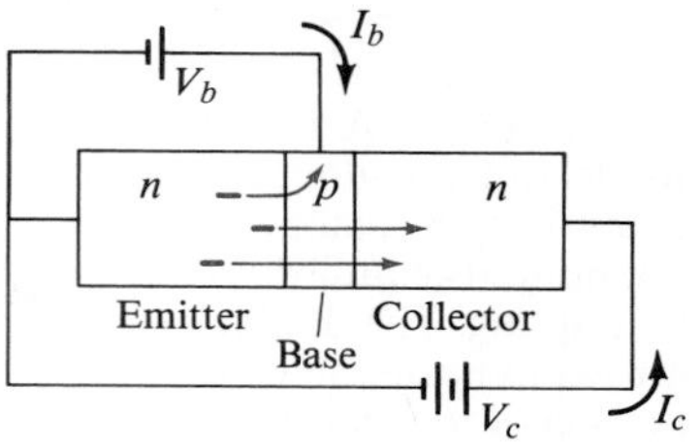

The really widespread application of transistors in machines ranging from wristwatches to supercomputers became possible with the invention of the **integrated circuit** or IC. This is a single small piece, or "chip," of silicon, into which hundreds of thousands of transistors and other components are built (Fig. 17.27). Among the advantages of building the IC as a single unit are that it is extraordinarily compact and that the same manufacturing process makes both the thousands of components and the equally numerous connections.

FIGURE 17.26 An *n-p-n* junction transistor. The EMF V_b drives electrons from the emitter into the base. A few of these exit from the base and make up the current shown as I_b, but the base is so thin that most of the electrons coast through to the collector and cause the current I_c. The two currents are proportional, but I_c can be some 100 times bigger. Thus small variations in I_b are amplified into large variations in I_c.

FIGURE 17.27 An integrated circuit made for use in a computer. This small chip can store 4 million bits of information, the equivalent of several hundred printed pages. It can read a single bit in 65 ns, or all 4 million in about a quarter of a second. (Courtesy IBM.)

17.7 Lattice Vibrations and Phonons (optional)

In the last three sections we have discussed the motion of the electrons in a solid, but have completely ignored the motion of the atomic cores, which we have treated as if they formed a rigid, immovable lattice. For many purposes, this is an excellent approximation. Nevertheless, we should consider, at least briefly, the possible motions of the atomic cores, and we shall find that they have an important effect on certain macroscopic properties of solids, most notably the specific heat.

We know that the atoms* of a solid are anchored by interatomic forces to their equilibrium positions in the rigid lattice of the solid. However, they can vibrate about these equilibrium positions, and if we wish to find the effects of these vibrations, we must calculate the possible energies of this motion. In our discussion of molecules, we saw that the vibrational motion of an atom about its

* For brevity, we shall use "atoms" to denote the atoms, atomic cores, or ions that make up the solid lattice.

FIGURE 17.28 A schematic
model of a one-dimensional solid
comprising five atoms, each
bound to its equilibrium position
by a spring. The double-headed
arrows indicate the vibrational
motion of the atomic cores about
their equilibrium positions.

equilibrium position is well approximated as simple harmonic motion, which, in turn, is equivalent to the motion of a mass on a spring. As a simple model of a solid, therefore, we imagine that each atom is attached to its equilibrium position by a spring, as shown schematically in Fig. 17.28. If we consider for a moment a one-dimensional solid, the potential energy of each atom would be just

$$U = \tfrac{1}{2}kx^2, \tag{17.5}$$

where k is the spring constant and x is the atom's displacement from equilibrium.

The energy (17.5) is the potential-energy function of the simple harmonic oscillator (SHO) discussed in Section 8.9. Therefore, the quantized energy levels for the vibration of any one atom are those found in (8.96):

$$E_{\text{vib}} = (n + \tfrac{1}{2})\hbar\omega_c \qquad (n = 1, 2, 3, \ldots), \tag{17.6}$$

where $\omega_c = \sqrt{k/m}$ is the frequency of oscillation of a classical particle with mass m on a spring with force constant k. In most solids, the spacing $\hbar\omega_c$ of these vibrational levels is of order 0.01 eV. (This is roughly the same result as derived for a molecule in Example 16.4 of Section 16.6.) Now, at room temperature the average energy of thermal motion of an atom is of order 0.1 eV (as we shall see shortly); that is, the average energy is large compared to the level spacing. This suggests (what proves to be true) that we could treat the vibrational motion of the atoms classically, at least at room temperatures or higher.

The potential energy (17.5) is appropriate for an atom in one dimension. The corresponding energy when the atom moves in three dimensions is

$$U = \tfrac{1}{2}kr^2 = \tfrac{1}{2}k(x^2 + y^2 + z^2), \tag{17.7}$$

where $\mathbf{r} = (x, y, z)$ is the particle's displacement from equilibrium. The total energy of vibration is

$$E_{\text{vib}} = K + U = \frac{1}{2m}(p_x^2 + p_y^2 + p_z^2) + \frac{1}{2}k(x^2 + y^2 + z^2). \tag{17.8}$$

The average value of this energy at temperature T is determined (in classical statistical mechanics) by the **equipartition theorem**. According to this theorem, each of the six terms in (17.8) should have an average value of $\tfrac{1}{2}k_B T$, where k_B denotes Boltzmann's constant,

$$k_{\text{B}} = 1.38 \times 10^{-23} \text{ J/K} = 8.62 \times 10^{-5} \text{ eV/K}. \tag{17.9}$$

Therefore (according to classical theory), the average vibrational energy of any one atom at temperature T should be

$$\overline{E}_{\text{vib}} = 3k_B T. \tag{17.10}$$

(Notice that at room temperature, $T \approx 300$ K, this gives an energy of order 0.1 eV, as stated earlier.) If the solid has N atoms in all, the total energy of the whole solid should be

$$E_{\text{tot}} = N\overline{E}_{\text{vib}} = 3Nk_{\text{B}}T. \tag{17.11}$$

Using this result, we can now calculate the specific heat of a solid.

The **molar specific heat** C of a substance is defined as the energy needed to raise 1 mole of the substance through 1 degree; that is, C is the derivative

$$C = \frac{dE_{tot}}{dT}.$$

If we consider a solid comprised of a single element, a mole contains N_A atoms (where N_A is Avogadro's constant). Therefore, according to (17.11) its total energy is $3N_A k_B T$, and, differentiating, we find for the molar specific heat:

$$C = \frac{dE_{tot}}{dT} = 3N_A k_B. \tag{17.12}$$

The combination of constants $N_A k_B$ is called the universal gas constant R:

$$R = N_A k_B = 8.31 \text{ J/(mol} \cdot \text{K)}. \tag{17.13}$$

Thus we can rewrite (17.12) as

$$C = 3R = 24.9 \text{ J/(mol} \cdot \text{K)}. \tag{17.14}$$

This result, called the **Dulong–Petit law,** states that the molar specific heats of elemental solids should all have the same value, $3R = 24.9$ J/(mole K). Given the relative simplicity of its derivation, the Dulong–Petit law is confirmed remarkably well by the observed specific heats of the first four solids listed in Table 17.2.

If we consider instead a solid comprised of two elements, such as NaCl, then a mole contains $2N_A$ atoms. Therefore, the same argument that led to (17.14) now gives *twice* that result: $C = 6R = 49.8$ J/(mol $\cdot$ K). This prediction is supported by the last two entries in Table 17.2.

There are two features of the Dulong–Petit law that caused difficulties for classical physics, but are easily explained by quantum mechanics. First, the result (17.14), which agrees so well with experiment, was derived from the vibrational energy (17.11) of the atoms. However, we know that the conduction electrons in a metal can move around more-or-less freely. The total energy of a metal should include the energy of these electrons, and since this energy would be expected to increase with T, the specific heat of a metal should be larger than the Dulong–Petit value. In fact, we can easily estimate the expected classical contribution of the conduction electrons: Let us suppose that each atom con-

TABLE 17.2

Molar specific heats, C, of various solids at room temperature, all in J/(mol $\cdot$ K). For the first four solids, each containing a single element, the Dulong–Petit law predicts that $C = 3R = 24.9$ J/(mol $\cdot$ K), at least at sufficiently high temperatures. For the last two, which contain two elements, the prediction is $C = 6R = 49.8$.

Solid	C (measured)	C (predicted)
Aluminum	24.2	
Copper	24.5	24.9
Silver	25.2	
Sulfur	24.3	
NaCl	49.9	49.8
KF	48.4	

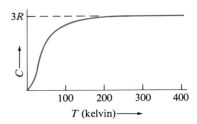

FIGURE 17.29 The molar specific heat of gold. At room temperature, C is equal to the classical value $3R$, but $C \to 0$ as $T \to 0$.

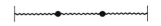

FIGURE 17.30 A one-dimensional "solid," comprising two identical atoms, joined to each other and to the walls by three identical springs.

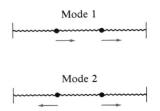

FIGURE 17.31 The two normal modes of the "solid" of Fig. 17.30. In the first mode the two atoms oscillate exactly in phase, and in the second they oscillate exactly out of phase.

tributes exactly one conduction electron, which behaves like a classical free particle. The energy of such a particle contains none of the potential-energy terms of (17.8). Therefore, according to the classical equipartition theorem each electron should have an average energy equal to half that of (17.10): $\bar{E}_{el} = 1.5k_BT$; and the total energy of the N conduction electrons should be $1.5Nk_BT$. Thus the result (17.11), $E_{tot} = 3Nk_BT$, should become $E_{tot} = 4.5Nk_BT$, and the specific heat (17.14) should be $C = 4.5R$ for a metal. Why, then, does the Dulong–Petit value, $C = 3R$, agree so well with experiment?

The answer to this puzzle is that electrons are quantum particles and obey the Pauli exclusion principle. At absolute zero, we know that all of the N conduction electrons are packed into the N lowest states in the conduction band, and that these occupied levels span 1 or 2 eV. As the temperature increases, each electron's energy should increase (according to classical theory) by an amount of order k_BT, which is about 0.02 eV at room temperature. But for most electrons, all of the levels to which this energy could lift an electron are *already filled.* (The only exception is for the small number of electrons within 0.02 eV of the highest occupied level.) Therefore, all the conduction electrons, except those few with the very highest energies, remain locked in the same states as at absolute zero; that is, their energy does not change with temperature. This is why the conduction electrons contribute so little to the specific heat of metals.*

The second classical puzzle connected with the Dulong–Petit law is that at low temperatures the specific heat of all solids drops *below* the predicted value, and, near absolute zero, $C \to 0$, as illustrated in Fig. 17.29. The reason for this behavior has already been mentioned: The allowed vibrational energies of the atoms are quantized. At absolute zero, all atoms would be in the vibrational ground state, and, as we raise the temperature, they remain locked in that same state, at least until k_BT is comparable with the level spacing. Thus, near absolute zero, the atoms' energies cannot change, and C is very small. Only when k_BT is appreciably larger than the vibrational spacing does C approach the classical value.

PHONONS

To conclude our discussion of lattice vibrations, we mention an important correction to the simple model discussed above. We have so far implied that the atoms vibrate independently of one another; that is, that the vibrations of any one atom about its equilibrium position are unaffected by the motion of any other atom. This is patently false, since the forces that restrain any one atom are the interatomic forces exerted by its neighbors, and no single atom can vibrate without forcing many others to vibrate as well.

To illustrate these ideas, we consider a one-dimensional classical "solid," comprising just two atoms, bound to each other and to the walls of its container by three identical springs, as in Fig. 17.30. Obviously, this has the property just discussed, that neither atom can vibrate without forcing the other to vibrate as well. The two simplest modes of vibration of this system are called **normal modes** and are illustrated in Fig. 17.31. In the first mode, the two atoms oscillate with equal amplitude, in the same direction, and exactly in phase. In the second, they oscillate with equal amplitude, but in opposite directions, exactly out of phase. Two important properties of these normal modes are (1) that they

---

* By the same argument, electrons in the lower bands contribute even less.

FIGURE 17.32   A one-dimensional solid comprising five atoms connected by springs. The arrows indicate a normal mode in which all of the atoms vibrate in phase, with various different amplitudes.

are independent, in the sense that either can occur without the other, and (2) that the most general possible motion is just a combination of these two simple motions. The two normal modes have different classical frequencies given by (Problem 17.46)

$$\omega_{c1} = \sqrt{\frac{k}{m}} \quad \text{and} \quad \omega_{c2} = \sqrt{\frac{3k}{m}}. \quad (17.15)$$

If we apply the Schrödinger equation to the same system, we get analogous results: The two atoms cannot move independently of one another, but we *can* build up the general solution in terms of solutions that are the quantum analogs of the two normal modes of Fig. 17.31.* In particular, the allowed energies of the system have the form

$$E = E_1 + E_2, \quad (17.16)$$

where E_1 denotes the allowed energies of mode 1, and E_2 those of mode 2. Moreover, the allowed energies of the two normal modes have precisely the familiar form for the quantum SHO:

$$E_1 = (n_1 + \tfrac{1}{2})\hbar\omega_{c1} \quad \text{and} \quad E_2 = (n_2 + \tfrac{1}{2})\hbar\omega_{c2} \quad (17.17)$$

where n_1 and n_2 are any two integers, 0, 1, 2, ...

To make our model a little more realistic, we could imagine replacing the two atoms of Fig. 17.30 by a large number N, bound in a long chain by identical springs. We have illustrated this (for the admittedly rather small number $N = 5$) in Fig. 17.32. Just as in the case $N = 2$, the individual atoms cannot vibrate independently; however, there are again certain normal modes — N of them in this case — which *are* independent. We have illustrated one simple normal mode in Fig. 17.32. In this mode — as in all normal modes — all of the atoms vibrate simultaneously at the same frequency.† On a macroscopic scale this kind of *collective oscillation* is just what we would describe as an *acoustic wave* in the solid. (Other names are *sound wave* and *elastic wave*.) We see that the microscopic oscillations of the atoms are, from a macroscopic point of view, just the acoustic, or elastic, vibrations of the solid.

If we apply the Schrödinger equation to the same system of N atoms, we find that every allowed energy is the sum of N terms, each of which is just the energy of one of the normal modes and has the familiar form

$$E = (n + \tfrac{1}{2})\hbar\omega_c \quad (17.18)$$

(where the classical frequency ω_c is, in general, different for different modes).

* In the language introduced in Chapter 9, we can describe the situation more precisely, as follows: In terms of the coordinates x_1 and x_2 of the two atoms, the Schrödinger equation does not *separate*, but in terms of the coordinates $y_1 = (x_1 + x_2)/2$ and $y_2 = x_2 - x_1$, which describe the normal modes of Fig. 17.31, the equation *does* separate (see Problem 17.47).

† This statement needs to be qualified a little: In certain normal modes, the amplitude of oscillation of some atoms can be zero; that is, there can be modes where not *all* of the atoms vibrate. However, it is easy to show that at least half the atoms vibrate in every mode.

The picture that emerges of the lattice vibrations is surprisingly similar to the simpleminded picture in which we treated the N atoms as if they could vibrate independently. In either view, the total energy of the lattice is the sum of N terms,* each with allowed values of the form (17.18) appropriate to an SHO. The main difference is that in our original picture these energies were supposed to be the energies of the individual vibrating atoms, whereas we now recognize that they are the energies of the collective, acoustical vibrations of the whole lattice.

An important consequence of (17.18) is that the energy of any lattice vibration can change only in multiples of $\hbar\omega_c$. This is reminiscent of electromagnetic waves, whose energy can only change by multiples of $\hbar\omega_{em}$ (as originally proposed by Planck — Chapter 5). In either case, there is a basic quantum of energy, $\hbar\omega$. In the electromagnetic case, we call these basic quanta *photons*. By analogy we call the quanta of acoustic vibrations in a solid **phonons** (a combination of "phone" and "photon").

The idea of phonons as the quanta of the lattice vibrations has proved surprisingly fruitful. For example, a conduction electron moving through a solid can lose energy when it collides with an atom of the lattice. (This process is in fact the origin of electrical resistance.) The energy lost by the electron goes into energy of the lattice vibrations, and we can view the process as the creation of a phonon by an electron, similar to the process in which a decelerating charge loses energy and creates a photon.

17.8 Superconductivity (optional)

To conclude this chapter we describe how some materials, when cooled below a certain critical temperature, become perfect conductors of electricity. Before we discuss these *superconductors,* we need to say a little more about ordinary conductors.

COLLISIONS AS THE CAUSE OF RESISTANCE

We saw in Section 17.5 that a conductor is characterized by a partially filled valence band. When an electric field is applied, the electrons in the highest occupied states can move into nearby unoccupied states, and a current flows. If these mobile electrons were completely free, they would continue to gain kinetic energy and produce an ever-increasing current. The reason this does *not* happen is that the electrons collide with the atoms of the lattice and lose energy. Each collision scatters an electron and randomizes its velocity, so that the process of acceleration has to start again. Thus the electrons undergo a regular succession of accelerations and randomizing collisions; on average, they move with a steady *drift velocity*† in the direction opposite to the applied field, and it is the drifting electrons that constitute the electric current.

The less frequently electrons collide with the lattice atoms, the larger the velocity that they can acquire between collisions. Thus solids in which the electrons suffer few collisions carry larger currents and hence have lower resistivity. In particular, if there were *no* collisions between the electrons and the lattice, the solid would have zero resistivity.

---

* This is for a one-dimensional solid. In three dimensions there are $3N$ terms.

† The name "drift velocity" suggests, correctly, that this average velocity is small. Typical values are less than 1 mm/s.

Although the discussion above was classical, its conclusion is very nearly correct. There is, however, one important difference that emerges if we treat the electrons properly, using quantum mechanics. Consider first an electron moving in a *perfectly regular* and *perfectly rigid* crystalline wire. If we were to solve the Schrödinger equation for this system, we would find the allowed energies and states of the electron. In some of these states the electron moves to the left, and in some it moves to the right. In the absence of an electric field, the N electrons occupy the N lowest states, and there are equal numbers traveling in any direction and hence no net current. When an electric field is applied to the right, some electrons can transfer to unoccupied states in which they move preferentially to the left. So far this account is very similar to the classical account, but we now find an important difference. The states of the electron result from solving the Schrödinger equation, which *includes* the effects of the crystalline lattice. Thus the presence of the lattice does *not* cause the electron to lose energy; if the lattice really were perfectly regular and rigid, there would be no energy-reducing collisions, the electron would move unscattered through the lattice, and the resistance would be zero.* In practice, of course, solids never are perfectly regular, nor are they perfectly rigid. Electrons can lose energy in collisions with the irregularities associated with crystal defects and impurities, and they can also lose energy in collisions with the lattice as it vibrates in its thermal motion.

The difference between the classical and quantum pictures of resistance is this: Classically, we would say that resistance comes from collisions of electrons with the *lattice;* according to quantum mechanics, resistance arises from collisions with *imperfections* in the lattice and the *vibrational motion* of the lattice. According to the classical picture, we would expect that an electron could travel only a few atomic diameters in the lattice before being scattered by one of the atoms. On the other hand, if we consider a nearly perfect metal at low temperature (to minimize vibration), then according to the quantum picture an electron should be able to travel many atomic diameters before being scattered by a lattice imperfection or vibration. Thus quantum mechanics predicts much lower resistivities than does the classical theory. The measured resistivities of pure metals at low temperature confirm the quantum view. In particular, it turns out that the electron's mean free path in a pure metal at low temperature can be hundreds of atomic diameters.

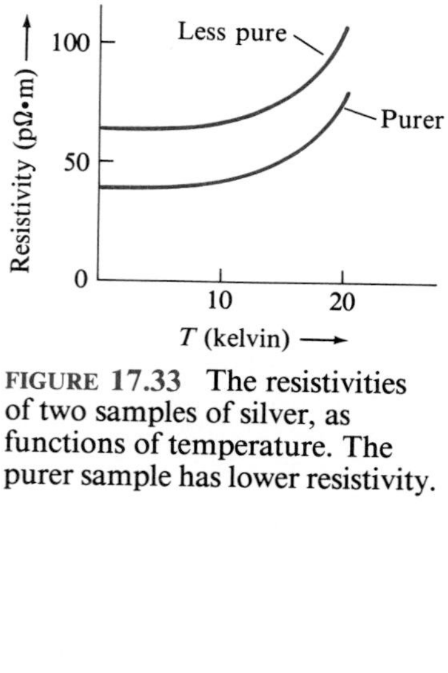

TEMPERATURE DEPENDENCE OF RESISTIVITY

As a conductor's temperature decreases, its thermal vibrations diminish, and its resistivity gets smaller.† However, even at absolute zero the electrons should still be scattered by the lattice imperfections and its zero-point vibrations. Thus as $T \to 0$, the resistivity should not approach zero. Instead, it should drop toward some nonzero value, which depends on the composition of the metal. This expected behavior is illustrated in Fig. 17.33, which shows the

FIGURE 17.33 The resistivities of two samples of silver, as functions of temperature. The purer sample has lower resistivity.

---

* To understand this surprising conclusion, it may help to draw a parallel with a light wave traveling through a perfect, transparent crystal. The presence of the crystal *does* affect the wave; for example, the wave speed may be less than in vacuum. Nevertheless, the wave travels unscattered and unattenuated through the crystal.

† By contrast the resistivity of a semiconductor *increases* as T drops, because the number of conduction electrons decreases.

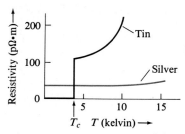

FIGURE 17.34 Resistivity of tin, as a function of temperature. At the critical temperature T_c the resistivity drops abruptly to zero. Below T_c the resistivity is exactly zero, and we say the material is superconducting. For comparison, the colored curve shows the behavior of the normal conductor silver.

resistivities of two different samples of silver. Both curves decrease as $T \rightarrow 0$, but the purer sample has less residual resistivity at $T = 0$, as we would expect.

SUPERCONDUCTIVITY

Although many conductors show the expected behavior of Fig. 17.33, the Dutch physicist Kamerlingh Onnes discovered in 1911 that some behave in a dramatically different and wholly unexpected manner. These materials behave normally down to a certain **critical temperature** T_c, which is different for different materials, but at T_c their resistivity drops abruptly to zero. For temperatures below T_c their resistivity is exactly zero, and we say that the material is in the **superconducting state**. Once a current is established in a superconducting circuit, it will continue indefinitely without any EMF to maintain it. The abrupt transition to the superconducting state is illustrated in Fig. 17.34, which shows the resistivity of tin as a function of T.

The known superconductors include many metals (tin, aluminum, lead, etc.), with critical temperatures all below about 10 K ($T_c = 9$ K for niobium). Several metal compounds (NbTi, PbMoS, etc.) are also superconductors and have somewhat higher critical temperatures, up to about 23 K (for Nb_3Ge). The potential applications of materials with no resistance are obvious and numerous. (For example, there would be no heat losses in power lines of zero resistance.) We shall describe a few successful applications shortly, but the need to keep the circuits at such low temperatures has meant that all such applications are expensive and specialized. For this reason, there was great excitement in 1986 when the physicists Bednorz and Müller at the IBM laboratory in Zurich discovered the first of a class of superconductors with higher critical temperatures. These new **high-temperature superconductors** are all layered crystals including planes of copper oxide, and some have critical temperatures higher than 100 K. Since liquid nitrogen has a temperature of 77 K and is relatively inexpensive, these new materials can be maintained in their superconducting state relatively easily and cheaply. Therefore, there has been a tremendous effort to find ways to manufacture the new superconductors conveniently and to discover still more of them with even higher critical temperatures. So far, practical applications are limited to uses involving thin films. Applications involving thicker cross sections (as in an ordinary wire) are not practical, because the new materials are very brittle and cease to be superconducting at rather low current densities.

MAGNETIC PROPERTIES OF SUPERCONDUCTORS

As if their electrical behavior were not striking enough, superconductors also have remarkable magnetic properties. Some of these magnetic properties are automatic consequences of the electrical properties. Since they have zero resistance, the electric field E in superconductors is always zero. (Otherwise, there would be an infinite current.) Now, from Faraday's law we know that any change in the magnetic field B necessarily produces an electric field E. Therefore, if E is always zero, it follows that B cannot change; that is, B must be constant in a superconductor.

Suppose, for example, that we place a superconductor in a field-free region (so that $B = 0$) and then bring up a magnet. Since B inside the superconductor was initially zero, and since it cannot change, B has to *remain* zero. The way that this happens is easy to see: By Faraday's law, the approaching magnet induces currents in the superconductor, and these currents induce a *B* field which exactly cancels that of the original magnet. This explains one version of the celebrated levitation demonstration shown in Fig. 17.35. Lowering a magnet toward a superconductor induces currents in the superconductor, and the magnetic field of these currents holds up the original magnet. Since the material is superconducting, the currents persist, and the magnet remains levitated indefinitely.

So far we have argued that the magnetic field in a superconductor cannot change. In 1933, Meissner discovered that the magnetic field in a superconductor is always *zero*. Thus, if a superconductor above its critical temperature T_c is put in a magnetic field and is then cooled below T_c, it ejects all *B* fields as it becomes superconducting. This **Meissner effect** is the basis of a second form of the levitation experiment of Fig. 17.35: Suppose that the magnet is placed *on* the superconductor above T_c and the system is then cooled. As the temperature passes T_c, currents appear in the superconductor (to eject the *B* field), and these currents lift the magnet up and hold it levitated above the superconductor.

FIGURE **17.35** A magnet remains levitated above one of the new superconductors in a dish of liquid nitrogen. The magnet can either be lowered toward the superconductor, which is already below T_c; or it can be placed on the superconductor, which is above T_c but is then cooled below it. In the first case the experiment demonstrates Faraday's law, in the second, the Meissner effect. (Courtesy Colorado Superconductors Inc.)

THEORIES OF SUPERCONDUCTIVITY

It was not until 1957 that a satisfactory theory of superconductivity was found. In that year the American physicists Bardeen, Cooper, and Schrieffer published what is now called the BCS theory. According to this theory the electrons in a superconductor form into pairs, called *Cooper pairs*. This comes about because any electron moving through the lattice attracts the atomic cores, causing a region of higher positive charge. A second electron can be attracted by this positive charge, and under the right conditions, this attraction can actually overcome the Coulomb repulsion of the original electron.

It turns out that once the Cooper pairs have formed, lattice imperfections and vibrations can only scatter them by dissociating them, which requires an energy of order 10^{-3} eV. Below a certain temperature (of order 10 K) thermal energies are insufficient to do this. Therefore, the Cooper pairs cannot be scattered by the lattice, and the resistivity of the solid is zero.

The BCS theory successfully accounts for the properties of all "low-temperature" superconductors. For example, a curious feature of these superconductors is that those metals that are the best *normal* conductors (such as copper and silver) exhibit *no* superconductivity. This is because the formation of Cooper pairs requires a relatively strong interaction between electrons and the lattice, and the good normal conductors are precisely the materials in which the electrons interact most weakly with the lattice.

The superconductivity of the new high-temperature superconductors is believed to be due to the formation of pairs, as in the BCS theory. However, in this case, the pairs are apparently *not* caused by the same electron–lattice interaction as in the BCS theory. Just what does cause the formation of pairs is not known and is the subject of intense research.

APPLICATIONS

The complete disappearance of resistance and the resulting absence of ohmic heating suggest several obvious uses for superconductors. Perhaps the most successful application to date has been to use superconducting wire to carry the large currents needed in high-field electromagnets. Such superconducting magnets are widely used in modern particle accelerators (Section 14.11). They are also used to produce the stable, large-volume magnetic fields needed for magnetic-resonance imaging in medicine.

Another application under study is the reduction of losses in power lines that deliver electricity from power plants. At present, the cost of refrigerating superconducting wires outweighs the potential saving, but new superconductors with higher critical temperatures (and hence lower refrigeration costs) may make such applications feasible in the future.

Another area of study is at a far smaller scale. The minaturization of computers is limited by the heat generated in the conductors linking the semiconductor elements. (A high density of these elements implies a high density of heat generation, resulting in excessive temperatures.) Use of superconducting connections could eliminate this problem.

All of the applications mentioned so far take advantage of the superconductor's zero resistance. A possible application of the magnetic properties of superconductors is to low friction bearings whose surfaces could be held apart by the Meissner effect, as in the levitation shown in Fig. 17.35.*

Still another class of applications of superconductivity involves the *Josephson effect*. This quantum phenomenon causes an alternating current to flow between two superconductors separated by a thin insulator when a small dc voltage is applied. Since the frequency of the alternating current is very sensitive to the applied voltage and to any applied magnetic field, the Josephson effect can be used to measure voltages accurately and to measure extremely small magnetic fields (the latter using the so-called *superconducting quantum interference device,* or *SQUID*). The Josephson effect is also the basis for several high-precision measurements of the fundamental constants.

IDEAS YOU SHOULD NOW UNDERSTAND FROM CHAPTER 17

Types of bonds: covalent, metallic, ionic, dipole–dipole

Cohesive energy

Crystalline lattices: cubic, FCC, BCC, HCP

Madelung constant

Amorphous solids

Long- and short-range order

Bands and band gaps

Conductors, insulators, and semiconductors

Holes

Impurity semiconductors; *n*- and *p*-type

The *n-p* diode and some applications

Transistors

[Optional sections: lattice vibrations; the Dulong–Petit law; normal modes; phonons; causes of resistance; superconductivity; critical temperature; the Meissner effect]

---

* An application that is frequently mentioned is the levitation of trains by superconducting magnets. However, this is not an application of the Meissner effect; rather, it is just another application of the superconductor's zero resistance, and there are magnetically levitated trains that operate with ordinary (nonsuperconducting) magnets.

SECTION 17.2 (BONDING OF SOLIDS)

17.1 • From the data in Table 17.1 find the cohesive energies of diamond and germanium in J/mol and kcal/mol.

17.2 • From the data in Table 17.1 find the cohesive energies of NaCl and LiF in J/mol and kcal/mol. (Be careful! How many atoms are contained in a mole?)

17.3 • The force between two electric dipoles (each of zero net charge) results from the imperfect cancellation of the various attractive and repulsive forces among the four charges involved. (a) Prove that the force between the two dipoles in Fig. 17.36(a) is repulsive. (b) What is it for the dipoles in Fig. 17.36(b)?

(a)   (b)

FIGURE 17.36 (Problem 17.3)

17.4 • • (a) Prove that the force between two dipoles with two like charges closest as in Fig. 17.37(a) is repulsive. (b) What is it if unlike charges are closest as in Fig. 17.37(b)?

(a)   (b)

FIGURE 17.37 (Problem 17.4)

17.5 • • The van der Waals force between two atoms arises because the fluctuating dipole moment of atom 1 induces a dipole moment in atom 2, and this induced moment is always oriented so as to experience an attractive force from atom 1. This point is illustrated in Fig. 17.2, for one orientation of the dipoles. (a) Sketch the electric field lines of dipole 1 in Fig. 17.2 and verify that the induced dipole 2 is oriented as shown. (b) Choose axes in the usual way (x across and y up the page) and put the origin at dipole 1. If atom 2 is moved onto the y axis, what is the orientation of its induced dipole moment, and what is the direction of the force on it? (c) Repeat for the case that atom 2 is put halfway between the x and y axes.

SECTION 17.3 (CRYSTALS AND NONCRYSTALS)

17.6 • One can think of a simple cubic lattice [Fig. 17.5(b)] as a stack of many unit cubes [Fig. 17.5(c)]. However, in counting atoms there is a danger of miscounting, since adjacent cubes *share* several atoms. (a) Consider any one atom in the lattice. How many cubes share this atom? (b) Consider any one cube. How many atoms does the cube have a share of? (c) Prove that in the lattice as a whole, there is one atom per unit cube.

17.7 • Consider the Na^+ ion at the center of the cube shown in Fig. 17.6(b). (a) In terms of the nearest-neighbor distance r_0, find the distance of any one of the next-nearest-neighbor ions. (b) How many of these ions are there? (c) Are they Na^+ or Cl^-?

17.8 • Consider the Na^+ ion at the center of the cube shown in Fig. 17.6(b). (a) In terms of the nearest-neighbor distance r_0, find the distance of any one of the third-nearest-neighbor ions. (b) How many of these ions are there? (c) Are they Na^+ or Cl^-?

17.9 • The density of solid NaCl is 2.16 g/cm³. Find the nearest-neighbor distance. (*Hint:* The number density of ions is 1 ion per r_0^3.)

17.10 • The KCl crystal has the same FCC structure as NaCl, and the nearest-neighbor distance is 0.315 nm. What is the density of KCl? (*Hint:* The number density of ions is 1 ion per r_0^3.)

17.11 • • In Example 17.1 [Eq. (17.2)] we estimated the potential energy of any one ion in solid NaCl. (a) Make a corresponding estimate of the PE of any one ion in KCl, whose crystal structure is the same as NaCl, but whose nearest-neighbor separation is 0.31 nm. (b) This estimate ignores the repulsive forces that come into play when the ions get very close. Thus your answer should be somewhat below the observed value of -7.2 eV. Use this observed value to find the atomic cohesive energy of KCl in eV/atom. (The ionization energy of K is 4.3 eV; the electron affinity of Cl is 3.6 eV.)

17.12 • • Consider the Na^+ ion at the center of the cube in Fig. 17.6(b). (a) How many fourth-nearest neighbors does it have? (b) Are they Na^+ or Cl^-? (c) Where are they? (d) What is their distance?

17.13 • • We saw in Example 17.1 that the PE of any one ion in NaCl has the form $U = -\alpha ke^2/r_0$, where $\alpha = 1.75$ is called the Madelung constant. The calculation of this constant in three dimensions is much more trouble than it is worth here, but the *one*-dimensional Madelung constant is fairly easy to calculate, as follows: Consider a long one-dimensional "crystal" of alternating Na^+ and Cl^- ions spaced a distance r_0 apart. Prove that the PE of any one ion is $U = -\alpha ke^2/r_0$ where $\alpha = 1.39$. [*Hint:* The series for $\ln(1 + z)$ in Appendix B (with $z = 1$) is useful in this calculation.]

17.14 •• Extend the reasoning of Problem 17.6 to prove that the number of atoms per unit cell of the BCC lattice is 2.

17.15 •• Extend the reasoning of Problem 17.6 to prove that the number of atoms per unit cell of the FCC lattice is 4.

17.16 •• Gold has the FCC structure, and its density is 19.3 g/cm³. (*a*) Use the result of Problem 17.15 to find the edge length of the unit cube. (*b*) What is the nearest-neighbor separation (center to center as usual)?

17.17 •• Copper has the FCC structure, and the center-to-center distance between nearest-neighbor atoms is 0.255 nm. Use this information and the result of Problem 17.15 to find the density of solid copper.

17.18 •• Iron has the BCC structure, and its density is 7.86 g/cm³. Use the result of Problem 17.14 to find the center-to-center distance between nearest-neighbor iron atoms.

17.19 •• In the CsCl crystal the Cs^+ ions and Cl^- ions are arranged on two identical simple cubic lattices. The Cl^- lattice is offset from the Cs^+ lattice, so that each Cl^- is at the exact center of a cube of Cs^+ ions, and vice versa. Thus a unit cell of CsCl looks just like the BCC cell in Fig. 17.7 (except that the center ion is Cl^-, whereas those at the corners are Cs^+, or vice versa). (*a*) The density of CsCl is 3.97 g/cm³. Use the result of Problem 17.14 to find the edge length of the unit cube of CsCl. (*b*) What is the nearest-neighbor distance?

17.20 •• (*a*) Imagine a large box full of identical rigid spheres packed in a simple cubic lattice and each touching all of its nearest neighbors. What fraction of the total volume is taken up by the spheres? (This fraction is called the *packing fraction*.) Answer the same question for (*b*) FCC and (*c*) BCC packing. (*d*) Which arrangement packs the most spheres per volume?

17.21 •• To estimate the tensile strength of the material in Example 17.2, we had to know the force needed to pull two atoms apart. To get an estimate of this force consider the semiclassical model of the H_2 molecule in Fig. 16.34 (Problem 16.11). Using the result $s = R_0/\sqrt{3}$ from Problem 16.11 and taking $R_0 = 0.4$ nm

as a representative lattice spacing, find the total force of one atom on the other.

17.22 •• Problem 17.21 suggests one way to estimate the force needed to separate two bonded atoms; here is another. The graph in Fig. 16.6 shows the energy of an Na^+–Cl^- pair as a function of their separation. Since the slope of this graph is the force between the ions, the maximum force needed to pull them apart is just the maximum slope of the graph. Use the graph to show that this force is of order 3×10^{-9} newton.

17.23 ••• In Example 17.1, Eq. (17.1), we wrote down the first three terms in the electrostatic PE of an ion in the NaCl lattice. What are the next two terms?

17.24 ••• In Example 17.1, Eq. (17.1), we said that the electrostatic PE of any one ion in NaCl is $U_{es} = -\alpha ke^2/r$, where $\alpha = 1.75$. This is the electrostatic energy of simple point charges, and we should have included a contribution due to the repulsion of the ions at close range. This repulsive contribution can be approximated by a term of the form $U_{rep} = +a/r^n$, where a and n are positive constants. Thus the total PE is

$$U(r) = U_{es}(r) + U_{rep}(r) = -\alpha \frac{ke^2}{r} + \frac{a}{r^n}. \quad (17.19)$$

(*a*) At the equilibrium separation $r = r_0$, this energy is a minimum. Use this fact to find the constant a, and hence show that the total PE of an ion at equilibrium is

$$U(r_0) = -\alpha \frac{ke^2}{r_0} \left(1 - \frac{1}{n}\right).$$

(*b*) This result differs from our rough estimate (17.2) only by the second term $1/n$. For NaCl, $r_0 = 0.281$ nm and $n = 8$. Use these values to find the total PE of an ion in NaCl. Compare with the rough answer of -9.0 eV found in Example 17.1 and the observed value of -7.92 eV.

SECTION **17.4** (ENERGY LEVELS OF ELECTRONS IN SOLIDS; BANDS)

17.25 • The density of solid silicon is 2.33 g/cm³. (*a*) Estimate the total number, N, of silicon atoms in a microcrystal 1 μm $\times$ 1 μm $\times$ 1 μm. (*b*) The valence band in silicon is about 1 eV wide. Given that there are $4N$ states in this band, make a rough estimate of the average separation between states in the valence band of this crystal. (*c*) Consider a silicon solar cell that is 1 cm $\times$ 1 cm $\times$ 1 mm; repeat both calculations for these dimensions.

17.26 • The density of solid germanium is 5.32 g/cm³. (*a*) Estimate the total number, N, of atoms in a microcrystal 1 μm $\times$ 1 μm $\times$ 1 μm. (*b*) The conduction band in germanium is about 2 eV wide. Given that there are $4N$ states in this band, make a rough estimate of the average separation between states in the conduction band of this crystal. (*c*) Consider a germanium crystal that is 1 cm $\times$ 1 cm $\times$ 1 cm; repeat both calculations for these dimensions.

17.27 • Imagine a crystal containing just one type of atom on a simple cubic lattice. If the lattice spacing is 0.2 nm and the whole crystal is 3 mm $\times$ 3 mm $\times$ 3 mm, how many states are there in the 2*p* band?

17.28 • Consider a solid containing N atoms and a band that evolved from an atomic level of angular momentum *l*. How many states does this band contain? Check that your answer gives $2N$ for an *s* band and $6N$ for a *p* band.

SECTIONS **17.5** AND **17.6** (CONDUCTORS AND INSULATORS AND SEMICONDUCTORS)

17.29 • Consider a single crystal of an insulator with a band gap of 3.5 eV. What colors of visible light will this crystal transmit?

17.30 • Crystalline sulfur absorbs blue light but no other colors; as a result, it is a pale yellow transparent solid. It is also an electrical insulator. From this information deduce the width of the band gap of crystalline sulfur, and describe the occupancy of the bands above and below the gap.

17.31 • Germanium is a semiconductor with a band gap of 0.8 eV. What colors of visible light does it transmit?

17.32 • Explain why many insulators are transparent to visible light, whereas most semiconductors are opaque to visible light but transparent to infrared.

17.33 • The longest wavelength of electromagnetic radiation that is absorbed in gallium arsenide is 890 nm. What is the band gap of this semiconductor?

17.34 • What is the longest-wavelength radiation that can produce a current in a photovoltaic cell which consists of a silicon *n-p* diode with band gap 1.1 eV?

17.35 • As a nuclear particle traverses a particle detector, it creates electron–hole pairs and loses energy. The average energy lost per pair created is 3.1 eV. (This is much bigger than the band gap of 1.1 eV because many of the excited electrons come from deep down in the valence band, and other processes — like inelastic scattering — take some energy.) A particle with initial kinetic energy of 2 MeV enters the detector and is brought to rest. (*a*) How many electron–hole pairs does it create? (*b*) What total charge flows around the circuit? (See Section 13.11 for more details.)

17.36 • (*a*) If a crystal of pure silicon is doped with atoms of phosphorus, what type of semiconductor results, *n*

or *p*? (*b*) What if the same crystal were doped with aluminum? (*c*) With antimony?

17.37 • (*a*) If a crystal of pure germanium is doped with atoms of boron, what type of semiconductor results, *n* or *p*? (*b*) What if the same crystal were doped with indium? (*c*) With arsenic?

17.38 •• Consider a hypothetical element that forms a solid with bands as shown in Fig. 17.38. (*a*) Suppose that the isolated atom has configuration $1s^2 2s^2$. If its equilibrium separation is $r_0 = a$, is the solid a conductor or insulator? What if $r_0 = b$? Answer both questions for the case that the atomic configuration is (*b*) $1s^2 2s^2 2p^1$ and (*c*) $1s^2 2s^2 2p^6$.

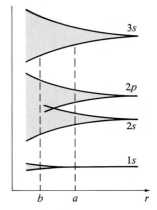

FIGURE 17.38 (Problem 17.38)

17.39 •• It can be shown that the current flowing through an ideal *n-p* diode is

$$I = I_0(e^{Ve/k_B T} - 1), \qquad (17.20)$$

where I_0 is a constant, characteristic of the diode, V is the applied voltage (positive in the forward direction, negative in the backward), k_B is Boltzmann's constant, and T is the absolute temperature. (*a*) Prove that for large negative voltages the current is $I = -I_0$. (This means there *is* a current in the reverse direction, but it is very small, as you will show now.) (*b*) Consider forward and backward voltages of 1 volt ($V = \pm 1$ volt). Taking $T = 293$ K (room temperature), write down the corresponding forward and backward currents and compute their ratio.

17.40 •• Using Eq. (17.20) (Problem 17.39) sketch the current I through an *n-p* junction as a function of the applied voltage V. (Show both positive and negative values of V. Your curve will show exponential growth on the right; in practice, unavoidable resistance eventually limits this growth to the familiar linear growth corresponding to Ohm's law.)

17.41 •• We saw that when germanium is doped with an atom of arsenic, the fifth valence electron of As is very weakly bound and hence is easily excited into the conduction band of the solid. The main reason for this extremely weak binding is that the electron is orbiting the As^+ ion (charge $+e$), not in a vacuum, but in solid Ge, which has a dielectric constant $\kappa \approx$ 16. (a) Consider an electron orbiting a charge $+e$ in a medium whose dielectric constant is κ. The Coulomb force on the electron is $F = ke^2/(\kappa r^2)$; that is, we have to replace the constant k used in vacuum by k/κ. Use either the Bohr model or the Schrödinger equation to argue that the allowed energies are $E_n = -E_R/(\kappa n)^2$. Note that with $\kappa \approx 16$ these energies are some 256 times smaller than those of hydrogen. (b) By what factor is the "Bohr radius" increased as compared to ordinary hydrogen?

17.42 •• Consider a current of 3 mA flowing through an LED whose band gap is about 2 eV. (a) How many photons are produced per second? (b) What is the power of the light produced? Notice that this small current produces a light that is easily visible. (For comparison, a typical flashlight bulb, which is very easily visible, consumes about a watt of power, of which about 50 mW is visible light). (c) Show in general that this power (in watts) is numerically just the product of the current (in amperes) and the band gap (in eV).

SECTION **17.7** (LATTICE VIBRATIONS AND PHONONS)

17.43 • The argument leading to the Dulong–Petit law treated the lattice vibrations classically. This would be expected to break down once the thermal energy ($\sim k_B T$) is comparable to, or less than, the quantized level spacing $\hbar \omega_c$. Given that the latter is of order 0.01 eV, estimate the temperature below which the Dulong–Petit law should fail. Compare your answer with Fig. 17.29.

17.44 •• The Dulong–Petit law predicts that the molar specific heats of elemental solids should all have the same value, $C = 3R$. On the other hand, the ordinary specific heats c, defined as the energy per degree per *unit mass,* can be quite different. (a) Prove that the specific heats, c, of elemental solids obeying the Dulong–Petit law are inversely proportional to the atomic masses. (b) Given the following data, check the accuracy of this prediction.

solid:	aluminum	silver	gold
c [cal/(g·K)]:	0.214	0.0558	0.0312

17.45 •• (a) Assuming their molar specific heats are given by the Dulong–Petit law, predict the specific heats, in cal/(g·K), of Al, Ag, and Au. (b) Find the percent discrepancies between your predictions and the measured values given in Problem 17.44.

17.46 •• Consider the one-dimensional "solid" of Fig. 17.30. Let x_1 and x_2 denote the displacements of the two atoms from their equilibrium positions, and let k be the force constant of the springs and m the mass of either atom. (a) Assuming that the atoms are vibrating in mode 1 of Fig. 17.31 ($x_1 = x_2$), write down Newton's second law for atom 1 and for atom 2. Show that the two equations are identical and that the angular frequency of this mode is $\omega_{c1} = \sqrt{k/m}$. (b) Do the same for mode 2 ($x_1 = -x_2$), but show that in this case the frequency is $\omega_{c2} = \sqrt{3k/m}$.

17.47 ••• Consider the one-dimensional "solid" of Fig. 17.30, in which the two atoms have mass m and the three springs have force constant k. Let x_1 and x_2 denote the displacements of the two atoms from their equilibrium positions. (a) Write down the equations of motion for x_1 and x_2 (that is, Newton's second law for each mass). Note that each equation involves *both* x_1 and x_2; this reflects the nonindependence of the motions of the two atoms. (b) Rewrite these two equations in terms of $y_1 = (x_1 + x_2)/2$ and $y_2 = x_2 - x_1$. Note that the equations of motion for y_1 and y_2 *are* independent. (c) What are the frequencies with which y_1 and y_2 can oscillate? (d) Describe the motion of the two atoms in the case that y_1 oscillates while y_2 remains zero. Repeat for the case that $y_1 = 0$ while y_2 oscillates. Compare with the normal modes of Fig. 17.31.

17.48 ••• We claimed that the level spacing $\hbar \omega_c$ for the vibrational motion of the atoms in a solid lattice is of order 0.01 eV. Verify this claim as follows: The PE of a single ion in NaCl is well approximated by Eq. (17.19) in Problem 17.24. (a) Differentiate this expression to find the force on the ion. (b) Use the fact that the force is zero at equilibrium ($r = r_0$) to eliminate the constant a. (c) Show that if $r = r_0 + x$, where x is a small displacement from equilibrium, the force has the form $F \approx -Kx$, where $K = \alpha ke^2(n - 1)/r_0^3$. (d) This means the ion oscillates like an SHO. Write down an expression for the classical frequency ω_c of a chlorine ion, and then estimate the spacing $\hbar \omega_c$ of its quantized levels.* (Values of α, n, and r_0 are given in Problem 17.24.)

* This argument uses several approximations: We have assumed that a single ion can oscillate while the others remain fixed —which we know is not really possible. Further, the expression (17.19) for $U(r)$ was derived for a solid in which *all* ions have moved to a common spacing r. Nevertheless, the estimate should, and does, give the right order of magnitude.

A P P E N D I C E S

Physical Constants

NAMED CONSTANTS

Most constants are given first with three significant figures, since this is almost always sufficiently accurate. The last two columns give the mantissa of the best known value and its uncertainty in parts per million. For example, in the first entry, the best known value of the atomic mass unit is

$$1 \text{ u} = 1.660540 \times 10^{-27} \text{ kg}$$

with an uncertainty of 0.6 part per million (so that the final zero is uncertain by about 1). All values were taken from E. R. Cohen and B. N. Taylor, *Reviews of Modern Physics,* vol. 59, p. 1139 (1987).

Name	Symbol and value	Best mantissa	Uncertainty (ppm)
Atomic mass unit	$1 \text{ u} = \frac{1}{12} m(^{12}\text{C atom})$	exact	
	$= 1.66 \times 10^{-27} \text{ kg}$	1660540	0.6
	$= 931.5 \text{ MeV}/c^2$	9314943	0.3
Avogadro's constant	$N_A = 6.02 \times 10^{23} \text{ particles/mol}$	6022137	0.6
Bohr magneton	$\mu_B = e\hbar/(2m_e)$		
	$= 5.79 \times 10^{-5} \text{ eV/T}$	57883826	0.09
	$= 9.27 \times 10^{-24} \text{ J/T (or A·m}^2)$	9274015	0.3
Bohr radius	$a_B = \hbar^2/(ke^2 m_e)$		
	$= 5.29 \times 10^{-11} \text{ m}$	52917725	0.04
Boltzmann's constant	$k_B = 8.62 \times 10^{-5} \text{ eV/K}$	861738	8
	$= 1.38 \times 10^{-23} \text{ J/K}$	138066	8
Coulomb force constant	$k = 1/(4\pi\epsilon_0) = \mu_0 c^2/(4\pi)$	exact	
	$= 8.99 \times 10^9 \text{ N·m}^2/\text{C}^2$	898755178 . . .	
Electron Compton wavelength	$\lambda_c = h/(m_e c)$		
	$= 2.43 \times 10^{-12} \text{ m}$	24263106	0.09
Electron volt	$1 \text{ eV} = 1.60 \times 10^{-19} \text{ J}$	16021773	0.3

Name	Symbol and value	Best mantissa	Uncertainty (ppm)
Elementary charge	$e = 1.60 \times 10^{-19}$ C	16021773	0.3
Fine-structure constant	$\alpha = ke^2/(\hbar c)$ $= 7.30 \times 10^{-3} \approx 1/137$	72973531	0.04
Gas constant	$R = 8.31$ J/(mol·K) $= 0.0821$ liter·atm/(mol·K)	831451 820578	8 8
Gravitational constant	$G = 6.67 \times 10^{-11}$ N·m²/kg²	66726	130
Mass of electron	$m_e = 5.49 \times 10^{-4}$ u $= 9.11 \times 10^{-31}$ kg $= 0.511$ MeV/c^2	54857990 9109390 5109991	0.02 0.6 0.3
Mass of proton	$m_p = 1.007$ u $= 1.673 \times 10^{-27}$ kg $= 938.3$ MeV/c^2	100727647 1672623 9382723	0.01 0.6 0.3
Mass of neutron	$m_n = 1.009$ u $= 1.675 \times 10^{-27}$ kg $= 939.6$ MeV/c^2	100866490 1674929 9395656	0.01 0.6 0.3
Nuclear magneton	$\mu_N = e\hbar/(2m_p)$ $= 3.15 \times 10^{-8}$ eV/T $= 5.05 \times 10^{-27}$ J/T	31524517 5050787	0.09 0.3
Permeability of space	$\mu_0 = 4\pi \times 10^{-7}$ N/A² $= 1.26 \times 10^{-6}$ N/A²	exact 125663706 . . .	
Permittivity of space	$\epsilon_0 = 1/(\mu_0 c^2)$ $= 8.85 \times 10^{-12}$ C²/(N·m²)	exact 885418781 . . .	
Planck's constants	$h = 6.63 \times 10^{-34}$ J·s $= 4.14 \times 10^{-15}$ eV·s	6626076 4135669	0.6 0.3
	$\hbar = h/(2\pi)$ $= 1.05 \times 10^{-34}$ J·s $= 6.58 \times 10^{-16}$ eV·s	10545727 6582122	0.6 0.3
Rydberg constant	$R = m_e k^2 e^4/(4\pi c \hbar^3)$ $= 1.10 \times 10^{-2}$ nm⁻¹	1097373153	0.001
Rydberg energy	$E_R = hcR = m_e k^2 e^4/(2\hbar^2)$ $= 13.6$ eV	13605698	0.3
Speed of light	$c = 3.00 \times 10^8$ m/s	299792458	exact

USEFUL COMBINATIONS

	Best mantissa	Uncertainty (ppm)

$$hc = 1240 \text{ eV}\cdot\text{nm} = 1240 \text{ MeV}\cdot\text{fm}$$ — 12398424 — 0.3

$$\hbar c = 197 \text{ eV}\cdot\text{nm} = 197 \text{ MeV}\cdot\text{fm}$$ — 19732705 — 0.3

$$ke^2 = 1.44 \text{ eV}\cdot\text{nm} = 1.44 \text{ MeV}\cdot\text{fm}$$ — 14399652 — 0.3

$$N_A \times (1 \text{ u}) = 1 \text{ gram}$$ — exact

$$k_B T = 0.026 \text{ eV at room temperature (300 K)}$$

CONVERSION FACTORS

Area:

$$1 \text{ barn} = 10^{-28} \text{ m}^2 \qquad \text{exact}$$

Energy:

$$1 \text{ cal} = 4.184 \text{ J} \qquad \text{exact}$$
$$1 \text{ eV} = 1.60 \times 10^{-19} \text{ J} \qquad 16021773 \quad 0.3$$

Length:

$$1 \text{ Å} = 1 \text{ angstrom}$$
$$= 10^{-10} \text{ m} \qquad \text{exact}$$

$$1 \text{ ft} = 30.48 \text{ cm} \qquad \text{exact}$$
$$1 \text{ in} = 2.54 \text{ cm} \qquad \text{exact}$$
$$1 \text{ mi} = 1609 \text{ m} \qquad 1609344 \quad \text{exact}$$

Mass:

$$1 \text{ lb(mass)} = 0.454 \text{ kg} \qquad 45359237 \quad \text{exact}$$
$$1 \text{ MeV}/c^2 = 1.07 \times 10^{-3} \text{ u} \qquad 10735438 \quad 0.3$$
$$= 1.78 \times 10^{-30} \text{ kg} \qquad 17826627 \quad 0.3$$

$$1 \text{ u} = \tfrac{1}{12} m(^{12}\text{C atom}) \qquad \text{exact}$$
$$= 931.5 \text{ MeV}/c^2 \qquad 9314943 \quad 0.3$$
$$= 1.66 \times 10^{-27} \text{ kg} \qquad 1660540 \quad 0.6$$

Momentum:

$$1 \text{ MeV}/c = 5.34 \times 10^{-22} \text{ kg}\cdot\text{m/s} \qquad 5344288 \quad 0.3$$

SI PREFIXES

T	tera	10^{12}	c	centi	10^{-2}	p	pico	10^{-12}
G	giga	10^{9}	m	milli	10^{-3}	f	femto	10^{-15}
M	mega	10^{6}	μ	micro	10^{-6}	a	atto	10^{-18}
k	kilo	10^{3}	n	nano	10^{-9}			

B

Useful Mathematical Relations

TRIGONOMETRIC IDENTITIES

$$\sin(\theta + \phi) = \sin\theta\cos\phi + \sin\phi\cos\theta$$

$$\cos(\theta + \phi) = \cos\theta\cos\phi - \sin\theta\sin\phi$$

$$\sin\theta\cos\phi = \frac{\sin(\theta + \phi) + \sin(\theta - \phi)}{2}$$

$$\cos\theta\cos\phi = \frac{\cos(\theta + \phi) + \cos(\theta - \phi)}{2}$$

$$\sin\theta\sin\phi = \frac{\cos(\theta - \phi) - \cos(\theta + \phi)}{2}$$

$$\cos^2\theta = \frac{1 + \cos 2\theta}{2}$$

$$\sin^2\theta = \frac{1 - \cos 2\theta}{2}$$

$$\cos\theta + \cos\phi = 2\cos\frac{\theta + \phi}{2}\cos\frac{\theta - \phi}{2}$$

$$\cos\theta - \cos\phi = 2\sin\frac{\theta + \phi}{2}\sin\frac{\phi - \theta}{2}$$

$$\sin\theta + \sin\phi = 2\sin\frac{\theta + \phi}{2}\cos\frac{\theta - \phi}{2}$$

$$\sin^2\theta + \cos^2\theta = 1$$

$$\sec^2\theta - \tan^2\theta = 1$$

LAW OF COSINES

$$C^2 = A^2 + B^2 - 2AB\cos\theta,$$

where A, B, C are the three sides of a triangle and θ is the angle opposite C.

COMPLEX NUMBERS

If $z = x + iy$, where x and y are real numbers, then

$$|z| = \sqrt{x^2 + y^2} = \text{absolute value (or modulus) of } z$$

and

$$z^* = x - iy = \text{complex conjugate of } z.$$

Hence

$$|z|^2 = z^* z.$$

COMPLEX TRIGONOMETRIC RELATIONS

$$e^{i\theta} = \cos\theta + i\sin\theta \qquad \textbf{(Euler's relation)}$$

$$\cos\theta = \frac{e^{i\theta} + e^{-i\theta}}{2}$$

$$\sin\theta = \frac{e^{i\theta} - e^{-i\theta}}{2i}$$

EXPANSIONS

Any function $f(z)$ that is analytic near the point $z = a$ can be expanded in a Taylor series,

$$f(z) = f(a) + f'(a)(z - a) + f''(a)\frac{(z - a)^2}{2!} + \cdots \qquad \textbf{(Taylor's series)}$$

for z near a. Important special cases of this expansion are:

$$e^z = 1 + z + \frac{z^2}{2!} + \frac{z^3}{3!} + \cdots$$

$$\ln(1 + z) = z - \frac{z^2}{2} + \frac{z^3}{3} - \frac{z^4}{4} + \cdots$$

$$\cos z = 1 - \frac{z^2}{2!} + \frac{z^4}{4!} - \cdots$$

$$\sin z = z - \frac{z^3}{3!} + \frac{z^5}{5!} - \cdots$$

$$(1 + z)^n = 1 + nz + \frac{n(n - 1)}{2!}z^2 + \frac{n(n - 1)(n - 2)}{3!}z^3 + \cdots \qquad \textbf{(binomial series)}$$

In all of these expansions z can be real or complex. The series for e^z, $\cos z$, and $\sin z$ converge for all values of z; that for $\ln(1 + z)$ converges if $|z| < 1$ (also for $z = 1$), but diverges if $|z| > 1$. If n is a positive integer, the binomial series terminates after $n + 1$ terms (all subsequent terms are zero) and is therefore a polynomial of degree n; if n is not a positive integer, the binomial series converges if $|z| < 1$, but diverges if $|z| > 1$.

When z is small—$|z| \ll 1$—one obtains a good approximation using just the first one or two terms of these five series. In particular, with $|z| \ll 1$ the binomial series reduces to the binomial approximation:

$$(1 + z)^n \approx 1 + nz \qquad \textbf{(binomial approximation).}$$

Integrals of the form

$$I_n = \int_0^\infty x^n e^{-\lambda x^2} \, dx,$$

where λ is a positive number, occur frequently in several branches of physics. When n is a positive integer, their value can be found from the following.

$$I_0 = \sqrt{\frac{\pi}{4\lambda}}$$

$$I_1 = \frac{1}{2\lambda}$$

$$I_2 = \sqrt{\frac{\pi}{16\lambda^3}}$$

and

$$I_n = -\frac{dI_{n-2}}{d\lambda}.$$

Notice that the integral $\int_{-\infty}^\infty x^n e^{-\lambda x^2} \, dx$ equals $2I_n$ when n is even, but is zero if n is odd.

Another common integral is the indefinite integral

$$J_n = \int x^n e^{-x/b} \, dx.$$

When n is a small integer, this is easily evaluated by parts. For example,

$$J_0 = -be^{-x/b}$$
$$J_1 = -(b^2 + bx)e^{-x/b}$$
$$J_2 = -(2b^3 + 2b^2 x + bx^2)e^{-x/b}.$$

In general,

$$J_{n+1} = b^2 \frac{\partial J_n}{\partial b}.$$

Note, in particular, that

$$\int_0^\infty e^{-x/b} \, dx = b$$

$$\int_0^\infty x e^{-x/b} \, dx = b^2$$

and

$$\int_0^\infty x^2 e^{-x/b} \, dx = 2b^3.$$

C

Alphabetical List of the Elements

LIST BY NAMES

Name	Symbol	Z	Name	Symbol	Z	Name	Symbol	Z
Actinium	Ac	89	Helium	He	2	Radium	Ra	88
Aluminum	Al	13	Holmium	Ho	67	Radon	Rn	86
Americium	Am	95	Hydrogen	H	1	Rhenium	Re	75
Antimony	Sb	51	Indium	In	49	Rhodium	Rh	45
Argon	Ar	18	Iodine	I	53	Rubidium	Rb	37
Arsenic	As	33	Iridium	Ir	77	Ruthenium	Ru	44
Astatine	At	85	Iron	Fe	26	Rutherfordium	Unq*	104
Barium	Ba	56	Krypton	Kr	36	Samarium	Sm	62
Berkelium	Bk	97	Lanthanum	La	57	Scandium	Sc	21
Beryllium	Be	4	Lawrencium	Lr	103	Selenium	Se	34
Bismuth	Bi	83	Lead	Pb	82	Silicon	Si	14
Boron	B	5	Lithium	Li	3	Silver	Ag	47
Bromine	Br	35	Lutetium	Lu	71	Sodium	Na	11
Cadmium	Cd	48	Magnesium	Mg	12	Strontium	Sr	38
Calcium	Ca	20	Manganese	Mn	25	Sulfur	S	16
Californium	Cf	98	Mendelevium	Md	101	Tantalum	Ta	73
Carbon	C	6	Mercury	Hg	80	Technetium	Tc	43
Cerium	Ce	58	Molybdenum	Mo	42	Tellurium	Te	52
Cesium	Cs	55	Neodymium	Nd	60	Terbium	Tb	65
Chlorine	Cl	17	Neon	Ne	10	Thallium	Tl	81
Chromium	Cr	24	Neptunium	Np	93	Thorium	Th	90
Cobalt	Co	27	Nickel	Ni	28	Thulium	Tm	69
Copper	Cu	29	Niobium	Nb	41	Tin	Sn	50
Curium	Cm	96	Nitrogen	N	7	Titanium	Ti	22
Dysprosium	Dy	66	Nobelium	No	102	Tungsten	W	74
Einsteinium	Es	99	Osmium	Os	76	Unnilhexium	Unh*	106
Erbium	Er	68	Oxygen	O	8	Unnilpentium	Unp*	105
Europium	Eu	63	Palladium	Pd	46	Unnilquadium	Unq*	104
Fermium	Fm	100	Phosphorus	P	15	Unnilseptium	Uns*	107
Fluorine	F	9	Platinum	Pt	78	Uranium	U	92
Francium	Fr	87	Plutonium	Pu	94	Vanadium	V	23
Gadolinium	Gd	64	Polonium	Po	84	Xenon	Xe	54
Gallium	Ga	31	Potassium	K	19	Ytterbium	Yb	70
Germanium	Ge	32	Praseodymium	Pr	59	Yttrium	Y	39
Gold	Au	79	Promethium	Pm	61	Zinc	Zn	30
Hafnium	Hf	72	Protactinium	Pa	91	Zirconium	Zr	40
Hahnium	Unp*	105						

* The names and symbols for elements 104 to 107 are disputed. The International Union of Pure and Applied Chemistry has recommended unnilquadium (Unq), unnilpentium (Unp), unnilhexium (Unh), and unnilseptium (Uns). In the United States, elements 104 and 105 are sometimes called rutherfordium (Rf) and hahnium (Ha).

LIST BY SYMBOLS

Symbol	Name	Z	Symbol	Name	Z	Symbol	Name	Z
Ac	Actinium	89	Ha	Hahnium*	105	Pu	Plutonium	94
Ag	Silver	47	He	Helium	2	Ra	Radium	88
Al	Aluminum	13	Hf	Hafnium	72	Rb	Rubidium	37
Am	Americium	95	Hg	Mercury	80	Re	Rhenium	75
Ar	Argon	18	I	Iodine	53	Rf	Rutherfordium*	104
As	Arsenic	33	In	Indium	49	Rh	Rhodium	45
At	Astatine	85	Ir	Iridium	77	Rn	Radon	86
Au	Gold	79	Ho	Holmium	67	Ru	Ruthenium	44
B	Boron	5	K	Potassium	19	S	Sulfur	16
Ba	Barium	56	Kr	Krypton	36	Sb	Antimony	51
Be	Beryllium	4	La	Lanthanum	57	Sc	Scandium	21
Bi	Bismuth	83	Li	Lithium	3	Se	Selenium	34
Bk	Berkelium	97	Lr	Lawrencium	103	Si	Silicon	14
Br	Bromine	35	Lu	Lutetium	71	Sm	Samarium	62
C	Carbon	6	Md	Mendelevium	101	Sn	Tin	50
Ca	Calcium	20	Mg	Magnesium	12	Sr	Strontium	38
Cd	Cadmium	48	Mn	Manganese	25	Ta	Tantalum	73
Ce	Cerium	58	Mo	Molybdenum	42	Tb	Terbium	65
Cf	Californium	98	N	Nitrogen	7	Tc	Technetium	43
Cl	Chlorine	17	Na	Sodium	11	Te	Tellurium	52
Cm	Curium	96	Nb	Niobium	41	Th	Thorium	90
Co	Cobalt	27	Nd	Neodymium	60	Ti	Titanium	22
Cr	Chromium	24	Ne	Neon	10	Tl	Thallium	81
Cs	Cesium	55	Ni	Nickel	28	Tm	Thulium	69
Cu	Copper	29	No	Nobelium	102	U	Uranium	92
Dy	Dysprosium	66	Np	Neptunium	93	Unh	Unnilhexium*	106
Er	Erbium	68	O	Oxygen	8	Unp	Unnilpentium*	105
Es	Einsteinium	99	Os	Osmium	76	Unq	Unnilquadium*	104
Eu	Europium	63	P	Phosphorus	15	Uns	Unnilseptium*	107
F	Fluorine	9	Pa	Protactinium	91	V	Vanadium	23
Fe	Iron	26	Pb	Lead	82	W	Tungsten	74
Fm	Fermium	100	Pd	Palladium	46	Xe	Xenon	54
Fr	Francium	87	Pm	Promethium	61	Y	Yttrium	39
Ga	Gallium	31	Po	Polonium	84	Yb	Ytterbium	70
Gd	Gadolinium	64	Pr	Praseodymium	59	Zn	Zinc	30
Ge	Germanium	32	Pt	Platinum	78	Zr	Zirconium	40
H	Hydrogen	1						

* The names and symbols for elements 104 to 107 are disputed. The International Union of Pure and Applied Chemistry has recommended unnilquadium (Unq), unnilpentium (Unp), unnilhexium (Unh), and unnilseptium (Uns). In the United States, elements 104 and 105 are sometimes called rutherfordium (Rf) and hahnium (Ha).

D

Atomic and Nuclear Data

In the following pages we list data for the ground states of all of the 252 known stable nuclei. Since more than 2000 unstable nuclei are known, we have included only a small selection of these. Specifically, we have included (1) all principal members of the natural radioactive series, (2) at least the two longest-lived unstable isotopes of each element, and (3) any unstable nuclei that are mentioned in the text or problems. Much more complete lists can be found in: J. K. Tuli, *Nuclear Wallet Cards** (National Nuclear Data Center, 1985), from which the data here are taken with permission; C. M. Lederer and V. S. Shirley, *Table of Isotopes,* 7th ed. (New York: Wiley, 1978); or the *Chart of the Nuclides* (General Electric Company).†

Data are listed by element, in order of atomic number Z; for each value of Z, they are given in order of mass number A. For each element, the first line lists the following:

1. Z = atomic number = number of protons in nucleus
2. Chemical symbol
3. $\overline{m}$ = chemical atomic mass in atomic mass units
4. Name of element

On the succeeding lines we list a selection of isotopes, including all those that are stable. Each line gives the following:

1. A = mass number (color = unstable)
2. m = mass of neutral atom in atomic mass units
3. j = quantum number for total angular momentum of nucleus
4. Percent abundance or half-life (and decay modes).

Notes:
1. The chemical atomic mass $\overline{m}$ is the average mass of the atoms in the

---

* This convenient booklet can be obtained from the National Nuclear Data Center, Brookhaven National Laboratory, Upton, NY 11973.

† This wall chart is available from General Electric Company, 175 Curtner Avenue, Mail Code 684, San Jose, CA 95125.

normal, natural mixture of isotopes on earth. For most unstable elements this average is not well defined.

2. All masses are given in atomic mass units, defined so that a neutral atom of carbon 12 has mass 12 u exactly.

3. The names of the elements beyond $Z = 103$ are disputed.

4. A mass number printed in color indicates that the nucleus is unstable. Be aware, however, that the distinction between stable and unstable nuclei is sometimes hard to draw, since an unstable nucleus whose half-life is sufficiently long may appear to be stable. In particular, as techniques for detecting nuclear decays improve, some nuclei that were considered stable are found to be unstable but with extremely long half-lives. Thus several nuclei that, until recently, were considered stable are listed here as unstable; for the same reason, some nuclei shown here as stable may eventually prove to be unstable.

5. The percent abundances are the percent of atoms of each isotope found in normal terrestrial sources.

6. In listing half-lives we have used the following abbreviations: ns = nanosecond, μs = microsecond, ms = millisecond, m = minute, d = day, y = year, ky = 10^3 y, My = 10^6 y, Gy = 10^9 y. For half-lives outside the range from ns to Gy we use exponential notation; for example, 6E−15s means 6×10^{-15} seconds and 7E+16y means 7×10^{16} years. The symbol $\geq$ means that the given value is a lower limit; for example, $\geq$4E+17y means that $t_{1/2}$ is at least 4×10^{17} years, and may be more.

7. A few very long-lived unstable atoms occur naturally with well-defined abundances. In such cases we list both abundance and half-life.

8. Decay modes are listed in order of decreasing importance. Those that contribute less than 1% are omitted. We use the following abbreviations: ec = electron capture, p = proton emission, sf = spontaneous fission, $2\beta^-$ = double β decay. (This rare decay mode involves emission of two electrons and two antineutrinos.) A few of the decay modes listed are known, or suspected, theoretically but have not actually been observed. If β^+ decay is possible, electron capture is always possible as well; but the converse is not true. Therefore, we use the following convention: β^+ means that β^+ decay and electron capture are both possible; ec means that electron capture is possible but not β^+ decay.

9. With one exception, all data refer to the nuclear ground state. The one exception is tantalum 180, one of whose excited states is very long lived ($t_{1/2} \geq 10^{15}$ years) and is a detectable constituent of the normal terrestrial mixture of isotopes.

Z	Symbol	$\overline{m}$	Name	
	A	m	j	% or $t_{1/2}$(mode)
0	n		Neutron	
	1	1.008 665	1/2	10.2m($\beta-$)
1	H	1.008	Hydrogen	
	1	1.007 825	1/2	99.985%
	2	2.014 102	1	0.015%
	3	3.016 049	1/2	12.3y($\beta-$)

Z	Symbol	$\overline{m}$	Name	
2	He	4.003	Helium	
	3	3.016 029	1/2	0.0001%
	4	4.002 603	0	99.9999%
	6	6.018 886	0	807ms($\beta-$)
	8	8.033 922	0	119ms($\beta-$)
3	Li	6.941	Lithium	
	6	6.015 121	1	7.5%
	7	7.016 003	3/2	92.5%
	8	8.022 486	2	838ms($\beta-$)
	9	9.026 789	3/2	178ms($\beta-$)
	11	11.043 908	1/2	8.7ms($\beta-$)

Z	Symbol	$\overline{m}$	Name	
	A	m	j	% or $t_{1/2}$(mode)

4	Be	9.012	Beryllium	
	7	7.016 928	3/2	53.3d(ec)
	9	9.012 182	3/2	100%
	10	10.013 534	0	1.6My($\beta-$)
	11	11.021 658	1/2	13.8s($\beta-$)
	12	12.026 921	0	24.4ms($\beta-$)
5	B	10.811	Boron	
	8	8.024 606	2	770ms($\beta+$)
	10	10.012 937	3	19.9%
	11	11.009 305	3/2	80.1%
	12	12.014 353	1	20.2ms($\beta-$)
6	C	12.011	Carbon	
	11	11.011 433	3/2	20.4m($\beta+$)
	12	12 exactly	0	98.90%
	13	13.003 355	1/2	1.10%
	14	14.003 242	0	5730y($\beta-$)
7	N	14.007	Nitrogen	
	11	11.026 742	1/2	6E$-$22s(p)
	12	12.018 613	1	11.0ms($\beta+$)
	13	13.005 739	1/2	9.96m($\beta+$)
	14	14.003 074	1	99.634%
	15	15.000 109	1/2	0.366%
	16	16.006 100	2	7.13s($\beta-$)
8	O	15.999	Oxygen	
	12	12.034 418	0	1E$-$21s(p)
	14	14.008 595	0	70.6s($\beta+$)
	15	15.003 065	1/2	122s($\beta+$)
	16	15.994 915	0	99.762%
	17	16.999 131	5/2	0.038%
	18	17.999 160	0	0.200%
9	F	18.998	Fluorine	
	17	17.002 095	5/2	64.5s($\beta+$)
	18	18.000 937	1	110m($\beta+$)
	19	18.998 403	1/2	100%
10	Ne	20.179	Neon	
	20	19.992 436	0	90.51%
	21	20.993 843	3/2	0.27%
	22	21.991 383	0	9.22%
	23	22.994 465	5/2	37.2s($\beta-$)
	24	23.993 612	0	3.38m($\beta-$)
11	Na	22.990	Sodium (Natrium)	
	22	21.994 434	3	2.60y($\beta+$)
	23	22.989 768	3/2	100%
	24	23.990 961	4	15.0h($\beta-$)
12	Mg	24.305	Magnesium	
	23	22.994 124	3/2	11.3s($\beta+$)
	24	23.985 042	0	78.99%
	25	24.985 837	5/2	10.00%
	26	25.982 594	0	11.01%
	27	26.984 341	1/2	9.46m($\beta-$)
	28	27.983 877	0	20.9h($\beta-$)
13	Al	26.982	Aluminum	
	26	25.986 892	5	720ky($\beta+$)
	27	26.981 539	5/2	100%
	29	28.980 446	5/2	6.56m($\beta-$)
14	Si	28.086	Silicon	
	28	27.976 927	0	92.23%
	29	28.976 495	1/2	4.67%
	30	29.973 770	0	3.10%
	31	30.975 362	3/2	2.62h($\beta-$)
	32	31.974 148	0	105y($\beta-$)
15	P	30.974	Phosphorus	
	29	28.981 802	1/2	4.14s(ec)
	31	30.973 762	1/2	100%
	32	31.973 907	1	14.3d($\beta-$)
	33	32.971 725	1/2	25.3d($\beta-$)
16	S	32.066	Sulfur	
	32	31.972 071	0	95.02%
	33	32.971 458	3/2	0.75%
	34	33.967 867	0	4.21%
	35	34.969 032	3/2	87.5d($\beta-$)
	36	35.967 081	0	0.02%
	38	37.971 163	0	2.84h($\beta-$)
17	Cl	35.453	Chlorine	
	35	34.968 853	3/2	75.77%
	36	35.968 307	2	301ky($\beta-,\beta+$)
	37	36.965 903	3/2	24.23%
	39	38.968 004	3/2	55.6m($\beta-$)
18	Ar	39.948	Argon	
	36	35.967 546	0	0.337%
	38	37.962 733	0	0.063%
	39	38.964 314	7/2	269y($\beta-$)
	40	39.962 384	0	99.600%
	42	41.963 049	0	32.9y($\beta-$)
19	K	39.098	Potassium (Kalium)	
	39	38.963 707	3/2	93.258%
	40	39.963 999	4	0.012% 1.28Gy($\beta-,\beta+$)
	41	40.961 825	3/2	6.730%
	43	42.960 717	3/2	22.3h($\beta-$)

Z	Symbol	$\overline{m}$	Name		
	A	**m**		**j**	**% or $t_{1/2}$(mode)**
20	Ca	40.078	Calcium		
	40	39.962 591		0	96.941%
	41	40.962 278		7/2	103ky(ec)
	42	41.958 618		0	0.647%
	43	42.958 766		7/2	0.135%
	44	43.955 481		0	2.086%
	45	44.956 185		7/2	164d($\beta-$)
	46	45.953 690		0	0.004%
	47	46.954 543		7/2	4.54d($\beta-$)
	48	47.952 533		0	0.187%
					$\geq$2E+16y(2$\beta-$)
21	Sc	44.956	Scandium		
	41	40.969 250		7/2	596ms($\beta+$)
	45	44.955 910		7/2	100%
	46	45.955 170		4	83.8d($\beta-$)
	47	46.952 409		7/2	3.34d($\beta-$)
22	Ti	47.88	Titanium		
	44	43.959 690		0	54.2y(ec)
	45	44.958 124		7/2	3.08h($\beta+$)
	46	45.952 629		0	8.0%
	47	46.951 764		5/2	7.3%
	48	47.947 947		0	73.8%
	49	48.947 871		7/2	5.5%
	50	49.944 792		0	5.4%
23	V	50.941	Vanadium		
	49	48.948 517		7/2	330d(ec)
	50	49.947 161		6	0.250%
					1.5E+17y($\beta+,\beta-$)
	51	50.943 962		7/2	99.750%
24	Cr	51.996	Chromium		
	48	47.954 033		0	21.6h($\beta+$)
	50	49.946 046		0	4.345%
	51	50.944 768		7/2	27.7d(ec)
	52	51.940 510		0	83.789%
	53	52.940 651		3/2	9.501%
	54	53.938 883		0	2.365%
25	Mn	54.938	Manganese		
	53	52.941 291		7/2	3.7My(ec)
	54	53.940 361		3	312d($\beta+$)
	55	54.938 047		5/2	100%
	56	55.938 907		3	2.58h($\beta-$)
26	Fe	55.847	Iron (Ferrum)		
	54	53.939 613		0	5.8%
	55	54.938 296		3/2	2.68y(ec)
	56	55.934 939		0	91.72%
	57	56.935 396		1/2	2.2%
	58	57.933 277		0	0.28%
	60	59.934 077		0	1.49My($\beta-$)
27	Co	58.933	Cobalt		
	56	55.939 841		4	78.8d($\beta+$)
	57	56.936 294		7/2	271d(ec)
	59	58.933 198		7/2	100%
	60	59.933 820		5	5.27y($\beta-$)
28	Ni	58.69	Nickel		
	56	55.942 134		0	6.10d($\beta+$)
	58	57.935 346		0	68.27%
	59	58.934 349		3/2	75ky($\beta+$)
	60	59.930 788		0	26.10%
	61	60.931 058		3/2	1.13%
	62	61.928 346		0	3.59%
	63	62.929 670		1/2	100y($\beta-$)
	64	63.927 968		0	0.91%
29	Cu	63.546	Copper		
	63	62.929 599		3/2	69.17%
	64	63.929 766		1	12.7h($\beta+,\beta-$)
	65	64.927 793		3/2	30.83%
	67	66.927 748		3/2	61.9h($\beta-$)
30	Zn	65.39	Zinc		
	64	63.929 145		0	48.6%
	65	64.929 243		5/2	244d($\beta+$)
	66	65.926 035		0	27.9%
	67	66.927 129		5/2	4.1%
	68	67.924 846		0	18.8%
	70	69.925 325		0	0.6%
	72	71.926 856		0	46.5h($\beta-$)
31	Ga	69.723	Gallium		
	67	66.928 204		3/2	3.26d(ec)
	69	68.925 580		3/2	60.1%
	71	70.924 701		3/2	39.9%
	72	71.926 365		3	14.1h($\beta-$)
32	Ge	72.59	Germanium		
	68	67.928 097		0	271d(ec)
	70	69.924 250		0	20.5%
	71	70.924 954		1/2	11.8d(ec)
	72	71.922 079		0	27.4%
	73	72.923 463		9/2	7.8%
	74	73.921 177		0	36.5%
	76	75.921 402		0	7.8%

Z	Symbol	$\overline{m}$	Name	
	A	m	j	% or $t_{1/2}$(mode)
33	As	74.922	Arsenic	
	73	72.923 827	3/2	80.3d(ec)
	74	73.923 928	2	17.8d($\beta+,\beta-$)
	75	74.921 594	3/2	100%
34	Se	78.96	Selenium	
	74	73.922 475	0	0.9%
	76	75.919 212	0	9.0%
	77	76.919 913	1/2	7.6%
	78	77.917 308	0	23.6%
	79	78.918 498	7/2	65ky($\beta-$)
	80	79.916 520	0	49.7%
	82	81.916 698	0	9.2%
				1.4E+20y($2\beta-$)
35	Br	79.904	Bromine	
	77	76.921 377	3/2	57.0h($\beta+$)
	79	78.918 336	3/2	50.69%
	81	80.916 289	3/2	49.31%
	82	81.916 802	5	35.3h($\beta-$)
36	Kr	83.80	Krypton	
	78	77.920 396	0	0.35%
	80	79.916 380	0	2.25%
	81	80.916 590	7/2	210ky(ec)
	82	81.913 483	0	11.6%
	83	82.914 135	9/2	11.5%
	84	83.911 507	0	57.0%
	85	84.912 532	9/2	10.7y($\beta-$)
	86	85.910 615	0	17.3%
	90	89.919 528	0	32.3s($\beta-$)
	92	91.926 270	0	1.85s($\beta-$)
37	Rb	85.468	Rubidium	
	83	82.915 143	5/2	86.2d(ec)
	85	84.911 794	5/2	72.165%
	87	86.909 187	3/2	27.835%
				48Gy($\beta-$)
38	Sr	87.62	Strontium	
	84	83.913 430	0	0.56%
	85	84.912 937	9/2	64.8d($\beta+$)
	86	85.909 267	0	9.86%
	87	86.908 884	9/2	7.00%
	88	87.905 619	0	82.58%
	90	89.907 738	0	28.6y($\beta-$)
39	Y	88.906	Yttrium	
	88	87.909 507	4	107d($\beta+$)
	89	88.905 849	1/2	100%
	91	90.907 303	1/2	58.5d($\beta-$)

Z	Symbol	$\overline{m}$	Name	
40	Zr	91.224	Zirconium	
	90	89.904 703	0	51.45%
	91	90.905 644	5/2	11.22%
	92	91.905 039	0	17.15%
	93	92.906 474	5/2	1.53My($\beta-$)
	94	93.906 315	0	17.38%
	96	95.908 275	0	2.80%
				$\geq$4E+17y($\beta-$)
41	Nb	92.906	Niobium	
	92	91.907 192	7	35My($\beta+$)
	93	92.906 377	9/2	100%
	94	93.907 281	6	20.3ky($\beta-$)
42	Mo	95.94	Molybdenum	
	92	91.906 809	0	14.84%
	93	92.906 813	5/2	3.5ky(ec)
	94	93.905 085	0	9.25%
	95	94.905 841	5/2	15.92%
	96	95.904 679	0	16.68%
	97	96.906 020	5/2	9.55%
	98	97.905 407	0	24.13%
	99	98.907 711	1/2	66.0h($\beta-$)
	100	99.907 476	0	9.63%
43	Tc		Technetium	
	97	96.906 364	9/2	2.6My(ec)
	98	97.907 215	6	4.2My($\beta-$)
44	Ru	101.07	Ruthenium	
	96	95.907 600	0	5.52%
	98	97.905 287	0	1.88%
	99	98.905 939	5/2	12.7%
	100	99.904 219	0	12.6%
	101	100.905 582	5/2	17.0%
	102	101.904 349	0	31.6%
	103	102.906 323	3/2	39.3d($\beta-$)
	104	103.905 424	0	18.7%
	106	105.907 322	0	372d($\beta-$)
45	Rh	102.906	Rhodium	
	101	100.906 159	1/2	3.3y(ec)
	102	101.906 814	6	2.9y($\beta+$)
	103	102.905 500	1/2	100%
46	Pd	106.42	Palladium	
	102	101.905 634	0	1.020%
	103	102.906 114	5/2	17.0d(ec)
	104	103.904 029	0	11.14%
	105	104.905 079	5/2	22.33%
	106	105.903 478	0	27.33%
	107	106.905 127	5/2	6.5My($\beta-$)
	108	107.903 895	0	26.46%
	110	109.905 167	0	11.72%

Z	Symbol	$\overline{m}$	Name		
	A	m		j	% or $t_{1/2}$(mode)
47	Ag	107.868	Silver (Argentum)		
	105	104.906 521		1/2	41.3d($\beta+$)
	107	106.905 092		1/2	51.839%
	109	108.904 756		1/2	48.161%
	111	110.905 295		1/2	7.45d($\beta-$)
48	Cd	112.41	Cadmium		
	106	105.906 461		0	1.25%
	108	107.904 176		0	0.89%
	109	108.904 953		5/2	463d(ec)
	110	109.903 005		0	12.49%
	111	110.904 182		1/2	12.80%
	112	111.902 757		0	24.13%
	113	112.904 400		1/2	12.22%
					9E+15y($\beta-$)
	114	113.903 357		0	28.73%
	116	115.904 755		0	7.49%
49	In	114.82	Indium		
	111	110.905 109		9/2	2.83d(ec)
	113	112.904 061		9/2	4.3%
	115	114.903 882		9/2	95.7%
					4E+14y($\beta-$)
50	Sn	118.710	Tin (Stannum)		
	112	111.904 827		0	0.97%
	114	113.902 784		0	0.65%
	115	114.903 347		1/2	0.36%
	116	115.901 747		0	14.53%
	117	116.902 956		1/2	7.68%
	118	117.901 609		0	24.22%
	119	118.903 311		1/2	8.58%
	120	119.902 199		0	32.59%
	122	121.903 440		0	4.63%
	123	122.905 722		11/2	129d($\beta-$)
	124	123.905 274		0	5.79%
	126	125.907 653		0	100ky($\beta-$)
51	Sb	121.75	Antimony (Stibium)		
	121	120.903 821		5/2	57.3%
	123	122.904 216		7/2	42.7%
	124	123.905 938		3	60.2d($\beta-$)
	125	124.905 252		7/2	2.73y($\beta-$)
52	Te	127.60	Tellurium		
	120	119.904 047		0	0.096%
	122	121.903 050		0	2.60%
	123	122.904 271		1/2	0.908%
					1.3E+13y(ec)
	124	123.902 818		0	4.816%
	125	124.904 429		1/2	7.14%
	126	125.903 310		0	18.95%
	128	127.904 463		0	31.69%
					$\geq$8E+24y(2$\beta-$)
	130	129.906 229		0	33.80%
					3E+21y(2$\beta-$)
53	I	126.904	Iodine		
	125	124.904 620		5/2	60.1d(ec)
	127	126.904 473		5/2	100%
	129	128.904 986		7/2	15.7My($\beta-$)
54	Xe	131.29	Xenon		
	124	123.905 894		0	0.10%
	126	125.904 281		0	0.09%
	127	126.905 182		1/2	36.4d(ec)
	128	127.903 531		0	1.91%
	129	128.904 780		1/2	26.4%
	130	129.903 509		0	4.1%
	131	130.905 072		3/2	21.2%
	132	131.904 144		0	26.9%
	133	132.905 889		3/2	5.24d($\beta-$)
	134	133.905 395		0	10.4%
	136	135.907 213		0	8.9%
55	Cs	132.905	Cesium		
	133	132.905 429		7/2	100%
	135	134.905 885		7/2	3My($\beta-$)
	137	136.907 074		7/2	30.2y($\beta-$)
56	Ba	137.33	Barium		
	130	129.906 281		0	0.106%
	132	131.905 043		0	0.101%
	133	132.905 988		1/2	10.7y(ec)
	134	133.904 485		0	2.417%
	135	134.905 665		3/2	6.592%
	136	135.904 553		0	7.854%
	137	136.905 812		3/2	11.23%
	138	137.905 233		0	71.70%
	140	139.910 581		0	12.7d($\beta-$)
	142	141.916 361		0	10.6m($\beta-$)
	143	142.920 483		5/2	14.5s($\beta-$)
	144	143.922 845		0	11.4s($\beta-$)
57	La	138.905	Lanthanum		
	137	136.906 463		7/2	60ky(ec)
	138	137.907 106		5	0.09%
					128Gy($\beta+,\beta-$)
	139	138.906 347		7/2	99.91%

Z	Symbol	$\overline{m}$	Name		
	A	m		j	% or $t_{1/2}$(mode)
58	Ce	140.12	Cerium		
	136	135.907 139		0	0.19%
	138	137.905 985		0	0.25%
	139	138.906 631		3/2	138d(ec)
	140	139.905 433		0	88.48%
	142	141.909 241		0	11.08%
					$\geq$5E+16y(2β−)
	144	143.913 643		0	284d(β−)
59	Pr	140.908	Praseodymium		
	141	140.907 647		5/2	100%
	142	141.910 039		2	19.1h(β−)
	143	142.910 814		7/2	13.6d(β−)
60	Nd	144.24	Neodymium		
	142	141.907 720		0	27.13%
	143	142.909 810		7/2	12.18%
	144	143.910 084		0	23.80%
					2E+15y(α)
	145	144.912 570		7/2	8.30%
					$\geq$6E+16y(α)
	146	145.913 113		0	17.19%
	148	147.916 889		0	5.76%
	150	149.920 887		0	5.64%
					$\geq$1E+18y(2β−)
61	Pm		Promethium		
	145	144.912 743		5/2	17.7y(ec)
	146	145.914 708		3	5.53y(β+,β−)
62	Sm	150.36	Samarium		
	144	143.911 998		0	3.1%
	147	146.914 894		7/2	15.0%
					106Gy(α)
	148	147.914 819		0	11.3%
					7E+15y(α)
	149	148.917 180		7/2	13.8%
					$\geq$2E+15y(α)
	150	149.917 272		0	7.4%
	152	151.919 729		0	26.7%
	154	153.922 205		0	22.7%
63	Eu	151.96	Europium		
	151	150.919 847		5/2	47.8%
	152	151.921 742		3	13.3y(β+,β−)
	153	152.921 225		5/2	52.2%
	154	153.922 975		3	8.8y(β−)

Z	Symbol	$\overline{m}$	Name		
64	Gd	157.25	Gadolinium		
	150	149.918 663		0	1.79My(α)
	152	151.919 787		0	0.20%
					1.1E+14y(α)
	154	153.920 861		0	2.18%
	155	154.922 617		3/2	14.80%
	156	155.922 118		0	20.47%
	157	156.923 956		3/2	15.65%
	158	157.924 099		0	24.84%
	160	159.927 049		0	21.86%
65	Tb	158.925	Terbium		
	157	156.924 023		3/2	150y(ec)
	158	157.925 411		3	150y(β+,β−)
	159	158.925 342		3/2	100%
66	Dy	162.50	Dysprosium		
	154	153.924 428		0	3My(α)
	156	155.924 277		0	0.06%
	158	157.924 403		0	0.10%
	159	158.925 735		3/2	144d(ec)
	160	159.925 193		0	2.34%
	161	160.926 930		5/2	18.9%
	162	161.926 796		0	25.5%
	163	162.928 728		5/2	24.9%
	164	163.929 171		0	28.2%
67	Ho	164.930	Holmium		
	163	162.928 731		7/2	10y(ec)
	165	164.930 319		7/2	100%
	166	165.932 280		0	26.8h(β−)
68	Er	167.26	Erbium		
	162	161.928 774		0	0.14%
	164	163.929 198		0	1.61%
	166	165.930 290		0	33.6%
	167	166.932 046		7/2	22.95%
	168	167.932 368		0	26.8%
	169	168.934 588		1/2	9.40d(β−)
	170	169.935 461		0	14.9%
	172	171.939 353		0	49.3h(β−)
69	Tm	168.934	Thulium		
	169	168.934 212		1/2	100%
	170	169.935 798		1	129d(β−)
	171	170.936 426		1/2	1.92y(β−)

Z	Symbol	$\overline{m}$	Name	
	A	m	j	% or $t_{1/2}$(mode)
70	Yb	173.04	Ytterbium	
	168	167.933 894	0	0.13%
	169	168.935 186	7/2	32.0d(ec)
	170	169.934 759	0	3.05%
	171	170.936 323	1/2	14.3%
	172	171.936 378	0	21.9%
	173	172.938 207	5/2	16.12%
	174	173.938 859	0	31.8%
	175	174.941 273	7/2	4.19d($\beta-$)
	176	175.942 564	0	12.7%
71	Lu	174.967	Lutetium	
	174	173.940 336	1	3.31y($\beta+$)
	175	174.940 770	7/2	97.41%
	176	175.942 679	7	2.59%
				36Gy($\beta-$)
72	Hf	178.49	Hafnium	
	174	173.940 044	0	0.162%
				2E+15y(α)
	176	175.941 405	0	5.206%
	177	176.943 217	7/2	18.606%
	178	177.943 696	0	27.297%
	179	178.945 812	9/2	13.629%
	180	179.946 546	0	35.100%
	182	181.950 550	0	9My($\beta-$)
73	Ta	180.948	Tantalum	
	179	178.945 930	7/2	665d(ec)
	180	179.947 461	1	8.1h(ec,$\beta-$)
	180*	179.947 496	9	0.012%
	metastable			$\geq$1E+15y(ec, $\beta-$)
	181	180.947 993	7/2	99.988%
	182	181.950 148	3	114d($\beta-$)
74	W	183.85	Tungsten (Wolfram)	
	180	179.946 701	0	0.13%
				$\geq$1E+15y(α)
	182	181.948 202	0	26.3%
	183	182.950 220	1/2	14.3%
	184	183.950 929	0	30.67%
				$\geq$3E+17y(α)
	186	185.954 356	0	28.6%
75	Re	186.207	Rhenium	
	183	182.950 817	5/2	70.0d(ec)
	185	184.952 951	5/2	37.40%
	187	186.955 745G	5/2	62.60%
				50Gy($\beta-$)

Z	Symbol	$\overline{m}$	Name	
	A	m	j	% or $t_{1/2}$(mode)
76	Os	190.2	Osmium	
	184	183.952 489	0	0.02%
				$\geq$1E+17y(α)
	186	185.953 830	0	1.58%
				2E+15y(α)
	187	186.955 741	1/2	1.6%
	188	187.955 829	0	13.3%
	189	188.958 137	3/2	16.1%
	190	189.958 436	0	26.4%
	192	191.961 468	0	41.0%
77	Ir	192.22	Iridium	
	189	188.958 712	3/2	13.2d(ec)
	191	190.960 584	3/2	37.3%
	192	191.962 580	4	73.8d($\beta-,\beta+$)
	193	192.962 917	3/2	62.7%
78	Pt	195.08	Platinum	
	190	189.959 916	0	0.01%
				600Gy(α)
	192	191.961 019	0	0.79%
	193	192.962 977	1/2	50y(ec)
	194	193.962 655	0	32.9%
	195	194.964 765	1/2	33.8%
	196	195.964 927	0	25.3%
	198	197.967 869	0	7.2%
79	Au	196.967	Gold (Aurum)	
	195	194.965 012	3/2	186d(ec)
	196	195.966 543	2	6.18d($\beta+,\beta-$)
	197	196.966 543	3/2	100%
80	Hg	200.59	Mercury (Hydrargyrum)	
	194	193.965 391	0	520y(ec)
	196	195.965 807	0	0.14%
	198	197.966 743	0	10.02%
	199	198.968 254	1/2	16.84%
	200	199.968 300	0	23.13%
	201	200.970 277	3/2	13.22%
	202	201.970 617	0	29.80%
	203	202.972 848	5/2	46.6d($\beta-$)
	204	203.973 466	0	6.85%
81	Tl	204.383	Thallium	
	202	201.972 085	2	12.2d($\beta+$)
	203	202.972 320	1/2	29.524%
	204	203.973 839	2	3.78y($\beta-$,ec)
	205	204.974 400	1/2	70.476%
	207	206.977 404	1/2	4.77m($\beta-$)
	208	207.981 998	5	3.05m($\beta-$)
	209	208.985 333	1/2	2.20m($\beta-$)

Z	Symbol	$\overline{m}$	Name		
	A	m		j	% or $t_{1/2}$(mode)
82	Pb	207.2	Lead (Plumbum)		
	204	203.973 020		0	1.4%
					$\geq$ 1.4E+17y
	205	204.974 458		5/2	15My(ec)
	206	205.974 440		0	24.1%
	207	206.975 871		1/2	22.1%
	208	207.976 627		0	52.4%
	209	208.981 065		9/2	3.25h($\beta-$)
	210	209.984 163		0	22.3y($\beta-$)
	211	210.988 735		9/2	36.1m($\beta-$)
	212	211.991 871		0	10.6h($\beta-$)
	214	213.999 798		0	26.8m($\beta-$)
83	Bi	208.980	Bismuth		
	207	206.978 446		9/2	32.2y($\beta+$)
	208	207.979 717		5	368ky($\beta+$)
	209	208.980 374		9/2	100%
	210	209.984 096		1	5.01d($\beta-$)
	211	210.987 255		9/2	2.14m(α)
	212	211.991 255		1	60.6m($\beta-,\alpha$)
	213	212.994 360		9/2	45.6m($\beta-,\alpha$)
	214	213.998 691		1	19.9m($\beta-$)
84	Po		Polonium		
	192	191.991 443		0	34ms(α)
	194	193.988 180		0	0.7s(α)
	196	195.985 539		0	5.5s(α)
	207	206.981 570		5/2	5.80h($\beta+$)
	208	207.981 222		0	2.90y(α)
	209	208.982 404		1/2	102y(α)
	210	209.982 848		0	138d(α)
	211	210.986 627		9/2	516ms(α)
	212	211.988 842		0	0.298μs(α)
	213	212.992 833		9/2	4.2μs(α)
	214	213.995 176		0	164μs(α)
	215	214.999 418		9/2	1.78ms(α)
	216	216.001 888		0	0.15s(α)
	218	218.008 966		0	3.11m(α)
85	At		Astatine		
	207	206.985 733		9/2	1.80h($\beta+,\alpha$)
	210	209.987 126		5	8.1h($\beta+$)
	211	210.987 470		9/2	7.21h(ec,α)
	213	212.992 911		9/2	0.11μs(α)
	215	214.998 638		9/2	0.10ms(α)
	217	217.004 695		9/2	32.3ms(α)

Z	Symbol	$\overline{m}$	Name		
86	Rn		Radon		
	210	209.989 669		0	2.4h($\alpha,\beta+$)
	211	210.990 575		1/2	14.6h($\beta+,\alpha$)
	212	211.990 680		0	24m(α)
	213	212.993 856		9/2	25.0ms(α)
	214	213.995 339		0	0.27μs(α)
	219	219.009 478		5/2	3.96s(α)
	220	220.011 368		0	55.6s(α)
	222	222.017 571		0	3.82d(α)
87	Fr		Francium		
	220	220.012 293		1	27.4s(α)
	221	221.014 230		5/2	4.9m(α)
	222	222.017 563		2	14.4m($\beta-$)
	223	223.019 733		3/2	21.8m($\beta-$)
88	Ra		Radium		
	206	206.003 800		0	0.4s(α)
	216	216.003 509		0	182ns(α)
	218	218.007 118		0	14μs(α)
	220	220.011 004		0	23ms(α)
	222	222.015 353		0	38.0s(α)
	223	223.018 501		1/2	11.4d(α)
	224	224.020 186		0	3.66d(α)
	225	225.023 604		3/2	14.8d($\beta-$)
	226	226.025 403		0	1600y(α)
	228	228.031 064		0	5.75y($\beta-$)
89	Ac		Actinium		
	225	225.023 204		3/2	10.0d(α)
	227	227.027 750		3/2	21.8y($\beta-,\alpha$)
	228	228.031 014		3	6.13h($\beta-$)
	231	231.038 551		1/2	7.5m($\beta-$)
90	Th	232.038	Thorium		
	227	227.027 703		3/2	18.7d(α)
	228	228.028 715		0	1.91y(α)
	229	229.031 755		5/2	7340y(α)
	230	230.033 128		0	75.4ky(α)
	231	231.036 298		5/2	25.5h($\beta-$)
	232	232.038 051		0	100%
					14.0Gy(α)
	234	234.043 593		0	24.1d($\beta-$)
91	Pa		Protactinium		
	228	228.030 974		3	22h($\beta+,\alpha$)
	231	231.035 880		3/2	32.7ky(α)
	233	233.040 242		3/2	27.0d($\beta-$)
	234	234.043 303		4	6.70h($\beta-$)

Z	Symbol	$\overline{m}$	Name		
	A	m	j	% or $t_{1/2}$(mode)	

Z	Symbol	$\overline{m}$ / A	m	j	% or $t_{1/2}$(mode)
92	U	238.029	Uranium		
	226	226.029 168	0	0.5s(α)	
	231	231.036 264	5/2	4.2d(ec)	
	232	232.037 129	0	68.9y(α)	
	233	233.039 628	5/2	159ky(α)	
	234	234.040 947	0	0.006% 245ky(α)	
	235	235.043 924	7/2	0.720% 704My(α)	
	236	236.045 563	0	23.4My(α)	
	238	238.050 785	0	99.274% 4.47Gy(α)	
	239	239.054 290	5/2	23.5m($\beta-$)	
93	Np		Neptunium		
	231	231.038 239	5/2	48.8m($\beta+,\alpha$)	
	236	236.046 559	6	115ky(ec,$\beta-$)	
	237	237.048 168	5/2	2.14My(α)	
	239	239.052 933	5/2	2.36d($\beta-$)	
94	Pu		Plutonium		
	239	239.052 158	1/2	24.1ky(α)	
	242	242.058 737	0	376ky(α)	
	244	244.064 198	0	80.8My(α)	
95	Am		Americium		
	241	241.056 824	5/2	432y(α)	
	243	243.061 375	5/2	7380y(α)	
96	Cm		Curium		
	240	240.055 503	0	27d(α)	
	247	247.070 347	9/2	15.6My(α)	
	248	248.072 343	0	340ky(α,sf)	
	250	250.078 352	0	7400y(sf,α,$\beta-$)	
97	Bk		Berkelium		
	247	247.070 300	3/2	1380y(α)	
	248	248.073 107		9y(α,$\beta-$)	
98	Cf		Californium		
	249	249.074 845	9/2	350y(α)	
	251	251.079 579	1/2	898y(α)	
99	Es		Einsteinium		
	252	252.082 945		472d(α,$\beta+$)	
	254	254.088 019	7	276d(α)	
100	Fm		Fermium		
	253	253.085 173	1/2	3.00d(ec,α)	
	257	257.095 099	9/2	100d(α)	
101	Md		Mendelevium		
	257	257.095 577	7/2	5.2h(ec,α)	
	258	258.098 572	8	55d(α)	
102	No		Nobelium		
	255	255.093 258	1/2	3.1m(α,$\beta+$)	
	258	258.098 143	0	1.2ms(sf)	
	259	259.100 932	9/2	60m(α,ec)	
103	Lr		Lawrencium		
	256	256.098 486		28s(α,$\beta+$)	
	260	260.105 314		180s(α)	
104	Rf/Unq		Rutherford/Unnilquadium		
	257	257.102 941		4.8s(α,$\beta+$)	
	261	261.108 685		65s(α,sf,ec)	
105	Ha/Unp		Hahnium/Unnilpentium		
	258	258.109 018		4s(α,$\beta+$)	
	262	262.113 763		34s(sf,α,ec)	
106	Unh		Unnilhexium		
	263	263.118 218		0.8s(sf,α)	
107	Uns		Unnilseptium		
	262	262.122 931		115ms(α)	
108					
	264	264.128 964	0		
	265	265.130 155			
109					
	266	266.137 638		5ms(α)	

GENERAL REFERENCES

There are several books that cover about the same material as ours at about the same level. Among these we recommend

N. ASHBY AND S. C. MILLER, *Principles of Modern Physics.* San Francisco: Holden-Day, 1970.

K. W. FORD, *Classical and Modern Physics,* vol. 3. Lexington, MA: Xerox, 1974.

K. S. KRANE, *Modern Physics.* New York: Wiley, 1983.

R. A. SERWAY, C. J. MOSES, AND C. A. MOYER, *Modern Physics.* Philadelphia: Saunders, 1989.

P. A. TIPLER, *Modern Physics.* New York: Worth, 1978.

Three books with about the same coverage but at a more adanced level are

J. J. BREHM AND W. J. MULLIN, *Introduction to the Structure of Matter.* New York: Wiley, 1989.

R. EISBERG AND R. RESNICK, *Quantum Physics of Atoms, Molecules, Solids, Nuclei, and Particles.* New York: Wiley, 1985.

R. L. SPROULL AND W. A. PHILLIPS, *Modern Physics.* New York: Wiley, 1980.

Some good references on the history of modern physics are

I. ASIMOV, *Asimov's Biographical Encyclopedia of Science and Technology,* 2nd ed. Garden City, NY: Doubleday, 1982.

H. BOORSE AND L. MOTZ, *The World of the Atom,* 2 vol. New York: Basic Books, 1966. An excellent selection of original writings from the Greeks to the 1960s.

E. SEGRÈ, *From X Rays to Quarks.* San Francisco: Freeman, 1980.

G. L. TRIGG, *Landmark Experiments in 20th Century Physics.* New York: Crane Russak, 1975.

An excellent list of references that is continually growing is the series of *Resource Letters* in the *American Journal of Physics.* You can find the latest of these by looking in the annual index. Many of them are accompanied by a volume that includes reprints of several of the best references. (These reprint books are listed as "RB" in the *Products Catalog* of the American Association of Physics Teachers, 5112 Berwyn Rd, College Park, MD 20740.) A resource letter on the history of physics (with a reprint book, number RB-50) is

S. G. BRUSH, "History of Physics," Resource Letter HP1, *American Journal of Physics,* vol. 55, pp. 683–691 (1987).

The internationally agreed values for the fundamental constants are updated at irregular intervals (on the order of 10 years). The latest values, which appear in several places, can be found in

Handbook of Chemistry and Physics. Boca Raton, FL: CRC Press, which is updated every year and is a rich source of information.

REFERENCES FOR PART I, RELATIVITY

Some fine books devoted entirely to relativity, and hence considerably more complete than our account, are

A. P. FRENCH, *Special Relativity.* New York: Norton, 1960.

N. D. MERMIN, *Space and Time in Special Relativity.* New York: McGraw-Hill, 1968.

E. F. TAYLOR AND J. A. WHEELER, *Spacetime Physics.* Freeman, San Francisco, 1966

A more advanced book is

W. RINDLER, *Essentials of Relativity.* New York: Springer-Verlag, 1977.

Tests of general relativity are described in

C. M. WILL, *Was Einstein Right? Putting General Relativity to the Test.* New York: Basic Books, 1986.

REFERENCES FOR PART II, QUANTUM MECHANICS

There are many books on quantum mechanics at a slightly higher level than our treatment. Two particularly fine ones are

R. P. FEYNMAN, R. B. LEIGHTON, AND M. SANDS, *The Feynman Lectures on Physics,* Vol. 3. Reading, MA: Addison-Wesley, 1965. Supposedly an introductory text, but packed with wonderful insights and a favorite with all physicists.

A. P. FRENCH AND E. F. TAYLOR, *An Introduction to Quantum Physics.* New York: Norton, 1978.

Historical books on quantum theory and atomic physics are

G. GAMOW, *The Thirty Years That Shook Physics.* Garden City, NY: Doubleday, 1966. One of some 20 nontechnical books by a master of popular writing, who was himself an active contributor to modern physics.

R. A. MILLIKAN, *The Electron.* Chicago: University of Chicago Press, 1963. A facsimile copy of Millikan's 1917 book, with a historical introduction.

G. P. THOMSON, *J. J. Thomson, Discoverer of the Electron.* Garden City, NY: Doubleday, 1964. Biography of "JJ" by his son, the physicist G. P. Thomson.

REFERENCES FOR PART III, SUBATOMIC PHYSICS

More advanced books on nuclear and particle physics are

D. GRIFFITHS, *Introduction to Elementary Particles.* New York: Harper & Row, 1987.

K. Gottfried and V. F. Weisskopf, *Concepts of Particle Physics.* New York: Oxford University Press, 1984.

K. S. Krane, *Introductory Nuclear Physics.* New York: Wiley, 1987.

E. Segrè, *Nuclei and Particles.* Reading, MA: Addison-Wesley, 1977.

Resource Letters in this area are

O. W. Greenberg, "Quarks," Resource Letter Q1, *American Journal of Physics,* vol. 50, pp. 1074–1089 (1982).

L. Lederman, "History of the Neutrino," Resource Letter Neu 1, *American Journal of Physics,* vol. 38, pp. 129–136 (1970).

J. Rosner, "New Particles," Resource Letter NP1, *American Journal of Physics,* vol. 48, pp. 90–103 (1980).

Tabulated data on nuclei and particles can be found in

J. Tuli, *Nuclear Wallet Cards,* 1985. A pocket-sized booklet with more data than almost anyone could need, available from the National Nuclear Data Center, Brookhaven National Laboratory, Upton, NY 11973.

Chart of the Nuclides. A wall chart of nuclear data, available from the General Electric Company, 175 Curtner Avenue, Mail Code 684, San Jose, CA 95125.

Particle Data Group, *Particle Properties Data Booklet,* 1990. Available from Technical Information Dept, MS 90-2125, Lawrence Berkeley Lab, Berkeley, CA 94720.

There are many nontechnical or semitechnical books on nuclear and particle physics. Three fine ones are

F. Close, M. Marten, and C. Sutton, *The Particle Explosion.* New York: Oxford University Press, 1987. This splendid "coffee-table" book with hundreds of beautiful pictures takes the story through the mid-1980s.

H. G. Graetzer and D. L. Anderson, *The Discovery of Nuclear Fission.* New York: Van Nostrand, 1971. This collection of reprints of original articles with helpful editorial comments gives a fascinating and readable account of the unfolding of one of the twentieth century's great discoveries.

S. Weinberg, *The Discovery of the Subatomic Particles.* New York: Freeman, 1983. An excellent account through about 1970.

REFERENCES FOR PART IV, RADIATION, MOLECULES, AND SOLIDS

More advanced books on lasers include

A. Siegman, *Lasers.* Mill Valley, CA: University Science Books, 1986.

O. Svelto, *Principles of Lasers.* New York: Plenum Press, 1976.

A Resource Letter on Lasers:

D. C. O'Shea and D. C. Peckham, "Lasers," Resource Letter L1, *American Journal of Physics,* vol. 49, pp. 915–925 (1981).

Three classic books on molecules are

G. Herzberg, *Molecular Spectra and Molecular Structure: I Spectra of Diatomic Molecules.* New York: Van Nostrand, 1950.

M. Karplus and R. N. Porter, *Atoms and Molecules: An Introduction for Students of Physical Chemistry.* Reading, MA: Addison-Wesley, 1970.

L. Pauling, *The Chemical Bond.* Ithaca, NY: Cornell University Press, 1967. A readable account by a pioneer of the subject.

A reasonably nontechnical book on solids is

A. Holden, *The Nature of Solids.* New York: Columbia University Press, 1968.

Three that are appreciably more advanced are

A. J. Dekker, *Solid State Physics.* Englewood Cliffs, NJ: Prentice-Hall, 1957.

C. Kittel, *Introduction to Solid State Physics,* 6th ed. New York: Wiley, 1986.

H. M. Rosenberg, *The Solid State.* Oxford: Oxford University Press, 1988.

Many properties of molecules and solids are tabulated in

Handbook of Chemistry and Physics. Boca Raton, FL: CRC Press.

The book of Herzberg mentioned above (and his three companion volumes) contains many tables of molecular properties; similarly, the book of Kittel gives many properties of solids.

Figure 3.10: Courtesy Dale Prull, University of Colorado.
Figure 5.2: After R. A. Millikan, *Physical Review*, vol. 7, p. 362 (1916). Fig. 5.8(b): Courtesy Cortlandt Pierpont and Curtis Haltiwanger, Molecular Structure Lab, University of Colorado. Fig. 5.9(b): Courtesy the Educational Development Center, Newton, MA. Fig. 5.10: After C. Ulrey, *Physical Review*, vol. 11, p. 401 (1918). Fig. 5.13: After A. H. Compton, *Physical Review*, vol. 22, p. 411 (1923).
Figure 6.6: After H. G. J. Moseley, *Philosophical Magazine*, vol. 27, p. 708 (1914). Fig. 6.8(b): Copied, with the permission of John Wiley & Sons, from Trajmar, Rice, and Kuperman, *Advances in Chemical Physics*, vol. 18, p. 32 (1970).
Figure 7.2: Courtesy AT&T Archives, Short Hills, NJ. Fig. 7.3(a) and (b): Courtesy the Educational Development Center, Newton, MA. Fig. 7.3(c): Courtesy Prof. C. G. Shull, MIT. Fig. 7.4(a): From *University Physics* by Sears, Zemansky, and Young (Addison-Wesley, 1982) p. 777; courtesy Addison-Wesley. Fig. 7.4(b): Courtesy Prof. C. Jönsson, Tübigen, Germany; first published in *Zeitschrift für Physik*, vol. 161, p. 454 (1961). Fig. 7.12: Adapted, with the permission of the McGraw-Hill Publishing Co., from *College Physics* by Weber, White, and Manning (McGraw-Hill, 1959), p. 265. Figs. 7.17 and 7.20: Courtesy Dale Prull, University of Colorado.
Figure 11.10: Courtesy Steven O'Neil, Joint Institute for Laboratory Astrophysics, Boulder, CO. Fig. 11.11(c): Data courtesy of Carl Lineberger, University of Colorado.
Figure 12.19: Redrawn, with the permission of Fred White and *Physical Review*, from White, Collins, and Rourke, *Physical Review*, vol. 101, p. 1791 (1956). Fig. 12.21: Copied, with the permission of the McGraw-Hill Publishing Co., from *Mass Spectrometry* by K. Biemann (McGraw-Hill, 1962), p. 295.
Figure 13.4: After G. J. Neary, *Proceedings of the Royal Society*, vol. A175, p. 79 (1940). Fig. 13.7(b): Courtesy Dale Prull. Fig. 13.10: From Blackett and Lees, *Proceedings of the Royal Society*, vol. A136, p. 338 (1932). Figs. 13.19 and 13.20: Courtesy Eloise Trabka, University of Colorado. Fig. 13.24: Courtesy the Cavendish Lab, Cambridge, England. Fig. 13.25: Courtesy the Carnegie Institution of Washington, Department of Terrestrial Magnetism. Fig. 13.27: Courtesy Los Alamos National Lab. Fig. 13.29: Courtesy University of Colorado Nuclear Physics Lab.
Figure 14.1: After Simons and Zuber, *Proceedings of the Royal Society*, vol. A159, p. 383 (1937). Fig. 14.4: From Powell, Fowler, and Perkins, *The Study of Elementary Particles by the*

Photographic Method (Pergamon, 1959) (a) p. 245, courtesy Science Photo Library, London, (b) p. 254, courtesy Professor Fowler, Bristol University. Fig. 14.5: Courtesy Professor Stevenson, University of California, Berkeley, CA. Fig. 14.6: Courtesy Lawrence Berkeley Lab, Berkeley, CA. Fig. 14.9: After Gottfried and Weisskopf, *Concepts of Particle Physics* (Oxford University Press, 1984). Fig. 14.11: Courtesy Brookhaven National Lab. Fig. 14.12: Redrawn, with the permission of Professor Ting and *Physical Review*, from Aubert et al., *Physical Review Letters*, vol. 33, p. 1405 (1974). Figs. 14.16 and 14.18: Courtesy DESY, Hamburg, Germany. Figs. 14.19 and 14.20: Courtesy CERN, Geneva, Switzerland. Fig. 14.21: Courtesy Stanford Linear Accelerator Center and the U.S. Department of Energy. Fig. 14.23: Courtesy Fermi National Lab.
Figures 15.2 and 15.3: Courtesy Dale Prull. Fig. 15.8: Redrawn, with the permission of University Science Books, from Siegmann, *Lasers* (University Science Books, 1986), Fig. 1.50, p. 60. Fig. 15.10: Courtesy Hughes Research Lab, Malibu, CA. Fig. 15.14: Courtesy Litton Guidance and Control Systems. Fig. 15.21: Copyright 1970, Hewlett-Packard Company; reproduced with permission.
Figure 16.2(c): Photo by Gopal Murti and David Prescott, University of Colorado; courtesy David Prescott. Fig. 16.31: Copied, with the permission of the McGraw-Hill Publishing Co., from G. M. Barrow, *Molecular Spectroscopy* (McGraw-Hill, 1962), p. 173.
Figure 17.4: Courtesy Digital Instruments, Inc., Santa Barbara, CA. Figs. 17.9 and 17.10: Courtesy U.S.X., Pittsburgh, PA. Fig. 17.27: Courtesy IBM. Fig. 17.35: Courtesy Colorado Superconductors, Inc., Fort Collins, CO.

The photographs of Nobel laureates and other eminent scientists were furnished by the following sources:

AIP/Niels Bohr Library: pp. 10, 15, 67, 82, 108, 112, 128, 140, 149, 155, 193, 254, 275, 360, 372, 397, 472, 503, 522

AIP/Meggers Gallery of Nobel Laureates: pp. 29, 169, 257, 317, 390, 409, 452, 463

Columbia University: p. 395

The Edgar Fahs Smith Collection; Dept. of Special Collections; Van Pelt-Dietrich Library; University of Pennsylvania: 285, 331

Harris and Ewing: p. 110

Town and Country Photographers: p. 351

ANSWERS TO ODD-NUMBERED PROBLEMS

CHAPTER 1

1.1. (a) $x = \frac{1}{2}g(\sin\theta - \mu\cos\theta)t^2$, $y = 0$;

$t = \sqrt{\dfrac{2l}{g(\sin\theta - \mu\cos\theta)}}$ (b) Main advantages of Oxy are: Motion is in x direction only; friction is in x direction; normal force is in y direction.

1.3. (a) $x = v_0 t\cos\theta - \frac{1}{2}gt^2\sin\phi$, $y = v_0 t\sin\theta - \frac{1}{2}gt^2\cos\phi$ (b) The main advantage of Oxy is that the condition for landing is just $y = 0$.

1.5. M continues on with speed v_0; m bounces back with speed $3v_0$.

1.7. (a) $c - v = 2.9976 \times 10^8$ m/s   (b) $c + v = 2.9982 \times 10^8$ m/s   (c) $\sqrt{c^2 - v^2} = 2.9979 \times 10^8$ m/s

1.9. (a) $u' = \sqrt{u^2 - v^2\sin^2\theta'} - v\cos\theta'$ (b) $u + v$ and $u - v$ (c) $\pm 5\%$

1.11.

x:	0.5	0.1	0.01	0.001
$(1 - x)^2$:	0.25	0.81	0.9801	0.998001
$1 - 2x$:	0	0.80	0.98	0.998
% difference:	100%	1%	0.01%	0.001%

difference $= x^2$

1.13. Tom wins by 250 s.

1.15. $\Delta N = 0.017$

CHAPTER 2

2.1. It loses 5.6×10^{-15} s.

2.3. (a) 1.4×10^6 m/s   (b) 1.1×10^7 m/s

2.5. $0.94c$

2.7. (a) $\tau = d(\gamma - 1)/(v\gamma)$; if $v \ll c$, this is approximately $dv/(2c^2) = 1.7$ ns.   (b) $\tau \approx dv/(2c^2) \to 0$ as $v \to 0$

2.9. 420 muons expected

2.11. $v = 0.6c$

2.13. If the ball's radius is r as measured in S, then in S' the ball is an oblate spheroid (a flattened sphere) with radius r in the $y'z'$ plane but radius r/γ in the x' direction.

2.15. (a) 60 cm   (b) 100 cm   (c) 91.7 cm   (d) 83.2 cm

2.17. $x' = 9.2$ c·s, $y' = z' = 0$, $t' = -1.15$ s

2.21. (a) $x_1 = 4500$ m, $y_1 = z_1 = 0$, $t_1 = 15$ μs
(b) $x_2 = 1500$ m, $y_2 = z_2 = 0$, $t_2 = 10$ μs
(c) $\Delta t = t_2 - t_1 = -5$ μs; $\Delta t' = t_2' - t_1' = 5$ μs

2.23. (a) $x_1' = y_1' = z_1' = t_1' = 0$; $x_2' = 5$ c·yr, $y_2' = z_2' = 0$, $t_2' = -3$ yr   (b) 5 c·yr   (c) No, they are not simultaneous as observed in S'.

2.25. $0.85c$

2.27.

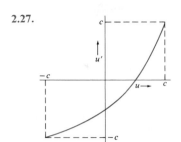

2.29. $v = 0.01c$ toward us

2.31. $v = 0.20c$

CHAPTER 3

3.1. The relativistic and nonrelativistic momenta, p_{rel} and p_{NR}, in units of 10^{-22} kg·m/s, are

β:	0.1	0.5	0.9	0.99
p_{rel}:	0.276	1.57	5.63	19.2
p_{NR}:	0.273	1.36	2.46	2.7

3.5. $u = 0.87c$

3.7. $u = 0.94c$

3.13. (a) $E = 5$ GeV   (b) $u = 0.8c$

3.15. $p = 1090$ MeV/$c = 5.81 \times 10^{-19}$ kg·m/s, $u = 0.76c = 2.27 \times 10^8$ m/s

3.17. $p = 12$ GeV/c, $u = 0.92c = 2.77 \times 10^8$ m/s

3.19. $K = 1.26 \times 10^{-12}$ J $= 7.9$ MeV

3.21. (a) 9×10^{-36} kg   (b) 1.5×10^{-10} (c) 1.5×10^{-9} g

3.23. (a) 184 MeV   (b) 1122 MeV

3.25. 249 MeV, 206 MeV/c, $0.83c$

3.27. $M = 2950$ MeV/c^2, $u = 0.65c$

3.29. $R = 4.7$ cm

3.33. $E_1 = E_2 = 67.5$ MeV

3.37. (b) 1.1×10^{-8}, 99.9999995%

3.39. $u = c\sqrt{1 - 1/(1 + K/mc^2)^2}$

K:	$0.001mc^2$	$0.01mc^2$	$0.1mc^2$	mc^2
u(rel):	$0.04469c$	$0.1404c$	$0.417c$	$0.87c$
u(NR):	$0.04472c$	$0.1414c$	$0.447c$	$1.41c$
% diff:	0.075%	0.75%	7.3%	63%

$K_{max} \approx 0.01mc^2$ (actually, $0.013mc^2$)

CHAPTER 4

4.1. Elements 1 to 42, 44 to 60, and 62 to 92 occur naturally. Elements 43 (tecnetium) and 61 (promethium) and elements 93, 94, . . . do not occur naturally and must be produced artificially.

4.3. 4.67 g (approximately—using the actual masses, one gets 4.63)

4.5. Hydrogen: ^{1}H(99.985%), ^{2}H(0.015%)
Helium: ^{3}He (0.0001%), ^{4}He (99.9999%)
Oxygen: ^{16}O (99.762%), ^{17}O (0.038%), ^{18}O (0.200%)
Aluminum: ^{27}Al (100%)

4.7. For example: titanium 50 and chromium 50; chromium 54 and iron 54; indium 115 and tin 115. An example of a triplet of isobars is tin 124, tellurium 124, and xenon 124.

4.9. 1 u $= 1.66054 \times 10^{-27}$ kg $= 931.494$ MeV/c^2

4.11. $\overline{m}$(C) = 12.0 u, $\overline{m}$(Cl) = 35.5 u, $\overline{m}$(Fe) = 55.9 u

4.13. (a) 4.0015 u (b) Correct answer is larger by 8.6×10^{-8} u. (c) Seven or eight significant figures

4.15. (a) 0.17 (b) 118 (c) 0.68 (d) 1.3

4.17. $-96,320$ C

4.19. $t = l/u$; $u_y = eEl/(mu)$

4.21. $u = 2\theta H/(QBl)$ and $m/e = QB^2l^2/(2\theta^2H)$, where H = heat

4.23. (b) $(1/t_d) + (1/t_u) = n(0.113$ s$^{-1})$ where $n = 2, 1, 3$, and 5 (c) $r = 6.94 \times 10^{-7}$ m (d) $Q = 2e, e, 3e$, and $5e$ with $e = 1.62 \times 10^{-19}$ C

4.25. $E_{max} = 3.5 \times 10^{12}$ V/m for Thomson, but 1.8×10^{21} V/m for Rutherford

4.27. (a) $R_{Ag} < 9$ fm or, a little better, $R_{Ag} + R_{He} \lesssim 9$ fm (b) $E \approx 6$ MeV

CHAPTER 5

5.1. From 1.8 to 3.1 eV; $E > 3.1$ eV for UV, and $E < 1.8$ eV for IR

5.3. 1.2×10^{-4} eV

5.5. (a) $\phi = 2.2$ eV (b) $V_s = 1.9$ volts

5.7. $h \approx 4.0 \times 10^{-15}$ eV·s $= 6.4 \times 10^{-34}$ J·s

5.9. $\theta = 4.6°$

5.11. $n = 3, 4$, and 5; $d = 0.11$ nm

5.13. $V_0 = 124$ kV

5.15. $\Delta\lambda = 0.0041$ nm, $\Delta\lambda/\lambda_0 = 5.8\%$

5.17. $E = 0.34$ MeV, $K_e = 0.66$ MeV

5.19. (a) $E = 10^5$ J, $p = 0.33 \times 10^{-3}$ kg·m/s (b) 0.33 m/s (c) 5.5×10^{-5} J. Almost all of the energy produces heat.

5.21. $\lambda_0 = 0.022$ nm

CHAPTER 6

6.1. $\lambda = 1.87$ μm; infrared

6.3. $ke^2 = 1.44$ eV·nm, $hc = 1240$ eV·nm

6.5. (a) 5.26×10^{-11} m $= 0.0526$ nm (b) 0.0527 nm (These were calculated using the three-figure data inside the front cover. The small difference is due to rounding errors. If one uses the more accurate data in Appendix A and rounds the answer to three significant figures, one finds 0.0529 nm either way.)

6.9. 365 nm $< \lambda \le 656$ nm; UV and visible

6.11. (a) $r_1 = 0.26$ pm, $E_1 = -2.8$ keV, (b) $\lambda = 0.59$ nm, soft X ray or far UV

6.13. $E_{ph} = 6.9$ keV, $\lambda = 0.18$ nm, X ray

6.17. (a) $r_e = rm_p/M$, $r_p = rm_e/M$, where $M = m_e + m_p$ (f) $r = n^2\hbar^2/(\mu ke^2)$ (g) -13.598 eV (correct), -13.606 eV (ignoring proton's motion)

6.19. Chlorine $(Z = 17)$

6.21. $r = 6.5 \times 10^{-13}$ m, which is much greater than R

6.23. (b) 3.3×10^{-5} eV

CHAPTER 7

7.1. $\lambda = 0.055$ nm

7.3. $K = 4.0 \times 10^{-9}$ eV

7.5. $\lambda_e = 1.0 \times 10^{-2}$ nm, $\lambda_\mu = \lambda_e/\sqrt{207} = 7.0 \times 10^{-4}$ nm

7.7. $\lambda = 6.4 \times 10^{-13}$ m

7.9.

Energy	Electron	Photon
1 keV	0.0388 nm	1.24 nm
1 MeV	870 fm	1240 fm
1 GeV	1.239 fm	1.240 fm

7.11. (a) 0.59 MeV (b) 1.23 MeV (c) 3.64 MeV

7.13. $d = 2.74$ nm

7.17. $v = 167$ cm/s, $\lambda = 0.52$ cm, $f = 318$ Hz

7.19. $k = 0.0114$ rad/nm, $\omega = 3.4 \times 10^{15}$ rad/s

7.21. $\lambda = 0.071$ nm, $k = 88.7$ rad/nm

7.23. (a) $\phi = \pi - kx_0$ (b) $\phi = \pi$ (c) Change x_0 by $-\lambda/2$

7.27. 5 min

7.29. (a) $\Delta k \ge 1.7$ rad/m (b) $2 \Delta t = 0.1$ s, $\Delta\omega \ge 10$ rad/s

7.31.

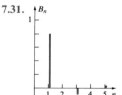 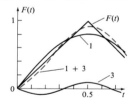

7.35. $\Delta v \ge 0.018c$

7.37. $K_{min} = 2.3 \times 10^{-56}$ J; it would travel 2.7×10^{-20} m—not very far!

7.39. Halving l halves Δx and doubles Δp_x; no change in their product

7.41. $K_{min} \approx 40$ MeV

7.43. $\Delta E \approx 30$ MeV

7.45. (a) $\Delta E \ge 3 \times 10^{-13}$ eV (b) $\Delta\lambda \approx 8 \times 10^{-11}$ nm, $\Delta\lambda/\lambda \approx 1.5 \times 10^{-13}$

CHAPTER 8

8.1. At $t = 0$: $y = 0, 1.76, 2.85, 2.85$, etc.; at $t = 0.05$ s $y = 0$ for all x; at $t = 0.07$ s: $y = 0, -1.04, -1.68, -1.68$, etc.

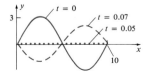

8.3. $y(x, t) = A \sin kx \cos \omega t$, where $A = 2$ cm, $k = \pi/10$ rad/cm, $\omega = 80\pi$ rad/s

8.5. The three pictures were *not* made at equally spaced times. (If they were, the height of the second would be $1/\sqrt{2}$ times the first.)

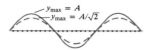

8.7. $y(x, t) = A \sin kx \cos \omega t$, where $A = 3$ cm, $k = \pi/40$ rad/cm, $\omega = 20$ rad/s

8.9. $x = r \cos \theta$, $y = r \sin \theta$, $r = \sqrt{x^2 + y^2}$, $\theta = \tan^{-1}(y/x)$ (defined to put x and y in the correct quadrant)

8.11. (a) $a = A \sin \phi$, $b = A \cos \phi$; $A = \sqrt{a^2 + b^2}$, $\phi = \tan^{-1}(a/b)$; (b) take $t' = t + \phi/\omega$

8.15. 8.2, 32.7, 73.5 MeV

8.17. $E_5 = 25E_1$, $E_6 = 36E_1$, $E_7 = 49E_1$

8.27. (a) $|\psi_4(x)|^2 = (2/a)\sin^2(4\pi x/a)$
(b) $x_{mp} = a/8, 3a/8, 5a/8$, and $7a/8$ (c) $P(0.50a \le x \le 0.51a) \approx 0 \approx P(0.75a \le x \le 0.76a)$

8.29. $x_{av} = a/2$

8.31. (a) $(c/a) - (1/2\pi) \sin(2\pi c/a)$ (b) 1 (this is the normalization integral) (c) 50% (d) 9%

8.33. (a)

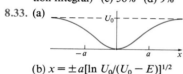

(b) $x = \pm a[\ln U_0/(U_0 - E)]^{1/2}$

8.35.

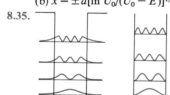

8.37.

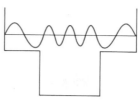

8.39.

8.41. (a) General solution $= Ae^x + Be^{-x}$; for the given starting conditions, $\psi(x) = e^x = 1.4918$ when $x = 0.4$; (b) and (c):

Steps:	1	2	3	5	10	50	100
$\psi(0.4)$:	1.40	1.44	1.456	1.4693	1.4802	1.4895	1.4906

8.43. (a) General solution $= Ae^{2x} + Be^x$; for the given starting conditions, $\psi(x) = 2e^x - e^{2x} = 0.7581$ when $x = 0.4$; (b) and (c):

Steps:	1	2	3	5	10	50	100
$\psi(0.4)$:	1	0.92	0.879	0.8383	0.8016	0.7674	0.7628

8.47. $A_1 = (4m\omega_c/\pi\hbar)^{1/4}$

8.49. (a) Turning points are $x = \pm b$. (b) 84% (c) 16%

CHAPTER 9

9.1.

	$\partial f/\partial x$	$\partial f/\partial y$
(a)	$2xy^3 + 4x^3y^2$	$3x^2y^2 + 2x^4y$
(b)	$3(x + y)^2$	$3(x + y)^2$
(c)	$\cos x \cos y$	$-\sin x \sin y$

9.3. (a) $\partial h/\partial x = $ (slope of hill if one walks due east); $\partial h/\partial y = $ (slope of hill if one walks due north) (b) He is ascending. (c) He is walking horizontally with the slope up on his right and down on his left.

9.5. (a) $12y(x + y^2)$ (b) $[4x + 2y + 4x(x + y)^2] \exp(x + y)^2$
(c) $\dfrac{x + y}{(x - y)^2}$

9.7. $\psi(x, y) = A \sin(n_x \pi x/a) \sin(n_y \pi x/b)$

9.9. Rectangular box ($E_o = \hbar^2\pi^2/2Ma^2$):

n_x, n_y:	1,1	2,1	1,2	2,2	3,1	1,3
E:	$2.21E_0$	$5.21E_0$	$5.84E_0$	$8.84E_0$	$10.21E_0$	$11.89E_0$
deg:	1	1	1	1	1	1

square box:

E:	$2E_0$	$5E_0$	$8E_0$	$10E_0$
deg:	1	2	1	2

9.11. All degeneracies arise because any permutation of n_x, n_y, n_z leaves E unchanged.

n_x n_y n_z	E	deg
1 1 4	$18E_0$	3
2 2 3	$17E_0$	3
1 2 3	$14E_0$	6
2 2 2	$12E_0$	1
1 1 3	$11E_0$	3
1 2 2	$9E_0$	3
1 1 2	$6E_0$	3
1 1 1	$3E_0$	1
	0	

9.15. If $P = (r, \phi)$, then $Q = (r, \phi + \pi)$.

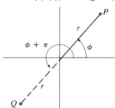

9.17. (c) $\dfrac{\partial^2 \psi}{\partial r^2} = \dfrac{\partial^2 \psi}{\partial x^2} \cos^2\phi + 2 \dfrac{\partial^2 \psi}{\partial x\, \partial y} \cos\phi \sin\phi + \dfrac{\partial^2 \psi}{\partial y^2} \sin^2\phi$

9.19. If $P = (x, y, z)$ or (r, θ, ϕ), then $Q = (-x, -y, -z)$ or $(r, \pi - \theta, \phi + \pi)$.

9.21. $\theta_{min} = 35.3°$

9.23. (a)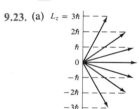

(b) 7 orientations

(c) $\theta_{min} = 30°$

9.25.

l:	1	2	3	4	10	100
L(correct):	$1.41\hbar$	$2.45\hbar$	$3.46\hbar$	$4.47\hbar$	$10.49\hbar$	$100.50\hbar$
L(Bohr):	$\hbar$	$2\hbar$	$3\hbar$	$4\hbar$	$10\hbar$	$100\hbar$
ratio:	1.41	1.22	1.15	1.12	1.05	1.005

As $l \to \infty$, the ratio L(correct)/L(Bohr) $\to 1$.

9.27. (c) General solution $\theta = A + B \ln[(1 + \cos\theta)/(1 - \cos\theta)]$

9.31. $(1/r)_{av} = 1/a_B$

9.35. $\displaystyle\int_0^\infty r^2 \exp(-2r/a_B)\, dr = a_B^3/4$

9.37. P(outside a_B) = 68%

9.43. $r_{mp} = n^2 a_B$

9.45. $r_{mp} = 1.89$ pm, $E = 10.7$ keV

9.47. $\lambda = 0.84$ nm

CHAPTER 10

10.1. (a) 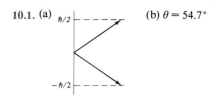 (b) $\theta = 54.7°$

10.3. (a) S_z has three possible values: $\hbar, 0, -\hbar$;

(b)

(c) $\theta_{min} = 45°$

10.5. $E = -E_R$, $n = 1$, $l = m = 0$, $m_s = \pm\frac{1}{2}$
$E = -E_R/4$, $n = 2$, $l = 0$, $m = 0$, $m_s = \pm\frac{1}{2}$
and $l = 1$, $m = 1, 0, -1$, $m_s = \pm\frac{1}{2}$

10.7. (a) $\mu = 1.26 \times 10^{-4}$ A·m^2
(b) $\Gamma = 1.88 \times 10^{-4}$ m·N
(c) $\Delta E = 3.77 \times 10^{-4}$ J

10.9. $i = 3.2 \times 10^{-4}$ A

10.11. $\mu_B = 9.27 \times 10^{-24}$ A·m^2 = 5.79×10^{-5} eV/T

10.13. (a)

$$\underline{\quad\quad} \; E_0 + 3\mu_B B$$
$$\underline{\quad\quad}$$
$$\underline{\quad\quad}$$
$$\underline{\quad\quad} \; E_0$$
$$\underline{\quad\quad}$$
$$\underline{\quad\quad}$$
$$\underline{\quad\quad} \; E_0 - 3\mu_B B$$

(b) Separation = 4.6×10^{-5} eV

10.15. (a)

$$\underline{\quad\quad} \; E_2 + 2\mu_B B$$
$$\underline{\quad\quad} \; E_2$$
$$\underline{\quad\quad} \; E_2 - 2\mu_B B$$

$$\underline{\quad\quad} \; E_1 + \mu_B B$$
$$\underline{\quad\quad} \; E_1$$
$$\underline{\quad\quad} \; E_1 - \mu_B B$$

(b) There are seven distinct conceivable photon energies.

10.17. (a) Separation of states = 8.1×10^{-5} eV
(b) $E_{ph} = 8.1 \times 10^{-5}$ eV; $\lambda = 1.5$ cm, which is microwave radiation

CHAPTER 11

11.1. $\mathscr{E} = 3.5 \times 10^{11}$ V/m

11.3. (a) $E = -91$ keV (b) $r = 650$ fm

11.5. (a) $U(r) = -ke^2/r$ (b) $E = -1.51$ eV

11.7. (a) 10 (b) 14 (c) $2(2l + 1)$

11.9. (a) Number of orientations = 4
(b) $_2$He: $1s$ ———•—•——— $_3$Li: $1s$ ———•—•—•———

11.11. (a)

Energy:	$2E_0$	$5E_0$	$8E_0$	$10E_0$
Degeneracy:	2	4	2	4

(b)

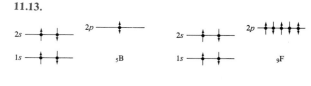

Six particles:

(c) Ten:

11.13.

11.15. With the $2s$ level full, the fifth electron of $_5$B must occupy the slightly higher $2p$ level and is therefore slightly easier to remove than any of the electrons in $_4$Be.

11.17. (a) $E \approx -8E_R$ ignoring the repulsion of the two electrons
(b) $U \approx ke^2/a_B = 2E_R$ (c) $E \approx -6E_R = -81.6$ eV

11.19. $_{30}$Zn: $1s^2 2s^2 2p^6 3s^2 3p^6 4s^2 3d^{10}$
$_{35}$Br: $\cdots\cdots\cdots\quad 4p^5$
$_{54}$Xe: $\cdots\cdots\cdot 4p^6 5s^2 4d^{10} 5p^6$
$_{85}$At: $\cdots\cdots\cdot 6s^2 4f^{14} 5d^{10} 6p^5$
$_{87}$Fr: $\cdots\cdots\cdot 6p^6 7s^1$

11.21. The 48 electrons in $_{48}$Cd fill all shells and subshells through $4d$. The last electron in $_{49}$In must occupy the $5p$ level, which is appreciably higher and hence leads to a lower ionization energy. Similarly, $_{80}$Hg just fills all subshells through $4f$ and $5d$, and the last electron in $_{81}$Tl must occupy the higher $6p$ level.

11.23. (a) $_{30}$Zn: $1s^2 2s^2 2p^6 3s^2 3p^6 4s^2 3d^{10}$
$_{80}$Hg: $\cdots\cdots\cdot 4p^6 5s^2 4d^{10} 5p^6 6s^2 4f^{14} 5d^{10}$
$_{37}$Rb: $\cdots\cdots\cdot\quad 5s^1$
$_{55}$Cs: $\cdots\cdots\cdot 5s^2 4d^{10} 5p^6 6s^1$
(b) $_{30}$Zn, zero; $_{80}$Hg, zero; $_{37}$Rb, total angular momentum = spin $(s = \frac{1}{2})$; $_{55}$Cs, total = spin $(s = \frac{1}{2})$.

11.25. If s were equal to $\frac{3}{2}$, the configurations would be
$_8$O: $1s^4 2s^4$
$_{10}$Ne: $1s^4 2s^4 2p^2$
$_{21}$Sc: $1s^4 2s^4 2p^{12} 3s^1$

11.27. $_3$Li $= [_2$He$]2s^1 = 1s^2 2s^1$
$_{10}$Ne $= [_2$He$]2s^2 2p^6 = \cdot 2s^2 2p^6$
$_{12}$Mg $= [_{10}$Ne$]3s^2 = \cdots 3s^2$
$_{19}$K $= [_{18}$Ar$]4s^1 = \cdots\cdot 3p^6 4s^1$
$_{28}$Ni $= [_{18}$Ar$]4s^2 3d^8 = \cdots\cdot 4s^2 3d^8$
$_{48}$Cd $= [_{36}$Kr$]5s^2 4d^{10} = \cdots\cdot 3d^{10} 4p^6 5s^2 4d^{10}$

11.29. Ga = gallium, $Z = 31$:
$1s^2 2s^2 2p^6 3s^2 3p^6 4s^2 3d^{10} 4p^1$
Xe = xenon, $Z = 54$:
$\cdots\cdots\cdot 4p^6 5s^2 4d^{10} 5p^6$
W = tungsten, $Z = 74$:
$\cdots\cdots\cdot 6s^2 4f^{14} 5d^4$
At = astatine, $Z = 85$:
$\cdots\cdots\cdot 5d^{10} 6p^5$
Md = mendelevium, $Z = 101$:
$\cdots\cdots\cdot 6p^6 7s^2 5f^{13}$

Unh= unnilhexium, $Z = 106$:
$\cdots\cdots\cdots\cdots\cdots\cdot 5f^{14} 6d^4$

11.31. (a) (IE of $_{56}$Ba) ≈ 5.29 eV
(b) (EA of $_{35}$Br) ≈ 3.33 eV
(c) (R of $_9$F) ≈ 0.055 nm

11.33. (EA of $_9$F) ≈ 3.88 eV

11.35. [Ne]$3s^1$, [Ne]$3p^1$, [Ne]$3d^1$, [Ne]$4s^1$

11.37. $1s^2$, $1s^1 2s^1$, $1s^1 2p^1$, $1s^1 3s^1$, $1s^1 3p^1$, $1s^1 3d^1$

CHAPTER 12

12.1. $_1^1$H$_0$, $_2^3$He$_1$, $_3^7$Li$_4$, $_{10}^{20}$Ne$_{10}$, $_{18}^{40}$Ar$_{22}$, $_{29}^{63}$Cu$_{34}$, $_{82}^{206}$Pb$_{124}$

12.3. (a) ^{70}Ge: isotopes: (^{70}Ge), ^{72}Ge, ^{73}Ge, ^{74}Ge, ^{76}Ge
isobars: ^{70}Zn (^{70}Ge)
isotones: ^{68}Zn, ^{69}Ga, (^{70}Ge)
(b) ^{27}Al: isotopes: (^{27}Al) (i.e., no isotopes)
isobars: (^{27}Al)
isotones: ^{26}Mg, (^{27}Al), ^{28}Si
(c) ^{88}Sr: isotopes: ^{84}Sr, ^{86}Sr, ^{87}Sr, (^{88}Sr)
isobars: (^{88}Sr)
isotones: ^{86}Kr, (^{88}Sr), ^{89}Y, ^{90}Zr, ^{92}Mo

12.5. m_{atom} (^{12}C) = 12 u exactly; fraction of mass in electrons = 2.75×10^{-4}

12.7. $(V_{atom}/V_{nuc}) \approx 10^{14}$; $(\rho_{atom}/\rho_{nuc}) \approx 10^{-14}$

12.9. $\lambda = 0.0026$ nm (nuclear transition); $\lambda = 690$ nm (atomic transition)

12.11. (a) Mass densities: $\rho_{nuc} = 3.3 \times 10^{17}$ kg/m^3; $\rho_{nuc}/\rho_{solid} \approx 4.2 \times 10^{13}$
(b) Charge densities: $\rho_{nuc} = 1.45 \times 10^{25}$ C/m^3; $\rho_{nuc}/\rho_{spark} \approx 2 \times 10^{29}$

12.13.

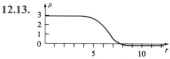

12.15. $a = 8.6 \times 10^{27}$ m/s$^2 \approx 10^{27}$ g

12.17. (a) $K_{min} \approx 3\pi^2 \hbar^2/(8mR_0^2 A^{2/3}) \approx (133$ MeV$)/A^{2/3}$
(b) $K_{min} \approx 31$ MeV for ^{9}Be, 15 MeV for ^{27}Al, 3.5 MeV for ^{238}U

12.19. $\Delta B = 2.4$ MeV (versus 1.7 MeV observed)

12.21. (a) $s =$ integer (b) $s =$ half-odd integer (c) Only proton–neutron model is consistent.

12.23. $U = -9.4$ MeV—much too small to offset the minimum KE of nearly 200 MeV.

12.25. (a) $B \approx 6.8 \times 10^{-3}$ T (b) $\Delta E \approx 8 \times 10^{-7}$ eV
(c) $\Delta E = 5.9 \times 10^{-6}$ eV. The rough estimate in (b) is an underestimate for several reasons, the simplest of which is that the proton's magnetic moment is actually nearly three times bigger than μ_N.

12.27. (a) $U = 4.4$ MeV in ^{12}C (b) $U = 28$ MeV in ^{208}Pb

12.29.

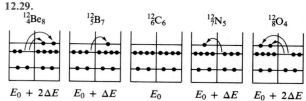

$^{12}_4$Be$_8$ $\quad$ $^{12}_5$B$_7$ $\quad$ $^{12}_6$C$_6$ $\quad$ $^{12}_7$N$_5$ $\quad$ $^{12}_8$O$_4$

$E_0 + 2\Delta E$ $\quad$ $E_0 + \Delta E$ $\quad$ E_0 $\quad$ $E_0 + \Delta E$ $\quad$ $E_0 + 2\Delta E$

12.31. (a) $m_{atom} = 4.002603$ u (b) $m_{nuc} = 4.001505$ u (c) No

12.33. $S_n = 8.2$ MeV

12.35. (a) $B = 1566.5$ MeV (b) $B = 1566.5$ MeV

12.37. $a_{Coul} = 0.807$ MeV (versus 0.711 MeV for best fit)

12.39. $B/A = 8.01$ MeV for ^{20}Ne, 8.74 MeV for ^{60}Ni, 7.41 MeV for ^{259}No. These follow the trend of Fig. 12.14, with a maximum near $A = 60$.

12.41. (b) Energy released = 6.7 MeV (versus 6.5 MeV observed)

12.43. (a) (Charge between r and $r + dr$) $= 3Qr^2dr/R^3$

12.45. (a) For $A = 25$, $Z = 11.9$, $N = 13.1$; for $A = 45$, $Z = 20.8$, $N = 24.2$; etc.

(b)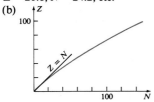

12.47. $j_{tot} = 0, \frac{7}{2}, 0, \frac{7}{2}, 0, \frac{7}{2}, 0, \frac{7}{2}, 0$, all in agreement with the values observed

12.49. $j_{tot} = \frac{1}{2}, 0, \frac{7}{2}$, all in agreement with the values observed

12.51. (a)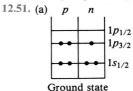

Ground state

(b) $j_{tot} = \frac{3}{2}$

(c) $j_{tot} = \frac{1}{2}$

First excited

12.53. For ^{19}F, Fig. 12.17 implies that $j = \frac{3}{2}$ (versus $j = \frac{1}{2}$ observed); for ^{121}Sb, Fig. 12.17 implies that $j = \frac{7}{2}$ (versus $j = \frac{5}{2}$ observed).

12.55. (Degeneracy of $j = l + \frac{1}{2}$ level) $= 2l + 2$
(degeneracy of $j = l - \frac{1}{2}$ level) $= 2l$

12.57. The separation of the two levels is proportional to $2l + 1$.

12.59. The COOH group has mass 45 u and, when it breaks off, leaves a fragment of mass 104 u.

12.61. (a) $R = \sqrt{2V_0 m/neB^2}$ (b) With $m = 4$ u, 12 u, 16 u, the three radii are the same, $R = \sqrt{8V_0(1\text{ u})/eB^2}$. (c) The differences are 0.03% and 0.02%.

CHAPTER 13

13.1. (a) 7.48×10^{11} (b) 9.49×10^8 (c) 9.01×10^5

13.3. (a) $r = 0.0101$ yr^{-1}, $\tau = 99.4$ yr (b) $r = 4.23 \times 10^{-3}$ μs^{-1}, $\tau = 237$ μs

13.5. (a) $r = 1.55 \times 10^{-10}$ yr^{-1} (b) $N = 2.53 \times 10^{21}$ (c) $R = 12,400$ decays/s.

13.11. $t_{av} = 1/r = \tau$

13.13. $N_B = N_0(1 - e^{-rt})$

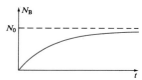

13.15. (a) $R_0 = 105$ decays/min (b) 19,400 yr

13.17. (a)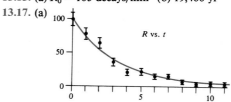

(b)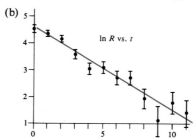

In (a), the $1/e$ time can be read as $\tau \approx 3$ hr. In (b), τ is the time for $\ln R$ to decrease by 1 (for example, from 4 to 3) and is $\tau = 3.25$ hr. In (a) one must draw an exponential curve, in (b) a straight line.

13.19. $K = 0.156$ MeV

13.21. $K = 0.783$ MeV

13.23. $(\Delta m_{atom})c^2 = 0.36$ MeV, enough for ec, but not for β^+

13.25. (a) $mc^2 = Am_n c^2 + Z(m_p - m_n)c^2 - a_{vol}A + a_{surf}A^{2/3} + a_{Coul}A^{-1/3}Z^2 + a_{sym}A^{-1}(2Z - A)^2 - \epsilon a_{pair}A^{-1/2}$

(b)

(d)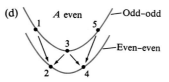

In the arrangement shown for A even, both of the nuclei 2 and 4 would be stable.

(e) ^{36}S and ^{36}Ar; ^{40}Ar and ^{40}Ca; ^{124}Sn, ^{124}Te, and ^{124}Xe

13.27. $K = -7.1, -6.4, -2.1, +5.5$, and $+25$ MeV; first three impossible, last two possible

13.29. (a) $K_\alpha = 6.4$ MeV (b) $K_\beta = -0.86$ MeV (impossible) (c) $K_\alpha = 6.2$ MeV, $K_\beta = 2.2$ MeV (both possible) (d) The thorium series has a branch at ^{212}Bi.

13.31. $^{235}U \rightarrow {}^{231}Th \rightarrow {}^{231}Pa \rightarrow {}^{227}Ac \rightarrow ({}^{227}Th \text{ or } {}^{223}Fr) \rightarrow$
$^{223}Ra \rightarrow {}^{219}Rn \rightarrow {}^{215}Po \rightarrow {}^{211}Pb \rightarrow {}^{211}Bi \rightarrow {}^{207}Tl \rightarrow {}^{207}Pb$

13.33. (a) $K = 6.147$ MeV (b) When the offspring is produced in an excited state there is less kinetic energy released. (c) 30, 60, 80 keV

13.37. Since $K_f < 0$, the reaction is impossible (at the incident energy given).

13.39. $m_n = 1.0067$ u compared with 1.0087 u

13.41. $\sigma = 0.6$ ft^2

13.43. (a) Impossible (energy) (b) Impossible (nucleon number) (c) Possible (d) Impossible (nucleon number) (e) Possible.

13.45. (a) 50 MeV (b) 53 MeV

13.47. (a) 23.8 MeV (b) 183 MeV

13.49. (a) six neutrons (b) Six neutrons (c) No; each still has four neutrons too many and will undergo β^- decay.

13.51. (a) $\Delta K/K = 4\mu/(1+\mu)^2$ (c) $\Delta K/K = 0.89$ for deuterium and 0.28 for carbon.

13.53. (a) 52 kg (b) 12 kg

13.55. (a) ^{13}N, ^{13}N, ^{14}N, ^{14}N, ^{15}N, ^{15}N
(c) In the CNO cycle the highest Coulomb barrier $(Z_1 Z_2 ke^2/r)$ is for the ^1H and ^{15}N with $Z_1 Z_2 = 1 \times 7 = 7$; in the p-p cycle the highest barrier is for the two ^3He nuclei, for which $Z_1 Z_2 = 2 \times 2 = 4$. The corresponding values of r are about the same. Therefore, (ratio of temperatures) $\approx$ (ratio of barrier heights) $\sim \frac{7}{4}$.

13.57. (a) $t_{1/2}(Th)/t_{1/2}(Ra) = 2.45 \times 10^{24}$, $ln[\cdots] = 56$ (b) 60

13.59. (a)

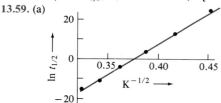

(b) Slope of graph = 315, predicted slope = $aZ = a86 = 340$. The agreement is improved if we take into account the variation of the second term in Eq. (13.80).

13.61. (a) ^{209}Pb, $K = 31.85$ MeV
(c) $P(^{14}C) = 2.7 \times 10^{-37}$, $P(^{14}C)/P(\alpha) = 2 \times 10^{-9}$

13.63. (a) $l_2 = 9.3$ cm (b) $l_{10} = 19.8$ cm

13.65. $p = 192$ MeV/$c = 1.02 \times 10^{-19}$ kg·m/s; $K = 4.95$ MeV $= 7.91 \times 10^{-13}$ J

13.67. (a) 6.3 μg (b) 1.0 g

13.69. 2.2 m

CHAPTER 14

14.1. (a) $E = \sqrt{(hc/\lambda)^2 + (mc^2)^2}$

14.7. (a) 50.8 ns (b) 40.3 ns (c) Yes, by a reasonable (though not enormous) margin

14.9. (a) $R = \infty$ for electromagnetic forces (b) $m_W \approx 100$ GeV/c^2

14.11. 2.4×10^{-24} s

14.17. $U_{nuc} \approx -100$ MeV, $U_{elec} = 1.44$ MeV

14.19. $(1 - v/c)$ is 0.73 for the μ decay, but 2.65×10^{-5} for the e decay. Ratio $\approx 3 \times 10^4$.

14.21. (b) $\tau^+ \rightarrow e^+ + \nu_e + \bar{\nu}_\tau$; $\tau^- \rightarrow \mu^- + \bar{\nu}_\mu + \nu_\tau$

14.23. (a) $\tau = 1.0 \times 10^{-20}$ s (b) $\tau = 1.8 \times 10^{-23}$ s

14.25. $B(\Lambda^\circ) = 1$, $B(K^\circ) = 0$

14.27. (a) $p_K = 150$ MeV/c, $p_\mu = 255$ MeV/c (b) $p_\nu = 250$ MeV/c (at 74.5° from the incoming K^+) (c) $E_\mu = 276$ MeV, $E_\nu = 250$ MeV (d) $E_K = 526$ MeV, $m_K = 504$ MeV/c^2, $u_K = 0.29c$

14.29. (a) $(\bar{u}d) + (uud) \rightarrow (u\bar{s}) + (dds)$; (b) $(dds) \rightarrow (udd) + (\bar{u}d)$; (c) $e^- + (udd) \rightarrow e^- + (udd)^*$; (d) $(udd)^* \rightarrow (\bar{u}d) + (uud)$

14.31. $(uds) = \Sigma^\circ$ with charge zero, and $(dds) = \Sigma^-$ with charge $-e$

14.33.

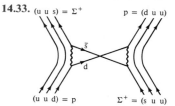

14.35. $R \approx 1.2 \times 10^{-18}$ m

14.37.

14.43. Mass of water required $\approx 2.6 \times 10^5$ metric tons. This has a volume of 2.6×10^5 m^3 = 26 m $\times$ 100 m $\times$ 100 m.

14.45. (a) 8.5×10^5 orbits (b) Distance = 5×10^6 km, total time = 17 s

14.47. (a) Diameter = 26 km (b) Period = 0.47 ms

CHAPTER 15

15.1. (a) $P = (2kq^2\omega^4 x_0^2 \sin^2 \omega t)/(3c^3)$

15.3. (a) $r = 10^{26}$ photons/s (b) No

15.5. (a) $P = 5.23 \times 10^{-4}$ eV/s (b) $P = 2.05 \times 10^5$ eV/s, $P_{elec}/P_{prot} = 3.9 \times 10^8$

15.7. (a) $P = 7.45 \times 10^{12}$ eV/s (b) $P_{tot} = 2.4$ MW

15.9. $P = 2k^3 q^4 Q^2/(3c^3 m^2 r^4)$ (b) $P \rightarrow 16P$ (c) $P \rightarrow 4P$

15.11. $P_{He} = 64 P_H = 1.9 \times 10^{13}$ eV/s

15.13. (a) $dE/dr = ke^2/(2r^2)$ (b) $dE/dt = -2(ke^2)^3/(3m^2 c^3 r^4)$ (c) $dr/dt = -4(ke^2)^2/(3m^2 c^3 r^2)$, $\Delta t = m^2 c^3 a_B^3/4(ke^2)^2 = 1.6 \times 10^{-11}$ s

15.15. (a) $3 \rightarrow 4$ and $4 \rightarrow 3$, but if all atoms are in the ground state no transitions will be observed. (b) $1 \rightarrow 5$ and $5 \rightarrow 1$; only $1 \rightarrow 5$ would occur in a cool gas. (c) $\lambda = 443$ nm

15.19. $r = 1.86 \times 10^8$ s^{-1}; $\tau = 5.4 \times 10^{-9}$ s

15.21. (a) Because of the Pauli principle, the spins have to be opposite ($s_{tot} = 0$) if both electrons have the same spatial wave function.

(b)

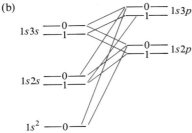

(c) Both $1s2s$ levels are metastable since all downward transitions are forbidden by $\Delta s_{tot} = 0$, or $\Delta l = 1$, or both.

15.23. The excited state has $j_{tot} = 9$ and all possible decays have $\Delta j_{tot} = 8$ or 9. Therefore, the excited state is very nearly stable.

15.25. (a) and (c):

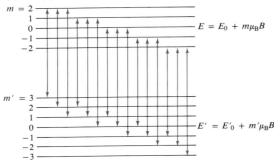

(b) In principle there are 35 conceivable transitions, but only 11 distinct energy differences.

(d) There are three distinct frequencies.

15.27. (a) 8.0×10^{17} photons per pulse (b) $f_1 = 6.4 \times 10^{-11}$ (c) $\delta\theta = 1.75 \times 10^{-5}$ rad (d) 1.4×10^4 m (e) $f_2 = 5.1 \times 10^{-9}$ (f) About 80

CHAPTER 16

16.1. (a) $F = 8.34 \times 10^{23}$ N (b) $a = 1.55 \times 10^{29}g$

16.3.

Molecule	eR_0 (C·m)	p/eR_0
KCl	4.32×10^{-29}	76%
LiF	2.56×10^{-29}	82%
NaBr	4.00×10^{-29}	76%
NaCl	3.84×10^{-29}	78%

16.5. 16 km/liter = 38 mi/gal

16.7. 9.2×10^6 J

16.9. 4.4×10^7 J

16.11. $U = ke^2\left(\dfrac{1}{R_0} + \dfrac{1}{s} - \dfrac{8}{\sqrt{s^2 + R_0^2}}\right)$; U is minimum at $s = R_0/\sqrt{3}$.

16.13. (a)

Molecule:	CsI	NO	N_2	NaBr
ΔE (eV):	0.8	13.0	14.5	1.7

(b) CsI and NaBr are ionic; NO and N_2 are not.

16.15.

Molecule:	KI	NaBr	LiF
R_c (nm):	1.13	0.81	0.72

16.17. $R_0 = 0.22$ nm; this is an overestimate (the actual value is 0.20 nm).

16.19. CaS, CaCl$_2$, Li$_2$O, Na$_3$P

16.21. (a) Both have valence 2. (b) $2eR_0 = 8.0 \times 10^{-29}$ C·m (c) 45%

16.23. H$_3$P, H$_2$Te, HBr

16.25. H—C≡C—H

16.27. (a) ClF, BrCl, ICl (b) $p/eR_0 = 12\%, 6\%, 11\%$

16.33. $k = 2.6 \times 10^3$ eV/nm^2 = 4.2×10^2 N/m

16.35. $k = 840$ eV/nm^2

16.39. (b) $\hbar\omega_c(D) = \hbar\omega_c(H)/\sqrt{2} = 0.26$ eV

16.41. (a) $k \approx 2 \times 10^2$ eV/nm^2 (b) $\frac{1}{2}\hbar\omega_c \approx 0.2$ eV (compare observed value of 0.15 eV)

16.43. (a) and (d):

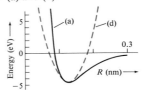

(b) $R_0 = 0.127$ nm, $B = 4.57$ eV (c) $k = 3030$ eV/nm^2

16.45. $R_0 = 0.14$ nm

16.47. (a) $E_{ph} = 4.79 \times 10^{-3}, 6.36 \times 10^{-3}, 7.95 \times 10^{-3}$ eV. These correspond to rotational transitions: $(2 \leftrightarrow 3)$, $(3 \leftrightarrow 4)$, and $(4 \leftrightarrow 5)$. (b) $R_0 = 0.162$ nm

16.49. (a) $E_{ph} = 4.79 \times 10^{-4}, 9.61 \times 10^{-4}, 14.36 \times 10^{-4}$ eV. These correspond to rotational transitions: $(0 \leftrightarrow 1)$, $(1 \leftrightarrow 2)$, $(2 \leftrightarrow 3)$. (b) $R_0 = 0.113$ nm

16.51. $k = 1.85 \times 10^3$ eV/nm^2, $R_0 = 0.16$ nm

CHAPTER 17

17.1. Diamond: 710 kJ/mol = 170 kcal/mol; for Ge, 371 kJ/mol = 88.7 kcal/mol

17.3. (b) Attractive

17.5.

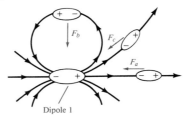

17.7. (a) $r_0\sqrt{2}$ (b) 12 ions (c) Na$^+$

17.9. $r_0 = 0.284$ nm

17.11. (a) PE $= -8.1$ eV (b) 3.25 eV/atom

17.17. $\rho(Cu) = 9.0$ g/cm^3

17.19. (a) Edge length $= 0.41$ nm   (b) $r_0 = 0.36$ nm

17.21. $F \approx 6 \times 10^{-9}$ N

17.23. PE $= \dfrac{ke^2}{r_0}\left(-6 + \dfrac{12}{\sqrt{2}} - \dfrac{8}{\sqrt{3}} + \dfrac{6}{2} - \dfrac{24}{\sqrt{5}} + \cdots\right)$

17.25. (a) $N = 5 \times 10^{10}$ atoms   (b) $\delta E = 5 \times 10^{-12}$ eV
(c) 5×10^{21} atoms and $\delta E = 5 \times 10^{-23}$ eV

17.27. 2.0×10^{22} states

17.29. All visible light is transmitted.

17.31. No visible light is transmitted.

17.33. $E_{gap} = 1.4$ eV

17.35. 6.5×10^5 pairs, 1.0×10^{-13} C

17.37. (a) p (b) p (c) n

17.39. (b) $I(\text{forward}) = 1.57 \times 10^{17}\, I_0$, $I(\text{back}) = -I_0$, ratio $=$
1.57×10^{17}

17.41. (b) It is increased by a factor of $\kappa \approx 16$.

17.43. $T \sim 120$ K. In Fig. 17.29, the curve drops away from $3R$
around $T \sim 200$ K.

17.45. Solid:

	Al	Ag	Au
c(predicted):	0.221	0.0553	0.0303 cal/(g·K)
Discrepancy:	3%	-1%	-3%

17.47. (a) $m\ddot{x}_1 = -2kx_1 + kx_2$; $m\ddot{x}_2 = kx_1 - 2kx_2$
(b) $m\ddot{y}_1 = -ky_1$; $m\ddot{y}_2 = -3ky_2$
(c) $\omega_{c1} = \sqrt{k/m}$; $\omega_{c2} = \sqrt{3k/m}$
(d) The two modes are precisely the normal modes of
Fig. 17.31.

Index

Except where noted otherwise, all references are to page numbers. Occasionally, if a topic is the subject of a whole section, chapter, or appendix, we give the section, chapter, or appendix number [for example, "Crystal, Sec. 17.3," or "Physical constants, App. A"]. When a reference is to a footnote or figure, we show this in parentheses, [for example, "Stellar aberration, 11 (Ftn.)"]. Finally, for references to the problems at the ends of chapters, we provide the problem number in parentheses after the page number [for example, "CNO cycle, 385 (Pr. 13.55)"].